G. R. Liu

MESH FREE METHODS

Moving beyond the Finite Element Method

CRC PRESS
Boca Raton London New York Washington, D.C.

Library of Congress Cataloging-in-Publication Data

Liu, Gui-Rong.
Mesh free methods : moving beyond the finite element method / by Gui-Rong Liu.
p. cm.
Includes bibliographical references and index.
ISBN 0-8493-1238-8 (alk. paper)
1. Engineering mathematics. 2. Numerical analysis. I. Title.

TA335 .L58 2002
620′.001′51—dc21 2002023347
CIP

International Standard Book Number 0-8493-1238-8
Library of Congress Card Number 2002023347
Printed in the United States of America 2 3 4 5 6 7 8 9 0
Printed on acid-free paper

Dedication

To Zuona

Yun, Kun, Run,

and my family

for the time and support they gave to me

Preface

Topics related to modeling and simulation play an increasingly important role in building an advanced engineering system in rapid and cost-effective ways. Since the invention by Sir Thomas Harriet (1560–1621), people have been using the finite difference method (FDM) to perform the task of modeling and simulation of engineering systems, in particular to solve partial differential equation systems. It works very well for problems of simple geometry. For decades, we have used techniques of finite element methods (FEM) to perform more-challenging tasks arising from increasing demands on flexibility, effectiveness, and accuracy in challenging problems with complex geometry. I still remember, during my university years, doing a homework assignment using FDM to calculate the temperature distribution in a rectangular plate. This simple problem demonstrated the power of numerical methods. About a year later, I created an FEM program to solve a nonlinear mechanics problem for a frame structural system, as my final year project. Since then, FEM has been one of my major tools in dealing with many engineering and academic problems. In the past two decades I have participated in and directed many engineering problems of very large scale with millions of degrees of freedom (DOFs). I thought, and many of my colleagues agreed, that with the advances of FEM and the computer, there were very few problems left to solve. Soon, I realized that I was wrong and for very simple reasons. When a class of problems is solved, people simply move on to solve a class of problems that are more complex and to demand results that are more accurate. In reality, problems can be as complex as we want them to be; hence, we can never claim that problems are solved. We solve problems that are idealized and simplified by us. Once the simplification is relaxed, new challenges arise. The older methods often cannot meet the demands of new problems of increasing complexity, and newer and more advanced methods are constantly born.

I heard about meshless methods in about 1993, while I was working at Northwestern University, but somehow I was reluctant to move into this new research area probably because I was quite happy with what I was doing using techniques of FEM. It was also partially because I was concentrating on the development of my strip element method (see the monograph by G. R. Liu and Xi, 2001). During 1995–1996, I handled a number of practical engineering problems for the defense industry using FEM packages, and encountered difficulties in solving mesh distortion-related problems. I struggled to use re-meshing techniques, but the solution was far from satisfactory. I then began to look for methods that can solve the mesh distortion problems encountered in my industrial research work. I immediately started to learn more about meshless methods.

I worked alone for about a year feeling as if I was walking in a maze of this new research area. I wished that I had a book on mesh free methods to guide me. I was excited for a time about the small progress I made, which motivated me to work day and night to write a proposal for a research grant from NSTB (a research funding agency of the Singapore government). I was lucky enough to secure the grant, which quickly enabled me to form a research team at the Centre for Advanced Computations in Engineering Science (ACES) working on element free methods. The research team at ACES is still working very hard in the area of mesh free methods. This book will cover many of the research outcomes from this research group.

This book provides systematic steps that lead the reader to understand mesh free methods; how they work; how to use and develop a mesh free method, as well as the problems associated with the element free methods. I experienced difficulties in the process of learning mesh free methods, because no single book was thus far available dedicated to the topic. I therefore hope my effort in writing this monograph can help researchers, engineers, and students who are interested in exploring mesh free methods.

My work in the area of mesh free methods has been profoundly influenced by the works of Professors Belytschko, Atluri, W. K. Liu, and many others working in this area. Without their significant contributions to this area, this book would not exist.

In preparing this book, a number of my colleagues and students have supported and contributed to its writing. I express my sincere thanks to all of them. Special thanks to Y. T. Gu, X. L. Chen, L. Liu, V. Tan, L. Yan, K. Y. Yang, M. B. Liu, Y. L. Wu, Z. H. Tu, J. G. Wang, X. M. Huang, Y. G. Wu, Z. P. Wu, K. Y. Dai, and X. Han. Many of these individuals have contributed examples to this book in addition to their hard work in carrying out a number of projects related to the mesh free methods covered in this book.

G. R. Liu

The Author

G. R. Liu received his Ph.D. from Tohoku University, Japan in 1991. He was a postdoctoral fellow at Northwestern University, U.S.A. He is currently the Director of the Centre for Advanced Computations in Engineering Science (ACES), National University of Singapore. He is also an associate professor at the Department of Mechanical Engineering, National University of Singapore. He has authored more than 250 technical publications including four books and 150 international journal papers. He is the recipient of the Outstanding University Researchers Awards (1998), for his development of the strip element method. He is also a recipient of the Defence Technology Prize (national award, 1999) for his contribution to development of underwater shock technology at Singapore. He won the Silver Award at CrayQuest 2000 (nationwide competition in 2000) for his development of mesh free methods. His research interests include computational mechanics, element free methods, nano-scale computation, vibration and wave propagation in composites, mechanics of composites and smart materials, inverse problems, and numerical analysis.

Contents

1

Introduction

1.1 Defining Mesh Free Methods

Designing advanced engineering systems requires the use of computer-aided design (CAD) tools. In such tools, computational simulation techniques are often used to model and investigate physical phenomena in an engineering system. The simulation requires solving the complex differential or partial differential equations that govern these phenomena.

Traditionally, such complex partial differential equations are largely solved using numerical methods, such as the finite element method (FEM) and the finite difference method (FDM). In these methods, the spatial domain where the partial differential governing equations are defined is often *discretized* into meshes.

A *mesh* is defined as any of the open spaces or interstices between the strands of a net that is formed by connecting nodes in a predefined manner. In FDM, the meshes used are also often called grids; in the finite volume method (FVM), the meshes are called volumes or cells; and in FEM, the meshes are called elements. The terminologies of grids, volumes, cells, and elements carry certain physical meanings as they are defined for different physical problems. However, all these grids, volumes, cells, and elements can be termed *meshes* according to the above definition of mesh. The key here is that a mesh must be predefined to provide a certain relationship between the nodes, which is the base of the formulation of these conventional numerical methods.

By using a properly predefined mesh and by applying a proper principle, complex differential or partial differential governing equations can be approximated by a set of algebraic equations for the mesh. The system of algebraic equations for the whole problem domain can be formed by assembling sets of algebraic equations for all the meshes.

The mesh free method, abbreviated *MFree* method in this book, is used to establish a system of algebraic equations for the whole problem domain without the use of a predefined mesh. MFree methods use a set of nodes scattered within the problem domain as well as sets of nodes scattered on the boundaries of the domain to *represent* (not discretize) the problem domain and its boundaries. These sets of scattered nodes do not form a mesh, which means that no information on the relationship between the nodes is required, at least for field variable interpolation.

There are a number of MFree methods, such as the element free Galerkin (EFG) method (Belytschko et al., 1994b), the meshless local Petrov–Galerkin (MLPG) method (Atluri and Zhu, 1998), the point interpolation method (PIM) (Liu, G. R. and Gu, 1999), the point assembly method (PAM) (Liu, G. R., 1999), the finite point method (Onate et al., 1996), the finite difference method with arbitrary irregular grids (Liszka and Orkisz, 1980; Jensen, 1980), smooth particle hydrodynamics (SPH) (Lucy, 1977; Gingold and Monaghan, 1977), reproducing kernel particle method (Liu, W. K. et al., 1993), which is an improved version of SPH, and so forth. They all share the same feature that predefined meshes are not used, at least for field variable interpolation. The names for the various MFree methods are still being debated. Because the methodology is still in a rapid development stage, new names of methods are constantly proposed. It may take some time before all the methods are properly categorized and unified to avoid confusion in the community.

In contrast to FEM, the term *element free method* is preferred, and in contrast to FDM, the term *finite difference method using arbitrary or irregular grids* is preferred. Some of the MFree methods have been assembled and termed *meshless methods* (see the excellent review paper by Belytschko et al., 1996b). The term mesh free method has also appeared in the literature in recent years. This book uses the term mesh free method or MFree method for the collection of all the different mesh free methods because the term carries more a positive sense than meshless method. A reason to prefer the abbreviation MFree is because the M can also stand for manpower and money. *M*esh free methods free *M*anpower from the chore of meshing and free substantial *M*oney for computer-aided engineering (CAE) projects. The question now is what qualifies as an MFree method?

The *minimum* requirement for an MFree method:

- That a predefined mesh is not necessary, at least in field variable interpolation.

The *ideal* requirement for an MFree method:

- That no mesh is necessary at all throughout the process of solving the problem of given arbitrary geometry governed by partial differential system equations subject to all kinds of boundary conditions.

The reality is that the MFree methods developed so far are not really ideal, and fail in one of the following categories:

- Methods that require background cells for the integration of system matrices derived from the weak form over the problem domain. These methods are not truly mesh free. EFG methods belong in this category. These methods are practical in many ways, as the creation of a background mesh is generally feasible and can always be automated using a triangular mesh for two-dimensional (2D) domains and a tetrahedral mesh for three-dimensional (3D) domains.
- Methods that require background cells locally for the integration of system matrices over the problem domain. MLPG methods belong in this category. These methods are said to be essentially mesh free because creating a local mesh is easier than creating a mesh for the entire problem domain; it is a simpler task that can be performed automatically without any predefinition for the local mesh.
- Methods that do not require a mesh at all, but that are less stable and less accurate. Collocation methods and finite difference methods using irregular grids belong to this category. Selection of nodes based on the type of a physical problem is still important for obtaining stable and accurate results (Song et al., 1999; Xu and Liu, 1999; Cheng and Liu, 1999). Automation of nodal selection and improving the stability of the solution are some of the challenges in these kinds of methods. This type of method has a very significant advantage: it is very easy to implement, because no integration is required.
- Particle methods that require a predefinition of particles for their volumes or masses. The algorithm will then carry out the analyses even if the problem domain undergoes extremely large deformation and separation. SPH methods

belong to this category. This type of method suffers from problems in the imposition of boundary conditions. In addition, predefining the particles still technically requires some kind of mesh.

1.2 Need for MFree Methods

FEM is robust and has been thoroughly developed for static and dynamic, linear or nonlinear stress analysis of solids, structures, as well as fluid flows. Most practical engineering problems related to solids and structures are currently solved using a large number of well-developed FEM packages that are commercially available. However, the following limitations of FEM are becoming increasingly evident:

1. Creation of a mesh for the problem domain is a prerequisite in using FEM packages. Usually the analyst spends the majority of his or her time in creating the mesh, and it becomes a major component of the cost of a simulation project because the cost of CPU (central processing unit) time is drastically decreasing. The concern is more the manpower time, and less the computer time. Therefore, ideally the meshing process would be fully performed by the computer without human intervention.
2. In stress calculations, the stresses obtained using FEM packages are discontinuous and less accurate.
3. When handling large deformation, considerable accuracy is lost because of the element distortions.
4. It is very difficult to simulate both crack growth with arbitrary and complex paths and phase transformations due to discontinuities that do not coincide with the original nodal lines.
5. It is very difficult to simulate the breakage of material into a large number of fragments as FEM is essentially based on continuum mechanics, in which the elements formulated cannot be broken. The elements can either be totally "eroded" or stay as a whole piece. This usually leads to a misrepresentation of the breakage path. Serious error can occur because the nature of the problem is nonlinear, and therefore the results are highly path dependent.
6. Re-mesh approaches have been proposed for handling these types of problems in FEM. In the re-mesh approach, the problem domain is re-meshed at steps during the simulation process to prevent the severe distortion of meshes and to allow the nodal lines to remain coincident with the discontinuity boundaries. For this purpose, complex, robust, and adaptive mesh generation processors have to be developed. However, these processors are only workable for 2D problems. There are no reliable processors available for creating hexahedral meshes for 3D problems due to the technical difficulty.
7. Adaptive processors require "mappings" of field variables between meshes in successive stages in solving the problem. This mapping process often leads to additional computation as well as a degradation of accuracy. In addition, for large 3D problems, the computational cost of re-meshing at each step becomes very high, even if an adaptive scheme is available.

8. FDM works very well for a large number of problems, especially for solving fluid dynamics problems. It suffers from a major disadvantage in that it relies on regularly distributed nodes. Therefore, studies have been conducted for a long time to develop methods using irregular grids. Efforts in this direction are still ongoing.

1.3 The Idea of MFree Methods

A close examination of these difficulties associated with FEM reveals the root of the problem: the need to use elements, which are the building block of FEM. A mesh with predefined "connectivity" is required to form the elements. As long as elements must be used, the problems mentioned above will not be easy to solve. Therefore, the idea of eliminating the elements and hence the mesh has evolved naturally. The concept of element free or mesh free methods has been proposed, in which the domain of the problem is represented by a set of arbitrarily distributed nodes.

The MFree method has great potential for solving the difficult problems mentioned above. Adaptive schemes can be easily developed, as there is no mesh, and hence no connectivity concept involved. Thus, there is no need to provide *a priori* any information about the relationship of the nodes. This provides flexibility in adding or deleting points/nodes whenever and wherever needed. For stress analysis of a solid domain, for example, there are often areas of stress concentration, even singularity. One can relatively freely add points in the stress concentration area without worrying about their relationship with the other existing nodes. In crack growth problems, nodes can be easily added around the crack tip to capture the stress concentration with desired accuracy. This nodal refinement can be moved with a propagation crack through a background arrangement of nodes associated with the global geometry. Adaptive meshing for a large variety of problems, 2D or 3D, including linear and nonlinear, static and dynamic stress analysis, can be very effectively treated in MFree methods in a relatively simple manner.

Because there is no need to create a mesh, and the nodes can be created by a computer in a fully automated manner, the time an engineer would spend on conventional mesh generation can be saved. This can translate to substantial cost and time savings in modeling and simulation projects.

There have been a number of MFree methods developed thus far. The major features of these methods are listed in Table 1.1. A software package MFree2D© with its own pre- and postprocessor has also been developed by G. R. Liu and co-workers (Liu, G. R. and Tu, 2001).

1.4 Outline of the Book

This book provides an introduction to MFree methods, and their application to various mechanics problems. This book covers the following types of problems:

- Mechanics for solids (2D solids)
- Mechanics for structures (beams, plates, and shells)
- Fluid mechanics (fluid flow, convection flow, and hydrodynamics)

TABLE 1.1

Some MFree Methods Developed Thus Far and Their Features

Method	Ref.	System Equation to be Solved	Method of Function Approximation	Chapter Covering the Method
Diffuse element method	Nayroles et al., 1992	Weak form	MLS approximation, Galerkin method	5, 6
Element free Galerkin (EFG) method	Belytschko et al., 1994b	Weak form	MLS approximation, Galerkin method	5, 6, 11, 12, 14–16
Meshless local Petrov–Galerkin (MLPG) method	Atluri and Zhu, 1998	Local weak form	MLS approximation, Petrov–Galerkin method	5, 7, 9–11, 14
Finite point method	Onate et al., 1996; Liszka and Orkisz, 1980; Jensen, 1980	Strong form	Finite differential representation (Taylor series), MLS approximation	5 (brief only)
Smooth particle hydrodynamics	Lucy, 1977; Gingold and Monaghan, 1977	Strong form	Integral representation	5, 9
Reproducing kernel particle method	Liu, W. K. et al., 1993	Strong form or weak form	Integral representation (RKPM)	5 (brief only)
hp-clouds	Oden and Abani, 1994; Armando and Oden, 1995	Weak form	Partition of unity, MLS	5 (introduction only)
Partition of unity FEM	Babuška and Melenk, 1995	Weak form	Partition of unity, MLS	5 (introduction only)
Point interpolation method	Liu, G. R. and Gu, 1999, 2000b, 2001a,b,c,d	Weak form and local weak form	Point interpolation	5, 8–14
Boundary node methods	Mukherjee and Mukherjee, 1997a,b	Weak form and local weak form	MLS	13 (brief only)
Boundary point interpolation methods	Liu, G. R. and Gu, 2000d; Gu and Liu, G. R., 2001a,e	Weak form and local weak form	Point interpolation	13, 14

The bulk of the material in the book is the result of the intensive research work by G. R. Liu and his research team in the past 6 years. Works of other researchers are also introduced. The significance of this book is as follows:

1. This is the first book published that comprehensively covers MFree methods.
2. The book covers, in a systematic manner, basic theories, principles, techniques, and procedures in solving mechanics problems using MFree methods. It will be very useful for researchers entering this new area of research on MFree methods, and for professionals and engineers developing computer codes for the next generation of computational mechanics.
3. Readers will benefit from the research outcome of G. R. Liu's research team and their work on an award-winning project on MFree methods founded by the Singapore government. Many materials in this book are the results of ongoing projects, and have not been previously published.

4. A software package named MFree2D© with its own pre- and postprocessors, which has been developed by G. R. Liu's team, is also introduced.
5. A large number of examples with illustrations are provided for validating, benchmarking, and demonstrating MFree methods. These examples can be useful reference materials for other researchers.

The book is written for senior university students, graduate students, researchers, and professionals in engineering and science. Mechanical engineers and practitioners and structural engineers and practitioners will also find the book useful. Knowledge of FEM is not required but would help a great deal in understanding many concepts and procedures of MFree methods. Basic knowledge of mechanics is also helpful in reading this book smoothly. A very brief introduction on mechanics is provided to prepare readers who are not familiar with the basics of mechanics.

The MFree method is a very new area of research. There exist many problems, problems that offer ample opportunities for research to develop the next generation of numerical methods. The method is also in a rapidly developing and growing stage. Different techniques are developed every day. This book addresses some of the current important issues, both positive and negative, related to MFree methods, which should prove beneficial to researchers, engineers, and students who are interested in venturing into this area of research.

The chapter-by-chapter description of this book is as follows:

Chapter 1: Provides a brief introduction to MFree methods including background and motivations that have led to the development of MFree methods, as well as this book.

Chapter 2: Describes MFree procedures by comparing them with conventional FEM. The basic steps involved in using MFree methods for solving engineering problems are listed. Some important common terms frequently used in MFree methods are defined.

Chapter 3: Presents a brief introduction to the basics of mechanics for solids, beams, and plates. Readers who are familiar with mechanics may skip this chapter.

Chapter 4: Briefly introduces some principles and weak forms that will be used for creating discretized system equations. Constrained weak forms that are useful for MFree formulation are also presented. Readers who are familiar with variational, energy, and weighted residual methods may skip this chapter.

Chapter 5: Provides a detailed description of the methods for constructing MFree shape functions. This is one of the core chapters of this book discussing one of the most important issues of MFree methods. A new classification of theories of function approximation methods is presented. Methods of approximation of functions are then systematically introduced. Issues related to constancy, reproduction, representation, compatibility, and convergence of an approximation are discussed. Traditional methods, SPH and MLS, and new methods of point interpolation are formulated in detail for the construction of MFree shape functions.

Chapter 6: Introduces the element free Galerkin method (EFG), one of the MFree methods that has created a great difference. Very detailed formulation based on the constrained Galerkin weak form is presented, and a number of techniques that are used for handling essential boundary conditions are discussed in detail. Issues related to the background integration are also examined.

Chapter 7: Introduces the meshless local Petrov–Galerkin (MLPG) method, which requires only local cells of a background mesh for integration and which is a significant advance. The constrained local Petrov–Galerkin method is used to formulate the MLPG method for both static and dynamic problems. Issues on types of local domains, effects of the dimension of these domains, handling local integration, and procedures in dealing with essential boundary conditions are discussed.

Chapter 8: Introduces the point interpolation methods (PIM), which are formulated based on both the Galerkin formulation and the local Petrov–Galerkin formulation. Different methods of PIM, both conforming and nonconforming, are formulated, examined, and benchmarked, issues related to the selection of parameters are discussed, and a large number of examples are provided.

Chapter 9: Introduces MFree methods for computational fluid dynamics problems. SPH, MLPG, and PIM are formulated and applied to solve a number of benchmarking and application problems. Issues related to iterative solution procedures for MFree methods in solving fluid flow problems are discussed.

Chapter 10: Introduces PIM methods developed for analysis of beams. Both thin beams governed by the Euler–Bernoulli beam theory and thick beams governed by the Timoshenko beam theory are used in the formulations. Shear-locking issues for beams in using MFree methods based on shear deformable beams are also discussed.

Chapter 11: Introduces MFree methods developed for analysis of plates. Both thin plates governed by the Mindlin plate theory and thick plates governed by the third-order shear deformation theory are used to develop the MFree methods. Three methods—EFG, PIM, and MLPG—are formulated. Shear-locking issues for plates are also discussed in great detail.

Chapter 12: Introduces MFree methods developed for analysis of shells. Both thin shells governed by the Kirchhoff–Love theory and thick shells are used to formulate the MFree methods. Two methods—EFG and PIM—are formulated for static and dynamic problems. Advantages of using MFree methods for shells are discussed. Shear- and membrane-locking issues for shells are also briefly discussed.

Chapter 13: Formulates two boundary-type MFree methods—BPIM and radial BPIM. Procedures of using the MFree concept for solving boundary integral equations are provided. A number of examples are presented to demonstrate the advantages of these methods in solving problems of infinite domain.

Chapter 14: Introduces methods that are formulated by coupling domain and boundary types of MFree methods. MFree methods that couple with the traditional FEM and boundary element method (BEM) are also formulated and benchmarked.

Chapter 15: Discusses a number of implementation issues in coding with MFree methods, including node generation, background cell generation, determination of nodal influence domain, node selection, error estimations, and adaptive procedures.

Chapter 16: Presents MFree2D©, which has been developed by G. R. Liu and coworkers with pre- and postprocessors for adaptive analysis of 2D solids. Its functions and usage are introduced.

2

Mesh Free Methods for Engineering Problems

In building a modern and advanced engineering system, engineers must undertake a very sophisticated process of modeling, simulation, visualization, analysis, designing, prototyping, testing, fabrication, and construction. The process is illustrated in the flowchart shown in Figure 2.1. The process is very often iterative in nature; that is, some of the procedures are repeated based on the results obtained at the current stage to achieve optimal performance for the system under construction.

This book deals with topics related mainly to modeling and simulation, as well as some issues related to visualization, which are underlined in Figure 2.1. Under these topics, we address the computational aspects, which are also underlined in Figure 2.1. The focus will be on physical, mathematical, and computational modeling and computational simulation. These topics play an increasingly important role in building an advanced engineering system in a rapid and cost-effective way. Many methods and computational techniques can be employed to deal with these topics. The book mainly focuses on the development and use of the mesh free (MFree) methods. This chapter addresses the overall procedures of modeling and simulation using MFree methods and the differences between the MFree method and other existing methods, especially the widely used finite element method (FEM, see Liu and Quek, 2002).

2.1 Physical Phenomena in Engineering

There are a large number of different physical phenomena in engineering systems, so many that it is not possible to model and simulate them all. In fact, only major phenomena, which significantly affect the performance of the system, need be modeled and simulated to provide a necessary and sufficient in-depth understanding of the system.

The physical problems covered in this book are in the areas of mechanics for solids, structures, and fluid flows, and mathematic models have been developed for the phenomena in these areas. Different types of differential or partial differential governing equations have also been derived for these phenomena. These phenomena can be simulated if a proper tool can be found to solve these equations. Similar to conventional FEM, the finite difference method, and finite volume methods, the MFree method is actually a tool for solving partial differential equations that govern different physical phenomena.

2.2 Solution Procedure

The procedure in FEM and the MFree method for solving engineering problems can in principle be outlined using the chart given in Figure 2.2. These two methods diverge at the stage of mesh creation. The fundamental difference between these two methods is the construction of the shape functions. In FEM, the shape functions are constructed using elements,

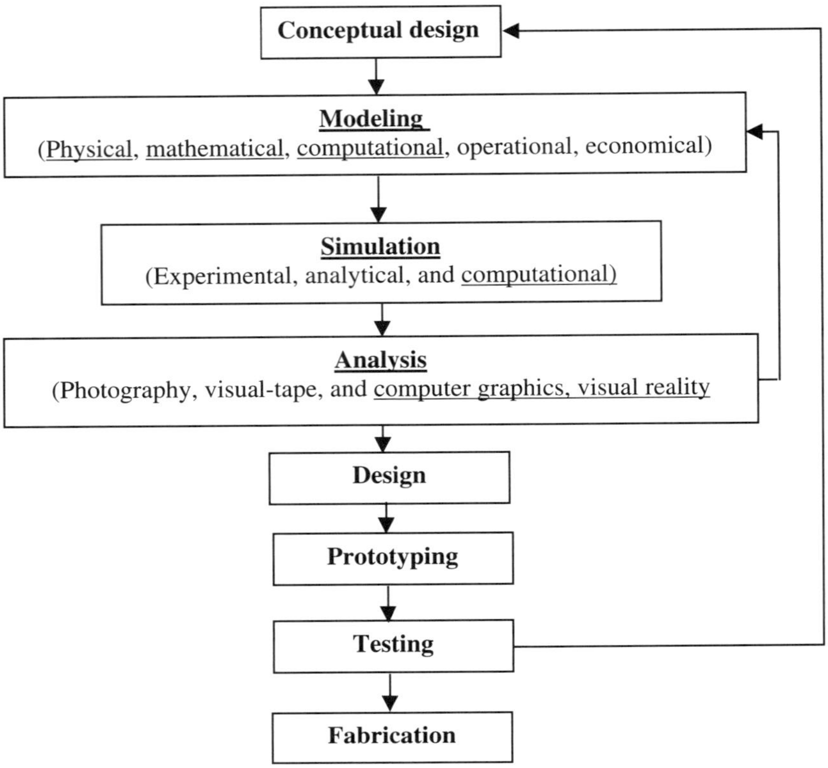

FIGURE 2.1
Processes that lead to building an engineered system.

and the shape functions will be the same for the element. In fact, if the natural coordinate systems are used, the shape functions in the natural coordinates are the same for all the elements of the same type. These shape functions are usually predetermined for different types of elements before the finite element analysis starts. In MFree methods, however, the shape functions constructed are usually only for a particular point of interest. The shape function changes as the location of the point of interest changes. The construction of the element free shape function is performed *during* the analysis, not *before* the analysis.

Once the global discretized system equation is established, the MFree method follows a procedure similar to FEM, except for some minor differences in the details of implementation. Therefore, many techniques developed over the past decades in FEM can be utilized in MFree methods with or without modifications.

The following sections present the basic procedures in MFree methods, by discussing the differences between FEM and the MFree method at major stages of analysis.

2.3 Modeling the Geometry

Real structures, components, or domains are in general very complex and have to be reduced to a manageable geometry. In FEM, curved parts of the geometry and its boundary can be modeled using curves and curved surfaces using high-order elements. However, it should be noted that the geometry is eventually represented by a collection of elements,

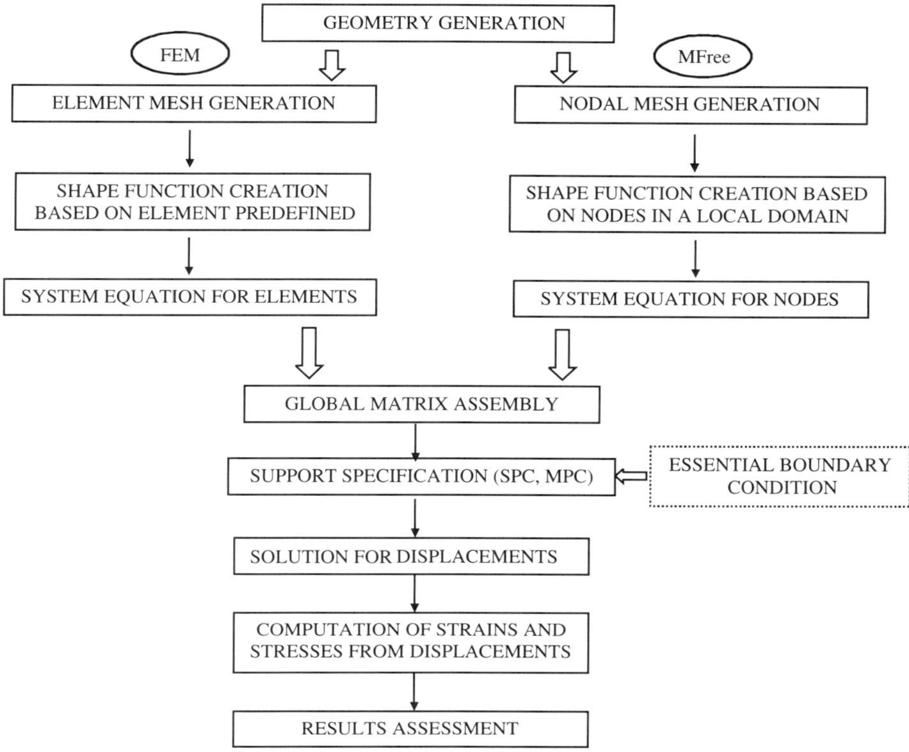

FIGURE 2.2
Flowchart for FEM and MFree method procedures.

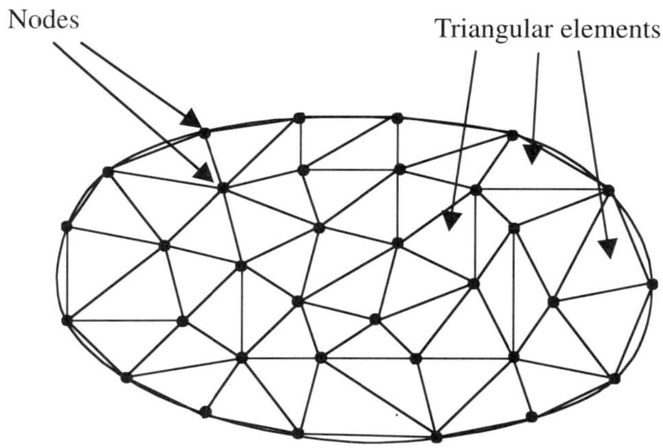

FIGURE 2.3
Smoothed boundary is represented in FEM by straight lines of the edges of triangular elements.

and the curves and curved surfaces are approximated by piecewise curves and surfaces of the elements. If linear elements are used, which is often the case in practical situations, these curves and surfaces are straight lines or flat surfaces. Figure 2.3 shows an example of a smooth boundary represented in the finite element model by straight lines of the edges of triangular elements. The accuracy of representation of the curved parts is controlled

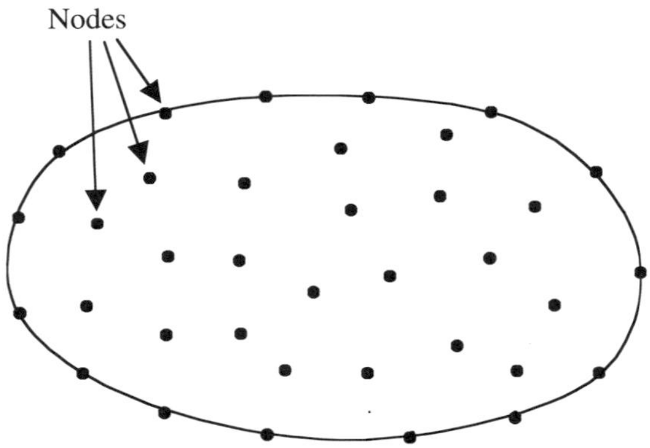

FIGURE 2.4
Smoothed boundary is represented in the mesh free method by nodes.

by the number of the elements and the order of the elements used. A finer mesh of elements can generally lead to more accurate results. However, because of the constraints on time and computational resources including hardware and software, it is always required to limit the number of elements. Therefore, fine details of the geometry need to be modeled only if very accurate results are required for those regions. The results of simulation have to be interpreted with these geometric approximations in mind. The analyst has to determine the distribution of the density of the mesh required to achieve a desired accuracy at important areas and regions of the problem domain.

In MFree methods, however, the boundary is *represented* (not discretized) by nodes, as shown in Figure 2.4. At any point between two nodes on the boundary, one can interpolate using MFree shape functions. Because the MFree shape functions are created using nodes in a *moving* local domain, the curved boundary can be approximated very accurately even if linear polynomial bases are used. It is common in MFree methods to use higher-order polynomials. Note that this geometric interpolation can be performed using the same technique for field variable interpolation in MFree methods.

Depending on the software used, there are many ways to create a properly simplified geometry in the computer. Points can be created simply by keying in the coordinates of the point. Lines/curves can be created by simply connecting points/nodes. Surfaces can be created by connecting/rotating/translating the existing lines/curves. Solids can be created by connecting/rotating/translating the existing surfaces. Points, lines/curves, surfaces, and solids can be translated/rotated/reflected to form new ones. Graphic interfaces are used for assisting the creation and manipulation of the objects. There are a number of CAD (computer-aided design) software packages used in engineering design that can produce files containing the geometry of the designed engineering system. These files can often be read by modeling software packages. Making use of the CAD files can save significant time in creating the geometry of the models. However, in many cases, the objects read directly from a CAD file may need to be modified and simplified before performing meshing. These tools for creating the geometry of the problem domain can be used for both the FEM and the MFree method.

Knowledge, experience, and engineering judgment are very important in modeling the geometry of a system. In many cases, finely detailed geometric features play only an aesthetic role, and will not affect the functionality or the performance of the engineering system very much. These features can be simply deleted, ignored, or simplified. This, however, may not be true for some cases, where a fine geometric change can give rise to a significant difference in the simulation results. Adaptive analysis is ideal for solving this problem objectively and independently of the judgment of the analyst. MFree methods provide more flexible ways for adaptive analyses.

Another very important issue is the simplification required by mathematic modeling. For example, a plate has three dimensions geometrically, but the plate in the plate theory of mechanics is represented mathematically only in two dimensions (the reason will be examined in the next chapter). Therefore, the geometry of a "mechanics" plate is a two-dimensional (2D) flat surface represented usually by the *neutral surface*. In FEM, *plate elements* are used in meshing the plate surfaces. A similar situation occurs in shells. A beam has also three dimensions geometrically. The beam in the beam theory of mechanics is represented mathematically only in one dimension. Therefore, the geometry of a "mechanics" beam is a one-dimensional (1D) straight or curved line. In FEM, *beam elements* have to be used to model the lines. A similar situation occurs in truss structures.

In MFree methods, beams, plates, and shells can all be represented using sets of arbitrarily distributed nodes. In the formulation of the MFree methods, corresponding theories used in the FEM must be used. The difference, again, lies mainly in the creation of the shape functions. This book presents these formulations in Chapter 5.

2.4 Node Generation

In FEM, meshing is performed to discretize the geometry created into small meshes called *elements* or *cells*, and many types of elements have been developed for different problems. The rationale behind domain discretization can be explained in a very rough and straightforward manner. We can expect that the solution for an engineering problem will be very complex, and will vary in a way that is usually unpredictable using functions defined globally across the whole problem domain. However, if the problem domain can be divided (*meshed*) into small elements using a set of *nodes* that are connected in a predefined manner using nodal lines, the solution within each element can be approximated very easily using simple functions such as polynomials, which are termed shape functions. The solutions for all the individual elements form the solution for the whole problem domain.

Mesh generation is a very important part of the *preprocess* in FEM, and it can be a very time-consuming task for the analyst. The domain has to be meshed properly into elements of specific shapes such as triangles and quadrilaterals. No overlapping and gaps are allowed. Information, such as the *element connectivity*, must also be created during the meshing for later simulation. It is ideal to have an entirely automated mesh generator; unfortunately, one is not available on the market. Semiautomatic preprocessors are available for most commercially available application software packages. There also exist packages designed mainly for meshing. Such packages can generate files of a mesh, which can be read by other modeling and simulation packages.

Triangulation is the most flexible way to create meshes of triangular elements. The process can be almost fully automated for 2D planes and even three-dimensional (3D) spaces. Therefore, it is used in most commercial preprocessors. The additional advantage of using triangles is the flexibility of modeling a complex geometry and its boundaries. The disadvantage is that the accuracy of the simulation results based on triangular elements is often much lower than that obtained using quadrilateral elements for the same density of nodes. Quadrilateral elements, however, are more difficult to generate automatically.

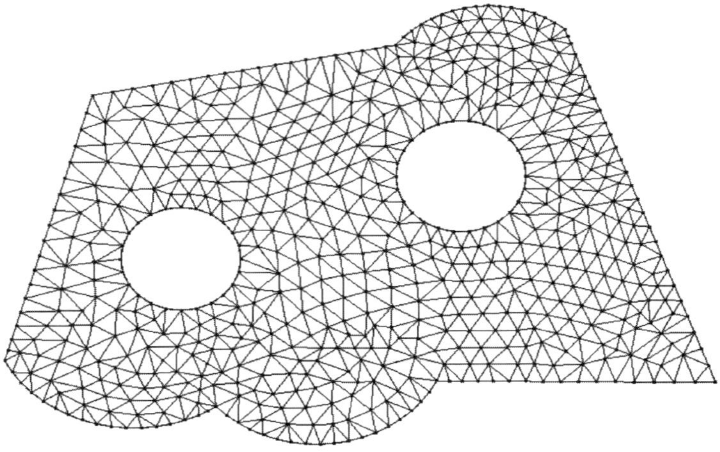

FIGURE 2.5
Mesh of triangular elements for a 2D problem domain.

An example of triangular meshes is shown in Figure 2.5, which is generated using MFree2D© (Liu, G. R. et al., 2000; Liu, G. R. and Tu, 2001).

In MFree methods, the problem domain is represented by a set of arbitrarily distributed nodes, as schematically illustrated in Figure 2.4. There is no need to use meshes or elements for field variable interpolation. Hence, there is no need to prescribe the relationship between the nodes. The nodes can be generated simply using triangulation algorithms that are routinely available for both 2D and 3D domains. The significance of MFree methods in terms of meshing is that the process of node generation can be fully automated without human intervention. The analysis can be performed in a fully adaptive manner, as in MFree2D. This can significantly save time for an analyst when creating a mesh for the problem domain.

All the MFree methods do not need a mesh of elements for field variable interpolation. However, some meshless methods require a background mesh of cells for integration of the system matrices, such as the element free Galerkin (EFG) methods (Belytschko et al., 1994b). Because the mesh is needed only for integration, any form of cells is acceptable as long as it provides sufficient accuracy in the integrations. The most convenient mesh to use is a mesh of triangular cells that can be generated automatically in MFree2D. There are MFree methods, such as the meshless local Petrov–Galerkin (MLPG) method, originally developed by Atluri and Zhu (1998), that require no mesh of elements for both field variable interpolation and background integration. These types of MFree methods are called essentially MFree methods, in that they require only a simple form of local mesh for integration of the system equations, which can be generated automatically with relative ease.

The procedure is often called *node generation*, which is usually performed using preprocessors. The preprocessor generates unique numbers for all the nodes for the solid or structure automatically. There are very few dedicated node generators available commercially; thus, we have to use preprocessors that have been developed for FEM. These processors are usually very sophisticated, and MFree node generation uses a very small portion of their capacity. All that is needed for MFree node generation is a small processor to generate triangular elements. We often use just the nodes of the triangular element mesh and discard the elements. In some MFree methods, we may also use the element mesh for background integration.

2.5 Shape Function Creation

In FEM, shape functions are created based on elements, and therefore, the computation of shape functions has been straightforward. In the early years of the development of FEM, much of the work involved the formulation of all different types of elements. All the shape functions of finite elements satisfy the Kronecker delta function property.

In MFree methods, however, the construction of shape functions has been and still is the central issue. This is because shape functions have to be computed with the use of predefined knowledge about the relationship of the nodes. This has posed the major challenge for MFree methods. The currently most widely used method for constructing MFree shape functions is the method of moving least squares (MLS) approximation. The application of MLS approximation has led to the development of many MFree methods and techniques. The major problem in MLS approximation is that the shape functions constructed do not possess the Kronecker delta function property.

The new promising method for constructing shape functions is the point interpolation method (PIM) originated by G. R. Liu and Gu (1999), because it produces shape functions that always have the Kronecker delta function property. Hence, PIM eliminates a number of difficult issues that have concerned many researchers in the area of MFree methods.

2.6 Property of Material or Media

In FEM, material properties can be defined for a group of elements or for each individual element if needed. For different phenomena to be simulated, different material properties are required. Inputting a material property into a preprocessor in both FEM and the MFree method is usually straightforward. All the analyst need do is key in the material property data and specify to which region of the geometry or which elements the data apply. Obtaining these properties, however, is not always simple. There are materials databases commercially available to choose from. Experiments are usually required to determine accurately the material properties to be used in the system. This is, however, beyond the scope of this text. In this book, we usually assume that the material property is known.

In MFree methods, material properties can be defined for subdomains of the problem. There are, however, some implementation issues related to the handling of the interfaces of different types of materials. There are challenging problems to be resolved, such as the method of field variable interpolation near the interface of different materials and the calculation of the stresses effectively and accurately near the interfaces without the use of a mesh. Some methods for dealing with interfaces of different materials are discussed in Chapter 6.

2.7 Boundary, Initial, and Loading Conditions

Boundary, initial, and loading conditions play a decisive role in the solution of a simulation. In FEM, inputting these conditions is not very difficult in most commercial preprocessors, and it is often interfaced with graphics. Users can specify these conditions either to the geometric identities (points, lines/curves, surfaces, and solids) or to the elements or grids. Again, to simulate this condition accurately for actual engineering systems requires experience, knowledge, and proper engineering judgment. There are standard procedures or techniques for the implementation of the boundary conditions, in the form of either single point or multipoint constraints. All the techniques developed in FEM are applicable (with some modification) to MFree methods.

In MFree methods using MLS approximations for constructing shape functions, special techniques are required to impose essential (displacement) boundary conditions, because the shape functions created do not satisfy the Kronecker delta conditions. MFree methods using shape functions created using PIM possess the Kronecker delta property. The imposition of the essential boundary conditions is the same as that in FEM.

In MFree methods, the shape functions are constructed concurrently with the process of assembling the global system equations. This provides an alternative way to implement essential boundary conditions by imposing these boundary conditions in the stage of shape function creation or stiffness matrix calculation. In this way, the final discretized system equation will not contain the degrees of freedom for the nodes with specified boundary conditions. Much work is still required in this direction of development.

2.8 Simulation

2.8.1 Discrete System Equations

Proper principles must be followed for *discretizing* the governing differential equations based on discretized domains. These principles differ from problem to problem. Both FEM and MFree methods use these principles; however, the procedure and preference of applying these principles can be different due to the different nature of these two methods.

In FEM, a set of discrete simultaneous system equations can be formulated using principles and shape functions created based on the element mesh generated. There are basically four principles used for establishing the simultaneous equations. The first is based on the principle of virtual work, such as Hamilton's principle, the minimum total potential energy principle, and so on. Traditional FEM is founded on these principles. The second is based on the residual methods, and is, in fact, a more general form of principle that can be used for deriving FEM equations both for solids and structures and for fluid flows, as long as the partial differential governing equations are provided. The third is based on the Taylor series, which has led to the formation of the traditional finite difference method (FDM). The fourth is based on the control of conservation laws on each finite volume (element) in the domain. The finite volume method (FVM) was established using this approach. The engineering practice so far shows that the first two principles are more *often* used for solid and structures, and the other two principles are more *often* used for fluid flow and heat transfer simulations. However, FEM has also been used to develop commercial packages for fluid flow and heat transfer problems, and FDM can be used for solids and structures.

It may be mentioned here without detailed illustration that the mathematical foundation of all these approaches is the *residual method*. A proper choice of the test and trial functions in the residual method can lead to FEM, FDM, or FVM formulation.

Many MFree methods can be formulated using the first three principles. In spatial discretization, the first two principles are used more often. Formulations based on the first two principles are termed *weak form*, and that based on the third principle is termed *strong form*. The discretized equation systems derived based on the weak form are more stable and can give much more accurate results. Therefore, this book mainly covers MFree methods of weak formulation. Some MFree methods favor a residual method of local form, especially in developing so-called truly meshless methods. This book covers methods using the local weak form in great detail. For the discretization of time, the third principle of Taylor series is often used. In summary, the use of the principles in MFree methods is very much the same as that in FEM.

2.8.2 Equation Solvers

A set of discretized system equations of a computation model is created and then fed to a *solver* to solve for the field variables. The process for FEM and the MFree methods is basically the same and places great demand on computer hardware. Different software packages use different algorithms according to the physical phenomenon to be simulated. There are two very important considerations when choosing algorithms for solving system equations. One is the storage required, and another is the central processing unit (CPU) time needed. Similar to FEM, MFree methods produce banded system matrices that can be handled in the same manner to reduce the storage and to maximize the efficiency of computation. In general, the bandwidth of these matrices produced by MFree methods is slightly larger than that of FEM. Techniques developed in FEM for reducing the bandwidth of the system matrices by optimizing the nodal arrangement are also applicable to MFree methods.

There are, in general, two categories of methods for solving simultaneous equations: direct methods and iterative methods. Often-used direct methods are the Gauss elimination method and the matrix decomposition method. These methods work well for relatively smaller systems. Direct methods operate on fully assembled system equations, and therefore demand larger storage. They can also be coded in such a way that the assembling is done only for those elements that are involved in the current stage of equation solving. This can significantly reduce the requirements for storage. All these techniques, which were developed for FEM, are applicable to MFree methods.

Iterative methods include the Gauss–Jacobi method, the Gauss–Seidel method, the successive overrelaxation method (SOR), generalized conjugate residual methods, the line relaxation method, and so on. These methods work well for relatively larger systems. Iterative methods are often coded in such a way to avoid full assembly of the system matrices to save significantly on storage. The performance in terms of the rate of convergence of these methods is usually very much problem dependent. In general, they perform better for large systems, especially for 3D problems.

For nonlinear problems, another iterative loop is needed. The nonlinear equation has to be properly formulated into a linear equation in the iteration. For time-dependent problems, time stepping is additionally required. There are generally two approaches to time stepping: implicit and explicit. Implicit approaches are usually more stable numerically but less efficient computationally than explicit approaches. Moreover, contact algorithms can be much more easily developed using explicit FEM. Again, these techniques developed in FEM are applicable to MFree methods with some modifications.

MFree methods usually demand additional CPU time, as the creation of the shape functions is more time-consuming and is performed during the computation. For meshless methods based on the MLS method, the less efficient use of CPU time is also caused by the extra effort needed in imposing essential boundary conditions. For meshless methods based on the local Petrov–Galerkin method, the lower efficiency is also attributed to the asymmetry of the stiffness matrix created. These issues are covered in great detail in relevant chapters.

Note that, today, people are more concerned about the time engineers spend on a project, and much less concerned about the CPU time, because the CPU time has become cheaper and cheaper, while the cost for well-trained engineers has become more and more expensive. Therefore, the small extra demand on CPU time does not disadvantage MFree methods significantly. In addition, the results obtained using MFree methods are usually more accurate than those obtained using FEM, as there are no stress discontinuity problems existing on the interfaces between the finite elements. In terms of the ratio of accuracy to CPU cost, MFree methods are in general superior to FEM (Belytschko et al., 1996b).

2.9 Visualization

The results generated by FEM and the MFree method after solving the system equation are usually in the form of a vast volume of digital data. The results have to be visualized in such a way that they can be easily interpolated, analyzed, and presented. The visualization is performed by the postprocessor that comes with the software package. Most of these processors allow users to display 3D objects in many convenient and colorful ways on the screen. The object can be displayed in the form of wire frames, collections of elements, and collections of nodes. The user can rotate, translate, and zoom in/out on the objects. Field variables can be plotted on the object in the form of contours, fringes, wire frames, and deformations. There are usually tools available for users to produce isosurfaces and vector fields of variables. Tools to enhance the visual effects are also available, such as shading, lighting, and shrinking. Animation and movies can also be produced to simulate dynamic aspects. Output in the form of tables, text files, x–y plots are also routinely available.

Advanced visualization tools, such as visual reality, are available today. These advanced tools allow users to display objects and results in a much more realistic 3D way. The platform can be a goggle, immersive desk, or even an immersive room. When the object is immersed in a room, analysts can walk through the object, go to the exact location, and view and analyze the results.

The visualization techniques used in FEM are applicable to visualizing the results obtained using MFree methods. However, major modification may be needed because there is no element-related information in MFree output files. There are also differences in the retrieval of the results from the nodal values. In FEM, there are ways to calculate strains and stresses due to the discontinuity of the stresses on the interfaces of the elements. In MFree methods, in contrast, one needs to recreate MFree shape functions for the point of interest to calculate the results required. The shape functions and the interpolating process for retrieving the results can be the same for creating the system equations in MFree methods. In FEM, however, they can be different.

Note that the MFree techniques may also be useful in developing visualization tools, where interpolation is often used.

2.10 MFree Method Procedure

Now that the role of MFree methods in simulating engineering systems and the differences between MFree methods and FEM in treating various issues have been described, we here summarize the general procedure for MFree methods in solving mechanics problems. We use mechanics problems of solids and structures as an example to describe these basic steps. Some important terminology frequently used in the MFree methods is defined.

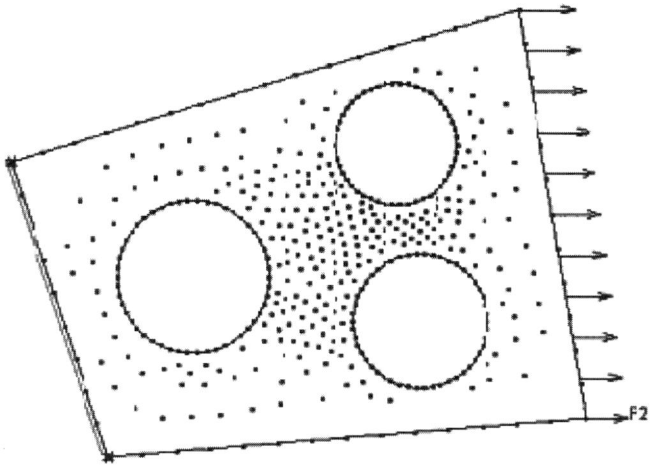

FIGURE 2.6
Example of an MFree model for 2D solids generated using MFree2D.

2.10.1 Basic Steps

STEP 1: DOMAIN REPRESENTATION The solid body of the structure is first modeled, and is represented using sets of nodes scattered in the problem domain and its boundary. Boundary conditions and loading conditions are then specified in the MFree model, as shown in Figure 2.6. The density of the nodes depends on the accuracy requirement of the analysis and the resources available. The nodal distribution is usually not uniform, and a denser distribution of nodes is often used in the area where the displacement gradient is larger. Because adoptive algorithms can be used in MFree methods, the density is eventually controlled automatically and adaptively in the code of the MFree methods. Therefore, we do not worry much about the distribution quality of the initial nodes used. In addition, as an MFree method, it should not demand too much for the pattern of nodal distribution. It should be workable within reason for arbitrarily distributed nodes.

Because the nodes will carry the value of the field variables in an MFree formulation, they are often called *field nodes*.

STEP 2: DISPLACEMENT INTERPOLATION Because a mesh of elements is not used in MFree methods, the field variable (say, a component of the displacement) u at any point at $\mathbf{x} = (x, y, z)$ within the problem domain is interpolated using the displacements at its nodes within the support domain of the point at $\mathbf{x}$, i.e.,

$$u(\mathbf{x}) = \sum_{i=1}^{n} \phi_i(\mathbf{x}) u_i = \mathbf{\Phi}(\mathbf{x}) \mathbf{U}_s \tag{2.1}$$

where n is the number of nodes included in a "small local domain" of the point at $\mathbf{x}$, u_i is the nodal field variable at the ith node in the small local domain, $\mathbf{U}_s$ is the vector that collects all the field variables at these nodes, and $\phi_i(\mathbf{x})$ is the shape function of the ith node determined using the nodes that are included in the small domain of $\mathbf{x}$. This small local domain is termed in this book as the *support domain* of $\mathbf{x}$. A support domain of a point $\mathbf{x}$ determines the number of nodes to be used to support or approximate the function value at $\mathbf{x}$. A support domain can be (but does not have to be) weighted using a weighted function, as shown in Figure 2.7. It can have different shapes and its dimension and shape can be different for different points of interest $\mathbf{x}$, as shown in Figure 2.8. The shapes most often used are circular or rectangular.

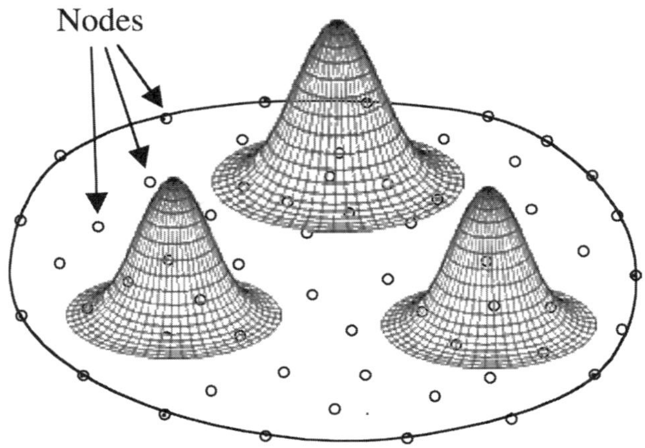

FIGURE 2.7
Domain representation of a 2D structure and nodes in a local weighted support domain.

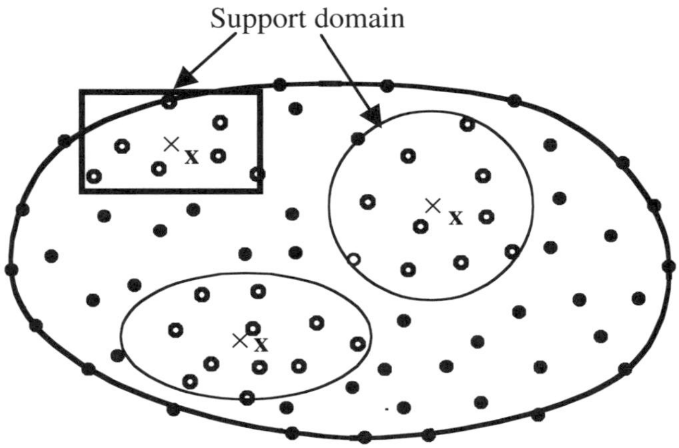

FIGURE 2.8
Support domain determines nodes (marked by **o**) that are used for approximation or interpolation of field variable at point **x**. A support domain can have different shapes and can be different from point to point. Most often used shapes are circular or rectangular.

The concept of support domain works well if the nodal density does not vary too drastically in the problem domain. However, in solving practical problems, such as problems with stress singularity, the nodal density can vary drastically. The use of a support domain based on the current point of interest can lead to an unbalanced selection of nodes for the construction of shape functions. In extreme situations, all the nodes used could be located on one side only, and the shape functions so constructed can result in serious error, due to extrapolation. To prevent this kind of problem, the concept of influence domain of a node should be used. MFree2D, introduced in Chapter 16, uses the approach of influence domain to select nodes for constructing shape functions. The concept of influence domain is explained in Section 2.10.4.

Here, we always use the concept of support domain to select the nodes for constructing shape functions, unless specifically noted otherwise, although we have different ways of selecting nodes for constructing shape functions. Section 2.10.2 presents a simple way to determine the dimension of the support domain.

Note also that this interpolation, defined in Equation 2.1, is performed for all the components of all the field variables in the same support domain. By taking a 3D solid mechanics problem as an example, the displacement is usually chosen as the field variable, and the displacement should have three components: displacements in the *x*, *y*, and *z* directions. The same shape function is used for all three displacement components in the support domain of the same point.

STEP 3: FORMATION OF SYSTEM EQUATIONS The discrete equations of an MFree method can be formulated using the shape functions and a strong or weak form system equation. These equations are often written in nodal matrix form and are assembled into the global system matrices for the entire problem domain.

The global system equations are a set of *algebraic equations* for static analysis, *eigenvalue equations* for free-vibration analysis, and *differential equations* with respect to time for general dynamic problems. The procedures for forming system equations are slightly different for different MFree methods. Hence, we discuss them in later chapters.

STEP 4: SOLVING THE GLOBAL MFree EQUATIONS Solving the set of global MFree equations, we obtain solutions for different types of problems.

1. **For static problems**, the displacements (or their parameters) at all the nodes in the entire problem domain are first obtained. The strain and stress in any element can then be retrieved. A standard linear algebraic equation solver, such as a Gauss elimination method, LU decomposition method, and iterative methods, can be used.
2. **For free-vibration and buckling problems**, eigenvalues and corresponding eigenvectors can be obtained using the standard eigenvalue equation solvers. The commonly used methods are the following:
   - Jacobi's method
   - Given's method and Householder's method
   - The bisection method (using Sturm sequences)
   - Inverse iteration
   - QR method
   - Subspace iteration
   - Lanczos' method
3. **For dynamics problems**, the time history of displacement, velocity, and acceleration are to be obtained. The following standard methods of solving dynamics equation systems can be used:
   - The modal superposition method may be a good choice for vibration types of problems and problems of far field response to low speed impact with many load cases.
   - For problems with a single load or few loads, the *direct integration method* can be used, which uses the FDM for time stepping with implicit and explicit approaches.
     - The implicit method is more efficient for relatively slow phenomena of vibration types of problems.
     - The explicit method is more efficient for very fast phenomena, such as impact and explosion.

   For computational fluid dynamics problems, the discretized system equations are basically nonlinear, and one needs an additional iteration loop to obtain the results.

2.10.2 Determination of the Dimension of a Support Domain

The accuracy of interpolation depends on the nodes in the support domain of the point of interest (which is often a quadrature point $\mathbf{x}_Q$ or the center of integration cells). Therefore, a suitable support domain should be chosen to ensure a proper area of coverage for interpolation. To define the support domain for a point $\mathbf{x}_Q$, the dimension of the support domain d_s is determined by

$$d_s = \alpha_s d_c \tag{2.2}$$

where α_s is the dimensionless size of the support domain and d_c is a characteristic length that relates to the nodal spacing near the point at $\mathbf{x}_Q$. If the nodes are uniformly distributed, d_c is simply the distance between two neighboring nodes. In the case where the nodes are non-uniformly distributed, d_c can be defined as an "average" nodal spacing in the support domain of $\mathbf{x}_Q$.

The physical meaning of the dimensionless size of the support domain α_s is very clear. It is simply the factor of the average nodal spacing. For example, $\alpha_s = 2.1$ means a support domain whose radius is 2.1 times the average nodal spacing. The actual number of nodes, n, can be determined by counting all the nodes in the support domain. The dimensionless size of the support domain α_s should be predetermined by the analyst, usually by carrying out numerical experiments for the same class of problems for which solutions already exist. Generally, an $\alpha_s = 2.0$ to 3.0 leads to good results.

Note that, if background cells are provided, support domains can also be defined based on the background cells.

2.10.3 Determination of the Average Nodal Spacing

For 1D cases, a simple method of defining an "average" nodal spacing is

$$d_c = \frac{D_s}{(n_{D_s} - 1)} \tag{2.3}$$

where D_s is an estimated d_s (the estimate does not have to be very accurate but should be known and a reasonably good estimate of d_s) and n_{D_s} is the number of nodes that are covered by a known domain with the dimension of D_s. By using Equation 2.3, it is very easy to determine the dimension of the support domain d_s for a point at x_Q in a domain with non-uniformly distributed nodes. The procedure is as follows:

1. Estimate d_s for the point at x_Q, which gives D_s.
2. Count nodes that are covered by D_s.
3. Use Equation 2.3 to calculate d_c.
4. Finally, calculate d_s using Equation 2.2 for a given (desired) dimensionless size of support domain α_s.

For 2D cases, a simple method of defining an "average" nodal spacing is

$$d_c = \frac{\sqrt{A_s}}{\sqrt{n_{A_s}} - 1} \tag{2.4}$$

where A_s is an estimated area that is covered by the support domain of dimension d_s (the estimate does not have to be very accurate but should be known and a reasonably good estimate), and n_{A_s} is the number of nodes that are covered by the estimated domain with the area of A_s. By using Equation 2.4 and the same procedure described for the 1D case, it is very easy to determine the dimension of the support domain d_s for a point at $\mathbf{x}_Q$ in a 2D domain with nonuniformly distributed nodes.

Similarly, for 3D cases, a simple method of defining an "average" nodal spacing is

$$d_c = \frac{\sqrt[3]{V_s}}{\sqrt[3]{n_{V_s}} - 1} \tag{2.5}$$

where V_s is an estimated volume that is covered by the support domain of dimension d_s, and n_{V_s} is the number of nodes that are covered by the estimated domain with the volume of V_s. By using Equation 2.5, and the same procedure described for the 1D case, we can determine the dimension of the support domain d_s for a point at $\mathbf{x}_Q$ in a 3D domain with non-uniformly distributed nodes.

2.10.4 Concept of the Influence Domain

Note that this book distinguishes between *support domain* and *influence domain*, terms that are often used in the MFree community to carry the same meaning as the support domain defined here. The influence domain in this book is defined as a domain that a node exerts an influence upon. It goes with a *node*, in contrast to the support domain, which goes with a *point* of interest **x** that can be, but does not necessarily have to be, at a node. The following explains in detail the concept of the influence domain.

Use of an influence domain is an alternative way to select nodes for interpolation, and it works well for domains with highly nonregularly distributed nodes. The influence domain is defined for each node in the problem domain, and it can be different from node to node to represent the area of influence of the node, as shown in Figure 2.9. Node 1 has

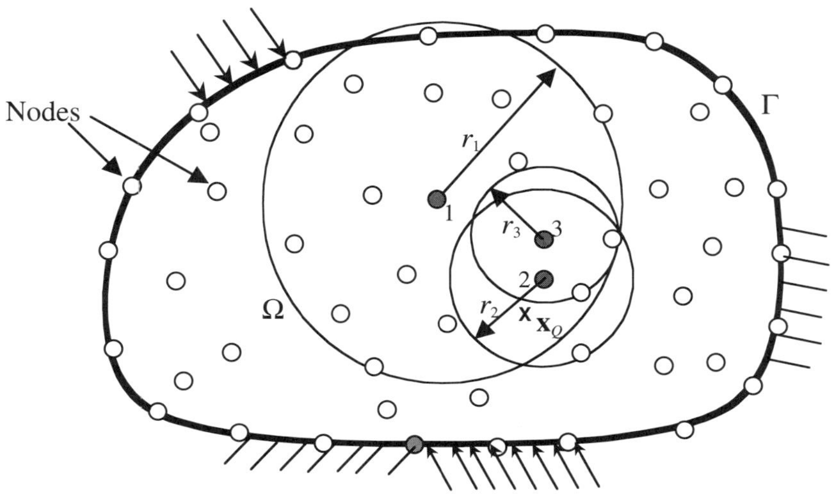

FIGURE 2.9
Influence domains of nodes. In constructing shape functions for point marked with **x** at point $\mathbf{x}_Q$, nodes whose influence domains covers **x** are to be used for construction of shape functions. For example, nodes 1 and 2 are included, but node 3 is not included.

an influence radius of r_1, and node 2 has an influence radius of r_2, etc. The node will be involved in the shape function construction for any point that is within its influence domain. For example, in constructing the shape functions for the point marked with **x** at point $\mathbf{x}_Q$ (see Figure 2.9), nodes 1 and 2 will be used, but node 3 will not be used. The fact that the dimension of the influence domain can be different from node to node allows some nodes to have further influence than others and prevents unbalanced nodal distribution for constructing shape functions. As shown in Figure 2.9, node 1 is included for constructing shape functions for the point marked with **x** at point $\mathbf{x}_Q$, but node 3 is not included, even though node 3 is closer to **x** compared with node 1.

The dimensions of the influence domain can be determined using a procedure similar to that described in Section 2.10.2. If background cells are provided, the influence domain can also be defined based on the background cells. MFree2D defines influence domains using information from triangular background cells.

2.10.5 Property of MFree Shape Functions

A compulsory condition that a shape function must satisfy is the *partition of unity*, that is,

$$\sum_{i=1}^{n} \phi_i(\mathbf{x}) = 1 \tag{2.6}$$

This is a necessary condition for the shape function to be able to produce any rigid motion of the problem domain.

There are also conditions that a shape function preferably satisfies. The first preferable condition is the *linear field reproduction* condition, that is,

$$\sum_{i=1}^{n} \phi_i(\mathbf{x}) x_i = \mathbf{x} \tag{2.7}$$

This condition is required for the shape function to pass the *standard patch test*, which has been used very often in testing finite elements. This condition is not compulsory because shape functions that fail to pass the patch test can still be used as long as a converged solution is produced. Many finite elements cannot pass the patch test but are widely used in FEM packages.

Another preferable condition is the *Kronecker delta function property*, that is,

$$\phi_i(\mathbf{x} = \mathbf{x}_j) = \begin{cases} 1 & i = j, \quad j = 1, 2, \ldots, n \\ 0 & i \neq j, \quad i, j = 1, 2, \ldots, n \end{cases} \tag{2.8}$$

This condition is preferred because a shape function that possesses this property permits use of a simple procedure to impose essential boundary conditions.

In element free methods, however, the shape functions created may or may not satisfy condition 2.8, depending on the method used for creating the shape functions. The methods of shape function creation are discussed in Chapter 5 in great detail, as they are the central issue for MFree methods.

2.11 Remarks

The similarities and differences between FEM and MFree methods are listed in Table 2.1.

TABLE 2.1

Differences between FEM and MFree Methods

	Items	FEM	MFree Method
1	Element mesh	Yes	No
2	Mesh creation and automation	Difficult due to the need for element connectivity	Relatively easy and no connectivity is required
3	Mesh automation and adaptive analysis	Difficult for 3D cases	Can always be done
4	Shape function creation	Element based	Node based
5	Shape function property	Satisfy Kronecker delta conditions; valid for all elements of the same type	May or may not satisfy Kronecker delta conditions depending on the method used; different from point to point
6	Discretized system stiffness matrix	Bonded, symmetrical	Bonded, may or may not be symmetrical depending on the method used
7	Imposition of essential boundary condition	Easy and standard	Special methods may be required; depends on the method used
8	Computation speed	Fast	1.1 to 50 times slower compared to the FEM depending on the method used
9	Retrieval of results	Special technique required	Standard routine
10	Accuracy	Accurate compared with FDM	Can be more accurate compared with FEM
11	Stage of development	Very well developed	Infancy, with many challenging problems
12	Commercial software package availability	Many	Very few and close to none

3

Mechanics of Solids and Structures

The basics of mechanics for solids and structural components are outlined in a concise and general manner in this chapter. Readers with experience in mechanics may skip the chapter, but it may prove helpful to skim the chapter to become familiar with the terminology conventions used in the book. Readers with less experience in mechanics may have difficulty fully understanding the details of all the concepts and equations presented here, but a global view of the basics of mechanics will help in the following chapters. We recommend such readers study this chapter without delving too deeply into the equations. Readers who are not interested in mechanics at all may simply skip this chapter, and read the chapters on MFree methods merely as a means to solve partial differential equations.

3.1 Basics

Solids and structures made of solids are stressed when they are subjected to *loads* or *forces*. The *stresses* are, in general, not uniform as the forces usually vary with coordinates. The stresses lead to *strains*, which can be observed as *deformation* or *displacement*. *Solid mechanics* and *structural mechanics* deal with the relationships between stresses and strains, displacements and forces, stresses (strains) and forces for given boundary conditions of solids and structures. These relationships are critical in modeling, simulating, and designing engineered structural systems.

Forces can be static and/or dynamic. *Statics* deals with mechanics of solids and structures subject to static loads. Solids and structures will experience vibration under the action of dynamic forces, which vary with time. In this case, the stress, strain, and displacement will be functions of time, and the principles and theories of *dynamics* apply. As statics is a special case of dynamics, the equations for statics can be derived by simply dropping out the dynamic terms in the dynamic equations. This book adopts the approach of deriving the dynamic equations first and obtaining the static equations directly from the derived dynamic equations.

Depending on the material property, solids can be *elastic*, that is, the deformation in the solids disappears fully if it is unloaded. There are also solids that are *plastic*, that is, the deformation in the solids cannot be fully recovered when it is unloaded. *Elasticity* deals with solids and structures of elastic materials and *plasticity* deals with those of plastic materials. This chapter is concerned mainly with the solids and structures of elastic materials. Methods to deal with plastic materials are treated as advanced topics and are covered in later relevant chapters.

Materials can be *anisotropic*, that is, the material property varies with direction. Deformation in an anisotropic material caused by a force applied in one direction may be different from that caused by the same force applied in another direction. Composite materials are often anisotropic. Many material constants have to be used to define the material properties of anisotropic materials. Many engineering materials are, however, *isotropic*, that is, their material properties are not direction dependent. Isotropic materials are a special case of anisotropic material. There are only two independent material constants for isotropic material, often known as Young's modulus and Poisson's ratio. This chapter is concerned mainly with solids and structures of isotropic materials. Topics related to solids and structures of anisotropic materials are covered in later chapters on a case-by-case basis.

Boundary conditions are another important consideration in mechanics. There are displacement (*essential*) and force (*natural*) boundary conditions for solids and structures. These terms of essential and natural boundary conditions are related to the use of weak forms to solve mechanics problems, and may be a little bit too difficult to understand at this stage. If we put it in a very simple way, the essential boundary conditions are the conditions that have to be satisfied by the trial functions before they are substituted into the weak form. It is *essential* because we have to consider them *during* the very first stage of constructing the trial functions. The natural boundary conditions are the conditions that will be *naturally* derived from the weak form procedure. In the stage of constructing the trial functions, one does not have to consider any of these natural conditions, because they will come out later naturally. For mechanics problems of solids and structures, the displacement boundary conditions are essential conditions, and the force boundary conditions are natural conditions.

Treatment of the essential boundary conditions is a very important topic in some MFree methods, as the essential boundary conditions may be satisfied when the trial function is constructed. Therefore, special treatments are required. We cover this in detail in Chapter 6. For now, we focus on the concepts of displacement, force, strain, and stress, as well as their relationships.

Structures are made of structural components, which are in turn made of solids. There are basically four types of structural components most often used: truss, beam, plate, and shell, as shown in Figure 3.1. The major purpose of using these structural components is to utilize material effectively to reduce the weight of the structure and the costs for building the structure. A practical structure can consist of different types of structural components including solid blocks. Theoretically, the principles and methodologies in solid mechanics can be applied to solve mechanics problems for all the structural components. However, it is often not efficient. Theories and formulations for taking geometric advantage of the structural components have, therefore, been developed. Formulations in MFree methods for trusses are very much the same as those in the finite element method (FEM), as a mesh is not necessary for trusses in FEM. All one needs is the geometric skeleton structure of the truss. The nodes for the FEM formulation are the same as the nodes of the skeleton structure. This book discusses in detail the formulation of MFree methods for beams, 2D solids, plates, and shells.

Mechanics for solids and structures is by itself a huge topic that needs more than a book to cover in detail. To provide a fairly clear and systematic discussion of this topic, we start with the formulation of 3D solids. Formulations for other types of structural components, beams, plates, shells, are then deduced from these general equations of 3D solids.

3.2 Equations for Three-Dimensional Solids

3.2.1 Stress and Strain

Consider a continuum of 3D elastic solids with a volume Ω and a surface boundary Γ, as shown in Figure 3.2. The surface of the solid is further divided into two types of surfaces. The surface on which the external forces are prescribed is denoted as Γ_t, and the surface

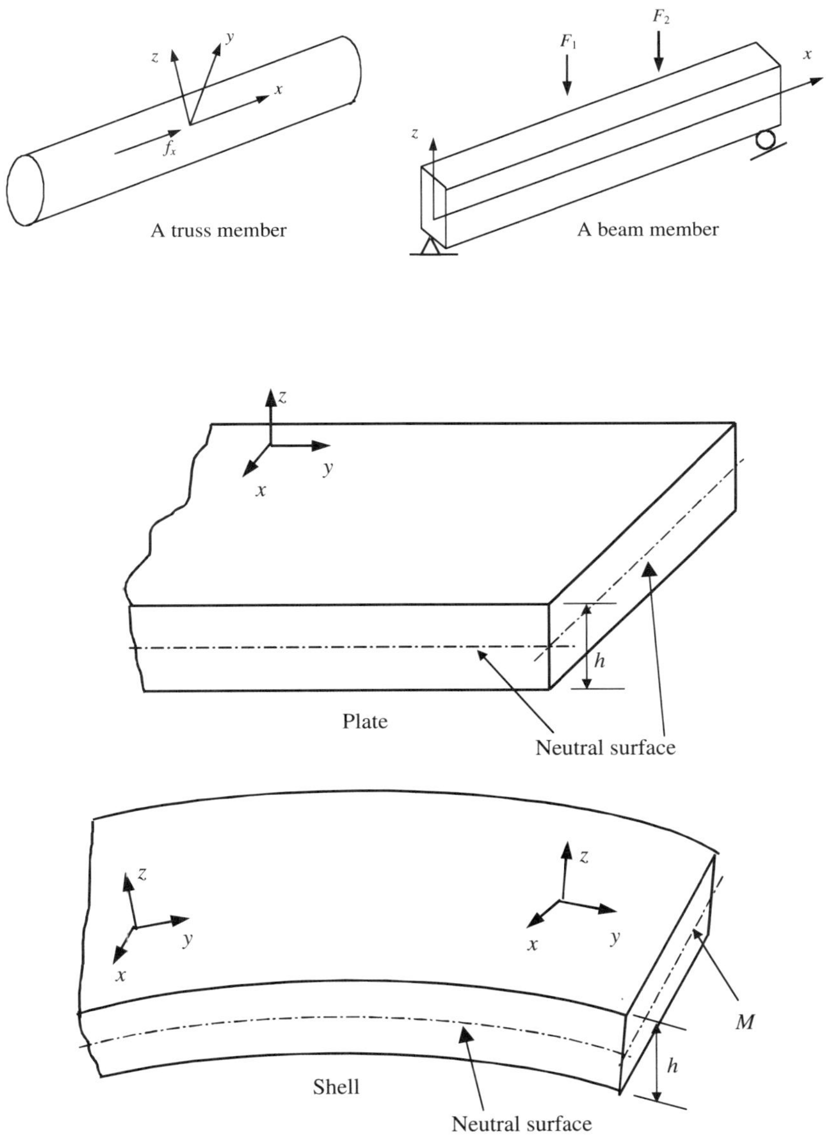

FIGURE 3.1
Types of structural components.

on which the displacements are prescribed is denoted Γ_u. The solid can also be loaded by body force **b** and surface traction (force) **t** in any distributed fashion in the volume of the solid.

At any point in the solid, there are, in general, six components of stress, as indicated on the surface of an "infinite" small cubicle shown in Figure 3.3. On each surface, there is a normal component of stress, and two components of shearing stress. The sign convention for the subscript is that the first letter represents the surface on which the stress is acting, and the second letter represents the direction of the stress. Note that there are also stresses acting on the other three hidden surfaces. As the normals of these surfaces are in the opposite direction to the corresponding coordinates, positive directions to the stresses

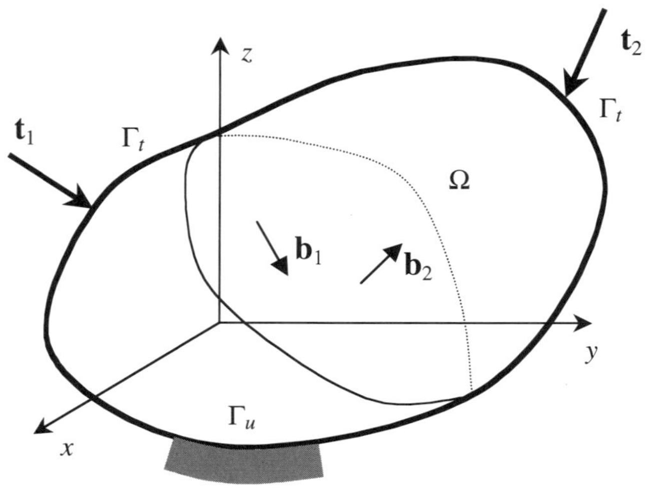

FIGURE 3.2
A continuum of solids subjected to forces.

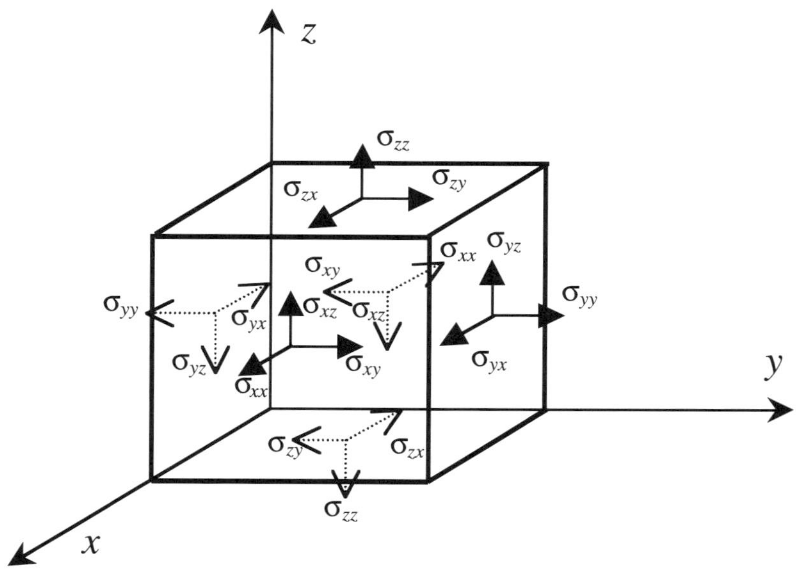

FIGURE 3.3
Stress components at a point in a solid.

should also be in the opposite direction of the coordinates. There are a total of nine stress components shown on the cubicle. By taking moments of force about the central axes of the cubicle at the state of equilibrium, it is easy to confirm that

$$\sigma_{xy} = \sigma_{yx}; \qquad \sigma_{xz} = \sigma_{zx}; \qquad \sigma_{zy} = \sigma_{yz} \tag{3.1}$$

Therefore, there are a total of six independent stress components at a point in solids. These stresses form a *stress tensor*. It is also often written in vector form:

$$\boldsymbol{\sigma}^T = \{\sigma_{xx} \;\; \sigma_{yy} \;\; \sigma_{zz} \;\; \sigma_{yz} \;\; \sigma_{xz} \;\; \sigma_{xy}\} \tag{3.2}$$

There are six corresponding strain components at any point in solids, which can also be written in a similar vector form:

$$\boldsymbol{\varepsilon}^T = \{\varepsilon_{xx} \;\; \varepsilon_{yy} \;\; \varepsilon_{zz} \;\; \varepsilon_{yz} \;\; \varepsilon_{xz} \;\; \varepsilon_{xy}\} \tag{3.3}$$

The strain is the rate of displacement per length, and therefore the components of strain can be obtained by derivatives of the displacements as follows:

$$\begin{aligned}
\varepsilon_{xx} &= \frac{\partial u}{\partial x} \\
\varepsilon_{yy} &= \frac{\partial v}{\partial y} \\
\varepsilon_{zz} &= \frac{\partial w}{\partial z} \\
\varepsilon_{xy} &= \frac{\partial u}{\partial y} + \frac{\partial v}{\partial x} \\
\varepsilon_{xz} &= \frac{\partial u}{\partial z} + \frac{\partial w}{\partial x} \\
\varepsilon_{yz} &= \frac{\partial v}{\partial z} + \frac{\partial w}{\partial y}
\end{aligned} \tag{3.4}$$

where u, v, w are the displacement components in the x, y, and z directions, respectively. The strain–displacement relation can be written in the following matrix form:

$$\boldsymbol{\varepsilon} = \mathbf{L}\mathbf{u} \tag{3.5}$$

where $\mathbf{u}$ is the displacement vector of the form:

$$\mathbf{u} = \begin{Bmatrix} u \\ v \\ w \end{Bmatrix} \tag{3.6}$$

and $\mathbf{L}$ is a differential operator matrix given by

$$\mathbf{L} = \begin{bmatrix} \partial/\partial x & 0 & 0 \\ 0 & \partial/\partial y & 0 \\ 0 & 0 & \partial/\partial z \\ 0 & \partial/\partial z & \partial/\partial y \\ \partial/\partial z & 0 & \partial/\partial x \\ \partial/\partial y & \partial/\partial x & 0 \end{bmatrix} \tag{3.7}$$

3.2.2 Constitutive Equations

The constitutive equation, which gives the relationship between the stress and strain in the material of a solid, is often termed Hooke's law. The generalized Hooke's law for general anisotropic materials can be given in the following matrix form:

$$\boldsymbol{\sigma} = \mathbf{c}\boldsymbol{\varepsilon} \tag{3.8}$$

where **c** is a matrix of material constants, which have to be obtained through experiments. The constitutive equation can written explicitly as

$$\begin{Bmatrix} \sigma_{xx} \\ \sigma_{yy} \\ \sigma_{zz} \\ \sigma_{yz} \\ \sigma_{xz} \\ \sigma_{xy} \end{Bmatrix} = \begin{bmatrix} c_{11} & c_{12} & c_{13} & c_{14} & c_{15} & c_{16} \\ & c_{22} & c_{23} & c_{24} & c_{25} & c_{26} \\ & & c_{33} & c_{34} & c_{35} & c_{36} \\ & & & c_{44} & c_{45} & c_{46} \\ & sy. & & & c_{55} & c_{56} \\ & & & & & c_{66} \end{bmatrix} \begin{Bmatrix} \varepsilon_{xx} \\ \varepsilon_{yy} \\ \varepsilon_{zz} \\ \varepsilon_{yz} \\ \varepsilon_{xz} \\ \varepsilon_{xy} \end{Bmatrix} \tag{3.9}$$

Note that the $c_{ij} = c_{ji}$. Thus, there are 21 independent material constants c_{ij}. For isotropic materials, **c** can be gradually reduced to

$$\mathbf{c} = \begin{bmatrix} c_{11} & c_{12} & c_{12} & 0 & 0 & 0 \\ & c_{11} & c_{12} & 0 & 0 & 0 \\ & & c_{11} & 0 & 0 & 0 \\ & & & \frac{c_{11}-c_{12}}{2} & 0 & 0 \\ & sy. & & & \frac{c_{11}-c_{12}}{2} & 0 \\ & & & & & \frac{c_{11}-c_{12}}{2} \end{bmatrix} \tag{3.10}$$

where

$$c_{11} = \frac{E(1-\nu)}{(1-2\nu)(1+\nu)}; \qquad c_{12} = \frac{E\nu}{(1-2\nu)(1+\nu)}; \qquad \frac{c_{11}-c_{12}}{2} = G \tag{3.11}$$

in which E, ν, and G are Young's modulus, Poisson's ratio, and shear modulus of the material. There are only two independent constants among these three constants. The relationship between these three constants is

$$G = \frac{E}{2(1+\nu)} \tag{3.12}$$

Given any two of these three constants, the other can then be calculated using the above equation.

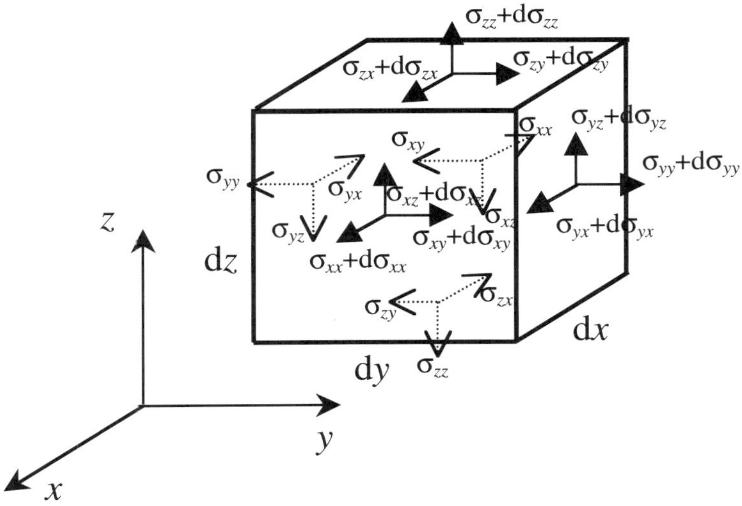

FIGURE 3.4
Stresses on an infinitesimal cube.

3.2.3 Dynamic Equilibrium Equations

Figure 3.4 shows an infinite small block in a solid. The equilibrium of forces on the block in the x direction gives

$$\begin{aligned}&(\sigma_{xx} + \mathrm{d}\sigma_{xx})\mathrm{d}y\,\mathrm{d}z - \sigma_{xx}\mathrm{d}y\,\mathrm{d}z + (\sigma_{yx} + \mathrm{d}\sigma_{yx})\mathrm{d}x\,\mathrm{d}z - \sigma_{yx}\mathrm{d}x\,\mathrm{d}z \\ &\quad + (\sigma_{zx} + \mathrm{d}\sigma_{zx})\mathrm{d}x\,\mathrm{d}y - \sigma_{zx}\mathrm{d}x\,\mathrm{d}y + b_x\mathrm{d}x\,\mathrm{d}y\,\mathrm{d}z = \rho\ddot{u}\,\mathrm{d}x\,\mathrm{d}y\,\mathrm{d}z \end{aligned} \tag{3.13}$$

where the term on the right-hand side is the initial force term, and b_x is the component in the x direction of the external body force applied at the center of the small block. Note that

$$\mathrm{d}\sigma_{xx} = \frac{\partial\sigma_{xx}}{\partial x}\mathrm{d}x, \qquad \mathrm{d}\sigma_{yx} = \frac{\partial\sigma_{yx}}{\partial y}\mathrm{d}y, \qquad \mathrm{d}\sigma_{zx} = \frac{\partial\sigma_{zx}}{\partial z}\mathrm{d}z \tag{3.14}$$

Equation 3.13 becomes one of the equilibrium equations:

$$\frac{\partial\sigma_{xx}}{\partial x} + \frac{\partial\sigma_{yx}}{\partial y} + \frac{\partial\sigma_{zx}}{\partial z} + b_x = \rho\ddot{u} \tag{3.15}$$

Similarly, the equilibrium of forces in the y and z directions results in the other two equilibrium equations:

$$\frac{\partial\sigma_{xy}}{\partial x} + \frac{\partial\sigma_{yy}}{\partial y} + \frac{\partial\sigma_{zy}}{\partial z} + b_y = \rho\ddot{v} \tag{3.16}$$

$$\frac{\partial\sigma_{xz}}{\partial x} + \frac{\partial\sigma_{yz}}{\partial y} + \frac{\partial\sigma_{zz}}{\partial z} + b_z = \rho\ddot{w} \tag{3.17}$$

The equilibrium equations, Equations 3.15 to 3.17, can be written in a concise matrix form:

$$\mathbf{L}^T\boldsymbol{\sigma} + \mathbf{b} = \rho\ddot{\mathbf{u}} \tag{3.18}$$

where **b** is the vector of external body forces in the x, y, and z directions:

$$\mathbf{b} = \begin{Bmatrix} b_x \\ b_y \\ b_z \end{Bmatrix} \tag{3.19}$$

By using Equations 3.5 and 3.8, the equilibrium equation 3.18 can be further written in terms of displacement:

$$\mathbf{L}^T\mathbf{c}\mathbf{L}\mathbf{u} + \mathbf{b} = \rho\ddot{\mathbf{u}} \tag{3.20}$$

This is the general form of dynamic equilibrium equation for elastic solids. If the loads applied on the solid are static, the concern is then only the static status of the solid, and the static equilibrium equation can be obtained simply by dropping the dynamic terms in Equation 3.20.

$$\mathbf{L}^T\mathbf{c}\mathbf{L}\mathbf{u} + \mathbf{b} = 0 \tag{3.21}$$

Equations obtained in this section are applicable to 3D elastic solids. Theoretically, these equations can be applied to all other types of structures, such as trusses, beams, plates, and shells, because they are all made of 3D solids. However, treating all the structural components as 3D solids makes computation very expensive, and practically impossible. Therefore, theories for taking geometric advantage of different types of solids and structural components have been developed. Application of these theories in a proper manner can reduce analytical and computational effort drastically. A brief description about these theories is given in the following sections.

Equations 3.15 through 3.17 are also often written in the following form using a recursive indicator:

$$\sigma_{ij,j} + b_i = \rho\ddot{u}_i \tag{3.22}$$

where $i, j = 1, 2, 3$, represent, respectively, x, y, and z, and $u_1 = u$, $u_2 = v$, $u_3 = w$. When $i = 1, 2, 3$, Equation 3.22 becomes, respectively, Equations 3.15 through 3.17. Equation 3.22 is convenient for deriving weak form system equations using weighted residual methods.

3.3 Equations for Two-Dimensional Solids

3.3.1 Stress and Strain

For 2D solids, let us assume, for example, that all the dependent variables are independent of the z axis, and all the external loads are independent of the z coordinate, and are applied only in the x–y plane. There are two types of 2D solids. One is a *plane stress* solid, and the other is a *plane strain* solid. Plane stress solids are solids whose thickness in the z direction

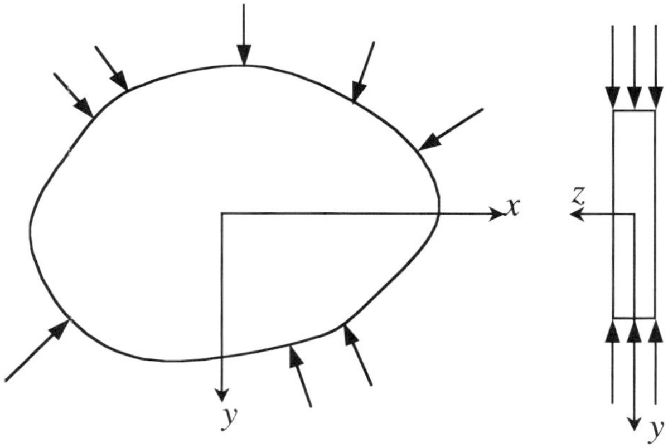

FIGURE 3.5
A plane stress problem.

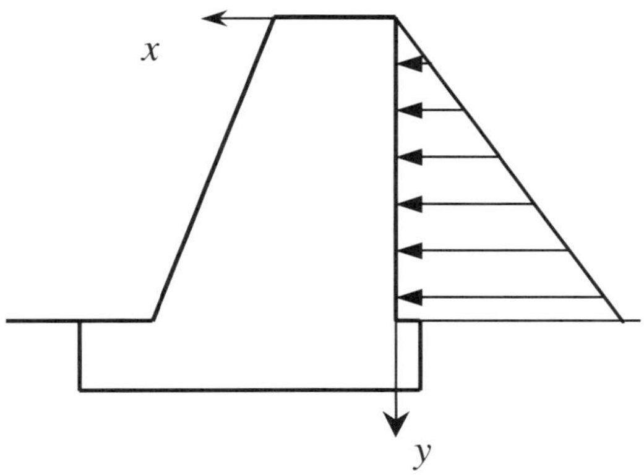

FIGURE 3.6
A plane strain problem.

is very small compared with dimensions in the x and y directions. As external forces are applied only in the x–y plane, and stresses in the z direction (σ_{zz}, σ_{xz}, σ_{yz}) are all zero, as shown in Figure 3.5. We have, therefore, only three in-plane stresses (σ_{xx}, σ_{yy}, σ_{xy}) to deal with.

Plane strain solids are those solids whose thickness in the z direction is very large compared with their dimensions in the x and y directions. As external forces are applied evenly along the z axis, the movement in the z direction at any point is constrained. The strain components in the z direction (ε_{zz}, ε_{xz}, ε_{yz}) are, therefore, all zero, as shown in Figure 3.6. We have only three in-plane stresses (ε_{xx}, ε_{yy}, ε_{xy}) to handle.

The system equations for 2D solids can be obtained by simply omitting terms related to the z direction in the system equations for 3D solids. Equations for isotropic materials are given as follows.

The stress components are

$$\boldsymbol{\sigma} = \begin{Bmatrix} \sigma_{xx} \\ \sigma_{yy} \\ \sigma_{xy} \end{Bmatrix} \tag{3.23}$$

There are three corresponding strain components at any point in 2D solids, which can also be written in a similar vector form:

$$\boldsymbol{\varepsilon} = \begin{Bmatrix} \varepsilon_{xx} \\ \varepsilon_{yy} \\ \varepsilon_{xy} \end{Bmatrix} \tag{3.24}$$

The strain–displacement is

$$\begin{aligned} \varepsilon_{xx} &= \frac{\partial u}{\partial x} \\ \varepsilon_{yy} &= \frac{\partial v}{\partial y} \\ \varepsilon_{xy} &= \frac{\partial u}{\partial y} + \frac{\partial v}{\partial x} \end{aligned} \tag{3.25}$$

where u, v are the displacement components in the x, y directions, respectively. The strain–displacement relation can also be written in the following matrix form:

$$\boldsymbol{\varepsilon} = \mathbf{L}\mathbf{u} \tag{3.26}$$

where the displacement vector has the form of

$$\mathbf{u} = \begin{Bmatrix} u \\ v \end{Bmatrix} \tag{3.27}$$

and differential operator matrix is given by

$$\mathbf{L} = \begin{bmatrix} \frac{\partial}{\partial x} & 0 \\ 0 & \frac{\partial}{\partial y} \\ \frac{\partial}{\partial y} & \frac{\partial}{\partial x} \end{bmatrix} \tag{3.28}$$

3.3.2 Constitutive Equations

Hooke's law for 2D solids has the following matrix form:

$$\boldsymbol{\sigma} = \mathbf{c}\boldsymbol{\varepsilon} \tag{3.29}$$

where $\mathbf{c}$ is a matrix of material constants, which have to be obtained through experiments. For isotropic materials, we have

$$\mathbf{c} = \frac{E}{1-\nu^2}\begin{bmatrix} 1 & \nu & 0 \\ \nu & 1 & 0 \\ 0 & 0 & (1-\nu)/2 \end{bmatrix} \quad \text{(Plane stress)} \tag{3.30}$$

For plane strain problems, the matrix of material constants **c** can be obtained by simply replacing E and ν, respectively, with $E/(1-\nu^2)$ and $\nu/(1-\nu)$, which leads to

$$\mathbf{c} = \frac{E(1-\nu)}{(1+\nu)(1-2\nu)}\begin{bmatrix} 1 & \frac{\nu}{1-\nu} & 0 \\ \frac{\nu}{1-\nu} & 1 & 0 \\ 0 & 0 & \frac{1-2\nu}{2(1-\nu)} \end{bmatrix} \quad \text{(Plane strain)} \tag{3.31}$$

3.3.3 Dynamic Equilibrium Equations

The dynamic equilibrium equations for 2D solids can be easily obtained by removing the terms related to the z coordinate from Equations 3.15 and 3.16.

$$\frac{\partial \sigma_{xx}}{\partial x} + \frac{\partial \sigma_{yx}}{\partial y} + b_x = \rho \ddot{u} \tag{3.32}$$

$$\frac{\partial \sigma_{xy}}{\partial x} + \frac{\partial \sigma_{yy}}{\partial y} + b_y = \rho \ddot{v} \tag{3.33}$$

These equilibrium equations can be written in the same concise matrix form:

$$\mathbf{L}^T\boldsymbol{\sigma} + \mathbf{b} = \rho\ddot{\mathbf{u}} \tag{3.34}$$

where **b** is the external force vector given by

$$\mathbf{b} = \begin{Bmatrix} b_x \\ b_y \end{Bmatrix} \tag{3.35}$$

Equation 3.34 has exactly the same form as Equation 3.18. For static problems, the equilibrium equations can be written as

$$\mathbf{L}^T\boldsymbol{\sigma} + \mathbf{b} = 0 \tag{3.36}$$

Equations 3.34 and 3.36 are much easier to solve compared with their counterpart equations for 3D solids.

3.4 Equations for Truss Members

A truss member is a solid whose dimension in one direction is much larger than in the other two directions, as shown in Figure 3.7. The force is applied only in the x direction. Therefore, a truss member is actually a 1D solid. The equations for 1D solids can be obtained by further omitting the stress related to the y direction, σ_{yy}, σ_{xy} from the 2D plane stress case.

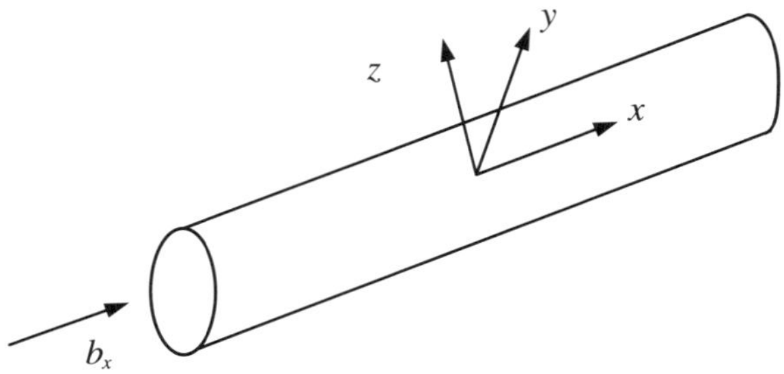

FIGURE 3.7
A truss member.

3.4.1 Stress and Strain

The stress in a truss member is σ_{xx}, which is often simplified as σ_x. The corresponding strain in a truss member is ε_{xx}, which is often simplified as ε_x. The strain–displacement relationship is given simply by

$$\varepsilon_x = \frac{\partial u}{\partial x} \tag{3.37}$$

3.4.2 Constitutive Equations

Hooke's law for 1D solids has the following simple form:

$$\sigma = E\varepsilon \tag{3.38}$$

This is actually the original Hooke's law. The Young's modulus E can be obtained using a simple tensile test.

3.4.3 Dynamic Equilibrium Equations

The dynamic equilibrium equation for 1D solids is

$$\frac{\partial \sigma_x}{\partial x} + b_x = \rho \ddot{u} \tag{3.39}$$

The static equilibrium equation for 1D solids is

$$\frac{\partial \sigma_x}{\partial x} + b_x = 0 \tag{3.40}$$

3.5 Equations for Beams

A beam possesses dimensional characteristics similar to a truss member, as shown in Figure 3.8. The difference is that the forces applied on a beam are transversal, that is, the direction of the force is perpendicular to the axis of the beam. Therefore, a beam experiences bending resulting in deflection in the y direction, which is a function of x.

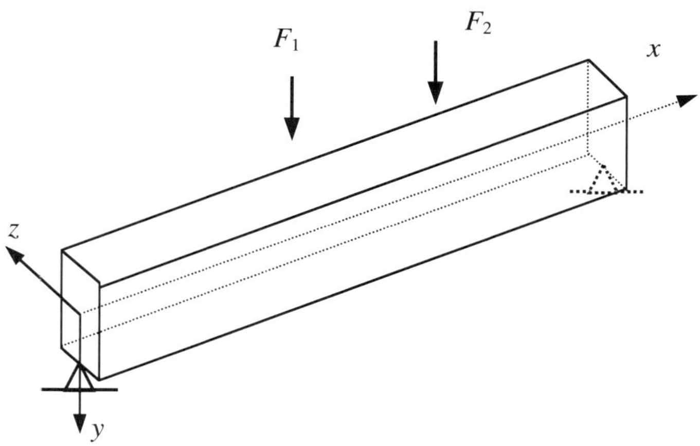

FIGURE 3.8
A simply supported beam.

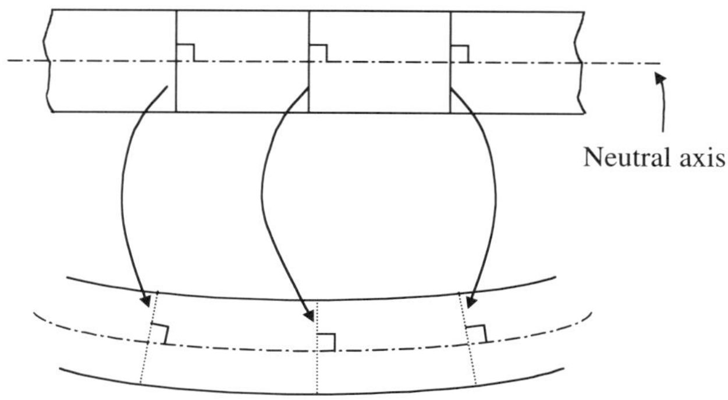

FIGURE 3.9
Euler–Bernoulli assumption.

3.5.1 Stress and Strain

The stresses on the cross section of a beam are normal stress σ_{xx} and shear σ_{xy}. There are several theories for analyzing beam deflections. These theories can be basically divided into two major categories: theory for *thin* beams and theory for *thick* beams. This book deals with both beams. The thin beam theory is often called the *Euler–Bernoulli* beam theory. The Euler–Bernoulli beam theory assumes that plane cross sections, which are normal to the undeformed neutral axis, remain plane after bending and remain normal to the deformed axis, as shown in Figure 3.9. With this "normal stays normal" assumption, one can first have

$$\varepsilon_{xy} = 0 \tag{3.41}$$

Second, the axial displacement, u, at a distance y from the neutral axis can be expressed by

$$u = y\theta \tag{3.42}$$

where θ is the rotation in the x–y plane. The rotation can be obtained from the deflection of the neutral axis of the beam, v, in the y direction.

$$\theta = \frac{\partial v}{\partial x} \tag{3.43}$$

The relationship between the normal strain and the deflection can be given by

$$\varepsilon_{xx} = \frac{\partial u}{\partial x} = y\frac{\partial^2 v}{\partial x^2} = Lv \tag{3.44}$$

where L is the differential operator given by

$$L = y\frac{\partial^2}{\partial x^2} \tag{3.45}$$

3.5.2 Constitutive Equations

The original Hooke's law is applicable to beams:

$$\sigma_{xx} = E\varepsilon_{xx} \tag{3.46}$$

3.5.3 Moments and Shear Forces

Figure 3.10 shows a small representative cell of length dx of the beam. The beam cell is subjected to external force b_y, moment M_z, shear force V, and inertial force $\rho A\ddot{v}$, where ρ is the density of the materials and A is the area of the cross section. The moment on the cross section at x results from the distributed normal stress σ_{xx}, as shown in Figure 3.11. The normal stress can be calculated by substituting Equation 3.44 into Equation 3.46:

$$\sigma_{xx} = ELv \tag{3.47}$$

It is seen from the above equation that the normal stress σ_{xx} varies linearly in the vertical direction on the cross section of the beam. The moments resulting from the normal stress

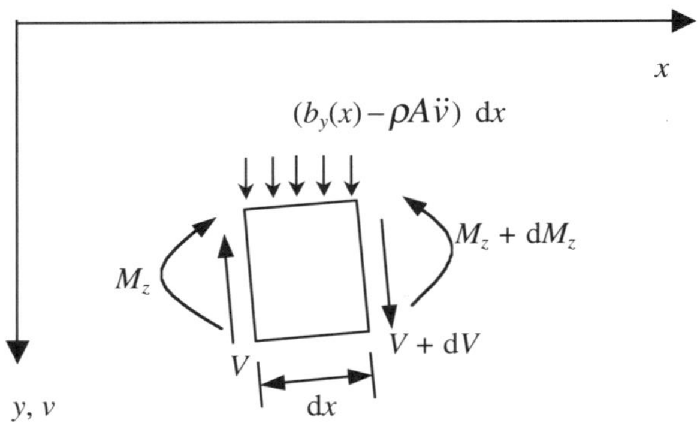

FIGURE 3.10
Isolated beam cell of length dx.

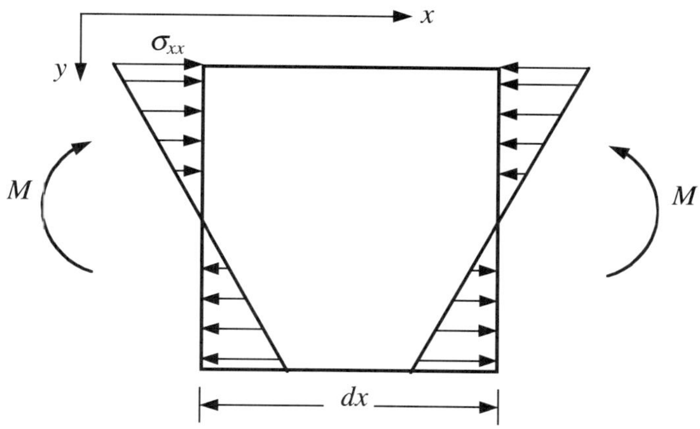

FIGURE 3.11
Stress and moment.

on the cross section can be calculated by the following integration over the area of the cross section.

$$M_z = \int_A -\sigma_{xx} y \, \mathrm{d}A = -E\left(\int_A y^2 \mathrm{d}A\right) Lv = EI_z Lv = -EI_z \frac{\partial^2 v}{\partial x^2} \tag{3.48}$$

where I_z is the second moment of area (or moment of inertia) of the cross section with respect to the z axis, which can be calculated for a given shape of the cross section using the following equation:

$$I_z = \int_A y^2 \mathrm{d}A \tag{3.49}$$

Considering the force equilibrium of the small beam cell in the y direction, we have

$$\mathrm{d}V + (b_y(x) - \rho A\ddot{v})\mathrm{d}x = 0 \tag{3.50}$$

or

$$\frac{\mathrm{d}V}{\mathrm{d}x} = b_y(x) - \rho A\ddot{v} \tag{3.51}$$

Considering the moment equilibrium of the small beam cell with respect to any point at the right surface of the cell, we have

$$\mathrm{d}M_z - V\mathrm{d}x + \frac{1}{2}(b_y - \rho A\ddot{v})(\mathrm{d}x)^2 = 0 \tag{3.52}$$

Neglecting the second-order small term leads to

$$\frac{\mathrm{d}M_z}{\mathrm{d}x} = V \tag{3.53}$$

Substituting Equation 3.48 into Equation 3.53 gives

$$V = -EI_z \frac{\partial^3 v}{\partial x^3} \tag{3.54}$$

3.5.4 Dynamic Equilibrium Equations

The dynamic equilibrium equation for beams can be obtained simply by substituting Equation 3.54 into Equation 3.51, which gives

$$EI_z \frac{\partial^4 v}{\partial x^4} + \rho A \ddot{v} = b_y \tag{3.55}$$

The static equilibrium equation for beams can be obtained simply by dropping the dynamic term in Equation 3.55:

$$EI_z \frac{\partial^4 v}{\partial x^4} = b_y \tag{3.56}$$

3.5.5 Equations for Thick Beams

When the beam becomes thicker or deeper in relation to its span, the transverse shear force affects the deformation of the beam. The shear force forces the cross section of the beam to deform and renders the "normal to normal" assumption invalid. Hence, the Euler–Bernoulli beam theory can no longer be used. Instead, we need to use the Timoshenko beam theory. The Timoshenko beam theory accounts for transverse shear deformation. The general equation of static deformation of the Timoshenko beam theory is written as

$$\begin{aligned} \frac{\partial}{\partial x}\left[GA\kappa\left(\frac{\partial v}{\partial x} + \theta\right)\right] + b_y &= 0 \\ -\frac{\partial}{\partial x}\left(EI\frac{\partial \theta}{\partial x}\right) + GA\kappa\left(\frac{\partial v}{\partial x} + \theta\right) &= 0 \end{aligned} \tag{3.57}$$

where v is the deflection of the beam, A is the cross-sectional area, G is the shear modulus, and κ is the shear correction coefficient.

Note that there are two variables in the Timoshenko beam theory: deflection v and rotation θ. The relation in Equation 3.43 is no longer valid. The shear strain is calculated by

$$\varepsilon_{xy} = \frac{\partial v}{\partial x} + \theta \tag{3.58}$$

and the shear stress is given by

$$\sigma_{xy} = \kappa G \varepsilon_{xy} \tag{3.59}$$

3.6 Equations for Plates

A plate possesses a dimensional characteristic geometrically similar to a 2D solid, as shown in Figure 3.12. The difference is that the forces applied on a plate are in the direction perpendicular to the plane of the plate. A plate can also be viewed as a 2D analog of a beam. Therefore, a plate experiences bending resulting in deflection w in the z direction, which is a function of x and y. The stress σ_{zz} in a plate is assumed to be zero.

Similar to beams, there are several theories for analyzing deflection in plates. These theories can also be basically divided into two major categories: theory for *thin* plates and theory for *thick* plates. This book addresses the following:

- Thin plate theory, often called *classic* plate theory (CPT) and also known as *Kirchhoff* plate theory
- The *first-order shear deformation theory* (FSDT) known as *Mindlin* plate theory
- The *third-order shear deformation theory* (TSDT)

3.6.1 Thin Plates

Stress and Strain

CPT assumes that normals to the neutral surface of the undeformed plate remain straight and normal to the neutral surface during deformation or bending. This assumption is often called the Kirchhoff assumption. The Kirchhoff assumption results in, first,

$$\varepsilon_{xz} = 0, \qquad \varepsilon_{yz} = 0 \tag{3.60}$$

Therefore, we need only worry about three strains—ε_{xx}, ε_{yy}, and ε_{xy}—and they are all in the plane of the plate. Therefore, it can share the same constitutive equation of 2D solids

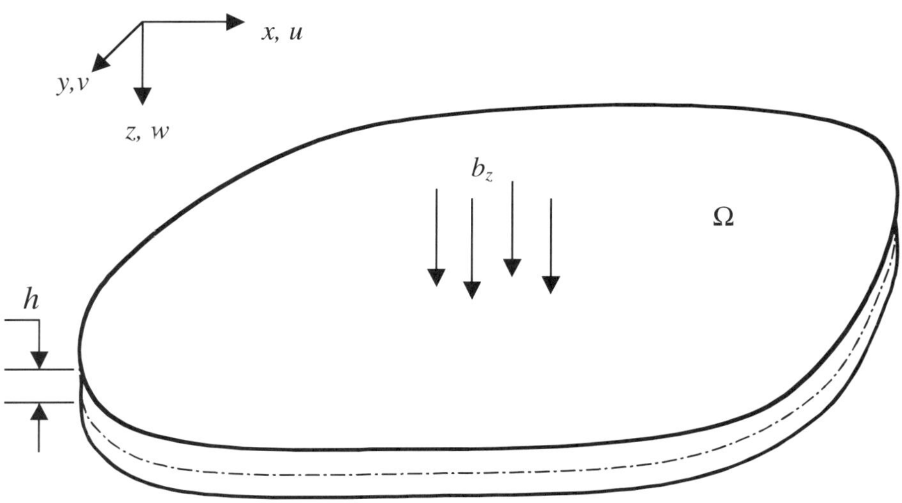

FIGURE 3.12
Plate subjected to transverse load.

of plane stress, Equation 3.29. Second, the displacements parallel to the undeformed middle surface, u and v, at a distance z from the neutral surface can be expressed by

$$u = -z\frac{\partial w}{\partial x} \tag{3.61}$$

$$v = -z\frac{\partial w}{\partial y} \tag{3.62}$$

where w is the deflection of the middle plane of the plate in the z direction. Using Equations 3.61 and 3.62, we obtain a simple relationship:

$$\mathbf{u} = \begin{Bmatrix} u \\ v \\ w \end{Bmatrix} = \begin{Bmatrix} -z\frac{\partial}{\partial x} & -z\frac{\partial}{\partial y} & 1 \end{Bmatrix}^T w = \mathbf{L}_u w \tag{3.63}$$

It is clearly shown here that all three displacement components are expressed in terms of the deflection w, due to the Kirchhoff assumption. The relationship between the three components of strain and the deflection can be given by

$$\varepsilon_{xx} = \frac{\partial u}{\partial x} = -z\frac{\partial^2 w}{\partial x^2} \tag{3.64}$$

$$\varepsilon_{yy} = \frac{\partial v}{\partial y} = -z\frac{\partial^2 w}{\partial y^2} \tag{3.65}$$

$$\varepsilon_{xy} = \frac{\partial u}{\partial y} + \frac{\partial v}{\partial x} = -2z\frac{\partial^2 w}{\partial x \partial y} \tag{3.66}$$

or in matrix form:

$$\boldsymbol{\varepsilon} = z\mathbf{L}w \tag{3.67}$$

where $\boldsymbol{\varepsilon}$ is the vector of in-plane strains defined by Equation 3.24, and $\mathbf{L}$ is the differential operator matrix given by

$$\mathbf{L} = \begin{bmatrix} -\frac{\partial^2}{\partial x^2} \\ -\frac{\partial^2}{\partial y^2} \\ -2\frac{\partial^2}{\partial x \partial y} \end{bmatrix} \quad \text{for CPT} \tag{3.68}$$

We define pseudo-strain as

$$\boldsymbol{\varepsilon}_p = \mathbf{L}w \tag{3.69}$$

that is, independent of the coordinate z.

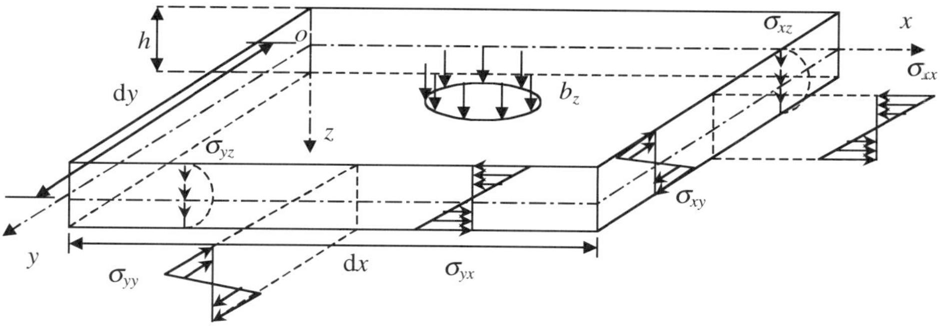

FIGURE 3.13
Stresses on the cross sections of an isolated plate cell.

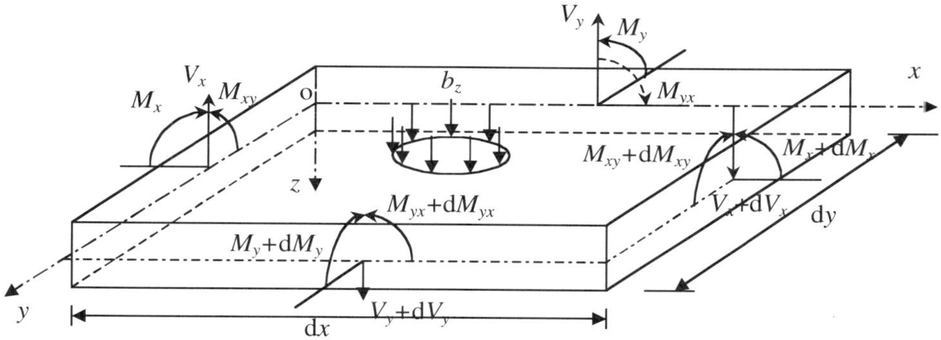

FIGURE 3.14
Shear forces and moments in an isolated plate cell of dx × dy.

Moments and Shear Forces

Figure 3.13 shows the stresses on the cross sections of a small representative cell of dx × dy from a plate of thickness h. The plate cell is subjected to external force b_z, and inertial force $\rho h\ddot{w}$, where ρ is the density of the materials. Figure 3.14 shows the moments M_x, M_y, M_z, and M_{xy}, and shear forces V_x and V_y. The moments and shear forces result from the distributed in-plane (normal and shear) stresses σ_{xx}, σ_{yy}, and σ_{xy} shown in Figure 3.14. The stresses can be obtained by substituting Equation 3.67 into Equation 3.29:

$$\boldsymbol{\sigma} = z\,\mathbf{c}\,\mathbf{L}w \tag{3.70}$$

It is seen from the above equation that the in-plane stresses vary linearly in the vertical direction on the cross sections of the plate. The moments on the cross section can be calculated by carrying out the following integration:

$$\begin{Bmatrix} M_x \\ M_y \\ M_{xy} \end{Bmatrix} = \int_{-h/2}^{h/2} \boldsymbol{\sigma} z\,\mathrm{d}z = \left(\int_{-h/2}^{h/2} \mathbf{c}z^2\mathrm{d}z\right)\mathbf{L}w = \left(\int_{-h/2}^{h/2} \mathbf{c}z^2\mathrm{d}z\right)\boldsymbol{\varepsilon}_p = \mathbf{D}\boldsymbol{\varepsilon}_p \tag{3.71}$$

where **D** is a matrix of constants related to the material property and the thickness of the plate. For inhomogeneous plates such as laminated plates, **c** can be a function of z, **D** depends on the configuration of the laminated plate, and it will have the following general form:

$$\mathbf{D} = \int_{-h/2}^{h/2} \mathbf{c}z^2\mathrm{d}z = \begin{bmatrix} D_{11} & D_{12} & D_{16} \\ D_{12} & D_{22} & D_{26} \\ D_{16} & D_{26} & D_{66} \end{bmatrix} \tag{3.72}$$

where D_{ij} ($i, j = 1, 2, 6$) are constants that can be determined once the material and the layer configuration of the laminate are defined. For homogeneous plates, we simply have

$$\mathbf{D} = \int_{-h/2}^{h/2} \mathbf{c}z^2\mathrm{d}z = \frac{h^3}{12}\mathbf{c} \tag{3.73}$$

Using Equation 3.30, we obtain

$$\mathbf{D} = \begin{bmatrix} D_{11} & D_{12} & D_{16} \\ D_{12} & D_{22} & D_{26} \\ D_{16} & D_{26} & D_{66} \end{bmatrix} = \frac{Eh^3}{12(1-\nu^2)} \begin{bmatrix} 1 & \nu & 0 \\ \nu & 1 & 0 \\ 0 & 0 & (1-\nu)/2 \end{bmatrix} \tag{3.74}$$

We define now pseudo-stress as

$$\boldsymbol{\sigma}_p = \begin{Bmatrix} M_x \\ M_y \\ M_{xy} \end{Bmatrix} \tag{3.75}$$

which is also independent of coordinate z.

Constitutive Equation for Thin Plate

By using Equations 3.71 and 3.75, the generalized Hooke's law for a thin plate becomes

$$\boldsymbol{\sigma}_p = \mathbf{D}\boldsymbol{\varepsilon}_p \tag{3.76}$$

Using Equations 3.68, 3.69, 3.74, and 3.75, we can have explicit expressions for the moments in homogeneous plates of isotropic materials:

$$M_x = -\left(D_{11}\frac{\partial^2 w}{\partial x^2} + D_{12}\frac{\partial^2 w}{\partial y^2}\right) \tag{3.77}$$

$$M_y = -\left(D_{12}\frac{\partial^2 w}{\partial x^2} + D_{22}\frac{\partial^2 w}{\partial y^2}\right) \tag{3.78}$$

$$M_{xy} = -2D_{66}\frac{\partial^2 w}{\partial x \partial y} \tag{3.79}$$

In deriving the system equations, we consider first the equilibrium of the small plate cell in the z direction (see Figure 3.14), and note that

$$\mathrm{d}V_x = \frac{\partial V_x}{\partial x}\mathrm{d}x \qquad \text{and} \qquad \mathrm{d}V_y = \frac{\partial V_y}{\partial y}\mathrm{d}y$$

We have

$$\left(\frac{\partial V_x}{\partial x}\mathrm{d}x\right)\mathrm{d}y + \left(\frac{\partial V_y}{\partial y}\mathrm{d}y\right)\mathrm{d}x + (b_z - \rho h\ddot{w})\mathrm{d}x\,\mathrm{d}y = 0 \tag{3.80}$$

or

$$\frac{\partial V_x}{\partial x} + \frac{\partial V_y}{\partial y} + b_z = \rho h\ddot{w} \tag{3.81}$$

Consider then the moment equilibrium of the plate cell with respect to the x axis, and neglecting the second-order small term leads to a formula for shear force V_x given in terms of moments.

$$V_x = \frac{\partial M_x}{\partial x} + \frac{\partial M_{xy}}{\partial y} \tag{3.82}$$

Considering, finally, the moment equilibrium of the plate cell with respect to the y axis and neglecting the second-order small term give

$$V_y = \frac{\partial M_{xy}}{\partial x} + \frac{\partial M_y}{\partial y} \tag{3.83}$$

Dynamic Equilibrium Equations

The dynamic equilibrium equation for plates can be obtained by substituting Equation 3.71 into Equations 3.82 and 3.83, and then into Equation 3.81. For homogeneous and isotropic plates we have

$$D\left(\frac{\partial^4 w}{\partial x^4} + 2\frac{\partial^4 w}{\partial x^2 \partial y^2} + \frac{\partial^4 w}{\partial y^4}\right) + \rho h\ddot{w} = b_z \tag{3.84}$$

where the bending stiffness of the plate is defined by

$$D = \frac{Eh^3}{12(1 - \nu^2)} \tag{3.85}$$

A more general form of partial differential equation for governing symmetric laminates of anisotropic materials can be derived in a similar manner, and is given by (Reddy, 1993)

$$\begin{aligned}&\frac{\partial^2}{\partial x^2}\left(D_{11}\frac{\partial^2 w}{\partial x^2} + D_{12}\frac{\partial^2 w}{\partial y^2}\right) + \frac{\partial^2}{\partial y^2}\left(D_{12}\frac{\partial^2 w}{\partial x^2} + D_{22}\frac{\partial^2 w}{\partial y^2}\right)\\ &\quad + 2\frac{\partial^2}{\partial x \partial y}\left(2D_{66}\frac{\partial^2 w}{\partial x \partial y}\right) - b_z(x, y) + I_0\frac{\partial^2 w}{\partial t^2} - I_2\frac{\partial^2}{\partial t^2}\left(\frac{\partial w}{\partial x} + \frac{\partial w}{\partial y}\right) = 0\end{aligned} \tag{3.86}$$

where D_{ij} needs to be obtained using Equation 3.73. In Equation 3.86, I_0 is the mass per unit area of the plate defined by

$$I_0 = \int_{-h/2}^{h/2} \rho \mathrm{d}z \tag{3.87}$$

For plates of homogeneous materials, we have

$$I_0 = \int_{-h/2}^{h/2} \rho \mathrm{d}z = \rho h \tag{3.88}$$

I_2 is the mass moments of inertia given by

$$I_2 = \int_{-h/2}^{h/2} \rho z^2 \mathrm{d}z \tag{3.89}$$

For plates of homogeneous materials, we have

$$I_2 = \int_{-h/2}^{h/2} \rho z^2 \mathrm{d}z = \rho h^3/12 \tag{3.90}$$

The static equilibrium equation for plates of isotropic material can be obtained by dropping the dynamic term in Equation 3.84:

$$D\left(\frac{\partial^4 w}{\partial x^4} + 2\frac{\partial^4 w}{\partial x^2 \partial y^2} + \frac{\partial^4 w}{\partial y^4}\right) = b_z \tag{3.91}$$

The static equilibrium equation for symmetric laminates of anisotropic materials can be obtained by dropping the dynamic term in Equation 3.86:

$$\frac{\partial^2}{\partial x^2}\left(D_{11}\frac{\partial^2 w}{\partial x^2} + D_{12}\frac{\partial^2 w}{\partial y^2}\right) + \frac{\partial^2}{\partial y^2}\left(D_{12}\frac{\partial^2 w}{\partial x^2} + D_{22}\frac{\partial^2 w}{\partial y^2}\right) + 2\frac{\partial^2}{\partial x \partial y}\left(2D_{66}\frac{\partial^2 w}{\partial x \partial y}\right) - b_z(x, y) = 0 \tag{3.92}$$

3.6.2 Mindlin Plates

The Mindlin plate theory is applied for thick plates, as the shear deformation and rotary inertia effects can be included. The Mindlin plate theory is also known as the first-order shear deformation theory (FSDT). The Mindlin theory does not demand that the cross section be perpendicular to the neutral plane after deformation. The situation is very similar to that of the Timoshenko beam shown in Figure 3.9, but is extended into one more dimension. Thus, we usually have $\varepsilon_{xz} \neq 0$, $\varepsilon_{yz} \neq 0$. Therefore, we will have five components of strains and stress to handle.

The displacements u and v, which are parallel to the undeformed neutral surface, at a distance z from the neutral plan, can be expressed by

$$u = z\varphi_x \tag{3.93}$$

$$v = z\varphi_y \tag{3.94}$$

where φ_x and $-\varphi_y$ denote rotations of the cross section of the plate about the y and x axes, respectively. The deflection of the plate is still represented by the deflection at the neutral plane of the plate, and is denoted by w. The vector of the displacements can the expressed as

$$\begin{Bmatrix} u \\ v \\ w \end{Bmatrix} = \underbrace{\begin{bmatrix} 0 & z & 0 \\ 0 & 0 & z \\ 1 & 0 & 0 \end{bmatrix}}_{\mathbf{L}_u} \underbrace{\begin{Bmatrix} w \\ \varphi_x \\ \varphi_y \end{Bmatrix}}_{\mathbf{u}} = \mathbf{L}_u\mathbf{u} \tag{3.95}$$

where

$$\mathbf{u} = \begin{Bmatrix} w \\ \varphi_x \\ \varphi_y \end{Bmatrix} \tag{3.96}$$

is the vector of three independent field variables for Mindlin plates.

By using Equations 3.5 and 3.7 for general solids, the linear strains in the Mindlin plate are as follows:

$$\begin{Bmatrix} \varepsilon_{xx} \\ \varepsilon_{yy} \\ \varepsilon_{xy} \\ \varepsilon_{xz} \\ \varepsilon_{yz} \end{Bmatrix} = \underbrace{\begin{bmatrix} 0 & z\dfrac{\partial}{\partial x} & 0 \\ 0 & 0 & z\dfrac{\partial}{\partial y} \\ 0 & z\dfrac{\partial}{\partial y} & z\dfrac{\partial}{\partial x} \\ \dfrac{\partial}{\partial x} & 1 & 0 \\ \dfrac{\partial}{\partial y} & 0 & 1 \end{bmatrix}}_{\mathbf{L}} \begin{Bmatrix} w \\ \varphi_x \\ \varphi_y \end{Bmatrix} = \mathbf{L}\mathbf{u} \tag{3.97}$$

The stresses can then be obtained using Equations 3.8 to 3.11, by simply removing the components σ_{zz}. For isotropic materials, we have

$$\begin{Bmatrix} \sigma_{xx} \\ \sigma_{yy} \\ \sigma_{xy} \\ \sigma_{xz} \\ \sigma_{yz} \end{Bmatrix} = \underbrace{\frac{E}{1-\nu^2}\begin{bmatrix} 1 & \nu & 0 & 0 & 0 \\ \nu & 1 & 0 & 0 & 0 \\ 0 & 0 & (1-\nu)/2 & 0 & 0 \\ 0 & 0 & 0 & \kappa(1-\nu)/2 & 0 \\ 0 & 0 & 0 & 0 & \kappa(1-\nu)/2 \end{bmatrix}}_{\mathbf{D}} \begin{Bmatrix} \varepsilon_{xx} \\ \varepsilon_{yy} \\ \gamma_{xy} \\ \gamma_{xz} \\ \gamma_{yz} \end{Bmatrix} = \mathbf{D}\boldsymbol{\varepsilon} \tag{3.98}$$

where ν is Poisson's ratio, κ is the shear effectiveness factor, and $\kappa = \frac{5}{6}$ for Mindlin plates. This stress–strain relationship is also applicable to the plate theory of third-order shear deformation simply by assigning $\kappa = 1$.

The strain (or stress) components given in Equation 3.97 can be divided into in-plane strains defined by Equation 3.24 and off-plane (transverse) strains. The division is useful in some of the energy formulations when we need to account for in-plane and off-plane strain energies separately. The in-plane strains are given by

$$\boldsymbol{\varepsilon} = \mathbf{L}\boldsymbol{\varphi} \tag{3.99}$$

where

$$\mathbf{L} = \begin{bmatrix} z\frac{\partial}{\partial x} & 0 \\ 0 & z\frac{\partial}{\partial y} \\ z\frac{\partial}{\partial x} & z\frac{\partial}{\partial y} \end{bmatrix} \tag{3.100}$$

$$\boldsymbol{\varphi} = \begin{Bmatrix} \varphi_x \\ \varphi_y \end{Bmatrix} \tag{3.101}$$

The transverse shear strains ε_{xz} and ε_{yz} can be obtained from Equation 3.97 as

$$\boldsymbol{\gamma} = \begin{Bmatrix} \varepsilon_{xz} \\ \varepsilon_{yz} \end{Bmatrix} = \begin{Bmatrix} \varphi_x + \frac{\partial w}{\partial x} \\ \varphi_y + \frac{\partial w}{\partial y} \end{Bmatrix} \tag{3.102}$$

Note that if the transverse shear strains are negligible, the above equation will lead to

$$\varphi_y = -\frac{\partial w}{\partial y} \tag{3.103}$$

$$\varphi_x = -\frac{\partial w}{\partial x} \tag{3.104}$$

The FSDT plate theory becomes the CPT.

The transverse average shear stress $\boldsymbol{\tau}$ relates to the transverse shear strain in the form:

$$= \begin{Bmatrix} \sigma_{xz} \\ \sigma_{yz} \end{Bmatrix} = \kappa \underbrace{\begin{bmatrix} G & 0 \\ 0 & G \end{bmatrix}}_{\mathbf{D}_s} \boldsymbol{\gamma} = \kappa\mathbf{D} \tag{3.105}$$

where G is the shear modulus.

3.6.3 Third-Order Theory of Plates

The Mindlin plate theory or the FSDT has some problems in the solution, such that the transverse shear forces obtained are not zero on the plate surfaces, which contradicts the actual situation. A more accurate theory for thick plates, termed third-order shear deformation theory (TSDT) (see a very good description by Reddy, 1984), has been proposed and widely used. Here, we briefly describe TSDT following the formulation of C. M. Wang et al. (2000), where readers can find a more-detailed description. Based on TSDT, the displacement field of the plate can be expressed as

$$\begin{Bmatrix} u \\ v \\ w \end{Bmatrix} = \underbrace{\begin{bmatrix} -\alpha z^3 \frac{\partial}{\partial x} & z - \alpha z^3 & 0 \\ -\alpha z^3 \frac{\partial}{\partial y} & 0 & z - az^3 \\ 1 & 0 & 0 \end{bmatrix}}_{\mathbf{L}_u} \underbrace{\begin{Bmatrix} w \\ \varphi_x \\ \varphi_y \end{Bmatrix}}_{\mathbf{u}} = \mathbf{L}_u \mathbf{u} \tag{3.106}$$

where $\alpha = 4/(3h^2)$, h is the thickness of the plate, $\mathbf{u}$ is the same as Equation 3.96, w is the transverse deflection of the neutral plane of the plate, and φ_x and $-\varphi_y$ denote rotations of the cross section of the plate about the y and x axes, respectively.

It can be easily seen that by setting $\alpha = 0$, Equation 3.106 becomes Equation 3.95, and the displacement field based on FSDT can be obtained. Furthermore, if we have $\alpha = 0$, $\varphi_x = -\partial w/\partial x$, and $\varphi_y = -\partial w_0/\partial y$, the displacement field based on CPT can be obtained. The independent variables are w, φ_x, and φ_y for both plates based on FSDT and TSDT. Note that for the classic (Kirchhoff) plate theory, there is only one independent variable w, as shown in the first part of this section.

The linear strains for plates of TSDT are as follows:

$$\begin{Bmatrix} \varepsilon_{xx} \\ \varepsilon_{yy} \\ \varepsilon_{xy} \\ \varepsilon_{xz} \\ \varepsilon_{yz} \end{Bmatrix} = \underbrace{\begin{bmatrix} -\alpha z^3 \frac{\partial^2}{\partial x^2} & (z - \alpha z^3)\frac{\partial}{\partial x} & 0 \\ -\alpha z^3 \frac{\partial^2}{\partial y^2} & 0 & (z - \alpha z^3)\frac{\partial}{\partial y} \\ -2\alpha z^3 \frac{\partial^2}{\partial x \partial y} & (z - \alpha z^3)\frac{\partial}{\partial y} & (z - \alpha z^3)\frac{\partial}{\partial x} \\ (1 - \beta z^2)\frac{\partial}{\partial x} & (1 - \beta z^2) & 0 \\ (1 - \beta z^2)\frac{\partial}{\partial y} & 0 & (1 - \beta z^2) \end{bmatrix}}_{\mathbf{L}} \begin{Bmatrix} w \\ \varphi_x \\ \varphi_y \end{Bmatrix} = \mathbf{Lu} \tag{3.107}$$

where $\beta = 3\alpha$. Again, if we set $\alpha = 0$, the foregoing equation becomes Equation 3.97, that is, for plates of FSDT.

The stress–strain relationship is the same as that for the FSDT plates given in Equation 3.98 with $\kappa = 1$.

3.7 Remarks

This chapter presents some basic formulations and concepts of the mechanics of solids and simple structural components. It is not intended to cover all the issues in mechanics, but, rather, to list equations that are used in the later chapters and to provide concise explanations of some of the most important terms of mechanics. Understanding the materials covered in this chapter should be sufficient for the reader to study the later chapters for most of the mechanics problems of solids and structures. We omitted mechanics for shells, as these formulations are more advanced. They are not easy to explain and would have led to confusion. Shells will be analyzed in the book using MFree methods as an advanced topic. Therefore, all information related to shells is contained in Chapter 12, so that readers who are not interested in shells can skip that chapter entirely. We also omitted materials related to fluid mechanics, but cover them in Chapter 9 together with the MFree treatment, so that all information on the mechanics of fluids can be grouped together.

4

Principles for Weak Forms

Principles used for creating weak forms for MFree methods are outlined in a concise, practical, and easy-to-understand manner in this chapter. We introduce the principles, and explain the procedures for their use. The principles used in MFree methods are similar to those used in the conventional finite element method (FEM). However, MFree methods use these principles in much more innovative ways. Therefore, this chapter covers these principles in a new systematic way for easy application to MFree methods. Readers with experience in FEM may skip the chapter, but it may prove helpful to skim the chapter to become familiar with the terminology conventions used in the book. Readers with less experience with FEM may have difficulty fully understanding the details of all the procedures and equations presented here, but a global view of the general procedure and principles will help in the following chapters. We recommend that such readers study this chapter without delving too deeply into the equations. Subsequently, it may be a good idea to review this chapter again after reading the following chapters to understand the details of all the equations presented in this chapter.

4.1 Strong Forms vs. Weak Forms

The partial differential system equations developed in Chapter 3 are *strong form* system equations, such as Equations 3.20 through 3.22 for the mechanics of solids. *Weak form*, in contrast to strong form, requires weaker consistency on the assumed function of dependent variables (u, v, w in this case). Obtaining the *exact* solution for a strong form system equation is ideal but, unfortunately, it is usually very difficult for practical engineering problems that are very complex. The finite difference method (FDM), which uses finite differential representation (Taylor series) of a function in a local domain, can be used to solve system equations of strong form to obtain an *approximated* solution. However, FDM requires a regular mesh of grids, and can usually work only for problems with regular geometry and simple boundary conditions. One of the MFree methods for solving strong form system equations to obtain an approximate solution is to use arbitrarily distributed grids based on Taylor series expansions, and least square or moving least squares (MLS) approximations. The formulation is very simple but less stable, and the accuracy of the results often depends on the selection of the nodes for constructing the differential equations. There are also MFree methods for solving strong form system equations using an integral representation of field variable functions, such as the smooth particle hydrodynamics (SPH) methods. This formulation can deal well with dynamic problems of infinite domain, such as problems in astrophysics (Monaghan, 1992). SPH has proved stable for arbitrarily distributed nodes. This is due to the use of integral representation of field functions, which pass the differentiation operations on the field function to the weight function. Therefore, it reduces the requirement on the order of consistency on the approximated field function, and is actually very similar to that of weak form formulations. The difference is that the weak form operation is implemented in the stage of shape function construction, and not in the stage of creating the system equation. The major problem with the SPH methods is the treatment of boundaries of the problem domain and the boundary conditions.

Formulation based on weak forms can produce a stable set of algebraic system equations and gives discretized system equations that produce much more accurate results. Therefore, weak form is often preferred by many to obtain an approximated solution. This book uses weak form formulations to form discretized system equations for our MFree methods for mechanics problems of solids, structures, as well as fluids.

The consistency requirement on the approximation functions for field variables is very different for strong form and weak form formulation. For a 2kth-order differential governing system equation, the strong form formulation assumes the field variable possesses a consistency of 2kth order. The weak form formulation, however, requires a consistency of only kth order.

There are basically two major categories of often-used principles for constructing weak forms. They are variational methods and weighted residual methods. There are different forms of variational methods. Galerkin formulation may be the most widely used approach for establishing system equations. It is, of course, applicable to deriving MFree equations. Hamilton's principle is often employed to produce approximated system equations for dynamic problems, and is also applicable to MFree methods. The minimum total potential energy principle has been a convenient tool for deriving discrete system equations for FEM and also for many other types of approximation methods. The weighted residual method is a more general and powerful mathematical tool that can be used for creating discretized system equations for many types of engineering problems. It has been and is still used for developing new MFree methods. All these approaches are adopted in this book for creating discretized system equations for various types of MFree methods.

4.2 Hamilton's Principle

Hamilton's principle is one of the variational principles; it states, "Of all possible time histories of consistent displacement states which satisfy*

(a) the compatibility conditions (4.1)

(b) the essential (displacement or kinematical) boundary conditions (4.2)

(c) the conditions at initial (t_1) and final time (t_2) (4.3)

the history corresponding to the actual solution makes the *Lagrangian functional* a minimum." These conditions are also called admissible conditions. Mathematically, Hamilton's principle states

$$\delta \int_{t_1}^{t_2} L \, \mathrm{d}t = 0 \tag{4.4}$$

where L is the Lagrangian functional. For a system of solids and structures it can be defined as

$$L = T - \Pi_s + W_f \tag{4.5}$$

*See Section 5.1 for the definitions of consistency and compatibility.

where T is the kinetic energy, Π is the strain energy, and the W_f is the work done by the external forces.

The kinetic energy is defined by

$$T = \frac{1}{2}\int_{\Omega} \rho \dot{\mathbf{u}}^T \dot{\mathbf{u}} \, d\Omega \tag{4.6}$$

where Ω stands for the whole volume of the solid. For solids and structures of elastic materials, the strain energy of the system can be expressed as

$$\Pi_s = \frac{1}{2}\int_{\Omega} \boldsymbol{\varepsilon}^T \boldsymbol{\sigma} \, d\Omega \tag{4.7}$$

The work done by the external forces can be obtained by

$$W_f = \int_{\Omega} \mathbf{u}^T \mathbf{b} \, d\Omega + \int_{\Gamma_t} \mathbf{u}^T \bar{\mathbf{t}} \, d\Gamma \tag{4.8}$$

where Γ_t stands for the boundary of the solids on which traction forces $\bar{\mathbf{t}}$ are prescribed (see Figure 3.2).

The procedure for solving a problem using Hamilton's principle is described as follows:

1. Construct a shape function to approximate the field functions using their values at nodes in the problem domain. The approximated field function should be consistent and admissible, i.e., it satisfies the conditions of Equations 4.1 to 4.3.
2. Calculate the kinetic energy T, the strain energy Π, and the work done by the external forces W_f in terms of the approximated field functions.
3. Form the Lagrangian functional L using Equation 4.5, and then substitute it into Equation 4.4. Use integration by parts and carry out the spatial and temporal integrations that lead to a set of differential equations with respect only to time.
4. Solve the set of differential equations with respect to time, using standard procedures to obtain the dynamic field.
5. For static problems, drop the time-related terms, and step 3 leads to a set of algebraic equations, which can be solved using a standard algebraic equation solver to obtain the static field.

At step 1, the shape function is constructed in FEM using a mesh of elements. In MFree methods, techniques without using elements are required.

Note that in order to use Hamilton's principle, a functional has to be obtained, which may not be possible for all problems. The advantage of using Hamilton's principle is that we do not have to know the strong form of the system equation.

4.3 Constrained Hamilton's Principle

There are cases when the approximated field function does not satisfy the condition, Equations 4.1 and 4.2, on parts of the problem domain including parts of the boundaries, discrete curves, and points of locations. Hamilton's principle has to be modified or constrained for such situations.

Consider the following set of k conditions that the approximated field function cannot satisfy:

$$\mathbf{C}(\mathbf{u}) = \begin{Bmatrix} C_1(\mathbf{u}) \\ C_2(\mathbf{u}) \\ \mathrm{M} \\ C_k(\mathbf{u}) \end{Bmatrix} = 0 \tag{4.9}$$

where $\mathbf{C}$ is a given matrix of coefficients.

Our purpose now is to seek the stationary point of the Lagrangian functional subjected to the constraint of Equation 4.9. There are basically two methods often used to modify the functional that accommodates these constraints. They are the method of Lagrange multipliers and the penalty method.

4.3.1 Method of Lagrange Multipliers

In the method of Lagrange multipliers, the modified Lagrangian is written as follows:

$$\tilde{L} = L + \int_{\Omega} \boldsymbol{\lambda}^T \mathbf{C}(\mathbf{u}) \mathrm{d}\Omega \tag{4.10}$$

where $\boldsymbol{\lambda}$ is a vector of the Lagrange multipliers given by

$$\boldsymbol{\lambda}^T = \{\lambda_1 \ \lambda_2 \ \ldots \ \lambda_k\} \tag{4.11}$$

These Lagrange multipliers are unknown functions of independent coordinates in the domain Ω. The modified Hamilton's principle then seeks the stationary of the following functional:

$$\delta \int_{t_1}^{t_2} \tilde{L} \, \mathrm{d}t = 0 \tag{4.12}$$

Note that since the Lagrange multipliers are unknown functions, the total number of unknown field functions of the whole system is then increased. In the process of seeking discretized system equations, these Lagrange multipliers must also be approximated in a manner similar to the field functions. Therefore, the total number of *nodal unknowns* in the discretized system equation will also be increased. The method of Lagrange multipliers will, however, rigorously enforce the constraints. The penalty method, introduced in the following subsection, does not increase the number of unknowns.

4.3.2 Penalty Method

In examining the constraint equations, Equation 4.9, we construct the following functional:

$$\mathbf{C}^T \boldsymbol{\alpha} \mathbf{C} = \alpha_1 C_1^2 + \alpha_2 C_2^2 + \cdots + \alpha_k C_k^2 \tag{4.13}$$

where $\boldsymbol{\alpha}$ is a diagonal matrix given by

$$\boldsymbol{\alpha} = \begin{bmatrix} \alpha_1 & 0 & 0 & 0 \\ 0 & \alpha_2 & 0 & 0 \\ 0 & 0 & O & 0 \\ 0 & 0 & 0 & \alpha_k \end{bmatrix} \tag{4.14}$$

where $\alpha_1, \alpha_2, \ldots, \alpha_k$ are penalty factors. They can be given as functions of coordinates, but usually they are assigned positive constant numbers. In any case, $\mathbf{C}^T\boldsymbol{\alpha}\mathbf{C}$ will always be non-negative, and hence zero is the minimum of the functional $\mathbf{C}^T\boldsymbol{\alpha}\mathbf{C}$. It would be zero only if all the conditions in Equation 4.9 are fully satisfied. Therefore, the following stationary condition of the functional $\mathbf{C}^T\boldsymbol{\alpha}\mathbf{C}$ guarantees the best satisfaction of the constraint equations, Equation 4.9:

$$\delta(\mathbf{C}^T\boldsymbol{\alpha}\mathbf{C}) = 0 \tag{4.15}$$

Performing the variation using the chain rule, we have

$$\delta(\mathbf{C}^T\boldsymbol{\alpha}\mathbf{C}) = 2\mathbf{C}^T\boldsymbol{\alpha}\delta\mathbf{C} = 2\delta\mathbf{C}^T\boldsymbol{\alpha}\mathbf{C} = 0 \tag{4.16}$$

which leads to the following minimization condition:

$$\boldsymbol{\alpha}\mathbf{C}(\mathbf{u}) = \begin{Bmatrix} \alpha_1 C_1(\mathbf{u}) \\ \alpha_2 C_2(\mathbf{u}) \\ \mathrm{M} \\ \alpha_k C_k(\mathbf{u}) \end{Bmatrix} = 0 \tag{4.17}$$

If $\alpha_i = 0$, the essential boundary condition is not enforced at all, because any C_i will still satisfy the ith equation in Equation 4.17. If α_i goes to infinity, the essential boundary condition is fully enforced, because C_i must be zero in order to satisfy Equation 4.17.

The above analysis shows that $\mathbf{C}^T\boldsymbol{\alpha}\mathbf{C}$ should be the right functional for our constraint. The modified Lagrangian is then written as follows:

$$\tilde{L} = L + \frac{1}{2}\int_\Omega \mathbf{C}^T(\mathbf{u})\boldsymbol{\alpha}\mathbf{C}(\mathbf{u})\mathrm{d}\Omega \tag{4.18}$$

The $\frac{1}{2}$ serves only to counter the 2 that will be produced in the later variational operation. The important difference between the penalty factor α_i and the Lagrange multiplier λ is that the penalty factor is a given constant (no variation is allowed), whereas the Lagrange multiplier is a variable.

Because α is a known constant, there will be no increase in the number of unknowns in the system. The question is how to choose the penalty factor. To impose the constraint fully, the penalty factor must be infinite, which is not possible in practical numerical analysis. Therefore, in the penalty method these constraints cannot be satisfied exactly, but only approximately. In general, the use of a larger penalty factor will enforce the constraint. The problem is that if the penalty factor is too small, the constraints will not be properly enforced, but if it is too large, numerical problems will be encountered. A compromise should be reached. Some kind of formula that is universally applicable would be useful. To find such a formula, one may need to determine the factors that affect the selection of the penalty factor in the actual event of solving the discretized system equations.

Note that use of the penalty method is a routine operation even in the FEM for imposing essential boundary conditions including single-point and multi-point constraints.

4.3.3 Determination of Penalty Factor

Because discretization errors can be comparable in magnitude to the errors due to the poor satisfaction of the constraint, Zienkiewicz (1989) has suggested using the following formula for FEM analysis:

$$\alpha = \text{constant}(1/h)^n \tag{4.19}$$

where h is the characteristic length, which can be the ratio of the element size to the dimension of the problem domain, and n is the order of the elements.

In extending this formula to MFree methods, we suggest that h be the ratio of the nodal spacing to the dimension of the problem domain, and $n = 1$. The constant in Equation 4.19 should relate to the material property of the solid or structure. It can be 10^{10} times Young's modulus.

This book prefers the following simple method for determining the penalty factor:

$$\alpha = 1.0 \times 10^{4\text{–}13} \times \max(\text{diagonal elements in the stiffness matrix}) \tag{4.20}$$

In most of the examples reported in later chapters using penalty methods, the foregoing equation is adopted.

It has also been suggested to use

$$\alpha = 1.0 \times 10^{5\text{–}8} \times \text{Young's modulus} \tag{4.21}$$

for some examples, which works well.

Note that trials may be needed to choose a proper penalty factor. More discussion on this is provided in Example 6.12 in Section 6.4.5, when we deal with nonlinear problems.

4.4 Galerkin Weak Form

The Galerkin weak form can be directly derived using Hamilton's principle for problems of solid mechanics. By using Equations 4.5 to 4.8, the Lagrangian L becomes

$$L = -\frac{1}{2}\int_\Omega \boldsymbol{\varepsilon}^T\boldsymbol{\sigma}\,\mathrm{d}\Omega + \int_\Omega \mathbf{u}^T\mathbf{b}\,\mathrm{d}\Omega + \int_{\Gamma_t} \mathbf{u}^T\bar{\mathbf{t}}\,\mathrm{d}\Gamma + \frac{1}{2}\int_\Omega \rho\dot{\mathbf{u}}^T\dot{\mathbf{u}}\,\mathrm{d}\Omega \tag{4.22}$$

Substituting the above into Equation 4.4, we have

$$\delta\int_{t_1}^{t_2}\left[-\frac{1}{2}\int_\Omega \boldsymbol{\varepsilon}^T\boldsymbol{\sigma}\,\mathrm{d}\Omega + \int_\Omega \mathbf{u}^T\mathbf{b}\,\mathrm{d}\Omega + \int_{\Gamma_t} \mathbf{u}^T\bar{\mathbf{t}}\,\mathrm{d}\Gamma + \frac{1}{2}\int_\Omega \rho\dot{\mathbf{u}}^T\dot{\mathbf{u}}\,\mathrm{d}\Omega\right]\mathrm{d}t = 0 \tag{4.23}$$

Moving the variation operation into the integral operations, we obtain

$$\int_{t_1}^{t_2}\left[-\frac{1}{2}\int_{\Omega}\delta(\boldsymbol{\varepsilon}^T\boldsymbol{\sigma})\mathrm{d}\Omega+\int_{\Omega}\delta\mathbf{u}^T\mathbf{b}\mathrm{d}\Omega+\int_{\Gamma_t}\delta\mathbf{u}^T\bar{\mathbf{t}}\mathrm{d}\Gamma+\frac{1}{2}\int_{\Omega}\delta(\rho\dot{\mathbf{u}}^T\dot{\mathbf{u}})\mathrm{d}\Omega\right]\mathrm{d}t = 0 \quad (4.24)$$

Changing the order of operation does not affect the results because all these operations are linear, and therefore it does not matter which one operates first. The integrand in the first integral term can be written as follows using the chain rule of variation:

$$\delta(\boldsymbol{\varepsilon}^T\boldsymbol{\sigma}) = \delta\boldsymbol{\varepsilon}^T\boldsymbol{\sigma}+\boldsymbol{\varepsilon}^T\delta\boldsymbol{\sigma} \quad (4.25)$$

Because the two terms in the foregoing equation are all scalars, their transposes are still themselves. For the last term in Equation 4.25, we should have

$$\boldsymbol{\varepsilon}^T\delta\boldsymbol{\sigma} = (\boldsymbol{\varepsilon}^T\delta\boldsymbol{\sigma})^T = \delta\boldsymbol{\sigma}^T\boldsymbol{\varepsilon} \quad (4.26)$$

Using the constitutive equation of Equation 3.8 and the symmetric property of the matrix of material constant **c**, we should have

$$\delta\boldsymbol{\sigma}^T\boldsymbol{\varepsilon} = \delta(\mathbf{c}\boldsymbol{\varepsilon})^T\boldsymbol{\varepsilon} = \delta\boldsymbol{\varepsilon}^T\mathbf{c}^T\boldsymbol{\varepsilon} = \delta\boldsymbol{\varepsilon}^T\mathbf{c}\boldsymbol{\varepsilon} = \delta\boldsymbol{\varepsilon}^T\boldsymbol{\sigma} \quad (4.27)$$

Therefore, Equation 4.25 becomes

$$\delta(\boldsymbol{\varepsilon}^T\boldsymbol{\sigma}) = 2\delta\boldsymbol{\varepsilon}^T\boldsymbol{\sigma} \quad (4.28)$$

Let's now look at the last term in Equation 4.24. We first move the time integration into the spatial integration; we should have

$$\int_{t_1}^{t_2}\left[\frac{1}{2}\int_{\Omega}\delta(\rho\dot{\mathbf{u}}^T\dot{\mathbf{u}})\mathrm{d}\Omega\right]\mathrm{d}t = \frac{1}{2}\int_{\Omega}\left[\int_{t_1}^{t_2}\delta(\rho\dot{\mathbf{u}}^T\dot{\mathbf{u}})\mathrm{d}t\right]\mathrm{d}\Omega \quad (4.29)$$

Again, this change of the order of integral operation does not affect the results because they are linear operations. The time integration in Equation 4.29 can be changed as follows, using the chain rule of variation and then the scalar property:

$$\int_{t_1}^{t_2}\delta(\rho\dot{\mathbf{u}}^T\dot{\mathbf{u}})\mathrm{d}t = \rho\int_{t_1}^{t_2}[\delta\dot{\mathbf{u}}^T\dot{\mathbf{u}}+\dot{\mathbf{u}}^T\delta\dot{\mathbf{u}}]\mathrm{d}t = 2\rho\int_{t_1}^{t_2}[\delta\dot{\mathbf{u}}^T\dot{\mathbf{u}}]\mathrm{d}t \quad (4.30)$$

Note that the variation and time differentiation are also exchangeable, i.e.,

$$\int_{t_1}^{t_2}[\delta\dot{\mathbf{u}}^T\dot{\mathbf{u}}]\mathrm{d}t = \int_{t_1}^{t_2}\left[\frac{\mathrm{d}\delta\mathbf{u}^T}{\mathrm{d}t}\frac{\mathrm{d}\mathbf{u}}{\mathrm{d}t}\right]\mathrm{d}t \quad (4.31)$$

Integrating by parts with respect to the time, we have

$$\int_{t_1}^{t_2}\left[\frac{\mathrm{d}\delta\mathbf{u}^T}{\mathrm{d}t}\frac{\mathrm{d}\mathbf{u}}{\mathrm{d}t}\right]\mathrm{d}t = \int_{t_1}^{t_2}\left[-\delta\mathbf{u}^T\frac{\mathrm{d}^2\mathbf{u}}{\mathrm{d}t^2}\right]\mathrm{d}t+\delta\mathbf{u}^T\frac{\mathrm{d}\mathbf{u}}{\mathrm{d}t}\bigg|_{t_1}^{t_2} \quad (4.32)$$

Invoking now the condition given in Equation 4.3, we know that the **u** has already satisfied the conditions at the initial (t_1) and the final time (t_2). Therefore, $\delta\mathbf{u}^T$ has to be zero at t_1 and t_2 (no variation can exist for any given constant value). Therefore, the last term in Equation 4.32 vanishes, which gives

$$\int_{t_1}^{t_2}\left[\frac{\mathrm{d}\delta\mathbf{u}^T}{\mathrm{d}t}\frac{\mathrm{d}\mathbf{u}}{\mathrm{d}t}\right]\mathrm{d}t = \int_{t_1}^{t_2}\left[-\delta\mathbf{u}^T\frac{\mathrm{d}^2\mathbf{u}}{\mathrm{d}t^2}\right]\mathrm{d}t \tag{4.33}$$

Therefore, Equation 4.29 becomes

$$\int_{t_1}^{t_2}\left[\frac{1}{2}\int_\Omega \delta(\rho\dot{\mathbf{u}}^T\dot{\mathbf{u}})\mathrm{d}\Omega\right]\mathrm{d}t = \int_\Omega\left[\rho\int_{t_1}^{t_2}\left[-\delta\mathbf{u}^T\frac{\mathrm{d}^2\mathbf{u}}{\mathrm{d}t^2}\right]\mathrm{d}t\right]\mathrm{d}\Omega \tag{4.34}$$

We now switch the order of integration to obtain

$$\int_{t_1}^{t_2}\left[\frac{1}{2}\int_\Omega \delta(\rho\dot{\mathbf{u}}^T\dot{\mathbf{u}})\mathrm{d}\Omega\right]\mathrm{d}t = -\int_{t_1}^{t_2}\left[\rho\int_\Omega[\delta\mathbf{u}^T\ddot{\mathbf{u}}]\mathrm{d}\Omega\right]\mathrm{d}t \tag{4.35}$$

Equation 4.24 now becomes

$$\int_{t_1}^{t_2}\left[-\int_\Omega \delta\boldsymbol{\varepsilon}^T\boldsymbol{\sigma}\mathrm{d}\Omega + \int_\Omega \delta\mathbf{u}^T\mathbf{b}\mathrm{d}\Omega + \int_{\Gamma_t}\delta\mathbf{u}^T\bar{\mathbf{t}}\mathrm{d}\Gamma - \int_\Omega \rho\delta\mathbf{u}^T\ddot{\mathbf{u}}\mathrm{d}\Omega\right]\mathrm{d}t = 0 \tag{4.36}$$

To satisfy the above equation for all possible choices of **u**, the integrand of the time integration has to vanish, which leads to

$$\int_\Omega \delta\boldsymbol{\varepsilon}^T\boldsymbol{\sigma}\,\mathrm{d}\Omega - \int_\Omega \delta\mathbf{u}^T\mathbf{b}\mathrm{d}\Omega - \int_{\Gamma_t}\delta\mathbf{u}^T\bar{\mathbf{t}}\mathrm{d}\Gamma + \int_\Omega \rho\delta\mathbf{u}^T\ddot{\mathbf{u}}\mathrm{d}\Omega = 0 \tag{4.37}$$

This is the well-known Galerkin weak form. For static problems, it simply reduces to

$$\int_\Omega \delta\boldsymbol{\varepsilon}^T\boldsymbol{\sigma}\,\mathrm{d}\Omega - \int_\Omega \delta\mathbf{u}^T\mathbf{b}\mathrm{d}\Omega - \int_{\Gamma_t}\delta\mathbf{u}^T\bar{\mathbf{t}}\mathrm{d}\Gamma = 0 \tag{4.38}$$

Equation 4.37 can also be viewed as the principle of *virtual work*, which states that if a solid body is in equilibrium, the total virtual work performed by all the stresses in the body and all the external forces applied on the body should vanish when the body is subjected to a *virtual displacement*. The virtual work can be viewed as an alternative statement of the equilibrium equation. In the situation given in Equation 4.37, we can view that the solid is subjected to a virtual displacement of $d\mathbf{u}$. The first term in Equation 4.37 is the virtual work done by the internal stress in problem domain Ω; the second term is the virtual work done by the external body force; the third term is the virtual work done by the external tractions on the natural boundaries; and the last term is the virtual work done by the inertial forces. Therefore, using the principle of virtual work, we can actually write out Equation 4.37 or 4.38 straight away.

The Galerkin weak form can be formulated in many other ways. One alternative is to formulate it via the method of weighted residuals, which is introduced in the following section.

By using the stress–strain relation of Equation 3.8, and then the strain–displacement relation of Equation 3.5, Equations 4.5 and 4.38 can be explicitly expressed as follows in terms of displacement vector **u**:

$$\int_{\Omega}\delta(\mathbf{Lu})^T\mathbf{c}(\mathbf{Lu})\mathrm{d}\Omega-\int_{\Omega}\delta\mathbf{u}^T\mathbf{b}\mathrm{d}\Omega-\int_{\Gamma_t}\delta\mathbf{u}^T\bar{\mathbf{t}}\mathrm{d}\Gamma+\int_{\Omega}\rho\delta\mathbf{u}^T\ddot{\mathbf{u}}\mathrm{d}\Omega=0 \tag{4.39}$$

This is the Galerkin weak form written in terms of displacement, and therefore it is convenient to use, as the displacement is to be approximated in FEM or MFree methods. For static problems, Equation 4.39 simply reduces to

$$\int_{\Omega}\delta(\mathbf{Lu})^T\mathbf{c}(\mathbf{Lu})\mathrm{d}\Omega-\int_{\Omega}\delta\mathbf{u}^T\mathbf{b}\mathrm{d}\Omega-\int_{\Gamma_t}\delta\mathbf{u}^T\bar{\mathbf{t}}\mathrm{d}\Gamma=0 \tag{4.40}$$

The above two equations of Galerkin weak form are very handy in application to problems of solid mechanics, because one does not need to perform integration by parts. The discretized system equation can be derived very easily using approximated displacements. The Galerkin procedure used in MFree methods is as follows:

1. Approximate the displacement at a point using MFree shape functions and nodal displacements at the nodes surrounding the point. The approximation should be consistent and has to satisfy Equations 4.1 and 4.2.
2. Substitute the approximated displacements into Equation 4.37, and factor out the variation of the nodal displacements, leading to a set of differential equations with respect only to time.
3. Solve the set of differential equations with respect to time, using standard procedures to obtain the dynamic field.
4. For static problems, use Equation 4.38 instead, which leads to a set of algebraic equations that can be solved using standard algebraic equation solvers to obtain the static field.

This Galerkin procedure is applied repeatedly in the following chapters for all kinds of mechanics problems of solids and structures.

Note that in using the Galerkin procedure one need not know the strong form of the system equation.

4.5 Constrained Galerkin Weak Form

For cases when the approximated field function does not satisfy the condition, Equation 4.1 or 4.2, on parts of the problem domain including parts of the boundaries and discrete points of locations, we should use the constrained Hamilton's principle to derive the constrained Galerkin weak form. The procedure is the same as in Section 4.4, except the modified Lagrange $\tilde{L}$ is used for the formulation. The following presents the final expressions.

4.5.1 Galerkin Weak Form with Lagrange Multipliers

For dynamic problems,

$$\int_\Omega \delta(\mathbf{Lu})^T \mathbf{c}(\mathbf{Lu})d\Omega - \int_\Omega \delta\mathbf{u}^T\mathbf{b}d\Omega - \int_{\Gamma_t} \delta\mathbf{u}^T\bar{\mathbf{t}}d\Gamma$$
$$-\int_\Omega \delta\boldsymbol{\lambda}^T\mathbf{C}(\mathbf{u})d\Omega - \int_\Omega \boldsymbol{\lambda}^T\delta\mathbf{C}(\mathbf{u})d\Omega + \int_\Omega \rho\delta\mathbf{u}^T\ddot{\mathbf{u}}d\Omega = 0 \tag{4.41}$$

For static problems, it simply reduces to

$$\int_\Omega \delta(\mathbf{Lu})^T \mathbf{c}(\mathbf{Lu})d\Omega - \int_\Omega \delta\mathbf{u}^T\mathbf{b}d\Omega - \int_{\Gamma_t} \delta\mathbf{u}^T\bar{\mathbf{t}}d\Gamma - \int_\Omega \delta\boldsymbol{\lambda}^T\mathbf{C}(\mathbf{u})d\Omega - \int_\Omega \boldsymbol{\lambda}^T\delta\mathbf{C}(\mathbf{u})d\Omega = 0 \tag{4.42}$$

4.5.2 Galerkin Weak Form with Penalty Factors

For dynamic problems,

$$\int_\Omega \delta(\mathbf{Lu})^T \mathbf{c}(\mathbf{Lu})d\Omega - \int_\Omega \delta\mathbf{u}^T\mathbf{b}d\Omega - \int_{\Gamma_t} \delta\mathbf{u}^T\bar{\mathbf{t}}d\Gamma - \int_\Omega \delta\mathbf{C}(\mathbf{u})^T\boldsymbol{\alpha}\mathbf{C}(\mathbf{u})d\Omega + \int_\Omega \rho\delta\mathbf{u}^T\ddot{\mathbf{u}}d\Omega = 0 \tag{4.43}$$

For static problems, it simply reduces to

$$\int_\Omega \delta(\mathbf{Lu})^T \mathbf{c}(\mathbf{Lu})d\Omega - \int_\Omega \delta\mathbf{u}^T\mathbf{b}d\Omega - \int_{\Gamma_t} \delta\mathbf{u}^T\bar{\mathbf{t}}d\Gamma - \int_\Omega \delta\mathbf{C}(\mathbf{u})^T\boldsymbol{\alpha}\mathbf{C}(\mathbf{u})d\Omega = 0 \tag{4.44}$$

4.6 Minimum Total Potential Energy Principle

Another very often used energy principle in FEM is the minimum total potential energy principle. This principle states that for a structure system that is at equilibrium, the total potential energy in the system must be stationary for variations of admissible displacements. This principle works in a very straightforward manner in the following three simple steps:

1. Approximate the field function (displacement) in terms of the nodal variables, say, **d**, which is a vector that collects all the nodal displacements in the problem domain.
2. Express the total potential energy in terms of the nodal variables **d**. For solids and structures of elastic materials, the total potential energy can be expressed as

$$\Pi = \Pi_s - W_f \tag{4.45}$$

3. Use the following stationary conditions to create a set of discretized system equations:

$$\frac{\partial \Pi}{\partial \mathbf{d}} = \begin{Bmatrix} \frac{\partial \Pi}{\partial d_1} \\ \frac{\partial \Pi}{\partial d_2} \\ \text{M} \end{Bmatrix} = 0 \tag{4.46}$$

The number of the equation created is exactly the total number of the nodal variables.

4.7 Weighted Residual Method

The weighted residual method is a very simple but powerful mathematical tool to obtain approximated system equations of weak form. The concept of the weighted residual method is straightforward and applicable, in principle, for most of the partial differential equations that govern engineering problems, including mechanics of solids, structures, and fluids.

Let us consider the general form of the partial differential equations for solid mechanics. It can be rewritten in a general concise functional form:

$$\mathbf{f}(\mathbf{u}(x, y, z)) = 0 \tag{4.47}$$

For example, for solid mechanics problems formulated in Section 3.2.3, we have

$$\mathbf{f}(\mathbf{u}(x, y, z)) = \mathbf{L}^T\mathbf{c}\mathbf{L}\mathbf{u} + \mathbf{b} - \rho\ddot{\mathbf{u}} \tag{4.48}$$

In general, it is difficult to obtain the exact solution $\mathbf{u}(x, y, z)$ that satisfies Equation 4.47. We therefore seek a $\mathbf{u}(x, y, z)$ that satisfies Equation 4.47 in a weighted integral sense over a quadrature domain:

$$\int_\Omega \widehat{\mathbf{W}}^T \mathbf{f}(\mathbf{u}(x, y, z))\mathrm{d}\Omega = 0 \tag{4.49}$$

where $\widehat{\mathbf{W}}$ is a vector of weight functions corresponding to the equations in $\mathbf{f}$. We hope that $\mathbf{u}(x, y, z)$ is a good approximation of the exact solution. This is the idea of the weighted residual approach. It is, indeed, very simple.

Although the method is simple and, in principle, it is applicable to most partial differential equations, different ways of implementation will lead to solutions of different accuracies. Variational methods and finite difference methods can all be derived using the weighted residual method by properly choosing the weight functions.

Note that function $\mathbf{f}$ can be either static or dynamic. The weighted residual procedure will produce these two types of discrete systems accordingly.

The procedure for solving a problem using the weighted residual method is as follows:

1. Construct a shape function to approximate the field function using the field variables at the nodes in the domain with a certain order of consistency.
2. Construct weight functions. When the Dirac delta function is used, the weighted residual method leads to the collocation method. If the shape function constructed in step 1 is also used as the weight function, the weighted residual method leads to the Galerkin weak form, which was derived in the previous section. In cases where the weight function constructed is different from the shape function used, the method is often termed the Petrov–Galerkin method, which is used in this book frequently.
3. Substitute the shape and weight functions into Equation 4.49 using integration by parts, and carry out the integrations using boundary conditions, which leads to a set of differential equations with respect only to time.
4. Solve the set of differential equations using standard procedures to obtain the dynamic field.
5. For static problems, step 3 leads to a set of algebraic equations, which can be solved using standard algebraic equation solvers to obtain the static field.

In the strong form used in Equation 4.49, there may be higher-order differentiations, say, an order of $2k$. However, the approximated field function needs a consistency of only kth order, because the operation of integration by parts transfers another kth order of differentiation to the weight function. This reduction of the requirement of order of consistency on the approximated field function is very important to create equation systems that provide accurate solutions. On the other hand, due to the translation of the differentiation to the weight function, the weight function used has to be differentiable up to the kth order. Other than this, there is no compulsory requirement on the weight function used. We are free to use weight functions that give us convenience in the formulation. The choice of weight function will also affect to a certain degree the accuracy of the results.

Note that to use the weighted residual method, the strong form of the system equations needs to be known, but we do not have to know whether there is a functional available for the problem.

It is very important to mention here that the quadrature domain Ω used in Equation 4.49 does not have to be the entire domain of the problem. When a local quadrature domain is used, the weighted residual method can be termed a *local weighted residual method*. This local weighted residual method is used very often in later chapters, in the form of the local Petrov–Galerkin method where the test and trial functions are chosen independently.

Note also that if the approximated field function contains the exact solution, the weighted residual method will produce the exact solutions as long as there is no numerical error in the computation. This important feature is useful in testing MFree methods developed based on the weighted residual method.

4.8 Weighted Residual Method with Constraints

When we must satisfy constraint equations, such as Equation 4.9, in addition to the strong form of governing system equations, such as Equation 4.47, the weighted residual method can be written by simply adding the weighted residual of the *penalized* constraint equations:

$$\int_{\Omega} \widehat{\mathbf{W}}^T \mathbf{f}(\mathbf{u}(x,y,z))\mathrm{d}\Omega + \int_{\Omega} \widehat{\mathbf{W}}^T \boldsymbol{\alpha}\mathbf{C}(\mathbf{u})\mathrm{d}\Omega = 0 \tag{4.50}$$

where α is a diagonal matrix given by Equation 4.14. Note that $\widehat{\mathbf{W}}$ in the first integral is a vector of weight functions corresponding to the equations in $\mathbf{f}$, and $\widehat{\mathbf{W}}$ in the second integral is a vector of weight functions corresponding to the equations in $\mathbf{C}$. All these weight functions can be, but do not have to be, the same.

4.9 Points to Note

NOTE 1 In both the method of Lagrange multipliers and the penalty method, the additional functionals need not be the domain type. If the unsatisfied conditions are on a boundary or curves in the domain, the integrals should be changed to curve integrals; if they are at points, the integrals should be changed to a summation over those points.

NOTE 2 In using the method of Lagrange multipliers, we form the constrained Lagrangians by adding a functional for the constraint to the original Lagrangians, as shown in Equation 4.10. Variation leads to a set of simultaneous system equations that is enlarged with Lagrange multipliers as additional unknowns. This procedure works for static problems. For dynamic problems, however, one may need special ways to deal with the *enlarged* set of dynamic system equations, as both the stiffness and mass matrices are nonpositive. A more practical way to obtain a set of well-behaved system equations is to formulate separately the system equations with the original Lagrangian and the constraint equations using their own weak form to obtain two separate discretized sets of equations. The discretized constrained equations are decomposed to obtain a set of orthogonal vectors that are then used to produce a *condensed* set of system equations using orthogonal transformation techniques, which at the same time ensures the satisfaction of the constraints. We demonstrate this procedure in Section 11.1 in detail when we deal with free-vibration problems using the EFG method for plates.

4.10 Remarks

We make the following two remarks:

- The weak forms to be used in MFree methods are the same as those used in FEM. In FEM, one seldom uses the constrained principles and weak forms. In MFree methods, there are many methods that use the constrained weak forms.
- The procedures used in applying these weak forms in MFree methods will be slightly different from those in FEM, because of the difference in the forms of the shape functions. The integration domain may no longer be the union of the element, and it may overlap depending on the MFree method used.

Because of the uniqueness of the MFree method in creating shape functions, the development of efficient methods to create MFree shape functions is the most important issue in MFree methods. Once a robust and efficient method is developed to approximate admissible field functions, a weak form presented in this chapter can then be used to derive the discretized system equations. The next chapter is therefore devoted entirely to methods for creating MFree shape functions, and the properties of the MFree shape functions created.

In following chapters, we use Hamilton's principle, the Galerkin weak form, the minimum total potential energy principle, and the local Petrov–Galerkin method to formulate various MFree methods for mechanics problems of solids and structures. The local Petrov–Galerkin method is used to formulate various MFree methods for mechanics problems of fluids.

5

MFree Shape Function Construction

5.1 Overview

Creation of MFree shape functions is the central and most important issue in MFree methods. The challenge is how to create shape functions using only nodes scattered arbitrarily in a domain without any predefined mesh to provide connectivity of the nodes. Development of more effective methods for constructing shape functions is thus one of the hottest areas of research in the area of MFree methods. A good method of shape function construction should satisfy the following basic requirements:

1. The nodal distribution can be arbitrary within reason, at least more flexible than that in the finite element method (FEM) (*arbitrary nodal distribution*).
2. The algorithm must be stable (*stability*).
3. The shape function constructed should satisfy a certain order of consistency (*consistency*).
4. The domain for field variable approximation/interpolation (termed the support domain or influence domain or smoothing domain) should be small compared with the entire problem domain (*compact support*).
5. The algorithm should be computationally efficient. It should be of the same order of complexity as that of FEM (*efficiency*).
6. Ideally, the shape function should possess the Kronecker delta function property (*delta function property*).
7. Ideally, the field approximation using the shape function should be compatible throughout the problem domain (*compatibility*).

Satisfaction of the above requirements ensures both easy implementation of the MFree method and accuracy of the numerical solutions. The first requirement is obvious. The stability (the second requirement) of an algorithm should always be checked, because there could be uncertainties caused by the arbitrariness in the distribution of nodes. The consistency condition (requirement 3) are essential for the convergence of the numerical results, when the nodal spacing is reduced. Satisfaction of the compact condition (requirement 4) leads to a banded system matrix that can be handled with good computational efficiency. The domain for field variable approximation/interpolation should be kept as small as possible to ensure a narrow bandwidth in the discretized system matrices. Requirement 5 prevents unacceptably expensive shape function constructions, because a too costly procedure will eventually become impractical, no matter how good it is. The sixth requirement eases imposition of essential boundary conditions, to place a limit on the extra effort needed for handling the essential boundary conditions. This requirement is not rigid because one can use special measures to impose essential boundary conditions, of course, at additional expense. The last requirement removes the need for enforcing compatibility in using the (global) Galerkin weak forms for establishing the discrete equation systems. The compatibility requirement is unnecessary if the local weight residual weak form is employed. It requires only the reproduction or consistency of the shape function to achieve convergence of the solution. (See Section 5.11 for more details.)

A number of ways to construct shape functions have been proposed. This book classifies these methods into three major categories:

1. Finite integral representation methods, which include:
   - i. Smoothed particle hydrodynamics (SPH) method
   - ii. Reproducing kernel particle method (RKPM)
   - iii. General kernel reproduction method (GKR)
2. Finite series representation methods, which include:
   - a. Moving least squares (MLS) methods:
     - i. MLS approximation
     - ii. Modified MLS approximation
   - b. Point interpolation methods (PIM):
     - i. Polynomial PIM
     - ii. Radial PIMs
   - c. Partition of unit (PU) methods:
     - i. Partition of unity finite element (PUFE)
     - ii. *hp*-clouds
   - d. Finite element methods:
     - i. Element-based interpolations
3. Finite differential representation methods, which include:
   - i. Finite difference method (regular grids)
   - ii. Finite point method (irregular grids)

Figure 5.1 shows these methods schematically. Finite integral representation methods are relatively young, but have found a special place in MFree methods with the development of smoothed particle hydrodynamics (SPH). The function is represented using its information in a local domain (smoothing domain or influence domain) via an integral form, as illustrated in Figure 5.1. Consistency is achieved by properly choosing the weight function.

Finite series representation methods have a long history of development. They are well developed in FEM, and are very active now in the area of MFree methods. Consistency is ensured by the use of the basis functions. The inclusion of special terms in the basis can also improve the accuracy of the results for certain classes of problems.

Finite difference representation methods have also been used for a long time. Convergence of the representation is ensured via the theory of the Taylor series. Finite difference representation methods are usually used for establishing system equations based on strong formulation, where we may, but usually do not, construct shape functions. The following sections detail the first two types of methods, which are widely used for creating shape functions for MFree methods.

Note that the terms "consistency" and "compatibility" are used distinctively in this book. Consistency is the capability of the field function approximation method to reproduce the fields of lowest orders of complete polynomials at any point in the problem domain.

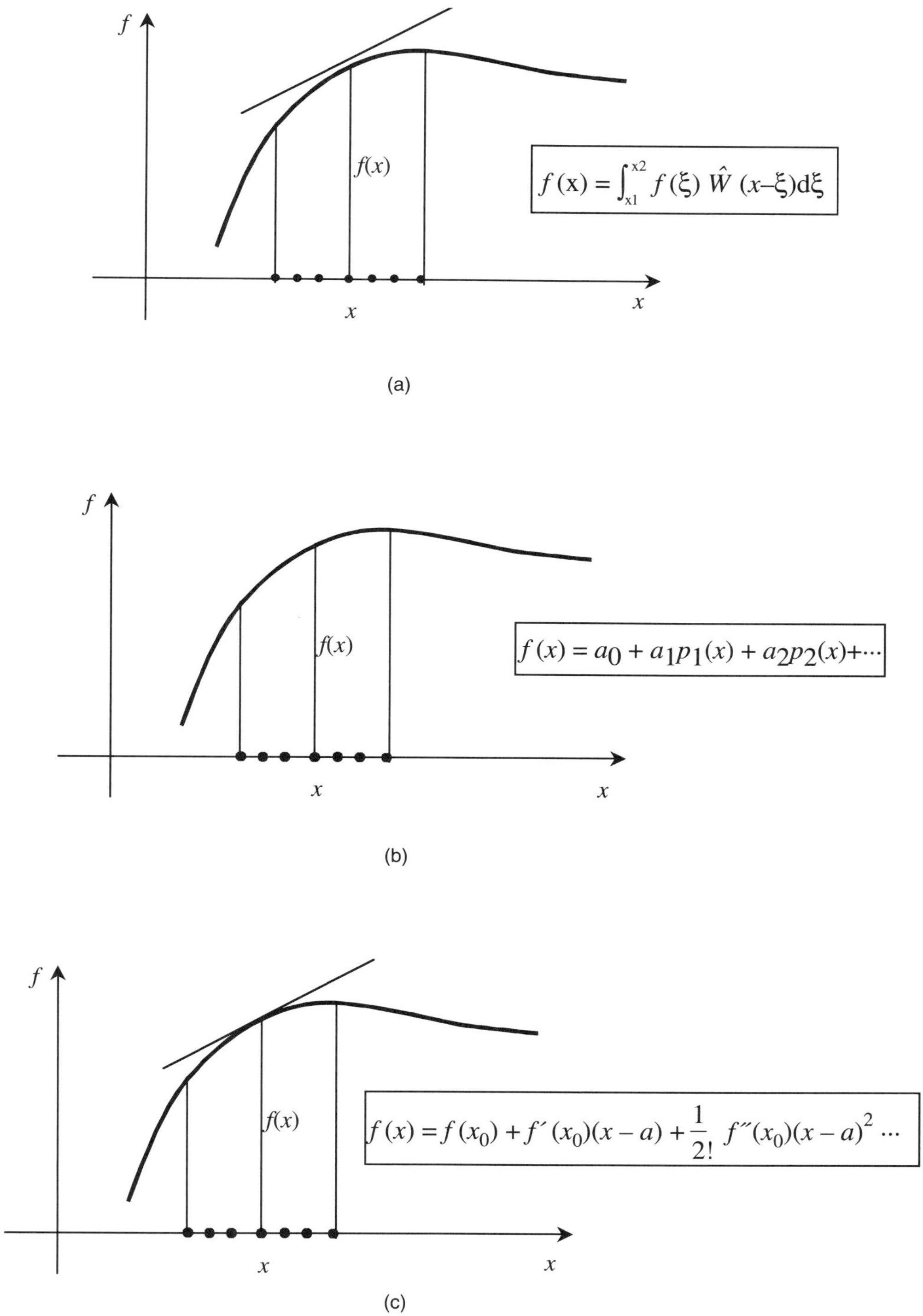

FIGURE 5.1
Methods of function representation at x using the information in its vicinity. (a) Finite integral representation. $\widehat{W}$: weight or smoothing function. (b) Finite series representation. $p_i(x)$ are basis functions. (c) Finite differential representation, where derivatives of function are used.

If the method can reproduce polynomials of up to the kth order, the method is said to have kth-order consistency. Compatibility refers to the continuity of the approximation on the boundaries between subdomains, based on which shape the functions are constructed. Both consistency and compatibility affect the accuracy and convergence of the numerical results. As will be seen later, the MLS shape functions are both consistent and compatible; the PIM shape functions, however, are consistent but not compatible. The SPH shape functions are compatible but not consistent near the boundary of the problem domain.

5.2 Smoothed Particle Hydrodynamics Approach

The SPH method uses integral representation of a function. Consider a function of $u(\mathbf{x})$ at any point of $\mathbf{x} = (x, y, z)$. Its integral representation can be given by

$$u(\mathbf{x}) = \int_{-\infty}^{+\infty} u(\boldsymbol{\xi})\delta(\mathbf{x}-\boldsymbol{\xi})\mathrm{d}\boldsymbol{\xi} \tag{5.1}$$

where $\delta(\mathbf{x})$ is the Dirac delta function. Note that this integral representation of a function is *exact*, but it is difficult to use for numerical analysis.

In SPH (Lucy, 1977; Gingold and Monaghan, 1977; Monaghan, 1982), $u(\mathbf{x})$ is *approximated* by the following finite integral form of representation:

$$u^h(\mathbf{x}) = \int_{\Omega} u(\boldsymbol{\xi})\widehat{W}(\mathbf{x}-\boldsymbol{\xi}, h)\mathrm{d}\boldsymbol{\xi} \tag{5.2}$$

where $u^h(\mathbf{x})$ represents the *approximation* of function $u(\mathbf{x})$, $\widehat{W}(\mathbf{x} - \boldsymbol{\xi}, h)$ is a kernel or weight or smoothing function, and h is termed the *smoothing length* in SPH. The smoothing length controls the size of the compact *support domain* Ω, which is often termed the *influence domain* or *smoothing domain* in SPH. The presentation of a function in the integral form of Equation 5.2 can be viewed as an approximation of the integral function representation given in Equation 5.1 over a finite domain.

In contrast to the finite differential representation of a function, this approximated integral presentation can be termed finite integral representation. The finite integral representation is conventionally termed *kernel approximation*. A finite integral representation is valid and converges when the weight function satisfies certain conditions (Monaghan, 1982):

1. $\widehat{W}(\mathbf{x}-\boldsymbol{\xi}, h) > 0$ over Ω (Positivity) (5.3)

2. $\widehat{W}(\mathbf{x}-\boldsymbol{\xi}, h) = 0$ outside Ω (Compact) (5.4)

3. $\int_{\Omega} \widehat{W}(\mathbf{x}-\boldsymbol{\xi}, h)\mathrm{d}\boldsymbol{\xi} = 1$ (Unity) (5.5)

4. $\widehat{W}$ is a monotonically decreasing function (Decay) (5.6)

5. $\widehat{W}(s, h) \to \delta(s)$ as $h \to 0$ (Delta function behavior) (5.7)

The first condition, *positivity*, is not necessary mathematically as a function representation requirement, but is important to ensure a meaningful (or stable) presentation of some physical phenomena. For example, in fluid dynamics problems, one of the field variables could be the density of the media, which can never be negative. There are different versions of SPH that do not always satisfy this condition, such as the reproducing kernel particle method (RKPM) (W. K. Liu et al., 1993), which ensures higher-order reproduction of the function and the derivatives of the function.

The second condition, *compact*, is important to the SPH method because it enables the approximation to be generated from a local representation of nodes; i.e., $u^h(\mathbf{x})$ will depend only on the values of u at nodes (particles) that are in the smoothing domain Ω in which $\widehat{W}$ is nonzero.

The third condition, *unity*, assures the zeroth-order consistency (C^0) of the integral form representation of the continuum function. Note that this does not necessarily guarantee the C^0 consistency of the discrete form of approximation.

Condition 4 is, again, not a mathematical requirement, but is imposed based on the physical considerations for the SPH method that a force exerted by a particle on another particle decreases with the increase of the distance between the two particles. This condition is violated in some MFree methods, such as the corrected kernel produced by RKPM, especially near the boundary of the problem domain.

Condition 5 is redundant, as a function that satisfies conditions 1 to 4 would naturally satisfy condition 5. In addition, the smoothing length h never goes to zero in practical computation. Condition 5 exists to allow us to observe explicitly that the method is converging to its exact form (Equation 5.1).

In summary, conditions 2 and 3 (compact and unity) are the minimum requirements for constructing a weight function for MFree methods based on finite integral representation.

The discretized form of $u^h(\mathbf{x})$ is obtained when nodal quadrature or so-called particle approximation is applied to evaluate the integral in Equation 5.2. The integral is approximated by the following summation for all the particles in the smoothing domain Ω:

$$u^h(\mathbf{x}) = \sum_I \widehat{W}(\mathbf{x} - \mathbf{x}_I) u_I \Delta V_I \tag{5.8}$$

where ΔV_I represents the volume of particle I. One difficulty in the application of the SPH method to many engineering problems is how to automatically calculate the particle volume for an arbitrary body or medium without using a mesh. In solving problems of fluid flow, the volumes of particles are treated as field variables, and updated automatically in the solution process. We still, of course, need an initial definition for these particles that represents the continuum media. The clear advantage of SPH is that, once the initial particles are defined, the subsequent updating is handled by the SPH formulation, which can naturally simulate many extreme situations, such as explosion and penetration. Some applications are covered in Chapter 7.

Equation 5.8 can be written in the following form, which is similar (in form) to the finite element formulation:

$$u^h(\mathbf{x}) = \sum_I \phi_I(\mathbf{x}) u_I \tag{5.9}$$

where $\phi_I(\mathbf{x})$ are the SPH shape functions given by

$$\phi_I(\mathbf{x}) = \widehat{W}(\mathbf{x} - \mathbf{x}_I) \Delta V_I \tag{5.10}$$

Note that, despite the similarity in form, the SPH shape function behaves very differently from the FE shape functions. The FE shape functions satisfy the Kronecker delta function defined by Equation 2.8, but the SPH shape functions do not. From Equation 5.10, it can be seen that the shape function depends only on the weight function (assume the uniform particle distribution). It is very difficult to construct a weight function that satisfies conditions 1 to 4 and the Kronecker delta function property at the same time.

Because of the lack of the delta function property, we have, in general, $u_I \neq u^h(\mathbf{x}_I)$. Therefore, u_I is termed a *nodal parameter* at node *I*, which is, in general, not the nodal value of the field variable at the node. The shape functions defined by Equation 5.10 do not provide *interpolants* because the approximated function does not pass through the parameter values at the nodes used for constructing the shape functions. Equation 5.9 is not an *interpolation* of a function, and it should be termed an *approximation* of a function. Because of this special property of the SPH shape function, the true value of the field variable should be retrieved using Equation 5.9 again, after obtaining the nodal parameters u_I at all the field nodes (particles).

5.2.1 Choice of Weight Function

Weight functions play an important role in MFree methods. They should be constructed according to the conditions given in Section 5.2. Most MFree weight functions are bell-shaped. The following is a list of commonly used weight functions.

The cubic spline weight function (W1):

$$\widehat{W}(\mathbf{x}-\mathbf{x}_I) \equiv \widehat{W}(\bar{d}) = \begin{cases} \frac{2}{3} - 4\bar{d}^2 + 4\bar{d}^3 & \text{for } \bar{d} \leq \frac{1}{2} \\ \frac{4}{3} - 4\bar{d} + 4\bar{d}^2 - \frac{4}{3}\bar{d}^3 & \text{for } \frac{1}{2} < \bar{d} \leq 1 \\ 0 & \text{for } \bar{d} > 1 \end{cases} \tag{5.11}$$

The quartic spline weight function (W2):

$$\widehat{W}(\mathbf{x}-\mathbf{x}_I) \equiv \widehat{W}(\bar{d}) = \begin{cases} 1 - 6\bar{d}^2 + 8\bar{d}^3 - 3\bar{d}^4 & \text{for } \bar{d} \leq 1 \\ 0 & \text{for } \bar{d} > 1 \end{cases} \tag{5.12}$$

The exponential weight function (W3):

$$\widehat{W}(\mathbf{x}-\mathbf{x}_I) \equiv \widehat{W}(\bar{d}) = \begin{cases} e^{-(\bar{d}/\alpha)^2} & \bar{d} \leq 1 \\ 0 & \bar{d} > 1 \end{cases} \tag{5.13}$$

where α is constant. We often use $\alpha = 0.3$.

In Equations 5.11 to 5.13,

$$\bar{d} = \frac{|\mathbf{x}-\mathbf{x}_I|}{d_W} = \frac{d}{d_W} \tag{5.14}$$

where d_W is directly related to the smoothing length h for the SPH. It defines the dimension of the domain where $\widehat{W} \neq 0$. In general, d_W can be different from point to point.

By following a general procedure for constructing weight (smoothing) functions (Liu, G. R. et al., 2002a), a new quartic weight (smoothing) function is constructed (W4):

$$\widehat{W}(\mathbf{x}-\mathbf{x}_I) \equiv \widehat{W}(\bar{d}) = \begin{cases} \frac{2}{3} - \frac{9}{2}\bar{d}^2 + \frac{19}{3}\bar{d}^3 - \frac{5}{2}\bar{d}^4 & \text{for } \bar{d} \leq 1 \\ 0 & \text{for } \bar{d} > 1 \end{cases} \tag{5.15}$$

A parabolic weight function also exists but is used less frequently. The formulation is given in Equation 8.48.

Figure 5.2 plots all four weight functions given by Equations 5.11 through 5.13, and 5.15. It can be clearly seen that the quartic weight function (W4) given in Equation 5.15 has a shape very similar to the piecewise cubic spline weight function (W1) given in Equation 5.11,

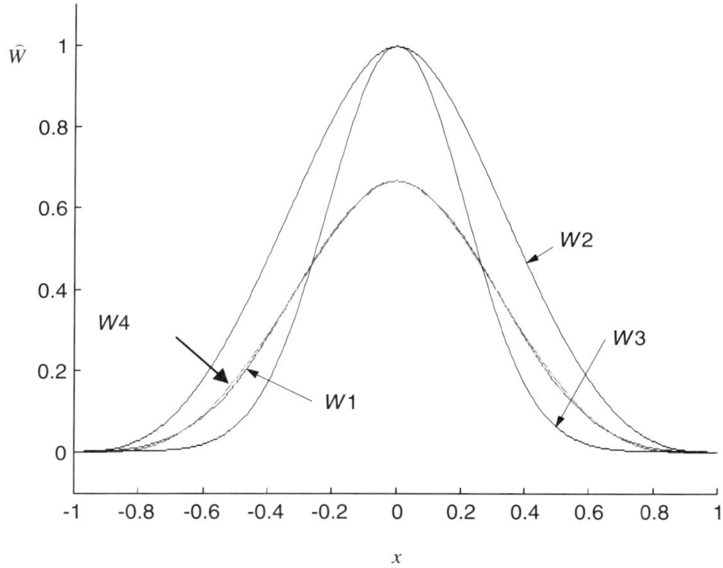

FIGURE 5.2
Weight functions. W1: cubic spline weight function; W2: quartic spline weight function; W3: exponential weight function ($\alpha = 0.3$); W4: quartic weight function.

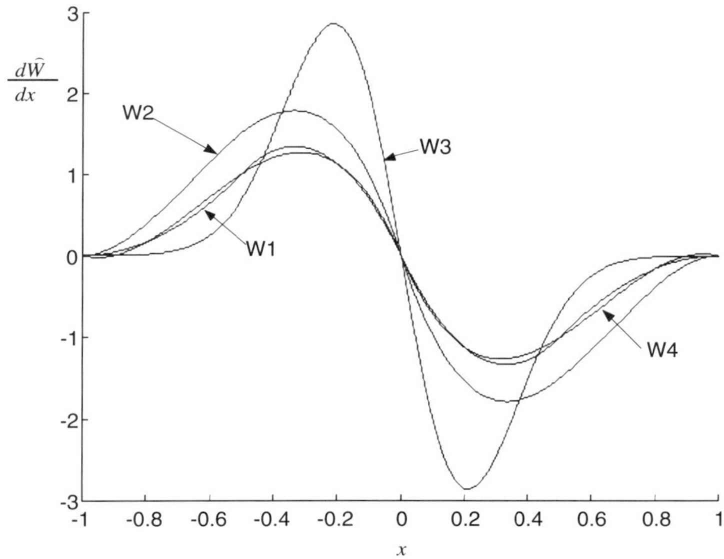

FIGURE 5.3
The first derivative of weight functions. W1: cubic spline weight function; W2: quartic spline weight function; W3: exponential weight function ($\alpha = 0.3$); W4: quartic weight function.

which has been tested and works very well for many applications. The new quartic weight function W4, however, has a simple form of one single piece, and possesses second-order reproducing capacity. Figure 5.3 plots the first derivative of all four weight functions. It is shown that the first derivatives of all four are smooth. It can be clearly seen that the first derivative of W4 still behaves similarly to that of the piecewise cubic spline function (W1).

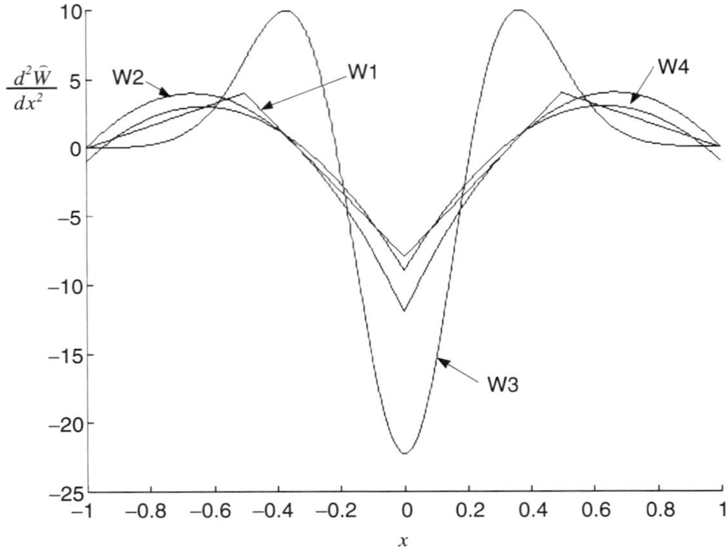

FIGURE 5.4
The second derivative of weight functions. W1: cubic spline weight function; W2: quartic spline weight function; W3: exponential weight function ($\alpha = 0.3$); W4: quartic weight function.

Figure 5.4 plots the second derivative of all four weight functions and shows that the second derivative of the cubic spline (W1) is no longer smooth. The second derivative of the new quartic weight function (W4) is still smooth but does not equal zero on the boundary. W4 is for readers who prefer the performance of W1 but at the same time want a simple one-piece formulation.

Note that the weight functions shown in Equations 5.11, 5.12, 5.13, and 5.15 need to be scaled to satisfy the condition of unity defined by Equation 5.5 for problems of different dimensions, if they are used in such finite integral representation methods as SPH. This is to ensure the consistency of function representation, as is seen in the next subsections. The scaling is immaterial if the shape functions are used in the finite series representation methods such as EFG, MLPG, PIMs, which are discussed in later sections.

In SPH methods, the following SPH weight function is often used (for 1D problems):

$$\widehat{W}(\mathbf{x}-\mathbf{x}_I, h) \equiv \widehat{W}(\bar{d}, h) = \frac{2}{3h}\begin{cases} 1-\frac{2}{3}\bar{d}^2+\frac{3}{4}\bar{d}^3 & \text{for } \bar{d}\le 1 \\ \frac{1}{4}(2-\bar{d})^3 & \text{for } 1<\bar{d}<2 \\ 0 & \text{for } \bar{d}\ge 2 \end{cases} \tag{5.16}$$

where $\bar{d} = d/h$ and h is the smoothing length. This SPH weight function is actually exactly the same as the cubic spline function given in Equation 5.11, but different in form and in the dimension of the smoothing domain.

5.2.2 Consistency

Similar to conventional FEM, an MFree method must converge, meaning that the numerical solution obtained by the MFree method must approach the exact solution when the nodal spacing approaches zero. For an MFree method to converge, the shape functions used have to satisfy a certain degree of consistency. The degree of consistency is often measured by the order of the polynomial field functions that can be exactly represented by the approximation using the shape function. If the approximation can produce a constant field function exactly, the approximation is then said to have zero-order consistency, or C^0 consistency. In general, if the approximation can produce a polynomial of up to kth order exactly, the approximation is said to have kth-order consistency, or C^k consistency. This notation convention is the same as that in conventional FEM. In finite element formulations, we also use the term *completeness*, meaning that the approximation of C^k consistency has to be completely consistent for all the lower orders from 0 to $k - 1$. In using polynomial shape functions, the C^k completeness is guaranteed by the use of all the polynomial terms completely up to the kth order. In this book, when we require C^k consistency, we imply also all the consistencies from C^0 to C^k.

In solving any partial differential equation (PDE) based on the weak form formulation, such as the Galerkin approach, there is a minimum consistency requirement for ensuring the convergence of the solution from the discretized equation system. The minimum consistency requirement depends on the order of the PDE. For a PDE of order $2k$, the minimum requirement of the consistency is C^k for Galerkin formulation. This is equivalent to the requirement of representing the polynomial of all orders up to the kth order. The reason is that an approximation that can exactly represent the polynomial of all the orders up to the kth order can represent any smooth function and all its derivatives up to the kth order with arbitrary accuracy as the nodal distance approaches zero (Belytscho et al., 1996b).

Representing a polynomial exactly is also said to reproduce the polynomial. Therefore, the reproducing concept is directly related to the concept of consistency. Let us now examine the consistency of the SPH approximation.

SPH approximation starts from the integral approximation (Equation 5.2). If we want the approximation to be of C^0 consistency, we need to require it to reproduce a constant c. Assume the field is given by $u\,(x) = c$. Substituting it into Equation 5.2, we obtain

$$u^h(\mathbf{x}) = \int_\Omega c\widehat{W}(\mathbf{x}-\boldsymbol{\xi}, h)\,\mathrm{d}\boldsymbol{\xi} = c \tag{5.17}$$

or

$$\int_\Omega \widehat{W}(\mathbf{x}-\boldsymbol{\xi}, h)\,\mathrm{d}\boldsymbol{\xi} = 1 \tag{5.18}$$

This is the condition of unity given in Equation 5.5 that a weight function has to satisfy. It is now clear that the condition of unity (Equation 5.5) ensures the lowest C^0 consistency for the SPH approximation.

Let us examine now whether the SPH approximation possesses C^1 consistency. Assume a linear field given by

$$u(\boldsymbol{\xi}) = c_0 + \mathbf{c}_1^T\boldsymbol{\xi} \tag{5.19}$$

where $\mathbf{c}_1$ and $\boldsymbol{\xi}$ are, respectively, a constant vector and the Cartesian coordinate vector. For 2D cases, they should be

$$\mathbf{c}_1 = \begin{Bmatrix} c_{1x} \\ c_{1y} \end{Bmatrix} \tag{5.20}$$

and

$$\boldsymbol{\xi} = \begin{Bmatrix} \xi \\ \eta \end{Bmatrix} \tag{5.21}$$

For 3D cases, they are

$$\mathbf{c}_1 = \begin{Bmatrix} c_{1x} \\ c_{1y} \\ c_{1z} \end{Bmatrix} \tag{5.22}$$

and

$$\boldsymbol{\xi} = \begin{Bmatrix} \xi \\ \eta \\ \zeta \end{Bmatrix} \tag{5.23}$$

Substituting the above equation into Equation 5.2, we obtain

$$u^h(\mathbf{x}) = \int_\Omega (c_0 + \mathbf{c}_1^T\boldsymbol{\xi})\widehat{W}(\mathbf{x} - \boldsymbol{\xi}, h)\mathrm{d}\boldsymbol{\xi} \tag{5.24}$$

If the approximation possesses C^1 consistency, we should have

$$u^h(\mathbf{x}) = c_0 + \mathbf{c}_1^T\mathbf{x} \tag{5.25}$$

Equating the right-hand sides of the above two equations, we have

$$c_0 + \mathbf{c}_1^T\mathbf{x} = \int_\Omega (c_0 + \mathbf{c}_1^T\boldsymbol{\xi})\widehat{W}(\mathbf{x} - \boldsymbol{\xi}, h)\mathrm{d}\boldsymbol{\xi} \tag{5.26}$$

or

$$c_0 + \mathbf{c}_1^T\mathbf{x} = c_0\int_\Omega \widehat{W}(\mathbf{x} - \boldsymbol{\xi}, h)\mathrm{d}\boldsymbol{\xi} + \mathbf{c}_1^T\int_\Omega \boldsymbol{\xi}\widehat{W}(\mathbf{x} - \boldsymbol{\xi}, h)\mathrm{d}\boldsymbol{\xi} \tag{5.27}$$

Using the condition Equation 5.5, the above equation becomes

$$\mathbf{x} = \int_\Omega \boldsymbol{\xi}\widehat{W}(\mathbf{x} - \boldsymbol{\xi}, h)\mathrm{d}\boldsymbol{\xi} \tag{5.28}$$

This gives the condition that the weight function has to satisfy for C^1 consistency. Therefore, in general, SPH does not possess C^1 consistency, if the weight function satisfies only the conditions given in Equations 5.3 to 5.7.

To examine further the conditions required for the weight function to achieve an approximation of C^1 consistency, we first multiply with **x** on both sides of Equation 5.18, which gives

$$\mathbf{x} = \int_{\Omega_I} \mathbf{x}\widehat{W}(\mathbf{x} - \boldsymbol{\xi}, h)\mathrm{d}\boldsymbol{\xi} \tag{5.29}$$

To show this more clearly, we then subtract Equation 5.28 from 5.29 to obtain

$$0 = \int_{\Omega_I} (\mathbf{x} - \boldsymbol{\xi})\widehat{W}(\mathbf{x} - \boldsymbol{\xi}, h)\mathrm{d}\boldsymbol{\xi} \tag{5.30}$$

The above integral is the first moment of the weight function. Therefore, the condition that the weight function must satisfy for C^1 consistency is that the first moment of the

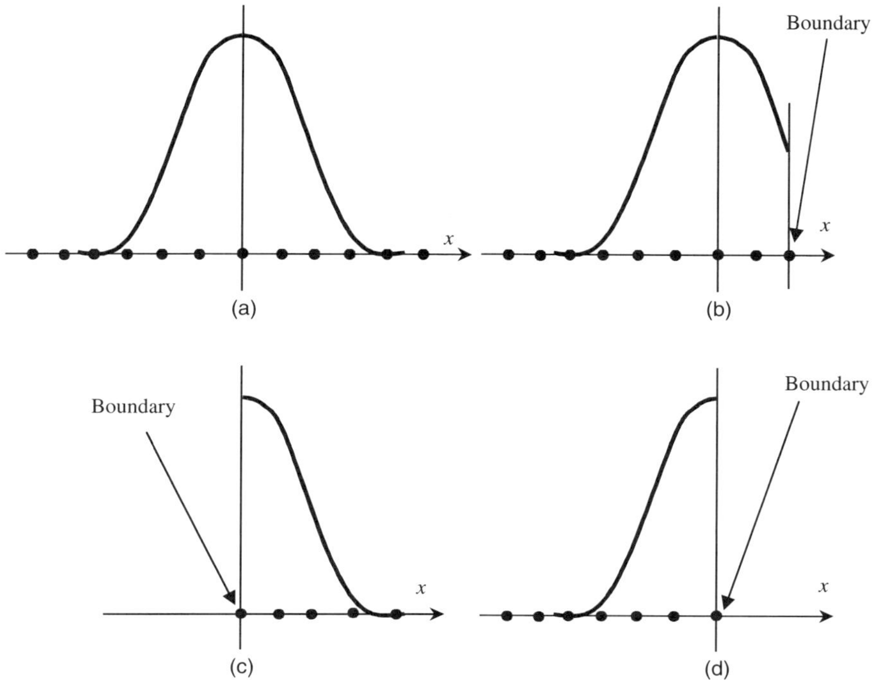

FIGURE 5.5
SPH weight functions for 1D case. (a) For an interior point, the weight function can be symmetric and the first moment vanishes. For a point near the boundary (b), or on the boundaries (c, d), the weight functions are not symmetrical.

weight function has to vanish. This condition can be satisfied if the weight function is symmetric about the origin. For an infinite problem domain, this symmetric condition is not very difficult to meet, and all the weight functions listed in Section 5.2.1 satisfy this condition. The problem occurs on and near the boundary, where it is not easy to construct a symmetric weight function. Figure 5.5 shows an example of the 1D situation. Figure 5.5a shows a weight function for an interior point, where a symmetric function can be easily defined to have linear consistency. For points near the boundary (Figure 5.5b) and on the boundary (Figure 5.5c and d), it is difficult to maintain the symmetry for linear consistency. Special treatments are required to enforce the linear consistency. This is discussed further in the following section.

Following the same procedure, it is easy to prove that the two quartic weight functions given by Equations 5.12 and 5.15 possess C^2 consistency, when the entire smoothing domain is located within the problem domain. The details can be found in a paper by G. R. Liu et al., 2002a.

5.3 Reproducing Kernel Particle Method

W. K. Liu et al. (1995) have developed a method that ensures the certain degree of consistency of the finite integral approximation and named it the reproducing kernel particle method (RKPM). W. K. Liu's advancement is achieved by adding a correction function to the kernel in Equation 5.2. This correction function is particularly useful in improving the SPH approximation near the boundaries as well as making it linearly or C^1 consistent near the boundary. The finite integral representation of a function with the correction function can be given by

$$u^h(\mathbf{x}) = \int_{\Omega} u(\boldsymbol{\xi})C(\mathbf{x}, \boldsymbol{\xi})\widehat{W}(\mathbf{x}-\boldsymbol{\xi}, h)d\boldsymbol{\xi} \tag{5.31}$$

where $C(\mathbf{x}, \boldsymbol{\xi})$ is the correction function. An example of the correction function in one dimension is

$$C(\mathbf{x}, \boldsymbol{\xi}) = c_1(\mathbf{x}) + c_2(\mathbf{x})(\boldsymbol{\xi}-\mathbf{x}) \tag{5.32}$$

where $c_1(\mathbf{x})$ and $c_2(\mathbf{x})$ are coefficients. The coefficients are found (Liu, W. K. et al., 1995) by enforcing the corrected kernel to reproduce the function:

$$c_1(\mathbf{x}) = \frac{m_2(\mathbf{x})}{(m_0(\mathbf{x})m_2(\mathbf{x}) - m_1^2(\mathbf{x}))} \tag{5.33}$$

$$c_2(\mathbf{x}) = \frac{m_1(\mathbf{x})}{(m_0(\mathbf{x})m_2(\mathbf{x}) - m_1^2(\mathbf{x}))} \tag{5.34}$$

where m_0, m_1, and m_2 are the moments of W, defined by

$$m_0(\mathbf{x}) = \int \widehat{W}(\mathbf{x}-\boldsymbol{\xi})d\boldsymbol{\xi} \tag{5.35}$$

$$m_1(\mathbf{x}) = \int \boldsymbol{\xi}\widehat{W}(\mathbf{x}-\boldsymbol{\xi})d\boldsymbol{\xi} \tag{5.36}$$

$$m_2(\mathbf{x}) = \int \boldsymbol{\xi}^2\widehat{W}(\mathbf{x}-\boldsymbol{\xi})d\boldsymbol{\xi} \tag{5.37}$$

If the integral in Equation 5.31 is discretized, then a function $u(\mathbf{x})$ can be approximated using the surrounding particles:

$$u^h(\mathbf{x}) = \sum_I C(\mathbf{x}, \mathbf{x}_I)\widehat{W}(\mathbf{x}-\mathbf{x}_I)u_I\Delta V_I = \sum_I \phi_I(\mathbf{x})u_I \tag{5.38}$$

where $\phi_I(\mathbf{x})$ are the RKPM shape functions given by

$$\phi_I(\mathbf{x}) = C(\mathbf{x}, \mathbf{x}_I)\widehat{W}(\mathbf{x}-\mathbf{x}_I)\Delta V_I \tag{5.39}$$

Note that the corrected weight function may not satisfy the conditions of Equations 5.3 and 5.6 listed in the previous subsection.

The RKPM method has been advanced and applied successfully to solve many problems of solids, structures, acoustics, fluids, etc. Readers are referred to publications by the group led by W. K. Liu (Liu, W. K. et al., 1993, 1995, 1996, 1997a,b,c; Chen, J. S. et al., 1996; Uras et al., 1997).

5.4 Moving Least Squares Approximation

Moving least squares (MLS), originated by mathematicians for data fitting and surface construction, is often termed local regression and loss (Lancaster and Salkauskas, 1981; Cleveland, 1993). It can be categorized as a method of finite series representation of functions. An excellent description of the MLS method can be found in a paper by Lancaster and Salkauskas (1981). The MLS method is now a widely used alternative for constructing MFree shape functions for approximation. Nayroles et al. (1992) were the first to use MLS approximation to construct shape functions for their diffuse element method (DEM) for mechanics problems. DEM was modified by Belytschko et al. (1994b), who named it the element free Galerkin (EFG) method, where the MLS approximation is also employed. The invention of DEM and the advances in EFG have created great impact on the development of MFree methods. The MLS approximation has two major features that make it popular: (1) the approximated field function is continuous and smooth in the entire problem domain; and (2) it is capable of producing an approximation with the desired order of consistency. The procedure of constructing shape functions for MFree methods using MLS approximation is detailed in this section.

5.4.1 MLS Procedure

Let $u(\mathbf{x})$ be the function of the field variable defined in the domain Ω. The approximation of $u(\mathbf{x})$ at point $\mathbf{x}$ is denoted $u^h(\mathbf{x})$. MLS approximation first writes the field function in the form:

$$u^h(\mathbf{x}) = \sum_j^m p_j(\mathbf{x})a_j(\mathbf{x}) \equiv \mathbf{p}^T(\mathbf{x})\mathbf{a}(\mathbf{x}) \tag{5.40}$$

where m is the number of terms of monomials (polynomial basis), and $\mathbf{a}(\mathbf{x})$ is a vector of coefficients given by

$$\mathbf{a}^T(x) = \{a_0(x) \ \ a_1(x) \ \ \ldots \ \ a_m(x)\} \tag{5.41}$$

which are functions of x.

In Equation 5.40, $\mathbf{p}(\mathbf{x})$ is a vector of basis functions that consists most often of monomials of the lowest orders to ensure minimum completeness. Enhancement functions can, however, be added to achieve better efficiency or to produce stress fields of special characteristics, such as singularity at the crack tip and stress discontinuity at interfaces of different types of materials. Here we discuss the use of the pure polynomial basis, which is very similar to that used in FEM (Kardestuncer, 1987; Zienkiewicz, 1989; Rao, 1999). In 1D space, a complete polynomial basis of order m is given by

$$\mathbf{p}^T(x) = \{p_0(x), p_1(x), \ldots, p_m(x)\} = \{1, x, x^2, \ldots, x^m\} \tag{5.42}$$

and in 2D space,

$$\mathbf{p}^T(\mathbf{x}) = \mathbf{p}^T(x, y) = \{1, x, y, xy, x^2, y^2, \ldots, x^m, y^m\} \tag{5.43}$$

In this case, the Pascal triangle shown in Figure 5.6 can be utilized to build $\mathbf{p}^T(\mathbf{x})$, and the number of nodes in the support domain can be chosen accordingly.

In 3D space, we have

$$\mathbf{p}^T(\mathbf{x}) = \mathbf{p}^T(x, y, z) = \{1, x, y, z, xy, yz, zx, x^2, y^2, z^2, \ldots, x^m, y^m, z^m\} \tag{5.44}$$

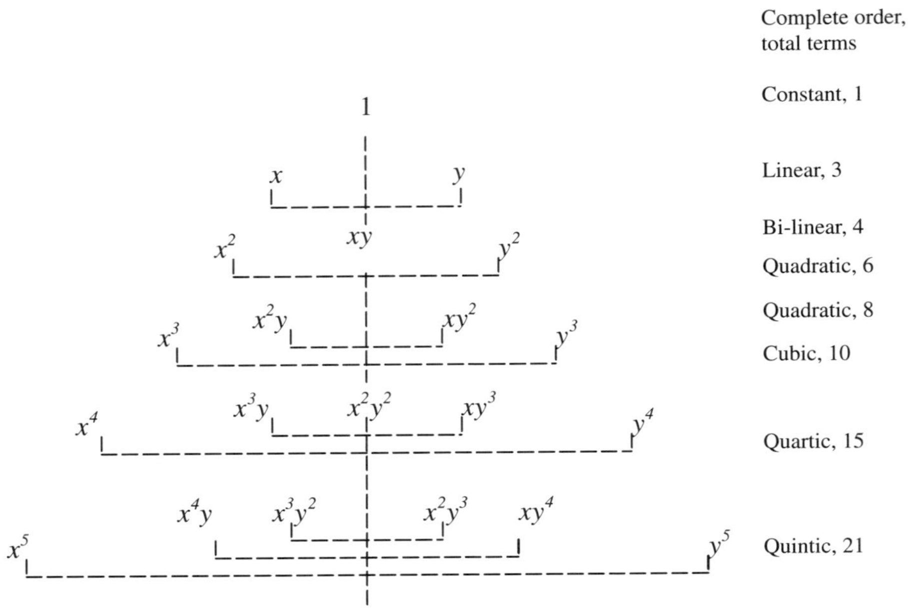

FIGURE 5.6
Pascal triangle of monomials, 2D case.

In this case, the Pascal pyramid shown in Figure 5.7 can be employed to build $\mathbf{p}^T(\mathbf{x})$. The vector of coefficients $\mathbf{a}(\mathbf{x})$ in Equation 5.40 is determined using the function values at a set of nodes that are included in the *support domain* of $\mathbf{x}$. A support domain of a point $\mathbf{x}$ determines the number of nodes that are used locally to approximate the function value at $\mathbf{x}$, as shown in Figure 2.8.

We explained in Section 2.10 that we have an alternative way of selecting nodes for constructing shape functions using the concept of an influence domain.

Given a set of n nodal values for the field function $u_1, u_2, \ldots, u_n$, at n nodes $\mathbf{x}_1, \mathbf{x}_2, \ldots, \mathbf{x}_n$ that are in the *support domain*, Equation 5.40 is then used to calculate the approximated values of the field function at these nodes:

$$u^h(\mathbf{x}, \mathbf{x}_I) = \mathbf{p}^T(\mathbf{x}_I)\mathbf{a}(\mathbf{x}), \quad I = 1, 2, \ldots, n \tag{5.45}$$

Note that $\mathbf{a}(\mathbf{x})$ here is an arbitrary function of $\mathbf{x}$. A functional of weighted residual is constructed using the approximated values of the field function and the *nodal parameters*, $u_I = u(\mathbf{x}_I)$, that are shown in Figure 5.8, i.e.,

$$\begin{aligned} J &= \sum_I^n \widehat{W}(\mathbf{x}-\mathbf{x}_I)[u^h(\mathbf{x}, \mathbf{x}_I) - u(\mathbf{x}_I)]^2 \\ &= \sum_I^n \widehat{W}(\mathbf{x}-\mathbf{x}_I)[\mathbf{p}^T(\mathbf{x}_I)\mathbf{a}(\mathbf{x}) - u_I]^2 \end{aligned} \tag{5.46}$$

where $\widehat{W}(\mathbf{x} - \mathbf{x}_I)$ is a weight function, and u_I is the nodal parameter of the field variable at node I.

Note that the weight function used in Equation 5.46 has a different mathematical mission than that used for finite integral representation methods, such as that in Equation 5.2. The weight function used in Equation 5.46 plays two important roles in constructing MLS

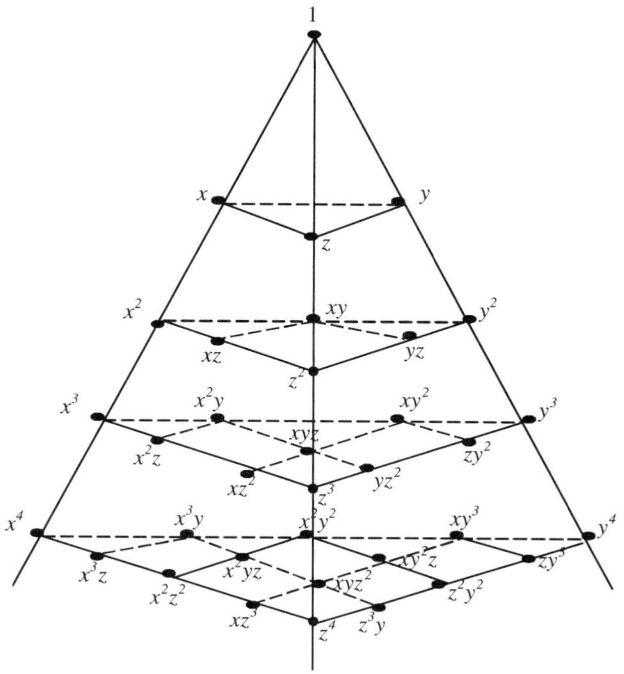

FIGURE 5.7
Pascal pyramid of monomials, 3D case.

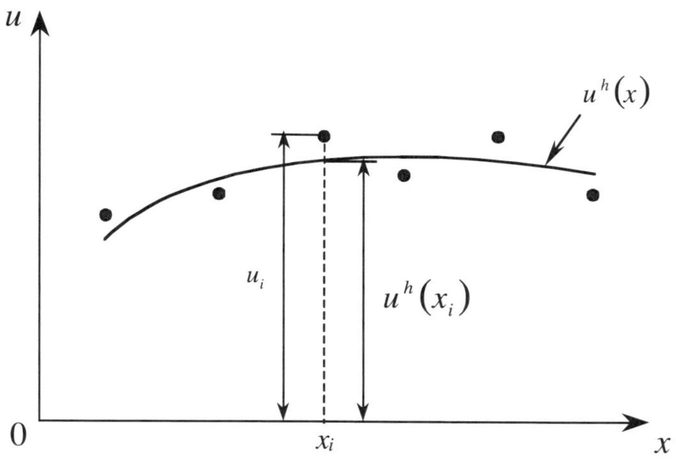

FIGURE 5.8
The approximation function $u^h(x)$ and the nodal parameters u_I in the MLS approximation.

shape functions. The first is to provide weightings for the residuals at different nodes in the support domain. We usually prefer nodes farther from x to have small weights. The second role is to ensure that nodes leave or enter the support domain in a gradual (smooth) manner when x moves. The second role of the weight function is very important, because it makes sure that the MLS shape functions to be constructed satisfy the compatibility condition, Equation 4.1. The weight function is not responsible for the consistency of the shape functions to be created later. The weight function used in Equation 5.46 can theoretically be any function as long as it satisfies the conditions of Equations 5.3, 5.4, and 5.6. Equation 5.4 also ensures a local support feature that leads to a banded system matrix. Equation 5.6 gives more weighting to the nodes that are closer to **x** where the field variable is to be approximated. Any weight functions shown in Equations 5.11, 5.12, 5.13, and 5.15 can be and have been used in MLS approximation. The scaling to meet the condition of unity that is required in SPH is not necessary here for MLS approximation.

In MLS approximation, at an arbitrary point **x**, **a**(**x**) is chosen to minimize the weighted residual. The minimization condition requires

$$\frac{\partial J}{\partial \mathbf{a}} = 0 \tag{5.47}$$

which results in the following linear equation system:

$$\mathbf{A}(\mathbf{x})\mathbf{a}(\mathbf{x}) = \mathbf{B}(\mathbf{x})\mathbf{U}_s \tag{5.48}$$

where **A** is called the (weighted) moment matrix given by

$$\mathbf{A}(\mathbf{x}) = \sum_I^n \widehat{W}_I(\mathbf{x})\mathbf{p}(\mathbf{x}_I)\mathbf{p}^T(\mathbf{x}_I) \tag{5.49}$$

where

$$\widehat{W}_I(\mathbf{x}) \equiv \widehat{W}(\mathbf{x} - \mathbf{x}_I) \tag{5.50}$$

In Equation 5.48, matrix **B** has the form of

$$\mathbf{B}(\mathbf{x}) = [\mathbf{B}_1, \mathbf{B}_2, \ldots, \mathbf{B}_n] \tag{5.51}$$

$$\mathbf{B}_I = \widehat{W}_I(\mathbf{x})\mathbf{p}(\mathbf{x}_I) \tag{5.52}$$

and $\mathbf{U}_s$ is the vector that collects the nodal parameters of the field variables for all the nodes in the support domain:

$$\mathbf{U}_s = \{u_1, u_2, \ldots, u_n\}^T \tag{5.53}$$

Solving Equation 5.48 for **a**(**x**), we obtain

$$\mathbf{a}(\mathbf{x}) = \mathbf{A}^{-1}(\mathbf{x})\mathbf{B}(\mathbf{x})\mathbf{U}_s \tag{5.54}$$

Substituting the above equation back into Equation 5.46 leads to

$$u^h(\mathbf{x}) = \sum_I^n \sum_j^m p_j(\mathbf{x})(\mathbf{A}^{-1}(\mathbf{x})\mathbf{B}(\mathbf{x}))_{jI} u_I \tag{5.55}$$

or

$$u^h(\mathbf{x}) = \sum_I^n \phi_I(\mathbf{x}) u_I \tag{5.56}$$

where the MLS shape function $\phi_I(\mathbf{x})$ is defined by

$$\phi_I(\mathbf{x}) = \sum_j^m p_j(\mathbf{x})(\mathbf{A}^{-1}(\mathbf{x})\mathbf{B}(\mathbf{x}))_{jI} = \mathbf{p}^T\mathbf{A}^{-1}\mathbf{B}_I \tag{5.57}$$

Note that m is the number of terms of polynomial basis $p(\mathbf{x})$, which is usually much smaller than n, which is the number of nodes used in the support domain for constructing the shape function. The requirement of $n >> m$ prevents the singularity of the weighted moment matrix, so that $\mathbf{A}^{-1}$ exists.

Equation 5.56 can also be written in the following matrix form:

$$u^h(\mathbf{x}) = \mathbf{\Phi}(\mathbf{x})\mathbf{U}_s \tag{5.58}$$

where $\mathbf{\Phi}(\mathbf{x})$ is the vector of MLS shape functions corresponding to n nodes in the support domain:

$$\mathbf{\Phi}([\mathbf{x}]) = [\phi_1(\mathbf{x}), \phi_2(\mathbf{x}), \ldots, \phi_n(\mathbf{x})] \tag{5.59}$$

To determine the spatial derivatives of the function of the field variable, which are required for deriving the discretized system equations, it is necessary to obtain the derivatives of the MLS shape functions. For convenience, to obtain the partial derivatives of shape functions, Equation 5.59 is first rewritten as follows using Equation 5.57:

$$\mathbf{\Phi}(\mathbf{x}) = \boldsymbol{\gamma}^T(\mathbf{x})\mathbf{B}(\mathbf{x}) \tag{5.60}$$

where $\boldsymbol{\gamma}(\mathbf{x})$ is determined by

$$\mathbf{A}(\mathbf{x})\boldsymbol{\gamma}(\mathbf{x}) = \mathbf{p}(\mathbf{x}) \tag{5.61}$$

The partial derivatives of $\boldsymbol{\gamma}(\mathbf{x})$ can be obtained as follows:

$$\mathbf{A}\boldsymbol{\gamma}_{,i} = \mathbf{p}_{,i} - \mathbf{A}_{,i}\boldsymbol{\gamma} \tag{5.62}$$

$$\mathbf{A}\boldsymbol{\gamma}_{,ij} = \mathbf{p}_{,ij} - (\mathbf{A}_{,i}\boldsymbol{\gamma}_{,j} + \mathbf{A}_{,j}\boldsymbol{\gamma}_{,i} + \mathbf{A}_{,ij}\boldsymbol{\gamma}) \tag{5.63}$$

$$\mathbf{A}\boldsymbol{\gamma}_{,ijk} = \mathbf{p}_{,ijk} - (\mathbf{A}_{,i}\boldsymbol{\gamma}_{,jk} + \mathbf{A}_{,j}\boldsymbol{\gamma}_{,ik} + \mathbf{A}_{,k}\boldsymbol{\gamma}_{,ij} + \mathbf{A}_{,ij}\boldsymbol{\gamma}_{,k} + \mathbf{A}_{,ik}\boldsymbol{\gamma}_{,j} + \mathbf{A}_{,jk}\boldsymbol{\gamma}_{,i} + \mathbf{A}_{,ijk}\boldsymbol{\gamma}) \tag{5.64}$$

where i, j, and k denote coordinates x and y. A comma designates a partial derivative with respect to the indicated spatial variable. The partial derivatives of shape function $\mathbf{\Phi}$ can then be obtained as follows:

$$\mathbf{\Phi}_{,i} = \boldsymbol{\gamma}^T_{,i}\mathbf{B} + \boldsymbol{\gamma}^T\mathbf{B}_{,i} \tag{5.65}$$

$$\mathbf{\Phi}_{,ij} = \boldsymbol{\gamma}^T_{,ij}\mathbf{B} + \boldsymbol{\gamma}^T_{,i}\mathbf{B}_{,j} + \boldsymbol{\gamma}^T_{,j}\mathbf{B}_{,i} + \boldsymbol{\gamma}^T\mathbf{B}_{,ij} \tag{5.66}$$

$$\mathbf{\Phi}_{,ijk} = \boldsymbol{\gamma}^T_{,ijk}\mathbf{B} + \boldsymbol{\gamma}^T_{,ij}\mathbf{B}_{,k} + \boldsymbol{\gamma}^T_{,ik}\mathbf{B}_{,j} + \boldsymbol{\gamma}^T_{,jk}\mathbf{B}_{,i} + \boldsymbol{\gamma}^T_{,i}\mathbf{B}_{,jk} + \boldsymbol{\gamma}^T_{,j}\mathbf{B}_{,ik} + \boldsymbol{\gamma}^T_{,k}\mathbf{B}_{,ij} + \boldsymbol{\gamma}^T\mathbf{B}_{,ijk} \tag{5.67}$$

It should be noted that MLS shape functions do not satisfy the Kronecker delta criterion $\phi_I(\mathbf{x}_J) \neq \delta_{IJ}$ that results in $u^h(\mathbf{x}_I) \neq u_I$; i.e., the nodal parameters u_I are not the nodal values of $u^h(\mathbf{x}_I)$. Therefore, they are not interpolants, but rather approximates of a function. Figure 5.8 gives a 1D example of the MLS approximation. The approximation of the displacement at the Ith node $u^h(\mathbf{x}_I)$ depends not only on the nodal parameter u_I but also on the nodal parameters u_1 through u_n, parameters that correspond to all the nodes within the support domain of node I. This is expressed in the sum given in Equation 5.56. This property makes the imposition of essential boundary conditions more complicated than that in the FEM.

A plot of a typical 1D MLS weight function and shape function is given in Figure 5.9. The shape function is for the node at $x = 0$ and is obtained using five nodes evenly distributed in the support domain of [−1, 1]. The quartic spline weight function (W2) is used. It can be seen that the MLS shape function attains a maximum value that is considerably less than 1. For this plot, the quartic weight function (Equation 5.12) is used with $d_W = 0.45\alpha_s$, where $\alpha_s = 2.5$.

Note that the dimension of the support domain d_s in MLS approximation is determined by the dimension of the weight function d_W. Therefore, $d_W = d_s$. The procedure for determining d_s has already been covered in Sections 2.10.2 and 2.10.3. These methods can be used here to determine d_W for both uniformly and nonuniformly distributed nodes in 1D, 2D, and 3D domains.

5.4.2 Consistency

The consistency of the MLS approximation depends on the complete order of the monomial employed in Equation 5.42 or 5.43. If the complete order of the monomial is k, the MLS shape function will possess C^k consistency. To demonstrate, we follow the argument of Krongauz and Belytschko (1996).

Note that J in Equation 5.46 is positive definite, because the weight function is chosen positive (Equation 5.3). Therefore, its minimum is non-negative. Consider a field given by

$$u(\mathbf{x}) = \sum_j^k p_j(\mathbf{x})\alpha_j(\mathbf{x}), \quad k \leq m \tag{5.68}$$

Such a given field can always be written in the form of

$$u(\mathbf{x}) = \sum_j^m p_j(\mathbf{x})\alpha_j(\mathbf{x}) \tag{5.69}$$

by simply assigning $\alpha_j = 0$ for $k < j$. Then, if we let $a_i(\mathbf{x}) = \alpha_i$, J will vanish and it will necessarily be a minimum, which leads to

$$u^h(\mathbf{x}) = \sum_j^k p_j(\mathbf{x})\alpha_j(\mathbf{x}) = u(\mathbf{x}) \tag{5.70}$$

This proves that any field given by Equation 5.68 will be exactly represented or reproduced by the MLS approximation. This proof procedure also implies that any function in the basis is reproduced exactly. This feature of MLS approximation is, in fact, very easy to understand by intuition: the MLS approximation seeks a set of coefficient $\mathbf{a}(\mathbf{x})$ that can

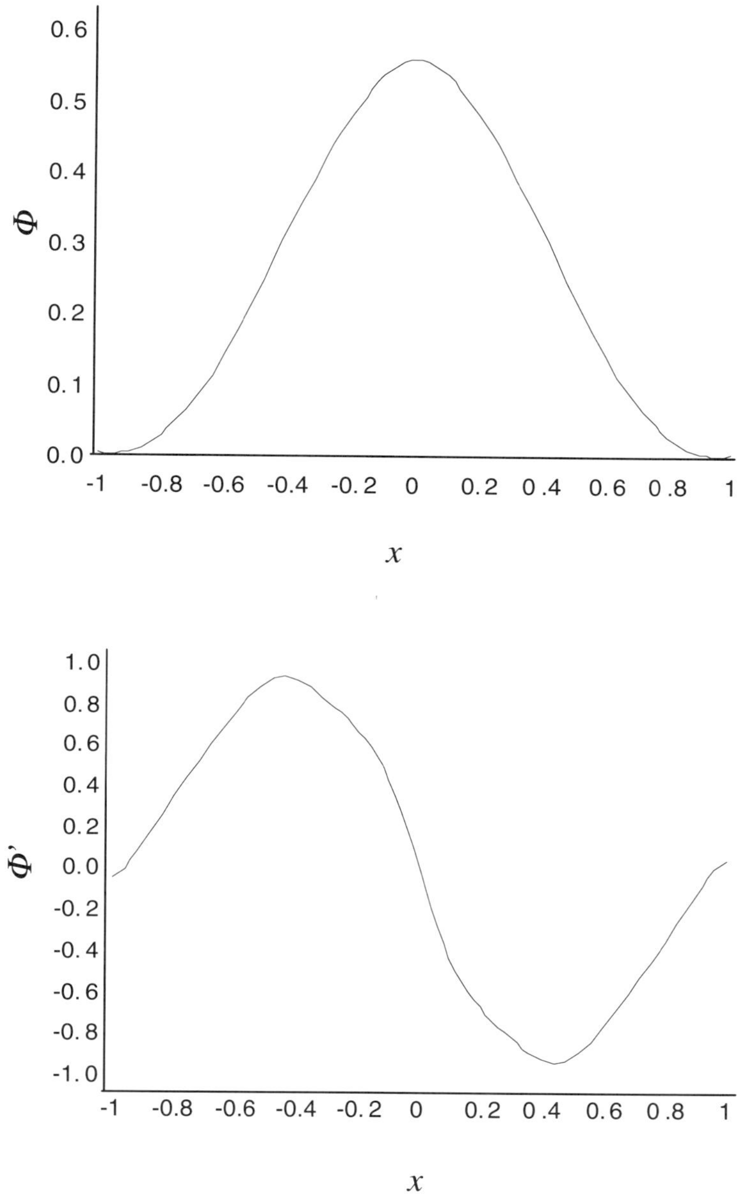

FIGURE 5.9
MLS shape function in 1D space for the node at $x = 0$ obtained using five nodes evenly distributed in the support domain of [–1, 1]. Quartic spline weight function (W2) is used. (a) MLS shape function; (b) derivative of the shape function. Note that the MLS shape function does not possess the Kronecker delta function property.

produce a function of $u(\mathbf{x}) = \sum_j^m p_j(\mathbf{x})\alpha_j(\mathbf{x})$ with a minimum distance norm to the actual function. If the actual function is in the basis of $p_j(\mathbf{x})$, MLS approximation will simply produce the basis because the distance norm is zero, which is, of course, the minimum.

The proof of the consistency of MLS approximation is valid for proving another important feature of MLS approximation: any function that appears in the basis can be reproduced exactly. To have the MLS approximation exhibit linear consistency, all one need do is include the constant and linear monomials into the basis. Making use of this feature further, one can develop shape functions for simulating a singular stress field at a crack tip by including singular functions into the basis (see Belytschko et al., 1994a, 1996a; Fleming et al., 1997). However, one has to make sure that the weighted moment matrix computed using Equation 5.49 is still invertible, when these additional basis functions are included.

5.4.3 Continuous Moving Least Square Approximation

Belytschko et al. (1996b) have also shown the relation between the approximations of RKPM and MLS. Their interesting procedure starts with the construction of the continuous form of the MLS approximation. The continuous counterpart of the Equation 5.46 can be written as

$$J(\mathbf{x}) = \int_\Omega \widehat{W}(\mathbf{x}-\boldsymbol{\xi})[u^h(\mathbf{x}, \boldsymbol{\xi}) - u(\boldsymbol{\xi})]^2 \mathrm{d}\boldsymbol{\xi} \tag{5.71}$$

where $\widehat{W}(\mathbf{x}-\boldsymbol{\xi})$ is a weight function of compact support. The approximation of the function has the form:

$$u^h(\mathbf{x}, \boldsymbol{\xi}) = \sum_i^m p_i(\boldsymbol{\xi}) a_i(\mathbf{x}) \tag{5.72}$$

The condition for minimizing $J(\mathbf{x})$ leads to the following equation for solving $a_j(\mathbf{x})$:

$$\frac{J(\mathbf{x})}{a_i(\mathbf{x})} = 0 \tag{5.73}$$

or

$$2\int_\Omega \left[\widehat{W}(\mathbf{x}-\boldsymbol{\xi})\left(\sum_i^m p_i(\boldsymbol{\xi})a_i(\mathbf{x}) - u(\boldsymbol{\xi})\right)\sum_j^m p_j(\boldsymbol{\xi})\mathrm{d}a_j(\mathbf{x})\right]\mathrm{d}\boldsymbol{\xi} = 0 \tag{5.74}$$

Because the foregoing equation has to be satisfied for all $\mathrm{d}a_j(\mathbf{x})$, we obtain the following equation for solving $a_j(\mathbf{x})$:

$$\sum_j^m \bar{A}_{ij} a_j(\mathbf{x}) = \int_\Omega [\widehat{W}(\mathbf{x}-\boldsymbol{\xi}) p_i(\boldsymbol{\xi}) u(\boldsymbol{\xi})]\mathrm{d}\boldsymbol{\xi} \tag{5.75}$$

where

$$\bar{A}_{ij}(\mathbf{x}) = \int_\Omega [\widehat{W}(\mathbf{x}-\boldsymbol{\xi}) p_i(\boldsymbol{\xi}) p_j(\boldsymbol{\xi})]\mathrm{d}\boldsymbol{\xi} \tag{5.76}$$

which is the continuous counterpart of the discrete moment matrix $\mathbf{A}(\mathbf{x})$ given in Equation 5.49. Solving Equation 5.75 for $a_j(\mathbf{x})$, we have

$$a_j(\mathbf{x}) = \bar{A}_{ij}^{-1}(\mathbf{x}) \int_\Omega [\widehat{W}(\mathbf{x}-\boldsymbol{\xi}) p_i(\boldsymbol{\xi}) u(\boldsymbol{\xi})]\mathrm{d}\boldsymbol{\xi} \tag{5.77}$$

Substituting the above equation into Equation 5.72, we obtain

$$u^h(\mathbf{x}, \boldsymbol{\xi}) = p_j(\boldsymbol{\xi})\bar{A}_{ij}^{-1}(\mathbf{x})\int_{\Omega}[\widehat{W}(\mathbf{x}-\boldsymbol{\xi}')p_i(\boldsymbol{\xi}')u(\boldsymbol{\xi}')]\mathrm{d}\boldsymbol{\xi}' \tag{5.78}$$

Note that in the above equation $\boldsymbol{\xi}'$ is used for the integral variable. The approximation at $\mathbf{x}$ is then obtained by letting $\mathbf{x} = \boldsymbol{\xi}$.

$$u^h(\mathbf{x}) = \int_{\Omega}[p_j(\mathbf{x})\bar{A}_{ij}^{-1}(\mathbf{x})p_i(\boldsymbol{\xi}')\widehat{W}(\mathbf{x}-\boldsymbol{\xi}')u(\boldsymbol{\xi}')]\mathrm{d}\boldsymbol{\xi}' \tag{5.79}$$

Comparison with Equation 5.2 reveals the similarity between the SPH and MLS approximations. Defining

$$C(\mathbf{x}, \boldsymbol{\xi}') = p_j(\mathbf{x})\bar{A}_{ij}^{-1}(\mathbf{x})p_i(\boldsymbol{\xi}') \tag{5.80}$$

produces the additional term for ensuring the consistency. This term is similar to the correction function used by W. K. Liu et al. (1995) for restoring consistency in SPH.

5.5 Point Interpolation Method

As the name implies, the point interpolation method (PIM) obtains its approximation by letting the interpolation function pass through the function values at each scattered node within the defined domain of support. PIM can be categorized as a finite series representation method. The basic procedure for PIM is given as follows.

Consider a function $u(\mathbf{x})$ defined in the problem domain Ω. The domain is represented by a set of nodes scattered in the problem domain and on the boundary. The PIM interpolates $u(\mathbf{x})$ using the nodal values at the nodes of the support domain of a point of interest at $\mathbf{x}_Q$. The PIM formulation starts with the following finite series representation:

$$u^h(\mathbf{x}, \mathbf{x}_Q) = \sum_{i=1}^{n} B_i(\mathbf{x})a_i(\mathbf{x}_Q) \tag{5.81}$$

where $B_i(\mathbf{x})$ are the basis functions defined in the Cartesian coordinate space $\mathbf{x}^T = [x, y, z]$, n is the number of nodes in the support domain of point $\mathbf{x}_Q$, and $a_i(\mathbf{x}_Q)$ is the coefficient for the basis function $B_i(\mathbf{x})$ corresponding to the given point $\mathbf{x}_Q$. There are two types of PIM that have been developed so far using different forms of basis functions. PIM using polynomial basis functions was originally developed by Liu and Gu (1999, 2001c). PIM using radial basis functions was developed by Wang and Liu (2000). The following subsection details the procedure of constructing polynomial PIM shape functions.

5.5.1 Polynomial PIM

The formulation of polynomial PIM starts with the following assumption:

$$u^h(\mathbf{x}, \mathbf{x}_Q) = \sum_{i=1}^{n} p_i(\mathbf{x})a_i(\mathbf{x}_Q) = \mathbf{p}^T(\mathbf{x})\mathbf{a}(\mathbf{x}_Q) \tag{5.82}$$

where $p_i(\mathbf{x})$ is the basis function of monomials in the space coordinates $\mathbf{x}^T = \{x, y, z\}$, n is the number of nodes in the support domain of point $\mathbf{x}_Q$, and $a_i(\mathbf{x}_Q)$ is the coefficient for the monomial $p_i(\mathbf{x})$ corresponding to the given point $\mathbf{x}_Q$. Vector $\mathbf{a}$ is defined as

$$\mathbf{a}^T(\mathbf{x}_Q) = \{a_1, a_2, a_3, \ldots, a_n\} \tag{5.83}$$

The $p_i(\mathbf{x})$ in Equation 5.82 is built utilizing the Pascal triangle shown in Figure 5.6 or 5.7, so that the basis is complete. A basis in one dimension has the form:

$$\mathbf{p}^T(x) = \{1, x, x^2, x^3, x^4, \ldots, x^n\} \tag{5.84}$$

A basis in the 2D domain is provided by

$$\mathbf{p}^T(\mathbf{x}) = \mathbf{p}^T(x, y) = \{1, x, y, xy, x^2, y^2, \ldots, x^n, y^n\} \tag{5.85}$$

and that in the 3D domain can be written as

$$\mathbf{p}^T(\mathbf{x}) = \mathbf{p}^T(x, y, z) = \{1, x, y, z, xy, yz, zx, x^2, y^2, z^2, \ldots, x^n, y^n, z^n\} \tag{5.86}$$

As a general rule, all the terms in the basis should be selected symmetrically from the Pascal triangle shown in Figure 5.6 or 5.7. Some higher-order terms can be selectively included in the polynomial basis if there is a need.

The coefficients a_i in Equation 5.82 can be determined by enforcing that Equation 5.82 be satisfied at the n nodes in the support domain of point $\mathbf{x}_Q$. At node i we can have equation

$$u_i = \mathbf{p}^T(\mathbf{x}_i)\mathbf{a} \quad i = 1 \sim n \tag{5.87}$$

where u_i is the nodal value of u at $\mathbf{x} = \mathbf{x}_i$. Equation 5.87 can be written in the following matrix form:

$$\mathbf{U}_s = \mathbf{P}_Q\,\mathbf{a} \tag{5.88}$$

where $\mathbf{U}_s$ is the vector that collects the values of field variables at all the n nodes in the support domain.

$$\mathbf{U}_s = \begin{Bmatrix} u_1 \\ u_2 \\ \mathrm{M} \\ u_n \end{Bmatrix} \tag{5.89}$$

and $\mathbf{P}_Q$ is called the *moment matrix* given by

$$\mathbf{P}_Q = \begin{bmatrix} \mathbf{p}^T(\mathbf{x}_1) \\ \mathbf{p}^T(\mathbf{x}_2) \\ \mathrm{M} \\ \mathbf{p}^T(\mathbf{x}_n) \end{bmatrix} \tag{5.90}$$

or in detail (for 2D cases):

$$\left\{\begin{array}{cccccccccc} 1 & x_1 & y_1 & x_1y_1 & x_1^2 & y_1^2 & x_1^2y_1 & x_1y_1^2 & x_1^3 & \text{K} \\ 1 & x_2 & y_2 & x_2y_2 & x_2^2 & y_2^2 & x_2^2y_2 & x_2y_2^2 & x_2^3 & \text{K} \\ \text{M} & \text{M} & \text{M} & \text{M} & \text{M} & \text{M} & \text{M} & \text{M} & \text{M} & \\ 1 & x_n & y_n & x_ny_n & x_n^2 & y_n^2 & x_n^2y_n & x_ny_n^2 & x_n^3 & \text{K} \end{array}\right. \tag{5.91}$$

Using Equation 5.88 and assuming that the inverse of the moment matrix $\mathbf{P}_Q$ exists,* we can then have

$$\mathbf{a} = \mathbf{P}_Q^{-1}\mathbf{U}_s \tag{5.92}$$

Substituting Equation 5.92 into Equation 5.82, we obtain

$$u^h(\mathbf{x}) = \sum_{i=1}^{n} \phi_i(\mathbf{x})u_i \tag{5.93}$$

or in matrix form:

$$u^h(\mathbf{x}) = \mathbf{\Phi}(\mathbf{x})\mathbf{U}_s \tag{5.94}$$

where $\mathbf{\Phi}(x)$ is a matrix of PIM shape functions $\phi_i(\mathbf{x})$ defined by

$$\mathbf{\Phi}(\mathbf{x}) = \mathbf{p}^T(\mathbf{x})\mathbf{P}_Q^{-1} = [\phi_1(\mathbf{x}),\ \phi_2(\mathbf{x}),\ \phi_3(\mathbf{x}),\ldots,\ \phi_n(\mathbf{x})] \tag{5.95}$$

It is possible that $\mathbf{P}_Q^{-1}$ in Equation 5.95 may not exist in some situations, which leads to a breakdown of the PIM method. Several techniques have been proposed to overcome the singularity issue, including methods of moving nodes, coordinate transformation, matrix triangulation algorithm, and use of radial function basis. A later part of this chapter will discuss these methods in detail. For now, we assume that the moment matrix is invertible.

Note that derivatives of the PIM shape functions can be obtained very easily, as all the functions involved are polynomials. The lth derivative of the shape functions are simply given by

$$\mathbf{\Phi}^{(l)}(\mathbf{x}) = [\mathbf{p}^{(l)}(\mathbf{x})]^T\mathbf{P}_Q^{-1} \tag{5.96}$$

Note that no weight function is used in constructing PIM shape functions. The dimension of the support domain d_s can be determined following the procedure described in Sections 2.10.2 and 2.10.3 for both uniformly and non-uniformly distributed nodes in 1D, 2D, and 3D domains.

* Special techniques are required, if otherwise.

5.5.2 Consistency

The consistency of the PIM shape function depends on the complete orders of the monomial $p_i(\mathbf{x})$ used in Equation 5.82, and hence also depends on the number of nodes in the support domain. If the complete order of the monomial is n, the shape functions will possess C^n consistency. To demonstrate, we consider a field given by

$$f(\mathbf{x}) = \sum_{j}^{k} p_j(\mathbf{x})\alpha_j, \quad k \leq n \tag{5.97}$$

where $p_j(\mathbf{x})$ are monomials that are included in Equation 5.82. Such a field can always be written using Equation 5.82 using all the basis terms including those in Equation 5.97:

$$f(\mathbf{x}) = \sum_{j}^{n} p_j(\mathbf{x})\alpha_j = \mathbf{p}^T(\mathbf{x})\boldsymbol{\alpha} \tag{5.98}$$

where

$$\boldsymbol{\alpha}^T = [\alpha_1, \alpha_2, \ldots, \alpha_k, 0, \ldots, 0] \tag{5.99}$$

Using n nodes in the support domain of $\mathbf{x}$, we can obtain the vector of nodal function value $\mathbf{U}_s$ as

$$\mathbf{U}_s = \begin{Bmatrix} f_1 \\ f_2 \\ \mathrm{M} \\ f_k \\ f_{k+1} \\ \mathrm{M} \\ f_n \end{Bmatrix} = \begin{bmatrix} p_1(x_1) & p_2(x_1) & \mathrm{L} & p_k(x_1) & p_{k+1}(x_1) & \mathrm{L} & p_n(x_1) \\ p_1(x_2) & p_2(x_2) & \mathrm{L} & p_k(x_2) & p_{k+1}(x_2) & \mathrm{L} & p_n(x_2) \\ \mathrm{M} & \mathrm{M} & \mathrm{L} & \mathrm{M} & \mathrm{M} & & \mathrm{M} \\ p_1(x_k) & p_2(x_k) & \mathrm{L} & p_k(x_k) & p_{k+1}(x_k) & \mathrm{L} & p_n(x_k) \\ p_1(x_{k+1}) & p_2(x_{k+1}) & \mathrm{L} & p_k(x_{k+1}) & p_{k+1}(x_{k+1}) & \mathrm{L} & p_n(x_{k+1}) \\ \mathrm{M} & \mathrm{M} & \mathrm{L} & \mathrm{M} & \mathrm{M} & & \mathrm{M} \\ p_1(x_n) & p_2(x_n) & \mathrm{L} & p_k(x_n) & p_{k+1}(x_n) & \mathrm{L} & p_n(x_n) \end{bmatrix} \begin{Bmatrix} \alpha_1 \\ \alpha_2 \\ \mathrm{M} \\ \alpha_k \\ 0 \\ \mathrm{M} \\ 0 \end{Bmatrix}$$

$$= \mathbf{P}_Q\mathbf{a} \tag{5.100}$$

Substituting Equation 5.100 into Equation 5.94, we have the approximation:

$$u^h(\mathbf{x}) = \mathbf{p}^T(\mathbf{x})\mathbf{P}_Q^{-1}\mathbf{U}_s = \mathbf{p}^T(\mathbf{x})\mathbf{P}_Q^{-1}\mathbf{P}_Q\boldsymbol{\alpha} = \mathbf{p}^T(\mathbf{x})\boldsymbol{\alpha} = \sum_{j}^{k} p_j(\mathbf{x})\alpha_j \tag{5.101}$$

which is exactly what is given in Equation 5.97. This proves that any field given by Equation 5.97 will be exactly reproduced by PIM, as long as the given field function is included in the basis functions used for constructing the PIM shape function. This feature of PIM shape function is, in fact, very easy to understand by intuition: any function given in the form of $f(\mathbf{x}) = \sum_j^k p_j(\mathbf{x})\alpha_j$ can be produced exactly by letting $a_j = \alpha_j$ $(j = 1, 2, \ldots, k)$ and $a_j = 0$ $(j = k+1, \ldots, n)$. This can always be done as long as the moment matrix $\mathbf{P}_Q$ is invertible so as to ensure the uniqueness of the solution for $\mathbf{a}$.

The proof of the consistency of PIM is valid for proving another important feature of PIM: that any function that appears in the basis can be reproduced exactly. This important property can be useful for creating fields of special features. For PIM to exhibit linear consistency, all one need do is include the constant and linear monomials into the basis. This feature of PIM can be used to compute accurate results for problems by including terms in the basis of PIM that are good approximations of the solution of the problem.

5.5.3 Properties of PIM Shape Functions

As long as the moment matrix is invertible, the PIM shape functions $\phi_i(\mathbf{x})$ constructed depend uniquely on the distribution of the scattered nodes. The PIM shape functions possess the following characteristics:

1. Shape functions are linearly independent in the support domain. This is because basis functions are linearly independent and $\mathbf{P}_Q^{-1}$ is assumed to exist. The existence of $\mathbf{P}_Q^{-1}$ implies that the shape functions are equivalent to the basis functions in function space, as shown in Equation 5.95, and hence are linearly independent.
2. Shape functions possess the Kronecker delta function property, that is,

$$\phi_i(\mathbf{x} = \mathbf{x}_j) = \begin{cases} 1 & i = j,\ j = 1, 2, \ldots, n \\ 0 & i \neq j,\ i, j = 1, 2, \ldots, n \end{cases} \tag{5.102}$$

This can be proved easily as follows. Because the PIM shape functions $\phi_i(\mathbf{x})$ are linearly independent, any vector of length n should be uniquely produced by linear combination of these n shape functions. Letting

$$\mathbf{U}_s = \{0, 0, \ldots, u_i, \ldots, 0\}^T \tag{5.103}$$

and substituting the above equation into Equation 5.93, we have at $\mathbf{x} = \mathbf{x}_j$, that

$$u^h(\mathbf{x}_j) = \sum_{1}^{n} \phi_i(\mathbf{x}_j)u_i = \phi_i(\mathbf{x}_j)u_i \tag{5.104}$$

When $i = j$, we obtain

$$u_i = \phi_i(\mathbf{x}_i)u_i \tag{5.105}$$

which leads to

$$\phi_i(\mathbf{x}_i) = 1 \tag{5.106}$$

This proves the first row of Equation 5.102. When $i \neq j$, we have

$$u_j = 0 = \phi_i(\mathbf{x}_j)u_i \tag{5.107}$$

which requires

$$\phi_i(\mathbf{x}_j) = 0 \tag{5.108}$$

This proves that PIM shape functions possess the Kronecker delta function property, Equation 5.102.
3. $\phi_i(\mathbf{x})$ is the partition of unity

$$\sum_{i=1}^{n} \phi_i(\mathbf{x}) = 1 \tag{5.109}$$

if the constant is included in the basis. This can be proved easily from the reproduction feature of PIM. Letting $u(\mathbf{x}) = c$, where c is a constant, we should have

$$\mathbf{U}_s = c\{1, 1,\ldots,1\}^T \tag{5.110}$$

Substituting the above equation into Equation 5.93, we obtain

$$u(\mathbf{x}) = c = \sum_1^n \phi_i(\mathbf{x})u_i = c\sum_1^n \phi_i(\mathbf{x}) \tag{5.111}$$

which gives Equation 5.109. This shows that the partition of unity of PIM shape functions in the support domain allows a constant field or rigid body movement to be reproduced. Note that Equation 5.109 does not require $0 \leq \phi_i(\mathbf{x}) \leq 1$.

4. $\phi_i(\mathbf{x})$ possesses reproducing properties

$$\sum_{i=1}^n \phi_i(\mathbf{x})x_i = \mathbf{x} \tag{5.112}$$

if the first-order monomial is included in the basis. This can be proved easily from the reproduction feature of PIM in exactly the same manner used for proving property 3. Letting $u(\mathbf{x}) = \mathbf{x}$, we should have

$$\mathbf{U}_s = \{\mathbf{x}_1, \mathbf{x}_2,\ldots,\mathbf{x}_n\}^T \tag{5.113}$$

Substituting the above equation into Equation 5.93, we obtain

$$u^h(\mathbf{x}) = \mathbf{x} = \sum_1^n \phi_i(\mathbf{x})\mathbf{x}_i \tag{5.114}$$

which is Equation 5.112.

5. $\phi_i(\mathbf{x})$ has derivatives of simple polynomial form, which is obvious.
6. $\phi_i(\mathbf{x})$ is of compact support as long as it is constructed using the nodes in a compact support domain.
7. There is no need for weight functions in constructing PIM shape functions, or the weight function used is a unit.
8. PIM shape function is not compatible. This is because the bell-shape weight function is not used in constructing the PIM shape function. The approximation using PIM shape function can be discontinuous when $\mathbf{x}_Q$ moves and the support domain updates its nodes.

Property 2 is important for handling essential boundary conditions. Properties 3 and 4 are essential for a PIM MFree method to pass the patch test, which is a conventional test used for decades in FEM for validation of elements. Property 5 will be very important in developing new truly MFree methods without using a background mesh of cells for integration. Property 6 leads to sparse and banded discretized system matrices. Property 7 eliminates the question of how to choose a weight function in constructing shape functions. Property 8 implies that a constrained energy weak form should usually be used for deriving discrete system equations. More discussion on this issue is in Section 5.11.

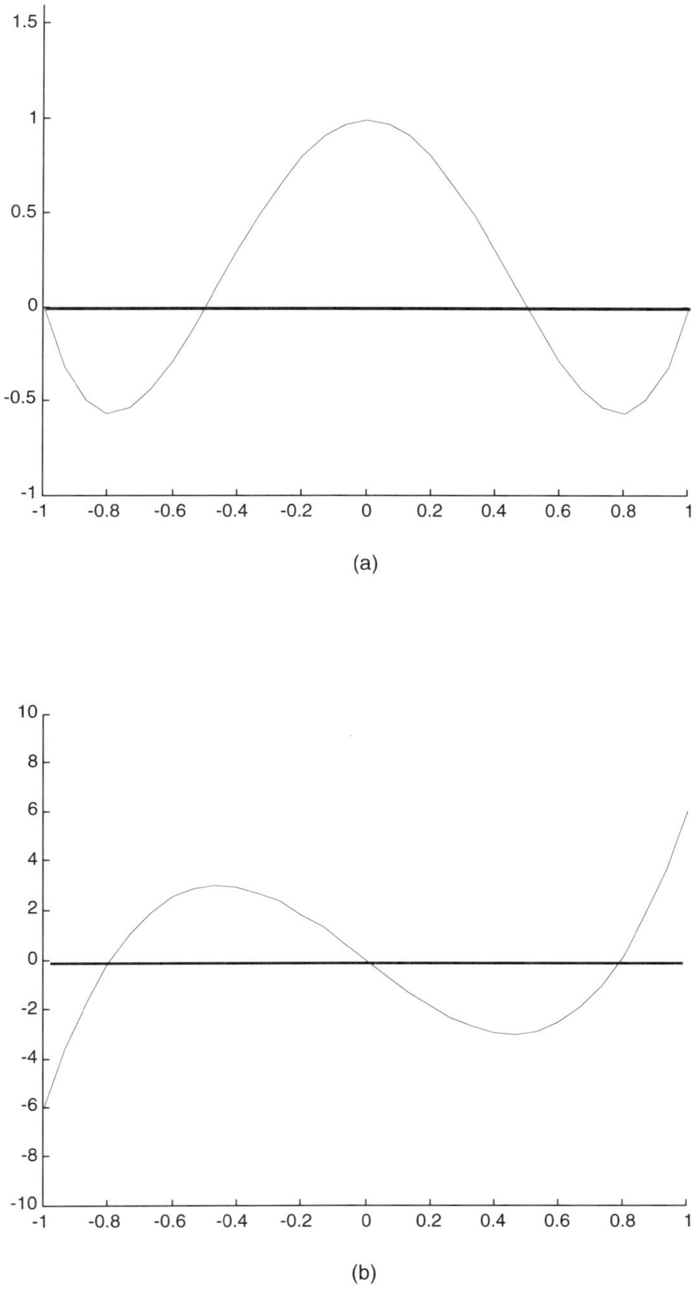

FIGURE 5.10
Polynomial PIM shape function in 1D space for the node at $x = 0$ obtained using five nodes evenly distributed in the support domain of [–1, 1]. (a) Shape function; (b) derivative of the shape function. Note that the PIM shape function possesses the Kronecker delta function property.

The PIM shape function is ideal for MFree methods in many ways, as it possesses the above excellent properties. Polynomial PIM formulation is the simplest and performs the best. Figure 5.10a shows a PIM shape function in 1D space for a node at $\mathbf{x} = 0$ obtained using five nodes evenly distributed in the support domain of [–1, 1]. It is clearly seen that the PIM

TABLE 5.1

Comparisons between PIM Interpolation and MLS Approximation

	Basis Function	Number of Basis Functions (m) and Number of Nodes (n)	Interpolation Coefficients	Delta Function Property	Compatibility
Point interpolation	Polynomial or radio functions or both	$m = n$	Constant	Yes	No
MLS approximation	Polynomial	$m \neq n$	Function	No	Yes

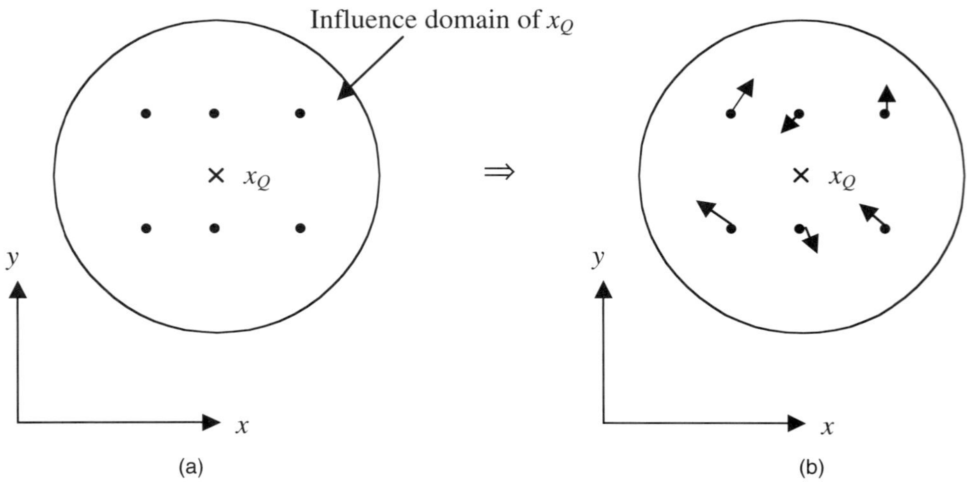

FIGURE 5.11
Node distribution in a domain of support of point $\mathbf{x}_Q$. (a) Six nodes in two parallel lines that lead to a singular moment matrix. (b) Moving nodes by a small distance randomly results in a nonsingular moment matrix.

shape function possesses the Kronecker delta function property. That is, $\phi(0.0) = 1.0$, $\phi(-1.0) = \phi(-0.5) = \phi(0.5) = \phi(1.0) = 0.0$. Figure 5.10b plots the first derivative of the shape function.

5.5.4 Difference between PIM Interpolation and MLS Approximation

PIM interpolation and MLS approximation are compared in Table 5.1. As the table shows, the main difference between PIM and MLS approximation is that the number of polynomial terms used in PIM is the same as the number of the nodes used in the support domain. The coefficients in PIM are constants, and those in MLS are functions. Most importantly, the PIM shape functions possess the Kronecker delta function property. The Kronecker delta function property allows essential boundary conditions to be easily treated in the same way as they are in conventional FEM. However, PIM interpolation is not, in general, compatible, while MLS approximation is compatible.

5.5.5 Methods to Avoid Singular Moment Matrix

As shown above, PIM shape functions possess many excellent properties that are very useful for MFree methods. However, the process of constructing PIM shape functions can break down as a result of the singularity of the moment matrix $\mathbf{P}_Q$. Figure 5.11 shows six nodes in the support domain of a point $\mathbf{x}_Q$. These six nodes sit in two lines parallel to the

x axis. When these six nodes are used, the polynomial basis can be of complete second order with respect to both the x and y coordinates:

$$\mathbf{p}^T(\mathbf{x}) = \{1, x, y, xy, x^2, y^2\} \tag{5.115}$$

However, these six nodes, as shown in Figure 5.11a, cannot possibly represent a second-order polynomial in the y direction, as there are only two distinct y coordinate values in all these six nodes. Therefore, the inverse of the moment matrix $\mathbf{P}_Q$ using these six nodes does not exist.

After selection of nodes and basis functions, the matrix $\mathbf{P}_Q$ is completely determined by the position of the scattered nodes in a given coordinate system; i.e., the existence of the inverse matrix $\mathbf{P}_Q^{-1}$ depends on the node distribution as well as on the coordinate system. Using polynomial basis functions is known to be difficult to guarantee the existence of $\mathbf{P}_Q^{-1}$ for a set of arbitrarily scattered nodes. However, the excellent properties of PIM shape functions warrant the effort needed to overcome the singular moment matrix problem. A number of methods for handling the singular moment matrix have been proposed by G. R. Liu and co-workers. These methods can be used to obtain an invertible moment matrix and are as follows.

1. The simplest method proposed to obtain a nonsingular moment matrix is to move or shift the nodes in the support domain by a small distance randomly in terms of both direction and the amount of shift, as shown in Figure 5.11b. The method is simple and effective for most problems. However, there is still a chance that the moment matrix will be singular, which may sometimes lead to a badly conditioned $\mathbf{P}_Q$.
2. The use of radial basis functions in constructing PIM shape functions is a method that always works and that guarantees the existence of the inverse of the moment matrix, if the guidelines for choosing shape parameters of the radial basis functions are followed. The drawback is that it is more expensive as more nodes are required to obtain accurate results comparable with those of polynomial PIM.
3. The use of radial basis functions with polynomial terms is an approach that improves the accuracy of the radial PIM, especially for patch tests. It can also reduce the sensitivity of the results on the shape parameters of radial basis functions. The computational cost is still more expensive as compared with polynomial PIM.
4. Polynomial PIM with coordinate transformation is an approach that has been developed by making use of the fact that the singularity of the moment matrix depends on the coordinate system where the moment matrix is formed. Singularity is avoided by rotating the local coordinate system. This method works for many cases, but does not provide a full proof for the problem.
5. The matrix triangularization method is very efficient and works well for most situations. It opens a new window of opportunity to fully solve the singularity problem of the moment matrix.

All the above methods have room for improvement. The author is very confident that development of a full proof method to create PIM shape functions that are accurate and efficient is not too far distant, especially in the direction of the fifth method. Alternatively, improvement in computational efficiency for radial PIM can also solve the problem.

The following sections detail the procedures and formulation for methods 2 through 5 listed above.

5.6 Radial PIM

5.6.1 Rationale for Radial Basis Functions

The advantage of using a polynomial basis is its simplicity and high accuracy, as will be evident in later chapters. It can reproduce a shape function of any order by increasing the number of nodes for interpolation. The major drawback of polynomial PIM is that singular moment matrix $\mathbf{P}_Q$ may sometimes occur. To create a nonsingular moment matrix, radial basis functions are introduced in PIM formulation for constructing shape functions (Wang and Liu, 2000). PIM using radial basis function (RBF) is termed radial PIM (RPIM).

5.6.2 PIM Formation Using Radial Basis Functions

In RPIM, we choose radial functions as the basis in Equation 5.81, i.e.,

$$u^h(\mathbf{x}, \mathbf{x}_Q) = \sum_{i=1}^{n} R_i(\mathbf{x})a_i(\mathbf{x}_Q) = \mathbf{R}^T(\mathbf{x})\mathbf{a}(\mathbf{x}_Q) \tag{5.116}$$

where vector $\mathbf{a}$ is defined in Equation 5.83, and R_i is a radial basis function with r the distance between point $\mathbf{x}$ and $\mathbf{x}_i$. For a 2D problem, r is defined as

$$r = [(x-x_i)^2 - (y-y_i)^2]^{1/2} \tag{5.117}$$

The vector $\mathbf{R}$ has the form:

$$\mathbf{R}^T(\mathbf{x}) = [R_1(\mathbf{x}), R_2(\mathbf{x}),\ldots,R_n(\mathbf{x})] \tag{5.118}$$

There are a number of forms of radial basis functions used by the mathematics community. Table 5.2 lists the four most often used forms of radial functions with some shape parameters that can be tuned for better performance. A classical form is the multiquadric (MQ) basis proposed by Hardy (1990). This form has been widely used in surface fitting and in constructing approximate solutions for partial differential equations (Kansa, 1990a,b; Powell,

TABLE 5.2
Typical Conventional Form of Radial Basis Functions

Item	Name	Expression	Shape Parameters
1	Multiquadrics (MQ)	$R_i(x, y) = (r_i^2 + C^2)^q = [(x-x_i)^2 + (y-y_i)^2 + C^2]^q$	C, q
2	Gaussian (EXP)	$R_i(x, y) = \exp(-cr_i^2) = \exp\{-c[(x-x_i)^2 + (y-y_i)^2]\}$	c
3	Thin plate spline (TPS)	$R_i(x, y) = r_i^\eta = [(x-x_i)^2 + (y-y_i)^2]^\eta$	η
4	Logarithmic RBF	$R_i(r_i) = r_i^\eta \log r_i$	η

1992; Coleman, 1996; Sharan et al., 1997; Fasshauer, 1997; Wendland, 1998; among others). The MQ basis function shown in Table 5.2 is a general form of the original MQ RBF that was suggested by Wang and Liu (2000, 2001a,b). When $q = \pm 0.5$, it reduces to the original MQ RBF proposed by Hardy. When $q = -0.5$, it reduces to the reciprocal MQ RBF. The general form of the MQ radial function has two shape parameters, C and q, which control the shape of the functions. These parameters can be tuned for different problems for better performance. The second form of radial function given in Table 5.2 is called the Gaussian radial function, or EXP, as it is an exponential function of the distance (Powell, 1992). The EXP radial basis function has only one shape parameter c, which controls the decay rate of the function. The third radial function in Table 5.2 is called the thin plate spline (TPS) function. The TPS is, in fact, a special case of the MQ radial function. The fourth form of radial basis function is the logarithmic RBF. This book will use and test the first three forms of radial functions, but will focus more on the first two forms of radial functions (MQ and EXP) and will give preference to the general form of the MQ radial basis function for reasons to be given in later chapters.

Enforcing $u(\mathbf{x})$ approximated by Equation 5.82 to pass through each scattered node in the support domain that is formed for the point of interest $\mathbf{x}$, we have the moment matrix of RBF:

$$\mathbf{R}_Q = \begin{bmatrix} R_1(r_1) & R_2(r_1) & \mathrm{L} & R_n(r_1) \\ R_1(r_2) & R_2(r_2) & \mathrm{L} & R_n(r_2) \\ \mathrm{M} & \mathrm{M} & \mathrm{O} & \mathrm{M} \\ R_1(r_n) & R_2(r_n) & \mathrm{L} & R_n(r_n) \end{bmatrix} \tag{5.119}$$

where

$$r_k = [(x_k - x_i)^2 + (y_k - y_i)^2]^{1/2} \tag{5.120}$$

Because the distance is directionless, we should have

$$R_i(r_j) = R_j(r_i) \tag{5.121}$$

Therefore, the moment matrix $\mathbf{R}_Q$ is symmetric.

The vectors of coefficients $\mathbf{a}$ in Equation 5.116 are determined by enforcing that the interpolation passes through all the n nodes within the support domain. The interpolation at the kth point has the form:

$$u_k = u(x_k, y_k) = \sum_{i=1}^{n} a_i R_i(x_k, y_k) \quad k = 1, 2, \ldots, n \tag{5.122}$$

or in matrix form:

$$\mathbf{U}_s = \mathbf{R}_Q \mathbf{a} \tag{5.123}$$

where $\mathbf{U}_s$ is the vector that collects all the field nodal variables at the n nodes in the support domain. A unique solution for vectors of coefficients $\mathbf{a}$ is obtained if the inverse of $\mathbf{R}_Q$ exists:

$$\mathbf{a} = \mathbf{R}_Q^{-1} \mathbf{U}_s \tag{5.124}$$

Substituting the foregoing equation into Equation 5.116 leads to

$$u^h(\mathbf{x}) = \mathbf{R}^T(\mathbf{x})\mathbf{R}_Q^{-1}\mathbf{U}_s = \mathbf{\Phi}(x)\mathbf{U}_s \tag{5.125}$$

where the matrix of shape functions $\mathbf{\Phi}(\mathbf{x})$ with n shape functions is:

$$\begin{aligned}\mathbf{\Phi}(\mathbf{x}) &= [R_1(\mathbf{x}), R_2(\mathbf{x}),\ldots,R_k(\mathbf{x}),\ldots,R_n(\mathbf{x})]\mathbf{R}_Q^{-1} \\ &= [\phi_1(\mathbf{x}), \phi_2(\mathbf{x}),\ldots,\phi_k(\mathbf{x}),\ldots,\phi_n(\mathbf{x})]\end{aligned} \tag{5.126}$$

in which $\phi_k(\mathbf{x})$ is the shape function for the kth node given by

$$\phi_k(\mathbf{x}) = \sum_{i=1}^{n} R_i(\mathbf{x})S_{ik}^a \tag{5.127}$$

where S_{ik}^a is the (i, k) element of matrix $\mathbf{R}_Q^{-1}$, which is a constant matrix for given locations of the n nodes in the support domain.

The derivatives of shape functions can be easily obtained as

$$\begin{aligned}\frac{\partial \phi_k}{\partial x} &= \sum_{i=1}^{n} \frac{\partial R_i}{\partial x} S_{ik}^a \\ \frac{\partial \phi_k}{\partial y} &= \sum_{i=1}^{n} \frac{\partial R_i}{\partial y} S_{ik}^a\end{aligned} \tag{5.128}$$

For the MQ basis function shown in Table 5.2, the partial derivatives for the MQ radial functions can be easily obtained using the following simple formulae:

$$\begin{aligned}\frac{\partial R_i}{\partial x} &= 2q(r_i^2 + C^2)^{q-1}(x - x_i) \\ \frac{\partial R_i}{\partial y} &= 2q(r_i^2 + C^2)^{q-1}(y - y_i)\end{aligned} \tag{5.129}$$

For the EXP radial function, the partial derivatives can also be obtained easily as follows:

$$\begin{aligned}\frac{\partial R_i}{\partial x} &= -2cR_i(x, y)(x - x_i) \\ \frac{\partial R_i}{\partial y} &= -2cR_i(x, y)(y - y_i)\end{aligned} \tag{5.130}$$

5.6.3 Nonsingular Moment Matrix

Note that the only difference between polynomial PIM and RPIM is the difference in the basis functions. Mathematicians have proved that the radial moment matrix $\mathbf{R}_Q$ is always invertible for arbitrary scattered nodes (Powell, 1992; Schaback, 1994; Wendland, 1998), as long as we avoid using some specific shape parameters, which are known. Therefore, $\mathbf{R}_Q$ can always be symmetric and invertible. The existence of $\mathbf{R}_Q^{-1}$ is the major advantage of using the radial basis over the polynomial basis.

5.6.4 Consistency

The radial PIM shape function is not consistent by the definition of consistency in this book. Mathematicians have also found that approximations of any continuous function using radial basis functions converge. Thus, there is no concern about the convergence issue.

Wang and Liu (2001b) found that the use of pure radial functions in the basis of PIM will not pass the standard patch test, which has been widely used in the FEM community for testing the performance of finite elements. This is because the radial function cannot produce the linear polynomials exactly, although it can approach polynomials in desired accuracy when the nodes are refined. The consistency of the radial shape functions can be installed by adding polynomial basis functions (see Section 5.7). For RPIM to pass the patch test, Wang and Liu (2001b) suggested using radial functions with polynomial terms of up to linear orders so as to construct shape functions with C^1 consistency. Adding polynomial terms to RBFs was proposed by Powell in 1992 for function approximation.

The radial basis functions have a wide range of shape parameters that may be tuned for better performance. However, if these parameters are not turned properly, the accuracy of the results suffers. Therefore, careful investigation of the effects of these shape parameters is necessary to provide proper guidelines for use of RPIM shape functions for different types of problems. Reliance of the accuracy on the shape parameters can be significantly reduced by adding polynomial basis functions, which are discussed later.

5.6.5 Radial Functions with Dimensionless Shape Parameters

The conventional forms of radial functions listed in Table 5.2 have been used by many researchers including the research group of the author. We found recently that it is very difficult to standardize the shape parameters of the radial basis functions. We therefore proposed a set of radial basis functions that have dimensionless parameters by performing some minor modification to the conventional radial basis functions. Some of the new forms of radial basis functions are listed in Table 5.3.

The MQ function with dimensionless shape parameters has the form:

$$R_i(x, y) = (r_i^2 + (\alpha_C d_c)^2)^q \quad \alpha_C \geq 0 \tag{5.131}$$

where α_C is the dimensionless shape parameter and d_c is the characteristic length that is

TABLE 5.3

Radial Basis Functions with Dimensionless Shape Parameters

Item	Name	Expression[a]	Shape Parameters	Parameter Relations[b]
1	Multiquadrics (MQ)	$R_i(x, y) = (r_i^2 + (\alpha_C d_c)^2)^q$	$\alpha_C \geq 0$, q	$\alpha_C = C/d_c$, $q = q$
2	Gaussian (EXP)	$R_i(x, y) = \exp\left[-\alpha_c\left(\frac{r_i}{d_c}\right)^2\right]$	α_c	$\alpha_c = c d_c$
3	Thin plate spline (TPS)[c]	$R_i(x, y) = r_i^\eta$	η	$\eta = \eta$
4	Logarithmic RBF[c]	$R_i(r_i) = r_i^\eta \log r_i$	η	$\eta = \eta$

[a] d_c is a characteristic length that is related to the nodal spacing in the local domain of the point of interest $\mathbf{x}_Q$. d_c is usually the average nodal spacing for all the nodes in the local domain (see Section 2.10.3 for a method to calculate d_c).

[b] The last column gives the relationship between the original parameters and the dimensionless parameters.

[c] The shape parameters in the TPS and logarithm RBFs are already dimensionless and, therefore, no change is needed.

usually the average nodal spacing for all the n nodes in the support domain (see Section 2.10.3 for how to determine d_c).

The first-order partial derivatives are obtained as follows:

$$\frac{\partial R_i}{\partial x} = 2q(r_i^2 + (\alpha_C d_c)^2)^{q-1}(x - x_i) \tag{5.132}$$

$$\frac{\partial R_i}{\partial y} = 2q(r_i^2 + (\alpha_C d_c)^2)^{q-1}(y - y_i) \tag{5.133}$$

$$R_{i,xx} = 2q[r_i^2 + (\alpha_C d_c)^2]^{q-1} + 4q(q-1)[r_i^2 + (\alpha_C d_c)^2]^{q-2}(x - x_i)^2 \tag{5.134}$$

$$R_{i,xy} = 4q(q-1)[r_i^2 + (\alpha_C d_c)^2]^{q-2}(x - x_i)(y - y_i) \tag{5.135}$$

$$R_{i,yy} = 2q[r_i^2 + (\alpha_C d_c)^2]^{q-1} + 4q(q-1)[r_i^2 + (\alpha_C d_c)^2]^{q-2}(y - y_i)^2 \tag{5.136}$$

The EXP radial function with dimensionless shape parameters can be written as

$$R_i(x, y) = \exp\left(-\alpha_c\left(\frac{r_i}{d_c}\right)^2\right) \tag{5.137}$$

where α_c is the dimensionless shape parameter. The first-order partial derivatives of the EXP radial basis functions are obtained as follows:

$$\frac{\partial R_i}{\partial x} = -\frac{2\alpha_c}{d_c^2} R_i(x, y)(x - x_i) \tag{5.138}$$

$$\frac{\partial R_i}{\partial y} = -\frac{2\alpha_c}{d_c^2} R_i(x, y)(y - y_i) \tag{5.139}$$

$$R_{i,xx} = \left[-2\left(\frac{\alpha_c}{d_c^2}\right) + 4\left(\frac{\alpha_c}{d_c^2}\right)^2 (x - x_i)^2\right] R_i(x, y) \tag{5.140}$$

$$R_{i,xy} = 4\left(\frac{\alpha_c}{d_c^2}\right)^2 R_i(x, y)(x - x_i)(y - y_i) \tag{5.141}$$

$$R_{i,yy} = \left[-2\left(\frac{\alpha_c}{d_c^2}\right) + 4\left(\frac{\alpha_c}{d_c^2}\right)^2 (y - y_i)^2\right] R_i(x, y) \tag{5.142}$$

Many results presented in this and later chapters were obtained before we had these new RBFs with dimensionless shape parameters, except results from some of our very recent work. It is simply not possible to repeat all the earlier examples using the new sets of radial functions. The book recommends that this new set of RBFs be used in the future, so that we can slowly build up a basis for comparison.

5.7 Radial PIM with Polynomial Reproduction

5.7.1 Rationale for Polynomials

RPIM with pure radial functions is not consistent and has a problem passing the standard patch tests, meaning that it fails to reconstruct the linear field exactly. The purpose of adding polynomials into the basis functions is to ensure the consistency of radial PIM shape functions. Adding polynomial terms up to the linear order can ensure the reproduction of the linear field (C^1 consistency) and hence help to pass the standard patch tests. This was our original motivation for adding polynomials to the radial basis for solving solid mechanics problems. Our study later found that, in general, adding polynomials can always improve the accuracy of the results. Another additional bonus of this formulation is that we have much more freedom in choosing shape parameters.

5.7.2 Formulation Using Radial-Polynomial Basis

By using the n nodes in the support domain, RPIM with polynomial basis functions approximates the field variable in the form:

$$u^h(\mathbf{x}) = \sum_{i=1}^{n} R_i(\mathbf{x})a_i + \sum_{j=1}^{m} p_i(\mathbf{x})b_j = \mathbf{R}^T(\mathbf{x})\mathbf{a} + \mathbf{p}^T(\mathbf{x})\mathbf{b} \tag{5.143}$$

where a_i is the coefficient for the radial basis $R_i(\mathbf{x})$ that is listed in Table 5.2, and b_j is the coefficient for the polynomial basis $p_j(\mathbf{x})$ that has the same form as the basis used in polynomial PIM. The number of radial basis functions n is determined by the number of the nodes in the support domain, and the number of polynomial basis m can be chosen based on the reproduction requirement. We often use a minimum number of terms of polynomial basis and more terms of radial basis ($m < n$) for better stability. To pass the patch test for 2D cases, one needs only three terms of polynomial basis.

The vector $\mathbf{a}$ in Equation 5.143 is defined as

$$\mathbf{a}^T = \{a_1, a_2, \ldots, a_n\} \tag{5.144}$$

and the vector $\mathbf{b}$ is defined as

$$\mathbf{b}^T = \{b_1, b_2, \ldots, b_m\} \tag{5.145}$$

The radial basis vector $\mathbf{R}$ in Equation 5.143 is defined as

$$\mathbf{R}^T(\mathbf{x}) = [R_1(\mathbf{x}), R_2(\mathbf{x}), \ldots, R_n(\mathbf{x})] \tag{5.146}$$

and the polynomial basis vector is written as

$$\mathbf{p}^T(\mathbf{x}) = [p_1(\mathbf{x}), p_2(\mathbf{x}), \ldots, p_m(\mathbf{x})] \tag{5.147}$$

The coefficients a_i and b_j in Equation 5.143 are determined by enforcing that the interpolation passes through all n nodes within the support domain. The interpolation at the kth point has the form:

$$u_k = u(x_k, y_k) = \sum_{i=1}^{n} a_i R_i(x_k, y_k) + \sum_{j=1}^{m} b_j p_j(x_k, y_k), \quad k = 1, 2, \ldots, n \tag{5.148}$$

or in matrix form:

$$\mathbf{U}_s = \mathbf{R}_Q \mathbf{a} + \mathbf{P}_m \mathbf{b} \tag{5.149}$$

where $\mathbf{U}_s$ is the vector that collects all the field nodal variables at the n nodes in the support domain. The polynomial term has to satisfy an extra requirement that guarantees unique approximation (Golberg et al., 1999) of a function, and the following constraints are usually imposed:

$$\sum_{i=1}^{n} p_j(x_i, y_i) a_i = 0 \quad j = 1, 2, \ldots, m \tag{5.150}$$

or in matrix form:

$$\mathbf{P}_m^T \mathbf{a} = \mathbf{0} \tag{5.151}$$

which is a set of homogeneous equations. Combination of Equations 5.149 and 5.151 gives

$$\begin{bmatrix} \mathbf{R}_Q & \mathbf{P}_m \\ \mathbf{P}_m^T & \mathbf{0} \end{bmatrix} \begin{Bmatrix} \mathbf{a} \\ \mathbf{b} \end{Bmatrix} = \begin{Bmatrix} \mathbf{U}_s \\ \mathbf{0} \end{Bmatrix} \tag{5.152}$$

or

$$\mathbf{G} \begin{Bmatrix} \mathbf{a} \\ \mathbf{b} \end{Bmatrix} = \begin{Bmatrix} \mathbf{U}_s \\ \mathbf{0} \end{Bmatrix} \tag{5.153}$$

The moment matrix corresponding to the radial function basis $\mathbf{R}_Q$ has been given by Equation 5.119, and the moment matrix $\mathbf{P}_m$ is an $n \times m$ matrix given by

$$\mathbf{P}_m = \begin{bmatrix} P_1(x_1, y_1) & P_2(x_1, y_1) & \mathrm{L} & P_m(x_1, y_1) \\ P_1(x_2, y_2) & P_2(x_2, y_2) & \mathrm{L} & P_m(x_2, y_2) \\ \mathrm{M} & \mathrm{M} & \mathrm{M} & \mathrm{M} \\ P_1(x_n, y_n) & P_2(x_n, y_n) & \mathrm{L} & P_m(x_n, y_n) \end{bmatrix}_{n \times m} \tag{5.154}$$

Because matrix $\mathbf{R}_Q$ is symmetric, matrix $\mathbf{G}$ will also be symmetric. A unique solution for vectors of coefficients $\mathbf{a}$ and $\mathbf{b}$ is obtained if the inverse of $\mathbf{G}$ exists:

$$\begin{Bmatrix} \mathbf{a} \\ \mathbf{b} \end{Bmatrix} = \mathbf{G}^{-1} \begin{Bmatrix} \mathbf{U}_s \\ \mathbf{0} \end{Bmatrix} \tag{5.155}$$

We choose not to prove the existence of the inverse of **G**. Instead, we will try to change the equations in a more efficient form, and then take up the existence issue.

Making use of the fact that Equation 5.151 is homogeneous, Equation 5.152 can be solved in the following more efficient procedure. Starting from Equations 5.149, and using the nonsingular property of matrix $\mathbf{R}_Q$, we have

$$\mathbf{a} = \mathbf{R}_Q^{-1}\mathbf{U}_s - \mathbf{R}_Q^{-1}\mathbf{P}_m\mathbf{b} \tag{5.156}$$

Substitution of the above expression into Equation 5.151 gives

$$\mathbf{b} = \mathbf{S}_b\mathbf{U}_s \tag{5.157}$$

where

$$\mathbf{S}_b = [\mathbf{P}_m^T\mathbf{R}_Q^{-1}\mathbf{P}_m]^{-1}\mathbf{P}_m^T\mathbf{R}_Q^{-1} \tag{5.158}$$

where $\mathbf{P}_m^T\,\mathbf{R}_Q^{-1}\,\mathbf{P}_m$ is termed a transformed moment matrix. Note also that $\mathbf{P}_m^T\,\mathbf{R}_Q^{-1}$ needs to be computed only once. Substituting Equation 5.157 back into Equation 5.156, we obtain

$$\mathbf{a} = \mathbf{S}_a\mathbf{U}_s \tag{5.159}$$

where

$$\mathbf{S}_a = \mathbf{R}_Q^{-1}[1 - \mathbf{P}_m\mathbf{S}_b] = \mathbf{R}_Q^{-1} - \mathbf{R}_Q^{-1}\mathbf{P}_m\mathbf{S}_b \tag{5.160}$$

Note that $\mathbf{R}_Q^{-1}\,\mathbf{P}_m$ can be obtained simply by transposing $\mathbf{P}_m^T\,\mathbf{R}_Q^{-1}$, which has been already computed.

Finally, the interpolation Equation 5.143 is expressed as

$$u(\mathbf{x}) = [\mathbf{R}^T(\mathbf{x})\mathbf{S}_a + \mathbf{p}^T(\mathbf{x})\mathbf{S}_b]\mathbf{U}_s = \mathbf{\Phi}(\mathbf{x})\mathbf{U}_s \tag{5.161}$$

where the matrix of shape functions $\mathbf{\Phi}(\mathbf{x})$ with n shape functions:

$$\mathbf{\Phi}(\mathbf{x}) = [\mathbf{R}^T(\mathbf{x})\mathbf{S}_a + \mathbf{p}^T(\mathbf{x})\mathbf{S}_b] = [\phi_1(\mathbf{x}),\ \phi_2(\mathbf{x}), \ldots,\ \phi_i(\mathbf{x}), \ldots,\ \phi_n(\mathbf{x})] \tag{5.162}$$

in which $\phi_i(\mathbf{x})$ is the shape function for the ith node given by

$$\phi_k(\mathbf{x}) = \sum_{i=1}^{n} R_i(\mathbf{x})S_{ik}^a + \sum_{j=1}^{m} p_j(\mathbf{x})S_{jk}^b \tag{5.163}$$

where S_{ik}^a is the (i, k) element of matrix $\mathbf{S}_a$, and S_{jk}^b is the (j, k) element of matrix $\mathbf{S}_b$, which are constant matrices for given locations of the n nodes in the support domain.

The derivatives of shape functions can be easily obtained as

$$\begin{aligned}\frac{\partial\phi_k}{\partial x} &= \sum_{i=1}^{n}\frac{\partial R_i}{\partial x}S_{ik}^a + \sum_{j=1}^{m}\frac{\partial p_j}{\partial x}S_{jk}^b\\ \frac{\partial\phi_k}{\partial y} &= \sum_{i=1}^{n}\frac{\partial R_i}{\partial y}S_{ik}^a + \sum_{j=1}^{m}\frac{\partial p_j}{\partial y}S_{jk}^b\end{aligned} \tag{5.164}$$

For the MQ basis function shown in Table 5.2, the partial derivatives for the MQ radial functions can be easily obtained using the following simple formulae:

$$\begin{aligned}\frac{\partial R_i}{\partial x} &= 2q(r_i^2 + C^2)^{q-1}(x - x_i)\\ \frac{\partial R_i}{\partial y} &= 2q(r_i^2 + C^2)^{q-1}(y - y_i)\end{aligned} \tag{5.165}$$

For the EXP radial function, the partial derivatives can also be obtained easily as follows:

$$\begin{aligned}\frac{\partial R_i}{\partial x} &= -2cR_i(x, y)(x - x_i)\\ \frac{\partial R_i}{\partial y} &= -2cR_i(x, y)(y - y_i)\end{aligned} \tag{5.166}$$

5.7.3 Singularity Issue of the Transformed Moment Matrix

Note that the transformed moment matrix $\mathbf{P}_m^T \mathbf{R}_Q^{-1} \mathbf{P}_m$ in Equation 5.158 may not be invertible and the whole procedure will fail. We therefore need to examine this possible situation.

Because $\mathbf{R}_Q$ is symmetric and invertible (with a full rank), the transformed moment matrix $\mathbf{P}_m^T \mathbf{R}_Q^{-1} \mathbf{P}_m$ in Equation 5.158 can be made at least symmetric. If the columns in $\mathbf{P}_m$ are independent (with a rank of m), we can easily prove that $\mathbf{P}_m^T \mathbf{R}_Q^{-1} \mathbf{P}_m$ is invertible by simply invoking the full rank property of $\mathbf{R}_Q$. To have all the columns of $\mathbf{P}_m$ be independent could be a problem in theory, but it is very easy to achieve in practice. We simply try to use a minimum number of terms of polynomial bases, so that $n \gg m$. This will ensure that $\mathbf{P}_m$ has a rank of m for all *practical* situations. It is very rare to have a case where the rank of $\mathbf{P}_m$ is less than m. It is, of course, possible to artificially make a case for $\mathbf{P}_m$ to have a rank of less than m by arranging the nodes in the support domain in a certain way (e.g., all the nodes are in a straight line). Note that the singularity issue is avoided in the weight moment matrix given in Equation 5.49 in MLS approximation using exactly the same strategy of having $n \gg m$. It is not a guarantee, but it is workable as a practical strategy. Therefore, $\mathbf{P}_m^T \mathbf{R}_Q^{-1} \mathbf{P}_m$ can be assumed, in general and in practical cases, invertible.

In using radial functions, one needs to investigate the effects of the shape parameters and to fine-tune these parameters for better performance. In the following examples, these parameters will be examined in detail via curve and surface fitting. For convenience, notations of MQ-PIM, EXP-PIM, and TPS-PIM refer to radial PIM using MQ, EXP, and TPS radial bases, respectively.

The following are some examples of shape functions constructed using the radial basis functions listed in Table 5.2.

Example 5.1 Sample RPIM Shape Functions

Examples of the RPIM shape functions are computed using the formulation given above. The radial basis functions listed in Table 5.2 are used in the computation. Figure 5.12 shows a typical shape function of MQ-PIM in 1D space. Shape parameters used in MQ-PIM are $C = 1.0$, $q = 0.5$. Five nodes evenly distributed in the support domain of [–1, 1] are used for computing the shape function for the node at $x = 0$. Figure 5.12a shows the shape function, and Figure 5.12b shows the derivative of the shape function.

Figure 5.13 shows a typical shape function of EXP-PIM with a shape parameter of $c = 0.3$. All the other conditions are the same as those used in computing Figure 5.12.

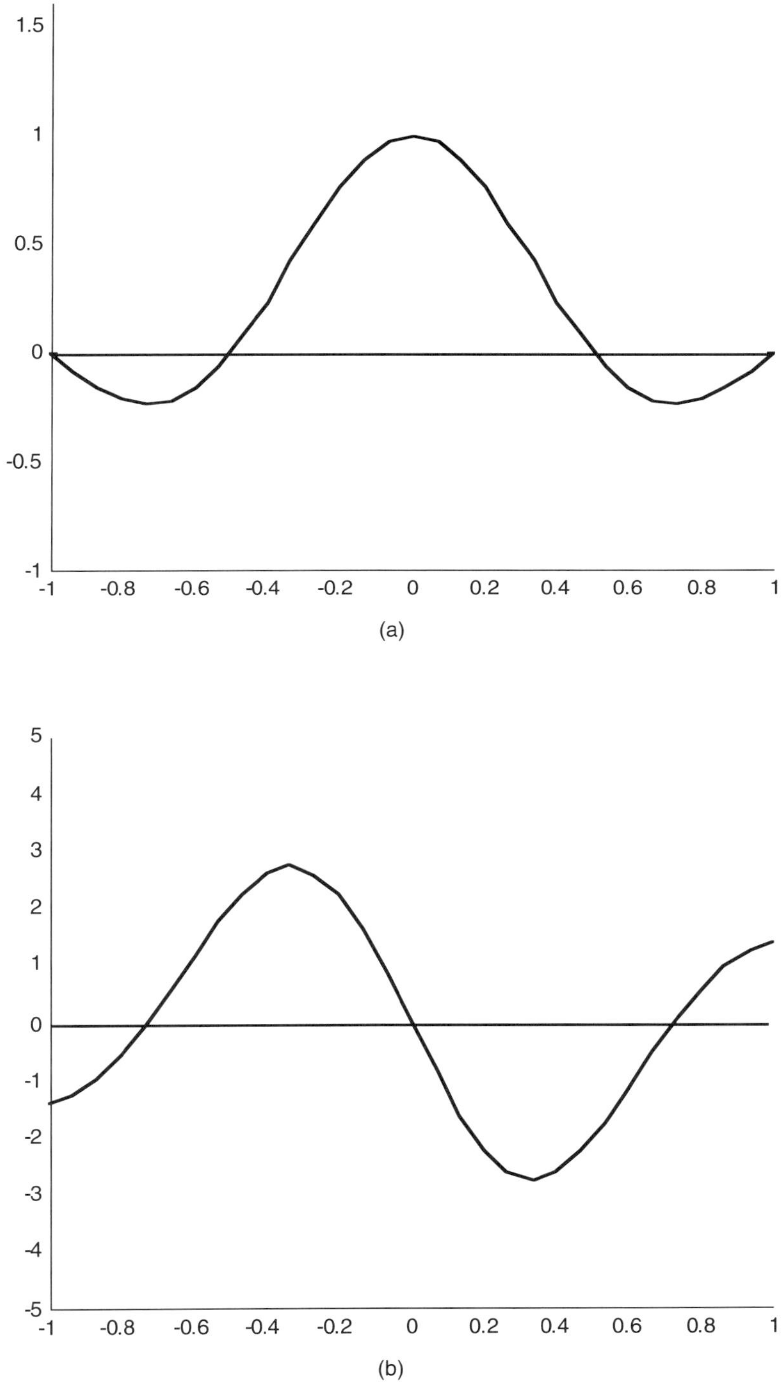

FIGURE 5.12
MQ-PIM ($C = 1.0$, $q = 0.5$) shape function in 1D space for the node at $x = 0$ obtained using five nodes evenly distributed in the support domain of $[-1, 1]$. (a) Shape function; (b) derivative of the shape function. Note that the PIM shape function possesses the Kronecker delta function property.

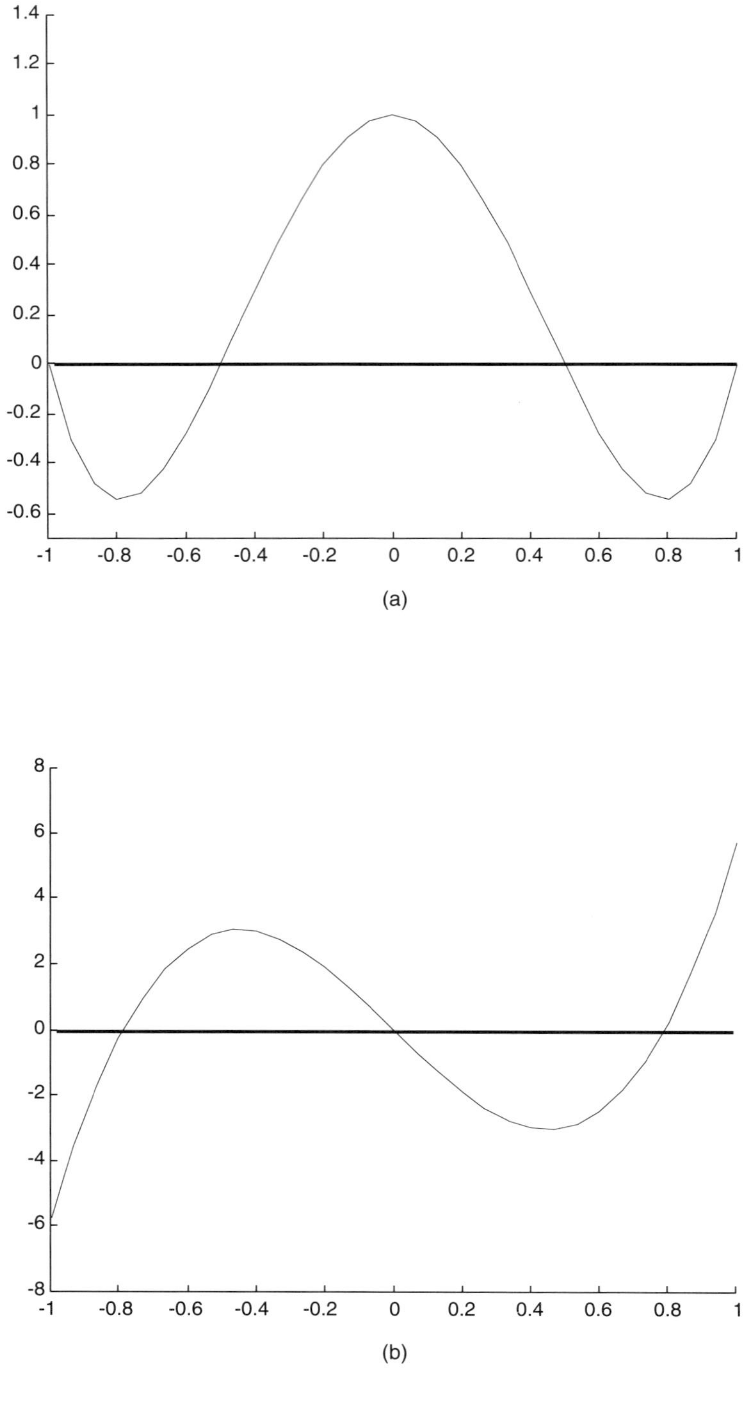

FIGURE 5.13
EXP-PIM ($c = 0.3$) shape function in 1D space for the node at $x = 0$ obtained using five nodes evenly distributed in the support domain of $[-1, 1]$. (a) Shape function; (b) derivative of the shape function. Note that the PIM shape function possesses the Kronecker delta function property.

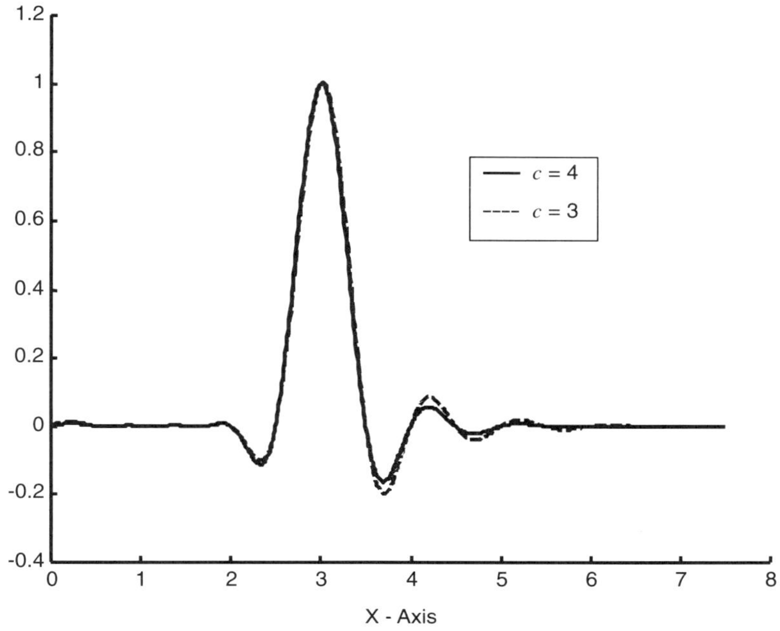

FIGURE 5.14
Effect of shape parameter c on PIM shape functions with EXP radial function basis.

Note that these shape functions possess the Kronecker delta function property given in Equation 5.102. As long as the polynomial basis $[1, x, y]$ ($m = 3$) is included in the basis, the shape functions so developed also satisfy Equation 5.112.

Example 5.2 Effects of Shape Parameters of RBFs on Shape Function

This example examines how the shape parameters affect the shape functions. To isolate the effects of the shape parameters of the radial functions, polynomial terms are not included in the basis function; that is, pure RPIM shape functions are used. Figure 5.14 shows the shape function of EXP-PIM at the seventh node computed using 16 nodes for a 1D interpolation. Different shape parameters c are used in the computation. Figure 5.14 shows that c affects the decay rate of the oscillations behind the first dominant peak. The larger the value of c, the faster the decay. Figure 5.15 shows the PIM shape function at the seventh node computed using 16 nodes for a 1D interpolation using an MQ radial basis function. The shape parameter C is fixed at $C = 1.42$ and the shape parameters q are investigated for two cases: 0.8 and 1.12. Figure 5.15 shows that q affects the decay rate of the oscillations behind the first dominant peak. The smaller the value of q, the faster the decay.

5.8 Polynomial PIM with Coordinate Transformation

This section introduces a method of coordinate transformation to produce an invertible moment matrix $\mathbf{P}_Q$. The method was originally proposed by Wang and Liu in 2000.

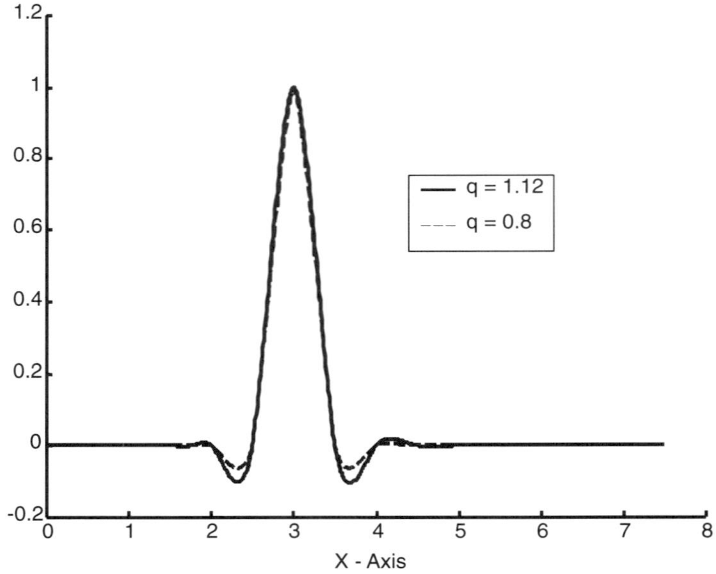

FIGURE 5.15
Effect of shape parameter q on PIM shape functions with MQ radial basis function ($C = 1.42$).

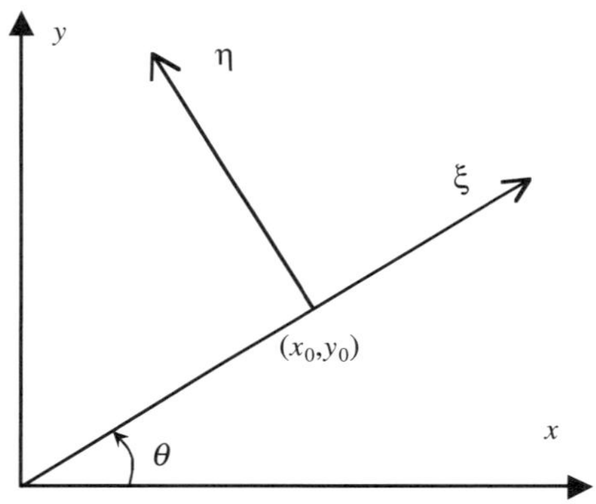

FIGURE 5.16
A local coordinate system (ξ, η) defined in a global coordinate system (x, y). The origin of (ξ, η) is located at (x_0, y_0).

5.8.1 Coordinate Transformation

We introduce a transformation between the global coordinate system (x, y) and the local coordinate system (ξ, η), as shown in Figure 5.16. This transformation can be performed using

$$\begin{aligned} \xi &= (x - x_0)\cos\theta + (y - y_0)\sin\theta \\ \eta &= -(x - x_0)\sin\theta + (y - y_0)\cos\theta \end{aligned} \tag{5.167}$$

where (x_0, y_0) is the origin of the local coordinate system (ξ, η), which is defined in a global coordinate system (x, y), and θ is the rotation angle for local coordination system (ξ, η) with respect to the global coordinate system.

The inverse transformation is expressed as

$$\begin{aligned} x &= x_0 + (\xi\cos\theta - \eta\sin\theta) \\ y &= y_0 + (\xi\sin\theta + \eta\cos\theta) \end{aligned} \tag{5.168}$$

In the local coordinates with a proper rotation angle θ, n polynomial basis terms $p_i(\boldsymbol{\xi})$ ($i = 0, 1,\ldots,n-1$) are chosen, and polynomial interpolation is performed to produce a nonsingular moment matrix $\mathbf{P}_0$ for point (0, 0), which corresponds to (x_0, y_0) in the global coordinate system. It is then inverted to obtain $\mathbf{P}_0^{-1}$, and the shape functions can be computed using (see Equation 5.95)

$$\boldsymbol{\Phi}(\boldsymbol{\xi}) = \mathbf{p}^T(\boldsymbol{\xi})\mathbf{P}_0^{-1} = [\phi_1(\boldsymbol{\xi}), \phi_2(\boldsymbol{\xi}), \phi_3(\boldsymbol{\xi}),\ldots,\phi_n(\boldsymbol{\xi})] \tag{5.169}$$

Because the coordinate transformation is very simple, and involves only rotation, the derivatives of the shape functions can be obtained efficiently using

$$\begin{Bmatrix} \left.\dfrac{\partial\phi_i}{\partial x}\right|_{(x_0,y_0)} \\ \left.\dfrac{\partial\phi_i}{\partial y}\right|_{(x_0,y_0)} \end{Bmatrix} = \begin{bmatrix} \cos\theta & -\sin\theta \\ \sin\theta & \cos\theta \end{bmatrix} \begin{Bmatrix} \left.\dfrac{\partial\phi_i}{\partial \xi}\right|_{(0,0)} \\ \left.\dfrac{\partial\phi_i}{\partial \eta}\right|_{(0,0)} \end{Bmatrix} \tag{5.170}$$

where

$$\begin{aligned} \left.\frac{\partial\phi_i}{\partial\xi}\right|_{(0,0)} &= (0,1,0,\ldots,0)\mathbf{P}_0^{-1} \\ \left.\frac{\partial\phi_i}{\partial\eta}\right|_{(0,0)} &= (0,0,1,0,\ldots,0)\mathbf{P}_0^{-1} \end{aligned} \tag{5.171}$$

5.8.2 Choice of Rotation Angle

The choice of the rotation angle determines the success of the method of coordinate transformation. A study was conducted to reveal the property of the moment matrix $\mathbf{P}_0$ in relation to the rotation angle. We define

$$f(\theta) = |\mathbf{P}_0| \tag{5.172}$$

where $|\mathbf{P}_0|$ is the determinant of $\mathbf{P}_0$. If any θ is not a root of $f(\theta)$, $\mathbf{P}_0^{-1}$ exists and the PIM shape function can be constructed. An example of six nodes, shown in Figure 5.17, is analyzed. Without coordinate transformation, $\mathbf{P}_Q^{-1}$ using these six nodes will not exist. Figure 5.18 shows the function of the determinant of the moment matrix $f(\theta)$ with respect to rotation angle θ. It is seen that the moment matrix is singular at four rotation angles marked with circles. Use of any rotational angle other than these four leads to a nonsingular moment matrix.

Note that how to choose the rotation angle is still an open question. One possible method is to choose a random number. However, this will not give 100% proof for a nonsingular moment matrix for more general cases. Note that the choice of the rotation angle depends on the nodal arrangement in the support domain. This further complicates the issue of choosing a proper rotation angle.

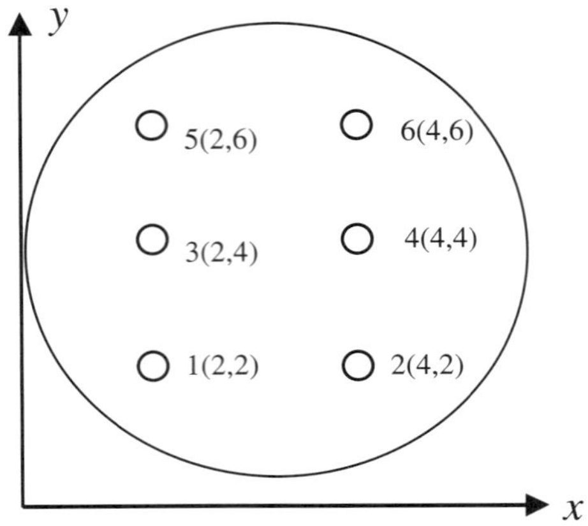

FIGURE 5.17
Six nodes in a support domain distributed in two parallel lines of $x = 2.0$ and 4.0.

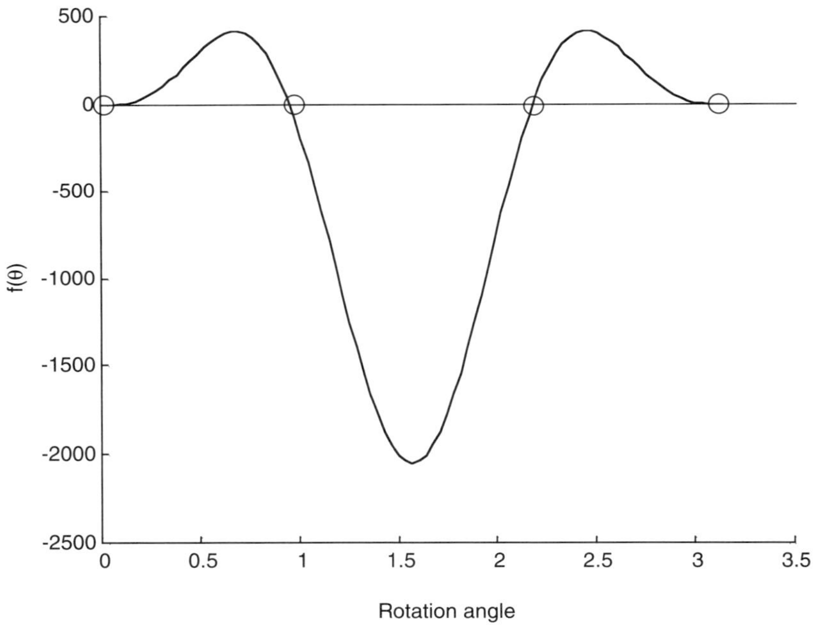

FIGURE 5.18
Determinant of the moment matrix formed in the local coordinate system via rotation. The moment matrix is singular at four rotation angles marked with circles.

5.9 Matrix Triangularization Algorithm

This section presents a matrix triangularization algorithm (MTA) for constructing PIM shape functions using polynomial bases. The MTA has been newly developed by Liu and Gu (2001d), after a close examination of the problem of singular moment matrix, which reveals the following important key facts:

1. The singularity is caused by improper enclosure of nodes in the support domain and improper selection of monomials in the basis. This means that the problem can be avoided if an algorithm can be developed to properly select the nodes and terms in the polynomial basis. For the case of nodal distribution shown in Figure 5.11, as examined in Section 5.5.5, the moment matrix using these six nodes is invertible. In this example, if the monomial x^2y is used to replace the monomial y^2 in the basis, the moment matrix will be invertible.
2. In MFree methods, we have total freedom to decide which nodes in the support domain are to be included, as long as a simple guideline is followed, which means that we do not have to choose all the nodes in the support domain. The guideline is that nodes near the point of interest, which is usually the center of the support domain, should have higher priority for inclusion. Also, lower orders of polynomial terms should have higher priority for inclusion into the basis. The only problem now is how to choose them without prior knowledge on the nodal distribution.
3. In solving an actual problem using MFree methods that use PIM shape functions, a singular moment matrix is not very often encountered. This implies that special treatment for including or excluding nodes is required but only for very rare situations.

MTA is an automatic procedure to ensure a proper node enclosure and a proper basis selection. The moment matrix is first triangularized to obtain the row and column ranks. In the triangularization process, we can also obtain information about which nodes need to be excluded from the support domain and which monomials need to be removed from the basis. Using MTA can ensure successful construction of PIM shape functions without preknowledge of the nodal distribution. The validity of the present MTA has been demonstrated using point interpolation examples and surface fittings. It has been confirmed that MTA ensures stability and flexibility in the point interpolation procedure.

5.9.1 MTA Procedure

MTA was invented as an automatic procedure that would work without too much increase in computational cost. Of course, manual manipulation is not allowed. The detailed procedure of the present MTA algorithm is outlined as follows:

1. **Normalize coordinates**. Suppose that there are n nodes included in the support domain of a quadrature point $\mathbf{x}_Q$. The coordinates of all the points in the support domain are first normalized with respect to $\mathbf{x}_Q$.
2. **Form $\mathbf{P}_Q$**. Using Equation 5.91, we can form an $n \times n$ moment matrix $\mathbf{P}_Q$. The rows of $\mathbf{P}_Q$ correspond to the nodes in the support domain, and its columns associate with the monomials in the polynomial basis.
3. **Compute rank via triangularization**. The moment matrix $\mathbf{P}_Q$ is then triangularized to determine the row rank, r. If $r = n$, $\mathbf{P}_Q$ is a full rank matrix. Therefore, shape functions can be directly obtained using the triangularized matrix, no computation done so far has been wasted, and no additional computation is required. If $r < n$, there is a rank deficiency in $\mathbf{P}_Q$. The reason for the rank deficiency is that some rows of $\mathbf{P}_Q(n - r$ rows) are linearly related with other rows. The $n - r$ rows and columns should be removed from $\mathbf{P}_Q$ to form a new full rank moment matrix. The removal procedure is as follows.

4. **Remove rows**. In the row triangularization process, the exchanges of rows are recorded. From the diagonal elements of the triangularized matrix, we can determine which rows (total $n - r$ rows) should be removed from $\mathbf{P}_Q$. Because the rows of $\mathbf{P}_Q$ relate to the nodes, we can then determine which nodes should be excluded from the support domain.
5. **Remove columns**. To determine the columns of $\mathbf{P}_Q$ that should be removed, $\mathbf{P}_Q^T$ is triangularized to obtain the column rank of $\mathbf{P}_Q$, and the information of exchanging rows of $\mathbf{P}_Q^T$ (columns of $\mathbf{P}_Q$) is also recorded. From matrix theory, the row rank and the column rank of a matrix are exactly the same. Therefore, the column rank of $\mathbf{P}_Q$ is also r. From the triangularized $\mathbf{P}_Q^T$, we can determine which columns ($n - r$ columns) should be removed from $\mathbf{P}_Q$. Because the columns of $\mathbf{P}_Q$ relate to the monomials in the basis, we can therefore determine which monomials (total $n - r$ terms) should be removed from the basis.
6. **Compute shape functions**. After the above operation, the $n \times n$ matrix $\mathbf{P}_Q$ is changed to a new $r \times r$ matrix that has full rank. The shape functions can finally be obtained easily using the triangularized matrix.

Note, in the case of rank deficiency, that the present MTA requires performing triangularization at least twice. Therefore, at least one triangularization is wasted in such a case. Fortunately, it is very rare to have rank deficiency cases for arbitrarily distributed nodes.

5.9.2 Normalization of the Support Domain

It should be noted here that nodes farther away from the quadrature point $\mathbf{x}$ and higher-order monomials in the basis functions should be excluded first in the case of $r < n$. This ensures the accuracy of the interpolations and the consistency of the basis. It can be achieved by using a local coordinate, $\tilde{\mathbf{x}}$, with the quadrature point $\mathbf{x}$ as its origin, and normalization of the support domain to obtain $\bar{\mathbf{x}}$. The operations are as follows:

Translate the axis to obtain the local coordinate $\tilde{\mathbf{x}}$

$$\tilde{\mathbf{x}}_i = \mathbf{x}_i - \mathbf{x}_Q \tag{5.173}$$

Normalize the support domain

$$\bar{\mathbf{x}}_i = \tilde{\mathbf{x}}_i / d_{\max} \tag{5.174}$$

where $d_{\max}$ is the maximum value of the coordinates $\tilde{\mathbf{x}}$ of the nodes in the support domain, i.e.,

$$d_{\max} = \max(x_i, y_i), \quad i = 1, 2, \ldots, n \tag{5.175}$$

Alternatively, we can simply use the dimension of support domain, d_s.

After these operations, we can observe the following properties in $\mathbf{P}_Q$:

1. The absolute coordinate values of all nodes in the support domain are not greater than one, i.e., $x_i \leq 1$ and $y_i \leq 1$.

2. The absolute coordinate values of a node farther away to the point $\mathbf{x}_Q$ is greater than that of the nodes at the vicinity of $\mathbf{x}_Q$.
3. In a basis function, the value of a higher-order monomial is less than that of lower-order ones.

Therefore, in the same column of $\mathbf{P}_Q$, the element of a nearer node is always of greater value than that of a farther node. In the same row of $\mathbf{P}_Q$, the element of a higher-order monomial is always of smaller value than that of a lower-order monomial.

Note that this normalization process also helps improve the conditioning of the moment matrix.

A Gauss elimination using a pivot of subdominant elements is used to compute the row rank to ensure that farther nodes are excluded from the support domain first. The subdominant elements are relatively smaller compared with the dominant element in the rows.

A Gauss elimination using a global pivot of dominant elements in computing the column rank will choose the pivots according the sequence from low-order monomials to high-order monomials. It can ensure the removal of higher-order monomials from the basis first.

Note that in examining the row rank, the Gauss elimination using a pivot of subdominant elements is performed. This could cause numerical problems, as the pivot is not the largest. However, this is not a great concern, because the pivot element is still one of the largest elements.

5.9.3 MTA Flowchart

The flowchart of the triangularization algorithm can be given briefly as follows:

(1) Determine the support domain for a point $\mathbf{x}$ to obtain an $n \times n$ moment matrix $\mathbf{P}_Q$.

(2) Triangularize $\mathbf{P}_Q$ to get row rank r and record the exchanges of rows.

(3) If $r < n$, then

  (a) Determine the nodes to be excluded from the support domain;

  (b) Triangularize $\mathbf{P}_Q^T$ to obtain the column rank of $\mathbf{P}_Q$;

  (c) Determine the monomials to be removed from the basis;

  (d) Remove corresponding rows and columns from $\mathbf{P}_Q$ to obtain $\tilde{\mathbf{P}}_Q$;

  (e) Go to (4);

  Else go to (4).

  End if

(4) Compute PIM shape functions from the triangularized matrix.

5.9.4 Test Examples

Example 5.3 Interpolation Using 6 Nodes in Parallel Lines

To show how MTA works, an example of the six nodes interpolation, as shown in Figure 5.19, is presented here in detail. These six nodes are included in the support domain of a quadrature point at $\mathbf{x}_Q = (0.5, 0.5)$. The basis given in Equation 5.115, which is of complete second order with respect to both the x and y coordinates, is used.

STEP 1 Normalization of the support domain

Using a local coordinate with the quadrature point $\mathbf{x}_Q$ as its origin, perform the normalization of the support domain. The new coordinate values $\bar{\mathbf{x}}$ of these nodes are listed in Table 5.4.

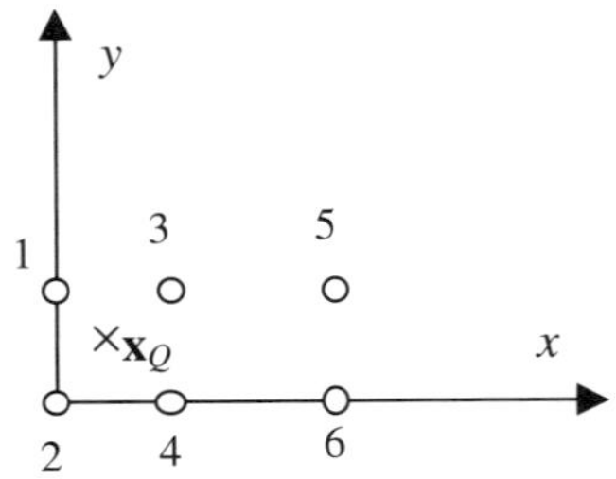

Nodes	x	y
1	0.0	1.0
2	0.0	0.0
3	1.0	1.0
4	1.0	0.0
5	2.0	1.0
6	2.0	0.0
$\mathbf{x}_Q$	0.5	0.5

FIGURE 5.19
Six nodes in parallel lines in the support domain of $\mathbf{x}_Q$ and its coordinates.

TABLE 5.4
Local Coordinates of the Six Nodes in Figure 5.19 after Normalization

Nodes	$\bar{x}_i$	$\bar{y}_i$
1	−0.33333	0.33333
2	−0.33333	−0.33333
3	0.33333	0.33333
4	0.33333	−0.33333
5	1.0	0.33333
6	1.0	−0.33333
$\bar{\mathbf{x}}$	0.	0.

STEP 2 Construction of the moment matrix $\mathbf{P}_Q$

Using Equation 5.115, the moment matrix $\mathbf{P}_Q$ is constructed as follows:

$$\mathbf{P}_Q = \begin{array}{c} \begin{matrix} 1 & x_i & y_i & x_iy_i & x_i^2 & y_i^2 \end{matrix} \\ \begin{bmatrix} 1.0 & -0.333 & 0.333 & -0.111 & 0.111 & 0.111 \\ 1.0 & -0.333 & -0.333 & 0.111 & 0.111 & 0.111 \\ 1.0 & 0.333 & 0.333 & 0.111 & 0.111 & 0.111 \\ 1.0 & 0.333 & -0.333 & -0.111 & 0.111 & 0.111 \\ 1.0 & 1.0 & 0.333 & 0.333 & 1.0 & 0.111 \\ 1.0 & 1.0 & -0.333 & -0.333 & 1.0 & 0.111 \end{bmatrix} \end{array} \begin{matrix} \leftarrow \text{node 1} \\ \leftarrow \text{node 2} \\ \leftarrow \text{node 3} \\ \leftarrow \text{node 4} \\ \leftarrow \text{node 5} \\ \leftarrow \text{node 6} \end{matrix} \quad (5.176)$$

It can be found that the rows of $\mathbf{P}_Q$ correspond to these six nodes in the support domain, and its columns associate with the monomials in the basis.

STEP 3 Row rank and row removal

Using the Gauss elimination with a pivot of subdominant elements, the moment matrix $\mathbf{P}_Q$ is triangularized to determine the row rank. The triangularized moment matrix $\mathbf{P}'_Q$ is

$$\mathbf{P}'_Q = \begin{bmatrix} -0.111 & -0.333 & 0.333 & 1.0 & 0.111 & 0.111 \\ 0.0 & 0.222 & 0.0 & 2.0 & -0.667 & 0.222 \\ 0.0 & 0.0 & 0.667 & 0.0 & 0.667 & 0.0 \\ 0.0 & 0.0 & 0.0 & -0.889 & 2.667 & -8.0 \\ 0.0 & 0.0 & 0.0 & 0.0 & 1.333 & 0.0 \\ 0.0 & 0.0 & 0.0 & 0.0 & 0.0 & 0.0 \end{bmatrix} \quad (5.177)$$

It can be found that the sixth row of $\mathbf{P}_Q'$ is zero and the row rank is $r = 5$, which means there is a rank deficiency in $\mathbf{P}_Q$. The sixth row of $\mathbf{P}_Q'$ is linearly related with at least one of the other five rows. One row and column should be removed from $\mathbf{P}_Q$ to form a new full rank moment matrix. In the row triangularization process, the exchanges of rows are recorded. The exchange history is given as

$$\text{IRE} = \{1, 2, 3, 4, 5, 6\} \tag{5.178}$$

This means that the sixth row of $\mathbf{P}_Q'$ corresponds to the sixth row of $\mathbf{P}_Q$. Therefore, the sixth row should be removed from $\mathbf{P}_Q$. Because the rows of $\mathbf{P}_Q$ relate to the nodes, we can then determine that node 6 should be excluded from the support domain.

From the above computational results, one can observe that the Gauss elimination using pivoting of subdominant elements will choose first the row of a near node as the pivot in the elimination process, and then choose the row of a farther node. It ensures that farther nodes are excluded from the support domain first. In this six-node interpolation, node 6, which is farther from the quadrature point $\mathbf{x}$, has been chosen automatically by MTA to be excluded from the support domain. This ensures accurate interpolations.

STEP 4 Column rank and column removal

The Gauss elimination using a global pivot of dominant elements is now performed to compute the column rank of $\mathbf{P}_Q$, i.e., the row rank of $\mathbf{P}_Q^T$. From matrix theory, the row rank and the column rank of a matrix are the same. Therefore, the column rank of $\mathbf{P}_Q$ is also $r = 5$. Similar to the procedure of the row removal, from the result of column rank it can be determined that the sixth column should be removed from $\mathbf{P}_Q$. Because the columns of $\mathbf{P}_Q$ relate to the monomials in the basis, one can therefore determine that the monomial y^2 should be removed from the basis.

It can be observed that there are only two different y values in all six nodes and interpolation of a second-order function of y will fail. Therefore, the monomial y^2 should be removed from the basis. Gauss elimination using a pivot of dominant elements can ensure that high-order monomials are removed from the basis so that the basis is of consistency. It should be noted that all these removals are performed purely mathematically and automatically without knowledge of the physical relations of the nodes.

STEP 5 Computation of the shape functions

After removing the sixth row and the sixth column, the original 6×6 matrix $\mathbf{P}_Q$ is changed to a new 5×5 matrix $\mathbf{P}_Q'$

$$\mathbf{P}_Q' = \begin{bmatrix} -0.111 & -0.333 & 0.333 & 1.0 & 0.111 & 0.111 \\ 0.0 & 0.222 & 0.0 & 2.0 & -0.667 & 0.222 \\ 0.0 & 0.0 & 0.667 & 0.0 & 0.667 & 0.0 \\ 0.0 & 0.0 & 0.0 & -0.889 & 2.667 & -8.0 \\ 0.0 & 0.0 & 0.0 & 0.0 & 1.333 & 0.0 \\ 0.0 & 0.0 & 0.0 & 0.0 & 0.0 & 0.0 \end{bmatrix} \tag{5.179}$$

It should be mentioned here that the $\mathbf{P}_Q'$ is a triangularized full rank matrix, and the shape functions can be obtained easily. The PIM shape functions of this six-node interpolation are obtained and given in Table 5.5.

Example 5.4 Interpolation Using 12 Nodes in Parallel Lines

MTA is now used for another case of singularity encountered when using 12 structured nodes arranged regularly in three parallel lines, as shown in Figure 5.20. These 12 nodes are located in parallel lines in both the x and y directions and are assumed to be within the support domain of point $\mathbf{x}_Q$. We initially choose the basis functions with 12 terms as

TABLE 5.5
Shape Functions of Six Parallel Nodes Obtained Using MTA and Evaluated at $\mathbf{x}_Q$

Nodes	1	2	3	4	5	Σ
ϕ_I	0.125	0.250	0.500	0.250	−0.125	1.000

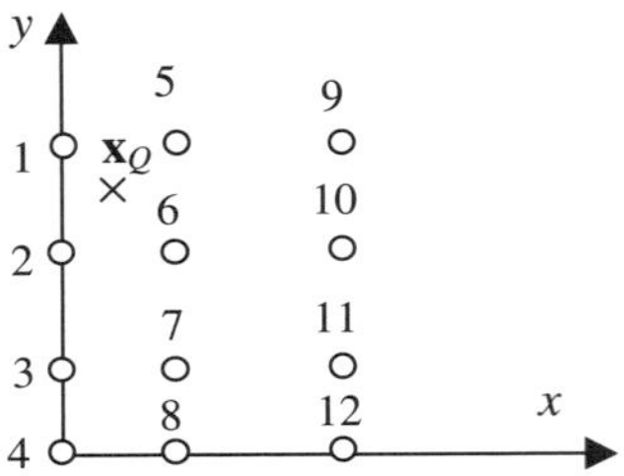

Nodes	x	y	Nodes	x	y
1	0.0	3.0	7	2.0	1.0
2	0.0	2.0	8	2.0	0.0
3	0.0	1.0	9	3.0	3.0
4	0.0	0.0	10	3.0	2.0
5	2.0	3.0	11	3.0	1.0
6	2.0	2.0	12	3.0	1.0
$\mathbf{x}_Q$	0.5	2.5			

FIGURE 5.20
Twelve nodes in parallel lines in the support domain of $\mathbf{x}_Q$ and its coordinates.

TABLE 5.6
Shape Functions of 12 Parallel Nodes Obtained Using MTA and Evaluated at $\mathbf{x}_Q$

Nodes	1	2	3	5	6	7	8	9	10	11	Σ
ϕ_I	0.1406	0.2812	−0.0468	0.2188	0.75	−0.2813	0.062	−0.0467	−0.0935	0.0157	1.000

$$\mathbf{p}(\mathbf{x}) = [1, x, y, xy, x^2, y^2, x^2y, xy^2, x^2y^2, x^3, y^3, x^3y]^T \tag{5.180}$$

By using these 12 nodes and the above basis functions, it is clear that point interpolation will fail, unless we do something about it. There are only three different x values, which cannot accommodate any third-order term of x. By using the present MTA, it is found that the rank of $\mathbf{P}_Q$ is $r = 10$. The nodes 4 and 12 are automatically excluded from the support domain, and the terms x^3 and x^3y are removed from the basis function. This result is exactly what we wanted. Hence, point interpolation can now be successful using the remaining nodes and terms of basis functions. The PIM shape functions are constructed and are listed in Table 5.6.

5.10 Comparison Study via Examples

Example 5.5 Comparison of Shape Functions Obtained Using Different Methods (1D Case)

Figure 5.21a shows a comparison of shape functions in 1D space obtained using four different methods: MLS, polynomial PIM, MQ-PIM, and EXP-PIM. The shape functions are obtained using five nodes at $x = -1.0, -0.5, 0.0, 0.5$, and 1.0 in the support domain of $[-1.0, 1.0]$. The shape function shown in Figure 5.21 is for the node at $x = 0.0$. It is noted that the shape functions obtained by the three PIM methods possess the Kronecker delta function property, that is, $\phi(0.0) = 1.0$, $\phi(-1.0) = \phi(-0.5) = \phi(0.5) = \phi(1.0) = 0.0$. The shape function obtained using MLS does not possess the Kronecker delta function property, that

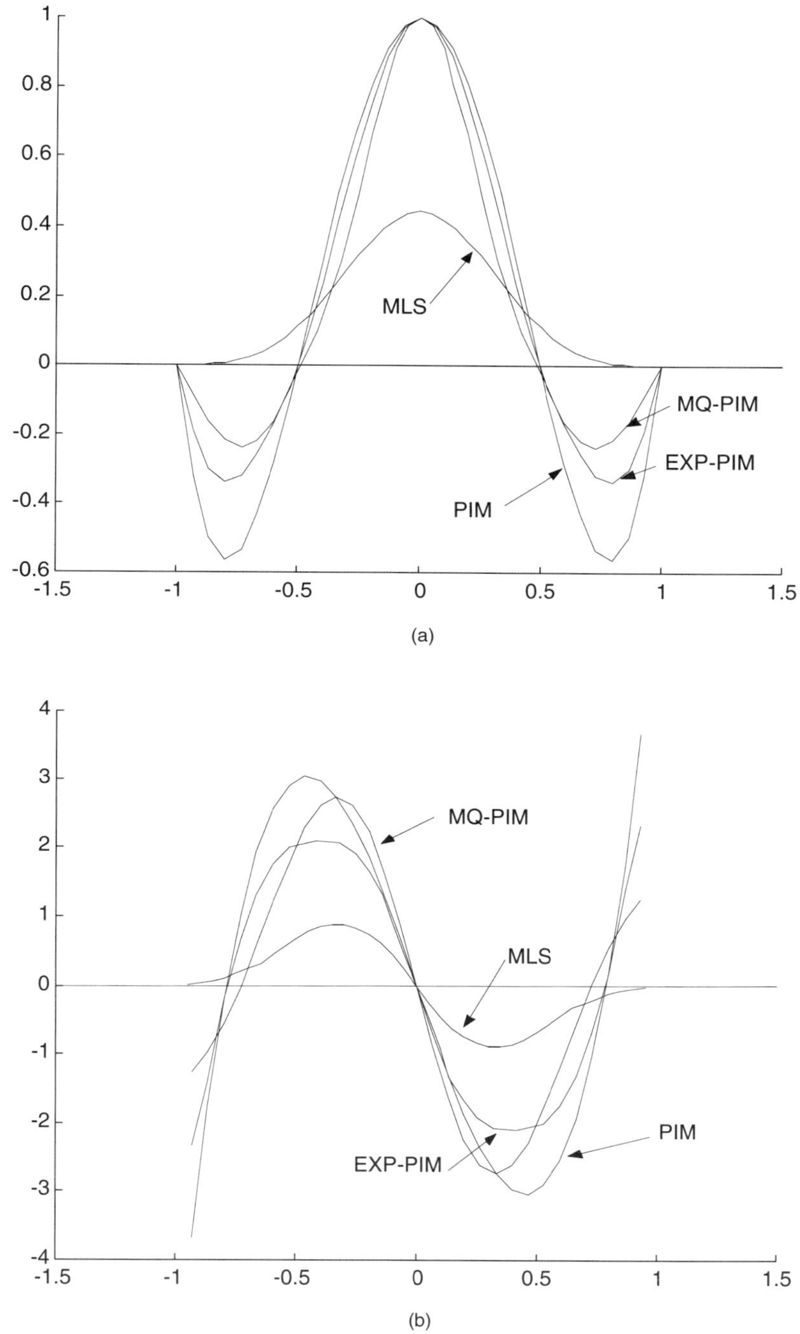

FIGURE 5.21
Comparison of shape functions obtained using four different methods: MLS, polynomial PIM, MQ-PIM, and EXP-PIM. (a) Shape functions; (b) derivatives of shape functions.

is, $\phi(0.0) \neq 1.0$, $\phi(-0.5) = \phi(0.5) \neq 0.0$. The MLS shape vanishes at $x = \pm 1.0$, because $x = \pm 1.0$ is on the boundary of the support domain. All the PIM shape functions vary more frequently than the MLS shape functions.

Figure 5.21b shows the first derivatives of shape functions. It is found that the derivatives of the MLS shape function vanish at two boundary points. This is because the quartic weight function is used in the construction of the shape function, whose first derivative also vanishes on the boundary (see Figure 5.4).

Example 5.6 Comparison of Shape Functions Obtained Using Different Methods (2D Case)

MFree shape functions are constructed in a domain of $(x, y) \in [-2, 2] \times [-2, 2]$ using 5×5 evenly distributed nodes in the domain. Figure 5.22 shows the shape function and its first derivative with respect to x computed using polynomial PIM. Figure 5.23 shows the shape function and its first derivative with respect to x computed using RPIM (MQ, $C = 1.0$, $q = 0.5$) with linear polynomials. It is shown that the PIM shape functions satisfy the Kronecker delta function. Figure 5.24 shows the MLS shape function. It is clear that the MLS shape functions do not satisfy the Kronecker delta function.

Comparing Figures 5.22 and 5.23 with Figure 5.24 reveals that the PIM shape function and its derivatives are much more complex compared with the MLS shape functions. This finding needs to be considered when we perform numerical integrations in computing discrete system equations.

Example 5.7 Curve Fitting Using MFree Shape Functions

Analyses are conducted to study the effects of shape parameters of the radial base functions on the accuracy when the shape functions are used for curve fitting. 1D and 2D problems are investigated. Accuracy in both the fitted function itself and in its derivatives is examined. The fitting of functions is based on the nodal function value sets that are generated at regularly as well as at irregularly distributed nodes. As will soon be shown (in Example 5.9), linear functions in 1D and 2D spaces can be reproduced exactly when $m = 3$. In this example, functions to be fitted are chosen to be nonpolynomial. Because the radial basis functions are of major concern, polynomial bases will not be included in the basis (i.e., $m = 0$) for this particular study. Curve fitting for the following three 1D functions defined in the domain of [0, 7] is performed:

$$f_1(x) = \sin(x) \tag{5.181}$$

$$f_2(x) = \sin^2(x) + (0.5x - 1)\cos(x) \tag{5.182}$$

$$f_3(x) = \frac{x^2}{8 + x^5} \tag{5.183}$$

The first function is a typically oscillatory function of x. The second is a mixture of polynomials and trigonometry terms. The last is a fractional function that approaches zero when x approaches infinity. All are types of functions that are different from radial basis functions. The first derivatives of these functions can be easily obtained as follows:

$$f_1'(x) = \cos(x) \tag{5.184}$$

$$f_2'(x) = 2\sin(x)\cos(x) - (0.5x - 1)\sin(x) + 0.5\cos(x) \tag{5.185}$$

$$f_3'(x) = \frac{16x - 3x^6}{(8 + x^5)^2} \tag{5.186}$$

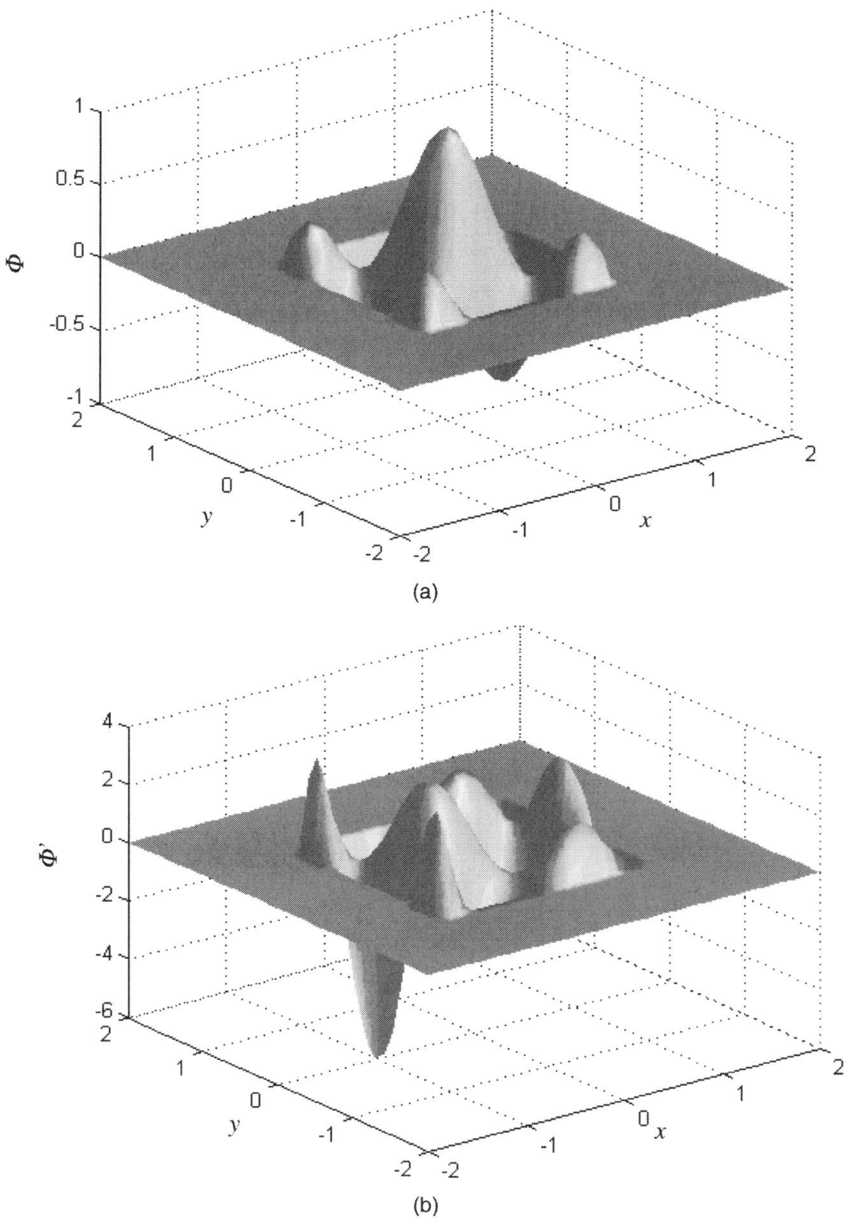

FIGURE 5.22
Polynomial PIM shape function and its derivative with respect to x. (a) PIM shape function; (b) derivative of the PIM shape function.

The curve fitting is carried out in the following procedure:

1. Create a set of nodes in the domain where the function is defined and is to be fitted.
2. For a given x where the function is to be fitted, choose n nodes in the support domain of x for the later construction of shape functions.

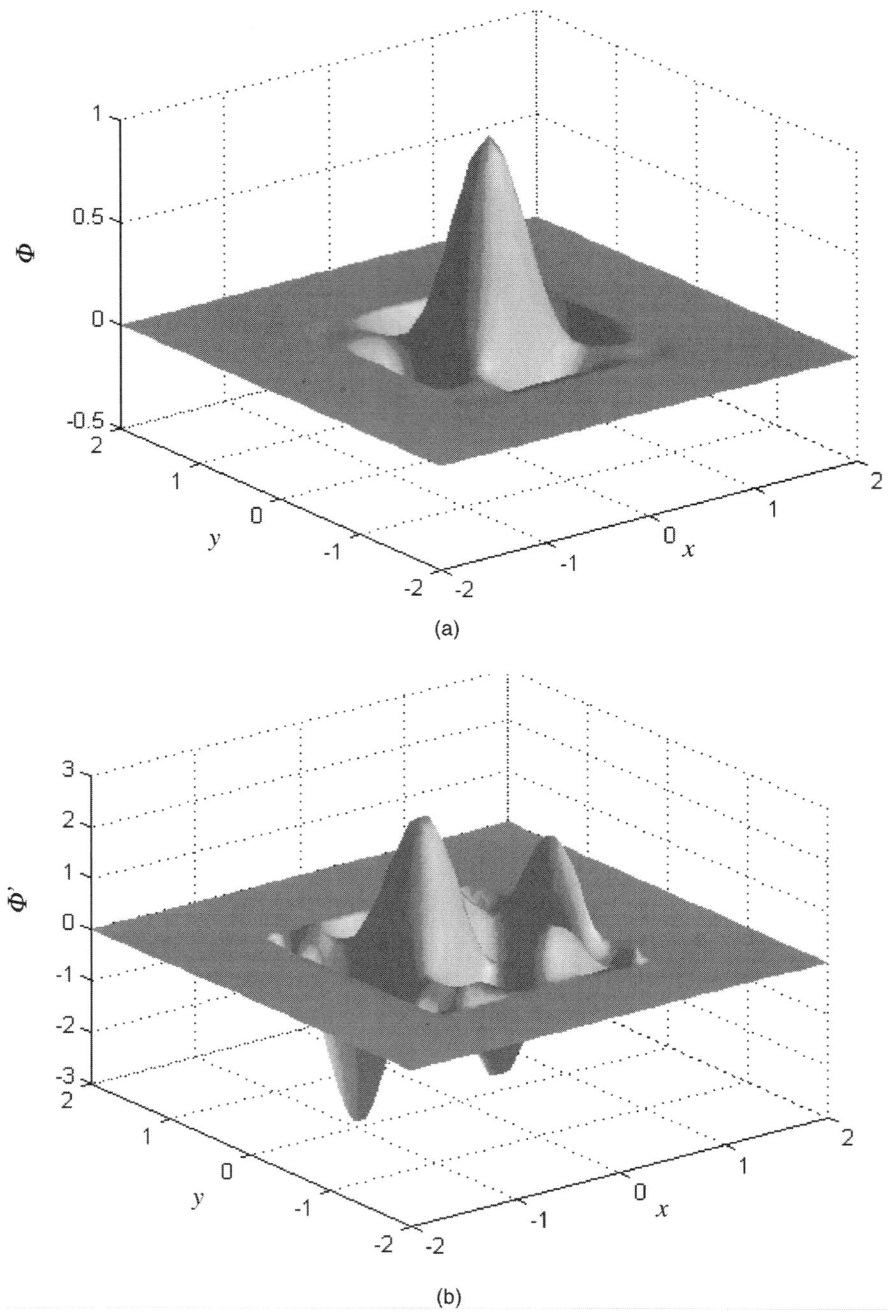

FIGURE 5.23
RPIM (MQ, $C = 1.0$, $q = 0.5$, with polynomial) shape function and its derivative with respect to x. (a) PIM shape function; (b) derivative of the PIM shape function.

3. Calculate the function values at these n nodes using Equation 5.181, 5.182, or 5.183.
4. Construct shape functions and their derivatives following the procedure discussed in the previous sections of this chapter.
5. Calculate the function value at x using these shape functions and their derivatives.

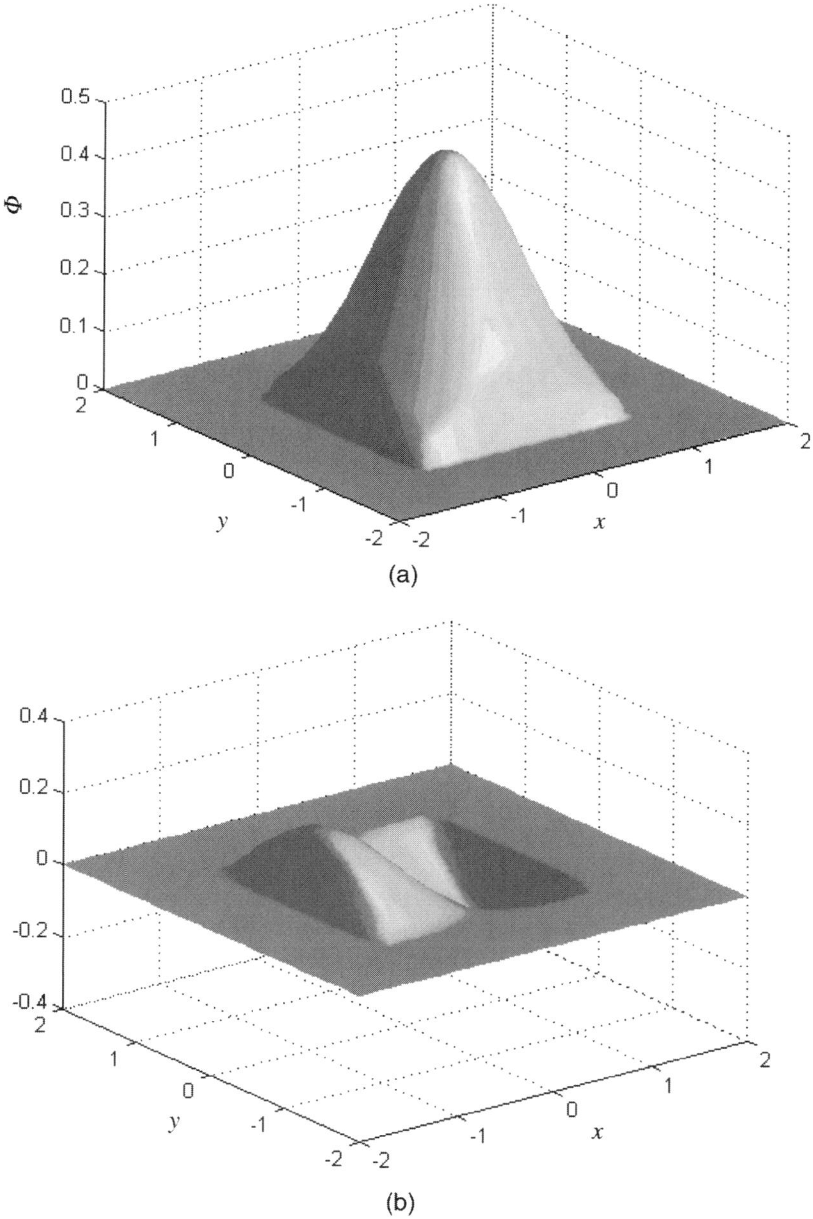

FIGURE 5.24
MLS shape function and its derivative with respect to x. (a) MLS shape function; (b) derivative of the MLS shape function.

The error of curve fitting at point i is defined as

$$e_i = \frac{f_i - \tilde{f}_i}{|\tilde{f}_i|} \tag{5.187}$$

where f_i is the true function value or its derivative at point i computed using Equation 5.184, 5.185, or 5.186, and $\tilde{f}_i$ is the fitted function value or the derivative of the fitted

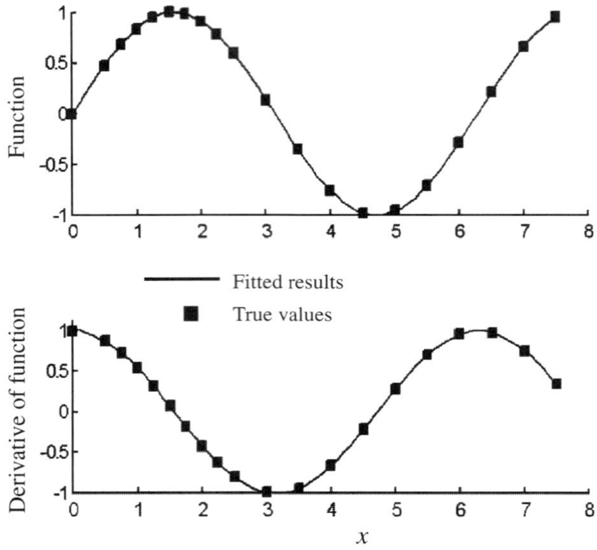

FIGURE 5.25
Comparison the original function value of $f = \sin(x)$ with the fitted function using a PIM shape function.

function at point i. The total fitting error in fitting the curve in the entire fitting domain is defined as

$$e_t = \sum_{i=1}^{n} |e_i| = \sum_{i=1}^{n} \left| \frac{f_i - \tilde{f}_i}{|\tilde{f}_i|} \right| \tag{5.188}$$

Figure 5.25 shows a typical result of curve fitting using PIM shape functions for $f_1(x) = \sin(x)$. Both the fitted function and the derivatives of the fitted function are shown in comparison with the original function values. It is found that on the plotted curves it is a perfect fitting, regardless of which shape function is used. Therefore, plots on error of fitting would be a much better gauge of the performance of different types of shape functions. Figure 5.26 shows the error in fitting $f_1 = \sin(x)$ using polynomial PIM shape functions created using five nodes in the vicinity of x. Both errors in fitted function and the derivative of the fitted function are shown. It is clearly shown that the error in the function itself is much smaller than that in the derivative of the fitted function. Figure 5.27 shows the same but is computed using MQ-PIM with shape parameters of $q = 0.5$, $C = 1.0$. Comparison with Figure 5.26 reveals that the accuracy in curve fitting by polynomial PIM shape functions is higher by about three orders than that using MQ-PIM shape functions. Figure 5.28 shows the same curve fitting error computed using EXP-PIM with a shape parameter of $c = 0.3$. It is seen that the accuracy of curve fitting using EXP-PIM is higher than that of MQ-PIM by about one order, but it is still lower than that using polynomial PIM by about two orders. Figure 5.29 shows the same results but is computed using MLS shape functions. The curve fitting error using MLS shape functions is the highest among all results obtained using these types of shape functions. Comparison with Figure 5.26 reveals that the accuracy in curve fitting by polynomial PIM shape functions is higher by about four orders than that using MLS shape functions.

Note from Figures 5.26 through 5.29 that larger errors often occur near the boundary of the domain. This is because at any point near the boundary the interpolation uses more

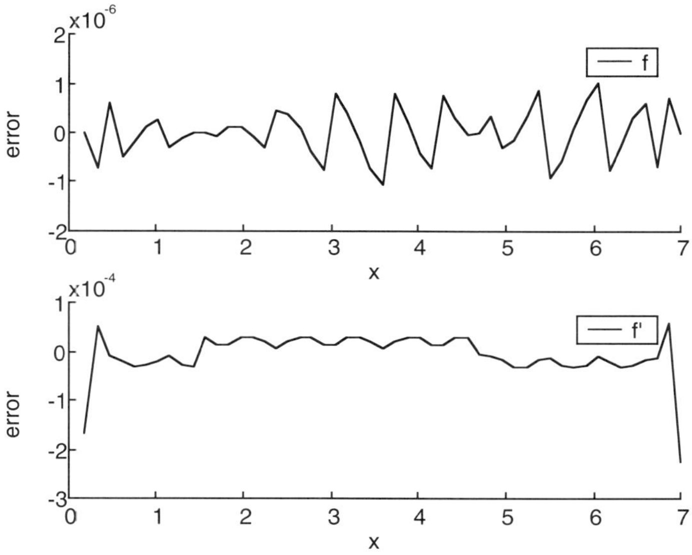

FIGURE 5.26
Error in fitting $f = \sin(x)$ using polynomial PIM shape functions created using five nodes in the vicinity of x. Both errors in fitted function and the derivative of the fitted function are shown.

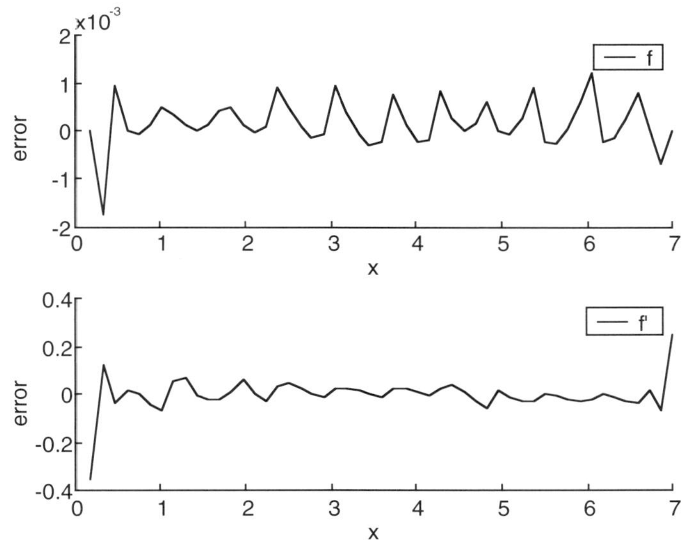

FIGURE 5.27
Error in fitting $f = \sin(x)$ using MQ-PIM ($q = 0.5$, $C = 1.0$) shape functions created using five nodes in the vicinity of x. Both errors in fitted function and the derivative of the fitted function are shown.

nodes on one side of the point. Note also that the fitted function is much more accurate on the function itself than on its derivatives, as one would expect.

Further note that the above comparison is performed under the same condition of nodal density and uses only five nodes in the creation of shape functions. The accuracy can be

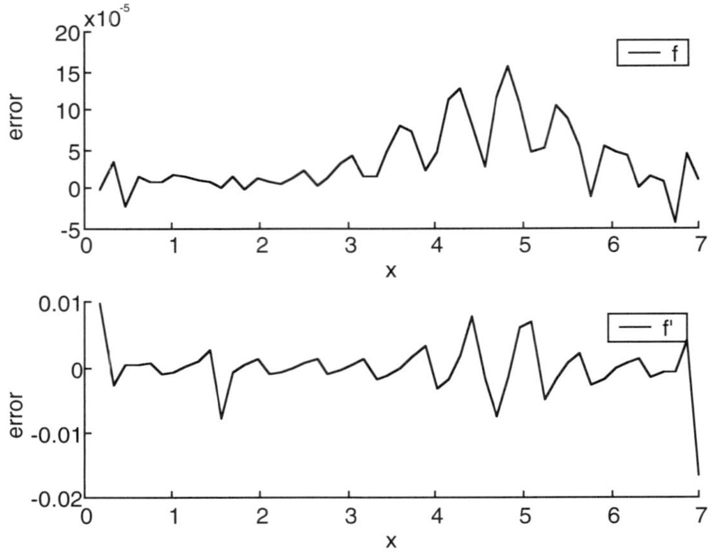

FIGURE 5.28
Error in fitting $f = \sin(x)$ using EXP-PIM ($c = 0.3$) shape functions created using five nodes in the vicinity of x. Both errors in fitted function and the derivative of the fitted function are shown.

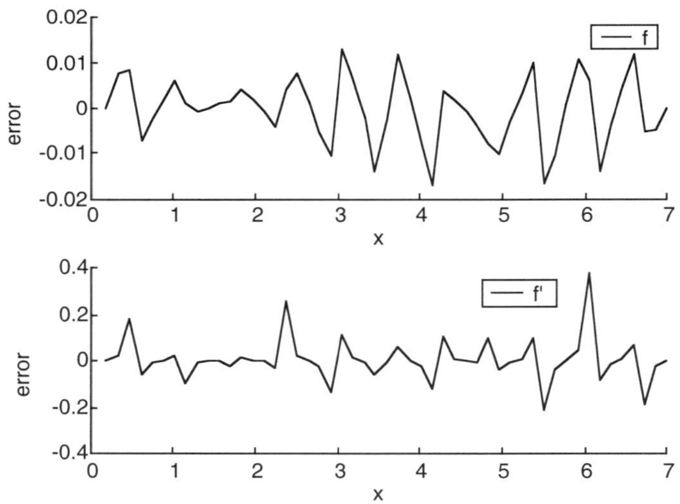

FIGURE 5.29
Error in fitting $f_1 = \sin(x)$ using MLS shape functions created using five nodes in the vicinity of x. Both errors in fitted function and the derivative of the fitted function are shown.

improved, of course, by increasing nodal density or by using more nodes in shape function creation. This applies to all types of shape functions. An intensive study has been conducted to examine the convergence of curve fitting using different shape functions. Figure 5.30 summarizes the results in terms of errors in fitted functions using different shape functions created using different dimensions of support domain as controlled by

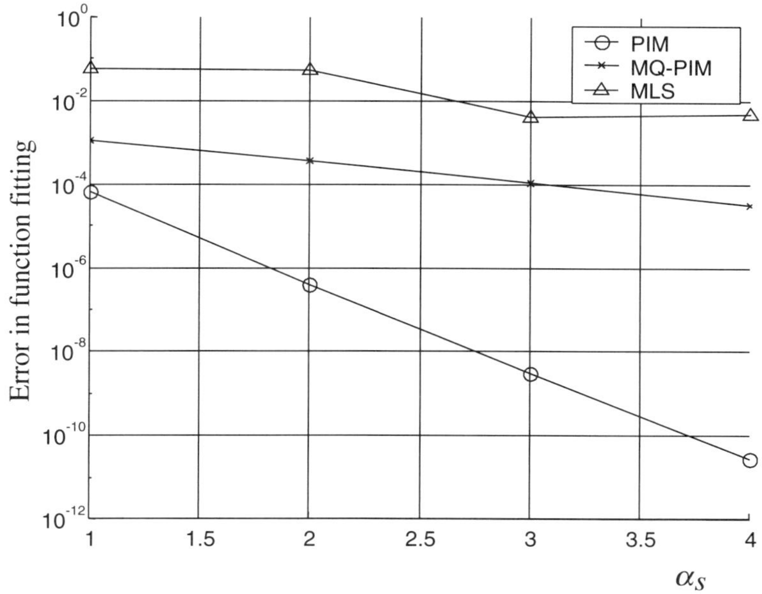

FIGURE 5.30
Error in curve fitting using different shape functions created using different dimensions of support domain. $\alpha_s = 2.1$, corresponding to the use of five nodes that is used in computing Figures 5.26 through 5.29.

Equation 2.2. The dimensionless size of the support domain of $\alpha_s = 2.1$ corresponds to the use of five nodes, as is used in computing Figures 5.26 through 5.29. It is found that polynomial PIM shape functions demonstrate much better convergence compared with RPIM and MLS shape functions. MLS has the lowest convergence rate.

The same study was also conducted for the second function defined by Equation 5.182. Similar findings as those for the first sine function have been obtained. Here we first present the results obtained using MQ-PIM shape functions. Figure 5.31 plots the error in the fitted function and its derivatives. We observe again very accurate fitting in the interior of the domain and less accuracy near the boundaries. It can be again seen that the fitted function is much more accurate on the function itself compared to its derivatives. Comparison with Figure 5.27 reveals that the fitting errors in function values are almost the same, but more accurate results are obtained for the derivative of the function. This may be attributed to the different shape parameters used. Figure 5.32 shows the error in the fitted function and its derivatives for the third function defined by Equation 5.186. Similar observations can be made.

Curve fitting using EXP-PIM was also examined. Figure 5.33 shows the fitted third function defined by Equation 5.186, together with the actual function values marked on the fitted curves. Figure 5.33b plots the errors in the fitted function and in its derivative. From Figure 5.33, similar conclusions can be drawn.

Example 5.8 Effects of Shape Parameters on the Condition Number of Moment Matrices and Curve Fitting

The properties of moment matrices $\mathbf{R}_Q$ and $\mathbf{G}$ are investigated when two polynomial terms ($m = 2$) are added to the basis for different shape parameters used in the radial function. Adding the polynomial term confirms that the property of matrix $\mathbf{G}$ is largely determined

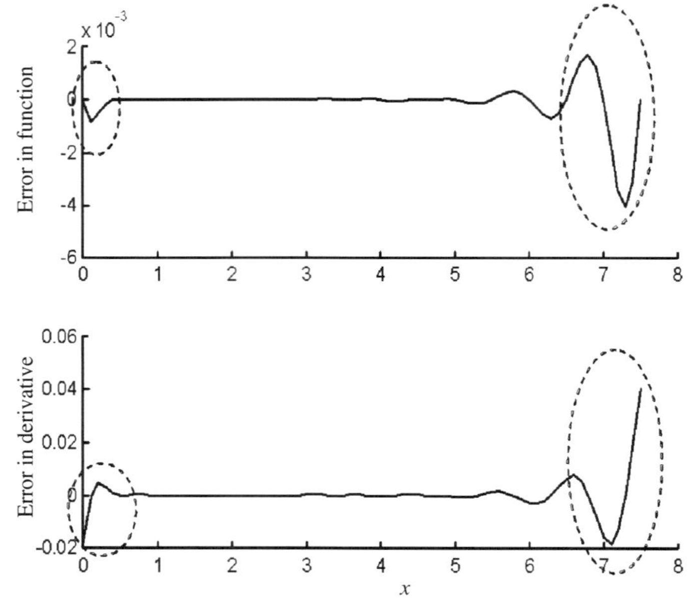

FIGURE 5.31
Distribution of error in the fitted function of $f = \sin^2(x) + (0.5x - 1)\cos(x)$ (MQ basis function with $q = 2.5$, $c = 0.1$).

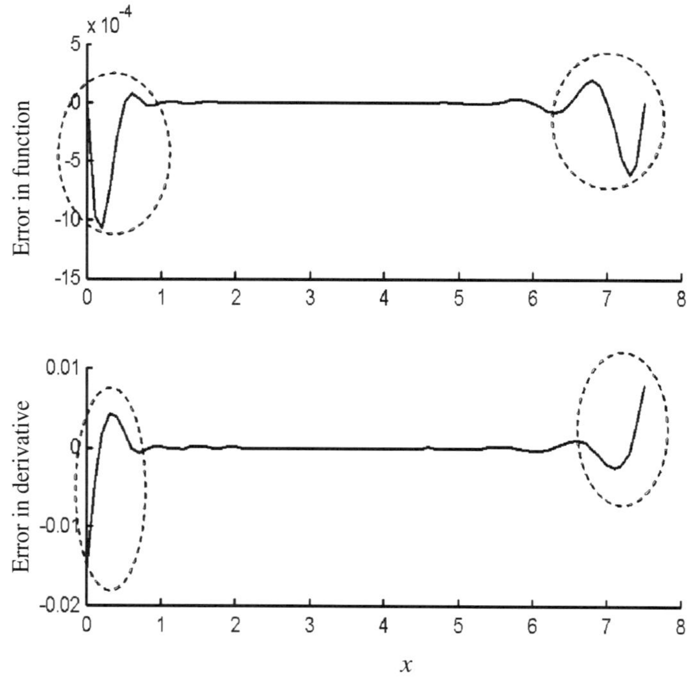

FIGURE 5.32
Distribution of error in the fitted function of $f = x^2/(8 + x^5)$ (MQ basis function with $q = 2.5$, $c = 0.1$).

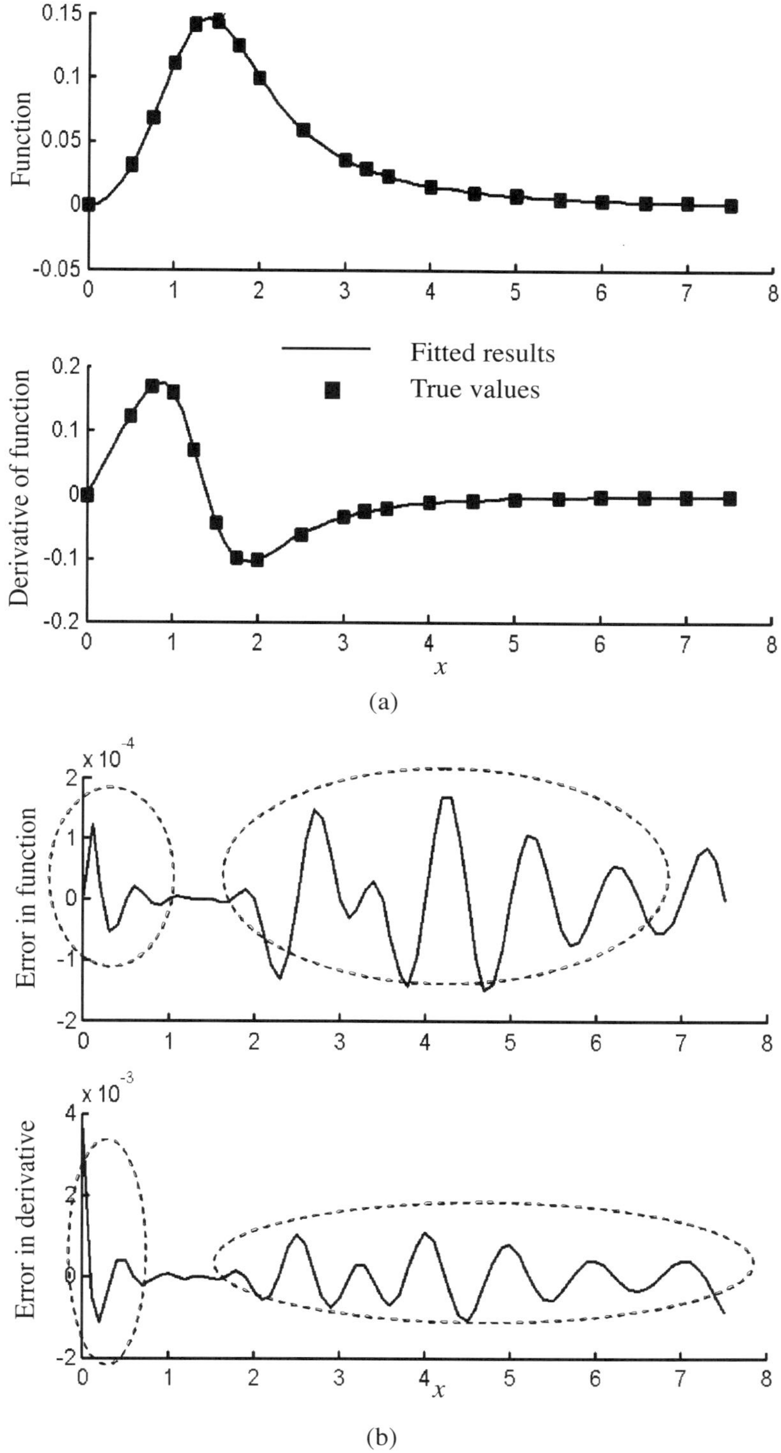

FIGURE 5.33
Interpolation of $f = x^2/(8 + x^5)$ using EXP radial function with $c = 2.0$. (a) Curve fitting of using EXP-FIM shape function; (b) error in function fitting.

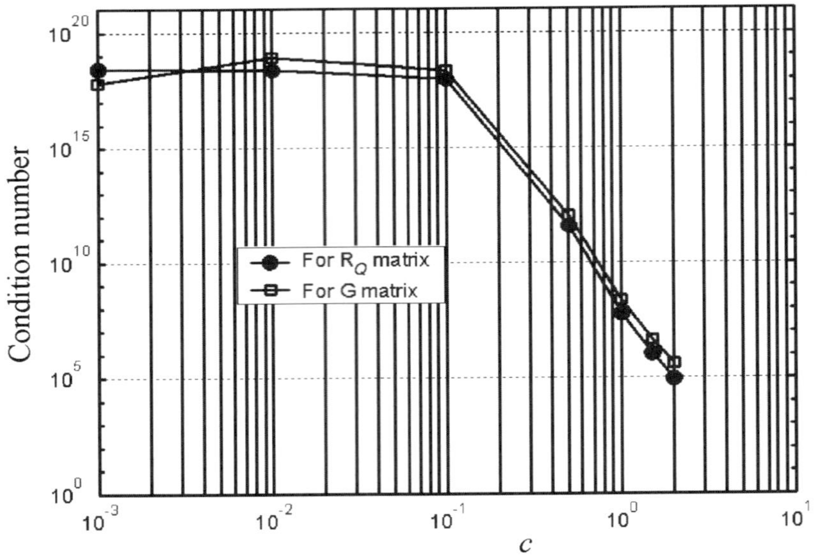

FIGURE 5.34
Effect of shape parameters c of the EXP radial function on condition number of moment matrices.

by that of $\mathbf{R}_Q$. In performing curve fittings, we set $m = 0$, to see the effects of parameters more clearly. We discuss the EXP radial function first, as it has only one parameter, c.

1. Condition numbers for both matrix $\mathbf{R}_Q$ and $\mathbf{G}$ are computed, and the results are shown in Figure 5.34. It is found that the condition number of matrix $\mathbf{R}_Q$ is very close to that of matrix $\mathbf{G}$. This implies that the condition of matrix $\mathbf{G}$ is largely determined by that of matrix $\mathbf{R}_Q$. Therefore, the discussion can be carried out based on only the condition of matrix $\mathbf{R}_Q$.
2. Shape parameter c significantly affects the condition number of matrix $\mathbf{R}_Q$ as shown in Figure 5.34. When c is smaller than 0.001, the condition number is too large for numerical computation. When $c = 0.001$, the fitted function is oscillatory but its derivative is very smooth, as shown in Figure 5.35. Although the inverse of matrix $\mathbf{R}_Q$ has been performed, the interpolation error is very large. Our study suggests that the parameter c should be larger than 0.001.

For MQ basis function, the following findings are obtained:

1. The condition number of the matrix $\mathbf{R}_Q$ is first investigated. It is found that both the parameters q and C affect the condition of matrix $\mathbf{R}_Q$. When q is an integer, matrix $\mathbf{R}_Q$ is ill-conditioned. Figure 5.36 summarizes the calculated condition numbers of matrix $\mathbf{R}_Q$ that vary with parameter q and C.
2. It is also confirmed that the condition number of matrix $\mathbf{G}$ is almost the same as that of $\mathbf{R}_Q$.
3. When $0 < q < 1$, where matrix $\mathbf{R}_Q$ has a smaller condition number, function fitting is smooth, but derivatives are oscillatory.
4. When $1 < q < 3.5$, where matrix $\mathbf{R}_Q$ has a slightly larger condition number, the fitted results on both the function and its derivatives are accurate with sufficient smoothness.

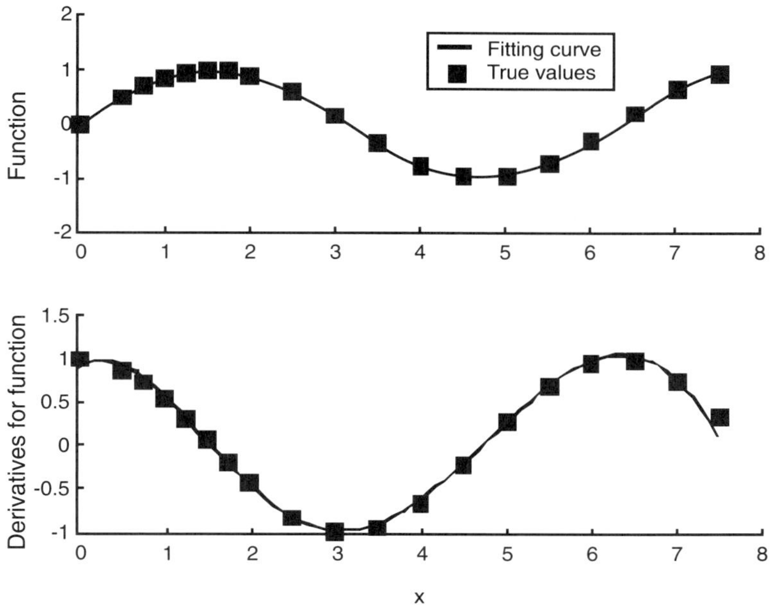

FIGURE 5.35
Sine function fitted by PIM using EXP radial basis function with a very small value of parameter c (=0.001); non-smooth-fitted function, but smooth derivatives.

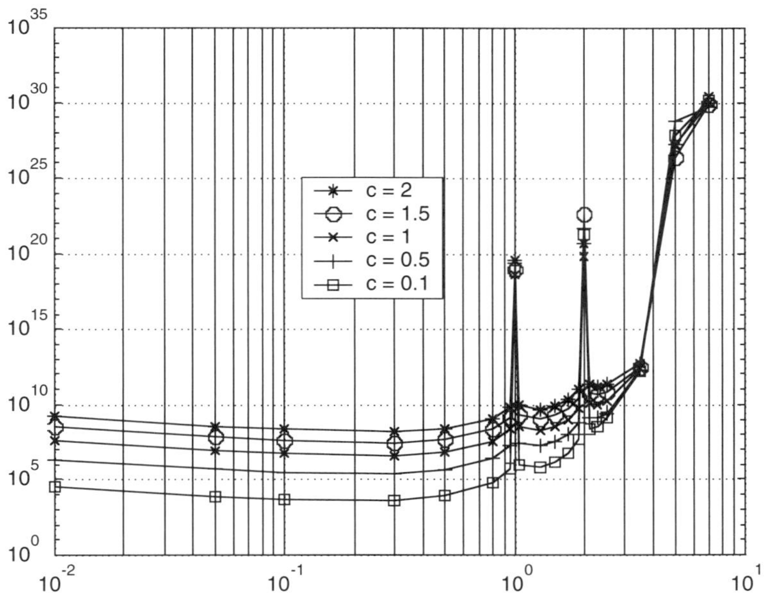

FIGURE 5.36
Effect of shape parameters q and C of the MQ radial basis function on condition number of matrix $\mathbf{R}_Q$.

5. When $3.5 < q < 10$, fitting results worsen if larger q is used. When $q = 10$, the error in the fitted function becomes unacceptably large.

In conclusion, an acceptable parameter q for the MQ radial function should be within $1 < q < 3.5$ for accurate curve fitting.

It has been confirmed that, in using RPIM shape functions for fitting nonpolynomial functions, adding polynomial terms does not improve the accuracy of the results significantly. This is shown in the following surface fitting examples.

Example 5.9 Surface Fitting Using MFree Shape Functions (Effects of Parameters)

In this example, two functions are used to examine the approximation quality of surfaces using PIM shape functions with polynomial and various radial basis functions. The results are compared with those using MLS approximation. The following two functions defined in 2D space are considered:

$$f_4(x, y) = x + y \tag{5.189}$$

$$f_5(x, y) = \sin x \cos y \tag{5.190}$$

Fitting these functions in a domain of $(x, y) \in [0, 1] \times [0, 1]$ is performed. Sixteen (4×4) evenly distributed points in the domain are used as field nodes, and all the nodes are used for constructing shape functions and performing interpolation. The approximate values at any point (x, y) within the domain are obtained through interpolations. In plotting the following figures, the shape function is first constructed, and values of functions at dense grids of 11×11 ($d_c = 0.1$) are calculated using Equation 5.125 or 5.161. In performing the MLS approximation, the cubic spline weight function is used. The errors of surface fitting are plotted to examine the quality of surface fitting using different shape functions.

In Figure 5.37, errors of surface fitting for the linear function $f_4(x, y)$ are plotted for four cases of different shape functions: (a) MLS shape function, (b) MQ-PIM shape function, (c) EXP-PIM shape function, and (d) TPS-PIM shape function. It is noted that the linear polynomial can be exactly reproduced by MLS approximation, but not by MQ-PIM, EXP-PIM, and TPS-PIM interpolation. It is clearly seen that if pure radial functions are used, MQ-PIM, EXP-PIM, and TPS-PIM cannot reproduce linear polynomials exactly. This implies that this type of PIM shape function will not pass the standard patch tests used in conventional FEM to examine element formulation.

The errors of surface fitting using RPIM shape functions with linear polynomial basis included are computed in the same way as above and are shown in Figure 5.38. It is found that all the RPIM shape functions, MQ-linear, EXP-linear, and TPS-linear, can exactly reproduce linear polynomials. This implies that RPIM shape functions with linear polynomial basis included will pass the patch tests.

The errors of surface fitting of function $f_5(x, y)$, which is a nonpolynomial, are plotted in Figure 5.39; they are computed using shape functions of (a) MLS, (b) MQ-PIM, (c) EXP-PIM, and (d) TPS-PIM. In these RPIM shape functions, no polynomial basis is included. From Figure 5.39, it is found that the errors are all on the order of 10^{-3}. In other words, PIM interpolations with radial basis functions have the same approximation quality as MLS approximation for functions other than polynomials. To investigate the effects of a polynomial basis in RPIM, the same calculation is performed using MQ-linear, EXP-linear, and TPS-linear PIM shape functions; the results are shown in Figure 5.40. Unlike the case for f_4, little improvement is obtained by including linear polynomial terms in the radial basis, as shown in Figure 5.40. Figure 5.40a gives the same result, but uses the polynomial shape function. It is clearly shown that when polynomial PIM is used, the error is much smaller—by as much as three orders. In summary:

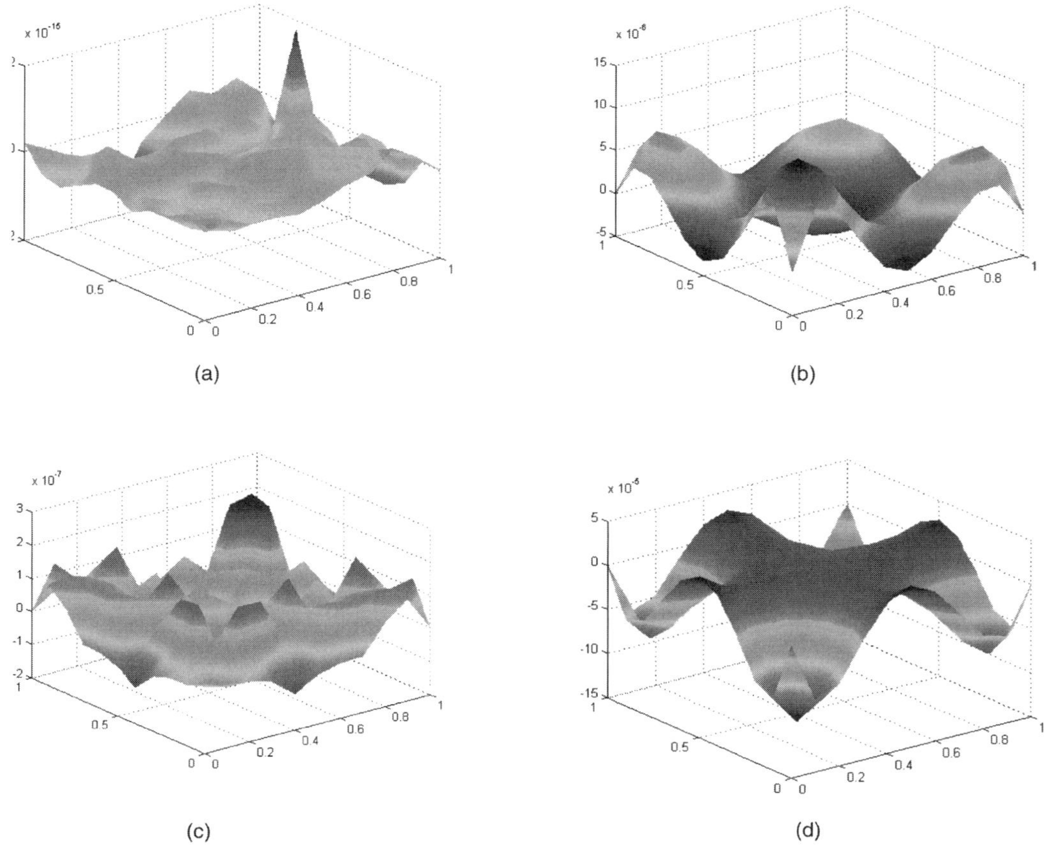

FIGURE 5.37
Error distribution of fitting function $f(x, y) = x + y$ using MLS and three radial PIM shape functions without polynomial terms. (a) MLS; (b) MQ-PIM ($C = 1.42$, $q = 1.03$); (c) EXP-PIM $c = 0.003$; (d) TPS-PIM ($\eta = 4.001$).

- Including a linear polynomial in a radial basis is not very effective in improving the accuracy in fitting functions of nonpolynomial type.
- Polynomial PIM is far superior to all the other shape functions in terms of accuracy in surface fitting.

Figure 5.41 plots the errors of surface fitting of MQ-PIM for parameter $q = \frac{1}{2}$ and $q = -\frac{1}{2}$. The results are compared because $q = \frac{1}{2}$ and $q = -\frac{1}{2}$ are the traditional parameters of the MQ radial function and are still widely used in the literature. Comparison with the results plotted in Figure 5.39 reveals that the parameter of $q = 1.03$ is slightly better in terms of accuracy. The parameter of $q = 1.03$ was first suggested by Wang and Liu, who found it performs much better for mechanics problems (see Chapter 8).

The effects of shape parameters of both MQ and EXP radial functions on surface fitting have been further investigated by J. G. Wang and G. R. Liu. They examined the condition numbers of the moment matrices that are formed in the process of computing the RPIM functions. The functions of nonpolynomial type defined by Equation 5.190 are chosen to be fitted in the domain of $(x, y) \in [0, 1] \times [0, 1]$. The findings of the investigation are summarized as follows.

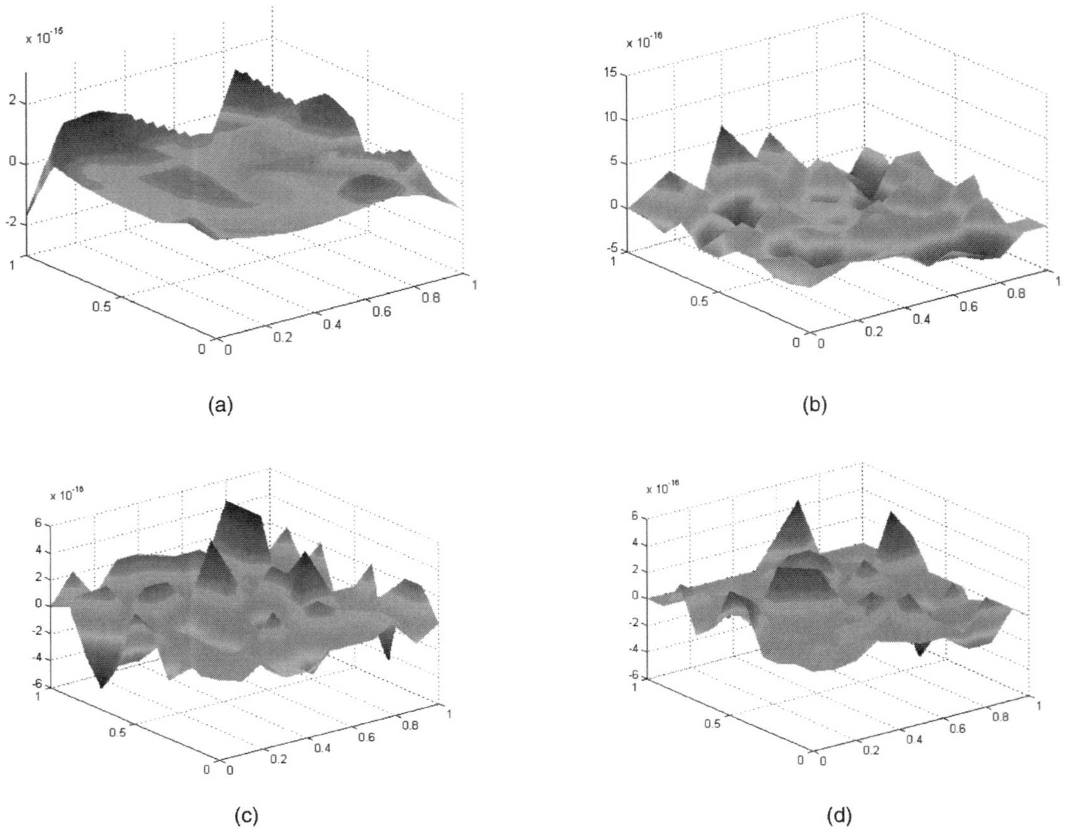

FIGURE 5.38
Error distribution of fitting function $f(x, y) = x + y$ using polynomial PIM and three types of radial PIM shape functions with linear polynomial basis. (a) Polynomial PIM; (b) MQ-PIM ($C = 1.42$, $q = 1.03$); (c) EXP-PIM $c = 0.003$; (d) TPS-PIM ($\eta = 4.001$).

For EXP Radial Basis Function

1. Condition numbers for both matrix $\mathbf{R}_Q$ and $\mathbf{G}$ are computed using the EXP radial basis function. Five different patterns of node distribution are used in the computation and the results are plotted in Figure 5.42. It is found that the condition number of matrix $\mathbf{R}_Q$ is very close to that of matrix $\mathbf{G}$. This implies again that the condition of matrix $\mathbf{G}$ is largely determined by that of matrix $\mathbf{R}_Q$. Therefore, the investigation can be carried out based only on the condition of matrix $\mathbf{R}_Q$. This finding is the same as that found in curve fitting.
2. The condition number of matrix $\mathbf{R}_Q$ decreases almost linearly with the shape parameter c, as shown in Figure 5.42a. A large value of c leads to a small condition number.
3. The nodal distribution pattern has little effect on the results.
4. The error in surface fitting is plotted in Figure 5.42b, which shows that the accuracy of surface fitting is low when the shape parameter c is larger. Therefore, one needs to use a small value of c for accurate surface fitting, when the condition number of matrix $\mathbf{R}_Q$ is large. On the other hand, too large a condition number of a matrix implies an ill-condition of the matrix, which leads to larger numerical

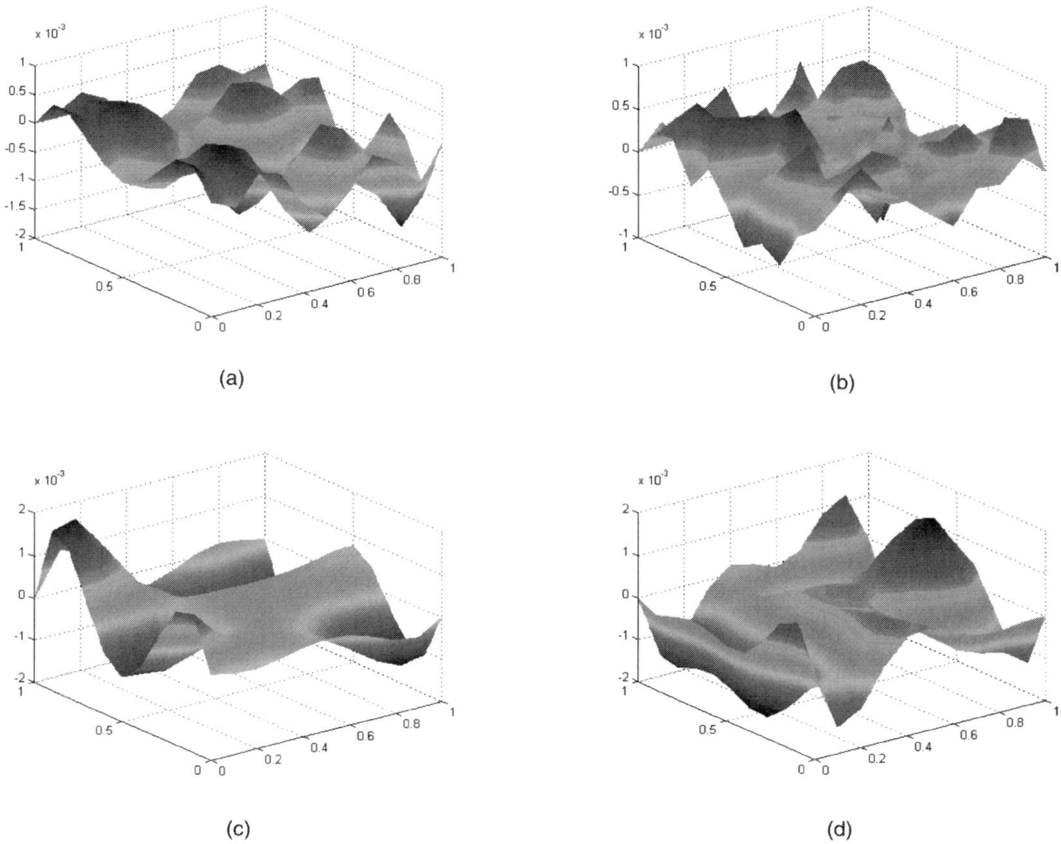

FIGURE 5.39
Error distribution of fitting function $f(x, y) = \sin x \cos y$ using polynomial PIM and three types of radial PIM shape functions without polynomial basis. (a) MLS; (b) MQ-PIM ($C = 1.42$, $q = 1.03$); (c) EXP-PIM $c = 0.003$; (d) TPS-PIM ($\eta = 4.001$).

errors. The preferred value of c is as small as possible that will produce a large condition number for the moment matrix, but that does not produce an ill-conditioned matrix $\mathbf{R}_Q$. A value between 0.1 to 0.4 seems to be a good choice for accurate surface fitting.

For MQ Radial Basis Function

The effects of the two parameters of the MQ basis function are shown in Figure 5.43 for the effects on the condition numbers on the moment matrix. Figure 5.44 illustrates the accuracy of surface fitting. The following may be noted:

1. Comparison of Figure 5.43a with Figure 5.43b reveals again that the condition number of matrix $\mathbf{R}_Q$ is very close to that of matrix $\mathbf{G}$. This reconfirms that the condition of matrix $\mathbf{G}$ is largely determined by that of matrix $\mathbf{R}_Q$.
2. The matrix is singular when q equals an integer.
3. The condition number changes little with shape parameter q, except when q is near 1, as shown in Figure 5.43a. Generally, the condition number for the MQ basis is much lower compared with the EXP basis functions, except in singular cases. This is one of the reasons we prefer MQ radial basis function.

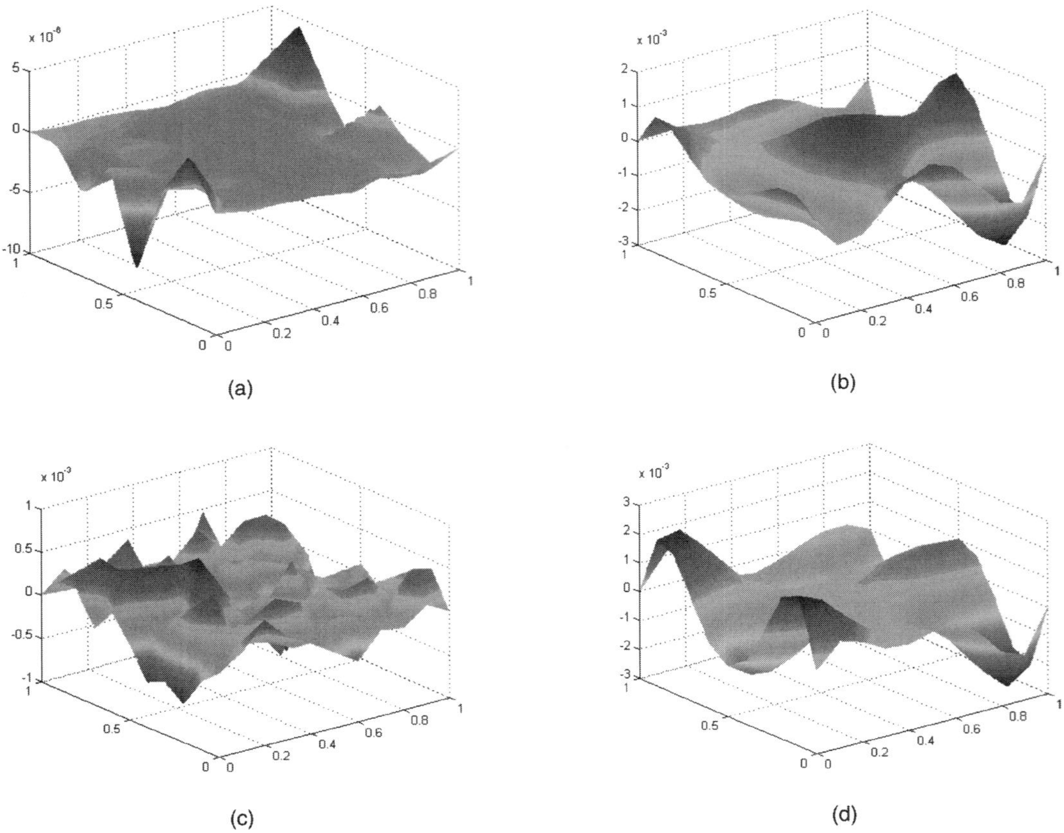

FIGURE 5.40
Error distribution of fitting function $f(x, y) = \sin x \cos y$ using polynomial PIM and three types of radial PIM shape functions with linear polynomial basis. (a) Polynomial PIM; (b) MQ-linear ($C = 1.42$, $q = 1.03$); (c) EXP-linear $c = 0.003$; (d) TPS-linear ($\eta = 4.001$).

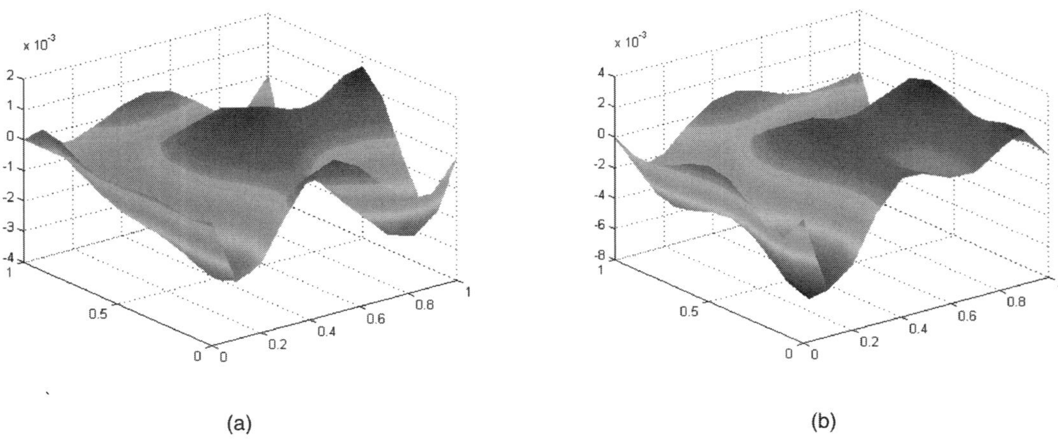

FIGURE 5.41
Error distribution of fitting function $f(x, y) = \sin x \cos y$ using traditional MQ-PIM shape functions. (a) MQ-PIM ($C = 1.42$, $q = 0.5$); (b) MQ-PIM ($C = 1.42$, $q = -0.5$).

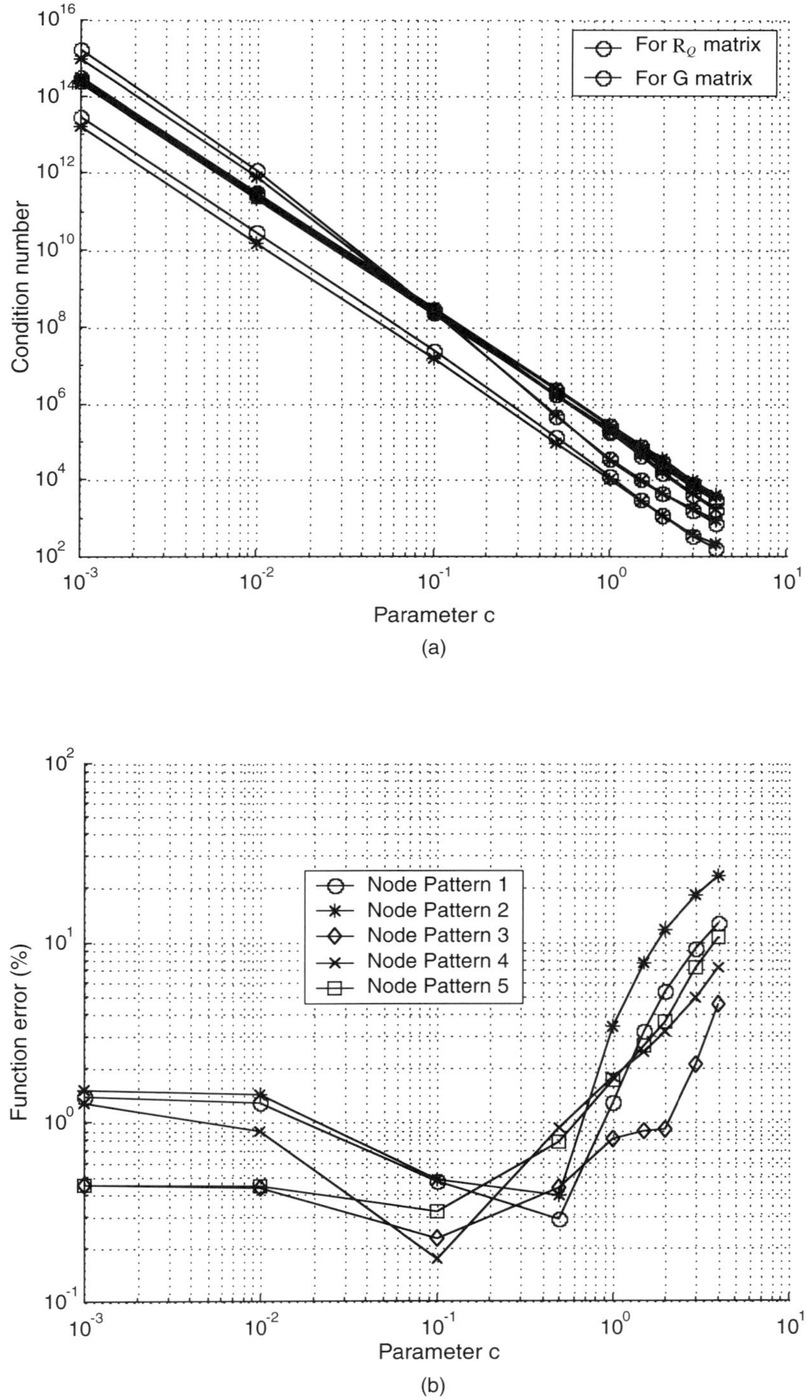

FIGURE 5.42
Effect of shape parameter c of EXP basis function. (a) Condition number obtained using different patterns of node distribution. (b) Surface fitting error for different patterns of node distribution. (From Wang, J. G. and Liu, G. R., *Comput. Methods Appl. Mech. Engrg.*, 191, 2611–2630, 2002. With permission.)

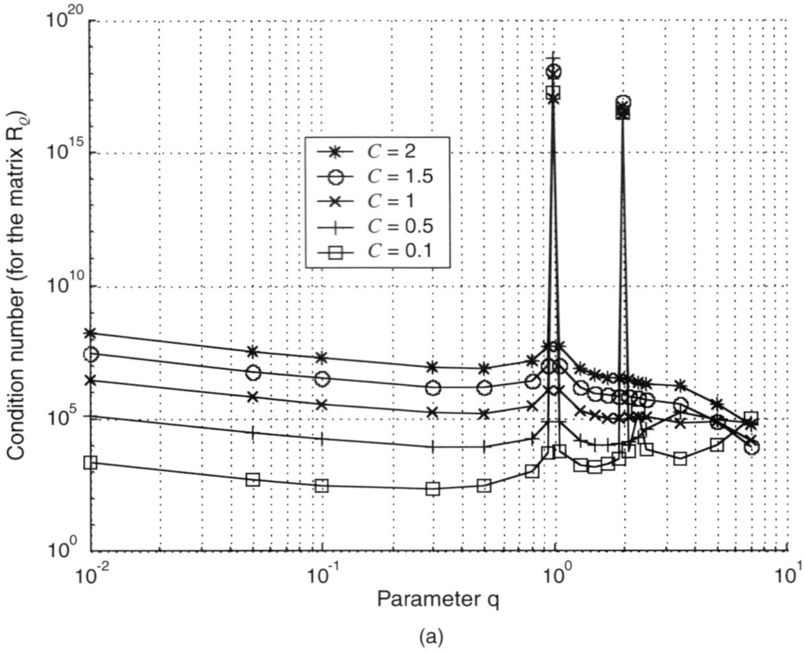

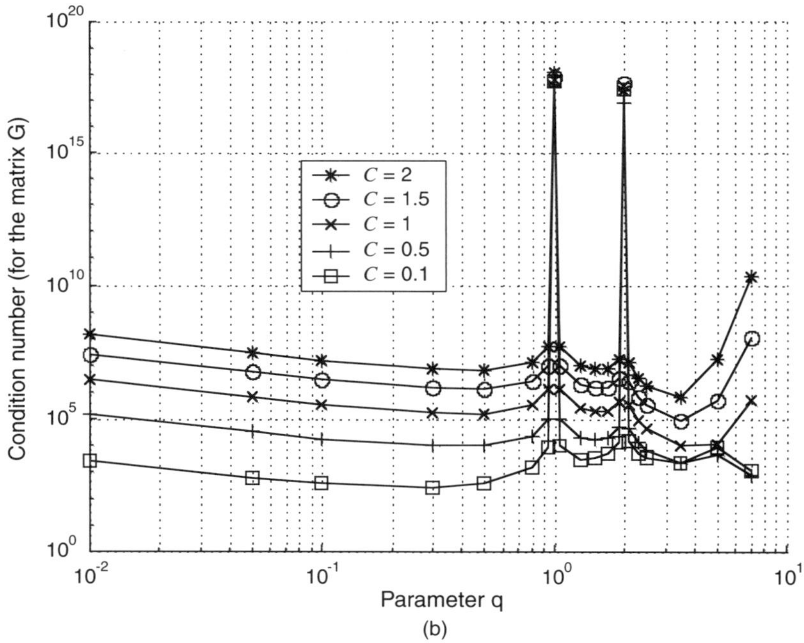

FIGURE 5.43
Effect of shape parameters of the MQ basis function. (a) Condition number of the moment matrix $\mathbf{R}_Q$ obtained using different parameters. (b) Condition number of the moment matrix **G** obtained using different parameters. (From Wang, J. G. and Liu, G. R., *Comput. Methods Appl. Mech. Engrg.*, 191, 2611–2630, 2002. With permission.)

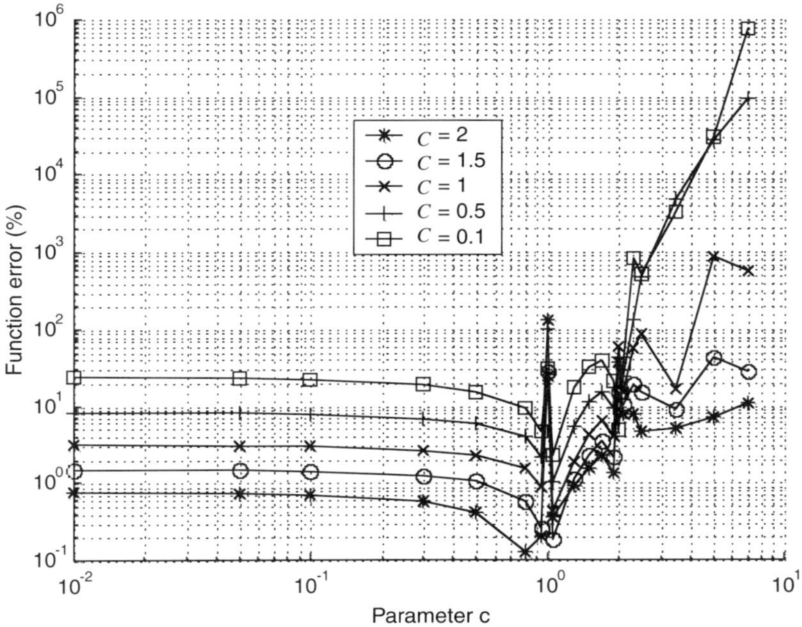

FIGURE 5.44
Effect of shape parameters on the error of surface fitting (MQ basis function). (From Wang, J. G. and Liu, G. R., *Comput. Methods Appl. Mech. Engrg.*, 191, 2611–2630, 2002. With permission.)

4. The shape parameter C has vital effect on the condition number. When C changes from 0.1 to 2.0, the condition number increases about 10^5 times. The accuracy is almost the same when $q < 1$ for fixed C.
5. Accuracy in the surface fitting increases when the q is near but not equal to 1.0, as shown in Figure 5.44. In terms of surface fitting accuracy, a larger C produces higher accuracy in surface fitting.

In summary, for both MQ and EXP basis functions, node distribution has little effect on surface fitting accuracy. This is very important as it implies that RPIM will be stable for irregular nodal distributions. The condition number of the matrix and surface fitting accuracy are heavily dependent on shape parameters. Parameters that lead to a large condition number in the moment matrix usually give more accurate results in surface fitting. However, too large a condition number will cause problems in computation. Therefore, a compromise is required. The preferred values of parameters are $q \approx 1.0$ (but not equal to 1.0, say, 0.98 and 1.03, etc.) and $C = 0.5$ to 2.0 (this is equivalent to $\alpha_C = C/d_c =$ 5 to 20, as $d_c = 0.1$) for the MQ basis, and $0.001 \leq c \leq 0.4$ (this is equivalent to $\alpha_c = cd_c =$ 0.0001 to 0.04) for the EXP basis.

Example 5.10 Surface Fitting Using MFree Shape Functions (Accuracy in Derivatives of the Fitted Surface)

In the following, the accuracy in the first and second derivatives of the fitted surface is considered. A comparison study is performed using MLS approximation, polynomial PIM, and RPIM (MQ with $C = 1.42$, $q = 1.03$, and linear polynomial terms). The function examined is

$$f_6(x, y) = \sin(x)\cos(y) + 0.5 \tag{5.191}$$

Support domains with $\alpha_s = 2.1$ are used; that is, the dimension of the support domain is 2.1 times the average nodal spacing. The fitting errors are plotted in Figure 5.45. The following remarks can be made:

1. All three interpolation techniques cannot exactly fit this surface, as the function is nonpolynomial.
2. The accuracy of the fitted surface is higher than that of the derivatives of the fitted surface. The higher the derivatives, the lower the accuracy.
3. Comparison of these three interpolation techniques reveals that the fitting accuracy of polynomial PIM is the highest, and the fitting accuracy of MLS is the lowest. The accuracies of the fitting surface and its derivatives using polynomial PIM are about one order higher than those using RPIM, and more than two orders higher than those using MLS approximation.

Example 5.11 Surface Fitting Using MFree Shape Functions (Effects of the Support Domain)

The accuracy of surface fitting also depends on the dimension of the support domain. Therefore, the fitting errors obtained using different dimensions of support domains of $\alpha_s = 1.1$, 1.6, 2.1, and 2.6 are plotted in Figure 5.46. Close study of this figure reveals the following:

1. When $\alpha_s = 1.1$, which means that about four nodes are used in the support domain, these three interpolation techniques have nearly the same fitting accuracy. In fact, polynomial PIM, RPIM (with linear polynomial terms), and MLS lead to the same shape functions in this case.
2. Polynomial PIM has higher accuracy in both the fitted surface and its derivatives, compared with RPIM and MLS. Polynomial PIM is stable and accurate in surface fitting. The fitting accuracy of PIM monotonously increases with the increase of the size of the support domain.
3. Although the fitting accuracy of RPIM also increases with the increase of the size of the support domain, the fitting accuracy is not improved in a smooth fashion. In addition, the parameters of the radial basis function chosen also affect the fitting results.
4. MLS with linear basis is the least accurate in surface fitting. Increase of the size of the support domain cannot significantly improve the fitting results.

5.11 Compatibility of MFree Function Approximation

In Chapter 4, we mentioned that in using energy principles the assumed or approximated field function has to be *compatible* (see Equation 4.1), meaning that the field function approximation needs to be continuous in the entire problem domain. In conventional FEM, the field function approximation in the problem domain is based on the *stationary element*.

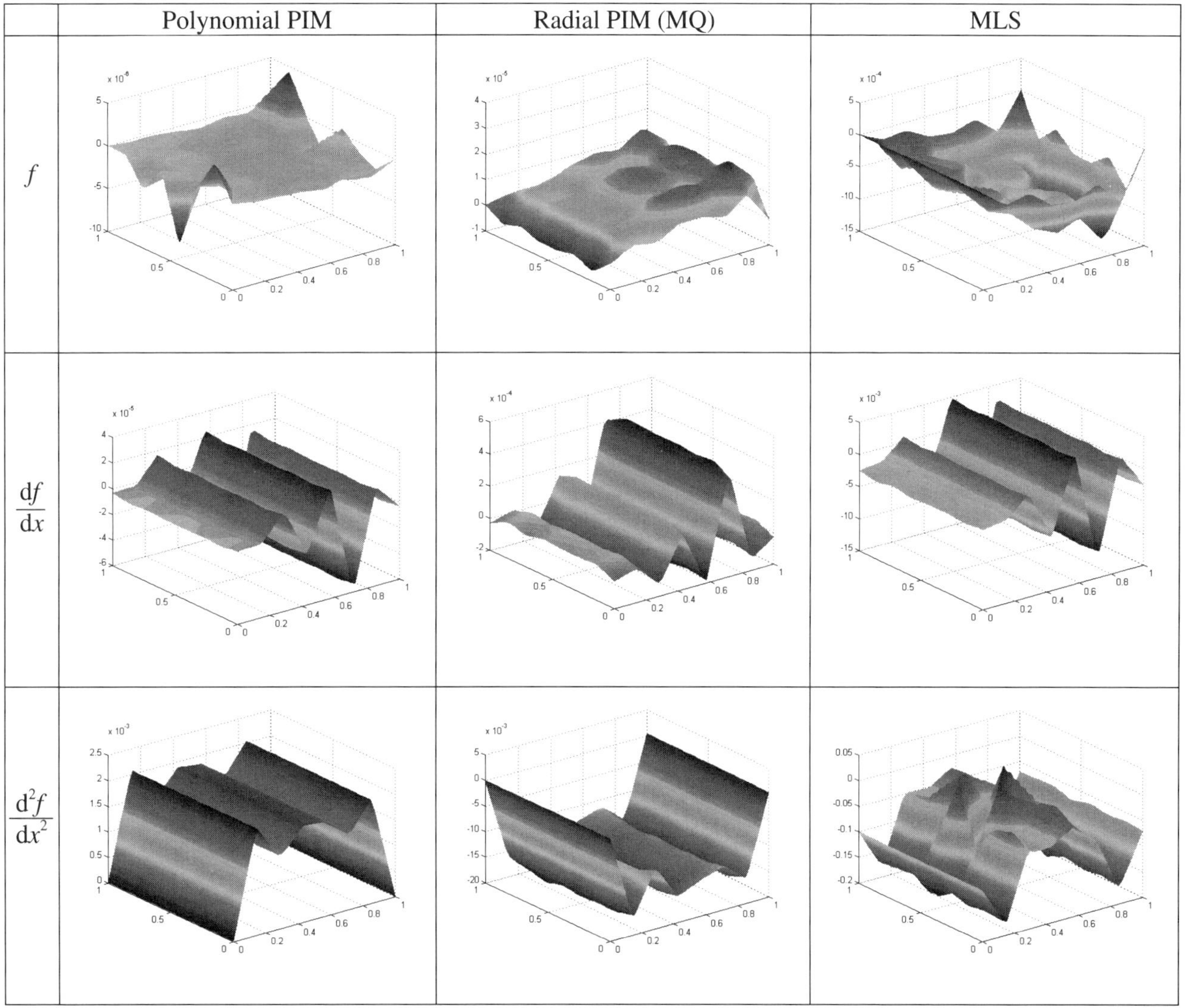

FIGURE 5.45
Error in fitted function and derivatives for surface of $f(x, y) = \sin x \cos y + 0.5$ using polynomial PIM, radial PIM (MQ, $C = 1.0$, $q = 0.5$, $m = 3$), and MLS shape functions.

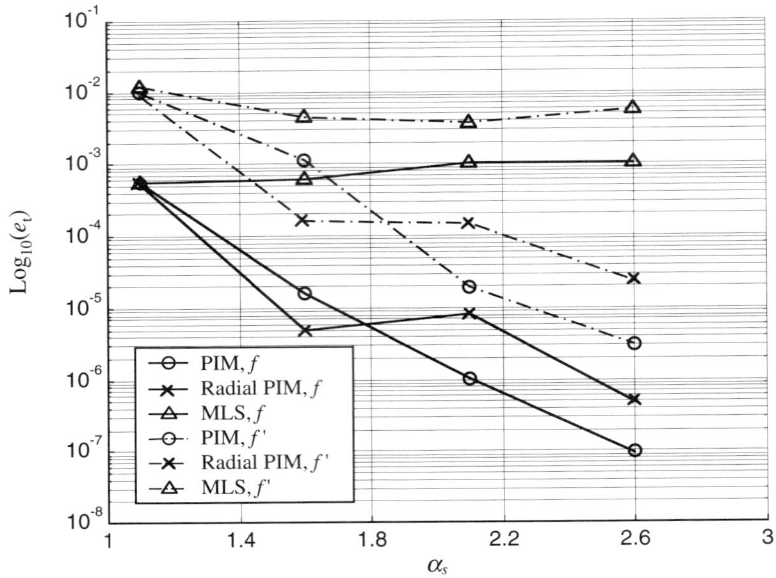

FIGURE 5.46
Error in fitted function and derivatives for surface of $f(x, y) = \sin x \cos y + 0.5$ using polynomial PIM, radial PIM (MQ, $C = 1.0$, $q = 0.5$, $m = 3$), and MLS shape functions. Effects of the dimension of the support domain.

The continuity of the field function approximation is ensured by either properly choosing shape functions of neighboring elements so that the order of the interpolation on the common boundary of the elements is the same or using so-called multi-point constraint equations (see, e.g., Liu and Quek, 2002) to enforce the compatibility. These types of elements are called *conforming* elements. It often happens that discontinuity is allowed to formulate so-called *nonconforming* finite elements. It has also been reported that some of the nonconforming finite elements perform better than conforming elements.

In MFree methods, the field function approximated is often *moving domain* based. The compatibility of field function approximation using MFree shape functions developed in this chapter may or may not always be satisfied. In using MLS approximation, compatibility is ensured by choosing weight functions that satisfy Equations 5.3, 5.4, and 5.6. The compatibility of field function approximation using the MLS shape function depends on the weight function used in Equation 5.46. Due to the use of the shape functions, the nodes can enter or leave the support domain in a smooth manner, which ensures compatibility while the point of interest is moving. The order of compatibility is determined by the smoothness of the weight function used. From Figures 5.2 to 5.4, it is seen that W1 and W2 provide at least second order compatibility, as their second derivatives are continuous in the support domain and vanish on the boundary of the support domain. W4 provides first order compatibility, as its first derivatives are continuous in the support domain and vanish on the boundary of the support domain. Although the second derivative of W4 is continuous in the support domain, it does not vanish on the boundary of the support domain (see Figure 5.4). As for W3, its derivatives of all orders are continuous within the support domain, but they are not exactly zero on the boundary of the support domain. Therefore, theoretically, W3 cannot provide compatibility of any order. However, the values of the function and its derivatives are very small on the boundary of the support domain. In practical numerical analyses, W3 provides very high order compatibility with a very small numerical error.

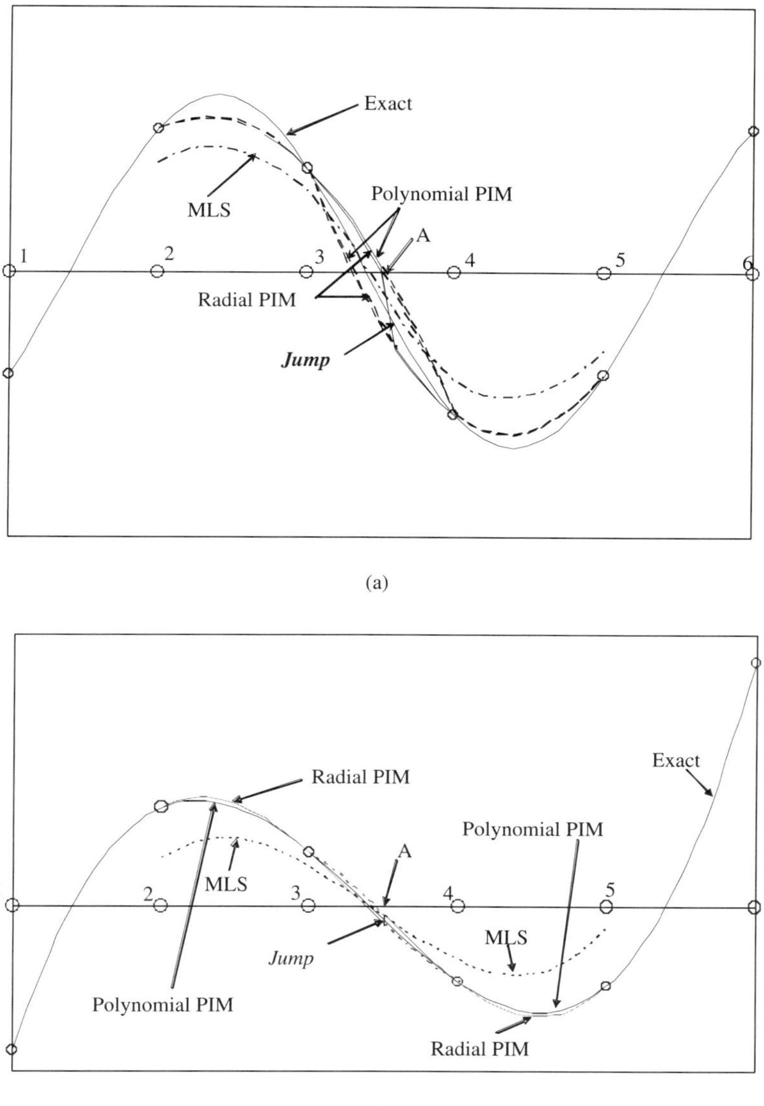

FIGURE 5.47
Field function approximation using MFree shape functions. Functions in the range of [0, 2.5] are approximated using 6 nodes. MLS, moving polynomial PIM, and moving radial PIM are used in function approximation. In MLS approximation the support domain is defined by $d_s = \alpha_s d_i$, where $\alpha_s = 1.9$, d_i is the space between two nodes. In PIMs, the nearest 4 nodes are used for interpolation (equivalent to $\alpha_s = 2.0$). (a) $f(x) = \sin[(x - 0.2)\pi]$; (b) $f(x) = (x - 0.2)(x - 1.2)(x - 2.2)$.

In using PIM shape functions, however, compatibility is not ensured, and the field function approximated could be discontinuous when nodes enter or leave the moving support domain. The nodes in the support domain are updated suddenly, meaning that when the nodes are entering or leaving the support domain, they are actually "jumping" into or out of the support domain. Therefore, the function approximated using the PIM shape functions can jump. Figure 5.47 shows an example of how a field function is approximated using MLS with quartic spline weight function (W2), polynomial PIM, and radial MQ-PIM shape functions, where we consider the approximation of a one-dimensional function using the function values at six nodes. Two functions are considered in this investigation: one is the sine function (Figure 5.47a) that is nonpolynomial, and another is a third order polynomial function. The following points may be noted from this figure.

1. MLS approximation is continuous, but does not pass through the function values at these nodes. There is no need to use the constrained energy principle to enforce the compatibility, but it is needed to enforce the essential boundary conditions.
2. All the PIM approximations pass through the field function values at the nodes due to the delta function property.
3. In approximating the sine functions that are nonpolynomial, the PIM and RPIM approximations are discontinuous at Point A ($x = 1.25$). This happens because, when the function is approximated between Points 2 and A, field nodes 1, 2, 3, 4, and 5 are used, and when the function is approximated between Points A and 5, field nodes 2, 3, 4, 5, and 6 are used for constructing the PIM shape functions. The consequences of this discontinuity of PIMs can create difficulties for PIMs to pass the standard patch tests and lead to a loss of accuracy. By simply applying the constrained energy principles (see Chapter 4), compatibility can be easily enforced in MFree methods. MFree methods using PIM shape functions with unconstrained energy principles are called nonconforming PIM or NPIM, and those using constrained energy principles in the formulation of discrete system equations are called conforming PIM or CPIM. The CPIM can, of course, always pass the standard patch test, and are discussed in more detail in Chapter 8.
4. In approximating the polynomial function of third order, the PIM approximation is continuous in the entire domain including at Point A, but the RPIM approximation is still discontinuous at Point A ($x = 1.25$). This is because of the reproduction feature of PIMs. The use of four nodes in interpolation allows the polynomial PIM to produce exactly the third order polynomial in the entire range. The RPIM can only reproduce the radial functions and not the polynomials, and therefore the RPIM approximation will still have a small jump at Point A. It is possible to remove the jump by adding sufficient polynomial terms into the radial basis. The reproduction feature of PIMs will allow the PIM methods to pass the standard patch tests when the local weighted residual formulation is used to create discrete system equations. We discuss this in more detail in Chapter 8.

We therefore conclude that in using either MLS or PIM shape functions, constrained energy principles (see Chapter 4) should be used to ensure either the compatibility of the trial function or essential boundary conditions. The use of unconstrained energy principles leads to the formulation of nonconformable MFree methods.

Note that for one-dimensional problems, the incompatibility problem can be avoided very easily by simply using one-piece PIM shape functions for the entire integration cell that is required in the integral form of energy principles. For the case shown in Figure 5.47, for example, one can simply use nodes 2, 3, 4, and 5 to construct PIM shape function for the entire integration cell of 3–4. For two- or three-dimensional domains, however, the use of one-piece PIM shape functions will not solve the incompatibility problem and *stitches* using the penalty method are needed to enforce the compatibility. This issue is discussed in detail in Chapter 8.

Note also that when the local residual weak form (see Chapter 4) is used to create the discretized system equations, the compatibility of the trail function is not a requirement. As long as the field approximation is consistent at any point in the quadrature domain, trail function is differentiable and the integrand is integrable. The PIM approximation satisfies all those requirements. Therefore, when PIM shape functions are used in the local residual weak form, no constraint is needed. Moreover, if it is required, compatibility can easily be achieved by using one-piece PIM shape functions for the entire local quadrature domain, which is usually very small. When using MLS approximation, the constrained local residual weak form has to be used to enforce the essential boundary conditions, as the MLS shape functions do not possess the Kronecker delta function property.

5.12 On the Concept of Reproduction

In the previous sections, we discussed the reproduction feature of the shape functions. We discuss this concept here again in comparison with the concepts of consistency and compatibility. The difference between reproduction and consistency is as follows:

- Reproduction is the capability of shape functions to reproduce functions that are in the basis function used to construct the shape functions.
- Consistency is the capability of the shape functions to reproduce the complete order of polynomials.

MLS and polynomial PIM shape functions are both reproductive and consistent. The radial PIM shape functions are reproductive, but if the basis function does not contain the polynomials, the shape function is not consistent. It can only reproduce functions that are combinations of radial basis functions in the basis. Any method that uses radial PIM shape functions will not pass the standard patch tests that require linear polynomial reproduction.

Shape functions that are reproductive do not necessarily guarantee the reproduction of a numerical method using the shape functions. The reproductivity of shape functions is the only *necessary* condition for a numerical method to be reproductive. The other important factor for a numerical method to be reproductive is the method or principles used to create the discrete system equations. If energy principles are used, the reproductivity of the numerical method primarily requires the field function approximation using the shape functions to be both reproductive and compatible. Other additional conditions should be accurate numerical implementation and imposition of essential boundary conditions. To have the method pass the standard patch tests, the field function approximation should be primarily both consistent (linear polynomial reproduction) and compatible. Compatibility can be imposed by using the constrained energy principles (see Chapter 4). If a numerical method based on energy principles is reproductive, it is also said to be *conforming*. For problems of solids and structures, it should provide the upper bound of the solution, and the displacement should converge to the exact solution from below when the nodal spacing approaches to zero.

If a local residual method, such as the local Petrov-Galerkin method, is used, the primary requirement for the resultant numerical method is the reproduction of the field function approximation using the shape functions. For problems of solids and structures, the displacement obtained using a reproductive method can converge to the exact solution

TABLE 5.7
Features of MFree Shape Functions

Shape Functions	Reproductivity	Consistency (Complete Order of Polynomial Reproductivity)	Compatibility in Field Function Approximation	Delta Function Property
SPH	No, on the boundary of the domain; Yes, in the interior of the domain	No, on the boundary of the domain; Yes, in the interior of the domain	Yes (for the continuous form of SPH)	No
RKPM	Yes	Yes	*	No
MLS	Yes	Yes	Yes	No
Polynomial PIM	Yes	Yes	No	Yes
Radial PIM	Yes	No	No	Yes
Radial PIM with polynomial basis	Yes	Yes	No	Yes

* *Note:* It is not clear how the correct function affects compatibility, but it is not an issue as long as the weak formulation is not used.

from both sides when the nodal spacing approaches zero. More discussion on this issue is given in Chapters 6, 7, and 8 using examples of standard patch tests.

Table 5.7 lists the features of MFree shape functions discussed in the previous sections of this chapter. Note that by combining these shape functions with the principles and methods of both weak forms (constrained or unconstrained) and strong forms, we can develop many different types of MFree methods. Such a combination offers tremendous opportunities for us to develop more effective and robust MFree methods. Understanding the features of the shape functions and the principles of weak and strong forms is crucial. The ultimate test of a numerical method being developed is the convergence test, and a conservative test would be standard patch. The chapters following present a number of MFree methods using some of the combinations of shape functions and principles of weak forms.

5.13 Other Methods

There are a number of other useful methods for constructing shape functions of high accuracy, such as the *hp*-clouds method and the partition of unity finite element method. We are unable to conduct a full-scale study and present examples in this book. The purpose of mentioning these here is merely to introduce an important area of development in MFree methods especially in the direction of adaptive analyses. Readers are referred to work by Duarte and Oden (1996) and Melenk and Babuska (1996), as well as their other recent publications.

5.14 Remarks

In FEM, the displacement field is expressed by displacements at *nodes* using *shape functions* defined over *elements*. The formation of the shape function has been relatively easy. The challenge in MFree methods is to efficiently construct stable, easy-to-use shape functions without using any predefined relations between nodes. Once the shape function is achieved, we will be more than half way toward our dream of MFree technology.

This chapter presents a number of ways to meet this challenge. There are still problems in achieving our ultimate goal, but we are reasonably close. In the following chapter, we apply these shape functions to different kinds of weak forms to produce discretized system equations for our mechanics problems.

6

Element Free Galerkin Method

The element free Galerkin (EFG) method is an MFree method developed by Belytschko et al. (1994b) based on the diffuse elements method (DEM) originated by Nayroles et al. (1992). The major features of the DEM and the EFG method are as follows:

1. Moving least square (MLS) approximation is employed for the construction of the shape function.
2. Galerkin weak form is employed to develop the discretized system equation.
3. Cells of the background mesh for integration are required to carry out the integration to calculate system matrices.

This chapter presents a very detailed procedure that leads to the EFG method. Detailed formulation and equations are provided. Technical issues, especially issues related to background integration, will also be examined. A typical benchmark problem of a cantilever beam is considered to illustrate the relationship between the density of the field nodes and the density of the global background mesh, as well as the number of integration sample points. The findings and remarks on the background integration are applicable to any MFree method that requires background integration.

Applications of the EFG method are presented for solving a number of academic and engineering problems including linear and nonlinear problems.

Note that the EFG is conforming due to the use of MLS shape functions that are consistent and compatible and the use of the constrained Galerkin approach to impose the essential boundary condition. This fact will be evidenced in the examples of patch tests.

6.1 EFG Formulation with Lagrange Multipliers

6.1.1 Formulation

A two-dimensional (2D) linear solid mechanics problem is used to present the procedure of the EFG method in formulating discretized system equations. The partial differential equation and boundary condition for a 2D solid mechanics problem can be written in the form:

$$\mathbf{L}^T\boldsymbol{\sigma} + \mathbf{b} = 0 \quad \text{Equilibrium equation in problem domain } \Omega \tag{6.1}$$

$$\mathbf{u} = \bar{\mathbf{u}} \quad \text{Boundary condition on essential boundary } \Gamma_u \tag{6.2}$$

which gives constraints to the field variable of displacement. The natural boundary conditions are given by

$$\boldsymbol{\sigma}\mathbf{n} = \bar{\mathbf{t}} \quad \text{on natural boundary } \Gamma_t \tag{6.3}$$

where

- $\mathbf{L}$ = differential operator defined by Equation 3.7 for three-dimensional (3D) solids and Equation 3.28 for 2D solids
- $\boldsymbol{\sigma}$ = the stress tensor defined by Equation 3.2 for 3D solids and Equation 3.23 for 2D solids
- $\mathbf{u}$ = the displacement vector given by Equation 3.6 for 3D cases and by Equation 3.27 for 2D cases
- $\mathbf{b}$ = the body force vector defined by Equation 3.19 for 3D cases and by Equation 3.35 for 2D cases
- $\bar{\mathbf{t}}$ = the prescribed traction displacement on the natural (stress) boundaries
- $\bar{\mathbf{u}}$ = the prescribed displacement on the essential (displacement) boundaries
- $\mathbf{n}$ = the vector of unit normal on the natural boundary

In the EFG method, the problem domain Ω is represented by a set of nodes scattered in the problem domain and on the boundaries of the domain. The MLS approximation procedure described in Section 5.4 is then used to approximate the displacement field $\mathbf{u}$ at a point of interest within the problem domain using the *nodal parameters* of displacement at the nodes that falls into the *support domain* of the point. Note that this book distinguishes the terms *support domain* and *influence domain*. These two domains can both be used to determine the nodes for MLS approximation, but they are based on slightly different concepts, as described in Section 2.10. The concept of the *nodal displacement parameter* comes from the fact that the displacement vector obtained by solving the discretized EFG system equation is not the actual displacement at the nodes. This is due to the lack of Kronecker delta function properties in the MLS shape function. We will pursue this issue in more detail later.

As discussed in Chapter 5 the MLS approximation is both consistent and compatible; the Galerkin procedure can then be used to establish a set of discretized system equations for the displacement parameters. The displacement at any point (including that at the nodes) is retrieved using MLS approximation and the nodal displacement parameters obtained after solving the discretized system equation. The strains at any point are also retrieved using the derivatives of the MLS shape functions and the nodal displacement parameters.

The constrained Galerkin weak form with Lagrange multipliers (see Chapter 4) for the problem stated by Equations 6.1 and 6.2 can be given by

$$\int_{\Omega} \delta(\mathbf{L}\mathbf{u})^T(\mathbf{c}\mathbf{L}\mathbf{u})d\Omega - \int_{\Omega} \delta\mathbf{u}^T\mathbf{b}d\Omega - \int_{\Gamma_t} \delta\mathbf{u}^T\bar{\mathbf{t}}d\Gamma - \int_{\Gamma_u} \delta\boldsymbol{\lambda}^T(\mathbf{u}-\bar{\mathbf{u}})d\Gamma - \int_{\Gamma_u} \delta\mathbf{u}^T\boldsymbol{\lambda}\, d\Gamma = 0 \tag{6.4}$$

Equation 6.4 is formed using Equation 4.41, by changing the area integrals for the constraint-related two terms into curve integrals because the constraints (essential boundary conditions) given in Equation 6.2 are defined only on the boundary. The last two terms in Equation 6.4 are produced by the method of Lagrange multipliers for handling essential boundary conditions for cases when $\mathbf{u}-\bar{\mathbf{u}} \neq 0$, which violates the condition of Equation 4.2. The Lagrange multipliers $\boldsymbol{\lambda}$ here can be viewed physically as *smart forces* that can force $\mathbf{u}-\bar{\mathbf{u}} = 0$. If the trial function $\mathbf{u}$ can be so chosen that $\mathbf{u}-\bar{\mathbf{u}} = 0$, the smart force will be zero, and the last two terms vanish completely.

The MLS approximation described in Chapter 5 is now used to express both the trial and test functions at any point of interest $\mathbf{x}$ using the nodes in the support domain of the point $\mathbf{x}$. For displacement comment u, we have

$$u^h(\mathbf{x}) = \sum_{I}^{n} \phi_I(\mathbf{x})u_I \tag{6.5}$$

where n is the number of nodes used in the support domain of the point at $\mathbf{x}$ for constructing the MLS shape function $\phi_I(\mathbf{x})$. The procedure of constructing $\phi_I(\mathbf{x})$ is detailed in Section 5.4, and the formulation of $\phi_I(\mathbf{x})$ is given in Equation 5.57.

For displacement component v, we should also have

$$v^h(\mathbf{x}) = \sum_I^n \phi_I(\mathbf{x}) v_I \tag{6.6}$$

Combining Equations 6.5 and 6.6, we obtain

$$\mathbf{u}^h = \begin{Bmatrix} u \\ v \end{Bmatrix}^h = \sum_I^n \underbrace{\begin{bmatrix} \phi_I & 0 \\ 0 & \phi_I \end{bmatrix}}_{\mathbf{\Phi}_I} \underbrace{\begin{Bmatrix} u_I \\ v_I \end{Bmatrix}}_{\mathbf{u}_I} = \sum_I^n \mathbf{\Phi}_I \mathbf{u}_I \tag{6.7}$$

where $\mathbf{\Phi}_I$ is the matrix of shape functions.

By using Equation 6.7, the product of $\mathbf{L}\mathbf{u}^h$ (which gives the strains) becomes

$$\mathbf{L}\mathbf{u}^h = \mathbf{L}\sum_I^n \mathbf{\Phi}_I \mathbf{u}_I = \sum_I^n \mathbf{L}\mathbf{\Phi}_I \mathbf{u}_I = \sum_I^n \begin{bmatrix} \frac{\partial}{\partial x} & 0 \\ 0 & \frac{\partial}{\partial y} \\ \frac{\partial}{\partial y} & \frac{\partial}{\partial x} \end{bmatrix} \begin{bmatrix} \phi_I & 0 \\ 0 & \phi_I \end{bmatrix} \mathbf{u}_I = \sum_I^n \underbrace{\begin{bmatrix} \phi_{I,x} & 0 \\ 0 & \phi_{I,y} \\ \phi_{I,y} & \phi_{I,x} \end{bmatrix}}_{\mathbf{B}_I} \mathbf{u}_I = \sum_I^n \mathbf{B}_I \mathbf{u}_I \tag{6.8}$$

where $\phi_{I,x}$ and $\phi_{I,y}$ represent the derivatives of the MLS shape function with respect to x and y, respectively, and $\mathbf{B}_I$ is the *strain matrix* for node I.

The use of MLS shape functions in Equations 6.5 and 6.6 leads to $\mathbf{u}^h - \bar{\mathbf{u}} \neq 0$ on the essential boundary, and the last two terms in Equation 6.4 are hence needed. The following explains why, using detailed formulation.

As described in previous chapters, the use of MLS approximation produces shape functions $\phi_I(\mathbf{x})$ that do not possess the Kronecker delta function property, i.e.,

$$\phi_I(\mathbf{x}_J) \neq \delta_{IJ} \tag{6.9}$$

This feature of the MLS shape function results in

$$u^h(\mathbf{x}_I) = \sum_I^n \phi_I(\mathbf{x}_I) u_I \neq u_I \tag{6.10}$$

$$v^h(\mathbf{x}_I) = \sum_I^n \phi_I(\mathbf{x}_I) v_I \neq v_I \tag{6.11}$$

This implies that the essential boundary condition Equation 6.2 cannot be exactly satisfied via enforcing

$$\mathbf{u}_I = \bar{\mathbf{u}}_I \quad \text{for node } I \text{ on } \Gamma_u \tag{6.12}$$

because what we really need to enforce is

$$\mathbf{u}(\mathbf{x}_I) = \bar{\mathbf{u}}_I \quad \text{for node } I \text{ on } \Gamma_u \tag{6.13}$$

Therefore, the fourth and fifth terms in Equation 6.4 are required to enforce the boundary conditions at the essential boundary Γ_u.

The Lagrange multiplier λ in Equation 6.4 should be an unknown function of the coordinates, which needs also to be interpolated using the nodes on the essential boundaries to obtain a discretized set of system equations, i.e.,

$$\lambda(\mathbf{x}) = \sum_{I}^{n_\lambda} N_I(s)\lambda_1 \quad \mathbf{x} \in \Gamma_u \tag{6.14}$$

where n_λ is the number of nodes used for this interpolation, s is the arc-length along the essential boundary, λ_I is the Lagrange multiplier at node I on the essential boundary, and $N_I(s)$ can be a Lagrange interpolant used in the conventional finite element method (FEM). The Lagrange interpolant of order n can be given in a general form of

$$N_k^n(s) = \frac{(s-s_0)(s-s_1)\cdots(s-s_{k-1})(s-s_{k+1})\cdots(s-s_n)}{(s_k-s_0)(s_k-s_1)\cdots(s_k-s_{k-1})(s_k-s_{k+1})\cdots(s_k-s_n)} \tag{6.15}$$

If we choose to use a first-order Lagrange interpolant, we have $n = 1$ and the Lagrange interpolants at point $s = s_0$ and $s = s_1$ will be

$$N_0(s) = \frac{(s-s_1)}{(s_0-s_1)}, \qquad N_1(s) = \frac{(s-s_0)}{(s_1-s_0)} \tag{6.16}$$

In this case, the Lagrange multiplier at s is interpolated using two nodes that are located before and after s. A higher-order Lagrange interpolant can also be used with more nodes on the boundary. By using Equation 6.14, the variation of Lagrange multiplier $\delta\lambda$ can be obtained by

$$\delta\lambda(\mathbf{x}) = \sum_{I}^{n_\lambda} N_I(s)\delta\lambda_I \quad \mathbf{x} \in \Gamma_u \tag{6.17}$$

The vector of Lagrange multipliers in Equation 6.4 can be written in matrix form:

$$\boldsymbol{\lambda} = \sum_{I=1}^{n_\lambda} \underbrace{\begin{bmatrix} N_I & \\ & N_I \end{bmatrix}}_{\mathbf{N}_I} \underbrace{\begin{Bmatrix} \lambda_{uI} \\ \lambda_{vI} \end{Bmatrix}}_{\boldsymbol{\lambda}_I} = \sum_{I=1}^{n_\lambda} \mathbf{N}_I \boldsymbol{\lambda}_I \tag{6.18}$$

where $\mathbf{N}_I$ is the Lagrange interpolant for node I on the essential boundary.
Substituting Equations 6.7 and 6.8 into Equation 6.4, we have

$$\int_\Omega \delta\left(\sum_I^n \mathbf{B}_I\mathbf{u}_I\right)^T\left(\mathbf{c}\sum_J^n \mathbf{B}_J\mathbf{u}_J\right)d\Omega - \int_\Omega \delta\left(\sum_I^n \mathbf{\Phi}_I\mathbf{u}_I\right)^T \mathbf{b}\,d\Omega - \int_{\Gamma_t} \delta\left(\sum_I^n \mathbf{\Phi}_I\mathbf{u}_I\right)^T \bar{\mathbf{t}}\,d\Gamma$$
$$- \int_{\Gamma_u} \delta\boldsymbol{\lambda}^T\left(\left(\sum_I^n \mathbf{\Phi}_I\mathbf{u}_I\right) - \bar{\mathbf{u}}\right)d\Gamma - \int_{\Gamma_u} \delta\left(\sum_I^n \mathbf{\Phi}_I\mathbf{u}_I\right)^T \boldsymbol{\lambda}\,d\Gamma = 0 \tag{6.19}$$

Notice that we use a different summation index for the term in the second bracket in the first integral term, to distinguish it from that in the first bracket. Let us first look at the first term in the above equation.

$$\int_\Omega \delta\left(\sum_I^n \mathbf{B}_I\mathbf{u}_I\right)^T\left(\mathbf{c}\sum_J^n \mathbf{B}_J\mathbf{u}_J\right)d\Omega = \int_\Omega \delta\left(\sum_I^n \mathbf{u}_I^T\mathbf{B}_I^T\right)\left(\mathbf{c}\sum_J^n \mathbf{B}_J\mathbf{u}_J\right)d\Omega \tag{6.20}$$

Note that the summation, variation, and integration are all linear operators, and therefore they are exchangeable. Hence, we can have

$$\int_\Omega \delta\left(\sum_I^n \mathbf{u}_I^T\mathbf{B}_I^T\right)\left(\mathbf{c}\sum_J^n \mathbf{B}_J\mathbf{u}_J\right)d\Omega = \sum_I^n\sum_J^n \delta\mathbf{u}_I^T \underbrace{\int_\Omega \mathbf{B}_I^T\mathbf{c}\mathbf{B}_J\,d\Omega}_{\mathbf{K}_{IJ}}\mathbf{u}_J$$
$$= \sum_I^{n_t}\sum_J^{n_t} \delta\mathbf{u}_I^T\mathbf{K}_{IJ}\mathbf{u}_J \tag{6.21}$$

In above equation, we made two changes. The first is the substitution of $\mathbf{K}_{IJ}$, which is a 2×2 matrix called in this book a *nodal stiffness matrix*. Note that the integration is over the entire problem domain, and all the nodes can be involved. Therefore, the summation limits have to be changed to n_t, which is the total number of nodes in the entire problem domain. Note also that the last summation in Equation 6.21 is nothing but a matrix assembly. To view this, we expand the summation and then group the components into matrix form as follows:

$$\sum_I^{n_t}\sum_J^{n_t} \delta\mathbf{u}_I^T\mathbf{K}_{IJ}\mathbf{u}_J = \delta\mathbf{u}_1^T\mathbf{K}_{11}\mathbf{u}_1 + \delta\mathbf{u}_1^T\mathbf{K}_{12}\mathbf{u}_2 + \cdots + \delta\mathbf{u}_1^T\mathbf{K}_{1n_t}\mathbf{u}_{n_t}$$
$$+ \delta\mathbf{u}_2^T\mathbf{K}_{21}\mathbf{u}_1 + \delta\mathbf{u}_2^T\mathbf{K}_{22}\mathbf{u}_2 + \cdots + \delta\mathbf{u}_2^T\mathbf{K}_{2n_t}\mathbf{u}_{n_t}$$
$$\mathrm{M}$$
$$+ \delta\mathbf{u}_{n_t}^T\mathbf{K}_{n_t1}\mathbf{u}_1 + \delta\mathbf{u}_{n_t}^T\mathbf{K}_{n_t2}\mathbf{u}_2 + \cdots + \delta\mathbf{u}_{n_t}^T\mathbf{K}_{n_tn_t}\mathbf{u}_{n_t}$$
$$= \delta\mathbf{U}^T\mathbf{K}\mathbf{U} \tag{6.22}$$

where **K** is the global stiffness matrix assembled using the nodal stiffness matrix in the form:

$$\mathbf{K} = \begin{bmatrix} \mathbf{K}_{11} & \mathbf{K}_{12} & \cdots & \mathbf{K}_{1n_t} \\ \mathbf{K}_{21} & \mathbf{K}_{22} & \cdots & \mathbf{K}_{2n_t} \\ \mathrm{M} & \mathrm{M} & \ddots & \mathrm{M} \\ \mathbf{K}_{n_t1} & \mathbf{K}_{n_t2} & \cdots & \mathbf{K}_{n_tn_t} \end{bmatrix} \tag{6.23}$$

The dimension matrix **K** should be $(2n_t) \times (2n_t)$, because $\mathbf{K}_{IJ}$ is 2×2.

Vector **U** is a *global displacement parameter vector* that collects the nodal parameter vectors of displacement at all nodes in the entire problem domain, which has the form:

$$\mathbf{U} = \begin{Bmatrix} \mathbf{u}_1 \\ \mathbf{u}_2 \\ \mathrm{M} \\ \mathbf{u}_{n_t} \end{Bmatrix} \tag{6.24}$$

where $\mathbf{u}_I$ is the *nodal displacement parameter vector* at node I, i.e.,

$$\mathbf{u}_I = \begin{Bmatrix} u_I \\ v_I \end{Bmatrix} \tag{6.25}$$

The length of vector **U** should be $(2n_t)$.

Next, let us examine the second term in Equation 6.19.

$$\int_{\Omega} \delta\left(\sum_{I}^{n} \mathbf{\Phi}_I \mathbf{u}_I\right)^T \mathbf{b}\,\mathrm{d}\Omega = \sum_{I}^{n} \delta \mathbf{u}_I^T \underbrace{\int_{\Omega} \mathbf{\Phi}_I^T \mathbf{b}\,\mathrm{d}\Omega}_{\mathbf{f}_I} = \sum_{I}^{n_t} \delta \mathbf{u}_I^T \mathbf{f}_I \tag{6.26}$$

In the above equation, we made two changes. The first is a substitution of

$$\mathbf{f}_I = \int_{\Omega} \mathbf{\Phi}_I^T \mathbf{b}\,\mathrm{d}\Omega \tag{6.27}$$

where $\mathbf{f}_I$ is called the *nodal force vector*. Note again that the integration is over the entire problem domain, and therefore the summations have to be changed for all the nodes in the problem domain, which is the second change in Equation 6.26. The last summation in Equation 6.26 can be expanded and then formed to a product of matrices as follows:

$$\sum_{I}^{n_t} \delta \mathbf{u}_I^T \mathbf{f}_I = \delta \mathbf{u}_1^T \mathbf{f}_1 + \delta \mathbf{u}_2^T \mathbf{f}_2 + \cdots + \delta \mathbf{u}_{n_t}^T \mathbf{f}_{n_t} = \delta \mathbf{U}^T \mathbf{F} \tag{6.28}$$

Vector **F** in Equation 6.28 is the *global force vector,* which collects force vectors at all the nodes in the problem domain and has the form:

$$\mathbf{F} = \begin{Bmatrix} \mathbf{f}_1 \\ \mathbf{f}_2 \\ \mathrm{M} \\ \mathbf{f}_{n_t} \end{Bmatrix} \tag{6.29}$$

where $\mathbf{f}_I$ is the nodal force vector at node I calculated using Equation 6.27, and consists of two components arranged as

$$\mathbf{f}_I = \begin{Bmatrix} f_{xI} \\ f_{yI} \end{Bmatrix} \tag{6.30}$$

where f_{xI} and f_{yI} are two components of nodal force in the x and y directions. The length of vector **F** should be $(2n_t)$.

The treatment for the third term in Equation 6.19 is exactly the same as that for the second term, except that the body force vector is replaced by the traction vector on the natural boundary, and the area integration is accordingly changed to curve integration. Therefore, the additional nodal force vector can be given as

$$\mathbf{f}_I = \int_{\Gamma_t} \mathbf{\Phi}_I^T \bar{\mathbf{t}}\, d\Gamma \tag{6.31}$$

The force vector therefore receives contributions from both the external body force and the external force applied on the natural boundaries.

Before we examine the fifth term, let us look at the last term in Equation 6.19.

$$\begin{aligned} \int_{\Gamma_u} \delta\left(\sum_I^n \mathbf{\Phi}_I \mathbf{u}_I\right)^T \boldsymbol{\lambda}\, d\Gamma &= \int_{\Gamma_u} \delta\left(\sum_I^n \mathbf{\Phi}_I \mathbf{u}_I\right)^T \left(\sum_J^{n_\lambda} \mathbf{N}_J \boldsymbol{\lambda}_J\right) d\Gamma \\ &= \sum_I^n \sum_J^{n_\lambda} \delta \mathbf{u}_I^T \underbrace{\int_{\Gamma_u} \mathbf{\Phi}_I^T \mathbf{N}_J\, d\Gamma}_{-\mathbf{G}_{IJ}} \boldsymbol{\lambda}_J \\ &= -\sum_I^{n_t} \sum_J^{n_{\lambda t}} \delta \mathbf{u}_I^T \mathbf{G}_{IJ} \boldsymbol{\lambda}_J = -\delta \mathbf{U}^T \mathbf{G} \boldsymbol{\lambda} \end{aligned} \tag{6.32}$$

where $n_{\lambda t}$ is the total number of nodes on the essential boundary, and **G** is also a global matrix formed by assembling its *nodal matrix* $\mathbf{G}_{IJ}$. Note that the dimension of matrix $\mathbf{G}_{IJ}$ is also 2×2, but it concerns only the nodes on the essential boundaries. The dimension of matrix **G** should be $(2n_t) \times (2n_{\lambda t})$.

Finally, let us examine the fourth term in Equation 6.19. Using Equations 6.7 and 6.18, we have

$$\begin{aligned}
&\int_{\Gamma_u} \delta\boldsymbol{\lambda}^T \left(\left(\sum_J^n \boldsymbol{\Phi}_J \mathbf{u}_J \right) - \bar{\mathbf{u}} \right) d\Gamma \\
&= \int_{\Gamma_u} \delta \left(\sum_I^{n_\lambda} \mathbf{N}_I \boldsymbol{\lambda}_I \right)^T \sum_J^n \boldsymbol{\Phi}_J \mathbf{u}_J d\Gamma - \int_{\Gamma_u} \delta \left(\sum_I^{n_\lambda} \mathbf{N}_I \boldsymbol{\lambda}_I \right)^T \bar{\mathbf{u}} d\Gamma \\
&= \sum_I^{n_\lambda} \sum_J^n \delta\boldsymbol{\lambda}_I^T \underbrace{\int_{\Gamma_u} \mathbf{N}_I^T \boldsymbol{\Phi}_J d\Gamma}_{-\mathbf{G}_{IJ}^T} \mathbf{u}_J - \sum_I^{n_\lambda} \delta\boldsymbol{\lambda}_I^T \underbrace{\int_{\Gamma_u} \mathbf{N}_I^T \bar{\mathbf{u}} d\Gamma}_{-\mathbf{q}_I} \\
&= -\sum_I^{n_{\lambda t}} \sum_J^{n_t} \delta\boldsymbol{\lambda}_I^T \mathbf{G}_{IJ}^T \mathbf{u}_J + \sum_I^{n_{\lambda t}} \delta\boldsymbol{\lambda}_I^T \mathbf{q}_I \\
&= -\delta\boldsymbol{\lambda}^T \mathbf{G}^T \mathbf{U} + \delta\boldsymbol{\lambda}^T \mathbf{q}
\end{aligned} \tag{6.33}$$

where matrix **G** is defined by Equations 6.32 and 6.40. The vector **q** in Equation 6.33 is of the length ($2n_{\lambda t}$), and is assembled using nodal vector $\mathbf{q}_I$.

Finally, summarizing Equations 6.22, 6.26, 6.28, 6.32, and 6.33, we obtain

$$\begin{aligned}
&\underbrace{\int_{\Omega} \delta \left(\sum_I^n \mathbf{B}_I \mathbf{u}_I \right)^T \left(\mathbf{c} \sum_J^n \mathbf{B}_J \mathbf{u}_J \right) d\Omega}_{\delta\mathbf{U}^T\mathbf{K}\mathbf{U}} - \underbrace{\int_{\Omega} \delta \left(\sum_I^n \boldsymbol{\Phi}_I \mathbf{u}_I \right)^T \mathbf{b} d\Omega - \int_{\Gamma_t} \delta \left(\sum_I^n \boldsymbol{\Phi}_I \mathbf{u}_I \right)^T \bar{\mathbf{t}} d\Gamma}_{\delta\mathbf{U}^T\mathbf{F}} \\
&- \underbrace{\int_{\Gamma_u} \delta\boldsymbol{\lambda}^T \left(\left(\sum_I^n \boldsymbol{\Phi}_I \mathbf{u}_I \right) - \bar{\mathbf{u}} \right) d\Gamma}_{\delta\boldsymbol{\lambda}^T[\mathbf{G}^T\mathbf{U}-\mathbf{q}]} - \underbrace{\int_{\Gamma_u} \delta \left(\sum_I^n \boldsymbol{\Phi}_I \mathbf{u}_I \right)^T \boldsymbol{\lambda} d\Gamma}_{\delta\mathbf{U}^T\mathbf{G}\boldsymbol{\lambda}} = 0
\end{aligned} \tag{6.34}$$

which is

$$\delta\mathbf{U}^T[\mathbf{K}\mathbf{U} + \mathbf{G}\boldsymbol{\lambda} - \mathbf{F}] + \delta\boldsymbol{\lambda}^T[\mathbf{G}^T\mathbf{U} - \mathbf{q}] = 0 \tag{6.35}$$

Because $\delta\mathbf{U}$ and $\delta\boldsymbol{\lambda}$ are arbitrary, the above equation can be satisfied only if

$$\begin{aligned}
\mathbf{K}\mathbf{U} + \mathbf{G}\boldsymbol{\lambda} - \mathbf{F} &= 0 \\
\mathbf{G}^T\mathbf{U} - \mathbf{q} &= 0
\end{aligned} \tag{6.36}$$

The above two equations can be written in the following matrix form:

$$\begin{bmatrix} \mathbf{K} & \mathbf{G} \\ \mathbf{G}^T & 0 \end{bmatrix} \begin{Bmatrix} \mathbf{U} \\ \boldsymbol{\lambda} \end{Bmatrix} = \begin{Bmatrix} \mathbf{F} \\ \mathbf{q} \end{Bmatrix} \tag{6.37}$$

This is the final discrete system equation for the entire problem domain. We now summarize all the nodal matrices and vectors that form Equation 6.37 for easy reference later.

$$\mathbf{K}_{IJ} = \int_{\Omega} \mathbf{B}_I^T \mathbf{c} \mathbf{B}_J \mathrm{d}\Omega \tag{6.38}$$

$$\mathbf{B}_I = \mathbf{L}\mathbf{\Phi}_I = \begin{bmatrix} \phi_{I,x} & 0 \\ 0 & \phi_{I,y} \\ \phi_{I,y} & \phi_{I,x} \end{bmatrix} \tag{6.39}$$

$$\mathbf{G}_{IJ} = -\int_{\Gamma_u} \mathbf{N}_I^T \mathbf{\Phi}_J \mathrm{d}\Gamma \tag{6.40}$$

$$\mathbf{\Phi}_I = \begin{bmatrix} \phi_I & 0 \\ 0 & \phi_I \end{bmatrix} \tag{6.41}$$

$$\mathbf{N}_I = \begin{bmatrix} N_I & 0 \\ 0 & N_I \end{bmatrix} \tag{6.42}$$

$$\mathbf{f}_I = \int_{\Omega} \mathbf{\Phi}_I^T \mathbf{b} \mathrm{d}\Omega + \int_{\Gamma_t} \mathbf{\Phi}_I^T \bar{\mathbf{t}} \, \mathrm{d}\Gamma \tag{6.43}$$

$$\mathbf{q}_I = -\int_{\Gamma_u} \mathbf{N}_I^T \bar{\mathbf{u}} \mathrm{d}\Gamma \tag{6.44}$$

The *nodal stiffness matrix* $\mathbf{K}_{IJ}$ defined in Equation 6.38 can in fact be viewed as the basic component for assembling the global stiffness matrix of EFG. It is clearly shown that MFree methods operate on nodes, in contrast to the operation on elements in the conventional FEM, where the *element stiffness matrix* is the basic component for assembling the global stiffness matrix.

By using the symmetric property of $\mathbf{c}$ matrix, it is obvious that

$$[\mathbf{K}_{II}]^T = \int_{\Omega} [\mathbf{B}_I^T \mathbf{c} \mathbf{B}_I]^T \mathrm{d}\Omega = \int_{\Omega} [\mathbf{B}_I^T \mathbf{c}^T \mathbf{B}_I] \mathrm{d}\Omega = \int_{\Omega} [\mathbf{B}_I^T \mathbf{c} \mathbf{B}_I] \mathrm{d}\Omega = \mathbf{K}_{II} \tag{6.45}$$

It is easy also to confirm that

$$[\mathbf{K}_{IJ}]^T = \int_{\Omega} [\mathbf{B}_I^T \mathbf{c} \mathbf{B}_J]^T \mathrm{d}\Omega = \int_{\Omega} [\mathbf{B}_J^T \mathbf{c}^T \mathbf{B}_I] \mathrm{d}\Omega = \int_{\Omega} [\mathbf{B}_J^T \mathbf{c} \mathbf{B}_I] \mathrm{d}\Omega = \mathbf{K}_{JI} \tag{6.46}$$

Observation of Equation 6.23 reveals that $\mathbf{K}$ is symmetric.

From Equation 6.38, it is shown that the nodal stiffness matrix $\mathbf{K}_{IJ}$ contains the stiffness coefficients between nodes I and J evaluated at a point in the problem domain. It is a function of coordinates, and needs to be integrated over the entire problem domain. It has to be assembled to the global stiffness matrix $\mathbf{K}$, as long as the nodes I and J are covered by the support domain of at least one quadrature point. If nodes I and J are far apart and if they do not share the same support domain of any quadrature point, $\mathbf{K}_{IJ}$ vanishes, and thus there is no need to compute. Therefore, as long as the support domain is compact and does not cover too wide a problem domain, many $\mathbf{K}_{IJ}$ in Equation 6.23 will be zero, and the global stiffness matrix $\mathbf{K}$ will be sparse. If the nodes are properly numbered, $\mathbf{K}$ will also be banded. We discuss briefly in Chapter 15 how to reduce the bandwidth of $\mathbf{K}$.

From Equations 6.31, 6.38, 6.40, and 6.44 it is evident that there is a need to integrate over the problem domain and the curve integrations for both natural and essential boundaries. These integrations have to be carried out via numerical quadrature techniques, and the Gauss quadrature scheme is most often used. In using the numerical quadrature scheme, a mesh of cells is required for the integration. This mesh of cells is called a background mesh. The background mesh is similar to the mesh of elements in FEM and no overlap or gap is permitted. In contrast to the mesh in FEM, the background mesh in EFG is used merely for the integration of the system matrices, and not for field variable interpolation. In principle, the background mesh can be totally independent of the arrangement of nodes. The only consideration in designing the cells of the background mesh is to ensure the accuracy of integration for the system matrices.

The matrices in Equation 6.37 can be much larger than the stiffness matrix in FEM, because of the presence of matrix $\mathbf{G}$ produced by the essential boundary conditions. Depending on the number of nodes on the essential boundaries, solution efficiency can be drastically reduced. From Equation 6.37, it can also be clearly seen that the matrix of the enlarged system is still symmetric but no longer positive definite, which further reduces computational efficiency in solving the system equations.

In the later sections of this chapter, we introduce alternative methods for enforcing essential boundary conditions, which lead to system equations with matrices of the same size as in conventional FEM; the system matrix will also remain positive definite.

6.1.2 EFG Procedure

The solution procedure of the EFG method is similar to that in FEM. The geometry of the problem domain is first modeled, and a set of nodes is generated to represent the problem domain, as shown in Figure 6.1. The system matrices are assembled via two loops. The outer loop is for all the cells of the background mesh, and the inner loop is for all the Gauss quadrature points in a cell. The flowchart of the algorithm for stress analysis using the EFG method is presented in Figure 6.2.

6.1.3 Background Integration

This subsection deals with integration issues in MFree methods. The discussion here on integration issues is based on the work by G. R. Liu and Yan (1999).

In either FEM or EFG, numerical integration is a time-consuming process required for the computation of the stiffness matrix that is established based on the variational method. In FEM, the integration mesh is the same as the element mesh. To obtain accurate results, the element mesh must be sufficiently fine and a sufficient number of integration points have to be used. In EFG, however, the background mesh is required only in performing the integration of computing the stiffness matrix. Therefore, a background mesh of proper density needs to be designed to obtain an approximate solution of desired accuracy. However, this can only be done after performing a detailed investigation to reveal the relationship between the density of the field nodes and the density of the background

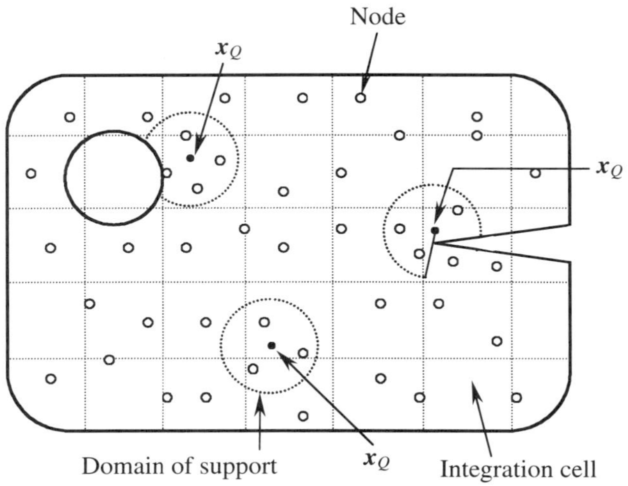

FIGURE 6.1
MFree model for EFG method with background mesh of cells for integration.

mesh. The first thing that needs to be addressed is the minimum number of integration points when numerical integration is adopted.

Zienkiewicz (1989) has shown for FEM that, if the number of independent relations provided by all integration points is less than the number of unknowns (displacements at all points in the element), the stiffness matrix **K** must be singular. This concept should also be applicable, in principle, to EFG. For a 2D problem, the number of unknown variables N_u should be

$$N_u = 2 \times n_t - n_f \tag{6.47}$$

where n_t and n_f are the node number in domain Ω and the number of constrained degrees of freedoms, respectively.

In evaluating the integrand at each quadrature (integration) point, three independent strain relations are used. Therefore, the number of independent equations used in all the quadrature points, N_Q, is

$$N_Q = 3 \times n_Q \tag{6.48}$$

where n_Q is the number of total quadrature points in domain Ω. Therefore, N_Q must be larger than N_u to avoid the singularity in the solution, and the minimum number of quadrature points must be greater than $N_u/3$. In other words, the total number of quadrature points n_Q should be at least two thirds of the total number of unconstrained field nodes in the problem domain, i.e.,

$$N_Q > N_u \approx 2n_t \quad \text{or} \quad n_Q > \frac{2}{3}n_t \quad \text{for 2D problems} \tag{6.49}$$

Note also that this rule is a necessary requirement, but not necessarily a sufficient requirement. The proper number of quadrature points is studied in the following section using benchmark problems.

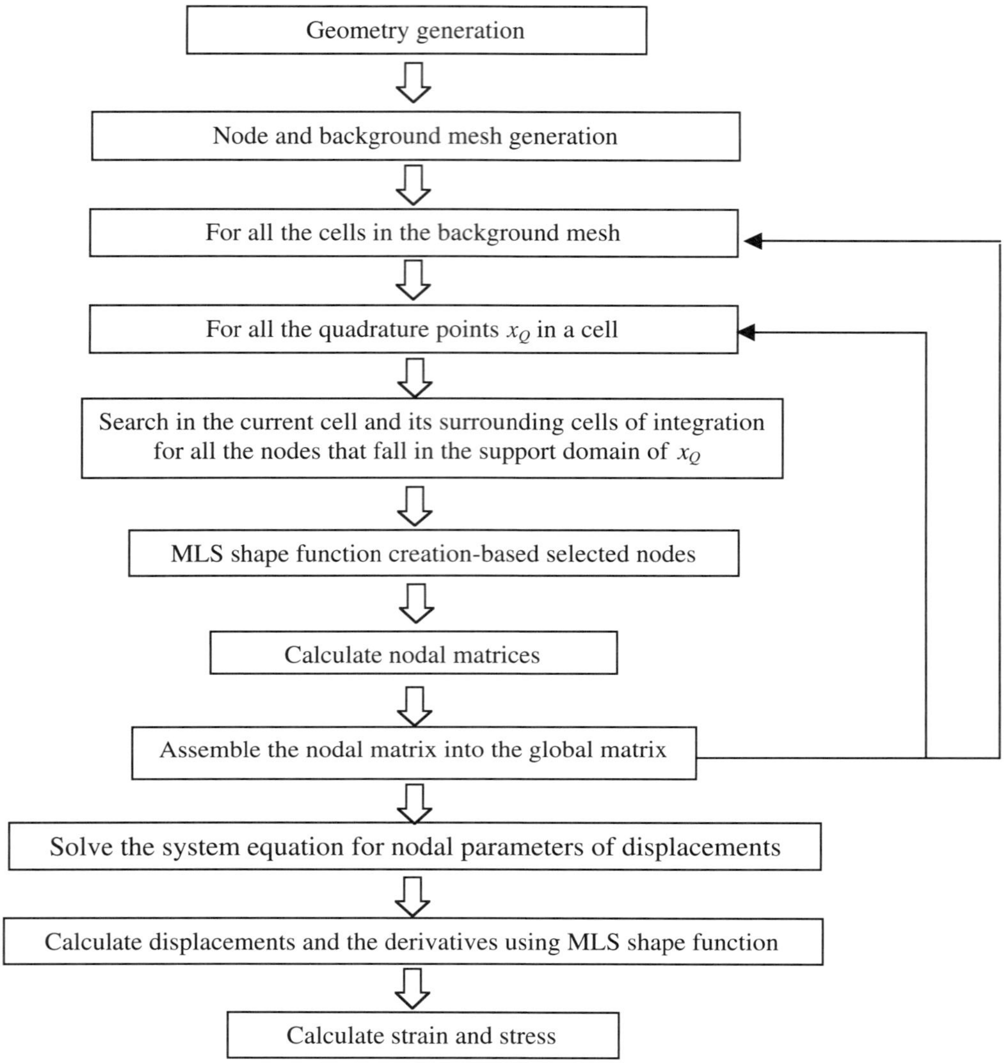

FIGURE 6.2
Flowchart of EFG method.

6.1.4 Numerical Examples

An EFG code has been developed based on the formulations provided above. Linear basis functions are employed in analyzing the following examples. The examples presented in this section are mainly designed for testing and benchmarking the EFG methods. In computing the system equations, the Gauss integration scheme often used in the conventional FEM is employed.

Example 6.1 Patch Test

For a numerical method to work for solid mechanics problems, the sufficient condition is to pass the standard patch test, which has been used very often in developing finite elements. The first numerical example, therefore, performs the standard patch test using the EFG method. The basic requirements for a patch are that the patch must have at least one

TABLE 6.1

Relative Error of Stresses at Node 5 in Patch Shown in Figure 6.3a

Coordinate of Node 5 (x, y)	Error in σ_{xx} (%)	Error in σ_{yy} (%)	Error in σ_{xy} (%)
(1.0, 1.0)	0.00	0.00	0.00
(1.1, 1.1)	−0.98	−0.77	−0.86
(1.9, 1.8)	194.59	142.13	164.46
(1.9, 0.1)	134.95	127.23	−130.43

Note: Lagrange multiplier is not used for imposing essential boundary conditions on the patch edges.

interior node and that a linear displacement is imposed on all the edges of the patch in the absence of body force. Satisfaction of the patch test then requires that the displacement at any interior node be given by the same linear function and that the strain and stress be constant in the entire patch. In our patch test, a square patch of dimension $L_x = 2$ by $L_y = 2$ is used. The displacements are prescribed on all outside boundaries by a linear function of x and y:

$$\bar{\mathbf{u}}(x, y) = \begin{Bmatrix} x \\ y \end{Bmatrix} \tag{6.50}$$

The patch is represented using a set of scattered nodes with some nodes in the interior of the patch. Both the regular and irregular nodal arrangements shown in Figure 6.3 are used for the test. The patch with regular nodal arrangement has eight boundary nodes and one interior node, as shown in Figure 6.3a, and the patch with irregular nodal arrangement has eight boundary nodes and three arbitrarily distributed interior nodes, as shown in Figure 6.3b. The numerical results show that the linear displacement field and the constant strain field have been reproduced within the patch to machine accuracy, as long as the numerical integration is accurate. This confirms that the EFG method exactly passes the patch test for both meshes, when an accurate numerical integration is performed. The issue of accurate integration is discussed in great detail in the next example.

Without the use of Lagrange multipliers, patch tests will fail, as reported by Belytschko et al. (1994b). The test was performed using the patch shown in Figure 6.3a for different locations of node 5. The relative errors of stresses are listed in Table 6.1. When node 5 is located at the center of the patch, there is no error in the results. This is, however, a very special case. When node 5 moves away from the center of the patch, error occurs. When node 5 is far from the center, meaning that the patch is highly irregular, the error is as large as almost 200%. This test clearly shows that the results can be very erroneous, if Lagrange multipliers are not used in enforcing the displacement (essential) boundary conditions.

Requirements for Galerkin Type Methods to Pass the Patch Test

The requirements for all the methods based on the Galerkin weak form to pass the patch test are as follows.

1. The shape functions are of at least linear consistency (see Chapter 5). This implies that the shape function is a linear field reproduction.
2. The field function approximation using the shape functions must be compatible (see Section 5.11). This is the condition given by Equation 4.1.

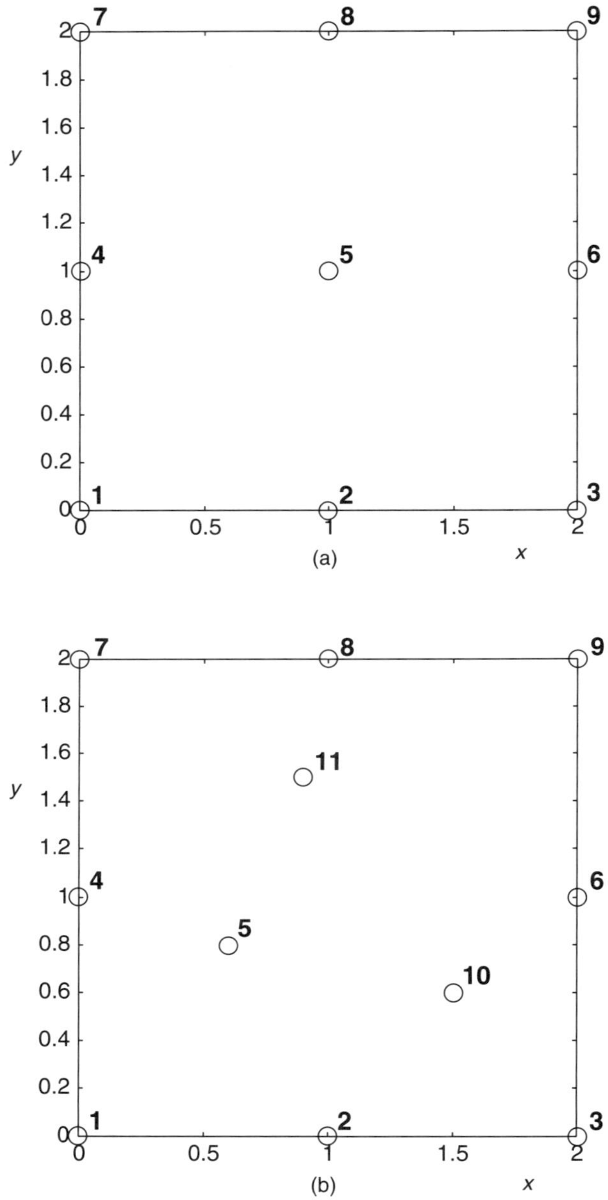

FIGURE 6.3
Nodal arrangement for patches (a) Regular nodal arrangement; (b) irregular nodal arrangement.

3. The essential boundary conditions (displacement constraints on the boundary of the patch) have to be accurately imposed. This is the condition given by Equation 4.2.

In addition, we require accurate numerical operations, such as integration, to form system equations in the process of testing.

The compatibility requirement is common to all the methods based on energy principles, because any possible gap or overlap (incompatibility) will affect the energy in the system and destroy the balance of the energy principle equation. The remedy is to use the constrained form of the energy principles that also include the energy caused by incompatible locations.

MLS shape functions can satisfy the first two requirements very easily as long as linear polynomial functions are included in the basis for constructing the shape functions. To satisfy the third condition, the constrained Galerkin form is required for constructing the system equations. We therefore conclude that the EFG method formulated in this chapter can always pass the standard patch test. In other words, the EFG is conforming. In its standard and accurate numerical implementation, it should provide the upper bound of the solution, and the displacement should converge to the exact solution from below when the nodal spacing approaches zero.

The PIM shape functions can satisfy the first requirement very easily as long as linear polynomial functions are included in the basis for constructing the shape functions. The PIM shape function can also satisfy the third requirement, as it possesses the Kronecker delta function property. To have the PIM approximation satisfy the second condition, however, the constrained Galerkin form is required in constructing the system equations. We discuss this issue in more detail in Chapter 8.

Example 6.2 Cantilever Beam (Numerical Integration)

Numerical study is conducted for a cantilever beam, which is often used for benchmarking numerical methods because the exact analytic solution for this problem is known. Our purpose here is to investigate issues related to background integration in the EFG method. There are a number of factors affecting the accuracy of the numerical results of the EFG method. These factors include the number of field nodes n, the background mesh density, and the order of Gauss integration. To provide a quantitative indication of how these factors affect the accuracy of results, a cantilever beam subjected to a load at the free end, as shown in Figure 6.4, is analyzed in detail using an EFG code. The exact solution of this problem is available as follows (Timoshenko and Goodier, 1977).

The displacement in the x direction is

$$u_x = -\frac{P_y}{6EI}\left[(6L-3x)x+(2+\nu)\left[y^2-\frac{D^2}{4}\right]\right] \tag{6.51}$$

where the moment of inertia I for a beam with rectangular cross section and unit thickness is given by

$$I = \frac{D^3}{12} \tag{6.52}$$

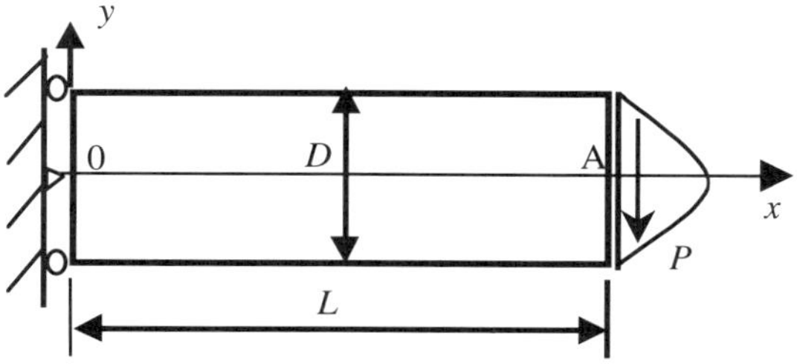

FIGURE 6.4
Cantilever beam loaded with an external force P distributed in a parabolic fashion at the end of the beam.

The displacement in the y direction is

$$u_y = \frac{P}{6EI}\left[3\nu y^2(L-x) + (4+5\nu)\frac{D^2x}{4} + (3L-x)x^2\right] \tag{6.53}$$

The normal stress on the cross section of the beam is

$$\sigma_x = -\frac{P(L-x)y}{I} \tag{6.54}$$

The normal stress in the y direction is

$$\sigma_y = 0 \tag{6.55}$$

The shear stress on the cross section of the beam is

$$\tau_{xy} = \frac{P}{2I}\left[\frac{D^2}{4} - y^2\right] \tag{6.56}$$

In this example, the parameters for this cantilever beam are taken as follows:

Loading: $P = -1000$ N
Young's modulus: $E = 3 \times 10^7$ N/m^2
Poisson's ratio: $\nu = 0.3$
Height of the beam: $D = 12$ m
Length of the beam: $L = 48$ m

The force P is distributed in a form of parabola at the right end of the beam:

$$t_{xy} = \frac{P}{2I}\left[\frac{D^2}{4} - y^2\right] \tag{6.57}$$

Strain energy error e is employed as an indicator of accuracy of the EFG numerical results:

$$e_e = \left\{\frac{1}{2}\int_\Omega (\varepsilon^{\text{num}} - \varepsilon^{\text{exact}}) D (\varepsilon^{\text{num}} - \varepsilon^{\text{exact}}) \, d\Omega\right\}^{1/2} \tag{6.58}$$

At the left boundary ($x = 0$) the displacements are prescribed using the analytical formulae (Equations 6.51 and 6.53):

$$u_x = -\frac{P(2+\nu)_y}{6EI}\left[y^2 - \frac{D^2}{4}\right] \tag{6.59}$$

$$u_y = \frac{P\nu L}{2EI} y^2 \tag{6.60}$$

On the right boundary ($x = L$), the applied external traction force is computed from the analytical formula Equation 6.57.

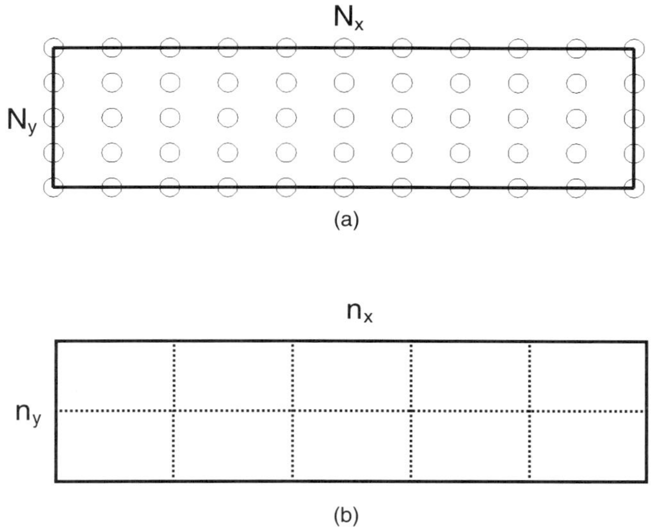

FIGURE 6.5
(a) Nodal arrangement; (b) background mesh.

TABLE 6.2
Strain Energy Error ($\times 10^{-2}$) Resulting from Using Different Number of Gauss Sampling Points and Number of Background Mesh (n_t = 55, N_u = 100)

Gauss Points (n_g)	Background Mesh ($n_x \times n_y$) 1 × 1	2 × 1	4 × 1	8 × 2	12 × 3
2 × 2	$e = \infty$, $N_Q = 12$	$e = \infty$, $N_Q = 24$	$e = \infty$, $N_Q = 48$	$e = 8.10$, $N_Q = 192$	$e = 3.12$, $N_Q = 432$
3 × 3	$e = \infty$, $N_Q = 27$	$e = \infty$, $N_Q = 54$	$e = \infty$, $N_Q = 108$	$e = 4.01$, $N_Q = 432$	$e = 2.95$, $N_Q = 972$
4 × 4	$e = \infty$, $N_Q = 48$	$e = \infty$, $N_Q = 96$	$e = 4.37$, $N_Q = 192$	$e = 3.62$, $N_Q = 768$	$e = 2.90$, $N_Q = 1728$
5 × 5	$e = \infty$, $N_Q = 75$	$e = 58.2$, $N_Q = 150$	$e = 4.63$, $N_Q = 300$	$e = 2.90$, $N_Q = 1200$	$e = 2.89$, $N_Q = 2700$
6 × 6	$e = \infty$, $N_Q = 108$	$e = 4.74$, $N_Q = 216$	$e = 3.74$, $N_Q = 432$	$e = 2.89$, $N_Q = 1728$	$e = 2.89$, $N_Q = 3888$
7 × 7	$e = \infty$, $N_Q = 147$	$e = 4.92$, $N_Q = 294$	$e = 2.96$, $N_Q = 588$	$e = 2.89$, $N_Q = 2352$	$e = 2.89$, $N_Q = 5292$
8 × 8	$e = \infty$, $N_Q = 192$	$e = 3.70$, $N_Q = 384$	$e = 2.99$, $N_Q = 768$	$e = 2.89$, $N_Q = 3072$	$e = 2.89$, $N_Q = 6912$
9 × 9	$e = \infty$, $N_Q = 243$	$e = 6.55$, $N_Q = 486$	$e = 2.93$, $N_Q = 972$	$e = 2.89$, $N_Q = 3888$	$e = 2.89$, $N_Q = 8748$
10 × 10	$e = 41.5$, $N_Q = 300$	$e = 9.52$, $N_Q = 600$	$e = 2.90$, $N_Q = 1200$	$e = 2.89$, $N_Q = 4800$	$e = 2.89$, $N_Q = 10800$

Notes: $N_Q = 3 \times n_Q = 3 \times (n_g \times n_x \times n_y)$.

For convenience of analysis, uniformly distributed nodes and background integration cells, as schematically shown in Figure 6.5, are used in the computation. N_x and N_y are the number of nodes with respect to the x and y directions. The density of the background mesh of cells is defined by $n_x \times n_y$, where n_x and n_y are, respectively, the number of background cells along the x and y directions. Table 6.2 summarizes the results of the error in terms of the energy norm defined by Equation 6.58 obtained by the EFG method using different numbers of Gauss integration points and different densities of background mesh of cells. In Table 6.2, n_g is the number of Gauss sampling points within a cell. The total number of quadrature points can be calculated using

$$n_Q = n_g \times n_x \times n_y \tag{6.61}$$

The total number of independent equations used in all the quadrature points, N_Q, can then be calculated using Equation 6.48, i.e.,

$$N_Q = 3 \times n_Q = 3 \times (n_g \times n_x \times n_y) \tag{6.62}$$

It is confirmed from Table 6.2 that, when $N_Q < N_u$, no stable solutions are obtained and even $N_Q > N_u$ does not necessarily guarantee an accurate solution. It can also be seen from Table 6.2 that an acceptably accurate result can be obtained using

$$N_Q > 4N_u \sim 5N_u \cong 9n_t \tag{6.63}$$

or

$$n_Q > 3n_t \tag{6.64}$$

Therefore, we may conclude as a rough rule-of-thumb that a sufficient number of all the Gauss points should be at least about three times the total number of nodes.

Efforts have been made to develop the EFG method without using a background mesh. Some of the so-called nodal integration approaches carry out the integration using only the field nodes without the use of additional Gauss points, to avoid using a background mesh of cells. In such cases, $n_Q \approx n_t$, which satisfies the minimum requirement of Equation 6.49. However, the above important finding of the $n_Q > 3n$ rule implies that any attempt at such a nodal integration scheme may suffer significant loss in accuracy, unless special measures, such as the use of stabilization terms (Beissel and Belytschko, 1996), can be taken to prevent that from happening. Note that the requirement of $n_Q > \frac{2}{3} n_t$ is only the minimum requirement to ensure a *nonsingular* system matrix; it does not guarantee the accuracy of the solution.

Figure 6.6 plots the exact and numerical solutions of EFG for beam deflection along the x axis. The plot shows excellent agreement between the exact solution and the numerical results for all the background meshes used. This fact reveals that the displacement is less sensitive to the background integration. A very coarse mesh can yield good displacement results. Figure 6.7 shows the distribution of stress σ_{xx} on the cross section of $x = L/2$ of the cantilever beam. Errors in stress between the exact solution and the numerical results are clearly evident. This fact implies that the stresses that are obtained using the derivatives of the displacement field are very sensitive to the way the integration is performed. A much finer mesh and more Gauss points have to be used for an accurate stress field. Figures 6.8 and 6.9 show, respectively, the stress components σ_{yy} and σ_{xy}. It is clearly shown again that the stresses are very sensitive to the cell density of the background integration, especially the shear stress σ_{xy}.

The total number of Gauss points depends on both the cell density of the background mesh and the number of Gauss points used in each cell, and these two have to be balanced. A finer background integral mesh can improve the accuracy, but there is a limit. On the other hand, more Gauss integration points give, in general, higher accuracy, but the background mesh should not be too coarse. By using different densities of background mesh and different numbers of Gauss sampling points, the displacement field and stress field are computed using EFG, and the accuracy is investigated. From Figures 6.7 through

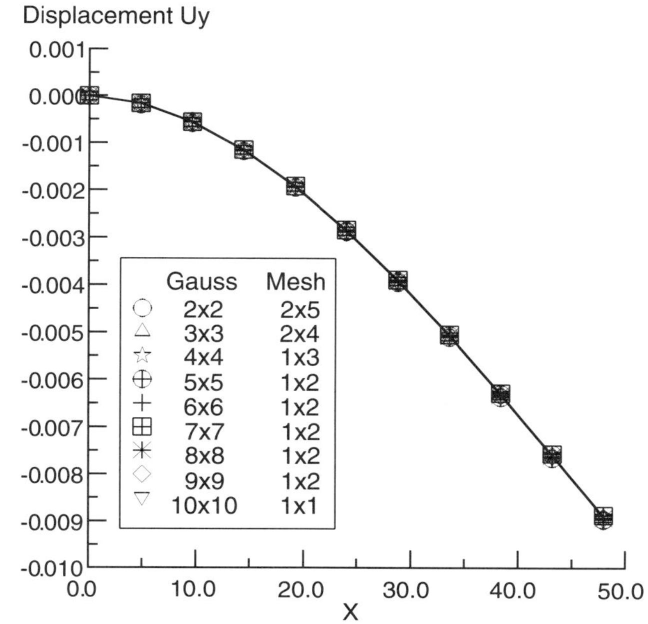

FIGURE 6.6
Deflection of the cantilever beam along $y = 0$.

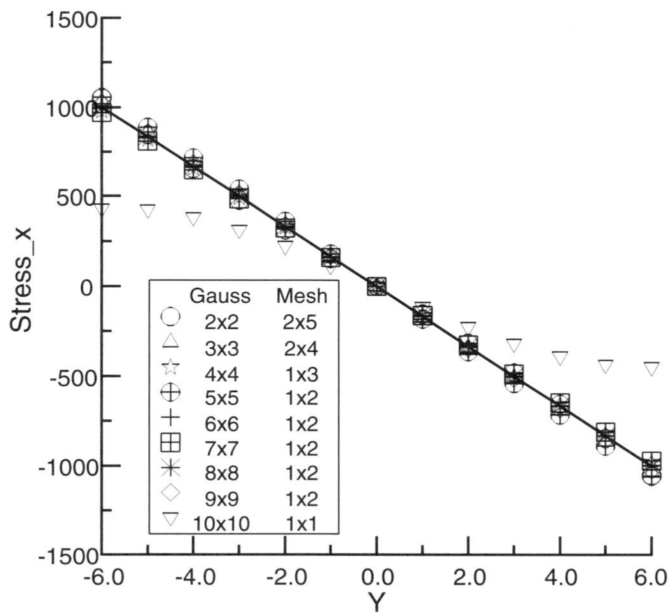

FIGURE 6.7
Distribution of stress σ_{xx} on the section of $x = L/2$ of the cantilever beam.

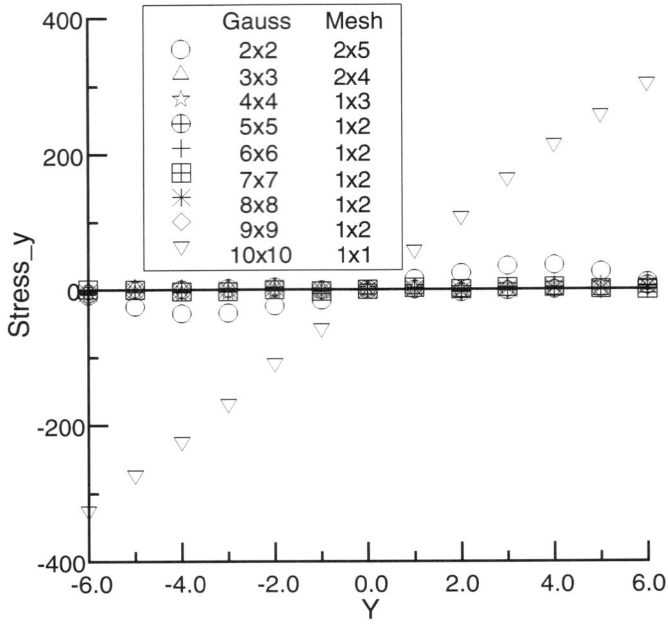

FIGURE 6.8
Distribution of stress σ_{yy} on the section of $x = L/2$ of the cantilever beam.

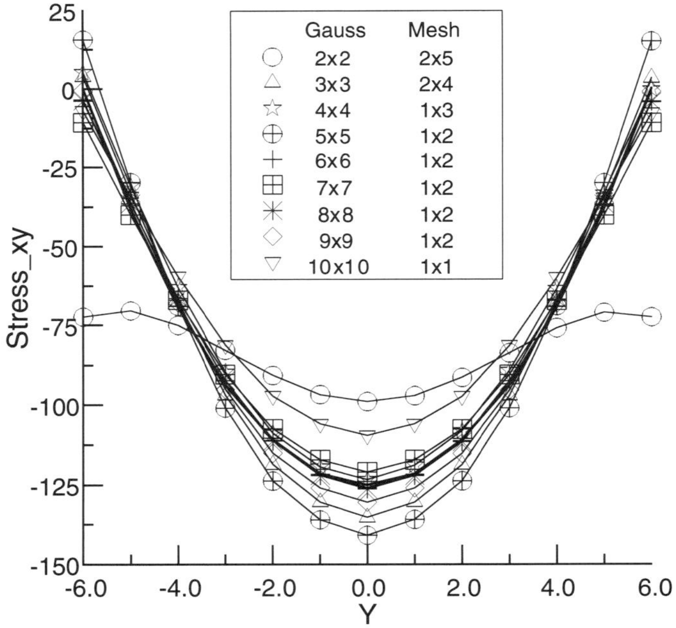

FIGURE 6.9
Distribution of stress τ_{xy} on the section of $x = L/2$ of the cantilever beam.

6.9, it can be observed that the cell density of the background mesh and the number of the Gauss points have to be balanced in order to obtain good results. Too fine a mesh without enough Gauss points or too many Gauss points with too coarse a mesh will not give accurate results. One should always avoid biases to either the number of Gauss points or the number of cells of the background mesh. The balance point should, in general, depend on the complexity of the field to be analyzed, and the basis used in the MLS approximation. Our study for solid mechanics problems has found that, when a linear basis function is used, the proper number of Gauss sampling points should be between $n_g = 2 \times 2$ and $n_g = 6 \times 6$. Once the number of the Gauss points is chosen, the density of the background mesh should then be calculated using the guideline of $n_Q > 3n_t$.

To investigate how the node number affects the accuracy of the result, the strain energy error is calculated for the cantilever beam using different nodal densities along the x and y directions. The background mesh is fixed at $n_x \times n_y = 12 \times 3$, and Gauss points of $n_g = 4 \times 4$ are used for the integration. Results are presented in Figure 6.10. It is clearly seen that increasing the number of nodes in the domain can improve the accuracy. For this particular problem of cantilever beam, increasing the nodes along the y direction improves the accuracy more efficiently than increasing the nodes along the x direction. Generally, when N_x is greater than eight, the results are sufficiently accurate. Further study on the effects of Gauss integration points is conducted by changing n_g from 2×2 to 10×10, and no significant change in strain energy error is observed.

Because both nodal density and background mesh density affect the accuracy of the stress field, the ratio of quadrature points to field nodes is defined as

$$\alpha_n = \frac{n_Q}{n_t} \tag{6.65}$$

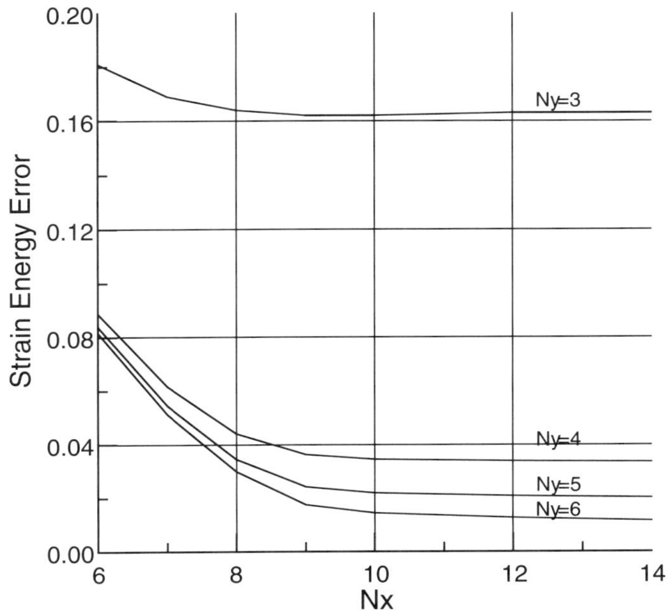

FIGURE 6.10
Effects of the nodal density on the strain energy error.

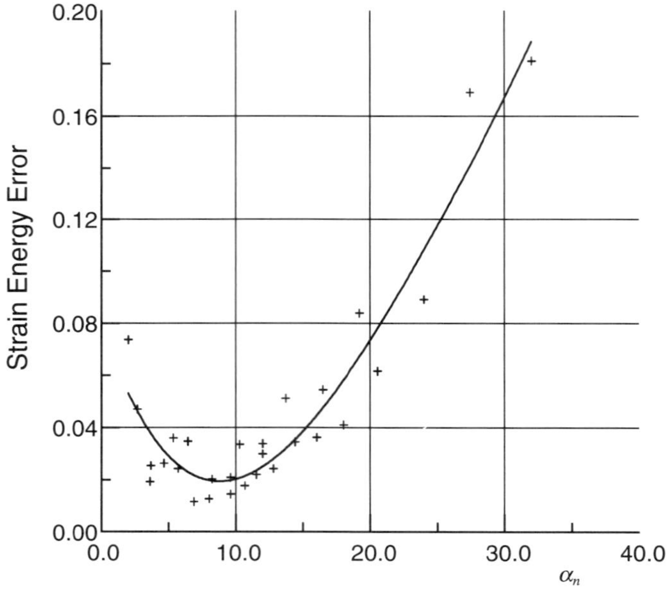

FIGURE 6.11
Relation between strain energy error and the ratio of quadrature points to field nodes α_n.

Further investigation has been conducted on the relationship between strain energy error and the ratio of quadrature points to nodes, α_n. The results are summarized in Figure 6.11. It is clearly shown that when α_n is around 7.0 to 9.0, the result obtained is most accurate. A reasonable result can be obtained using $\alpha_n > 3$. This confirms the finding from Table 6.2.

In the application of EFG for practical problems, the density of the field nodes should be determined by the gradient of the field variables. For most practical problems, the field nodes may not be evenly distributed. The sampling points should also be distributed accordingly in an uneven manner with α_n around 3.0 to 9.0.

Note that the Gauss point number suggested by Belytschko et al. (1994b) is $n_Q = n_t(\sqrt{n_c} + 2)^2$, where n_c is the number of nodes in a cell. When $n_c = 1$, we have $\alpha_n = 9$. For $n_c \geq 2$, this suggestion demands many more Gauss points.

6.1.5 Remarks

The EFG method is a meshless method that is based on global variational formulation—Galerkin variational principle. Therefore, although an element mesh is not required for field variable approximation over the problem domain, a global background mesh is still required to evaluate the integrals for calculating stiffness and mass matrices. Our numerical examination of the relationship between the density of field nodes and background mesh for 2D stress analysis problems shows the following:

- The minimum number of integration points must be greater than two thirds of the total number of the unfixed field nodes, i.e., $n_Q > \frac{2}{3} n_t$. This requirement of $n_Q > \frac{2}{3} n_t$ is only the minimum requirement to ensure a *nonsingular* system matrix; it does not guarantee the accuracy of the solution.
- The ratio of the integration points to the field nodes, α_n, is around 3 to 9, and economic results of acceptable accuracy can be obtained using $\alpha_n = 3$. This means

that a sufficient number of integration points should be about three times the number of the field nodes.

- The displacement field can be obtained rather accurately with a coarse background mesh of integration cells, whereas a finer background mesh is necessary for computing the stress field.
- Accuracy of the stress field can be improved efficiently by increasing the density of the field nodes, together with sufficient density of the background mesh.

6.2 EFG with Penalty Method

As described in the previous chapters, the use of MLS approximation produces shape functions that do not possess the Kronecker delta function property, i.e., $\phi_I(\mathbf{x}_J) \neq \delta_{IJ}$. This leads to $u^h(\mathbf{x}_J) = \sum_I^n \phi_I(\mathbf{x}_J)u_I \neq u_J$, which implies that one cannot impose essential boundary conditions in the same way as in conventional FEM. In the previous chapter, the essential boundary conditions are imposed by introducing Lagrange multipliers in the weak form. This method of Lagrange multipliers results in an enlarged nonpositive system matrix, as shown in Equation 6.37. The bandedness of the system matrix is also distorted. Therefore, it requires much more computational effort and resources in solving such system equations as Equation 6.37.

In this section, an alternative method—the penalty method—is introduced for the imposition of essential boundary conditions. The use of the penalty method produces equation systems of the same dimensions that conventional FEM produces for the same number of nodes, and the modified stiffness matrix is still positive definite. The problem with the penalty method lies in choosing a penalty factor that can be used universally for all problems. The penalty method has been used by many researchers; this section follows the formulation reported by G. R. Liu and Yang (1998).

6.2.1 Formulation

The penalty method has been frequently used in conventional FEM for enforcement of single or multipoint constraints (Zienkiewicz, 1989). In the EFG method, the essential boundary conditions needed to be enforced have the form:

$$\sum_{I}^{n} \phi_I(\mathbf{x})\mathbf{u}_I = \bar{\mathbf{u}}(\mathbf{x}) \quad \text{on } \Gamma_u \tag{6.66}$$

where $\bar{\mathbf{u}}(\mathbf{x})$ is the prescribed displacement on the essential boundary. Note that Equation 6.66 is but the continuous form of the so-called multipoint constraint (MPC) equations that are used very often in FEM analyses. Therefore, the penalty method can, of course, be applied to impose the essential boundary conditions in the MFree methods that use MLS shape functions for field variable approximation.

This section presents the formulation of a penalty method to impose essential boundary conditions in the EFG method. System equations for boundary value problems of both homogeneous and inhomogeneous materials are derived. The discrete system matrices derived using the penalty method from the constrained Galerkin weak form are positive definite (unless the essential boundary condition is improperly set) and banded, and the treatment of boundary conditions is as simple as it is in conventional FEM. Numerical examples are presented to demonstrate the procedure of enforcing the essential boundary conditions.

The present approach is also applied to problems with continuity conditions on the interfaces of multimaterial bodies, such as composite materials. Numerical examples demonstrate that the EFG method with penalty method is applicable in handling problems for composite materials, where the continuity conditions between different types of materials need to be enforced.

6.2.2 Penalty Method for Essential Boundary Conditions

Consider again the problem stated in Equations 6.1 and 6.2. Instead of using Lagrange multipliers, we introduce a penalty factor to penalize the difference between the displacement of MLS approximation and the prescribed displacement on the essential boundary. The constrained Galerkin weak form using the penalty method can then be posed as follows:

$$\int_{\Omega} \delta(\mathbf{Lu})^T \mathbf{c}(\mathbf{Lu})\mathrm{d}\Omega - \int_{\Omega} \delta\mathbf{u}^T \cdot \mathbf{b}\,\mathrm{d}\Omega - \int_{\Gamma_t} \delta\mathbf{u}^T \cdot \bar{\mathbf{t}}\,\mathrm{d}\Gamma - \delta\int_{\Gamma_u} \frac{1}{2}(\mathbf{u}-\bar{\mathbf{u}})^T \cdot \boldsymbol{\alpha} \cdot (\mathbf{u}-\bar{\mathbf{u}})\mathrm{d}\Gamma = 0 \tag{6.67}$$

Equation 6.67 is formed from Equation 4.43 by changing the area integrals for the constraint-related terms into curve integrals because the constraints (essential boundary conditions) given in Equation 6.2 are defined only on the boundary. Note that the difference between Equation 6.67 and Equation 6.4 is that the fourth and fifth terms in Equation 6.4 are replaced by the fourth term in Equation 6.67, where $\boldsymbol{\alpha} = \lceil \alpha_1\ \alpha_2\ \ldots\ \alpha_k \rfloor$ is a diagonal matrix of penalty factors, where $k = 2$ for 2D cases and $k = 3$ for 3D cases. The penalty factors α_i ($i = 1, \ldots, k$) can be a function of coordinates and they can be different from each other, but in practice we often assign them an identical constant of a large positive number, which can be chosen by following the method described in Section 4.3.3.

Substituting the expression of the MLS approximation for the displacement of Equation 6.7 into the weak form of Equation 6.67, and after similar derivations given in Section 6.1.1, we can arrive at the final system equation of

$$[\mathbf{K} + \mathbf{K}^{\alpha}]\mathbf{U} = \mathbf{F} + \mathbf{F}^{\alpha} \tag{6.68}$$

where **F** is the global external force vector assembled using the nodal force matrix defined by Equation 6.31, and **K** is the global stiffness matrix assembled using the nodal stiffness matrix given in Equation 6.38. The additional matrix $\mathbf{K}^{\alpha}$ is the global penalty matrix assembled using the nodal matrix defined by

$$\mathbf{K}_{IJ}^{\alpha} = \int_{\Gamma_u} \boldsymbol{\Phi}_I^T \boldsymbol{\alpha} \boldsymbol{\Phi}_J \,\mathrm{d}\Gamma \tag{6.69}$$

where $\boldsymbol{\Phi}_I$ is given by Equation 6.41.

The force vector $\mathbf{F}^{\alpha}$ is caused by the essential boundary condition, and its nodal vector has the form of

$$\mathbf{F}_I^{\alpha} = \int_{\Gamma_u} \boldsymbol{\Phi}_I^T \boldsymbol{\alpha} \bar{\mathbf{u}} \,\mathrm{d}\Gamma \tag{6.70}$$

Note that the integration is performed along the essential boundary, and hence matrix $\mathbf{K}^{\alpha}$ will have entries only for the nodes near the essential boundaries Γ_u, which are covered by the support domains of all the quadrature points on Γ_u.

Comparing Equation 6.68 with Equation 6.37, the advantages of the penalty method are obvious:

- The dimension and positive definite property of the matrix are preserved, as long as the penalty factors chosen are positive.
- The symmetry and the bandedness of the system matrix are preserved.

These advantages make the penalty method much more efficient and hence much more attractive compared with the Lagrange multipliers method. Detailed studies on implementation of the penalty method and computation of actual application problems have indicated the following minor disadvantages of the penalty method compared with the Lagrange multipliers method.

- It is necessary to choose penalty factors that are universally applicable for all kinds of problems. One hopes to use as large as possible penalty factors, but too large penalty factors often result in numerical problems, as we have experienced in the imposition of multipoint boundary condition in FEM. More discussion on this is provided in Section 4.3.3, Example 6.12 under Section 6.4.5, when we deal with nonlinear problems.
- The results obtained are in general less accurate, compared with the method of Lagrange multipliers.
- An essential boundary condition can never be precisely imposed. It is imposed only approximately.

Despite these minor disadvantages, the penalty method is much more favorable for many researchers. It is also implemented in MFree2D© (see Chapter 16). It can also be conveniently used for enforcing capability conditions in using PIM shape functions (see Chapter 8). Below is another application of the penalty method.

6.2.3 Penalty Method for Continuity Conditions

The domain of a problem can be composed of subdomains of different materials. The treatment of material discontinuity is straightforward in the conventional FEM, because we have elements to use. All one need do is have the element edges coincide with the interface of different materials. The properties of the material are used for elements that are located in the corresponding subdomain of the material. There is no special treatment required (this does not necessarily mean that such treatment is problem free in FEM).

In MFree methods, however, there is no mesh of elements, and hence the material interface cannot be defined based on elements. Special treatment is therefore needed. Cordes and Moran (1996) dealt with material discontinuity problems using the method of Lagrange multipliers. The conditions on the material interfaces were treated as a special essential boundary condition, and the approaches in the original EFG procedure discussed in Section 6.1 could then be followed to handle the material discontinuity. Krongauz and Belytschko (1998) have proposed another method to model material discontinuities in the EFG method by adding special shape functions that contain discontinuities in derivatives.

The penalty method introduced in this section can be an alternative for dealing with the material discontinuity problem. The detailed procedure was originally proposed by G. R. Liu and Yang in 1999, and is reported in detail in Yang's master's thesis. This section details the penalty method for handling the material discontinuity problem.

Multipoint constraints enforced using the penalty method are often used in FEM for modeling the different kinds of connections between subdomains of a structural system. The penalty method is also applicable for modeling similar situations in mechanics problems.

The penalty method is presented in this section to model two subdomains of different materials connected in a prescribed manner. A perfect connection is considered, but the approach is applicable for all kinds of connections, including partial connections.

Consider first an inhomogeneous medium consisting of two homogeneous bodies. On the boundary of the two homogeneous bodies, an interface is first defined by a set of nodes that belong to both materials. We then impose a *nonpenetration rule* for the influence domains of the nodes. The nonpenetration rule states that points contained in material 1 can only be influenced by the nodes in material 1 plus the interface nodes; points contained in material 2 can only be influenced by the nodes contained in material 2 plus the interface nodes. Our following treatment is based on this nonpenetration rule of influence domains. This rule confines the influence domain of a node within the subdomain of the material of the node.

Figure 6.12 illustrates the determination of the domains of influence for the nodes in problem domains of homogeneous and inhomogeneous materials. In Figure 6.12, circular domains of influence are employed. For the homogeneous case (Figure 6.12a), point a is contained in the influence domains of both nodes 4 and 5. Therefore, nodes 4 and 5 are considered neighbors of point a. Similarly, point b has neighbors of nodes 3 and 5, and

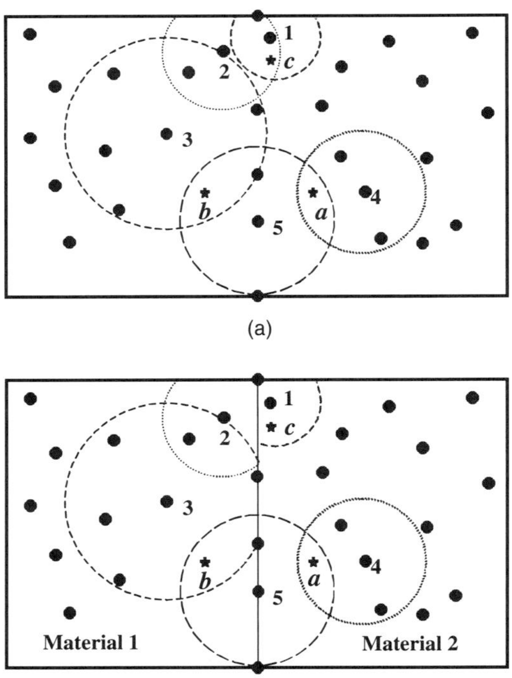

FIGURE 6.12
Determination of domains of influence. (a) Domains of influence in a homogeneous body; (b) domains of influence in an inhomogeneous body.

point c has neighbors of nodes 1 and 2. However, as the interface exists in the inhomogeneous materials in Figure 6.12b, the neighbors of each of the points a, b, and c may change, due to the blockage of the material interface. The influence domain for node 4 is the same as in homogeneous materials because the influence domain of node 4 does not intersect the interface, and, therefore, point a is still within its influence. The influence domain for node 5 is also the same as in homogeneous materials because its lies on the interface of both materials. Therefore, point a is still within the influence of node 5. The neighbors of point b still include nodes 3 and 5 since each pertains to material 1. However, point c is not included in the support domain of node 2 due to the nonpenetration of the influence domain of node 2. In this case, point c has only one neighbor, that is, node 1.

The connection of the subdomains is achieved by enforcing conditions of continuity. On the interface of the two materials Γ_s, we enforce the constraints of

$$\bar{\mathbf{u}}^{+} = \bar{\mathbf{u}}^{-} \tag{6.71}$$

where $\bar{\mathbf{u}}^{+}$ and $\bar{\mathbf{u}}^{-}$ correspond to the displacement in the two materials but on the interface of Ω^{+} and Ω^{-}, respectively. This constraint is then imposed using the penalty method in the constrained Galerkin weak form, i.e.,

$$\begin{aligned}&\int_{\Omega}\delta(\mathbf{Lu})^{T}(\mathbf{cLu})\mathrm{d}\Omega-\int_{\Omega}\delta\mathbf{u}^{T}\mathbf{b}\mathrm{d}\Omega-\int_{\Gamma_t}\delta\mathbf{u}^{T}\bar{\mathbf{t}}\,\mathrm{d}\Gamma\\&\quad-\delta\frac{1}{2}\int_{\Gamma_u}(\mathbf{u}-\bar{\mathbf{u}})^{T}\boldsymbol{\alpha}(\mathbf{u}-\bar{\mathbf{u}})\mathrm{d}\Gamma-\delta\frac{1}{2}\int_{\Gamma_s}(\bar{\mathbf{u}}^{+}-\bar{\mathbf{u}}^{-})^{T}\boldsymbol{\beta}(\bar{\mathbf{u}}^{+}-\bar{\mathbf{u}}^{-})\mathrm{d}\Gamma=0\end{aligned} \tag{6.72}$$

Note that the difference between Equations 6.72 and 6.67 is the additional term in Equation 6.72, where $\boldsymbol{\beta}$ is a diagonal matrix of penalty factors that have the same form of $\boldsymbol{\alpha}$, but the values in $\boldsymbol{\beta}$ may be different. The approximate values of $\bar{\mathbf{u}}^{+h}$ and $\bar{\mathbf{u}}^{-h}$ are expressed using MLS approximation (see Equation 6.7).

$$\bar{\mathbf{u}}^{+h}(\mathbf{x}) = \sum_{I}^{n^{+}}\boldsymbol{\Phi}_I^{+}(\mathbf{x})\mathbf{u}_I^{+} \tag{6.73}$$

and

$$\bar{\mathbf{u}}^{-h}(\mathbf{x}) = \sum_{I}^{n^{-}}\boldsymbol{\Phi}_I^{-}(\mathbf{x})\mathbf{u}_I^{-} \tag{6.74}$$

where n^{+} and n^{-} are the number of nodes, respectively. $\boldsymbol{\Phi}_I^{+}$ and $\boldsymbol{\Phi}_I^{-}$ are the matrices of shape functions created using the nodes that have influence on $\mathbf{x}$ in Ω^{+} and Ω^{-}, respectively.

Substituting Equations 6.5, 6.73, and 6.74 into Equation 6.72 leads to a set of discretized system equations:

$$[\mathbf{K}+\mathbf{K}^{\alpha}+\mathbf{K}^{\beta}]\mathbf{U}=\mathbf{F}+\mathbf{F}^{\alpha} \tag{6.75}$$

where

$$\mathbf{K}_{IJ}^{\beta} = \int_{\Gamma_s}[\boldsymbol{\Phi}_I^{+}-\boldsymbol{\Phi}_I^{-}]^{T}\cdot\boldsymbol{\beta}\cdot[\boldsymbol{\Phi}_J^{+}-\boldsymbol{\Phi}_J^{-}]\mathrm{d}\Gamma \tag{6.76}$$

Matrix $\mathbf{K}^{\beta}$ is the stiffness matrix for connecting the two different materials. Note that the integration is performed along the interfaces of two materials, and hence matrix $\mathbf{K}^{\beta}$ will have entries for the nodes near (not only on) the interfaces, which have influence on the quadrature points on the interface of different materials.

6.2.4 Numerical Examples

Numerical Examples for Treating Essential Boundary Conditions

Example 6.3 Patch Test

The first numerical example is the standard patch test. The same patch tests conducted in Example 6.1 are repeated here using the penalty method for imposing the linear displacement on the boundaries of the patches shown in Figure 6.3. The EFG method with penalty method exactly passes the test for both kinds of meshes to machine accuracy. In both cases, the maximum errors in the displacement are of order 10^{-13}; the stresses remain the same in the patch and the maximum errors are of order 10^{-11}. The displacements for the regular and irregular nodal arrangements are given in Tables 6.3 and 6.4.

Example 6.4 Timoshenko Beam

Consider a beam of characteristic length L and height D subjected to a parabolic traction at the free end, as shown in Figure 6.4, which was examined in Example 6.2. The beam is of unit thickness and the plane stress state is considered. The exact solution is given by

TABLE 6.3

Coordinates and Displacements Solved for the Patch Test with Regular Nodal Arrangement Using EFG with Penalty Method

Nodes	Coordinates (x, y)	Displacements Solved (u_x, u_y)
1	(0, 0)	(0.00000000000000, 0.00000000000000)
2	(1, 0)	(1.00000000000000, 0.00000000000000)
3	(2, 0)	(1.99999999999999, 0.00000000000000)
4	(0, 1)	(0.00000000000000, 1.00000000000000)
5	(1, 1)	(1.00000000000007, 0.99999999999996)
6	(2, 1)	(2.00000000000000, 1.00000000000000)
7	(0, 2)	(0.00000000000000, 2.00000000000000)
8	(1, 2)	(1.00000000000000, 2.00000000000000)
9	(2, 2)	(2.00000000000000, 2.00000000000000)

TABLE 6.4

Coordinates and Displacements Solved for the Patch Test with Irregular Nodal Arrangement Using EFG with Penalty Method

Nodes	Coordinates (x, y)	Displacements Solved (u_x, u_y)
1	(0, 0)	(0.00000000000000, –0.00000000000001)
2	(1, 0)	(1.00000000000000, 0.00000000000000)
3	(2, 0)	(2.00000000000000, 0.00000000000000)
4	(0, 1)	(0.00000000000000, 1.00000000000000)
5	(0.6, 0.8)	(0.59999999999995, 0.80000000000009)
6	(1, 2)	(1.00000000000000, 2.00000000000000)
7	(0, 2)	(0.00000000000000, 2.00000000000000)
8	(1, 2)	(1.00000000000000, 2.00000000000000)
9	(2, 2)	(2.00000000000000, 2.00000000000000)
10	(1.5, 0.6)	(1.49999999999997, 0.60000000000001)
11	(0.9, 1.5)	(0.90000000000000, 1.50000000000004)

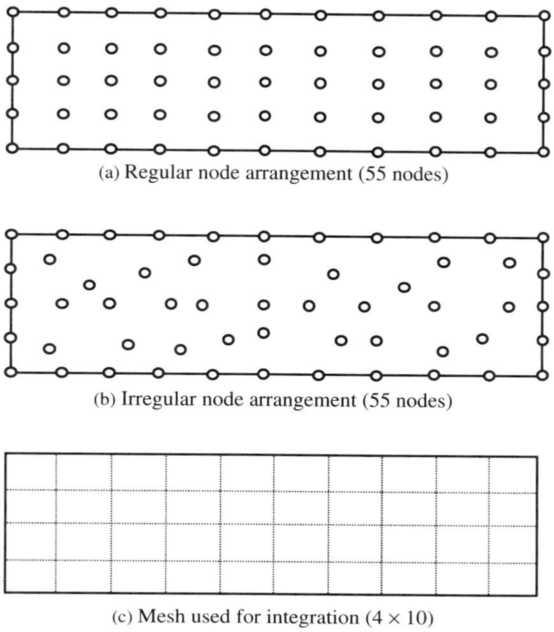

FIGURE 6.13
Nodal arrangements and background mesh for the cantilever beam. (a) Regular node arrangement; (b) irregular node arrangement; (c) mesh used for integration.

Timoshenko and Goodier (1970), and is listed in Equations 6.51 through 6.56. The parameters used in this section are as follows:

Loading: $P = 100$ N

Young's modulus: $E = 3 \times 10^6$ N/m^2

Poisson's ratio: $\nu = 0.3$

Height of the beam: $D = 1.2$ m

Length of the beam: $L = 4.8$ m

Both a regular and irregular arrangement of nodes and a regular background mesh of cells are used for numerical integrations to calculate the system equations, as shown in Figure 6.13. In each integration cell, a 4×4 Gauss quadrature scheme is used to evaluate the stiffness matrix. The linear basis function and cubic spline weight function are used in the MLS approximation. The dimension of the support domain α_s is chosen to be 3.5 so that the domain of support of each quadrature point contains at least 40 nodes to avoid the singularity of the moment matrix in constructing MLS shape functions.

Figure 6.14 plots the analytical solution based on 2D elasticity and the numerical solution using the present EFG method for the beam deflection along the x axis. The plot shows excellent agreement between the analytical and present numerical results for both regular and irregular nodal arrangements.

Figures 6.15 and 6.16 illustrate the comparison between the stresses calculated using the analytical solution and the present EFG with penalty method. The normal stress σ_x at the section of $x = L/2$ is shown in Figure 6.15, and the shear stress τ_{xy} is shown in Figure 6.16. Very good agreement is observed. It should be noted that the accuracy of the shear stress in the case of the irregular nodal arrangement is lower than that in the regular arrangement.

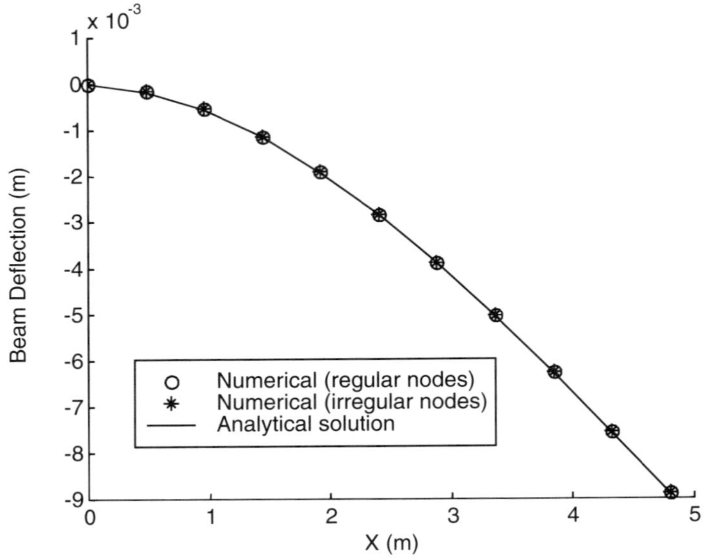

FIGURE 6.14
Analytical and present numerical solutions for the deflection of the cantilever beam.

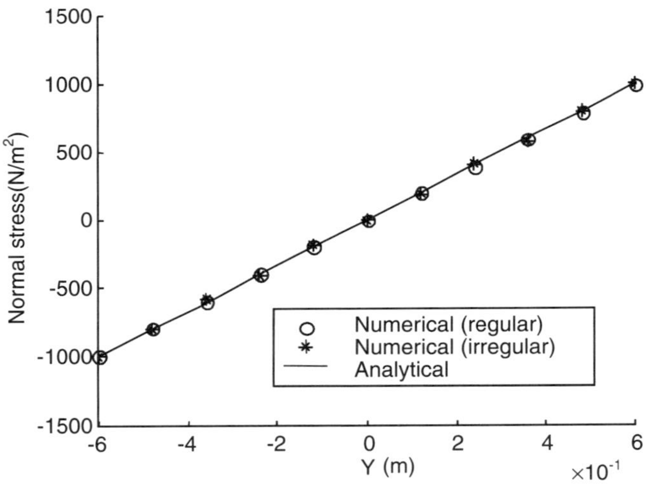

FIGURE 6.15
Analytical and present numerical solutions for the normal stress at the section $x = L/2$ of the cantilever beam.

Table 6.5 compares the numerical result for the vertical displacement at point A on the beam (see Figure 6.4) with the exact vertical displacement given in Equation 6.53. The calculation was performed for models with 10, 18, 55, and 189 nodes. This table shows that the numerical result approaches the exact solutions as the number of the nodes increases.

Figure 6.17 is a plot of the rate of convergence in L_2 energy error for the beam problem. The rate of convergence in energy is calculated using Equation 6.58. The value h was

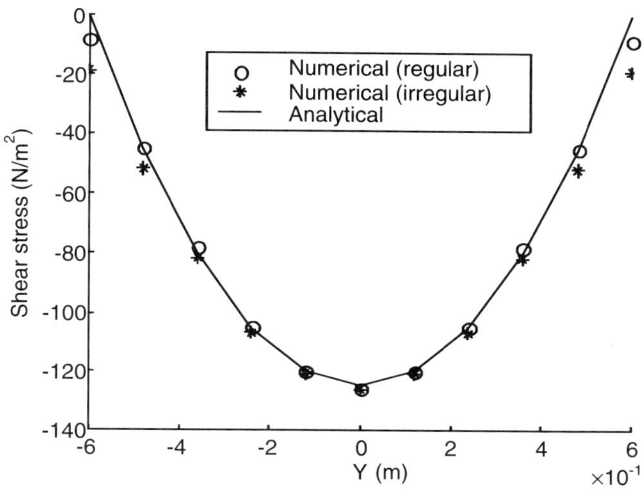

FIGURE 6.16
Analytical and numerical solutions for the shear stress at the section $x = L/2$ of the cantilever beam.

TABLE 6.5
Comparison of Vertical Deflection u_y(m) at End of Beam

Number of Nodes	Exact	EFG (Penalty)	%Error
10	–0.0089	–0.008099	9
18	–0.0089	–0.008511	4.4
55	–0.0089	–0.008883	0.2
189	–0.0089	–0.008898	0.02

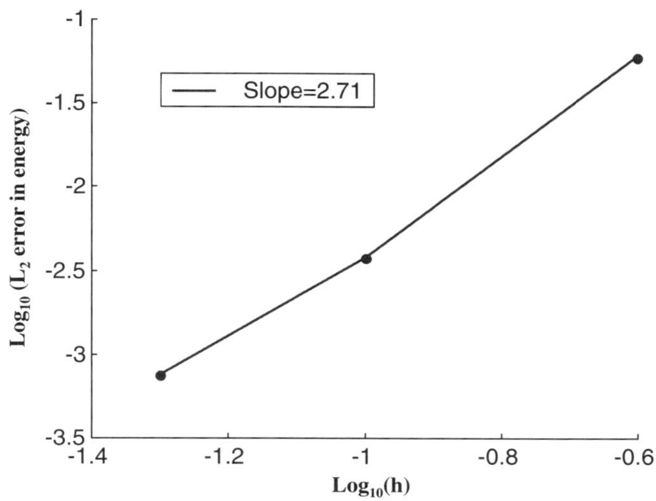

FIGURE 6.17
Rate of convergence in energy error tested on the cantilever beam.

TABLE 6.6

Computational Time Using EFG with Different Methods for Imposing Essential Boundary Conditions

	CPU Time (s)	
Nodes	EFG (Lagrange Multiplier)	EFG (Penalty)
55	1.1	0.6
189	35.4	3.5
561	115.2	13.8

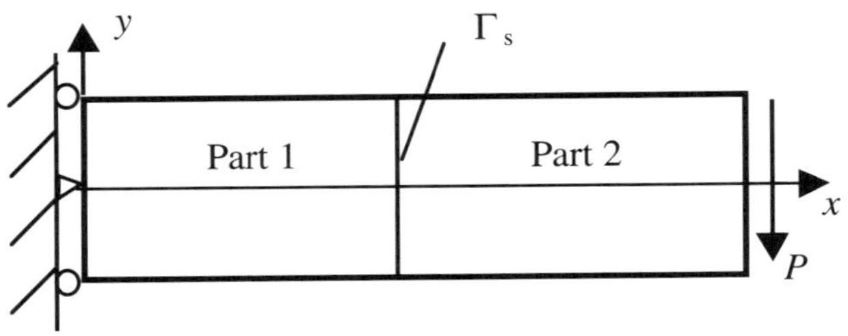

FIGURE 6.18
Cantilever beam made of two parts of different materials.

chosen to be the horizontal nodal spacing in the model. The dimensionless size of the support domain is $\alpha_s = 3.5$. The cubic spline weight function Equation 5.11 is used. The slope of the line plotted in Figure 6.17 is approximately 2.71, which is a greater rate of convergence than the linear finite elements, which should be around 1.0 (Hughes, 1987) for the same definition of error.

Computational Time

Because the penalty method does not increase the size of the system equation and the stiffness matrix is banded, computational efficiency can be improved greatly compared with the use of the method of Lagrange multipliers in EFG. Table 6.6 compares the CPU time of the penalty method vs. the method of Lagrange multipliers used in EFG for the cantilever beam. The computation is performed on the same HP workstation. The half-bandwidth technique is used to store the system matrices and solve the system equations. It can be seen that the penalty method is much faster than the method of Lagrange multipliers, especially for large numbers of field nodes.

Numerical Examples for Treating Continuity Conditions

Example 6.5 Cantilever Beam of Bi-Material

To validate the present method, the method is applied to a cantilever beam, comprising two parts of different materials connected at boundary Γ_s, as shown in Figure 6.18. We first assume that these two parts have the same material properties; therefore, this beam can be regarded as a homogeneous beam and the analytical solutions are valid. In the numerical analysis using EFG with penalty method, we still view the beam as two parts of different materials and do not allow the influence domains in both subdomains to go across the interface. The penalty method is applied on Γ_s to enforce the connectivity.

The parameters of the beam in this case are the same as those in Example 6.4. Figure 6.19 shows the comparison between the analytical and numerical results for the beam deflection.

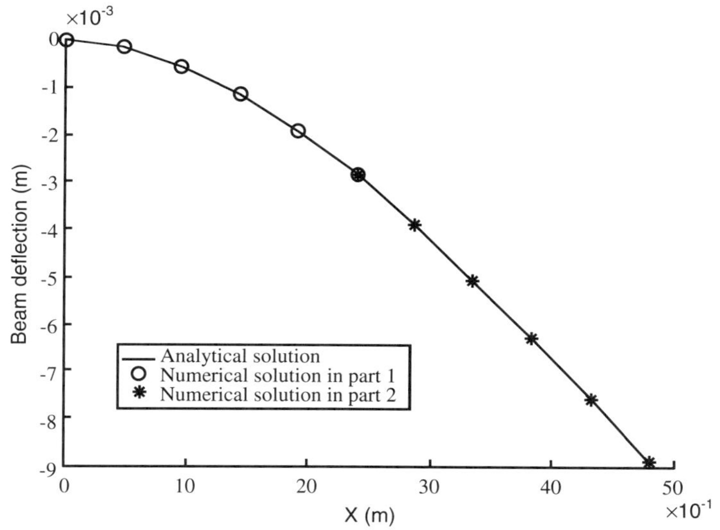

FIGURE 6.19
Analytical and numerical solutions for the deflection of the bi-material beam.

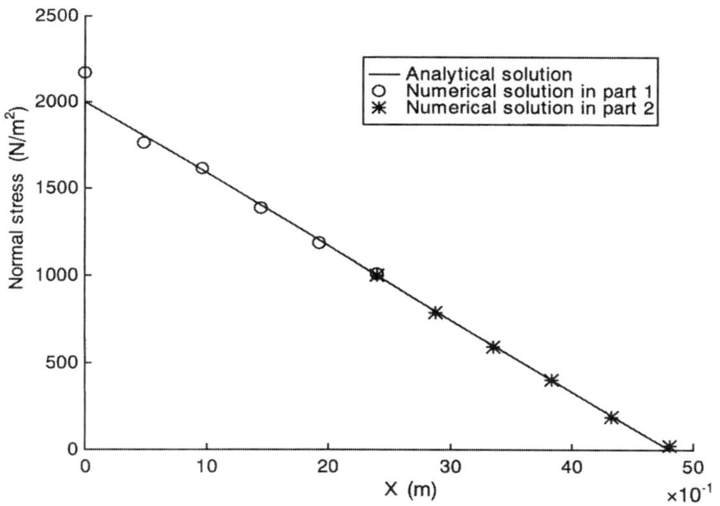

FIGURE 6.20
Analytical and numerical solutions for the normal stress at the upper surface ($y = D/2$) of the bi-material beam.

The solution for the normal stress σ_x on the beam sections at the upper surface of the beam is shown in Figure 6.20. These numerical solutions also exhibit good agreement between the MFree and the analytical results.

Example 6.6 Sandwich Composite Beam

In this example, a sandwich composite beam consisting of three layers of two materials, shown in Figure 6.21, is simulated. The two upper and lower surface layers are the same material thickness t_f and the thickness of the core material is denoted by t_c. The surface layer is stiffer than the core material, and all three layers are assumed to be perfectly connected together. The connection is enforced using the penalty method formulated in

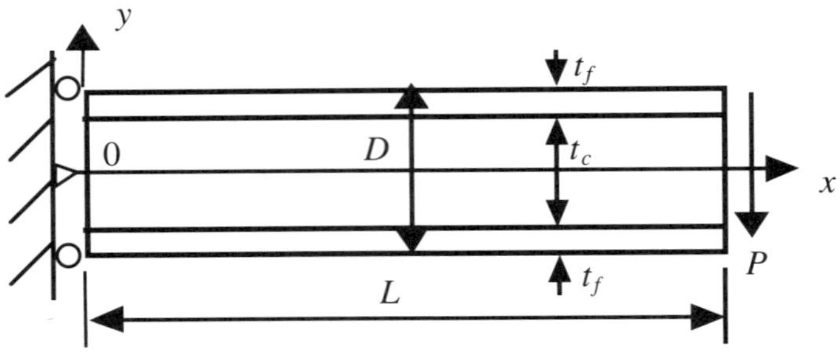

FIGURE 6.21
Sandwich composite cantilever beam.

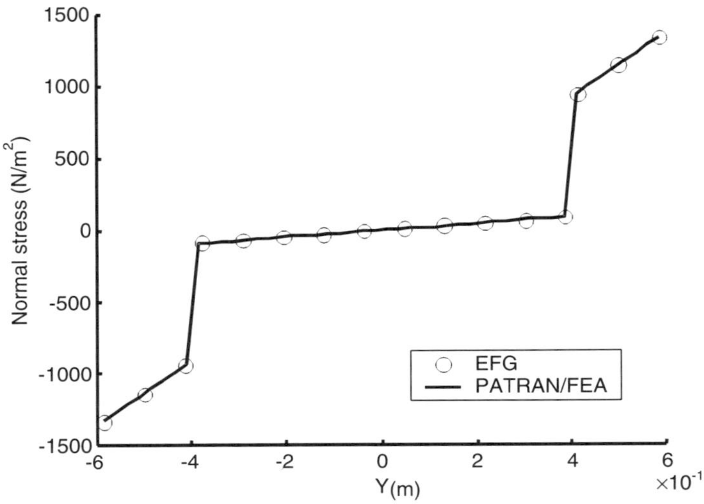

FIGURE 6.22
Numerical solutions for the normal stress at the section $x = L/2$ of the sandwich composite beam.

the above section. The parameters for this example are as follows:

Loading: $P = 100$ N
Young's modulus for the material of two surface layers: 1.67×10^9 N/m^2
Young's modulus for the core material: 1.67×10^8 N/m^2
Poisson's ratios for two materials: $\nu = 0.3$
Thickness of the two surface layers: $t_f = 3$
Thickness of the core layer: $t_c = 6$
Height of the beam: $D = 1.2$ m
Length of the beam: $L = 4.8$ m

The stresses calculated from the present method and PATRAN/FEA are compared in Figures 6.22 and 6.23. The normal stress σ_x at the section of $x = L/2$ is shown in Figure 6.22, and the shear stress τ_{xy} is shown in Figure 6.23. Very good agreement is observed. The discontinuity of the normal stress at the interface is clearly captured.

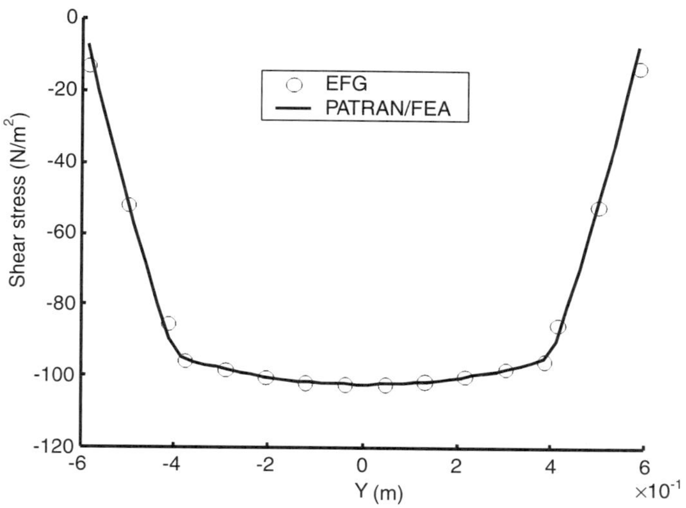

FIGURE 6.23
Numerical solutions for the shear stress at the section $x = L/2$ of the sandwich composite beam.

6.2.5 Remarks

In this section, the penalty method is used to impose the essential boundary and continuity conditions in the EFG method that uses MLS shape functions. It overcomes the drawbacks induced by the use of the method of Lagrange multipliers. The main advantage of the use of the penalty method is that it leads to a positive definite and banded stiffness matrix. The stiffness matrix also has a smaller dimension than those using Lagrangian multipliers, which improves computational efficiency. Numerical examples have demonstrated the performance of the penalty method.

The penalty method was also applied for the treatment of problems with material discontinuity. System equations for multibody problems are derived, and implemented in stress analysis in a composite beam. The numerical results agree well with the analytical and FEM solutions. This demonstrates the potential for application of the penalty method in MFree methods for analyzing structures of composite materials.

The methods for determining the penalty factor were given in Section 4.3.3. More discussion on this is provided in Example 6.12 in Section 6.4.5, when we deal with nonlinear problems.

6.3 Constrained Moving Least Square Method for EFG

As discussed, the root of the difficulty in imposing essential boundary conditions is the use of MLS approximation, which produces shape functions that do not satisfy the Kronecker delta function property. In areas that are far from the essential boundaries, the MLS shape function works just fine. The problems arise only for the nodes on the essential boundaries. If a procedure could be developed that produces MLS shape functions that possess the Kronecker delta function property only for the nodes on the essential boundaries, the problems would be overcome. The following presents an approach that uses constraints in the process of MLS approximation. The approach was originally proposed by G. R. Liu and Yang in 1999 and is termed the constrained moving least squares (CMLS) method. CMLS was first detailed in Yang's master's.

The basic idea of the CMLS is to impose the constraints of essential boundary conditions in the stage of shape function construction, so that the unconstrained Galerkin weak form can be used to produce a well-behaved equation system. The CMLS enforces the MLS approximation to pass through some desired values of the field variables at nodes at which the essential boundary condition is given. Thus, the approximation functions so obtained have the property of the delta function at the nodes only on the essential boundaries, and the treatment of the boundary conditions at the nodes can be as simple as in FEM. As shown later in this section, the system matrix established is banded, positive definite, and is of the same dimension as the equations produced by FEM.

6.3.1 Formulation

Let $u(\mathbf{x})$ be a function defined in the domain Ω and its approximation be $u^h(\mathbf{x})$. We are to approximate the function at a point of interest $\mathbf{x}$ using n nodes in the support domain of $\mathbf{x}$. The coordinates of the n nodes are defined by $\mathbf{x}_1,\ldots,\mathbf{x}_n$, where $\mathbf{x}_I = (x_I, y_I)$ in two dimensions. The nodal values of the function are denoted as

$$\mathbf{U}_s = \{u_1\ u_2\ \ldots\ u_n\}^T \tag{6.77}$$

The function at point $\mathbf{x}$ is approximated using m terms of monomials, i.e.,

$$u^h(\mathbf{x}) = \sum_J^m p_j(\mathbf{x})a_j(\mathbf{x}) \equiv \mathbf{p}^T(\mathbf{x})\mathbf{a}(\mathbf{x}) \tag{6.78}$$

where $\mathbf{p}(\mathbf{x})$ is a vector of monomial basis and $\mathbf{a}(\mathbf{x})$ is a vector of coefficients, which are the same as those in Equation 5.41.

Assume the approximation function $u^h(\mathbf{x})$ is required to be equal to the nodal values at $k(k \le n,\ k \le m)$ constrained nodes, which is written as a vector:

$$\mathbf{u}_b = \{\bar{u}_1\ \ \bar{u}_2\ \ \ldots\ \ \bar{u}_k\}^T \tag{6.79}$$

The nodal values of u_I at the remaining unconstrained nodes are written as

$$\mathbf{u}_m = \{u_{k+1}\ \ u_{k+2}\ \ \ldots\ \ u_n\}^T \tag{6.80}$$

Write Equation 6.78 in the form of subvectors as follows:

$$u^h(\mathbf{x}) = \mathbf{p}^T(\mathbf{x})\mathbf{a} = \{\mathbf{p}_b^T(\mathbf{x})\ \ \mathbf{p}_m^T(\mathbf{x})\}\begin{Bmatrix}\mathbf{a}_b\\ \mathbf{a}_m\end{Bmatrix} \tag{6.81}$$

where

$$\mathbf{p}_b^T(\mathbf{x}) = \{p_1(\mathbf{x})\ \ p_2(\mathbf{x})\ \ \ldots\ \ p_k(\mathbf{x})\} \tag{6.82}$$

$$\mathbf{p}_m^T(\mathbf{x}) = \{p_{k+1}(\mathbf{x})\ \ p_{k+2}(\mathbf{x})\ \ \ldots\ \ p_m(\mathbf{x})\} \tag{6.83}$$

and

$$\mathbf{a}_b = \{a_1 \ a_2 \ \ldots \ a_k\}^T \tag{6.84}$$

$$\mathbf{a}_m = \{a_{k+1} \ a_{k+2} \ \ldots \ a_m\}^T \tag{6.85}$$

We can then express k coefficients of $\mathbf{a}$, say, $\mathbf{a}_b = [a_1 \ a_2 \ \ldots \ a_k]^T$, in terms of the rest of the coefficients of $\mathbf{a}$, that is, $\mathbf{a}_m = [a_{k+1} \ \ldots \ a_m]^T$, from the constraint equations at k constrained nodes at the essential boundary. These constraint equations can be expressed as

$$\mathbf{u}_b = \{u^h(\mathbf{x}_1) \ u^h(\mathbf{x}_2) \ \ldots \ u^h(\mathbf{x}_k)\}^T \tag{6.86}$$

Substitute Equation 6.78 into Equation 6.86, and form subvectors as follows:

$$u^h(\mathbf{x}) = \mathbf{p}^T(\mathbf{x})\mathbf{a}(\mathbf{x}) = \{\mathbf{p}_b^T \ \mathbf{p}_m^T\}\begin{Bmatrix} \mathbf{a}_b \\ \mathbf{a}_m \end{Bmatrix} \tag{6.87}$$

we obtain

$$\mathbf{u}_b = \mathbf{P}_b\mathbf{a}_b + \mathbf{P}_m\mathbf{a}_m \tag{6.88}$$

where the moment matrix $\mathbf{P}_b$ is given by

$$\mathbf{P}_b = \begin{bmatrix} p_1(\bar{\mathbf{x}}_1) & p_2(\bar{\mathbf{x}}_1) & \cdots & p_k(\bar{\mathbf{x}}_1) \\ p_1(\bar{\mathbf{x}}_2) & p_2(\bar{\mathbf{x}}_2) & \cdots & p_k(\bar{\mathbf{x}}_2) \\ \mathrm{M} & \mathrm{M} & \ddots & \mathrm{M} \\ p_1(\bar{\mathbf{x}}_k) & p_2(\bar{\mathbf{x}}_k) & \cdots & p_k(\bar{\mathbf{x}}_k) \end{bmatrix} = \begin{bmatrix} \mathbf{p}_b^T(\bar{\mathbf{x}}_1) \\ \mathbf{p}_b^T(\bar{\mathbf{x}}_2) \\ \mathrm{M} \\ \mathbf{p}_b^T(\bar{\mathbf{x}}_k) \end{bmatrix} \tag{6.89}$$

$$\mathbf{P}_m = \begin{bmatrix} p_{k+1}(\bar{\mathbf{x}}_1) & \cdots & p_m(\bar{\mathbf{x}}_1) \\ p_{k+1}(\bar{\mathbf{x}}_2) & \cdots & p_m(\bar{\mathbf{x}}_2) \\ \mathrm{M} & \ddots & \mathrm{M} \\ p_{k+1}(\bar{\mathbf{x}}_k) & \cdots & p_m(\bar{\mathbf{x}}_k) \end{bmatrix} = \begin{bmatrix} \mathbf{p}_m^T(\bar{\mathbf{x}}_1) \\ \mathbf{p}_m^T(\bar{\mathbf{x}}_2) \\ \mathrm{M} \\ \mathbf{p}_m^T(\bar{\mathbf{x}}_k) \end{bmatrix} \tag{6.90}$$

We can then express k coefficients of $\mathbf{a}_b$ in terms of the rest of the coefficients of $\mathbf{a}_m$ using Equation 6.88, i.e.,

$$\mathbf{a}_b = \mathbf{C}_1 \cdot \mathbf{u}_b - \mathbf{C}_2 \cdot \mathbf{a}_m \tag{6.91}$$

where

$$\mathbf{C}_1 = \mathbf{P}_b^{-1} \tag{6.92}$$

$$\mathbf{C}_2 = \mathbf{P}_b^{-1} \cdot \mathbf{P}_m \tag{6.93}$$

The vector of coefficients, $\mathbf{a}_m$, is determined by the conventional MLS method over the "free" (unconstrained) nodes. At each free point $\mathbf{x}$ corresponding to $\mathbf{u}_m$, $\mathbf{a}_m$ are chosen to minimize the weighted residual (note that $\mathbf{a}_b$ is determined by Equation 6.91 and should be considered a constant in the weighted residual):

$$J = \sum_{I=k+1}^{n} \widehat{W}(\mathbf{x}-\mathbf{x}_I)[\mathbf{p}^T(\mathbf{x}_I)\mathbf{a}(\mathbf{x}) - u_I]^2 \tag{6.94}$$

where $n-k$ is the number of free nodes, $w(\mathbf{x}-\mathbf{x}_I)$ is a weight function, and u_I is the nodal value at node I ($I = k+1, k+2,\ldots,n$). The minimization of J with respect to coefficients $\mathbf{a}_m$ results in the following linear system:

$$\mathbf{A}_1(\mathbf{x})\cdot\mathbf{a}_b + \mathbf{A}_2(\mathbf{x})\cdot\mathbf{a}_m = \mathbf{B}(\mathbf{x})\mathbf{u}_m \tag{6.95}$$

where

$$\mathbf{A}_1(\mathbf{x}) = \sum_{I=k+1}^{n} \widehat{W}_I(\mathbf{x})\mathbf{p}_m(\mathbf{x}_I)\mathbf{p}_b^T(\mathbf{x}_I) \tag{6.96}$$

$$\mathbf{A}_2(\mathbf{x}) = \sum_{I=k+1}^{n} \widehat{W}_I(\mathbf{x})\mathbf{p}_m(\mathbf{x}_I)\mathbf{p}_m^T(\mathbf{x}_I) \tag{6.97}$$

$$\mathbf{B}(\mathbf{x}) = [\widehat{W}_1(\mathbf{x})\mathbf{p}_m(\mathbf{x}_1), \widehat{W}_2(\mathbf{x})\mathbf{p}_m(\mathbf{x}_2),\ldots,\widehat{W}_n(\mathbf{x})\mathbf{p}_m(\mathbf{x}_{n-k})] \tag{6.98}$$

and

$$\widehat{W}_I(\mathbf{x}) \equiv \widehat{W}(\mathbf{x}-\mathbf{x}_I) \tag{6.99}$$

From Equations 6.91 and 6.95, we can obtain $\mathbf{a}_b$ and $\mathbf{a}_m$

$$\mathbf{a}_b = \mathbf{E}_1\mathbf{u}_b + \mathbf{E}_2\mathbf{u}_m \tag{6.100}$$

and

$$\mathbf{a}_m = \mathbf{D}_1\mathbf{u}_b + \mathbf{D}_2\mathbf{u}_m \tag{6.101}$$

where

$$\mathbf{D}_1 = (\mathbf{A}_1\cdot\mathbf{C}_2 - \mathbf{A}_2)^{-1}\cdot\mathbf{A}_1\cdot\mathbf{C}_1 \tag{6.102}$$

$$\mathbf{D}_2 = -(\mathbf{A}_1\cdot\mathbf{C}_2 - \mathbf{A}_2)^{-1}\cdot\mathbf{B} \tag{6.103}$$

$$\mathbf{E}_1 = \mathbf{C}_1 - \mathbf{C}_2\cdot\mathbf{D}_1 \tag{6.104}$$

$$\mathbf{E}_2 = -\mathbf{C}_2\cdot\mathbf{D}_1 \tag{6.105}$$

The coefficient vector $\mathbf{a}$ can be expressed as

$$\mathbf{a} = \begin{Bmatrix}\mathbf{a}_b\\ \mathbf{a}_m\end{Bmatrix} = \begin{bmatrix}\mathbf{E}_1\\ \mathbf{D}_1\end{bmatrix}\mathbf{u}_b + \begin{bmatrix}\mathbf{E}_2\\ \mathbf{D}_2\end{bmatrix}\mathbf{u}_m = \begin{bmatrix}\mathbf{E}_1 & \mathbf{E}_2\\ \mathbf{D}_1 & \mathbf{D}_2\end{bmatrix}\begin{Bmatrix}\mathbf{u}_b\\ \mathbf{u}_m\end{Bmatrix} \tag{6.106}$$

Hence, we have

$$u^h(\mathbf{x}) = \mathbf{p}^T(\mathbf{x})\mathbf{a}(\mathbf{x}) = [\mathbf{p}_b^T\ \mathbf{p}_m^T]\begin{Bmatrix}\mathbf{a}_b\\ \mathbf{a}_m\end{Bmatrix} = [\mathbf{p}_b^T\ \mathbf{p}_m^T]\begin{bmatrix}\mathbf{E}_1 & \mathbf{E}_2\\ \mathbf{D}_1 & \mathbf{D}_2\end{bmatrix}\begin{Bmatrix}\mathbf{u}_b\\ \mathbf{u}_m\end{Bmatrix} \tag{6.107}$$

or

$$u^h(\mathbf{x}) = [\boldsymbol{\phi}_b \quad \boldsymbol{\phi}_m]\begin{Bmatrix} \mathbf{u}_b \\ \mathbf{u}_m \end{Bmatrix} = \boldsymbol{\Phi}\mathbf{U}_s \tag{6.108}$$

where

$$\boldsymbol{\phi}_b = \mathbf{p}_b^T\mathbf{E}_1 + \mathbf{p}_m^T\mathbf{D}_1 \tag{6.109}$$

and

$$\boldsymbol{\phi}_m = \mathbf{p}_b^T\mathbf{E}_2 + \mathbf{p}_m^T\mathbf{D}_2 \tag{6.110}$$

To determine the derivatives from the displacement Equation 6.108, it is necessary to obtain the shape function derivatives. The spatial derivatives of the shape functions are obtained by

$$\boldsymbol{\phi}_{,x} = [\boldsymbol{\phi}_{b,x} \quad \boldsymbol{\phi}_{m,x}] \tag{6.111}$$

where

$$\boldsymbol{\phi}_{b,x} = \mathbf{p}_{b,x}^T\mathbf{E}_1 + \mathbf{p}_b^T\mathbf{E}_{1,x} + \mathbf{p}_{m,x}^T\mathbf{D}_1 + \mathbf{p}_m^T\mathbf{D}_{1,x} \tag{6.112}$$

and

$$\boldsymbol{\phi}_{m,x} = \mathbf{p}_{b,x}^T\mathbf{E}_2 + \mathbf{p}_b^T\mathbf{E}_{2,x} + \mathbf{p}_{m,x}^T\mathbf{D}_2 + \mathbf{p}_m^T\mathbf{D}_{2,x} \tag{6.113}$$

6.3.2 Constrained Surfaces Generated by CMLS

To verify CMLS, two examples are presented, the first to show how the CMLS works in surface fitting with or without constraints on the boundary using the formulation developed in the previous subsection.

Example 6.7 Linear Constraint

The first example examines a flat surface fitted before and after a linear constraint is imposed, and the results are shown in Figures 6.24 and 6.25. The surface is parallel to the x–y plane before the application of a linear constraint to the nodes on the boundary (see Figure 6.24). After the imposition of the constraint, the flat surface follows these values at the boundary, as shown in Figure 6.25. This demonstrates that CMLS works well in imposing linear constraints on nodes.

Example 6.8 Parabolic Constraint

The second example examines the same situation, but a parabolic constraint is considered. The results are shown in Figures 6.26 and 6.27. These two examples demonstrate that the surfaces are enforced to pass through the constraint points, while the remaining part of the surface in the position far from the constraint points still possesses the property of a conventional MLS surface. This further confirms that the use of the CMLS algorithm works well in the imposition of constraints on boundary nodes.

It should be noted that the requirement for the order of the monomials in the basis used in CMLS is higher than that of MLS because some of the coefficients are determined by

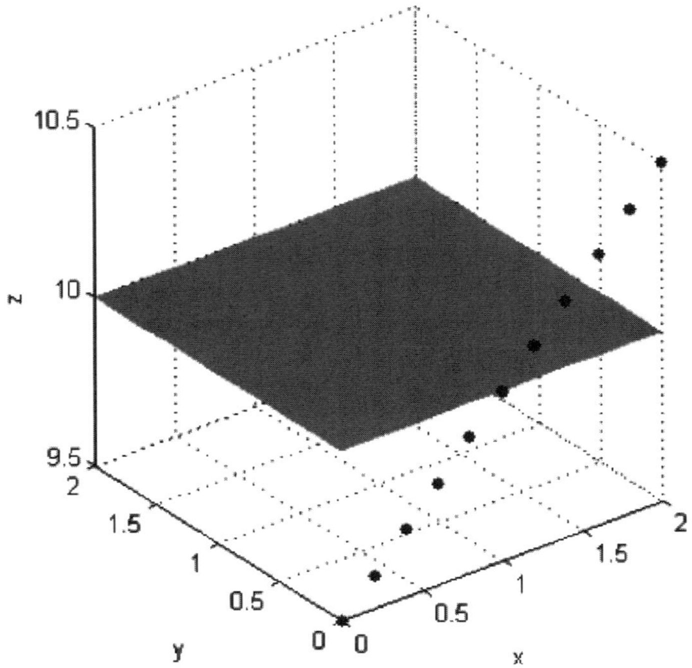

FIGURE 6.24
2D CMLS approximation function before the linear constraint is imposed.

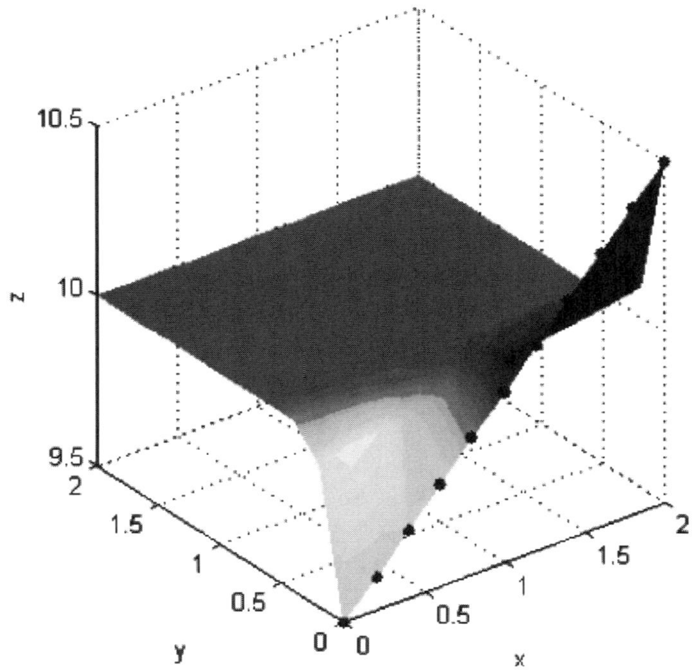

FIGURE 6.25
2D CMLS approximation function after the linear constraint is imposed.

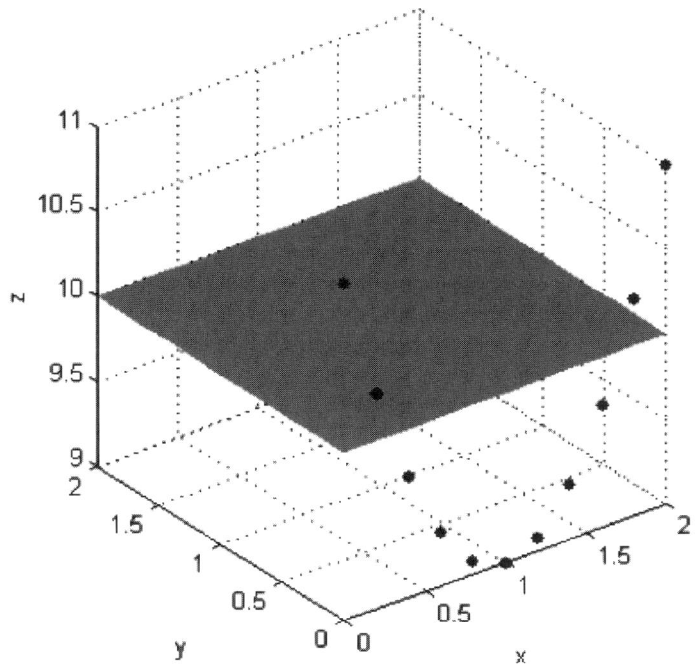

FIGURE 6.26
2D CMLS approximation function before the parabolic constraint is applied.

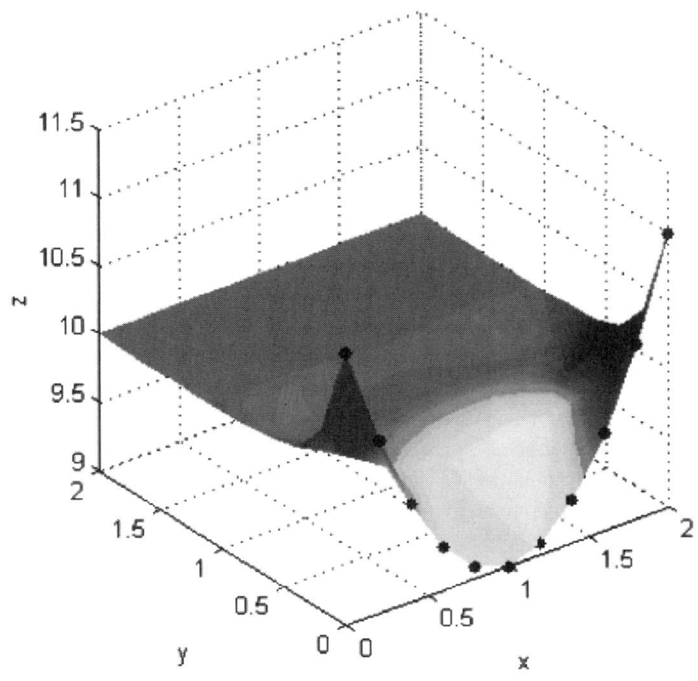

FIGURE 6.27
2D CMLS approximation function after the parabolic constraint is applied.

nodal constraints. In general, the greater the number of constrained points included in the support domain, the greater the number of terms of basis monomials of higher orders that should be used. The number of the polynomial coefficients determined by the nodal enforcement and the number by MLS should be balanced. For example, for a polynomial with six coefficients, it is best that three of the coefficients be determined by nodal enforcement and the rest by MLS. From our experience with 2D problems, the appropriate basis function should be polynomials of order 2 or 3. Including more free nodes can help to reduce the chance of producing singular moment matrices in the process of computing CMLS shape functions.

6.3.3 Weak Form and Discrete Equations

Consider again the mechanics problem stated in Equations 6.1 and 6.2, which was also dealt with in Section 6.2.1. Instead of using the penalty method, we now use CMLS to generate the shape functions that possess the Kronecker delta function property at the nodes on the essential boundaries. The prescribed displacement on the essential boundary can then be imposed directly, as in conventional FEM.

Using CMLS, the Galerkin weak form is as simple as in FEM, because the shape function created by CMLS has the same property as the shape functions of FEM at the essential boundary points. Based on Equation 6.67, we write

$$\int_{\Omega} \delta(\mathbf{Lu})^T(\mathbf{cLu})\mathrm{d}\Omega - \int_{\Omega} \delta \mathbf{u}^T \mathbf{b}\mathrm{d}\Omega - \int_{\Gamma_t} \delta \mathbf{u}^T \bar{\mathbf{t}}\,\mathrm{d}\Gamma = 0 \tag{6.114}$$

Note that the integration on the essential boundary has been removed, because the displacement function $\mathbf{u}$ is to be approximated using the CMLS approximation shown in Equation 6.108. Substituting the approximation of $\mathbf{u}$ into the weak form Equation 6.114 yields the discrete system equations:

$$\mathbf{KU} = \mathbf{F} \tag{6.115}$$

where $\mathbf{U}$ is a vector of nodal parameters of displacements for all the nodes in the problem domain, and $\mathbf{K}$ is the stiffness matrix assembled using the following nodal matrix:

$$\mathbf{K}_{ij} = \int_{\Omega} \mathbf{B}_i^T \mathbf{c} \mathbf{B}_j \mathrm{d}\Omega \tag{6.116}$$

where $\mathbf{B}_i$ and $\mathbf{c}$ have the same form as those in Equation 6.38, except that the shape functions used are different. The force vector $\mathbf{F}$ is assembled using the nodal force vector defined in Equation 6.31 but computed using the shape functions defined in this section.

The handling of essential boundary conditions at the boundary nodes is the same as in FEM. All one need do is impose the boundary condition directly to the nodal displacement in the final system equation. Using CMLS is very computationally efficient, especially for large systems, as the banded feature as well as the symmetry of the stiffness matrix is preserved. The drawback of this method is that the possibility of having a singular moment matrix is increased because fewer free nodes are used. Another drawback is that it is difficult to ensure the compatibility in the field function approximation, which leads to difficulty in passing the patch tests as discussed in the next section. Note that CMLS is used only for support domains that have at least one essential boundary point.

The Gauss quadrature scheme is still required to perform the integrations in computing system matrices. Ensuring accurate numerical integration when using CMLS is also more difficult compared with the use of MLS because the order of the CMLS shape functions created is usually higher.

6.3.4 Examples for Mechanics Problems

Example 6.9 Patch Test

The first numerical example of solid mechanics problem is the standard patch test. As square patch of $L_x = 2$ and $L_y = 2$ shown in Figure 6.28 is considered. The displacements are prescribed on all outside boundaries of the patch by a linear function defined by Equation 6.25. The nodal arrangement in this patch is also shown in Figure 6.28. A discrete system equation in the form of Equation 6.115 is established using the CMLS shape functions. Gauss quadrature is used to perform the integration. Table 6.7 shows the numerical results obtained using a background mesh of 40 × 40 quadrature cells with a 6 × 6 Gauss point each. The maximum errors of u_x and u_y are of order 10^{-7} and 10^{-6}, respectively.

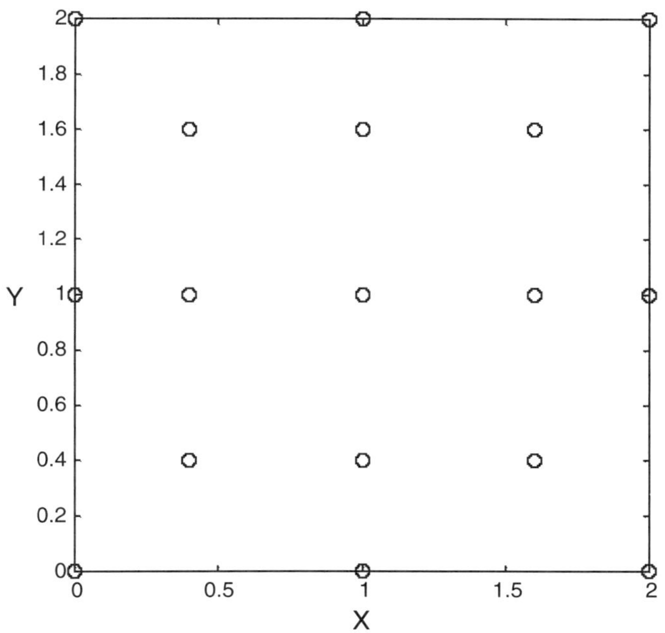

FIGURE 6.28
The nodal arrangement for the patch test.

TABLE 6.7
Numerical Results of the Displacements at Interior Nodes for the Patch Test

Node and Coordinates	u_x	u_y
9, (0.4,0.4)	0.40000024144004	0.40000024434450
10, (1,0.4)	1.00000008122310	0.39999996975363
11, (1.6,0.4)	1.59999956772840	0.39999902865713
12, (0.4,1)	0.40000058611168	1.00000023797684
13, (1,1)	1.00000009865193	0.99999996948019
14, (1.6,1)	1.59999981657464	0.99999978568420
15, (0.4,1.6)	0.39999990454283	1.60000015089939
16, (1,1.6)	0.99999979845197	1.59999992408260
17, (1.6,1.6)	1.59999922275895	1.59999878285197

TABLE 6.8

Maximum Error of the Displacements for a Patch Test for EFG-CMLS

Integration Cells (quadrature points)	Maximum Error of u_x	Maximum Error of u_y
10 × 10 (4 × 4)	2.6×10^{-3}	1.8×10^{-3}
20 × 20 (4 × 4)	1.4×10^{-4}	5.2×10^{-4}
40 × 40 (4 × 4)	1.5×10^{-5}	2.1×10^{-5}
40 × 40 (6 × 6)	8×10^{-7}	1.6×10^{-6}
8 × 8 (16 × 16)	5.2×10^{-5}	4×10^{-5}

Hence, the patch test is passed to the machine accuracy in this case. Selection of nodes often affects the results of the patch test. We tested a number of other patches of irregular internal nodes, and found that some cases had difficulty passing the patch test. We believe that the reason lies mainly in the compatibility of the field function approximation using CMLS. The nature of this problem is somewhat similar to that using PIM shape functions, which are discussed in detail in Chapter 8.

We found that the accuracy of the numerical results depends heavily on the accuracy of the numerical integration, as shown in Table 6.8. This is because the order of the shape functions created by CMLS is usually higher than those of MLS. Note that the Gauss quadrature is designed for integrating polynomial functions. The quadrature error may increase when it is applied to complex integrands like that given in Equation 6.116 because integrands are fractional functions, which in general cannot be represented exactly by polynomial functions. The Gauss quadrature, therefore, will not be able to produce exact results for the integration. An efficient and accurate numerical integration scheme should be used in EFG-CMLS to pass the patch test.

Example 6.10 Cantilever Beam

For benchmarking purposes, we consider again a beam of characteristic length L and height D subjected to a parabolic traction at the free end, as shown in Figure 6.4. The beam is considered to be of unit thickness, and the plane stress problem is considered. The exact solution is given by Equations 6.51 to 6.56 for displacements and stresses. The parameters used in this section are the same as in Example 6.4.

The arrangement of nodes and quadrature cells is shown in Figure 6.29. In each quadrature cell, 4 × 4 Gauss points are used. The solutions are obtained using a quadratic basis function with cubic spline weight function, and support domains of $\alpha_s = 2.5$ are used.

Figure 6.30 plots the analytical solution and the numerical solution using the present method for the beam deflection along the x axis. The plot shows excellent agreement between the analytical and present numerical results using CMLS.

Figure 6.31 illustrates the distribution of the normal stresses σ_x on the cross section at $x = L/2$ of the beam. Both the analytical solution and the present EFG-CMLS solution are plotted together for comparison. Very good agreement is observed between the stresses calculated by the analytical formulation and the present EFG-CMLS method. Figures 6.32 and 6.33 show the same comparison for, respectively, the normal stress σ_y and the shear stress τ_{xy} at the section of $x = L/2$ of the beam. Again, very good agreement is observed between the results calculated by the analytical formulation and the present EFG-CMLS method.

Table 6.9 compares the numerical and analytical results for the vertical displacement at point A on the beam (see Figure 6.4). The calculation was performed for models discretized with 18, 24, 55, and 189 nodes. This table shows that the numerical result converges as the number of nodes increases.

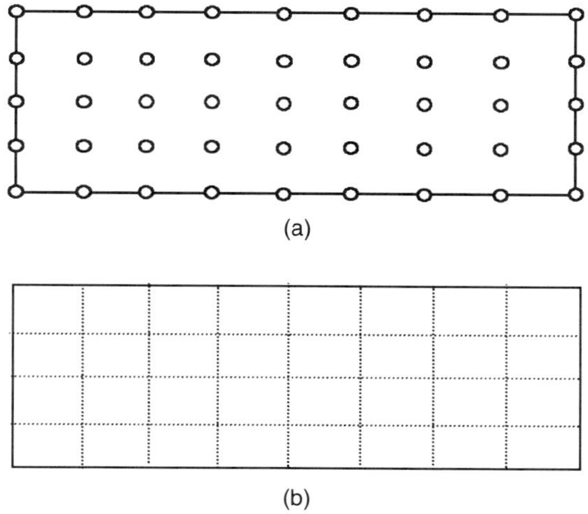

FIGURE 6.29
(a) Nodal arrangement; (b) mesh used for integration.

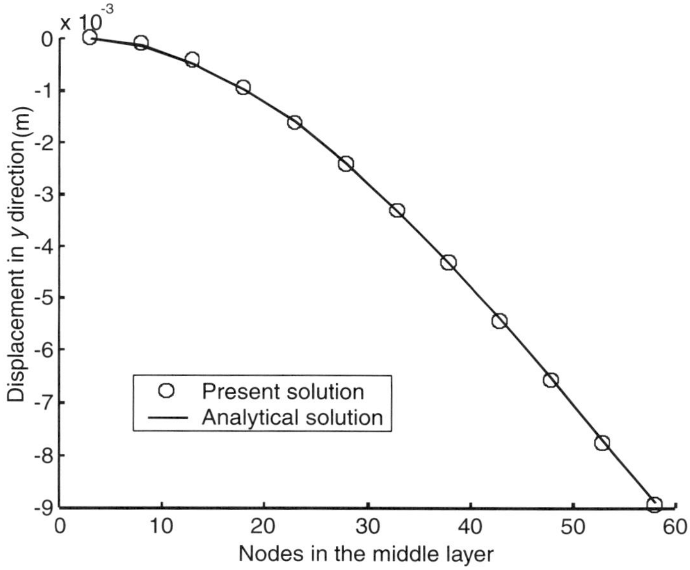

FIGURE 6.30
Analytical and EFG-CMLS numerical solutions for the deflection of the cantilever beam.

Example 6.11 Hole in an Infinite Plate

A plate with a circular hole subjected to a unidirectional tensile load in the x direction is considered, as shown in Figure 6.34. The plane stress condition is assumed. Due to symmetry, only the upper right quarter of the plate is modeled, as shown in Figure 6.35. Corresponding symmetric boundary conditions are applied on $x = 0$ and $y = 0$, i.e.,

$$u_x = 0, \qquad \sigma_{xy} = 0 \quad \text{when} \;\; x = 0 \tag{6.117}$$

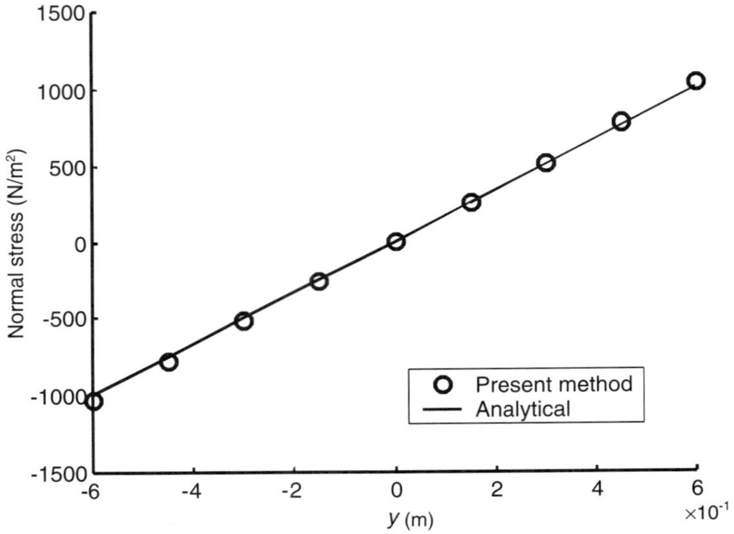

FIGURE 6.31
Analytical and the present EFG-CMLS numerical solutions for σ_x at the section of $x = L/2$ of the cantilever beam.

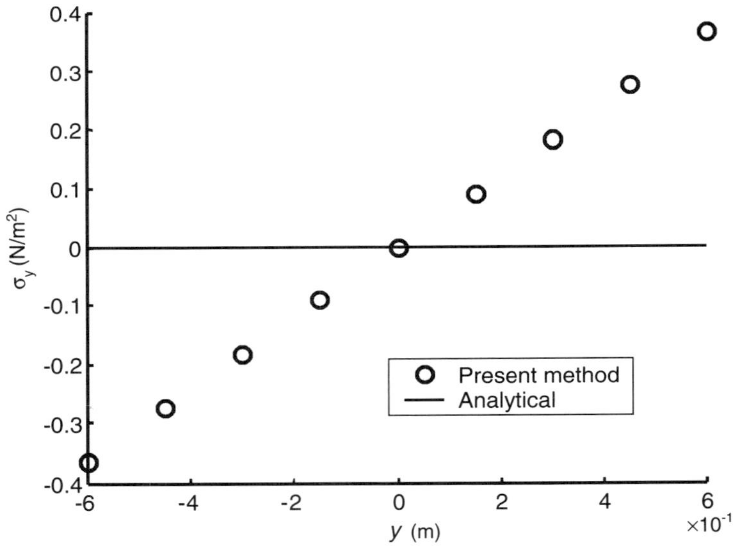

FIGURE 6.32
Analytical and the present EFG-CMLS numerical solutions for σ_y at the section of $x = L/2$ of the cantilever beam.

and

$$u_y = 0, \qquad \sigma_{xy} = 0 \quad \text{when} \;\; y = 0 \tag{6.118}$$

The boundary condition at the right edge is

$$\sigma_{xx} = p, \qquad \sigma_{yy} = \sigma_{xy} = 0 \quad \text{when} \;\; x = 5 \tag{6.119}$$

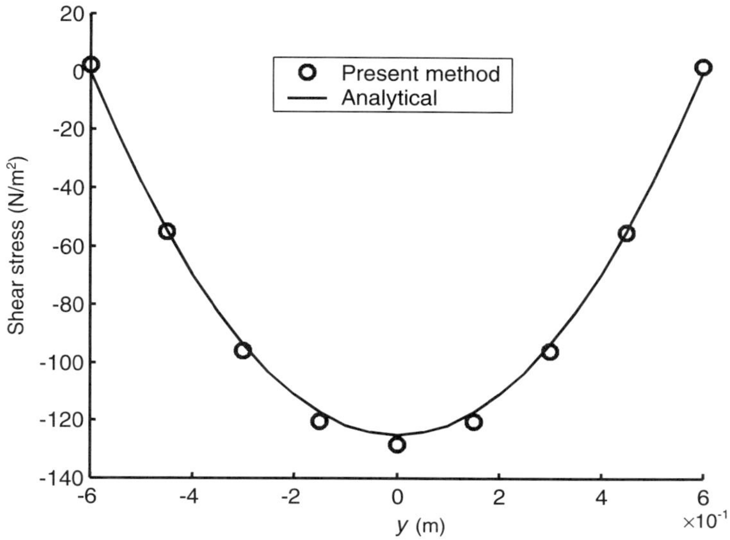

FIGURE 6.33
Analytical and EFG-CMLS numerical solutions for τ_{xy} at the section of $x = L/2$ of the cantilever beam.

TABLE 6.9

Comparison of Vertical Displacement at End of Beam

Number of Nodes	u_y (m) Exact	u_y (m) (EFG-CMLS)	%Error
18	–0.0089	–0.00792	11
24	–0.0089	–0.00837	6
55	–0.0089	–0.00887	0.34
189	–0.0089	–0.00891	0.1

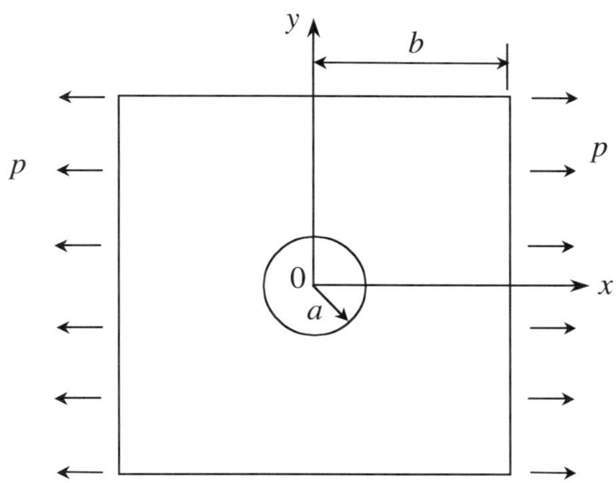

FIGURE 6.34
Plate with a hole subjected to a tensile load in the horizontal direction.

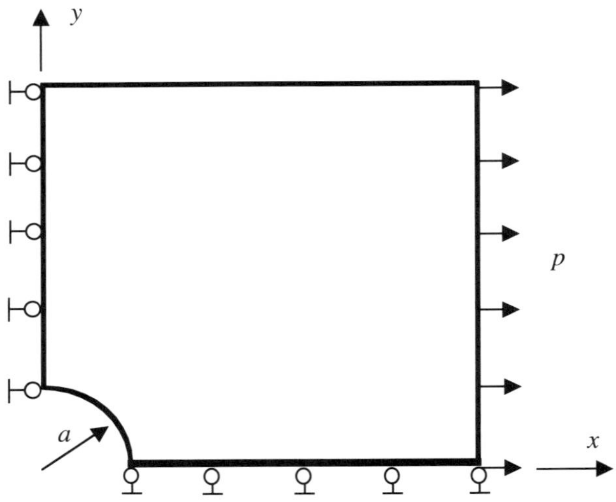

FIGURE 6.35
Quarter model of the plate with a hole subjected to a tensile load in the horizontal direction.

and the boundary condition at the upper edge is

$$\sigma_{xx} = 0, \qquad \sigma_{yy} = \sigma_{xy} = 0 \quad \text{when} \quad y = 5 \tag{6.120}$$

The parameters are listed as follows:

Loading: $p = 1$ N/m
Young's modulus: $E = 1.0 \times 10^3$ N/m^2
Poisson's ratio: $\nu = 0.3$
Height of the beam: $a = 1.0$ m
Length of the beam: $b = 5$ m

Symmetry conditions are imposed on the left and bottom edges, and the inner boundary of the hole is traction free. The tensile load in the x direction is imposed on the right edge. The exact solution for the stresses within the infinite plate is given by the following equations:

The displacement in the radial direction is given by

$$u_r = \frac{\sigma}{4\mu}\left\{ r\left[\frac{\kappa - 1}{2} + \cos 2\theta\right] + \frac{a^2}{r}[1 + (1 + \kappa)\cos 2\theta] - \frac{a^4}{r^3}\cos 2\theta \right\} \tag{6.121}$$

and the displacement in the tangent direction can be calculated using

$$u_\theta = \frac{\sigma}{4\mu}\left[(1 - \kappa)\frac{a^2}{r} - r - \frac{a^4}{r^3}\right]\sin 2\theta \tag{6.122}$$

where

$$\mu = \frac{E}{2(1 + \nu)} \qquad \kappa = \begin{cases} 3 - 4\nu & \text{plane strain} \\ \dfrac{3 - \nu}{1 + \nu} & \text{plane stress} \end{cases} \tag{6.123}$$

The normal stress in the x direction can be obtained using

$$\sigma_x(x, y) = 1 - \frac{a^2}{r^2}\left\{\frac{3}{2}\cos 2\theta + \cos 4\theta\right\} + \frac{3a^4}{2r^4}\cos 4\theta \tag{6.124}$$

The normal stress in the y direction is

$$\sigma_y(x, y) = -\frac{a^2}{r^2}\left\{\frac{1}{2}\cos 2\theta - \cos 4\theta\right\} - \frac{3a^4}{2r^4}\cos 4\theta \tag{6.125}$$

and the shear stress is given by

$$\tau_{xy}(x, y) = -\frac{a^2}{r^2}\left\{\frac{1}{2}\sin 2\theta - \sin 4\theta\right\} + \frac{3a^4}{2r^4}\sin 4\theta \tag{6.126}$$

where (r, θ) are the polar coordinates and θ is measured counterclockwise from the positive x axis. When the condition $b/a > 5$ is satisfied, the solution of a finite plate should be very close to that of an infinite plate. Therefore, the analytical results given in Equations 6.124 to 6.126 are employed as the reference results for comparison.

The CMLS method is used to perform the stress analysis. Two kinds of nodal arrangement are used, as shown in Figure 6.36. The results obtained using the present CMLS method for stress σ_x at $x = 0$ are plotted in Figure 6.37 together with the analytical results for the infinite plate. Figure 6.37 shows that the present CMLS method gives satisfactory results for the problem. The figure also shows that, as the number of the node increases, the results obtained are closer to the analytical solution.

6.3.5 Computational Time

The main advantage of the CMLS method is that it does not increase the number of the unknowns and leads to a banded stiffness matrix. Therefore, it is computationally cheaper than EFG with Lagrange multipliers. Table 6.10 compares the CPU time used in the EFG code, when the present EFG-CMLS method, MLS with Lagrange multipliers, and MLS with penalty method are used for imposing the essential boundary condition in solving the cantilever beam problem. The comparison was done on an HP UNIX workstation. The table shows that the CMLS method saves a significant amount of CPU time compared with MLS with Lagrange multipliers in EFG formulation. However, the CMLS method is

TABLE 6.10

CPU Time for EFG Code Using Different Method of Imposing Essential Boundary Conditions

	CPU Time (s)		
Nodes	**EFG (MLS + Lagrange Multiplier)**	**EFG (Penalty)**	**EFG (CMLS)**
55	1.1	0.6	0.7
189	35.4	3.5	9.3
561	115.2	13.8	30.8

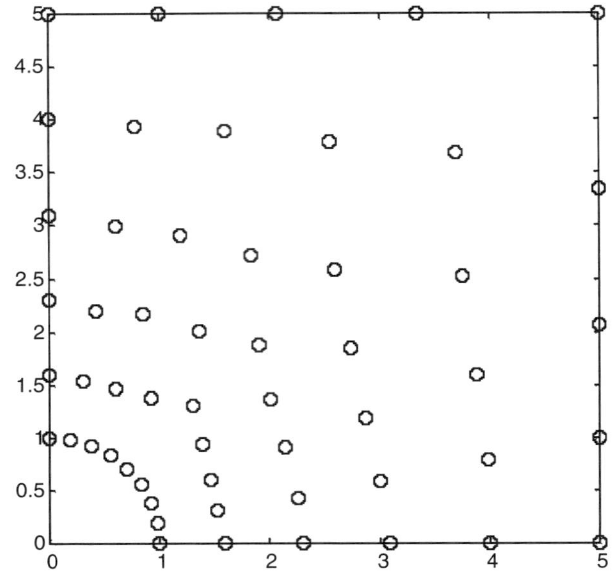

(a)

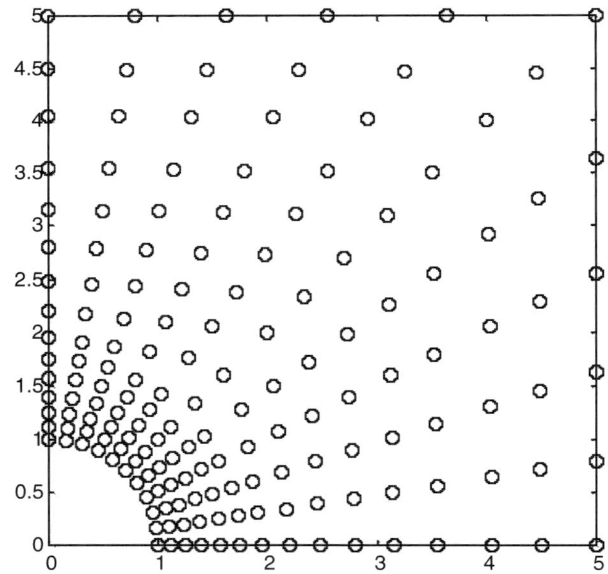

(b)

FIGURE 6.36
Nodal arrangement for the infinite plate with a central circular hole. (a) 54 nodes; (b) 165 nodes.

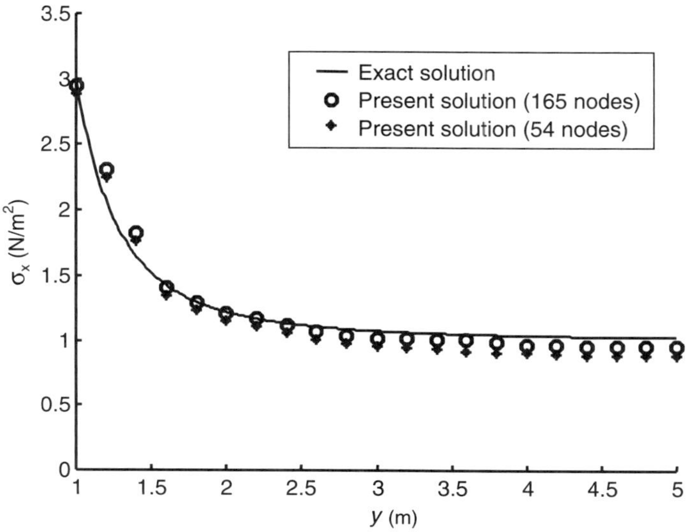

FIGURE 6.37
Comparison between the exact and present EFG-CMLS solution for stresses σ_x at $x = 0$.

less efficient than the penalty method. The reasons are (1) much time is used for computing the CMLS shape function compared with the MLS shape function, and (2) the number of integration points needs to increase because the CMLS shape functions are more complex than those of MLS. The advantage of CMLS over the penalty method may be that there is no need to choose the penalty factor. For large systems, the difference in CPU time between CMLS and the penalty method is expected to decrease as the majority of the CPU time is used for solving the system equation, which should be the same for both cases when CMLS and the penalty method are used.

6.3.6 Remarks

In this section, a technique called the constrained moving least squares method is introduced to construct the shape functions for MFree methods. CMLS enforces the approximation functions to pass through data points wherever necessary, while the usual MLS approximation is applied in other areas. CMLS treats the essential conditions at the stage of constructing shape functions, which simplifies the system equations. Hence, the system equations derived using the CMLS shape functions are positive definite and banded. The treatment of essential boundary conditions is as simple as in FEM. EFG-CMLS has problems passing the patch test, due perhaps to the incompatibility of CMLS approximation. More detailed study may be needed to fully realize the idea of CMLS. A method for solving the incompatibility issue for PIM shape functions is discussed in Chapter 8.

The idea of allowing certain conditions of a problem to be satisfied in the stage of constructing approximation functions should be explored further. MFree procedures for creating shape functions provide a lot of flexibility to meet different kinds of demands, not only on accuracy but also on constraints.

Before moving to the next section, it may be mentioned here that the imposition of essential boundary conditions can also be performed using finite elements attached to the MFree mesh on the portion of the essential boundaries (Krongauz and Belytschko, 1996). Coupling EFG with other numerical methods (see Chapter 13) is another alternative.

6.4 EFG for Nonlinear Elastic Problems

In engineering practice, problems are often nonlinear. Nonlinear problems in mechanics can be divided into two large categories: material and geometric nonlinear problems. There are also problems of both nonlinearities. In material nonlinear problems, the problems are often idealized as nonlinear elastic problems and elasto-plastic problems. The former assumes there is no residual stress when the material is unloaded, and the latter assumes that there are residual strains after the unloading but that it follows a simple "elasto-plastic" model. This section deals with elastic material nonlinear problems.

FEM has been the major computational tool to deal with nonlinear elastic problems and a large literature focuses on using conventional FEM to handle this kind of problem. However, very few publications exist on MFree methods. There is still a long distance to go to develop MFree packages that can perform the job of current FEM packages in solving nonlinear problems.

Methods and schemes developed in FEM can be modified and used in MFree methods. In dealing with nonlinear problems, the most common approach used in FEM is the incremental iteration approach, in which the loading on the structure is divided into a number of steps. At each step, an iterative method is used to obtain the displacement and stress–strain status. This incremental iteration approach is employed here for our EFG formulation.

This section introduces an autocorrector algorithm based on the EFG method, which can always ensure global equilibrium at each load step. This work was performed by Wang and Liu in 2000. In their formulation, a midpoint method is implemented for the constitutive law of nonlinear behaviors in both shear and volumetric deformation. The failure mode follows the Mohr–Coulomb law, which is widely accepted for frictional materials. This section is organized as follows. First, a nonlinear boundary-value problem and its autocorrection weak form in an incremental form are stated. An improved integration scheme for a nonlinear constitutive law, an *EB* model that is widely used in the community of foundation engineering, is employed to model the material. The numerical strategy is discussed in the iteration process during each incremental loading step. Finally, foundation problems are studied using the EFG formulation and compared with finite element results.

6.4.1 Basic Equations

The incremental method can be simply stated as follows.

Assume that the variables of stress, strain, displacement, force, and boundary conditions are all known at the beginning of a time interval $[t, t + \Delta t]$ and the force and boundary conditions are known at time $t + \Delta t$. Time t here refers to a point at loading history (not the physical time space). We are interested in determining the value of stress, strain, and displacement at time $t + \Delta t$. The problem is described by the following equations.

System Equations

- Equilibrium equation at any point in the problem domain:

$$\frac{\partial \sigma_{ij}}{\partial x_j} + b_i = 0 \quad \text{in } \Omega \tag{6.127}$$

where σ_{ij} is the stress tensor and b_i is the (external) body force. This equilibrium equation needs to be satisfied at all times. Based on the status at time t, we seek now the equilibrium status at time $t+\Delta t$. The incremental form of the equilibrium equations is expressed as

$$\frac{\partial \Delta\sigma_{ij}}{\partial x_j} + \Delta b_i = -\left(\frac{\partial \sigma_{ij}^t}{\partial x_j} + b_i^t\right) \quad \text{in } \Omega \tag{6.128}$$

where $\Delta\sigma_{ij}$ and Δb_i are, respectively, the stress and body force increments from time t to $t + \Delta t$. On the right-hand side of Equation 6.128, σ_{ij}^t is the stress tensor and b_i^t is the (external) body force at time t. The right-hand side is thus the unbalanced force at time t. If the equilibrium equation is precisely satisfied at time t, this unbalanced force term vanishes.

- The geometric relationship between the displacements u_i and the strains ε_{ij} in the incremental form can be written as

$$\Delta\varepsilon_{ij} = \frac{1}{2}\left(\frac{\partial \Delta u_i}{\partial x_j} + \frac{\partial \Delta u_j}{\partial x_i}\right) \quad \text{in } \Omega \tag{6.129}$$

where $\Delta\varepsilon_{ij}$ is the strain increments from time t to $t + \Delta t$.

- The constitutive law of nonlinear elastic materials in the incremental form can be expressed as

$$\mathrm{d}\sigma_{ij} = c_{ijkl}^{ep}\mathrm{d}\varepsilon_{ij} \quad \text{in } \Omega \tag{6.130}$$

where c_{ijkl}^{ep} is the tangential stress–strain matrix of material at the current stress level of σ_{ij}. In practice, the calculation is always done over a finite increment. A finite difference form of Equation 6.130 is required through an integration process using, for example, the Euler method. The finite difference form should be

$$\Delta\sigma_{ij} = c_{ijkl}^{ep^*}\Delta\varepsilon_{ij} \tag{6.131}$$

Here $c_{ijkl}^{ep^*}$ is the stress–strain matrix of material in finite difference form.

Boundary Condition

The boundary condition can have the same form as Equations 6.2 and 6.3. The incremental form is expressed as

$$\Delta u_i = \bar{u}_i - u_i^t = \Delta\bar{u}_i \quad \text{on } \Gamma_u \times [0, \infty) \tag{6.132}$$

$$\Delta\sigma_{ij}n_j + \sigma_{ij}^t n_j = \bar{T}_i^t + \Delta\bar{T}_i \quad \text{on } \Gamma_\sigma \times [0, \infty) \tag{6.133}$$

where $\mathbf{n} = \{n_1\ n_2\ n_3\}$ is the outward normal direction and n_i is its directional cosine.

6.4.2 Weak Form for Nonlinear Elastic Problems

In the present incremental formulation, two body forces are considered, i.e.,

1. Effective weight Δb_i^t
2. Pseudo-force $\left(\frac{\partial \sigma_{ij}^t}{\partial x_j} + b_i^t\right)$ due to unbalanced force at time t

Therefore, the constrained Galerkin weak form of the equilibrium equation over the time interval of $[t, t + \Delta t]$ can be written as

$$\begin{aligned}
&\int_{\Omega} \{\delta(\Delta\varepsilon)\}^T\{\Delta\sigma\}\mathrm{d}\Omega - \int_{\Omega}\{\delta(\Delta\bar{u})\}^T\{\Delta b\}\mathrm{d}\Omega - \int_{\Gamma_\sigma}\{\delta(\Delta\bar{u})\}^T\{\Delta\bar{T}\}\mathrm{d}s \\
&\quad + \int_{\Gamma_u}\{\delta(\Delta u - \Delta\bar{u})\}^T\mathbf{a}\{\Delta u - \Delta\bar{u}\}\mathrm{d}s \\
&= -\int_{\Omega}\{\delta(\Delta\varepsilon)\}^T\{\sigma^t\}\mathrm{d}\Omega + \int_{\Gamma_\sigma}\{\delta(\Delta\bar{u})\}^T\{\bar{T}^t\}\mathrm{d}\Omega + \int_{\Omega}\{\delta(\Delta\bar{u})\}^T\{b^t\}\mathrm{d}\Omega \\
&\quad - \int_{\Gamma_u}\{\delta(u^t - \bar{u}^t)\}\alpha\{u^t - \bar{u}^t\}\mathrm{d}s
\end{aligned} \tag{6.134}$$

The sum of all terms at the right-hand side is due to the unbalanced force at time t. Thus, it can automatically correct the unbalanced force at each incremental step. Equation 6.134 always maintains the global balance (in terms of energy) at any time and achieves the same accuracy at each time step. This incremental weak form is especially useful in nonlinear computation such as dissipation (Wang and Liu, 2001c).

Note that in Equation 6.134, the essential boundary condition is imposed using the penalty method, which is detailed in Section 6.2.

6.4.3 Discretization and Numerical Strategy

Substituting the expression of the MLS approximation for the displacement **u** Equation 6.7 into the weak form of Equation 6.134, and after a simple but lengthy derivation similar to what we performed in Section 6.1.1, we obtain the following discretized system of equations:

$$[\mathbf{K} + \mathbf{K}^{\alpha}]\mathbf{U} = \mathbf{F} \tag{6.135}$$

where **U** is a vector of nodal parameters of displacement for all the nodes in the problem domain. **K** is the global stiffness matrix assembled using the nodal stiffness matrix defined by

$$\mathbf{K}_{ij} = \int_{\Omega} \mathbf{B}_i^T \mathbf{c}_{ep}^* \mathbf{B}_j \,\mathrm{d}\Omega \tag{6.136}$$

where $\mathbf{B}_i$ is the strain matrix given by Equation 6.39, $\mathbf{c}_{ep}^*$ is the matrix of material constants that can be determined using a proper model of the materials, and $\boldsymbol{\alpha}$ is the diagonal matrix of penalty factors. For soil materials, for example, the matrix of material constants can be determined using the Duncan–Chang EB model (Duncan and Chang, 1970; Duncan et al., 1978).

The additional matrix $\mathbf{K}^{\alpha}$ is the penalty matrix assembled using the nodal matrix defined by Equation 6.69 and the force vector, and $\mathbf{F}^{\alpha}$ is caused by the essential boundary condition, and its nodal vector has the form of Equation 6.70.

Note that the integration is performed along the essential boundary, and hence matrix $\mathbf{K}^{\alpha}$ will have entries for the nodes near (not only on) the essential boundaries Γ_u.

The force vector **F** in Equation 6.135 is assembled using the nodal force vector defined by

$$\mathbf{f}_i = \int_{\Omega} \mathbf{\Phi}_i^T \mathbf{b}\, d\Omega + \int_{\Gamma_t} \mathbf{\Phi}_i^T \bar{\mathbf{t}}\, d\Gamma + \int_{\Gamma_u} \mathbf{\Phi}_i^T \boldsymbol{\alpha} \bar{\mathbf{u}}\, d\Gamma + \mathbf{f}_i^t \tag{6.137}$$

The nodal force vector shown in the foregoing equation consists of four components: body force, traction along force (natural) boundary, traction along the displacement (essential) boundary, and the unbalanced force $\mathbf{f}_i^t$ at load step t. The force vector $\mathbf{f}_i^t$ consists of the unbalanced force vectors at load step t for all the nodes, which are obtained by converting the stress status at load step t, as follows:

$$\mathbf{f}_i^t = \int_{\Omega} \begin{bmatrix} \frac{\partial \phi_i}{\partial x} \sigma_x^t + \frac{\partial \phi_i}{\partial y} \tau_{xy}^t \\ \frac{\partial \phi_i}{\partial y} \sigma_y^t + \frac{\partial \phi_i}{\partial x} \tau_{xy}^t \end{bmatrix} d\Omega \tag{6.138}$$

6.4.4 Numerical Procedure

Figure 6.38 shows the flowchart for material nonlinear analysis using the EFG method. The interaction procedure is basically the same as that using FEM. Variable replacement between consequent steps is important to nonlinear iteration. For our algorithm, the summation of all previous increments is adopted for any step including the iterating step. That is,

$$u^i = u^{i-1} + \Delta u^i \tag{6.139}$$

$$\varepsilon^i = \varepsilon^{i-1} + \Delta \varepsilon^i \tag{6.140}$$

$$\sigma^i = \sigma^{i-1} + \Delta \sigma^i \tag{6.141}$$

Modified Newton–Raphson with initial tangential modulus is suggested. For each construction step, the loading is divided into a number of steps. In each loading step, iteration is performed. The convergence criterion of the iteration is defined by the relative error of displacement as

$$e = \left| \frac{\|\Delta u^i\|}{\|\Delta u^{i-1}\|} - 1 \right| \tag{6.142}$$

where the norm of displacement is defined as

$$\|\Delta u\| = \sqrt{\Delta u_x^2 + \Delta u_y^2} \tag{6.143}$$

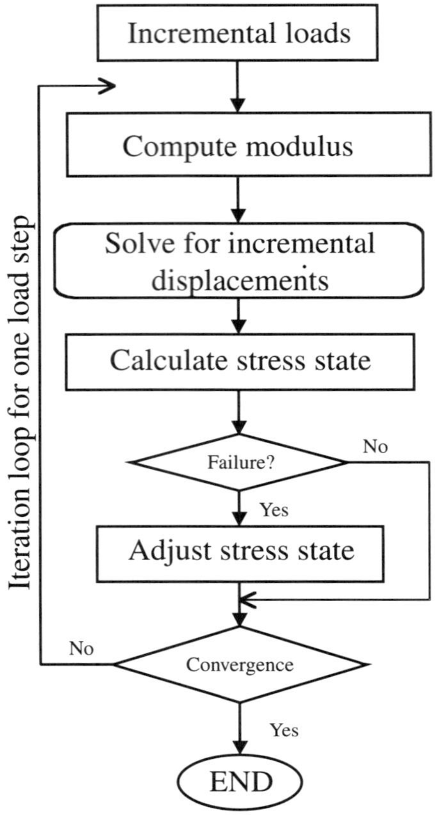

FIGURE 6.38
Flowchart for incremental algorithm of nonlinear analysis using EFG.

in which Δu_x and Δu_y are the incremental displacements. The relative energy error can also be defined in the same way by replacing incremental displacement with the incremental strain energy. The interaction is deemed convergent when the relative error reduces to a prescribed value.

6.4.5 Numerical Example

Example 6.12 Soil Foundation

A soil foundation is modeled as a 2D block subjected to a strip loading, as shown in Figure 6.39a. The load could be an embankment or road loading. The soil material of the foundation is modeled using the Duncan–Chang EB model. (Duncan and Chang, 1970; Duncan et al., 1978). Its model parameters are as follows: $K = 20.2$, $n = 1.0944$, $K_{ur} = 347.1$, $n_0 = 0.8304$, $K_b = 14.636$, $m = 1.0799$, $\sigma_d = 8$, $R_f = 0.7156$, $\phi = 30°$, and $C = 2.5$ kPa. The floating density $\gamma = 7$ kN/m^3. Elastic parameters of Young's modulus $E = 1000$ kPa and bulk modulus $B = 833.33$ kPa are used at the initial stage. Figure 6.40 shows a typical stress–strain curve obtained using the Duncan–Chang EB model. The detailed process of determining the matrix of material constants $\mathbf{c}^*_{ep}$ in Equation 6.136 can be found in papers by Duncan and co-workers (1970, 1978) using the above given parameters, as it is beyond the scope of this book.

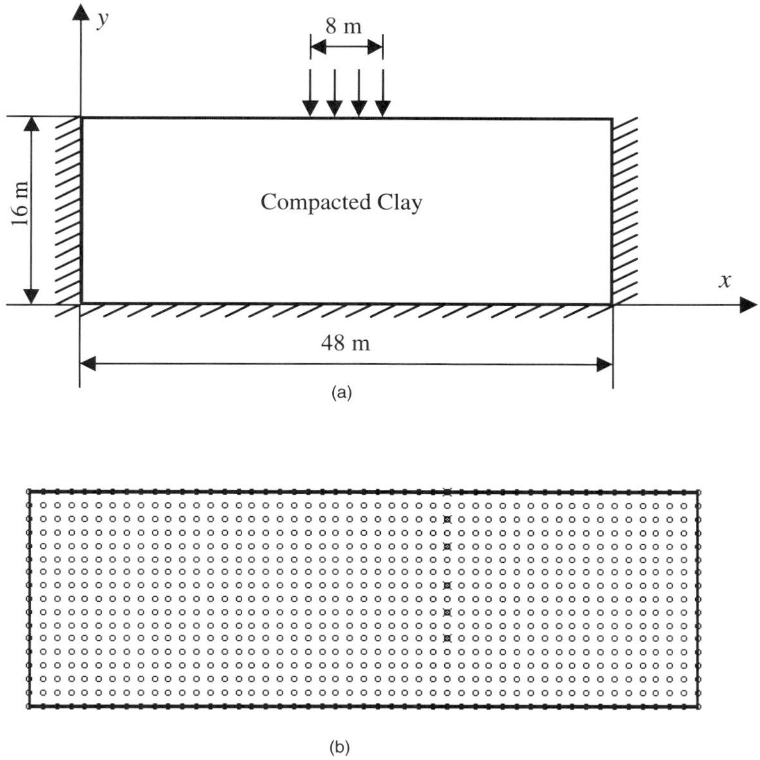

FIGURE 6.39
Nonlinear foundation model and the nodal arrangement. (a) Foundation problem; (b) meshless model.

All displacement boundaries are assumed to be fixed along the normal direction and free along the tangential direction. Regular and irregular nodal distributions are used in the computation. A regular nodal arrangement for the meshless model is shown in Figure 6.39b. For comparison, the same model is calculated using our in-house FEM, in which four-node isoparametric elements are employed.

Effect of Penalty Factor and Support Domain on Numerical Results

The penalty factor is an important parameter that affects the convergence rate and the numerical accuracy. It can be chosen differently for different portions of the essential boundary. Here, we choose to use a uniform penalty factor α for the entire essential boundary for convenient investigation of the effects of the penalty factor α on the results. Figure 6.41 shows the effect of the penalty factor on the convergence rate of the results of the displacement. It is found that within a wide range of α, the effect is not significant on the relative energy error, but it has some effect on the relative error of displacement. Our numerical examination also shows that too large an α will overemphasize the imposition of the essential boundary condition, and the equilibrium of the stress status in the domain is compromised. This causes some numerical problems such as stress oscillations along the essential boundary. When the penalty factor is too small, the essential boundary conditions are not properly imposed. A reasonable range would be from 10^5 to 10^8 times the Young's modulus at the current stage of stresses for nonlinear materials.

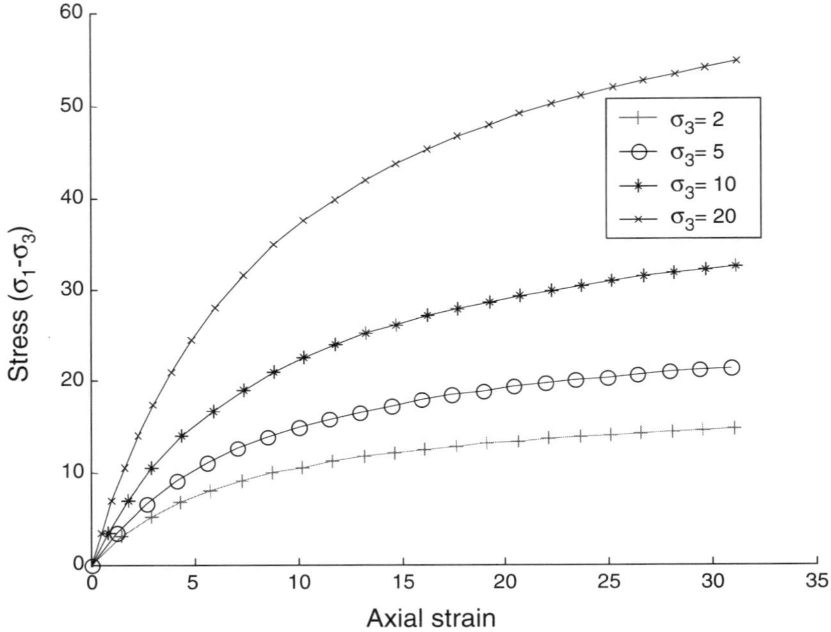

FIGURE 6.40
Stress–strain curves under different confining pressures.

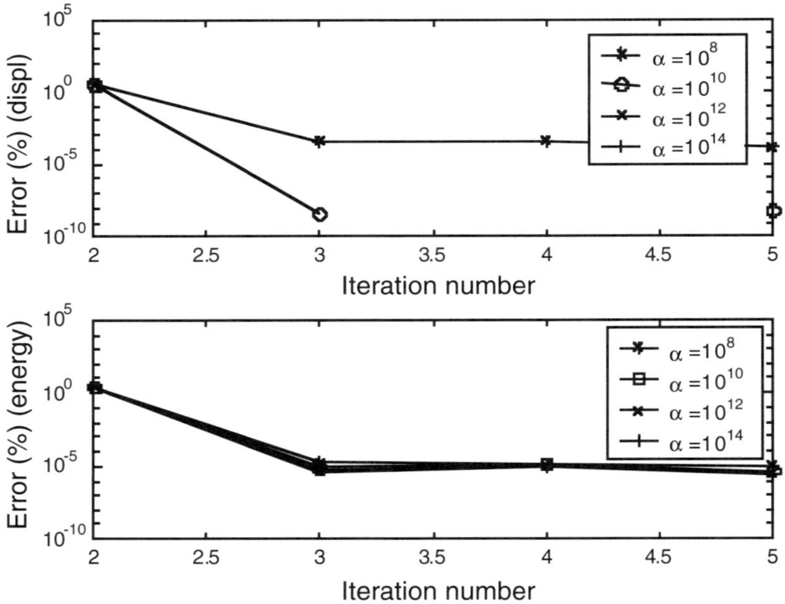

FIGURE 6.41
Effect of penalty parameter on convergence rate.

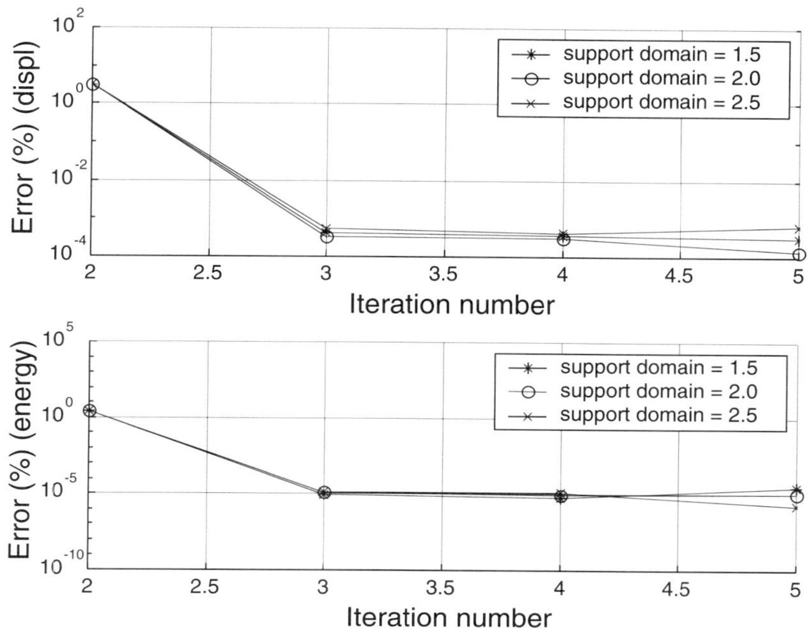

FIGURE 6.42
Effect of the dimension (α_s) of the support domain on convergence rate.

Another important factor for meshless methods is the support domain. Figure 6.42 shows the effect of the dimension (defined by α_s) of the support domain on the convergence rate when the support domain varies from 1.5 to 2.5 times the nodal distance between two neighboring nodes. Little change is observed, as long as it is larger than 1.5. It is our general observation that we can use a smaller support domain for nonlinear analysis. This could be because iteration is used in the analysis, which controls the accuracy of the results at the end.

The convergence rate of the EFG results is compared with FEM, as shown in Figure 6.43. It is shown that both EFG and FEM have good rates of convergence, and that the EFG has much lower relative errors especially for the energy error that relates to stresses, compared with the FEM. In other words, the meshless method requires less iteration to reach the same accuracy.

Results for a Foundation Subjected to Strip Loading

The process of simulation is typical for any analysis of problems of material nonlinearity. We divide the process into two stages. At the first stage, the foundation is loaded only by the gravity force of soil mass. The simulation at this stage is based on linear elasticity by ignoring the effects of material nonlinearity and can recover *in situ* stress in the foundation. We are not worried too much about accuracy at this stage, as we will later have an iteration to control the accuracy of the final results. After *in situ* stress is obtained, the analysis comes to the second stage, and a strip load of 100 kPa is then applied gradually (incrementally). This load is divided evenly into 25 loading steps. The soil is modeled using the Duncan EB model to obtain the matrix of material constants at the current stress status.

A typical history of the stress level computed at different nodes is shown in Figure 6.44. All of these nodes (from top to bottom: 815, 717, 619, 472, 374, and 276) are along a vertical line. The stress level increases faster when the node is near the loading surface (see node 815)

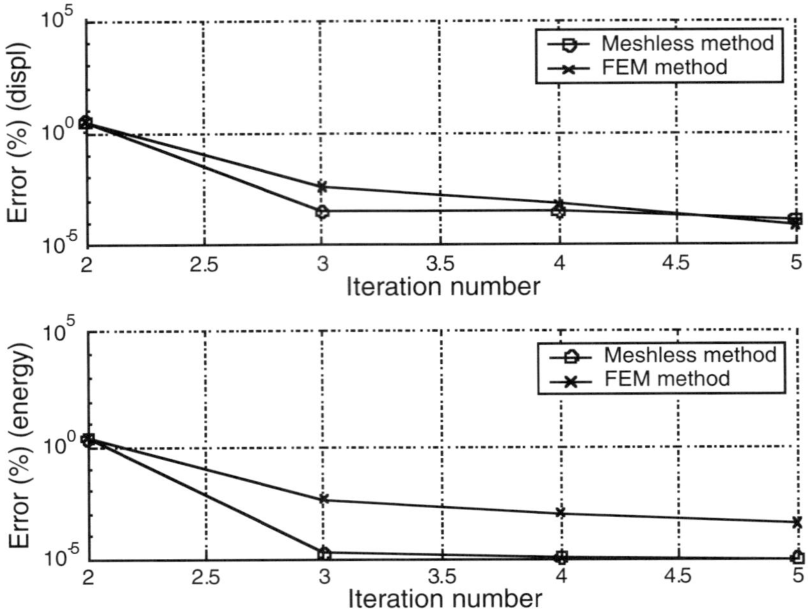

FIGURE 6.43
Comparison of convergence rates of the results between EFG meshless method and FEM.

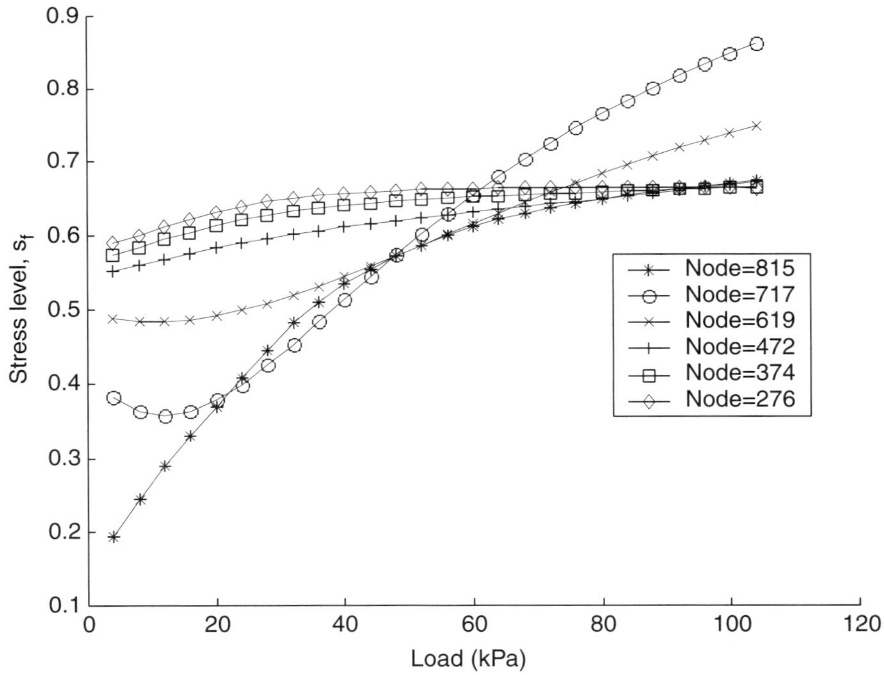

FIGURE 6.44
History of stress level at different nodes numbered from the top to bottom in the order of 815, 717, 619, 472, 374, and 276 along a vertical line marked in Figure 6.39b.

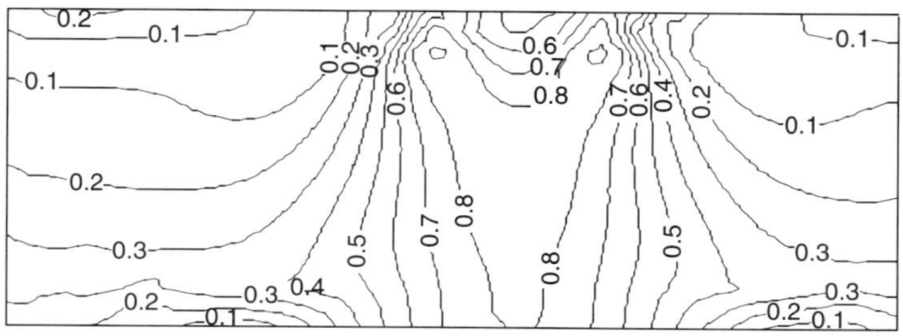

FIGURE 6.45
Stress level registered when the foundation is loaded to 100 kPa. Potential failure pattern with sliding zone is observed.

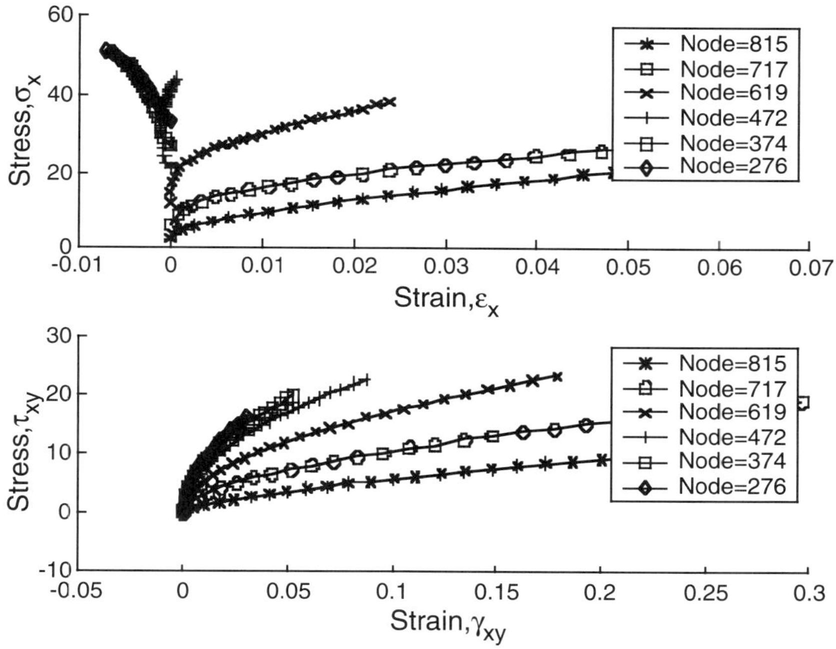

FIGURE 6.46
Stress–strain curves reproduced for the soil material at different nodes numbered from the top to bottom in the order of 815, 717, 619, 472, 374, and 276 along a vertical line marked in Figure 6.39b.

and little change is observed for the bottom nodes. The maximum stress level is registered at node 717 instead of node 815; thus, the possible failure point should be below the surface. Stress level changes significantly when nodes approach the edge of a potential sliding zone. When loading increases to 100 kPa, a potential failure pattern with sliding zone is formed, as shown in the stress level contour in Figure 6.45.

The computation also reveals the nonlinear property of soil material at different nodes, as shown in Figure 6.46. It is observed that the soil mass near the potential slide zone has strong nonlinearity, whereas that far from the slide lines shows slight nonlinearity.

The numerical results of the EFG method are further compared with FEM during the entire loading process. Figure 6.47 shows the stress history of σ_{xx} and σ_{yy} when the soil is progressively loaded. The results are obtained using both EFG and FEM. These results

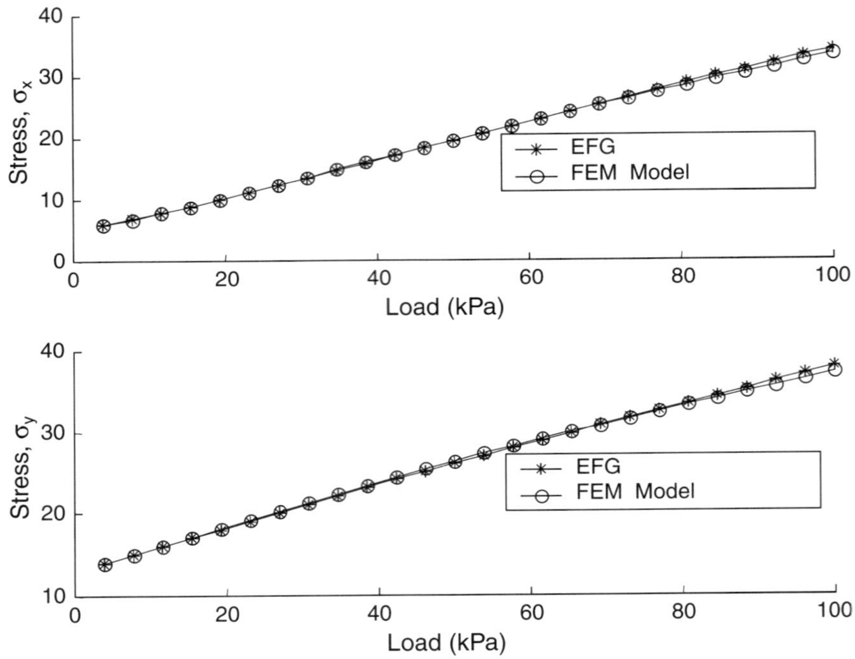

FIGURE 6.47
Comparison of stress history at node 717 obtained using EFG and FEM.

generally agree well at the initial stage of loading, but slight difference appears when loading increases to a certain stage. The stress level obtained by FEM is higher than that by EFG, as shown in Figure 6.48.

Spatial distribution of the displacement and strain is also examined. Figure 6.49 shows the displacement distribution on a horizontal line located 2 m (y = 14 m) below ground level. The loading is 100 kPa. The results are computed using both EFG and FEM, and very good agreement is observed. Figure 6.50 shows the distribution of the strains on a horizontal line located 2 m below ground level. Again, the EFG and FEM results agree very well.

6.4.6 Remarks

This section presents an example of using the EFG method for analyzing problems of material nonlinearity. The following points are noted.

- The EFG method is an effective method for nonlinear problems. A support domain of 1.5 to 2.5 times the local nodal distance is appropriate. The solution converges faster compared to that of FEM.
- Many numerical techniques that were developed in FEM can be utilized in the EFG method for solving nonlinear problems, with minor modifications. The major difference is in the field variable interpolation.
- Because of the lack of the Kronecker delta function property, efforts have to be made in the imposition of essential boundary conditions. In using the penalty method, the penalty factor should be 10^5 to 10^8 times the Young's modulus of the material near the essential boundary at the current stress status.

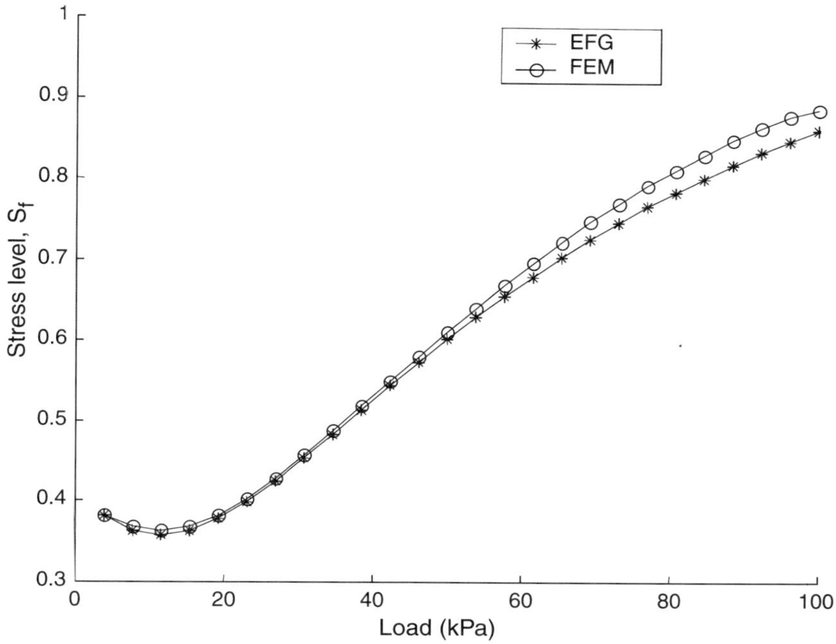

FIGURE 6.48
Comparison of stress level history at node 717 computed using EFG and FEM.

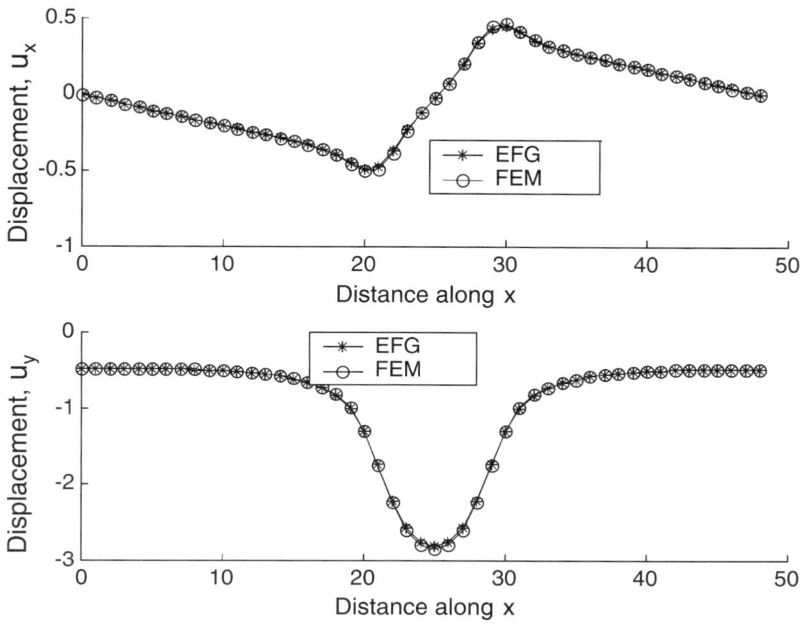

FIGURE 6.49
Displacement distribution on the horizontal line 2 m below the surface of the foundation (y = 14 m), when the load is 100 kPa. The results are computed using both EFG and FEM.

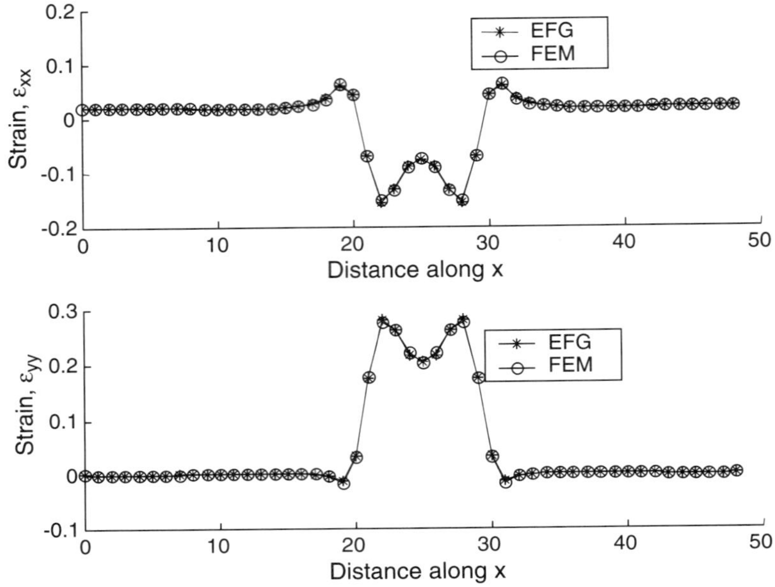

FIGURE 6.50
Strain distribution on the horizontal line 2 m below the surface of the foundation (y = 14 m), when the load is 100 kPa. The results are computed using both EFG and FEM.

6.5 Summary

This chapter presents the EFG method, which has been developed in the past decade. Works by Nayroles and co-workers and Belytschko and co-workers have, in fact, offered a new direction in the development of meshless methods. The works by Belytschko and co-workers on crack propagation problems are indeed impressive. Readers are referred to publications of Belytschko and co-workers. Following their work, a large number of researchers have also contributed significantly to the development of the EFG method. The first version of the commercial software package, MFree2D, has been developed by G. R. Liu and co-workers based on the EFG formulation. The MFree2D is, to the best knowledge of the author, the first commercialized software package on MFree methods that is fully packaged with pre- and postprocessor that is capable of carrying out adaptive analysis automatically.

The author feels that the development of EFG has indeed created a significant impact on the development of MFree methods. It is a significant advancement after the invention of the SPH method.

The challenges with the EFG method are (1) removal of the background cells for integration and (2) formulation of shape functions that possess Kronecker delta function property.

7

Meshless Local Petrov–Galerkin Method

The element free Galerkin (EFG) method still requires a mesh of background cells for integration in computing the system matrices, and so the method is not truly mesh free. Pursuit of truly MFree methods therefore continues.

The reason behind the need for background cells for integration is the use of the weak form for generating the discrete system equations. Is it possible to not use the weak form? The answer is yes. MFree methods that operate on strong forms, such as the finite point method (Liszka and Orkisz, 1980; Jensen, 1980; Onate et al., 1996; Cheng and Liu, G. R., 1999; Xu and Liu, G. R., 1999; Song et al., 1999), have been developed based on finite differential representation (Taylor series) of a function. However, these kinds of methods are generally not very stable, especially for arbitrary nodal distribution, and the results obtained are less accurate, especially when the derivatives of the field variables (strain or stress) are of interest. Efforts are still ongoing to stabilize these methods, especially in the direction of using radial functions.

Examination of the equation of the weighted residual form reveals that integration over the entire problem domain is required. This is because we are trying to satisfy the equation of the weighted residual form over the entire problem domain as one seamless piece. However, if we try to satisfy the equation point-by-point using information in a local domain of the point, the integration form can then be implemented locally by carrying out numerical integration over the local domain. The meshless local Petrov–Galerkin (MLPG) method originated by Atluri and Zhu (1998) uses the so-called local weak form of the Petrov–Galerkin residual formulation. MLPG has been fine-tuned, improved, and extended over the years by Atluri et al. (1999a,b), Ouatouati and Johnson (1999), G. R. Liu and Yan (2000), G. R. Liu and Gu (2000a, 2001e), Gu and G. R. Liu (2001c,f), and many others. This chapter details the MLPG method for two-dimensional (2D) solid mechanics problems.

In the MLPG implementation, MLS approximation is employed for constructing shape functions. Therefore, similar to the EFG method, there is an issue of imposition of essential boundary conditions. The original MLPG proposed by Atluri et al. (1999a,b) uses the penalty method. In the formulation by G. R. Liu and Yan (2000), a method called direct interpolation is used. This chapter formulates both methods.

A number of benchmark examples are presented to illustrate the effectiveness of the MLPG method. The effects of different parameters including the dimensions of different domains of MLPG on the accuracy of the results are also investigated via these examples.

Note that the point-by-point procedure in MLPG is also very similar to that of the collocation method. However, the MLPG is more stable due to the use of locally weighted residual integration. Note also that the MLPG is reproductive due to the combination of the MLS shape functions that is reproductive with the local Petrov–Galerkin approach. This fact will be evidenced in the examples of patch tests.

7.1 MLPG Formulation

We consider again a 2D solid mechanics problem, as shown in Figure 7.1, for illustrating the procedure for formulating the MLPG method. The problem domain is denoted by Ω, which is bounded by boundaries including essential (displacement) boundary Γ_u and natural (force or free) boundary Γ_t. The strong form of the problem has been given in Equations 6.1 to 6.3.

7.1.1 The Idea of MLPG

In the MLPG method, the problem domain is represented by a set of arbitrarily distributed nodes, as shown in same. The weighted residual method is used to create the discrete system equation. The weighted residual method is, of course, in integral form, and a background mesh of cells is still required for the integration. The major idea in MLPG is, however, that the implementation of the integral form of the weighted residual method is confined to a very small local subdomain of a node. This means that the weak form is satisfied at each node in the problem domain in a local integral sense. Therefore, the weak form is integrated over a "local quadrature domain" that is independent of other domains of other nodes. This is made possible by use of the Petrov–Galerkin formulation, in which one has the freedom to choose the weight and trial functions independently. If the Galerkin formulation is used, one has to use the same compatible function for the weight and trial functions in the same domain, which presents difficulties in confining the integration to a localized domain.

Because MLPG requires integrations only over a localized quadrature domain, what we need now is only a background mesh of cells for the local quadrature. The quadrature domain can be arbitrary in theory, but very simple regularly shaped subdomains, such as circles and rectangles for 2D problems and bricks and spheres for 3D problems, are often

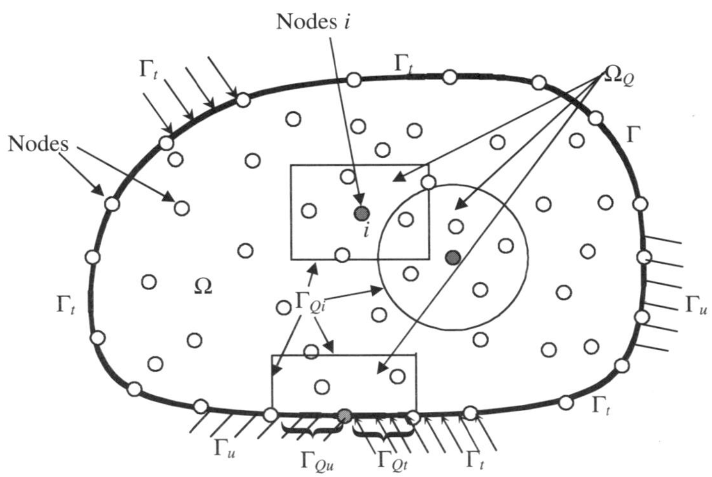

FIGURE 7.1
Domains and their boundaries. Problem domain Ω boundary bounded by its boundaries including essential (displacement) boundary Γ_u, natural (force or free) boundary Γ_t; quadrature domain of Ω_Q and its boundary including the interior boundary Γ_{Qi} that is located within the problem domain, the essential boundary Γ_{Qu} that intersects with Γ_u, and natural boundary Γ_{Qt} that intersects with Γ_t.

used for ease of implementation (see Figure 7.1 for the 2D case). Because of the simplicity of the quadrature domain, the creation of the local background mesh of cells is of course easier to perform compared to the creation of the background mesh for the entire problem domain, and can be automated. MLPG can be referred as a "truly" mesh-free method, or at least close to the ideal mesh-free method.

Note that the shape and dimensions of the quadrature domains do not have to be the same for all the nodes. Therefore, one is free to choose a proper shape for the local quadrature domain based on the local situation. This feature is important when the local quadrature domain encounters the global boundary of the problem. In addition, as long as the support domain is compact, MLPG will produce sparse system matrices. The major drawback of MLPG is the asymmetry of the system matrices due to the use of the Petrov–Galerkin formulation. Another drawback of MLPG is that the local background integration can be very tricky due to the complexity of the integrand produced by the Petrov–Galerkin approach, especially for domains that intersect with the boundary of the problem domain.

7.1.2 Formulation of MLPG

In MLPG (Atluri and Zhu, 1998), MLS approximation is employed to create shape functions for field variable approximation. Therefore, the same problem experienced in the EFG method will again surface in the imposition of essential boundary conditions. Atluri and Zhu (1998) use the penalty method to enforce the essential boundary condition. Following the formulation of Atluri and Zhu (1998), we use the strong form of Equation 3.22, which is equivalent to Equation 6.1:

$$\sigma_{ij,j} + b_i = 0 \tag{7.1}$$

The boundary conditions corresponding to Equations 6.2 and 6.3 are

$$\text{Essential boundary condition:} \quad u_i = \bar{u}_i \quad \text{on } \Gamma_u \tag{7.2}$$

$$\text{Natural boundary condition:} \quad \sigma_{ij}n_j = \bar{t}_i \quad \text{on } \Gamma_t \tag{7.3}$$

where $i, j = x, y$, and n_j is the jth component of the unit outward normal vector on the boundary.

The weak form for a node, say, node I, based on the local weighted residual method can be stated as (see Equation 4.50)

$$\int_{\Omega_Q} (\sigma_{ij,j} + b_i)\widehat{W}_I \, \mathrm{d}\Omega - \alpha \int_{\Gamma_{Qu}} (u_i - \bar{u}_i)\widehat{W}_I \, \mathrm{d}\Gamma = 0 \tag{7.4}$$

where $\widehat{W}$ is the weight or test function, and we use the same weight function for all the equations involved. Note that the second area integral in Equation 4.50 has been changed to a curve integral, because the constraint Equation 7.2 is on the essential boundary. Ω_Q is the domain of quadrature (integration) for node I, Γ_{Qu} is the part of the essential boundary that intersects with the quadrature domain Ω_Q (see Figure 7.1), and α is the penalty factor that we have seen in Chapter 6. Here we use the same penalty factor for all the displacement constraint equations (essential boundary conditions). The first term in Equation 7.4 is for the equilibrium requirement at node I, and the second term is only for the case when the essential boundary of the problem domain is part of the boundary of the local quadrature domain Ω_Q. If Ω_Q does not intersect with the essential boundary of the problem domain, the second term vanishes.

Using the divergence theorem, we obtain

$$\int_{\Gamma_Q} \sigma_{ij} n_j \widehat{W}_I \, \mathrm{d}\Gamma - \int_{\Omega_Q} \sigma_{ij} \widehat{W}_{I,j} \, \mathrm{d}\Omega + \int_{\Omega_Q} b_i \widehat{W}_I \, \mathrm{d}\Omega - \alpha \int_{\Gamma_{Qu}} (u_i - \bar{u}_i) \widehat{W}_I \, \mathrm{d}\Gamma = 0 \tag{7.5}$$

where $\Gamma_Q = \Gamma_{Qi} \cup \Gamma_{Qu} \cup \Gamma_{Qt}$, and Γ_{Qi} is the internal boundary of the quadrature domain, Γ_{Qt} is the part of the natural boundary that intersects with the quadrature domain, and Γ_{Qu} is the part of the essential boundary that intersects with the quadrature domain. When the quadrature domain is located entirely within the global domain, Γ_{Qt} and Γ_{Qu} vanish and $\Gamma_Q = \Gamma_{Qi}$. Unlike the Galerkin method, the Petrov–Galerkin method chooses the trial and test functions from different spaces. The weight function $\widehat{W}$ is purposely selected in such a way that it vanishes on Γ_{Qi}. Note that the weight functions mentioned in Chapter 5, e.g., the cubic or quartic spline, can be chosen to equal zero along the boundary of the internal quadrature domains; hence, they can be used as the weight functions for MLPG.

Equation 7.5 shows that the differentiation on the stresses is now transferred to the weight function. This reduces the constancy requirement when we approximate the trial displacement function that is used for obtaining the stresses.

Using a weight function $\widehat{W}$ that vanishes on Γ_{Qi}, we can then change the expression of Equation 7.5 to

$$\begin{aligned} &\int_{\Omega_Q} \sigma_{ij} \widehat{W}_{I,j} \, \mathrm{d}\Omega + \alpha \int_{\Gamma_{Qu}} u_i \widehat{W}_I \, \mathrm{d}\Gamma - \int_{\Gamma_{Qu}} \sigma_{ij} n_j \widehat{W}_I \, \mathrm{d}\Gamma \\ &\quad = \int_{\Gamma_{Qt}} \bar{t}_i \widehat{W}_I \, \mathrm{d}\Gamma + \alpha \int_{\Gamma_{Qu}} \bar{u}_i \widehat{W}_I \, \mathrm{d}\Gamma + \int_{\Omega_Q} b_i \widehat{W}_I \, \mathrm{d}\Omega \end{aligned} \tag{7.6}$$

When the quadrature domain is located entirely in the domain, integrals related to Γ_{Qu} and Γ_{Qt} vanish, and the Petrov–Galerkin form can be simplified as

$$\int_{\Omega_Q} \sigma_{ij} \widehat{W}_{I,j} \, \mathrm{d}\Omega = \int_{\Omega_Q} b_i \widehat{W}_I \, \mathrm{d}\Omega \tag{7.7}$$

Equation 7.7 is used to establish the discrete equations for all the nodes whose quadrature domain falls entirely within the problem domain. Equation 7.6 is used to establish the discrete equations for all the boundary nodes or the nodes whose quadrature domain intersects with the problem boundary.

Using Equation 7.6 or 7.7, and integrating over the quadrature domain, leads to discretized system equations for each node in the problem domain. This gives a set of algebraic equations for each node. By assembling all these sets of equations, a set of discretized system equations for the entire problem domain can then be obtained. Note that MLPG establishes algebraic equations based on nodes in the problem domain, which is in fact very similar to the finite difference procedure. It will be shown later that this feature can be used for imposition of the essential boundary conditions.

Equation 7.6 or 7.7 suggests that instead of solving the strong form of the system equation given in Equations 7.1 and 7.2, a relaxed weak form with integration over a small local quadrature domain is employed. This integration operation can "smear" out the numerical error and, therefore, make the discrete equation system much more accurate compared to the MFree procedures that operate directly on the strong forms of system equations. MLPG guarantees satisfaction of the equilibrium equation at a node in an integral sense over a quadrature domain, but it does not ensure satisfaction of the system equation of strong form exactly at the node. The size of the quadrature domain determines the "relaxing" extent of the differential equation of strong form. It will be shown later in the example problems that the quadrature domain needs to have sufficient dimension to produce an accurate and stable solution, and that too large a support domain does not necessarily provide significantly better results. This implies that the conventional weighted residual method integrated over the entire problem domain may be, to a certain extent, overkill.

In MLPG, Equation 7.6 or 7.7 is to be satisfied for all the local quadrature domains for each and every node in the entire problem domain, including the boundaries. This implies that the equilibrium equation and the boundary conditions are satisfied node by node in a weak sense of the local weighted residual. This node-by-node local formulation provides freedom in solving the system equation numerically.

Note that if the Kronecker delta function is used as the weight function, the method becomes an MFree method of strong formulation or collocation method.

Following now the MLS approximation procedure, one can generate the shape function for each node using the nodes in support domain Ω_s of a point (not necessarily a node). We will soon see that Ω_s is in general different from Ω_Q. The procedure is exactly the same as that in the EFG method. From Equation 6.7, we have

$$\mathbf{u}^h = \begin{Bmatrix} u \\ v \end{Bmatrix}^h = \sum_i^n \underbrace{\begin{bmatrix} \phi_i & 0 \\ 0 & \phi_i \end{bmatrix}}_{\mathbf{\Phi}_i} \underbrace{\begin{Bmatrix} u_i \\ v_i \end{Bmatrix}}_{\mathbf{u}_i} = \sum_i^n \mathbf{\Phi}_i \mathbf{u}_i \tag{7.8}$$

where n denotes the total number of nodes in the support domain Ω_s, $\phi_i(\mathbf{x})$ is the MLS shape function for node i that is created in the support domain Ω_s of point $\mathbf{x}$, u_i, and v_i are the nodal parameters of two displacement components in the x and y directions, $\mathbf{\Phi}_i$ is the matrix of shape functions given by

$$\mathbf{\Phi}_i = \begin{bmatrix} \phi_i & 0 \\ 0 & \phi_i \end{bmatrix} \tag{7.9}$$

and $\mathbf{u}_i$ is the nodal displacement for node i:

$$\mathbf{u}_i = \begin{Bmatrix} u_i \\ v_i \end{Bmatrix} \tag{7.10}$$

After using the divergence theorem that leads to Equation 7.6, we now change Equation 7.6 back to the following matrix form, to facilitate deriving the discrete system equations in matrix form:

$$\int_{\Omega_Q} \widehat{\mathbf{V}}_I^T \boldsymbol{\sigma}\, \mathrm{d}\Omega + \alpha \int_{\Gamma_{Qu}} \widehat{\mathbf{W}}_I \mathbf{u}\, \mathrm{d}\Gamma - \int_{\Gamma_{Qu}} \widehat{\mathbf{W}}_I \mathbf{t}\, \mathrm{d}\Gamma = \int_{\Gamma_{Qt}} \widehat{\mathbf{W}}_I \bar{\mathbf{t}}\, \mathrm{d}\Gamma + \alpha \int_{\Gamma_{Qu}} \widehat{\mathbf{W}}_I \bar{\mathbf{u}}\, \mathrm{d}\Gamma + \int_{\Omega_Q} \widehat{\mathbf{W}}_I \mathbf{b}\, \mathrm{d}\Omega \tag{7.11}$$

where $\widehat{\mathbf{V}}_I$ is a matrix that collects the derivatives of the weight functions in Equation 7.6, which has the form:

$$\widehat{\mathbf{V}}_I = \begin{bmatrix} \widehat{W}_{I,x} & 0 \\ 0 & \widehat{W}_{I,y} \\ \widehat{W}_{I,y} & \widehat{W}_{I,x} \end{bmatrix} \tag{7.12}$$

It is in fact the "strain" matrix caused by the weight (test) functions $\widehat{\mathbf{W}}$.

$\boldsymbol{\sigma}$ denotes the stress vector defined in Equation 3.23 for 2D solids. Using Equations 3.26, 3.29, and 7.8, we have

$$\boldsymbol{\sigma} = \mathbf{c}\boldsymbol{\varepsilon} = \mathbf{c}\mathbf{L}\mathbf{u}^h = \mathbf{c}\begin{bmatrix} \frac{\partial}{\partial x} & 0 \\ 0 & \frac{\partial}{\partial y} \\ \frac{\partial}{\partial y} & \frac{\partial}{\partial x} \end{bmatrix} \sum_j^n \boldsymbol{\Phi}_j \mathbf{u}_j = \mathbf{c}\sum_j^n \mathbf{B}_j \mathbf{u}_j \tag{7.13}$$

where

$$\mathbf{B}_J = \begin{bmatrix} \phi_{J,x} & 0 \\ 0 & \phi_{J,y} \\ \phi_{J,y} & \phi_{J,x} \end{bmatrix} \tag{7.14}$$

$\widehat{\mathbf{W}}$ is a matrix of weight functions given by

$$\widehat{\mathbf{W}}_I = \begin{bmatrix} \widehat{W}_I & 0 \\ 0 & \widehat{W}_I \end{bmatrix} \tag{7.15}$$

The tractions $\mathbf{t}$ of a point $\mathbf{x}$ can be written as

$$\mathbf{t} = \underbrace{\begin{bmatrix} n_x & 0 & n_y \\ 0 & n_y & n_x \end{bmatrix}}_{\mathbf{n}} \boldsymbol{\sigma} = \mathbf{n}\mathbf{c}\sum_j^n \mathbf{B}_j \mathbf{u}_j \tag{7.16}$$

in which (n_x, n_y) is the unit outward normal vector on the boundary:

$$\mathbf{n} = \begin{bmatrix} n_x & 0 & n_y \\ 0 & n_y & n_x \end{bmatrix} \tag{7.17}$$

Substitution of Equations 7.8 and 7.13 through 7.16 into Equation 7.11 leads to the following discrete systems of linear equations for the Ith node:

$$\begin{aligned} &\int_{\Omega_Q} \widehat{\mathbf{V}}_I^T \mathbf{c}\sum_{j=1}^n \mathbf{B}_j \mathbf{u}_j \, d\Omega + a \int_{\Gamma_Q} \widehat{\mathbf{W}}_I \sum_j^n \boldsymbol{\Phi}_j \mathbf{u}_j \, d\Gamma - \int_{\Gamma_Q} \widehat{\mathbf{W}}_I \mathbf{n}\mathbf{c}\sum_j^n \mathbf{B}_j \mathbf{u}_j \, d\Gamma \\ &= \int_{\Gamma_Q} \widehat{\mathbf{W}}_I \bar{\mathbf{t}} \, d\Gamma + \alpha \int_{\Gamma_Q} \widehat{\mathbf{W}}_I \bar{\mathbf{u}} \, d\Gamma + \int_{\Omega_Q} \widehat{\mathbf{W}}_I \mathbf{b} \, d\Omega \end{aligned} \tag{7.18}$$

The matrix form of Equation 7.18 can be assembled as

$$\sum_{j=1}^n \mathbf{K}_{Ij} \mathbf{u}_j = \mathbf{f}_I \tag{7.19}$$

where $\mathbf{K}_{Ij}$ is a 2×2 matrix called a *nodal stiffness matrix,* given by

$$\mathbf{K}_{Ij} = \int_{\Omega_Q} \widehat{\mathbf{V}}_{IW}^{T}\mathbf{c}\mathbf{B}_j \, d\Omega + \alpha \int_{\Gamma_Q} \widehat{\mathbf{W}}_I \mathbf{\Phi}_j \, d\Gamma - \int_{\Gamma_Q} \widehat{\mathbf{W}}_I \mathbf{n}\mathbf{c}\mathbf{B}_j \, d\Gamma \tag{7.20}$$

and $\mathbf{f}_I$ is the *nodal force vector* with contributions from body forces applied in the problem domain, tractions applied on the natural boundary, as well as the penalty force terms.

$$\mathbf{f}_I = \int_{\Omega_Q} \widehat{\mathbf{W}}_I \mathbf{b} \, d\Omega + \int_{\Gamma_{Qt}} \widehat{\mathbf{W}}_I \bar{\mathbf{t}} \, d\Gamma + \alpha \int_{\Gamma_{Qu}} \widehat{\mathbf{W}}_I \bar{\mathbf{u}} \, d\Gamma \tag{7.21}$$

Equation 7.19 presents two linear equations for node I. Using Equation 7.19 for all n_t nodes in the entire problem domain, two independent linear equations can be obtained for each node. Assemble all these $2n_t$ equations to obtain the final global system equations of

$$\mathbf{K}_{2n_t \times 2n_t} \mathbf{U}_{2n_t \times 1} = \mathbf{F}_{2n_t \times 1} \tag{7.22}$$

It can be easily seen that the system stiffness matrix $\mathbf{K}$ in the MLPG method is banded as long as the support domain is compact, but it is usually asymmetric. The asymmetry is caused by the use of the Petrov–Galerkin formulation, which employs different functions for the trial and test functions. It has been reported that the stiffness matrix $\mathbf{K}$ can be made symmetric (Atluri et al., 1999b).

7.1.3 Types of Domains

A number of subdomains are involved in practical implementation of the MLPG method. Each subdomain carries a different meaning, and some are similar to the domains used in the EFG method. The names of these domains are in fact very confusing in the current literature. Considering the terminology in the EFG method, we suggest the following systems of names for these subdomains.

All the subdomains are schematically drawn in Figure 7.2. The quadrature domain Ω_Q of node i at $\mathbf{x}_i$ is a domain for the integration in Equation 7.4. The weighted domain Ω_W is the domain where the weight (test) function is nonzero; i.e., $\widehat{W}_i(x) \neq 0$. Theoretically, the quadrature domain Ω_Q and the weighted domain Ω_W do not have to be the same, and Ω_W can be larger than Ω_Q. However, in practice we often use the same for both, i.e., $\Omega_Q = \Omega_W$, so that the curve integration along the interior boundary of Ω_Q will vanish, which simplifies the formulation and computation. Therefore, we always assume in this book that the quadrature domain is the weighted domain, unless specifically mentioned. The quadrature (or weight) domain can be theoretically arbitrary in shape. A circle or rectangular support domain is often used in practice for convenience.

For the local integration over Ω_Q, the Gauss quadrature may be used in the MLPG method. Therefore, for a mesh of local integration cells over its quadrature domain Ω_Q, a node is needed to employ the Gauss quadrature scheme. Because the quadrature domain is chosen to be simple, the creation of the local integral cells is not difficult. For each quadrature point $\mathbf{x}_Q$ in a cell, MLS interpolation is performed to compute the shape function and to obtain the integrand. A subdomain is then needed to choose the nodes for constructing the shape function. This subdomain carries exactly the same physical meaning of the support domain noted as Ω_s, which was defined in Chapter 5. The support domain Ω_s is independent of the quadrature domain Ω_Q (or the weighted domain Ω_W).

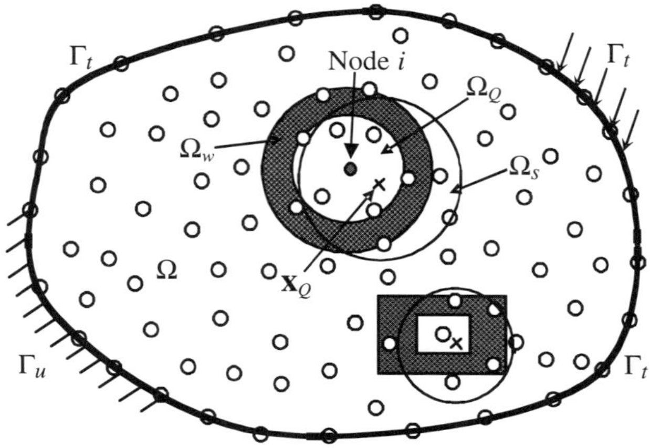

FIGURE 7.2
For node i, there are a number of subdomains: weighted domain Ω_W of a node at $\mathbf{x}_i$ is a domain in which $\widehat{W}_i(x) \neq 0$; quadrature domain Ω_Q is in Ω_W and often $\Omega_Q = \Omega_W$; and the support domain Ω_s for a quadrature point $\mathbf{x}_Q$. (From Gu, Y. T. and Liu, G. R., *Comput. Mech.*, 27, 188–198, 2001. With permission.)

Of course, the dimensions of these different domains will affect the results. These effects are addressed in later sections of this chapter via numerical examples.

7.1.4 Procedures for Essential Boundary Conditions

Enforcement of essential boundary conditions by the penalty method involves choice of penalty factor α. If α is chosen improperly, instability or erroneous results will sometimes occur. Alternatively, methods using an orthogonal transformation technique (Atluri et al., 1999b; Ouatouati and Johnson, 1999) have been proposed for imposition of essential boundary conditions. This section introduces a *method of direction interpolation* for the imposition of essential boundary conditions, which makes use of the special feature of the MLPG. This method was used by G. R. Liu and Yan (2000) to simplify the MLPG formulation.

As discussed above, the MLPG method establishes equations node by node, which makes it possible to use different sets of equations for the interior and boundary nodes. For node J located on the essential boundary, one can enforce the boundary condition using the equation of MLS approximation, i.e.,

$$u_J^h(\mathbf{x}) = \sum_{I=1}^{n} \phi_I(\mathbf{x}) u_I = \bar{u}_J \tag{7.23}$$

where $\bar{u}_J$ is the specified displacement at node J on the essential boundary. The foregoing equation is basically a linear algebraic equation for node J on the essential boundary. Therefore, for all the nodes on the essential boundary, there is no need to establish Equation 7.19. The essential boundary condition of Equation 7.23 is directly assembled into the global system equation. This treatment of the essential boundary condition is straightforward and very effective. It simplifies significantly the procedure of imposing essential boundary conditions, and the essential boundary conditions are satisfied exactly. Moreover, computation for all the nodes on the essential boundary has been simplified. This simple treatment is made possible because the MLPG method establishes discrete equations node by node.

Note also that this direct approach of imposing essential boundary conditions destroys the symmetry of the stiffness matrix. Fortunately, this does not create additional problems, because the stiffness matrix created using MLPG is not symmetric originally. If it were possible to apply this method in the EFG method, which produces symmetric matrices, it would not be used, because one probably loses more efficiency when the symmetry of the matrix is destroyed.

7.1.5 Numerical Investigation

Equations 7.20 and 7.21 require integration over the local quadrature domain and on the boundary that intersects with the quadrature domain. The integration has to be carried out via numerical quadrature schemes. In practice, the quadrature domain often needs to be further divided into cells, and the Gauss quadrature scheme is often used to evaluate the integration for each cell. Therefore, there will be a number of issues involved in the process, such as the number of cells and the number of the Gauss points to be used.

There exist, in general, difficulties in obtaining the exact (to machine accuracy) numerical integration in MFree methods (Atluri et al., 1999b; Liu, G. R. and Yan, 1999). Insufficiently accurate numerical integration may cause deterioration and rank deficiency in the numerical solution. In MLPG, the integration difficulty is more severe, because of the complexity of the integrand that results from the Petrov–Galerkin formulation. First, the shape functions constructed using MLS approximation have a complex feature, the shape functions have different forms in each small integration region, and the derivatives of the shape functions might have oscillations. Second, the overlapping of interpolation domains makes the integrand in the overlapping domain very complicated. To improve the accuracy of the numerical integration, the quadrature domain Ω_Q should be divided into small, regular partitions. In each small partition, more Gauss quadrature points should be used (Atluri et al., 1999b).

In this section, several numerical examples are employed to illustrate the implementation issues in the MLPG method using MLS approximation. The work was originally performed by G. R. Liu and Yan (2000). Rectangular quadrature domains Ω_Q are used, and the dimension of the quadrature domain for node I is defined as

$$(\alpha_Q \cdot d_{xI}) \times (\alpha_Q \cdot d_{yI}) \tag{7.24}$$

where α_Q is the dimensionless size of the rectangular quadrature domains, d_{xI} is the average nodal spacing in the horizontal direction between two neighboring nodes in the vicinity of node I, and d_{yI} is that in the vertical direction.

The support domain Ω_s used for constructing MLS shape functions is also a rectangle. The tensor product weight function for 2D problems is given by

$$\widehat{W}(\mathbf{x} - \mathbf{x}_I) = \widehat{W}(r_x) \cdot \widehat{W}(r_y) = \widehat{W}_x \cdot \widehat{W}_y \tag{7.25}$$

where $\widehat{W}(r_x)$ and $\widehat{W}(r_y)$ are any of the weight functions listed in Section 5.2.1 where $\bar{d}$ is replaced by r_x and r_y, which are given by

$$r_x = \frac{\|(x - x_I)\|}{x_{\max}} \tag{7.26}$$

$$r_y = \frac{\|(y - y_I)\|}{y_{\max}} \tag{7.27}$$

where x_{max} and y_{max} are, respectively, the dimensions of the rectangle in the x and y directions given by

$$x_{max} = \alpha_s d_{xI} \tag{7.28}$$

$$y_{max} = \alpha_s d_{yI} \tag{7.29}$$

where α_s is the dimensionless size of the support domain for computing the MLS shape functions.

The quadrature domain Ω_Q for constructing the weight (test) function is also rectangular, and any weight function given by Equations 5.11 through 5.13 and 5.15 can be used, where $\bar{d}$ is also replaced by r_x and r_y, which are defined by Equations 7.26 and 7.27. However, x_{max} and y_{max} are given by

$$x_{max} = \alpha_Q d_{xI} \tag{7.30}$$

$$y_{max} = \alpha_Q d_{yI} \tag{7.31}$$

where α_Q is the dimensionless size of the quadrature or weight domain. Note that in our implementation of MLPG, the weighted domain Ω_W coincides with the domain of quadrature Ω_Q; hence, the dimension of the quadrature domain is also the same as that of the weighted domain.

The quadrature domain is divided evenly by $n_c \times n_c$ cells, and 4×4 Gauss sampling points are used for each cell. To assess the accuracy, the relative error is defined as

$$r_e = \frac{\left\| \varepsilon^{exact} - \varepsilon^{num} \right\|}{\left\| \varepsilon^{exact} \right\|} \tag{7.32}$$

where the energy norm is defined as

$$\|\varepsilon\| = \left(\frac{1}{2} \int_{\Omega} \boldsymbol{\varepsilon}^T \mathbf{D} \boldsymbol{\varepsilon} \, d\Omega \right)^{1/2} \tag{7.33}$$

7.1.6 Examples

An MLPG code has been developed based on the above-mentioned formulation, and is used to conduct the following investigations. The direct approach is used for the imposition of essential boundary conditions. In the examples presented in this section, a rectangular support domain is used, and the dimension of the support domain is fixed at $\alpha_s = 3.5$, unless specified otherwise.

Example 7.1 Patch Test

Consider a standard patch test in a domain of dimension $[0,2] \times [0,2]$ with a linear displacement applied along its boundary: $u_x = x$, $u_y = y$. Satisfaction of the patch test requires that the value of u_x, u_y at any interior node be given by the same linear displacement function and that the derivative of the displacement be constant.

Three patterns of nodal arrangement shown in Figure 7.3 are considered: (a) 9 nodes with regular arrangement, (b) 9 nodes with a randomly distributed internal node, and

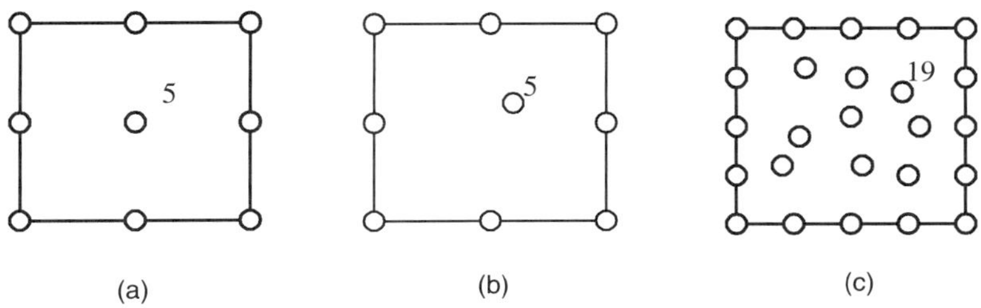

FIGURE 7.3
Nodal arrangement of patches for standard patch test: (a) 9 regular node patch; (b) 9 irregular node patch; (c) 25 irregular node patch.

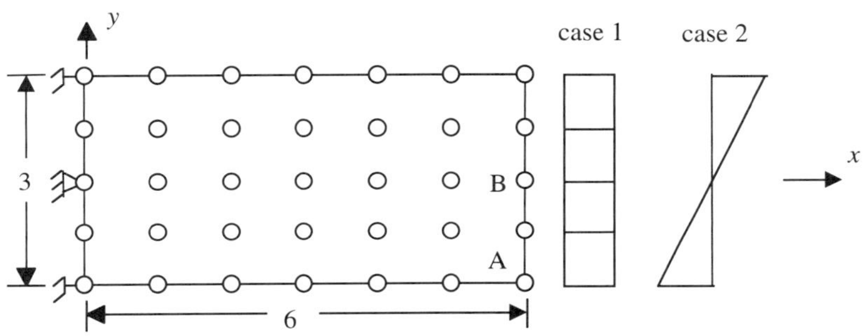

FIGURE 7.4
Nodal arrangement for high-order patch test.

(c) 25 nodes with irregular arrangement. The computational results have confirmed that MLPG can pass all the patch tests exactly (to machine accuracy) for both cubic and quartic spline weight functions with the given linear displacement boundary, as long as the integration is carried out accurately. Issues related to integration are discussed in detail in Example 7.3.

Example 7.2 High-Order Patch Test

A high-order patch of 3×6 shown in Figure 7.4 is used to study the effect of the order of polynomial basis used in MLS approximation and the quadrature domain on the numerical results of MLPG. The material properties for the patch are as follows:

Young's modulus: $E = 1$

Poisson's ratio: $\nu = 0.25$

The following two cases are examined.

CASE 1 A uniform axial stress with unit intensity is applied on the right end. The exact solution for this problem should be

$$\begin{aligned} u_x &= x \\ u_y &= -y/4 \end{aligned} \tag{7.34}$$

CASE 2 A linearly variable normal stress is applied on the right end.

TABLE 7.1

Displacement at the Right End of the High-Order Patch (number of nodes in the entire domain n_t: 5 ×7 =35; Gauss points: 4 ×4)

	u_x at Point A			u_y at Point B		
α_Q	Exact	Numerical	Error	Exact	Numerical	Error
1.0	−6.0	−6.418	6.97%	−12.0	−13.65	13.8%
1.5	−6.0	−5.942	−0.97%	−12.0	−11.84	−1.33%
2.0	−6.0	−5.959	−0.68%	−12.0	−11.86	−1.17%

α_Q: dimensionless size of the quadrature domain.

The exact solution for this problem should be

$$\begin{aligned} u_x &= 2xy/3 \\ u_y &= -(x^2 + y^2/4)/3 \end{aligned} \tag{7.35}$$

The linear basis is first used in MLS for creation of the shape functions. It is found that the MLPG can produce an exact solution to machine accuracy for case 1 and hence pass the patch test exactly. For case 2, however, it fails. Table 7.1 shows the relative error of displacement u_x and u_y at point A when the linear basis and quartic spline are used for case 2. It is shown that the MLPG results converge as the dimension of the quadrature domain α_Q increases, although it cannot produce the exact solution. Note that the exact solution of the displacement field for case 2 is quadratic, as shown in Equation 7.35. For the MLS shape function to reproduce the quadratic displacement field exactly, quadratic polynomial basis functions have to be used (see consistency issues with MLS discussed in Chapter 5). The patch is then attempted again using the quadratic basis, which confirms that MLPG has passed the patch test for case 2 as well.

Requirements for Local Petrov–Galerkin Methods to Pass the Patch Test

The requirements for all the methods based on the local Petrov–Galerkin residual method to pass the standard patch test may be given as follows:

1. The shape functions are of at least linear consistency (see Chapter 5). This implies that the shape function is capable of at least linear field reproduction.
2. The essential boundary conditions (displacement constraints on the boundary of the patch) have to be satisfied accurately. This is the condition given by Equation 4.2.

In addition, we require accurate numerical operations, such as integration to form system equations, in the process of testing.

Compared to the requirements for the Galerkin type of methods, local residual methods do not require compatibility on field function approximation (see the discussion given in Example 6.2). This can be easily understood intuitively using Equation 4.49.

We know that the exact solution of the standard patch test problem is a linear function (for the field variable of displacement). If a field variable approximation can produce the linear field at any point in the local quadrature domain of a node, which can satisfy Equation 4.57, Equation 4.49 is then naturally satisfied, which means that the local residual approach will produce the exact solution for the node. Because the local weighted residual method is applied independently to all the nodes in the problem domain, the linear field for the entire patch will be produced. The issue of incompatibility does not come into the picture. All we require is that the field variable approximation at any point in the quadrature domain be capable of linear field reproduction.

The MLS shape functions can satisfy the first requirement very easily as long as linear polynomial functions are included in the basis for constructing the shape functions. To satisfy the second requirement, the constrained local residual form is required in constructing the system equations or the direct interpolation method for the essential boundary nodes. We therefore conclude that the MLPG method formulated in this chapter can always pass the standard patch test, provided the numerical integration is accurate. In other words, the MLPG is productive.

PIM shape functions can satisfy both requirements very easily as long as linear polynomial functions are included in the basis for constructing the shape functions. We therefore can expect that PIM shape functions will work perfectly for local residual formulations and will not have any problem passing the standard patch tests. We discuss this issue in more detail in Chapter 8.

Example 7.3 Cantilever Beam

The cantilever beam described in Chapter 6 (Example 6.2) is tested again using the MLPG code. The beam is schematically drawn in Figure 6.4. The parameters for this example are as follows:

Loading: $P = -1000$ N

Young's modulus for the material: $E = 3.0 \times 10^7$ N/m^2

Poisson's ratio for two materials: $\nu = 0.3$

Height of the beam: $D = 12.0$ m

Length of the beam: $L = 48.0$ m

Three patterns with 55 (5 × 11), 189 (9 × 21), and 697 (17 × 41) regularly distributed nodes are employed to study the convergence of the MLPG method. Both linear basis and quartic spline are used. The results are presented for different numbers of integration cells n_c in each local quadrature domain Ω_Q (which is the same as the weighted domain Ω_W) and different dimensions of the quadrature domain.

Figures 7.5 and 7.6 plot the relationship between the relative error in strain defined in Equation 7.32 and the cell number n_c. Results in these two figures clearly show that

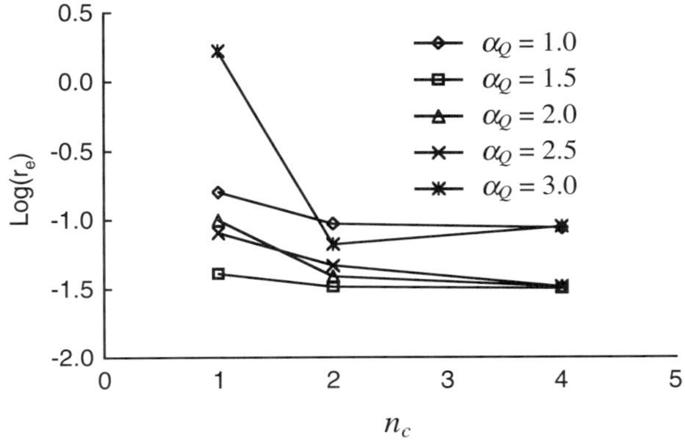

FIGURE 7.5
Rate of convergence in terms of relative error in strains computed in the cantilever beam using MLPG with MLS approximation (n_c = number of subdivision of the rectangular quadrature domain; α_Q = the dimension parameter of the rectangular quadrature domain; domain of support: $\alpha_s = 3.5$; total number of nodes: $n_t = 55$).

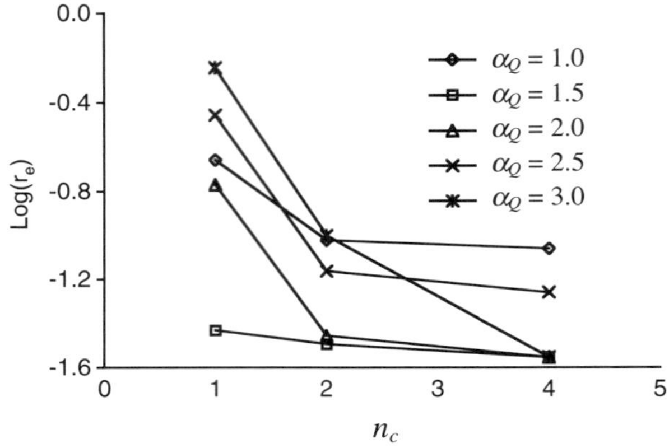

FIGURE 7.6
Same as Figure 7.5, but the number of field nodes n_t = 189.

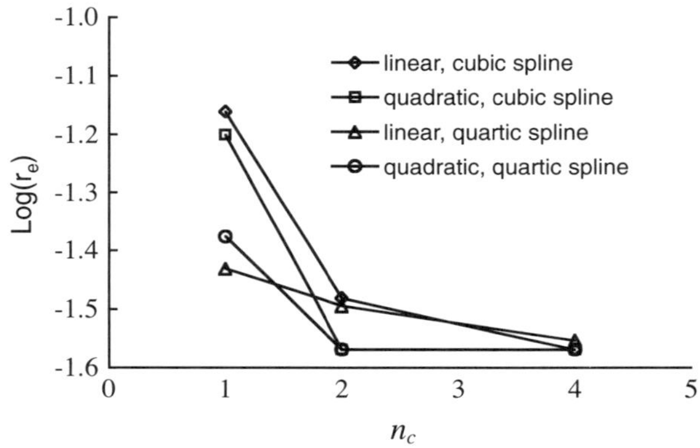

FIGURE 7.7
Rate of convergence in terms of relative error in strains computed in the cantilever beam using MLPG with MLS approximation. Effects of spline weight functions ($\alpha_Q = 1.5$; $n_t = 189$; $\alpha_s = 3.5$).

accuracy can be improved significantly by increasing the number of cells for local integration. These findings suggest the importance of subdivisions of the quadrature domain. Both Figures 7.5 and 7.6 also indicate an important fact: too large ($\alpha_Q = 3.0$) or too small ($\alpha_Q = 1.0$) a quadrature domain will give less accurate results. When the quadrature domain is too small, the area for the "smear" error is not large enough, and when the quadrature is too large, the error in the numerical integration will affect the accuracy of the results. Both Figures 7.5 and 7.6 show that a quadrature domain of $\alpha_Q = 1.5$ gives the most accurate results. This suggests that the dimension of the rectangular quadrature domain should be about 1.5 times the local nodal distance. This rule of 1.5 times nodal distance is widely used in MFree methods based on local weak forms.

The effects of the spline and basis functions on the relative error have also been investigated. The results are shown in Figure 7.7. It is found that there is no significant difference in accuracy using cubic and quartic spline weighted functions for a sufficient number of subdivisions of the quadrature domain. Using a quadratic basis function can somewhat increase the accuracy of the results, but it is not a clear indication.

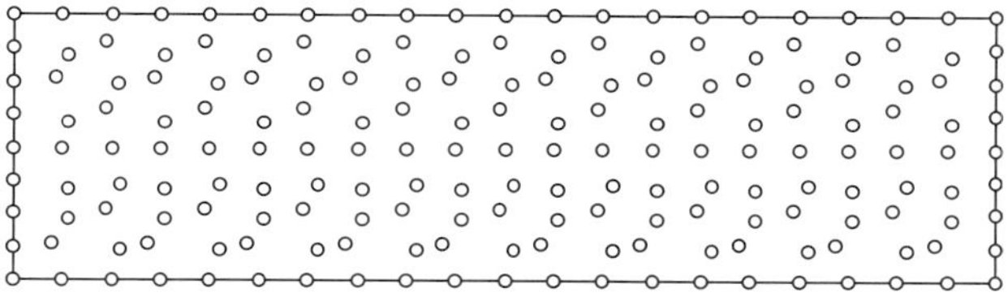

FIGURE 7.8
Irregular nodal arrangement for the cantilever beam (189 nodes).

TABLE 7.2
Relative Error for Irregular Node Arrangement

c_n	0.0	0.1	0.2	0.3	0.4
$r_e(\times 10^{-2})$	2.77	2.81	2.84	2.87	2.89

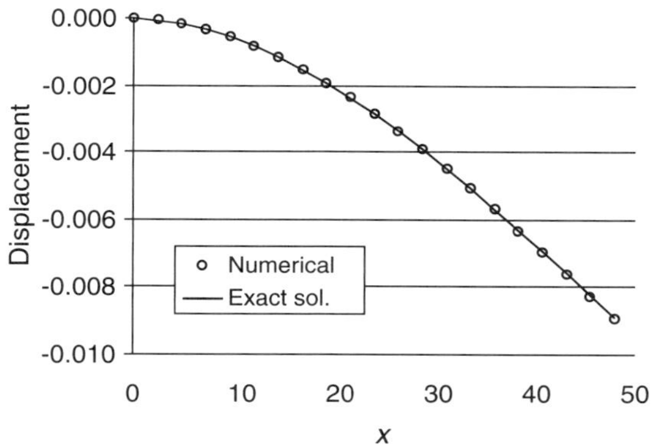

FIGURE 7.9
Deflection of the cantilever beam computed using MLPG with 189 irregular nodes. Comparison with the exact solution.

Irregularity of the nodal arrangement is also investigated. The irregularity of nodes is created by changing the coordinates of the interior nodes in the beam in the following manner:

$$\begin{aligned} x_I &= x_I \pm c_n d_{xI} \\ y_I &= y_I \pm c_n d_{yI} \end{aligned} \tag{7.36}$$

where c_n is the parameter that controls the irregularity of the nodes. For $c_n = 0.4$, some of the internal nodes are moved up to $0.4d_{xI}$ in the horizontal direction and $0.4d_{yI}$ in the vertical direction from its regular position. Parameter c_n is allowed to vary randomly in the range of 0.0 to 0.4. Figure 7.8 shows a typical irregular nodal arrangement. Table 7.2 shows the relative error defined in Equation 7.32 obtained using irregular node arrangements of different c_n values. It is seen that the irregularity of nodes has very little effect on the accuracy of the results. This fact reveals a very important feature of MLPG: it is stable for irregular nodal arrangements. The results based on $c_n = 0.4$ are obtained and plotted in Figure 7.9 for deflection of the beam, and in Figure 7.10 for the shear stress distribution at the central section of $x = 24$ m. Those results confirm that the effect of the nodal irregularity is very small, and can be neglected.

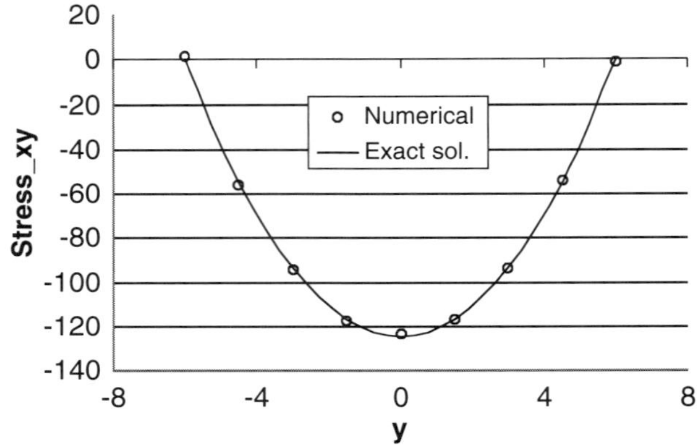

FIGURE 7.10
Shear stress distribution at central section of the cantilever beam computed using MLPG with 189 irregular nodes. Comparison with the exact solution.

Example 7.4 Infinite Plate with a Circular Hole

Example 6.11, examined using the EFG method, is now reexamined here using the MLPG method. The geometry of the plate is plotted in Figure 6.34. Because of the twofold symmetry, only a quarter of the plate shown in Figure 6.35 is modeled with symmetric boundary conditions applied on $x = 0$ and $y = 0$. The parameters are listed as follows:

Loading: $P = 1$ N/m

Young's modulus: $E = 1.0 \times 10^3$ N/m^2

Poisson's ratio: $v = 0.3$

Height of the beam: $a = 1.0$ m

Length of the beam: $b = 5$ m

The plate is subjected to a tension in the x direction at the edge of $x = 5$. The boundary condition at $x = 5$ is $\sigma_{xx} = p$, $\sigma_{yy} = \sigma_{xy} = 0$, and the boundary condition at $y = 5$ is free of all stresses. The analytical solution of displacement and the stress fields within the plate are provided by Equations 6.121 to 6.126 in the polar coordinates of (r, θ). A total of 165 nodes are used to represent the domain, as shown in Figure 6.36b. The stress components σ_{xx} obtained at $x = 0$ for $\alpha_s = 3.0$ are compared with the exact solution in Figure 7.11. Two types of cells of $n_c = 1$ and $n_c = 2$ are used. Figure 7.11 shows that finer cells ($n_c = 2$) give a more accurate result than coarse cells ($n_c = 1$). The results suggest again the importance of subdivision in the local quadrature domain. Figure 7.12 plots the results for different sizes of support domain α_Q, which implies a stable result with different α_Q.

Example 7.5 Half-Plane Problem

Stress analysis is carried out for a half plane of elastic solid subjected to a concentrated force, as shown in Figure 7.13. The results are compared with those obtained using the finite element method (FEM) at the section S-S' for the same distribution of nodes. We present the results of comparison only for stress, as it is much more critical than displacement. Figure 7.14 shows the distribution of the normal stress σ_x and the shear stress τ_{xy} along section S-S'. The results are obtained using different integral cells n_c and for given $\alpha_Q = 1.5$. Table 7.3 lists the normal stress at point A compared with those obtained using FEM.

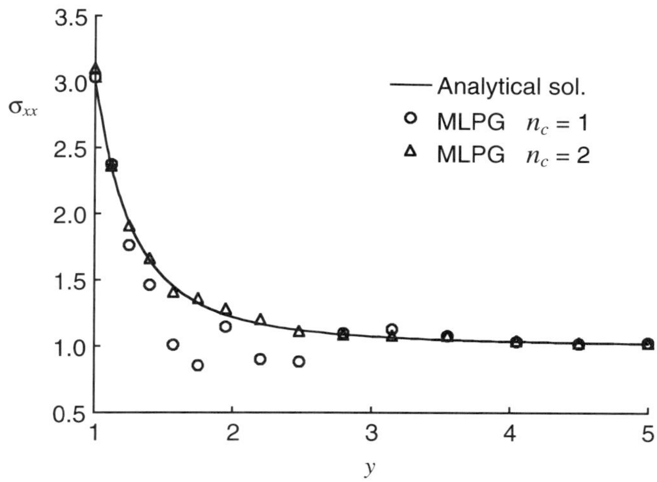

FIGURE 7.11
Comparison between the exact solution and MLPG with MLS approximation for σ_{xx} at $x = 0$ ($\alpha_s = 3.5$, $\alpha_Q = 3.0$).

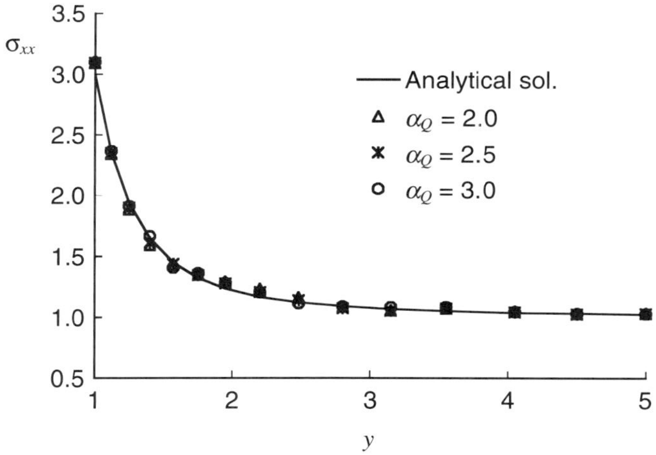

FIGURE 7.12
Comparison between the exact solution and MLPG with MLS approximation for σ_{xx} at $x = 0$ ($N = 165$, $\alpha_s = 3.5$, $n_c = 2$).

The results indicate that a sufficiently large dimension of quadrature domain together with a corresponding sufficient number of subdivisions of the quadrature domain for integration is necessary to obtain an accurate solution. Subdivision of the quadrature domain, however, leads to additional computation work. An acceptable combination of the number of subdivisions and the dimension of the quadrature domain is quite complicated, not very clear at this stage, and needs to be further investigated. The above examples seem to suggest that four subdivisions ($n_c = 2$) work well for rectangular quadrature domains of $\alpha_Q = 2.0$. It has been found for these examples that $n_c = \alpha_Q$ seems to be a necessary condition to obtain a reasonably accurate result. Because we find no reason to use a quadrature domain larger than $\alpha_Q = 2.0$, four subdivisions should be a good and economic choice for normal cases. If, for any reason, a large quadrature domain has to be

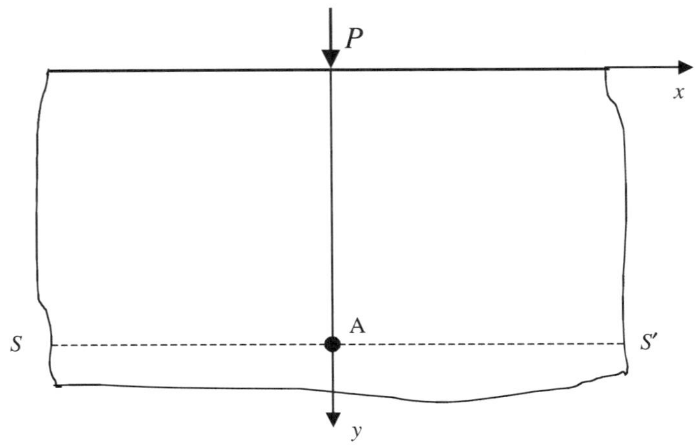

FIGURE 7.13
Half plane of elastic solid subjected to a vertical concentrated force *P*.

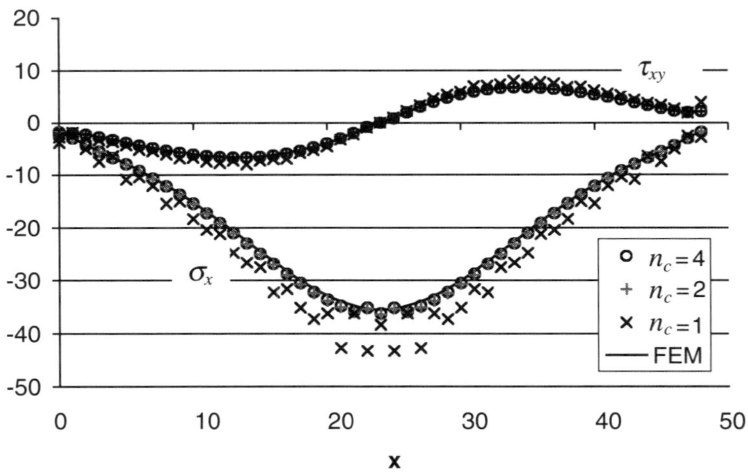

FIGURE 7.14
Stress distribution at section of S-S′ ($\alpha_Q = 1.5$).

TABLE 7.3

Comparison of Normal Stress at Point A with FEM Result (=35.4)

Number of Subdivision of Quadrature Domain $n_c \times n_c$	Dimension of Quadrature Domain (α_Q)		
	1.0	2.0	3.0
1 × 1	46.0	—	—
2 × 2	44.1	36.7	—
4 × 4	44.2	36.7	35.7

used, our suggestion would be

$$n_c = \text{round up to the nearest even number } (\alpha_Q) \tag{7.37}$$

The reason for the preference for an even number will be given in Sections 9.3 and 9.4. Discussion on local domain integration in Sections 9.3 and 9.4 also suggests four subdivisions for domains of standard squares that are obtained through coordinate transformation of irregular domains.

7.2 MLPG for Dynamic Problems

7.2.1 Statement of the Problem

Vibration analysis for structures is a very important field in computational mechanics. These dynamics problems are classically described by a linear partial differential equation associated with a set of boundary conditions and initial conditions. Exact analyses of these types of boundary and initial value problems are usually very difficult. Analytical solutions are available for very few problems with very simple geometry and boundary and initial conditions. Therefore, numerical techniques with different discretization schemes, such as FEM, have been widely used in solving practical vibration problems in science and engineering.

Examples presented in previous sections have demonstrated that the MLPG method, a truly mesh-free method, works well for static mechanics problems. It is, therefore, a natural extension to develop further MLPG for dynamic mechanics problems. Gu and G. R. Liu (2001c) have reported work on the development of MLPG for dynamic problems of 2D solids for both free-vibration and forced vibration analyses.

This section introduces their formulation. First, the local weak forms are presented using the weighted residual method locally and the strong form of partial differential dynamic system equations. MLS approximation is used to obtain the MLS shape functions, which are fed to the local Petrov–Galerkin formulation to derive a set of discretized dynamic system equations. In free-vibration analysis, the essential boundary conditions are formulated separately using the method of direct interpolation. The boundary conditions are then imposed utilizing orthogonal transform techniques to modify the discretized unconstrained dynamic system equations to obtain the eigenvalue equation. Frequencies and eigenmodes of free vibration are obtained by solving the eigenvalue equation. In forced vibration analysis, the penalty method is used to implement the essential conditions. Both the explicit time integration method (the central difference method) and the implicit time integration method (the Newmark method) are used to solve the forced vibration system equations.

Programs of the MLPG method have been developed, and a number of numerical examples of free-vibration and forced vibration analyses are presented to demonstrate the convergence, validity, and efficiency of the present methods. Some important parameters on the performance of the present MLPG method are also investigated in great detail.

The strong form of the initial/boundary value problem for small displacement elastodynamics can be given as follows (see Chapter 3):

$$\sigma_{ij,j} + b_i = \rho \ddot{u}_i + \eta_c \dot{u}_i \tag{7.38}$$

where ρ is the mass density, η_c is the damping coefficient, u_i is the displacement, $\ddot{u}_i = \partial^2 u_i / \partial t^2$ is the acceleration, $\dot{u}_i = \partial u_i / \partial t$ the velocity, σ_{ij} the stress tensor, b_i the body force tensor, and $(\,)_{,j}$ denotes the operation of $\partial/\partial x_j$. The auxiliary conditions are given as follows:

$$\text{Natural boundary condition:} \quad \sigma_{ij}n_j = \bar{t}_i \quad \text{on } \Gamma_t \tag{7.39}$$

$$\text{Essential boundary condition:} \quad u_i = \bar{u}_i \quad \text{on } \Gamma_u \tag{7.40}$$

$$\text{Displacement initial condition:} \quad u_i(\mathbf{x}, t_0) = \tilde{u}_i(\mathbf{x}) \quad \mathbf{x} \in \Omega \tag{7.41}$$

$$\text{Velocity initial condition:} \quad \dot{u}_i(\mathbf{x}, t_0) = \tilde{\dot{u}}_i(\mathbf{x}) \quad \mathbf{x} \in \Omega \tag{7.42}$$

in which the u_i, $\bar{t}_i$, u_0, and $\tilde{u}_i$ denote the prescribed displacements, tractions, initial displacements, and velocities, respectively. Note that the differences between the dynamic system equations for static and dynamic problems are (1) the inertial and damping terms in the equilibrium equations and (2) the additional equations of initial conditions.

7.2.2 Free-Vibration Analysis

The governing equation for an undamped free vibration can be written as follows:

$$\sigma_{ij,j} = \rho \ddot{u}_i \tag{7.43}$$

The boundary condition for the free vibration is reduced to only the essential boundary condition, Equation 7.40. In free-vibration analyses, the system is assumed to undergo harmonic motion, and the displacement $\mathbf{u}(\mathbf{x}, t)$ can be written in the form:

$$\mathbf{u}(\mathbf{x}, t) = \mathbf{u}(\mathbf{x})\sin(\omega t + \varphi) \tag{7.44}$$

where ω is the frequency of free vibration. Substituting Equation 7.44 into Equation 7.43 leads to the following equations of equilibrium for free vibration:

$$\sigma_{ij,j} + \omega^2 \rho u_i = 0 \tag{7.45}$$

It should be noted that the stresses, $\boldsymbol{\sigma}$, and displacements, $\mathbf{u}$, in Equation 7.45 are only functions of coordinate $\mathbf{x}$ for a given frequency ω.

A local weak form of Equation 7.45, over a local quadrature domain Ω_Q bounded by Γ_Q, can be obtained using the weighted residual method with integration over Ω_Q.

$$\int_{\Omega_Q} \widehat{W}(\sigma_{ij,j} + \omega^2 \rho u_i)\, \mathrm{d}\Omega = 0 \tag{7.46}$$

where $\widehat{W}$ is the weight (or test) function.

The first term on the left-hand side of Equation 7.46 can be integrated by parts to become

$$\int_{\Gamma_Q} \widehat{W}\sigma_{ij}n_j\, \mathrm{d}\Gamma - \int_{\Omega_Q} (\widehat{W}_{,j}\sigma_{ij} - \widehat{W}\omega^2 \rho u_i)\, \mathrm{d}\Omega = 0 \tag{7.47}$$

The local quadrature domain Ω_Q of a node i at $\mathbf{x}_i$ is the same as the weighted domain where $\widehat{W}(x) \neq 0$. The choice of Ω_Q is the same as that discussed in Section 7.1.2 for static problems. As shown in Figure 7.1, the boundary Γ_Q for the support domain Ω_Q is usually composed of three parts: the internal boundary Γ_{Qi} and the boundaries Γ_{Qu} and Γ_{Qt}, over which the essential and natural boundary conditions are specified. Imposing the natural boundary condition and noting that $\sigma_{ij}n_j = \partial u/\partial n \equiv t_i$ in Equation 7.47, and the fact that $\widehat{W}(x) = 0$ on Γ_{Qi}, we obtain

$$\int_{\Gamma_{Qu}} \widehat{\mathrm{W}} t_i \, \mathrm{d}\Gamma + \int_{\Gamma_{Qt}} \widehat{\mathrm{W}} \bar{t}_i \, \mathrm{d}\Gamma - \int_{\Omega_Q} (\widehat{\mathrm{W}}_{,j}\sigma_{ij} - \widehat{\mathrm{W}}\omega^2 \rho u_i) \, \mathrm{d}\Omega = 0 \tag{7.48}$$

For a support domain located entirely within the global domain, there is no intersection between Γ_Q and the global boundary Γ, and $\Gamma_{Qi}=\Gamma_Q$. This leads to the vanishing of integrals over Γ_{Qu} and Γ_{Qt}. Also note that for free vibration, we should have $\bar{t} = 0$ on Γ_t; the integrals over Γ_{Qt} vanish for all nodes in the free-vibration analysis. Equation 7.48 can be further expressed as follows. For nodes whose support domains do not intersect with the problem boundary:

$$\int_{\Omega_Q} (\widehat{W}_{,j}\sigma_{ij} - \widehat{W}\omega^2 \rho u_i) \mathrm{d}\Omega = 0 \tag{7.49}$$

For nodes whose support domains intersect with the problem boundary:

$$\int_{\Gamma_{Qu}} \widehat{W} t_i \, \mathrm{d}\Gamma - \int_{\Omega_Q} (\widehat{W}_{,j}\sigma_{ij} - \widehat{W}\omega^2 \rho u_i) \, \mathrm{d}\Omega = 0 \tag{7.50}$$

MLS approximation (Equation 7.8) is then used to approximate the field variables at any point in the support domain Ω_s of a node. Substituting Equation 7.8 and any weight functions given in Equations 5.11 through 5.13 and 5.15 into the local weak form Equation 7.49 or 7.50 for each and every node in the problem domain leads to the following discrete system equations. The procedure is exactly the same as in Section 7.2.1, except that the inertial term needs to be treated, which is trivial.

$$\mathbf{KU} - \omega^2 \mathbf{MU} = 0 \tag{7.51}$$

where the global stiffness matrix $\mathbf{K}$ is assembled using the nodal stiffness matrix $\mathbf{K}_{IJ}$. For nodes whose quadrature domains intersect with the problem boundary, we have

$$\mathbf{K}_{IJ} = \int_{\Omega_Q} \widehat{\mathbf{V}}_I^T \mathbf{c} \mathbf{B}_J \, \mathrm{d}\Omega - \int_{\Gamma_{Qu}} \widehat{\mathbf{W}}_I \mathbf{n} \mathbf{c} \mathbf{B}_J \, \mathrm{d}\Gamma \tag{7.52}$$

where $\widehat{\mathbf{V}}_I$, $\mathbf{B}_I$, $\widehat{\mathbf{W}}_I$, and $\mathbf{n}$ are defined, respectively, by Equations 7.12, 7.14, 7.15, and 7.17. For nodes whose quadrature domains do not intersect with the problem boundary, the nodal stiffness matrix is simplified as

$$\mathbf{K}_{IJ} = \int_{\Omega_Q} \widehat{\mathbf{V}}_I^T \mathbf{c} \mathbf{B}_J \, \mathrm{d}\Omega \tag{7.53}$$

The "mass" matrix $\mathbf{M}$ is obtained using

$$\mathbf{M}_{IJ} = \int_{\Omega_Q} \rho \widehat{\mathbf{W}}_I \mathbf{\Phi}_J \, \mathrm{d}\Omega \tag{7.54}$$

where $\mathbf{\Phi}_J$ is a matrix of the MLS shape function for node J, given by Equation 7.9.

For free-vibration analyses, Equation 7.51 can also be written in the following form of eigenvalue equation:

$$(\mathbf{K} - \lambda\mathbf{M})\mathbf{q} = 0 \tag{7.55}$$

where $\mathbf{q}$ is the eigenvector and λ is termed an eigenvalue that relates to the natural frequency in the form:

$$\lambda = \omega^2 \tag{7.56}$$

Equation 7.55 is the unconstrained eigenvalue equation that contains the rigid movement of the solid. To determine the frequencies, ω, and free-vibration modes for a constrained system, it is necessary to impose the essential boundary condition defined by Equation 7.40.

7.2.3 Imposition of Essential Boundary Conditions for Free Vibration

In the MLPG method that uses MLS shape functions, special effort has to be made to enforce essential boundary conditions for dynamic problems, because the shape functions constructed by MLS approximation lack the delta function property. In previous sections we have shown the penalty method (Atluri and Zhu, 1998) and the direct interpolation method (Liu, G. R. and Yan, 2000) for problems of static stress analyses. Here, we introduce the method using orthogonal transform techniques (Atluri et al., 1999b; Ouatouati and Johnson, 1999; Liu, G. R. and Chen, 2000, 2001) to establish a system equation of constrained free vibration.

Note the fact that for free-vibration analysis, the essential boundary conditions are always homogeneous, meaning that $\bar{u}_i = 0$ in Equation 7.40. Substituting Equation 7.8 into Equation 7.40, we find a set of algebraic linear equations of constraints

$$\mathbf{C}\mathbf{q} = 0 \tag{7.57}$$

where $\mathbf{C}$ is a flat matrix of $N_c \times N_t$ with many zero elements, N_c is the total number of constrained degrees of freedom, and N_t is the total number of degrees of freedom of the entire system. If the domain is represented by n_t nodes, $N_t = 2n_t$ for 2D solid mechanics problems. Using singular value decomposition (Strang, 1976), matrix $\mathbf{C}$ can be decomposed as

$$\mathbf{C} = \mathbf{U}_{Nc\times Nc}\begin{bmatrix}\Sigma_{N_r\times N_r} & 0\\ 0 & 0\end{bmatrix}_{N_c\times N_t}\mathbf{V}^T{}_{N_t\times N_t} \tag{7.58}$$

where $\mathbf{U}$ and $\mathbf{V}$ are orthogonal matrices, and $\mathbf{\Sigma}$ is a diagonal matrix with singular values of $\mathbf{C}$ for its diagonal terms. The matrix $\mathbf{V}$ can be written as

$$\mathbf{V}^T = \{\mathbf{V}_{N_t\times N_r}, \mathbf{V}_{N_t\times(N_t-N_r)}\}^T \tag{7.59}$$

where N_r is the rank of $\mathbf{C}$, namely, the number of independent constraints.

Performing coordinate transformation of

$$\mathbf{q} = \mathbf{V}_{N_t\times(N_t-N_r)}\tilde{\mathbf{q}} \tag{7.60}$$

the change of coordinates satisfies the constraint equation of Equation 7.57. Substituting Equation 7.60 into Equation 7.55 leads to

$$(\tilde{\mathbf{K}} - \omega^2\tilde{\mathbf{M}})\tilde{\mathbf{q}} = 0 \tag{7.61}$$

where the condensed stiffness matrix $\tilde{\mathbf{K}}$ given by

$$\tilde{\mathbf{K}}_{(N_t-N_r)\times(N_t-N_r)} = \mathbf{V}^T_{(N_t-N_r)\times N_t}\mathbf{K}_{N_t\times N_t}\mathbf{V}_{N_t\times(N_t-N_r)} \quad (7.62)$$

and the condensed stiffness mass matrix $\tilde{\mathbf{M}}$ becomes

$$\tilde{\mathbf{M}}_{(N_t-N_r)\times(N_t-N_r)} = \mathbf{V}^T_{(N_t-N_r)\times N_t}\mathbf{M}_{N_t\times N_t}\mathbf{V}_{N_t\times(N_t-N_r)} \quad (7.63)$$

Equation 7.61 is the eigenvalue equation for free vibration of a constrained solid. A detailed proof on this procedure is provided in Section 11.1.4.

It can be easily seen that the stiffness matrix **K** and the mass matrix **M** developed using the Petrov–Galerkin approach will be asymmetric. They will be banded as long as the support domain is compact.

A numerical integration is needed to evaluate the integration for computing both the stiffness and mass matrices, and Gauss quadrature can be used. Here we investigate also the effects of the dimensions of three local domains shown in Figure 7.2. In computing the stiffness matrix, it should be noted that for nodes whose quadrature domain intersects with the boundary of the problem domain, Equation 7.52 should be used. For interior nodes whose quadrature domains do not intersect with the boundary of the problem domain, Equation 7.53 should be used. In computing the mass matrix, Equation 7.54 is used for all the nodes.

Because the problem domains in the following examples are rectangles, rectangular local domains defined in Section 7.1.5 are used.

7.2.4 Numerical Examples

The MLPG method is used for free-vibration analysis of structures made of 2D solids.

Example 7.6 Cantilever Beam

The MLPG method is applied to analyze the free vibration of a cantilever beam, as shown in Figure 6.4. A plane stress problem is considered. The parameters for this example are as follows:

Young's modulus for the material: $E = 2.1 \times 10^4$ kgf/mm^2
Poisson's ratio for two materials: $\nu = 0.3$
Mass density: $\rho = 8.0 \times 10^{-10}$ kgfs^2/mm^4
Thickness of the beam: $t = 1$ mm
Height of the beam: $D = 10$ mm
Length of the beam: $L = 100$ mm

The problem has been analyzed by Nagashima (1999) using the node-by-node meshless (NBNM) method. The beam is first represented using a number of field nodes. Figure 7.15 shows two kinds of nodal arrangements: coarse (63 nodes) and fine (306 nodes). The effects of dimensions of the quadrature domain are investigated using different α_Q defined in Equations 7.30 and 7.31. It has been found that $\alpha_Q = 1.5$ to 2.5 can yield almost identical results in free-vibration analyses. This finding agrees well with that for static analyses presented in the previous section. From the static problems, we found that the quadrature domain used should be as small as possible, to reduce the burden in numerical integration. Therefore, $\alpha_Q = 1.5$ is used in the following free-vibration analyses.

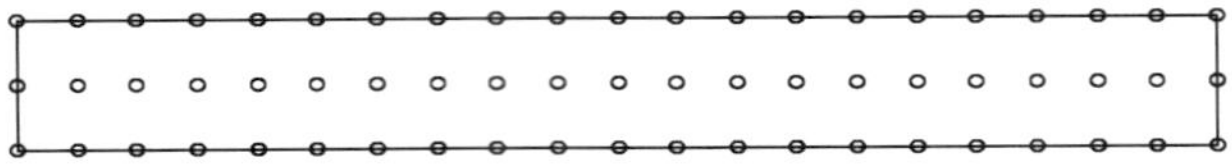

(a)

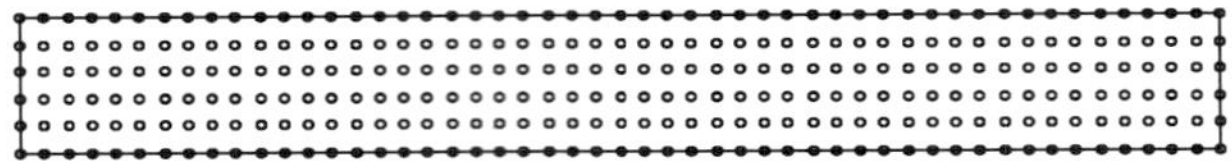

(b)

FIGURE 7.15
Nodal arrangement: (a) 63 nodes; (b) 306 nodes. (From Gu, Y. T. and Liu, G. R., *Comput. Mech.*, 27, 188–198, 2001. With permission.)

TABLE 7.4
Natural Frequencies (Hz) of a Cantilever Beam Computed Using MFree and FEM Models with Different Nodes ($\alpha_Q = 1.5$, $\alpha_s = 3.5$ for MLPG)

	Coarse MFree Model (63 nodes)			Fine MFree Model (306 nodes)		
Mode	MLPG	Nagashima (1999)	FEM (ABAQUS)	MLPG	Nagashima (1999)	FEM (ABAQUS)
1	919.47	926.10	870	824.44	844.19	830
2	5732.42	5484.11	5199	5070.32	5051.21	4979
3	12983.25	12831.88	12830	12894.73	12827.60	12826
4	14808.64	14201.32	13640	13188.12	13258.21	13111
5	26681.81	25290.04	24685	24044.43	23992.82	23818
6	38961.74	37350.18	37477	36596.15	36432.15	36308
7	40216.58	38320.59	38378	38723.90	38436.43	38436
8	55060.24	50818.64	51322	50389.01	49937.19	49958
9	64738.59	63283.70	63584	64413.89	63901.16	63917
10	68681.87	63994.48	65731	64937.83	64085.90	64348

Source: Gu, Y. T. and Liu, G. R., *Comput. Mech.*, 27, 188–198, 2001. With permission.

Frequency results of these two nodal arrangements obtained by MLPG are listed in Table 7.4. The results obtained by the FEM software ABAQUS and the NBNM method (Nagashima, 1999) are also listed in the table. The mesh used in the FEM mode has the same nodal arrangement. From this table, one can observe that the results by the present MLPG method are in good agreement with those obtained using FEM and the NBNM method. The convergence of the present method is also demonstrated in this table. As the number of nodes increases, results obtained by the present MLPG approach the FEM results (if we consider the FEM results as a reference). The lowest ten vibration modes obtained by the MLPG method are plotted in Figure 7.16. Comparison of the FEM results with Nagashima's (1999) results reveals that they are almost identical.

For beams with small slenderness ratios, the Euler–Bernoulli beam theory can be applied to obtain an analytical solution for their natural frequencies. The slenderness of a beam is expressed by the slenderness ratio, r/L, where $r = \sqrt{I/A}$ is the radius of gyration of the cross section, I is the moment of inertia of the cross section of the beam, and L is the length of the beam. To further benchmark the MLPG code developed, beams with two slenderness ratios, $r/L = 0.029$ ($L = 100$, $D = 10$, $t = 1.0$) and 0.144 ($L = 100$, $D = 50$, $t = 1.0$), are analyzed. The frequency results are listed in Table 7.5. For a beam of small slenderness ratio, one expects a very good prediction from the Euler–Bernoulli beam theory. Comparison with

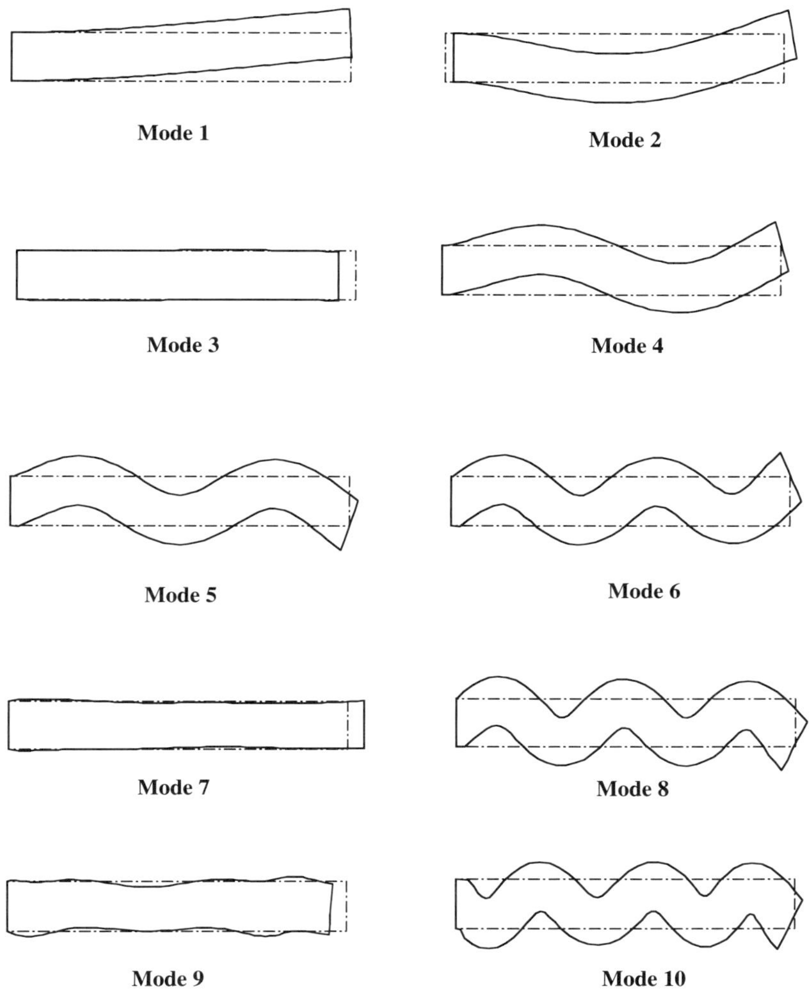

FIGURE 7.16
The lowest ten vibration modes of the cantilever beam. (From Gu, Y. T. and Liu, G. R., *Comput. Mech.*, 27, 188–198, 2001. With permission.)

TABLE 7.5

Natural Frequencies (Hz) of a Cantilever Beam with Different Slenderness (306 nodes are used in MLPG and FEM; $\alpha_Q = 1.5$, $n_s = 3.5$ are used for MLPG)

	***r/L* = 0.144**			***r/L* = 0.029**		
Modes	**MLPG**	**FEM (ABAQUS)**	**Euler Beam**	**MLPG**	**FEM (ABAQUS)**	**Euler Beam**
1	3565.81	3546.1	4138.23	824.44	830.19	827.65
Error with Euler beam (%)	−13.83	−14.31	—	−0.39	0.31	—
2	13025.06	12864	25933.86	5070.32	4979	5186.77
Error with Euler beam (%)	18.56	20.6	—	−2.24	−4.01	—

Source: Gu, Y. T. and Liu, G. R., *Comput. Mech.*, 27, 188–198, 2001. With permission.

the Euler–Bernoulli beam results reveals that, as the slenderness ratio r/L decreases, the natural frequencies of this 2D beam approach the values for an Euler–Bernoulli model. For the slender case of $r/L = 0.029$, the Euler–Bernoulli solution can be considered very close to the exact solution, and should be used as the reference. Table 7.5 shows that MLPG gives more accurate results compared with finite element results. It is well known that the fundamental (first) frequency of a 2D solid obtained by FEM should be larger than the exact value, meaning that the FEM results always give the upper bound of the exact results, and approach the exact results from the top. The MLPG result, however, does not guarantee an upper-bound solution. It approaches the exact solution from both sides. This is caused by the use of the local residual weak formulation.

Example 7.7 Cantilever Beam with Variable Cross Section

In this example, the present MLPG method is used in the free-vibration analysis of a cantilever beam with varying cross section, as shown in Figure 7.17. Results are obtained for the following parameters:

Young's modulus for the material: $E = 3.0 \times 10^7$ N/m^2

Poisson's ratio for two materials: $\nu = 0.3$

Mass density: $\rho = 1$ kg/m^3

Thickness of the beam: $t = 1$ m

Length of the beam: $L = 10$ m

Height of the beam: $D(0) = 5$ m and $D(10) = 3$ m

The nodal arrangement is shown in Figure 7.17. Results obtained by the MLPG method and the FEM software ABAQUS are listed in Table 7.6 for comparison. Results obtained by these two methods are in very good agreement.

Example 7.8 Shear Wall

MLPG is employed for free-vibration analysis of a shear wall with four openings, as shown in Figure 7.18. The shear wall is fully clamped on the bottom, and all the rest of the boundaries are free of external forces. The problem is considered a plane stress problem and a unit thickness is used. This problem has been studied using the boundary element

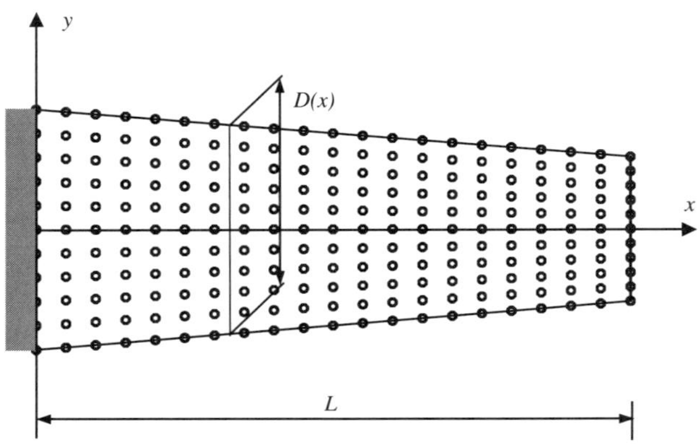

FIGURE 7.17
Cantilever beam of varying cross section. (From Gu, Y. T. and Liu, G. R., *Comput. Mech.*, 27, 188–198, 2001. With permission.)

TABLE 7.6

Natural Frequencies of the Cantilever Beam of Varying Cross Section ($\alpha_Q = 1.5$, $\alpha_s = 3.5$ for MLPG)

	ω (rad/s)				
Modes	**1**	**2**	**3**	**4**	**5**
MLPG method	263.21	923.03	953.45	1855.14	2589.78
FEM (ABAQUS)	262.09	918.93	951.86	1850.92	2578.63

Source: Gu, Y. T. and Liu, G. R., *Comput. Mech.*, 27, 188–198, 2001. With permission.

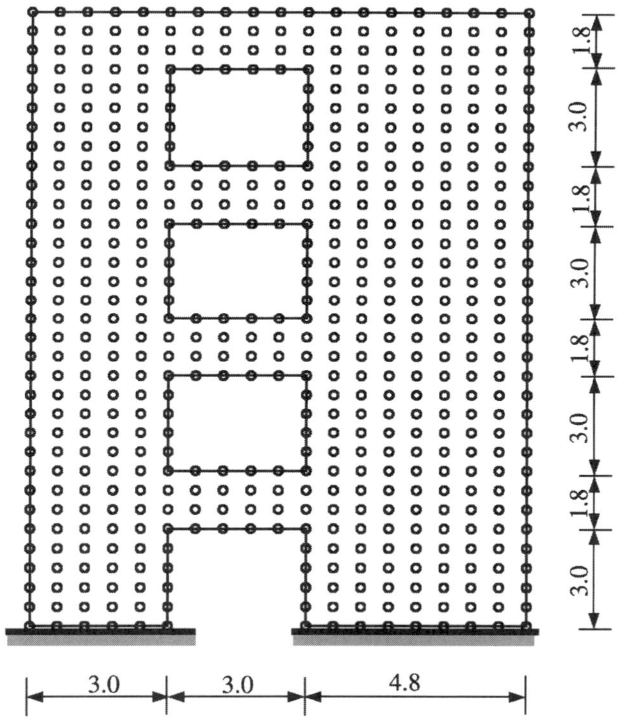

FIGURE 7.18
Shear wall with four openings (dimensions: m). (From Gu, Y. T. and Liu, G. R., *Comput. Mech.*, 27, 188–198, 2001. With permission.)

method (BEM) by some researchers (Brebbia et al., 1984) with parameters of $E = 1000$, $\nu = 0.2$, and $\rho = 1.0$. A total of 574 nodes are used to represent the problem domain shown in Figure 7.18. The problem is analyzed using MLPG and compared with the BEM and the FEM software ABAQUS. The natural frequencies of the lowest eight vibration modes are summarized and listed in Table 7.7. Results obtained by BEM and FEM are listed in the same table. Results obtained by the MLPG method are in very good agreement with those obtained using BEM and FEM.

7.2.5 Forced Vibration Analysis

The strong form of governing equation for forced vibration of 2D solids is given by Equation 7.38. The boundary conditions and initial conditions are given in Equations 7.39 to 7.42. The penalty method is used to enforce the essential boundary conditions. A local

TABLE 7.7

Natural Frequencies of a Shear Wall (α_Q=1.5, α_s=3.5 for MLPG)

	ω (rad/s)		
Mode	**MLPG Method**	**FEM (ABAQUS)**	**BEM (Brebbia et al., 1984)**
1	2.069	2.073	2.079
2	7.154	7.096	7.181
3	7.742	7.625	7.644
4	12.163	11.938	11.833
5	15.587	15.341	15.947
6	18.731	18.345	18.644
7	20.573	19.876	20.268
8	23.081	22.210	22.765

Source: Gu, Y. T. and Liu, G. R., *Comput. Mech.*, 27, 188–198, 2001. With permission.

weak form of the partial differential Equation 7.38, over a local quadrature domain Ω_Q bounded by Γ_Q, can be obtained using the weighted residual method locally

$$\int_{\Omega_Q} \widehat{W}(\sigma_{ij,j} + b_i - \rho\ddot{u}_i - \eta_c\dot{u}_i)\,\mathrm{d}\Omega - \alpha\int_{\Gamma_{Qu}} \widehat{W}(u_i - \bar{u}_i)\,\mathrm{d}\Gamma = 0 \tag{7.64}$$

The first term on the left-hand side of Equation 7.64 can be integrated by parts. Using the natural boundary condition defined by Equation 7.39, we obtain

$$\begin{aligned}&\int_{\Omega_Q} (\widehat{W}\rho\ddot{u}_i + \widehat{W}\eta_c\dot{u}_i + \widehat{W}_{,j}\sigma_{ij})\,\mathrm{d}\Omega - \int_{\Gamma_{Qi}} \widehat{W}t_i\,\mathrm{d}\Gamma - \int_{\Gamma_{Qu}} \widehat{W}t_i\,\mathrm{d}\Gamma + \alpha\int_{\Gamma_{Qu}} \widehat{W}u_i\,\mathrm{d}\Gamma \\ &= \int_{\Gamma_{Qt}} \widehat{W}\bar{t}_i\,\mathrm{d}\Gamma + \alpha\int_{\Gamma_{Qu}} \widehat{W}\bar{u}_i\,\mathrm{d}\Gamma + \int_{\Omega_Q} \widehat{W}b_i\,\mathrm{d}\Omega\end{aligned} \tag{7.65}$$

In the forced vibration analysis, **u** is a function of both the spatial coordinates and time. MLS approximation over the spatial domain is performed, and Equation 7.8 is rewritten as

$$u^h(\mathbf{x}, t) = \sum_I^n \mathbf{\Phi}_I(\mathbf{x})\mathbf{u}_I(t) \tag{7.66}$$

Substituting Equation 7.66 into the local weak form Equation 7.65 for all nodes leads to the following set of discrete equations:

$$\mathbf{M}\ddot{\mathbf{U}}(t) + \mathbf{C}\dot{\mathbf{U}}(t) + \mathbf{K}\mathbf{U}(t) = \mathbf{F}(t) \tag{7.67}$$

where the global mass matrix **M** is given by Equation 7.54. The global stiffness matrix **K** is obtained by assembling the nodal stiffness matrix defined by

$$\mathbf{K}_{IJ} = \int_{\Omega_Q} \widehat{\mathbf{V}}_I^T\mathbf{c}\mathbf{B}_J\,\mathrm{d}\Omega - \int_{\Gamma_{Qi}} \widehat{\mathbf{W}}_I\mathbf{n}\mathbf{c}\mathbf{B}_J\,\mathrm{d}\Gamma - \int_{\Gamma_{Qu}} \widehat{\mathbf{W}}_I\mathbf{n}\mathbf{c}\mathbf{B}_J\,\mathrm{d}\Gamma + \alpha\int_{\Gamma_{Qu}} \widehat{\mathbf{W}}_I\mathbf{\Phi}_J\,\mathrm{d}\Gamma \tag{7.68}$$

where $\widehat{\mathbf{V}}_I$, $\mathbf{B}_I$, $\widehat{\mathbf{W}}_I$, and **n** are defined, respectively, by Equations 7.12, 7.14, 7.15, and 7.17 and $\mathbf{\Phi}_I$ is a matrix of the MLS shape function for node I given by Equation 7.9.

The damping matrix **C** is obtained using

$$\mathbf{C}_{IJ} = \int_{\Omega_Q} \eta_c\widehat{\mathbf{W}}_I\mathbf{\Phi}_J\,\mathrm{d}\Omega \tag{7.69}$$

and the force vector **f** are defined as

$$\mathbf{f}_I(t) = \int_{\Gamma_{Qt}} \widehat{\mathbf{W}}_I \bar{\mathbf{t}}(t) \mathrm{d}\Gamma + \alpha \int_{\Gamma_{Qu}} \widehat{\mathbf{W}}_I \bar{\mathbf{u}} \mathrm{d}\Gamma + \int_{\Omega_Q} \widehat{\mathbf{W}}_I \mathbf{b}(t) \mathrm{d}\Omega \tag{7.70}$$

Again, Equation 7.68 will be reduced to Equation 7.53 for interior nodes whose boundary of quadrature domain does not intersect with the essential boundary.

7.2.6 Direct Analysis of Forced Vibration

The procedure of solving the discrete dynamic equation (Equation 7.67) is very much the same as that in standard FEM. There are two major approaches to solving Equation 7.67. One is the modal analysis approach, in which the natural frequencies and the vibration modes obtained in Section 7.2.2 are used to transform Equation 7.67 into a set of decoupled differential equations of second order with respect to time. These second-order differential equations can then be solved simply using the standard approach. The second approach is the methods of direct integration operating on Equation 7.67. The direct integration methods are utilized in this section. Several direct integration methods have been developed to solve the dynamic equation set similar to Equation 7.67, such as central difference method and the Newmark method (see, e.g., Petyt, 1990; Liu and Quek, 2002). Both the central difference and the Newmark methods are introduced here in a concise and easy-to-understand manner.

The Central Difference Method

The central difference method (CDM) consists of expressing the velocity and acceleration at time t in terms of the displacement at time $t - \Delta t$, t, and $t + \Delta t$ using central finite difference formulation:

$$\ddot{\mathbf{u}}(t) = \frac{1}{\Delta t^2}(\mathbf{u}(t-\Delta t) - 2\mathbf{u}(t) + \mathbf{u}(t+\Delta t)) \tag{7.71}$$

$$\dot{\mathbf{u}}(t) = \frac{1}{2\Delta t}(-\mathbf{u}(t-\Delta t) + \mathbf{u}(t+\Delta t)) \tag{7.72}$$

where Δt is a time step. The response at time $t + \Delta t$ is obtained by evaluating the equation of motion at time t. CDM is, therefore, an explicit method and is widely used in finite element packages for transient analysis.

CDM is conditionally stable, meaning that the solution is stable when the time step is sufficiently small. In FEM, the critical time step is calculated based on the size of the smallest element. The principle to be followed in calculating the critical time step is that the critical time should be smaller than the time the fastest wave propagates across the element. In the MFree method, there is no element, and there is a need for a new formula to compute the critical time. The time critical time step for CDM can be obtained from the maximum frequencies based on the dispersion relation (Belytschko et al., 2000):

$$\Delta t^{\mathrm{crit}} = \max_i \frac{2}{\omega_i}\left(\sqrt{\xi_i^2 + 1} - \xi_i\right) \tag{7.73}$$

where ω_i is the frequency and ξ_i the fraction of critical damping in this mode. For nonuniform arrangements of the nodes, the critical time step can be obtained by the eigenvalue inequality. The formula is

$$\Delta t^{\text{crit}} = \min \frac{2}{(\max_Q \lambda_{\max}^Q)^{1/2}} \tag{7.74}$$

where $\lambda_{\max}^Q$ is the maximum eigenvalue at the quadrature point $\mathbf{x}_Q$. The value of $\lambda_{\max}^Q$ depends on the size of local integration cells and the size of the interpolation domain.

The Newmark Method

The Newmark method is a generalization of the linear acceleration method. This latter method assumes that the acceleration varies linearly within the time interval of $(t, t+\Delta t)$. This gives

$$\ddot{\mathbf{u}}_{t+\Delta t} = \ddot{\mathbf{u}}_t + \frac{1}{\Delta t}(\ddot{\mathbf{u}}_{t+\Delta t} - \ddot{\mathbf{u}}_t)\tau \tag{7.75}$$

where $0 \le \tau \le \Delta t$, and

$$\dot{\mathbf{u}}_{t+\Delta t} = \dot{\mathbf{u}}_t + [(1-\delta)\ddot{\mathbf{u}}_t + \delta\ddot{\mathbf{u}}_{t+\Delta t}]\Delta t \tag{7.76}$$

$$\mathbf{u}_{t+\Delta t} = \mathbf{u}_t + \dot{\mathbf{u}}\Delta t + \left[\left(\frac{1}{2} - \beta\right)\ddot{\mathbf{u}}_t + \beta\ddot{\mathbf{u}}_{t+\Delta t}\right]\Delta t^2 \tag{7.77}$$

The response at time $t + \Delta t$ is obtained by evaluating the equation of motion at time $t + \Delta t$. The Newmark method is, therefore, an implicit method.

The Newmark method is unconditionally stable, meaning that the solution will always be stable regardless of the time step used, provided

$$\delta \ge 0.5 \quad \text{and} \quad \beta \ge \frac{1}{4}(\delta + 0.5)^2 \tag{7.78}$$

It has been found that $\delta = 0.5$ and $\beta = 0.25$ lead to acceptable results for most problems. Therefore, $\delta = 0.5$ and $\beta = 0.25$ are always used in this section for all the example problems. Note that although the Newmark method is unconditionally stable as long as Equation 7.78 is satisfied, and any time step used will produce stable results, the accuracy of the results is not guaranteed. A sufficiently small time step still must be used for accurate results.

7.2.7 Numerical Examples

Example 7.9 Cantilever Beam

For forced vibration analysis, a cantilever beam shown in Figure 6.4 is first examined using MLPG for benchmarking purposes. A plane stress problem is considered, and a unit thickness is used. The parameters used for this example are as follows:

Young's modulus for the material: $E = 3.0 \times 10^7$ N/m^2

Poisson's ratio for two materials: $\nu = 0.3$

Mass density: $\rho = 1$ kg/m^3

Length of the beam: $L = 48$ m

Height of the beam: $D = 12$ m

External excitation load: $P = -1000g(t)$

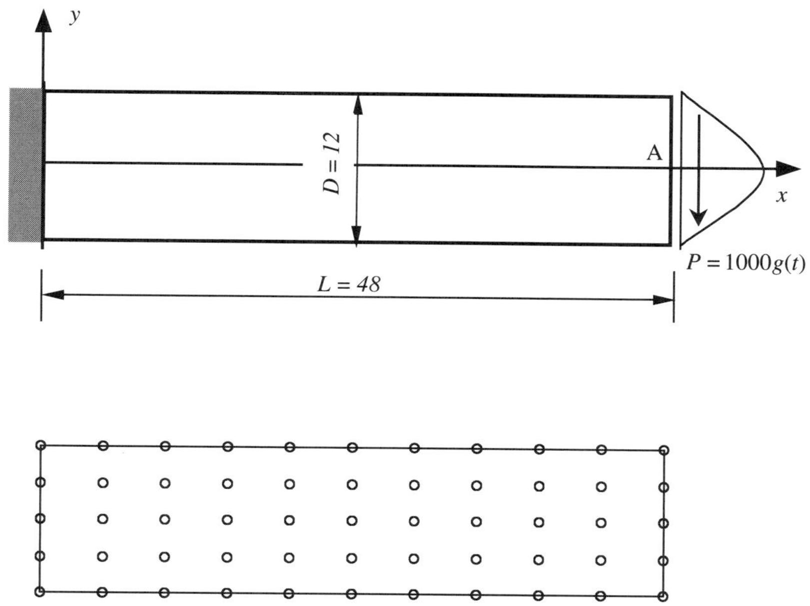

FIGURE 7.19
Configuration and nodal arrangement for the cantilever beam fixed at the left end and subjected to a dynamic force at the right end of the beam. (From Gu, Y. T. and Liu, G. R., *Comput. Mech.*, 27, 188–198, 2001. With permission.)

where $g(t)$ is a function of time. The external force is applied downward and distributed in a parabolic fashion on the right end of the beam. A total of 55 uniformly distributed nodes are used, as shown in Figure 7.19, to represent the problem domain. Displacements and stresses for all nodes are obtained. Detailed results of vertical displacement, u_y, on the middle node, A, of the free end of the beam are presented. For comparison, solutions for this problem are also obtained using the finite element software ABAQUS/Explicit.

Example 7.9a Simple Harmonic Loading

Consider first an external load of sinusoidal time function, i.e.,

$$g(t) = \sin(\omega_f t) \tag{7.79}$$

where ω_f is the frequency of the dynamic load, and $\omega_f = 27$ is used in this example. First, the effects of dimension parameter α_Q of the quadrature domain on the performance of the method for dynamic problems are investigated. Using Equation 7.74, the critical time calculated to have $\Delta t^{\text{crit}} \approx 1 \times 10^{-3}$.

The results of $\alpha_Q = 0.5$, 1.0, 1.5, and 2.0 are computed using the MLPG code. The displacements u_y at point A are plotted in Figures 7.20 and 7.21. These figures show that the results will be unstable for both CDM and the Newmark method when $\alpha_Q \leq 1.0$. Increasing α_Q is crucial to improve the accuracy and the stability for both CDM and the Newmark method. However, if the quadrature domain is too large, more subcells are needed to obtain accurate integrations, which will be computationally more expensive. Our study has found that $\alpha_Q = 1.5$ to 2.5 works for most problems of transient analyses. This finding is the same as that found for static and free-vibration analyses. In the following transient analyses $\alpha_Q = 1.5$ is employed.

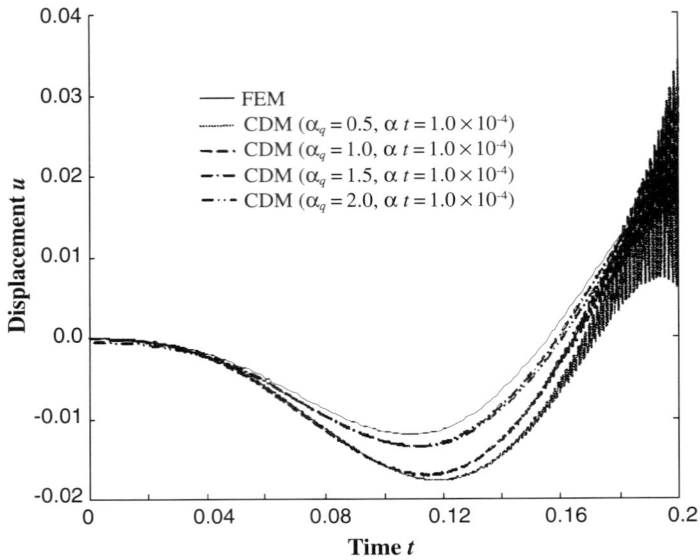

FIGURE 7.20
Displacement in the y direction at point A using CDM ($g(t) = \sin(\omega t)$). Results are unstable when $\alpha_Q \le 1.0$. (From Gu, Y. T. and Liu, G. R., *Comput. Mech.*, 27, 188–198, 2001. With permission.)

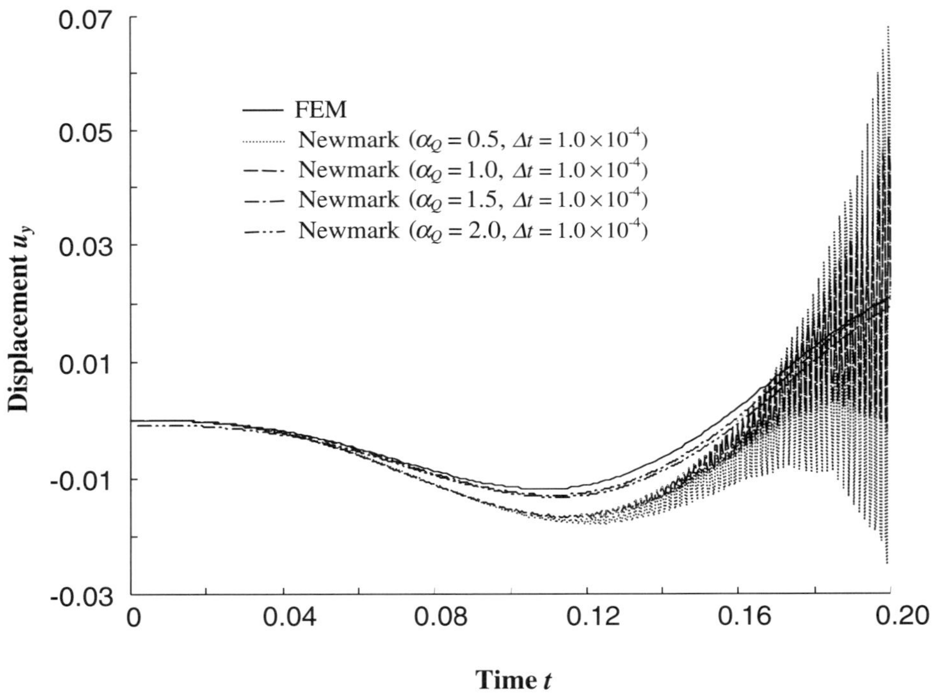

FIGURE 7.21
Displacement in the y direction at point A using the Newmark method ($\delta = 0.5$ and $\beta = 0.25$, with $g(t) = \sin(\omega t)$). Results are unstable when $\alpha_Q \le 1.0$. (From Gu, Y. T. and Liu, G. R., *Comput. Mech.*, 27, 188–198, 2001. With permission.)

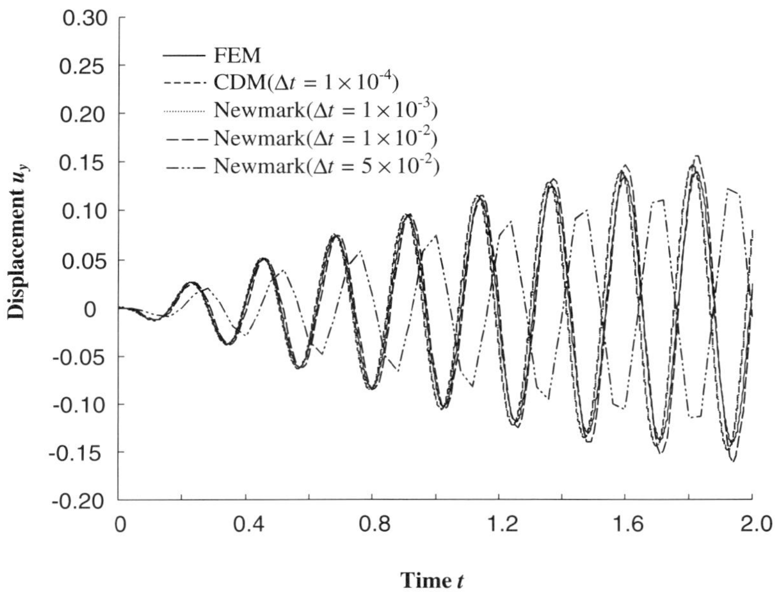

FIGURE 7.22
Displacement in the y direction at the point A ($g(t) = \sin(\omega t)$). (From Gu, Y. T. and Liu, G. R., *Comput. Mech.*, 27, 188–198, 2001. With permission.)

To investigate the property of two different direct time integration methods, CDM and the Newmark method, results of different time steps are obtained and plotted in Figure 7.22. It can be found that for $\Delta t = 1 \times 10^{-4}$ both methods obtain results in very good agreement with FEM. When $\Delta t \geq \Delta t^{crit}$ ($\Delta t^{crit} \approx 1 \times 10^{-3}$ according to Equation 7.74), the results obtained using CDM becomes unstable. However, the Newmark method is always stable for any time step used. It has been confirmed numerically that CDM is a conditionally stable method and that the Newmark method is an unconditionally stable method. A larger time step can be used in the Newmark method. A time step as large as $\Delta t = 1 \times 10^{-2}$ has been used, and very good results have been obtained using the Newmark method. However, it should be noted that the computational error would increase with the increase of time step in the Newmark method. For this example, it was found that the accuracy of the Newmark method would become unacceptable when the time step is too large, such as $\Delta t = 5 \times 10^{-2}$.

Many time steps are calculated to check the stability of the presented MLPG formulation. The Newmark method with $\Delta t = 5 \times 10^{-3}$ is used, and the damping coefficient, $c = 0.4$, is considered. Results for up to 20 s (about 100 natural vibration periods) are plotted in Figure 7.23. It can be found that a very stable result is obtained. After a long period of time, the forced vibration under the action of the sinusoidal dynamic loading becomes a steady sinusoidal vibration with the frequency of the external excitation ω_f. From the vibration theory (Meirovitch, 1980), a resonance will occur when $\omega_f = \omega_i$, where ω_i is the ith natural frequency. Figure 7.23 shows that the amplitude of vibration is very large (i.e., about 15 times the static displacement) because $\omega_f \approx \omega_1$. In addition, a beat vibration with the period T_b occurs when $\omega_f \approx \omega_1$. T_b can be obtained from Figure 7.23 as $T_b \approx 4.3$. From vibration theory, $T_b = 2\pi/|\omega_f - \omega_1|$, the first natural frequency of the system can be found as $\omega_1 = 28.3$, which is nearly the same as the result obtained in the free-vibration analysis by FEM, $\omega_1^{FEM} = 28$.

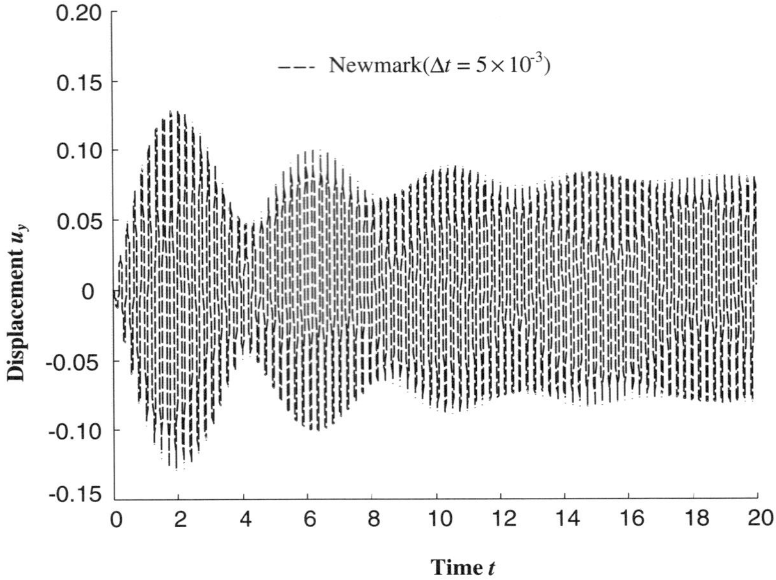

FIGURE 7.23
Displacement in the y direction at point A ($g(t) = \sin(\omega t)$). (From Gu, Y. T. and Liu, G. R., *Comput. Mech.*, 27, 188–198, 2001. With permission.)

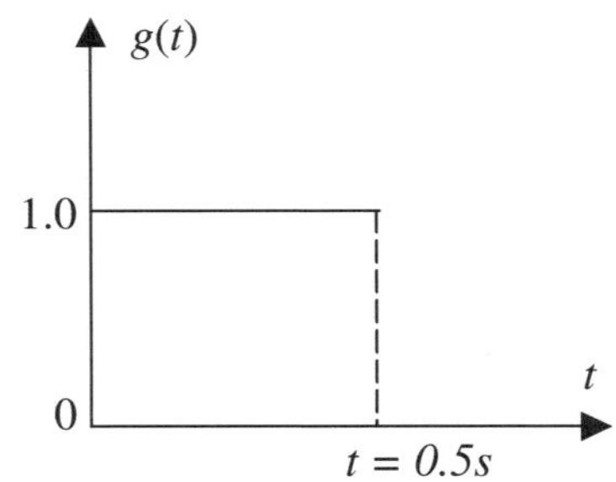

FIGURE 7.24
Rectangular pulse as the time function of the external force $g(t)$. (From Gu, Y. T. and Liu, G. R., *Comput. Mech.*, 27, 188–198, 2001. With permission.)

Example 7.9b Transient Loading

The transient response of a beam subjected to a suddenly loaded and suddenly vanished force $P = 1000g(t)$ is considered. The time function $g(t)$ is shown in Figure 7.24. The present MLPG method is used to obtain the transient response with and without damping. The Newmark method is utilized in this analysis. The result for a damping coefficient of $\eta_c = 0$ is plotted in Figure 7.25. For comparison, the result obtained by the finite element software ABAQUS/Explicit is shown in the same figure. Results obtained by the present MLPG method are in very good agreement with those obtained using FEM. Many time steps are calculated to check the stability of the presented MLPG formulation. The result for a damping coefficient $\eta_c = 0.4$ is plotted in Figure 7.26; which shows that the response declines with time because of damping. A very stable result is again obtained.

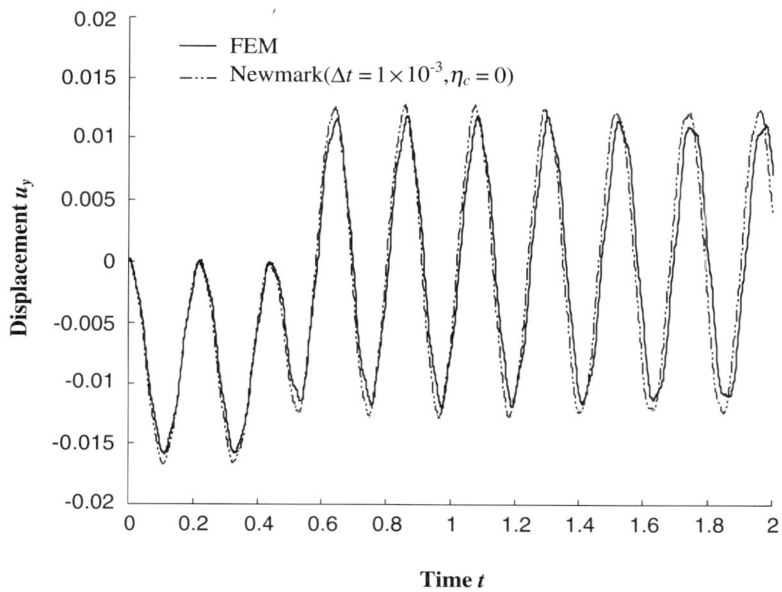

FIGURE 7.25
Transient displacement in the y direction at point A using the Newmark method ($\delta = 0.5$ and $\beta = 0.25$, η_c is the damping coefficient). (From Gu, Y. T. and Liu, G. R., *Comput. Mech.*, 27, 188–198, 2001. With permission.)

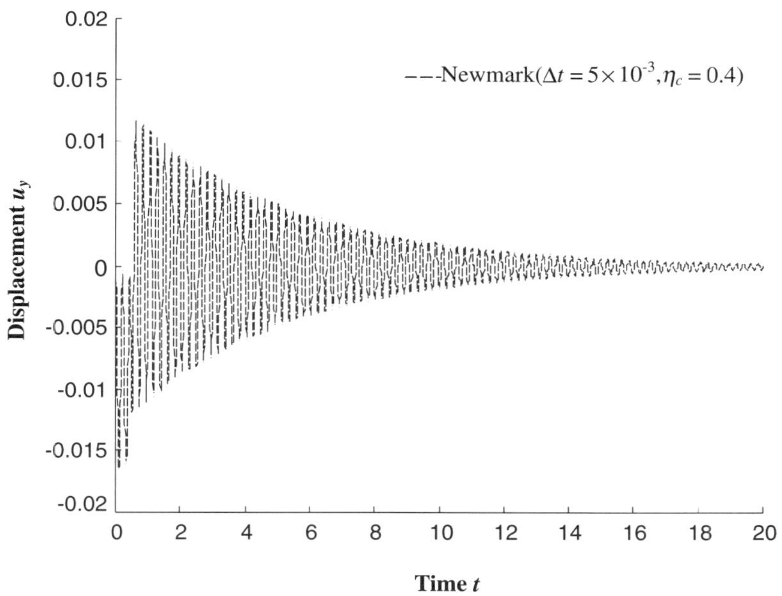

FIGURE 7.26
Transient displacement in the y direction at point A using the Newmark method ($\delta = 0.5$ and $\beta = 0.25$, η_c is the damping coefficient). (From Gu, Y. T. and Liu, G. R., *Comput. Mech.*, 27, 188–198, 2001. With permission.)

7.3 Remarks

MLPG formulations for static, free-vibration, and forced vibration analyses of 2D solids have been presented in this chapter. Some important parameters affecting the performance of the method for dynamic problems have been investigated; the important findings are summarized as follows:

1. MLPG is reproductive and works well for static and dynamic analysis of 2D solids.
2. The dimension of the quadrature domain is very important to the accuracy as well as the stability of the results. Larger α_Q will in general give more accurate and stable results, if the numerical integration can be carried out accurately. However, too large an α_Q often leads to difficulties in accurate numerical integration. A choice of α_Q=1.5 to 2.5 works for most problems, and α_Q= 1.5 is recommended as an economic choice.
3. The dimension parameter of the support domain is also very important to the accuracy as well as the stability of the results. A choice of α_s = 2.5 to 3.5 is good for most problems.
4. In the numerical integration in the quadrature domain, four subdivisions (2×2) works well for rectangular quadrature domains of α_Q= 2.0. For large quadrature domains, we suggest

$$n_c = \text{round up to nearest even number } (\alpha_Q) \tag{7.80}$$

5. MLPG is not as efficient as FEM in terms of computation time, because the system matrices produced are asymmetric, and the process of computing the MLS shape functions and their derivatives is more expensive than the FEM shape functions, which are usually given analytically. Improvement on these two issues for MLPG is very important.
6. Note that when α_Q approaches zero, MLPG becomes an MFree method of collocation. We can therefore conclude that the collocation method produces less accurate results especially for the derivatives of the field variables, at least if the shape function is constructed using the differential representation of field functions. The collocation method may work well if the shape function is constructed using integral representation of field functions, as the integration operation helps to smear out the error. However, this is nothing more than speculation. Further research is needed to draw a definitive conclusion on this issue.
7. One of the difficulties in MLPG is the integration for nodes near the boundaries of the problem domain, because the local quadrature domains for these nodes may intersect with the global boundary of the problem domain and create local quadrature domains of complex geometry. One method for solving this problem might be the use of a triangular mesh. We have also tried a simple trick, using very small regular quadrature domains for these nodes so that the boundary of their quadrature domains just touches but does not intersect with the global boundary. This simple trick works for some problems we have studied; however, we also found that the accuracy of the results for many problems could be affected by using this trick.

It is seen again that many numerical techniques used in conventional FEM can be used in MFree methods. Effectively making use of existing FEM techniques is very important in the process of developing MFree methods.

The idea of MLPG, proposed originally by Atluri and Zhu (1998), is a significant advance in the effort to develop truly MFree methods. This marks a new stage in the advance of MFree methods after the invention of EFG by Nayroles et al. (1992) and Belytschko et al. (1994b), and the earlier invention of the SPH method. After the invention of MLPG, many versions of methods have been proposed. The author regrets that it is not possible to mention all of them in this book.

The challenges with the MLPG method are (1) effective and accurate integration, (2) integration at complex boundaries, (3) effective production of symmetric system matrices, and (4) formulation of shape functions that possess the Kronecker delta function property.

8

Point Interpolation Methods

In Chapters 6 and 7, we introduced two recently developed MFree methods, the element free Galerkin (EFG) method and the meshless local Petrov–Galerkin (MLPG) method. These two methods are currently widely used and work well for a broad range of problems of continuum mechanics. Both methods use moving least squares (MLS) approximation for constructing shape functions, and hence they are accompanied by issues related to the imposition of essential boundary conditions. We discussed also a number of ways to tackle these issues, which require extra efforts both in formulation and in computation.

The point interpolation method (PIM) was proposed by G. R. Liu and Gu (1999, 2001a,b,c,d) to replace MLS approximation for creating shape functions. The major advantages of PIM are the excellent accuracy in function fitting and that the shape functions created possess the Kronecker delta function property, which allows simple imposition of essential boundary conditions as in the conventional finite element method (FEM). PIM was originally proposed by G. R. Liu and Gu in 1999, and has been modified, improved, and applied since to both Galerkin formulation and local Petrov–Galerkin formulation. In developing PIM, the battle has been how to overcome the problem related to the singular moment matrix, and how to make the point interpolation scheme stable numerically and flexible for arbitrarily distributed points. Two significant advances have been made over the past years. The first is the use of radial functions as the basis (Wang and Liu, 2000). The difficulty with PIM using radial basis functions is that it is less efficient compared with the original polynomial PIM. The second approach is the invention of a two-stage matrix regularization algorithm (MTA) (Liu, G. R. and Gu, 2001d), which can automatically exclude the nodes and the terms of the polynomial basis to create a nonsingular moment matrix. The two-stage matrix regularization scheme provides both numerical stability and accommodation of arbitrarily distributed nodes. The advantages of PIM with MTA are as follows:

1. The results obtained using PIM with MTA are more accurate than those using MLS approximation.
2. The shape functions constructed possess the Kronecker delta function property.
3. The shape functions constructed and their derivatives have very simple polynomial form, which allows analytical integration to compute the system matrix. This feature may eventually eliminate the need for numerical integration, and hence the background mesh, so as to develop a truly meshless methods based on local residual formulation.

This chapter introduces different versions of PIM for stress analysis of solids and structures. Formulations of PIM in both global Galerkin form and local Petrov–Galerkin form are presented. Various schemes are used for construction of shape functions. For convenience and clarity in description, these PIM methods are presented in Table 8.1. Note that

TABLE 8.1

Family of Point Interpolation Methods

Abbreviation	Full Name	Formulation	Feature*
NPIM	Nonconforming Point Interpolation Method	Galerkin formulation using polynomial PIM shape functions and various techniques for handling the singular moment matrix	Nonconforming
CPIM	Conforming Point Interpolation Method	Constrained Galerkin formulation using PIM shape function	Conforming
NRPIM	Nonconforming Radial Point Interpolation Method	Galerkin formulation using RPIM shape functions	Nonconforming
CRPIM	Conforming Radial Point Interpolation Method	Constrained Galerkin formulation using radial PIM shape function	Conforming
LPIM	Local Point Interpolation Method	Local Petrov–Galerkin formulation using polynomial PIM shape functions and various techniques for handling the singular moment matrix	Reproductive
LRPIM	Local Radial Point Interpolation Method	Local Petrov–Galerkin formulation using RPIM shape functions	Reproductive

* See discussion in Section 5.12.

because the PIM shape functions are not compatible, when the energy principles are used to formulate a PIM, it can be conforming or nonconforming. When the constrained energy principles are used to enforce compatibility, it will be conforming as long as the numerical implementation is accurate. A conforming PIM or CPIM should provide the upper bound of the solution, and the displacement should converge to the exact solution from below when the nodal spacing approaches zero. When the unconstrained energy principles are used, the PIM will be nonconforming. The displacement can converge to the exact solution from both sides when the nodal spacing approaches zero. When a PIM is formulated using the local Petrov–Galerkin method, it will always be reproductive. In this case, the displacement can converge to the exact solution from both sides when the nodal spacing approaches zero. Detailed discussions on this are given in the examples of patch tests.

8.1 Polynomial Point Interpolation Method

PIM was originated by G. R. Liu and Gu (1999) based on the Galerkin formulation using polynomial PIM shape functions. PIM has been improved over the past few years, especially the techniques for dealing with possible singularity in the moment matrix. This section details PIM formulations for mechanics problems for two-dimensional (2D) solids.

From Section 5.11 it is seen that field approximation using PIM shape function is not compatible at some locations. PIMs using Galerkian formulation are not conforming.

Conformability can be ensured by simply using a constrained Galerkian form that formulates conforming PIMs. Conforming and nonconforming PIMs using polynomial basis functions are discussed in this section.

8.1.1 Domain Discretization

In PIM, the problem domain is represented by properly scattered points as in any MFree method. PIM shape functions of delta function property are constructed based only on a group of field nodes arbitrarily distributed, using any of the schemes provided in Section 5.5.

Similar to the EFG method presented in Chapter 6, a background mesh of cells for integration will be used. The background cells are independent of field nodes.

8.1.2 Enclosure of Nodes

The support domain for a point of interest x_Q (which is usually a quadrature point in a cell of integration) can be defined using a local subdomain of simple shape, such as circles, ellipses, and rectangles for 2D problem domains and spheres, ellipsoids, and bricks for 3D problem domains. These support domains are usually formed with point x_Q at their geometric center. The number of nodes, n, can be determined by counting all the points in the support domain. The dimensions of the support domain should be so determined that it contains $n = 7$ to 30 (preferably 7 to 15) for 2D problems. Note that PIM allows the use of a wide range of different numbers of nodes for constructing shape functions for the point x_Q. However, using too many NPIM nodes (m) does not necessarily increase the accuracy but does lead to more computational work. In addition, more quadrature points are required to obtain accurate results, as the integrand becomes more complex when more nodes are used for constructing PIM shape functions. Therefore, a number of $n = 7$ to 15 is usually preferred for NPIM. In CPIM, however, the use of more nodes leads to significant improvement in the accuracy of the numerical results, which is very similar to the so-called p-convergence in conventional FEM.

In using NPIM, the enclosure of nodes is very much the same as in EFG as described above. The support domain is determined for each and every Gauss point. In CPIM, however, we use so-called one-piece shape functions for an integration cell to ensure the compatibility of the field function approximation with the cells. Therefore, the support domain is defined for the cell, meaning all the Gauss points in the cell share the same support domain and hence the same PIM shape functions. The support domain of a cell is determined in the manner similar to that discussed in the previous paragraph. The only difference is that the domain is centered by the geometrical center of the cell and not the Gauss point, as shown in Figure 8.1.

One can also fix the number of nodes needed. This can be achieved by using a larger support domain to "grab" a group of nodes more than the specified number. These nodes are then ranked by distance or some other criteria and the specified number of nodes selected from the rank list. In the case of using MTA (see Chapter 5), there is a chance of excluding nodes in the final construction of PIM shape functions. Therefore, one or two more nodes may be included initially. Note that for actual problems it is rare for MTA to exclude nodes, especially when the nodes are irregularly distributed.

When the density of the nodes varies drastically in space, the selection approach by distance based on the support domain may result in shape functions of extrapolation. This can happen for both PIM and MLS shape functions, but happens more often in PIM because PIM usually uses fewer nodes. In extreme situations, all the nodes selected may be located on one side of point x_Q. In such simulations, a more practical and efficient definition of an influence domain instead of a support domain for each individual node should be adopted (see Section 2.10 for definitions of influence domain and support domain).

The number of nodes used in the support domain for constructing MLS shape functions is usually about $n = 20$ to 50 (for 2D problems), which is much more than that used in PIM. The large number of nodes required in MLS serves to prevent a single moment matrix in the process of constructing MLS shape functions to ensure a successful construction. It is an expensive way to do this, because the global system matrix created using more nodes will have a larger bandwidth, in addition to the extra efforts needed for handling essential boundary conditions.

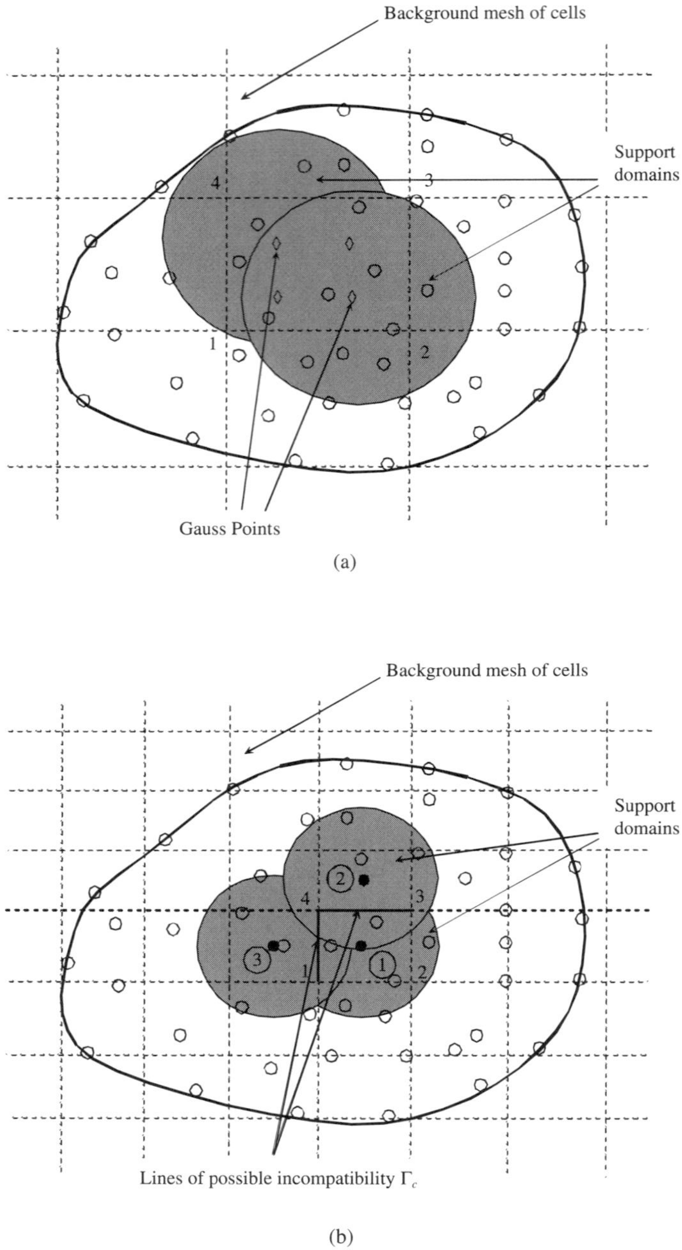

FIGURE 8.1
Support domains for PIMs. (a) Support domain for NPIM centered at the Gauss point. Each Gauss point has its own support domain and hence a set of shape functions. (b) Support domain in CPIM for an integration cell for constructing one-piece PIM shape functions. Possible incompatibility can occur between the interfaces of neighboring cells. Conforming PIM (CPIM) is formulated using the constrained Galerkin weak form.

8.1.3 Variational Form of Galerkin PIM

Consider a 2D problem of solid mechanics in domain Ω bounded by Γ. The strong form of system equation is given by Equations 6.1 to 6.3. When a PIM shape function is used, the unconstrained Galerkin weak form of the equilibrium equation can be simply posed as follows:

$$\int_{\Omega} \delta(\mathbf{Lu})^T(\mathbf{cLu})\,\mathrm{d}\Omega - \int_{\Omega} \delta\mathbf{u}^T\mathbf{b}\,\mathrm{d}\Omega - \int_{\Gamma_t} \delta\mathbf{u}^T\bar{\mathbf{t}}\,\mathrm{d}\Gamma - \frac{1}{2}\int_{\Gamma_c} \delta(\mathbf{u}^+ - \mathbf{u}^-)^T\boldsymbol{\alpha}(\mathbf{u}^+ - \mathbf{u}^-)\mathrm{d}\Gamma = 0 \quad (8.1)$$

Note that the difference between Equation 8.1 and Equation 6.67 is that there is no need for a term dealing with the essential boundary condition at Γ_u as the PIM shape function that possesses the Kronecker delta function property will be used. The condition of Equation 6.2 will be satisfied on the entire essential boundary as long as the nodal displacements are set to be the prescribed values at the nodes, which can be done by a simple procedure of row and column removal or some other methods as has been done in FEM.

However, because the PIM approximation is not compatible (see Section 5.11), enforcement of compatibility is needed on the incompatible curve Γ_c in the problem domain Ω to produce the conforming PIM (CPIM). If the last term in the left-hand side of Equation 8.1 is excluded, the formulation leads to nonconforming PIM (NPIM). In Equation 8.1, $\mathbf{u}^+$ and $\mathbf{u}^-$ are the displacements on the two sides of the incompatible interface Γ_c. In carrying out the integrations, a background mesh of cells is required as in the EFG method. The background mesh of cells can be independent of the nodes used for field variable interpolation. In each cell, Gauss quadrature is employed. The number of quadrature points depends largely on the nodal density. An investigation of the density of the nodes and the density of the Gauss points is detailed in Chapter 6, and the findings in Chapter 6 are applicable here as well for the Galerkin PIM. In formulating the NPIM, the PIM shape functions are formulated for each Gauss point using the support domain of the Gauss point. In formulating the CPIM, we deliberately construct so-called one-piece shape functions for the entire cell in the background integration mesh, meaning that all the Gauss points in a cell share the same shape function. The support domain is determined based on the geometrical center point of the integration cell. Therefore, the PIM shape function will be continuous within the entire cell, and the incompatible curves in the problem domain will be the interfaces of the integration cells. Figure 8.2 shows a rectangular domain with a triangular background mesh. Possible incompatibility can occur on edge 8-12 shared by cells 11 and 12.

Note that the alternative method for enforcing the compatibility condition in PIMs is to use the Lagrange multipliers method, as we did in EFG for imposing the essential boundary conditions. The pros and cons of using penalty methods vs. the Lagrange multipliers method are detailed in Chapter 4.

From Chapter 5, we have (see Equation 5.93)

$$u^h(\mathbf{x}) = \sum_{i=1}^{n} \phi_i(\mathbf{x})u_i \quad (8.2)$$

where ϕ_i is the PIM shape function at node i. Substituting the foregoing expression for all the displacement components of $\mathbf{u}$ into the weak form Equation 8.1 and following the exact procedure detailed in Section 6.1.1 yield the following global discrete system equations of PIM:

$$\mathbf{KU} = \mathbf{F} \quad (8.3)$$

where $\mathbf{U}$ is the displacement vector for all the nodes in the entire problem domain and $\mathbf{K}$ is the global stiffness matrix for the problem domain, which is assembled using the nodal stuffiness matrix defined by

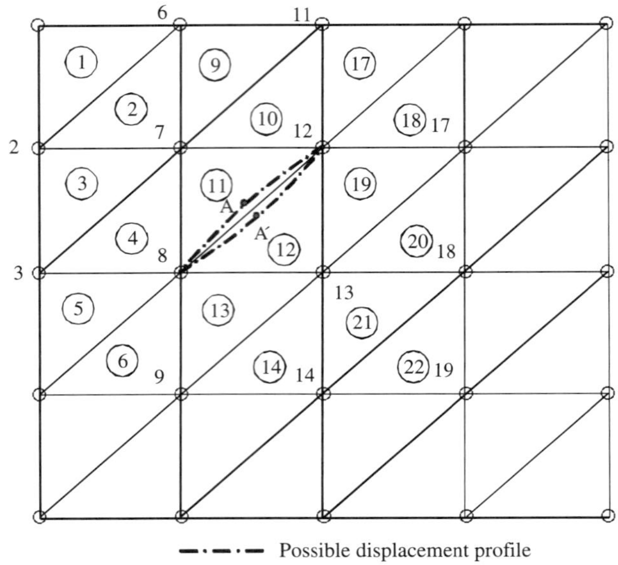

FIGURE 8.2
Compatibility on the interface of two neighboring integration cells with one-piece PIM shape functions used for cells. The gap on the interface 8–12 between cells 11 and 12 is observed in a 25-node patch test when 9 nodes were used in computing the PIM shape functions for cells. For background cell 11, nodes 1, 2, 3, 6, 7, 8, 11, 12, and 13 were used, and for cell 12, nodes 7, 8, 9, 12, 13, 14, 17, 18, and 19 were used. A stitch can be used to tie up points A and A′ using the penalty method to enforce the compatibility of the field function approximation of PIM and to pass the patch test.

$$\mathbf{K}_{ij} = \int_{\Omega} \mathbf{B}_i^T \mathbf{c} \mathbf{B}_j \, d\Omega + \int_{\Gamma_c} (\mathbf{\Phi}_i^+ - \mathbf{\Phi}_i^-)^T \boldsymbol{\alpha} (\mathbf{\Phi}_j^+ - \mathbf{\Phi}_j^-) d\Gamma = 0 \tag{8.4}$$

where **c** is a matrix of material constant given by Equation 3.30 for plane stress problems and Equation 3.31 for plane strain problems. The strain matrix is defined by

$$\mathbf{B}_i = \begin{bmatrix} \phi_{i,x} & 0 \\ 0 & \phi_{i,y} \\ \phi_{i,y} & \phi_{i,x} \end{bmatrix} \tag{8.5}$$

Therefore, the nodal stiffness matrix is 2×2. The global force vector **F** is assembled using the nodal force vector defined by

$$\mathbf{f}_i = \int_{\Gamma_i} \mathbf{\Phi}_i \bar{\mathbf{t}} d\Gamma + \int_{\Omega} \mathbf{\Phi}_i \mathbf{b} \, d\Omega \tag{8.6}$$

where $\mathbf{\Phi}_i$ is a matrix of the shape function at node i.

$$\mathbf{\Phi}_i = \begin{bmatrix} \phi_i & 0 \\ 0 & \phi_i \end{bmatrix} \tag{8.7}$$

The nodal force vector is therefore of dimension 2.

In Equation 8.4, $\mathbf{\Phi}_I^+$ is the matrix of shape functions constructed on the interface for an integral cell, and $\mathbf{\Phi}_I^-$ is the matrix of shape functions constructed on the same interface but for the neighboring integral cell. Note that the integration can be evaluated very easily as it needs to be carried out only on the edges of the elements. A conventional Gauss integral scheme works perfectly fine. Moreover, in practical computations, one does not even have to carry out the integration in Equation 8.4, but to enforce only one or a few points on the interface of two integration cells using the penalty method. For the case shown in Figure 8.2, we enforce the continuity of points A and A′. In other words, we can simply stitch a few points on the interfaces together. Therefore, the enforcement of compatibility is really a straightforward procedure. The benefits, however, are tremendous: allowing the field variable interpolation to be performed without the use of elements.

It may be noted that the displacement function $\mathbf{u}(x)$ approximated using PIM shape functions is always differentiable at any given point. In addition, the PIM shape function can be given in explicit forms of polynomials, and so the derivatives of the shape functions. The order of consistency depends on field nodes included in the support domain of the point. For 2D problems, the use of three nodes ensures C^1 consistency, and the use of six nodes ensures C^2 consistency (see Chapter 5 for proof).

The flowchart of nonconforming PIM is briefly as follows:

1. Loop over cells of domain
2. Loop over Gauss quadrature points x_Q in a cell
   a. Determine the domain of support of x_Q and find the nodes included in the support domain
   b. Compute $\phi_i(x_Q)$ and $\phi_{i,j}(x_Q)$ at the quadrature point
   c. Compute the nodal matrices, vectors
   d. Assemble the nodal contributions to the global matrices
3. End quadrature point loop
4. End cell loop

The above procedure is very much the same as that in EFG, except for the way in which the shape function is computed. For conforming PIM, the construction of PIM shape functions can be done outside the second loop over quadrature points.

8.1.4 Comparison of PIM, EFG, and FEM

PIM vs. FEM

The interpolation procedure in PIM is based on a group of arbitrarily distributed nodes. The interpolation procedure in FEM is the same as in PIM, but is based on elements. In both FEM and PIM, the number of monomials used in the basis functions, m, is the same as the number of nodes, n. Therefore, the interpolation functions have the property of the Kronecker delta function. This feature of PIM allows simple imposition of essential boundary conditions as in conventional FEM.

The interpolation at a quadrature sampling point in PIM is performed over the support domain of the point, which may overlap with the support domains of other sampling points. FEM defines the shape functions over predefined elements, and there is no overlapping in using nodes for computing shape functions. The finite element type interpolation can be regarded as *stationary cell interpolation.*

The interpolation in PIM (or MLS) can be regarded as *moving domain interpolation* in contrast to stationary cell interpolation. The difference lies not only between "moving" and "stationary," but also between "cell" and "domain." *Cell* requires a connectivity of the nodes that form the cell, whereas *domain* requires no connectivity for the nodes in the domain.

Integration and interpolation in FEM are based on the same mesh of elements defined by the connectivity of the nodes. In PIM, the background mesh can be completely independent of the field nodes. The separation of field variable interpolation and integration provides the freedom for the MFree methods of PIM.

PIM vs. EFG

The interpolation procedure of EFG is also based only on a group of arbitrarily distributed nodes in the support domain that moves with the point of interest. However, the basis number m is different from the number of nodes n in the support domain, $m < n$, and the MLS approximation is used to create shape functions (see Chapter 5). The MLS shape functions lack the delta function property, which leads to some complications in the imposition of essential boundary conditions, and the constrained Galerkin weak form must be used. The construction of MLS is also more complicated than the construction of PIM shape functions. In addition, a weight function is needed in the EFG method, and it takes extra effort to choose and compute the weight function. MLS approximation is continuous in the entire problem domain. PIM approximation is piece-wise continuous. Formulating conforming PIM requires the use of the constrained Galerkin weak form. In CPIM only one set of shape functions needs to be constructed for an integration cell.

The superior accuracy of PIM applied in curve and surface fitting compared with that of MLS approximation has been presented in great detail in Chapter 5 using a large number of examples.

8.1.5 Numerical Examples

Example 8.1 Patch Test

The first numerical example is the standard patch test, shown in Figure 8.3. Two patches of dimension $L_x = 2.0$ by $L_y = 2.0$ are tested. Figure 8.3a shows a patch with nine nodes of which one is an interior node. Figure 8.3b shows a patch of 15 nodes including seven

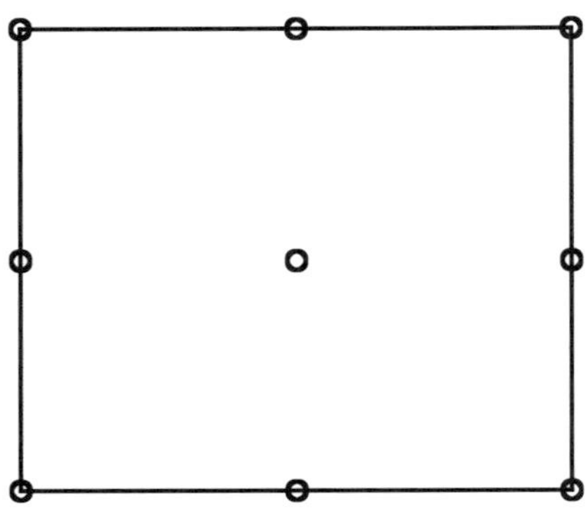

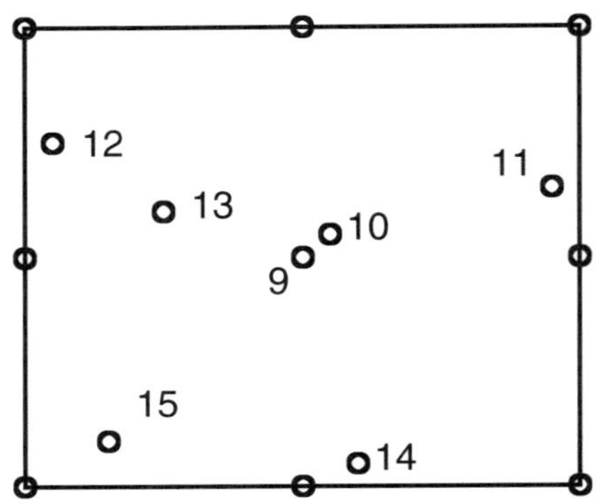

FIGURE 8.3
Nodal arrangement for patch tests: (a) patch with 9 nodes regularly distributed; (b) patch with 15 nodes irregularly distributed. (From Liu, G. R. and Gu, Y. T., *Int. J. Numer. Methods Eng.*, 50, 937–951, 2001. With permission.)

TABLE 8.2

NPIM Results at Interior Nodes Located Irregularly in the 15-Node Patch

Nodal Coordinates		u_x	u_y	σ_x	σ_y	τ_{xy}
9	(1.0, 0.0)	0.60000	0.0000	0.8571428571	0.8571428571	–3.94E-16
10	(1.1, 0.1)	0.66000	0.06000	0.8571428571	0.8571428571	–8.85E-17
11	(1.9, 0.3)	1.14000	0.18000	0.8571428571	0.8571428571	2.78E-16
12	(0.1, 0.5)	0.06000	0.30000	0.8571428571	0.8571428571	8.08E-16
13	(0.5, 0.2)	0.30000	0.12000	0.8571428571	0.8571428571	1.65E-15
14	(1.2, –0.9)	0.72000	–0.54000	0.8571428571	0.8571428571	–2.88E-15
15	(0.3, –0.8)	0.18000	–0.48000	0.8571428571	0.8571428571	–1.90E-15

Source: Liu, G. R. and Gu, Y. T., *Int. J. Numer. Methods Eng.*, 50, 937–951, 2001. With permission.

irregular interior nodes. A 2 × 2 rectangular background cell structure is used for integration in these patch tests.

In these patch tests, the displacements are prescribed on all outside boundaries by a linear function of x and y on the patches. The linear displacement functions are assumed as

$$\bar{\mathbf{u}}(x, y) = \begin{Bmatrix} 0.6x \\ 0.6y \end{Bmatrix} \tag{8.8}$$

The material constants are taken as $E = 1.0$ and $\nu = 0.3$. Satisfaction of the patch test requires that the displacement of any interior node be given by the same linear function and that the strains and stresses be constant in the entire patch.

In the 9-node patch test, a circular support domain with a radius of 1.2 is used for a Gauss point x_Q. Therefore, 4 to 6 nodes are usually used in the interpolation. The NPIM exactly passes the 9-node patch test. In the 15-node patch test, since the interior nodes are randomly distributed, 5 to 6 nodes are used in the interpolation depending on the distances between nodes and the Gauss point. The displacements and stresses of the 15-node patch test are computed and listed in Table 8.2. It is shown that the nonconforming PIM passes the patch test exactly to machine accuracy. In Figure 8.3b, nodes 9 and 10 are deliberately placed very close to each other. It is found that this does not affect the computational results. Note that manual selection of nodes in the support domain may be needed to avoid singularity in the moment matrix, unless MTA is used with PIM.

From the discussion given in Example 6.1, we can expect that the NPIM will not, in general, pass the patch test because the second requirement of compatibility is not met in NPIM. If, however, proper nodes are used to construct the PIM shape functions and properly arranged cells for the integration to prevent the occurrence of incompatibility, the NPIM can pass the patch test.

Note that passing the standard patch test is a *sufficient* requirement for a method to be able to converge to the true solution as the nodal spacing approaches zero. It is not a *necessary* requirement for a numerical method to converge. The ultimate test of a numerical method should be the convergence test. We have confirmed that the NPIM converges even for the standard patch test problem, meaning when the patch is refined, the result approaches the true solution. Further discussion is provided in Example 8.2.

The CPIM can always pass the patch tests provided there is:

1. No singularity in the moment matrix
2. Accurate numerical integration.

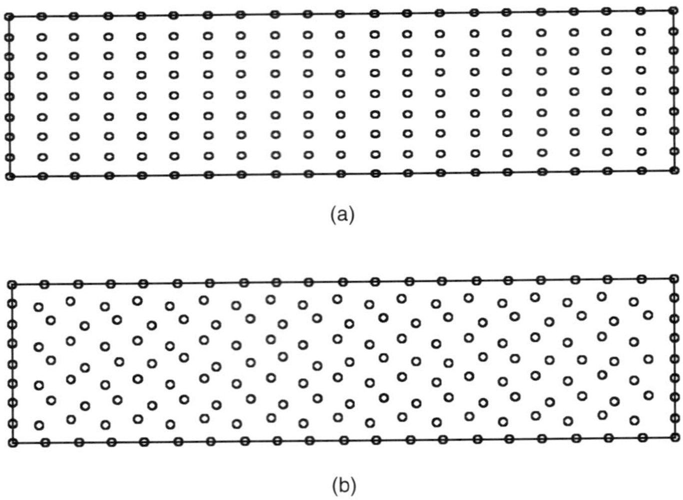

FIGURE 8.4
Nodal arrangement for stress analysis of the cantilever beam: (a) regular nodal arrangement (189 nodes); (b) irregular nodal arrangement (189 nodes). (From Liu, G. R. and Gu, Y. T., *Int. J. Numer. Methods Eng.*, 50, 937–951, 2001. With permission.)

Example 8.2 Cantilever Beam

PIM is then benchmarked using Example 6.2 to analyze stresses in a cantilever beam. The beam is of length L and height D as shown in Figure 6.4. The beam is subjected to a parabolic traction at the free end given in Equation 6.54. The beam has a unit thickness and a plane stress problem is considered. The analytical solution is available; it can be found in the textbook by Timoshenko and Goodier (1977) and is listed in Equations 6.48 to 6.53. The material properties and other parameters are taken as $E = 3.0 \times 10^7$, $\nu = 0.3$, $D = 12$, $L = 48$, and $P = -1000$. Both a regular nodal distribution and an irregular nodal distribution, as shown in Figure 8.4, are employed. For the regular nodal distribution, we move the nodes randomly by a small distance to avoid singularity in the moment matrix. A background mesh of 20×8 is used for integration. In each integration cell, 4×4 Gauss quadrature points are used to evaluate the stiffness matrix of the PIM. The computation is done using two different sizes of circular support domain with radii of $r_s = 3.6$ and $r_s = 4.8$.

Figures 8.5 and 8.6 show the deflection along the x axis of the beam computed using PIM together with the analytical solution for comparison. The plot shows an excellent agreement between the analytical and numerical results. The effects of the nodal irregularity are very small, as shown in both figures. A comparison of Figures 8.5 and 8.6 also reveals that the results obtained using two different sizes of support domains are very close. This implies that a support domain of $r_s = 3.6$ is large enough to obtain sufficiently accurate results.

Figure 8.7 illustrates the comparison between the shear stress calculated analytically and using PIM at the section of $x = L/2$. Again, very good agreement is observed between the analytical results and the NPIM results for both regular and irregular nodal distributions. It should be mentioned here that the irregularity of the nodes little affects the accuracy of the results, because of the accuracy in the numerical integration. When the node distribution is irregular, the shape function constructed becomes more complicated, leading to a complex integrand.

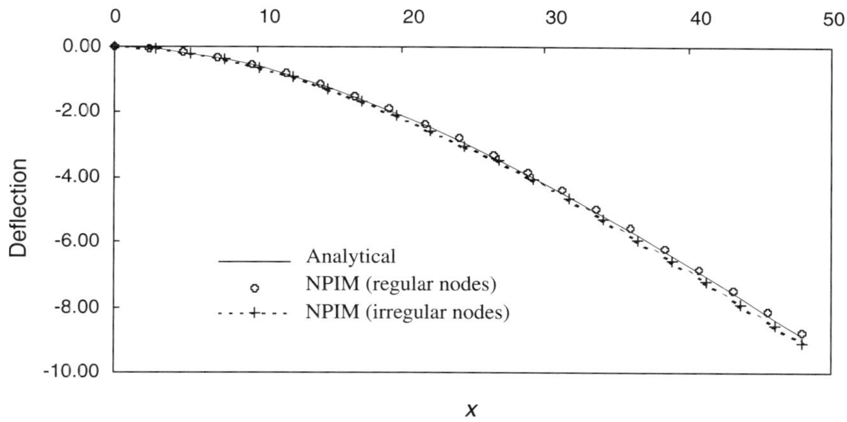

FIGURE 8.5
Deflection of the cantilever beam subject to a vertical force distributed at the free end. Computed using nonconforming Galerkin PIM with circular support domains of $r_s = 3.6$. (From Liu, G. R. and Gu, Y. T., *Int. J. Numer. Methods Eng.*, 50, 937–951, 2001. With permission.)

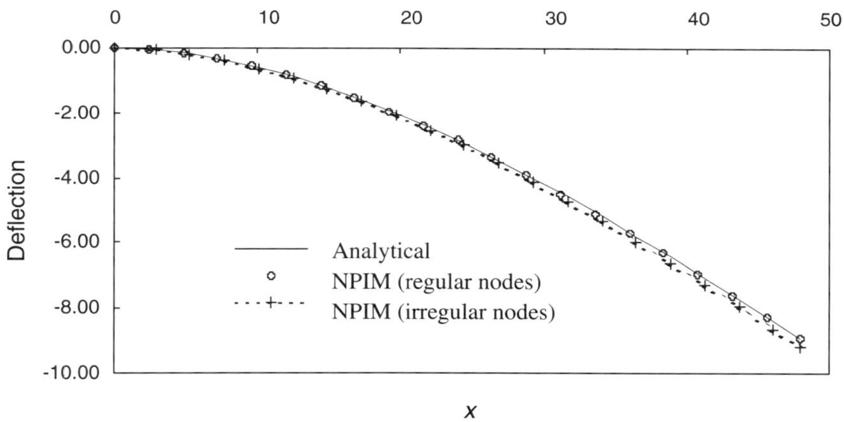

FIGURE 8.6
Deflection of the cantilever beam subject to a vertical force distributed at the free end. Computed using nonconforming Galerkin PIM with circular support domains of $r_s = 4.8$. (From Liu, G. R. and Gu, Y. T., *Int. J. Numer. Methods Eng.*, 50, 937–951, 2001. With permission.)

For more precise error analysis, we define the energy norm as an error indicator, because the accuracy in strain or stress is much more critical than that in displacement.

$$e_e = \left\{ \int_\Omega (\varepsilon^{\text{PIM}} - \varepsilon^{\text{EXACT}})^T c(\varepsilon^{\text{PIM}} - \varepsilon^{\text{EXACT}})\, \mathrm{d}\Omega \right\}^{1/2} \tag{8.9}$$

The convergence of NPIM is studied first, with regular nodal distribution. For easy comparison, the nodes coincide with the vertices of the background mesh. The investigation is conducted for $n = 4, 6, 8$, and 10. When $n = 4$ is used, then the NPIM is actually conforming and the results are the same as those of the conventional FEM. The convergence with mesh

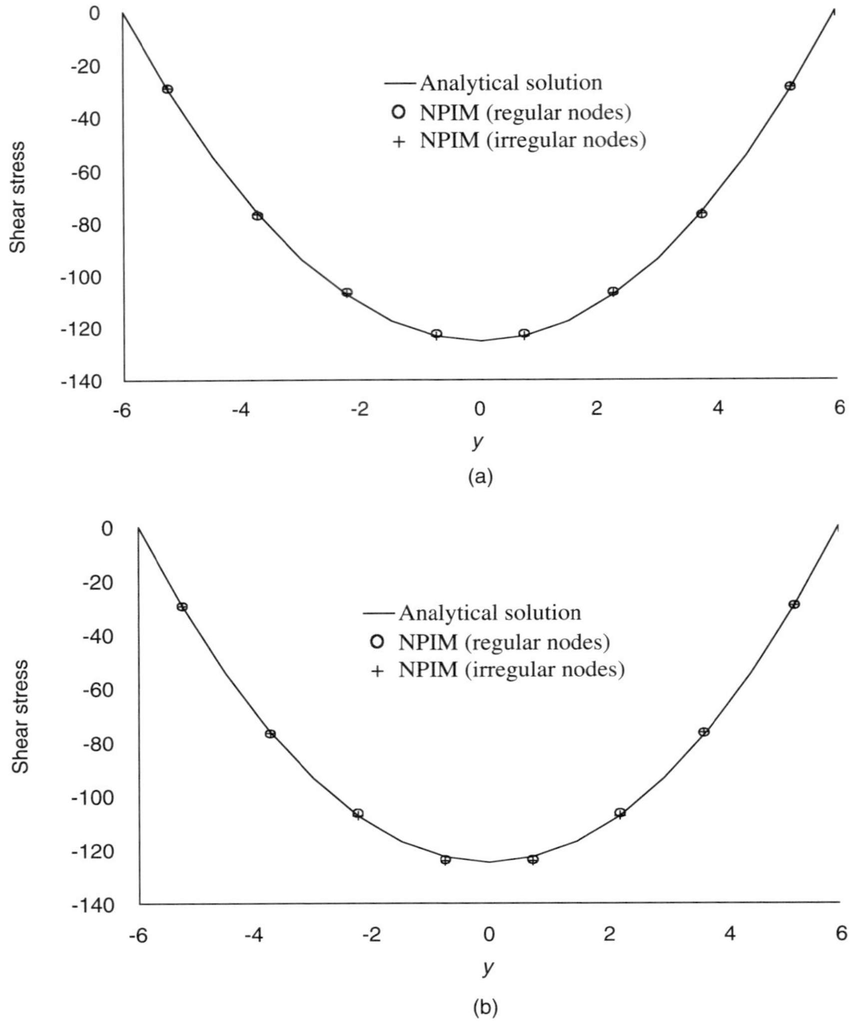

FIGURE 8.7
Shear stress σ_{xy} at the section $x = L/2$ of the beam ($r_s = 4.8$). (a) $r_s = 3.6$; (b) $r_s = 4.8$. (From Liu, G. R. and Gu, Y. T., *Int. J. Numer. Methods Eng.*, 50, 937–951, 2001. With permission.)

refinement is shown in Figure 8.9, where h is the nodal spacing. The convergence of NPIM is roughly the same as the four-node FEM shown in Figure 8.8. It is noted that the number of nodes used for interpolation, n, does not affect the accuracy significantly. This is because of the nonconformability of the NPIM.

The above convergence study of NPIM shows that one should use fewer points in constructing PIM shape functions, if NPIM is used. Accuracy should be achieved by increasing the nodal density, which is the so-called h-convergence.

Figure 8.9 shows a comparison of the rates of convergence of FEM, EFG, NPIM, and CPIM. The rate of convergence is denoted by R. In the FEM, four-node elements are used; in EFG, the support domain of $a_s = 2.5$ (about 25 nodes) is used for constructing MLS shape functions; in CPIM, one-piece shape functions are constructed using the nearest nine nodes surrounding the integration cell; in NPIM, the nearest nine nodes surrounding

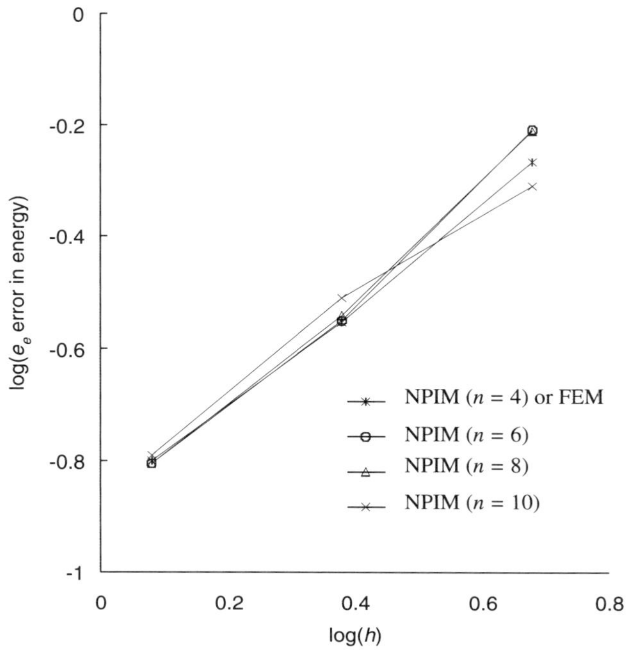

FIGURE 8.8
Convergence of PIM results in e_e norm of error. (From Liu, G. R. and Gu, Y. T., *Int. J. Numer. Methods Eng.*, 50, 937–951, 2001. With permission.)

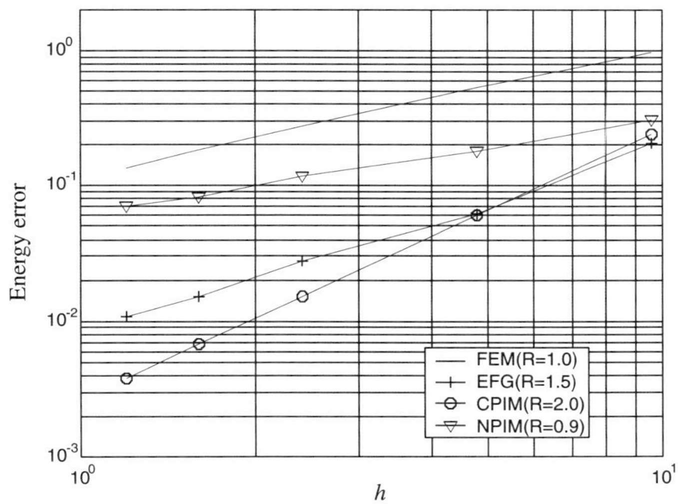

FIGURE 8.9
Comparison of the convergence of four different methods. The cantilever beam problem is used for this examination. R is the rate of convergence. In the FEM, 4-node elements are used; in EFG, the support domain of α_s = 2.5 (about 25 nodes) is used for constructing MLS shape functions; in CPIM, one-piece shape functions are constructed using the nearest 9 nodes surrounding the integration cell; in NPIM, the nearest 9 nodes surrounding the Gauss point are used. NPIM is more accurate than FEM, but converges a little slower. CPIM converges fastest. EFG performs between FEM and CPIM. When 16 nodes were used in CPIM, exact results were produced for the cantilever beam problem.

TABLE 8.3a

CPU Time(s) for Calculating the Cantilever Beam Using NPIM and EFG (Lagrange Multipliers)

No. of Nodes	EFG With Same n	EFG With Same Accuracy	PIM
55	3.3	8.4	2.1
189	67.2	95.1	8.2
561	1731.7	1818.4	32.1

Note: Tested on a DEC workstation.

Source: Liu, G. R. and Gu, Y. T., *Int. J. Numer. Methods Eng.*, 50, 937–951, 2001. With permission.

TABLE 8.3b

Comparison of CPU Time (Second) Used in NPIM and CPIM for the Cantilever Beam Problem

No. of Nodes	NPIM	CPIM	Difference (%)
55 nodes	0.651	0.697	6.6
189 nodes	1.492	1.695	11.6
697 nodes	8.621	9.478	9.0

Note: 9 nodes for interpolation in both NPIM and CPIM. Tested on a Datamini PC, 1.8G Intel Pentium 4.

the Gauss point are used. It has been found that NPIM is more accurate than FEM, but converges a little slower. It has also been seen that CPIM converges fastest with a convergence rate of 2.0. EFG performs between FEM and CPIM. We have also confirmed that the use of more nodes in CPIM can significantly reduce error in the analysis results. When $n = 16$ is used in CPIM for the cantilever beam problem, CPIM gives the exact solution. This fact clearly shows that CPIM can easily achieve p-convergence: One can easily use as many points as necessary without any manual operation. One can also very freely use different orders of PIM shape functions for different locations with compatibility ensured. The h-convergence will, of course, also work for CPIM, but may be less efficient compared to the p-convergence.

To study the efficiency, PIM is compared with the EFG method under the same conditions, but with Lagrange multipliers for imposing the essential boundary conditions. It is found that the computational error of EFG is larger than that of NPIM when the number of nodes n in the support domain is the same. To achieve the same accuracy, the n used in EFG must be larger than in NPIM. Therefore, the comparison is made for two situations: the same n and the same accuracy. The CPU time of NPIM and EFG is shown in Table 8.3a. It is found that NPIM uses much less CPU time than EFG. One may argue that this comparison is not fair, as the Lagrange multiplier method is used in EFG for imposing essential boundary conditions. If the penalty method is used instead, the computational time used by EFG with MLS will be much less. This is a valid argument, and it is true that when the penalty method is used, EFG will perform faster. A fairer comparison will be presented in later sections to demonstrate that methods using PIM shape functions still perform better than methods that use MLS shape functions with penalty methods for imposing essential boundary conditions.

Note that there is very little difference in CPU time between NPIM and CPIM. This can be evidenced from the test results shown in Table 8.3b. The cost of CPU time for imposing compatibility in PIM is only about 10%.

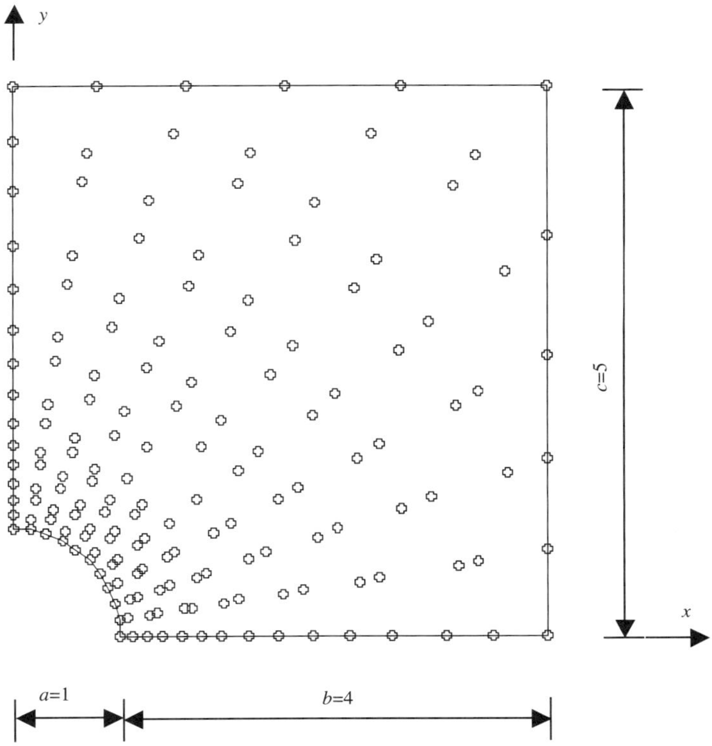

FIGURE 8.10
Nodes on plate with central hole subjected to unidirectional tensile load in the x direction. (From Liu, G. R. and Gu, Y. T., *Int. J. Numer. Methods Eng.*, 50, 937–951, 2001. With permission.)

Example 8.3 Hole in an Infinite Plate

Example 6.11, examined using the EFG method, is now reexamined here using the nonconforming Galerkin PIM method. The geometry of the plate is plotted in Figure 6.34. Because of the twofold symmetry, only a quarter of the plate is modeled with symmetric boundary conditions applied on $x = 0$ and $y = 0$. The parameters and the boundary conditions are exactly the same as those in Example 6.11. The displacement and the stress field within the plate are provided by Equations 5.104 to 5.109 in the polar coordinates (r, θ).

A circular support domain is used. The radius r_s of the circular support domain is defined as

$$r_s = \alpha_s \times d_I \tag{8.10}$$

where $\alpha_s = 2.0$ and d_I is the largest distance between two nodes that are on the boundary of the MFree model, as shown in Figure 8.10. A total of 165 nodes are used in the plate. Note that we deliberately have the nodes irregularly distributed. If the number of nodes in the support domain is more than 15, only 15 nodes with shorter distances to the Gauss quadrature point are used in the interpolation. It is found that for displacement, results obtained are almost identical with the analytical solution. As the stress is most critical, detailed results are presented here. The stress σ_x at $x = 0$ obtained using NPIM is shown in Figure 8.11. The result obtained by the FEM software PATRAN using the same nodes (regular mesh) as NPIM is shown in the same figure. It can be observed from this figure that NPIM yields very good results for the problem.

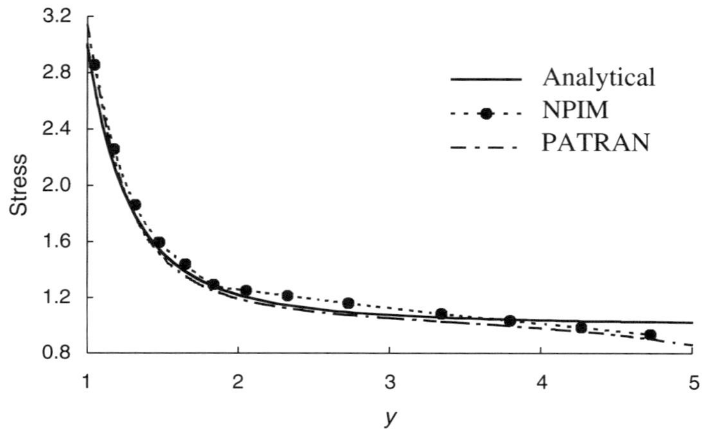

FIGURE 8.11
Stress distribution in a plate with a central hole subjected to a unidirectional tensile load (σ_x, at $x = 0$). (From Liu, G. R. and Gu, Y. T., *Int. J. Numer. Methods Eng.*, 50, 937–951, 2001. With permission.)

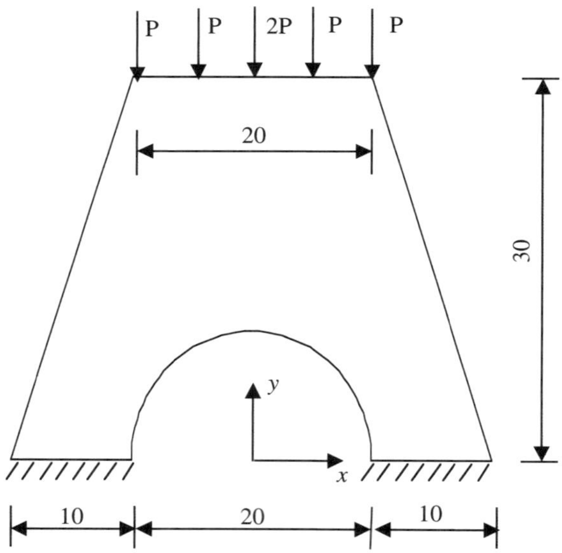

FIGURE 8.12
Bridge pier subjected to concentrated forces on the top (unit: m). (From Liu, G. R. and Gu, Y. T., *Int. J. Numer. Methods Eng.*, 50, 937–951, 2001. With permission.)

Example 8.4 Bridge Pier

In this example, NPIM is used in stress analysis of a bridge pier subjected to five concentrated forces on the top of the pier, as shown in Figure 8.12. The problem is solved as a plane strain problem. The parameters for this problem are as follows:

Loading: $p = 100$ kN

Young's modulus: $E = 40$ GPa

Poisson's ratio: $\nu = 0.15$

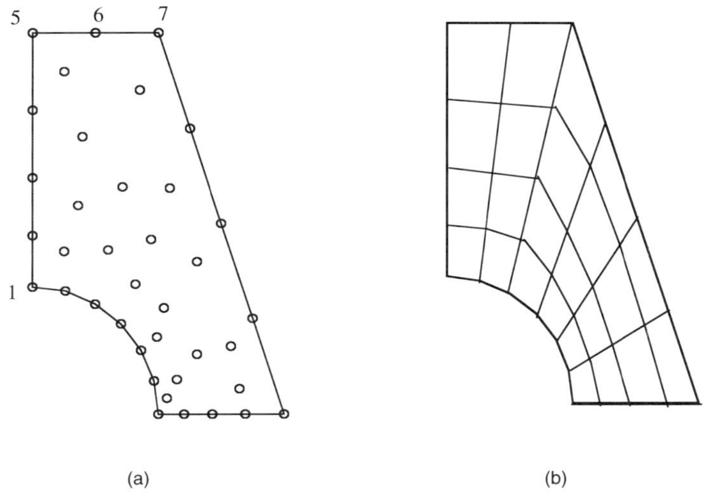

FIGURE 8.13
Models of the right half of the bridge pier: (a) nodal arrangement for PIM; (b) element mesh for FEM. (From Liu, G. R. and Gu, Y. T., *Int. J. Numer. Methods Eng.*, 50, 937–951, 2001. With permission.)

TABLE 8.4
Vertical Displacement of the Bridge Pier

Nodes	x	y	Displacement ($\times 10^{-5}$) NPIM	FEM
1	0.0	10.0	–1.6814	–1.7003
5	0.0	30.0	–2.6252	–2.6651
6	5.0	30.0	–2.2953	–2.2828
7	10.0	30.0	–2.2602	–2.2725

Source: Liu, G. R. and Gu, Y. T., *Int. J. Numer. Methods Eng.*, 50, 937–951, 2001. With permission.

As a result of symmetry, only the right half of the pier is modeled. The nodal arrangement is shown in Figure 8.13a. The problem is also analyzed by the FEM software PATRAN using the mesh shown in Figure 8.13b, which has the same number of nodes as NPIM.

The displacements at some nodes are listed in Table 8.4. The results obtained are in very good agreement with those obtained using FEM. The distribution of stress σ_y in the domain obtained by FEM and NPIM is shown in Figure 8.14. Thus, NPIM also obtains very good results for this problem.

8.1.6 Remarks

PIMs with Galerkin formulation have been presented. Because the PIM shape functions possess the Kronecker delta function property, difficulties in implementation of the essential boundary conditions are overcome. Numerical examples have demonstrated the effectiveness and simplicity of PIMs. PIM offers a simple and efficient numerical procedure to handle problems with industrial applications. The following section presents one of the application problems using PIM.

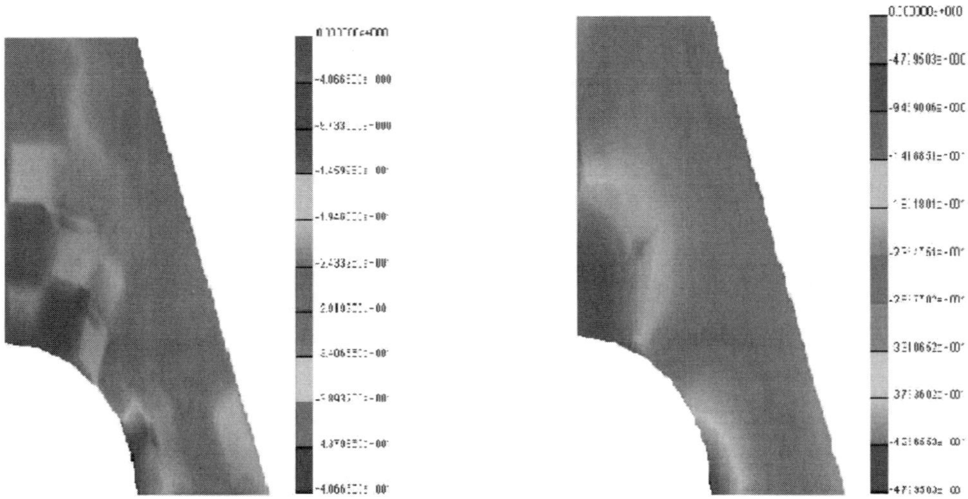

(a) (b)

FIGURE 8.14
Distribution of stresses σ_y in the bridge pier computed by FEM and NPIM: (a) FEM results; (b) NPIM results. (From Liu, G. R. and Gu, Y. T., *Int. J. Numer. Methods Eng.*, 50, 937–951, 2001. With permission.)

8.2 Application of PIM to Foundation Consolidation Problem

This section applies nonconforming PIM to a civil engineering problem of simulating the process of soil foundation consolidation, which is modeled as a 2D time-dependent problem. The work is performed by Wang et al. (2001) based on Biot's (1941) consolidation theory. A general form of Biot's consolidation theory is used, which can accommodate any constitutive law of materials. Spatial variables, displacement, and pore pressure are discretized using the same PIM shape functions. An incremental Galerkin weak form is used to create the discrete system equations, and the Crank–Nicholson method is used for the discretization of the time domain. Examples are presented and compared with closed-form solutions or FEM results.

8.2.1 Biot's Consolidation Theory and Its Weak Form

Biot's (1941) consolidation theory provides a macrolevel description for the interaction between soil and water in a saturated soil. The theory is composed of the following six sets of equations:

1. Equilibrium equation of soil–water mixture:

$$\frac{\partial \sigma_{ij}}{\partial x_j} + b_i = 0 \quad \text{in } \Omega \tag{8.11}$$

Its incremental form in time interval $[t, t + \Delta t]$ becomes

$$\frac{\partial \Delta\sigma_{ij}}{\partial x_j} + \Delta b_i = -\left(\frac{\partial \sigma_{ij}^t}{\partial x_j} + b_i^t\right) \quad \text{in } \Omega \tag{8.12}$$

2. Relationship of displacement and strain for soil skeleton:

$$\Delta\varepsilon_{ij} = \frac{1}{2}\left(\frac{\partial \Delta u_i}{\partial x_j} + \frac{\partial \Delta u_j}{\partial x_i}\right) \quad \text{in } \Omega \tag{8.13}$$

3. Constitutive law of soil skeleton in differential form:

$$\mathrm{d}\sigma'_{ij} = c^{ep}_{ijkl}\,\mathrm{d}\varepsilon_{kl} \quad \text{in } \Omega \tag{8.14}$$

4. Darcy's seepage law for pore water flow:

$$q_i = \frac{K_{ij}}{\gamma_w}\frac{\partial u}{\partial x_j} \quad \text{in } \Omega \tag{8.15}$$

5. Terzaghi's (Terzaghi and Peck, 1976) effective stress principle:

$$\sigma_{ij} = \sigma'_{ij} + u\delta_{ij} \tag{8.16}$$

6. Incompressibility of solid–water mixture or continuity equation:

$$\frac{\partial \varepsilon_v}{\partial t} = \frac{\partial q_i}{\partial x_i} \tag{8.17}$$

where σ_{ij}, σ'_{ij}, and u are the total stress tensor, the effective stress tensor, and the excess pore water pressure at time t and b_i is the body force. Δu_i is the displacement increment and $\Delta\sigma_{ij}$ and $\Delta\varepsilon_{ij}$ are, respectively, the total stress and strain increments in time interval $[t, t + \Delta t]$. The discharge of excess pore water is q_i in ith direction. γ_w is the density of water. c^{ep}_{ijkl} is the elastoplastic matrix of the soil skeleton, which is determined by constitutive laws. K_{ij} is the permeability tensor of the soil skeleton, which usually has nonzero components K_x in the x direction and K_y in the y direction, respectively. ε_v is the volume strain of the soil skeleton, which is expressed as

$$\varepsilon_v = \frac{\partial u_i}{\partial x_i} \tag{8.18}$$

Boundary conditions for this problem include two parts: boundary for the solid and boundary for the fluid (water).

For the soil skeleton boundary:

$$\begin{cases} u_i = \bar{u}_{i0} \\ \sigma'_{ij} n_j = \bar{T}_i \end{cases} \quad \text{on } \Gamma_u \times [0, \infty) \tag{8.19}$$

where $\mathbf{n} = \{n_1, n_2, n_3\}$ is the outward normal direction and n_i is its directional cosine.

For the fluid boundary:

$$\begin{cases} u = u_0 \\ q_i = q_{i0} \end{cases} \quad \text{on } \Gamma_u \times [0, \infty) \tag{8.20}$$

The initial conditions are

$$\begin{cases} u_i = 0 \\ u = 0 \end{cases} \quad \text{on } \Omega \times 0^- \tag{8.21}$$

Two components of body forces are acting on the soil skeleton:

1. Effective unit weight b_i' ($=b_i - \gamma_w$).
2. Seepage force induced by hydraulic gradient:

$$\left(-\frac{\partial u}{\partial x_i}\right)$$

In the time interval of $[t, t + \Delta t]$, the soil skeleton should satisfy the following Galerkin weak form:

$$\begin{aligned}
&\int_\Omega \{\delta(\Delta\varepsilon)\}^T\{\Delta\sigma'\}\,\mathrm{d}\Omega - \int_\Omega \left\{\delta\left(\frac{\partial \Delta u_i}{\partial x_i}\right)\right\}^T \{u\}^{t+\Delta t}\,\mathrm{d}\Omega - \int_\Omega \{\delta(\Delta\bar{u})\}^T\{\Delta b\}\,\mathrm{d}\Omega \\
&\quad + \int_{\Gamma_t} \{\delta(\Delta\bar{u})\}^T\{n\}u^{t+\Delta t}\,\mathrm{d}\Gamma - \int_{\Gamma_t}\{\delta(\bar{u})\}^T\{\Delta\bar{T}\}\,\mathrm{d}\Gamma \\
&\quad = -\int_\Omega \{\delta(\Delta\varepsilon)\}^T\{\sigma'^t\}\,\mathrm{d}\Omega + \int_{\Gamma_t}\{\delta(\bar{u})\}^T\{\bar{T}^t\}\,\mathrm{d}\Gamma - \int_\Omega \{\delta(\Delta\bar{u})\}^T\{b^t\}\,\mathrm{d}\Omega
\end{aligned} \tag{8.22}$$

The term at the right-hand side includes the unbalanced force at the previous time step. This force is included in every time step, so that Equation 8.22 can prevent error accumulation and maintain the equilibrium state at any time. Thus, the same accuracy is achieved at each time step.

Time integration is applied to continuity Equation 8.18, and the weak form for spatial variables (x, y) is expressed as

$$\begin{aligned}
&-\int_\Omega \{\delta u\}^T \left\{\frac{\partial \Delta u_i}{\partial x_i}\right\}\mathrm{d}\Omega \\
&\quad = \frac{1}{\gamma_w}\int_t^{t+\Delta t}\left[\int_{\Gamma_u}\{\delta u\}^T\{Ku\}\,\mathrm{d}\Gamma\right]\mathrm{d}t + \frac{1}{\gamma_w}\int_t^{t+\Delta t}\left[\int_\Omega \left\{\frac{\partial \delta u}{\partial x_i}\right\}^T\left\{K_i \frac{\partial u}{\partial x_i}\right\}\mathrm{d}\Omega\right]\mathrm{d}t
\end{aligned} \tag{8.23}$$

8.2.2 Discretization of Weak Form

The field variables of displacement increments $(\Delta u, \Delta v)$ and excess pore water pressure u at any time t are approximated using Equation 8.2; we then have

$$\begin{Bmatrix} \Delta u \\ \Delta v \\ u \end{Bmatrix} = \begin{bmatrix} \phi_1 & 0 & 0 & \cdots & \phi_n & 0 & 0 \\ 0 & \phi_1 & 0 & \cdots & 0 & \phi_n & 0 \\ 0 & 0 & \phi_1 & \cdots & 0 & 0 & \phi_n \end{bmatrix} \mathbf{U}_s \tag{8.24}$$

where the vector $\mathbf{U}_s$ collects field variables at all the nodes in the support domain, which are arranged as

$$\mathbf{U}_s = [\Delta u_1 \;\; \Delta v_1 u_1 \;\; \Delta u_2 \;\; \Delta v_2 u_2 \;\; \cdots \;\; \Delta u_n \;\; \Delta v_n u_n]^T \tag{8.25}$$

The time domain is integrated by the following equation:

$$\int_t^{t+\Delta t} f(x)\,dx = \Delta t[\theta f(t) + (1-\theta) f(t+\Delta t)] \tag{8.26}$$

Here $0 \le \theta \le 1$, and $f(x)$ represents a field function. By substituting Equation 8.24 into Equations 8.22 and 8.23, after a lengthy manipulation similar to that given in Section 6.1.1, the following system equations can be obtained:

$$\mathbf{KU} = \mathbf{F} \tag{8.27}$$

where $\mathbf{U}$ is the vector of nodal field variables that are arranged in the same way as in Equation 8.25, but for all the nodes in the entire problems domain. $\mathbf{K}$ is the global stiffness matrix assembled using the nodal stiffness matrix, which has the following form:

$$\mathbf{K}_{ij} = \begin{bmatrix} P_1 d_1 + P_3 d_2 + P_3 d_3 + P_9 d_4 & P_3 d_1 + P_9 d_2 + P_2 d_3 + P_5 d_4 & -\phi_j \frac{\partial \phi_i}{\partial x} \\ P_3 d_1 + P_2 d_2 + P_9 d_3 + P_5 d_4 & P_9 d_1 + P_5 d_2 + P_5 d_3 + P_4 d_4 & -\phi_j \frac{\partial \phi_i}{\partial y} \\ -\phi_i \frac{\partial \phi_j}{\partial x} & -\phi_i \frac{\partial \phi_j}{\partial y} & K_{33}^0 \end{bmatrix} \tag{8.28}$$

where P_i $(i = 1, 9)$ corresponds to the components of c_{ijkl}^{ep} in the material matrix $\mathbf{c}$ defined in Equation 8.14 that can be generally expressed as follows:

$$\mathbf{c} = \begin{bmatrix} P_1 & P_2 & P_3 \\ P_2 & P_4 & P_5 \\ P_3 & P_5 & P_9 \end{bmatrix} \quad \text{for plane problem} \tag{8.29}$$

For example, for linear elasticity with Young's modulus, E, and Poisson ratio, ν, matrix $\mathbf{c}$ is given by

$$\mathbf{c} = \frac{E}{1-\nu^2} \begin{bmatrix} 1 & \nu & 0 \\ \nu & 1 & 0 \\ 0 & 0 & (1-\nu)/2 \end{bmatrix} \quad \text{for plane stress} \tag{8.30}$$

Other coefficients in Equation 8.28 are obtained by

$$\begin{aligned} d_1 &= \frac{\partial \phi_i}{\partial x}\frac{\partial \phi_j}{\partial x} \qquad d_2 = \frac{\partial \phi_i}{\partial y}\frac{\partial \phi_j}{\partial x} \\ d_3 &= \frac{\partial \phi_i}{\partial x}\frac{\partial \phi_j}{\partial y} \qquad d_4 = \frac{\partial \phi_i}{\partial y}\frac{\partial \phi_j}{\partial y} \\ K_{33}^0 &= -(1-\theta)\frac{\Delta t}{\gamma_w}(K_x d_1 + K_y d_4) \end{aligned} \tag{8.31}$$

The global force vector **F** is assembled using its nodal counterparts that have the form:

$$\mathbf{f}_i = \int_{\Gamma_t} \mathbf{\Phi}_i \bar{\mathbf{t}}\, \mathrm{d}\Gamma + \int_{\Omega} \mathbf{\Phi}_i \mathbf{b}\, \mathrm{d}\Omega + \int_{\Omega} \mathbf{f}_{oi}\, \mathrm{d}\Omega \tag{8.32}$$

The force vector consists of three terms: external load on the natural boundary, effective body force, and unbalanced force at the previous time step, which can be obtained using

$$\begin{Bmatrix} f_{0x} \\ f_{0y} \\ f_{0u} \end{Bmatrix} = \begin{Bmatrix} \dfrac{\partial \phi_i}{\partial x}\sigma'_x + \dfrac{\partial \phi_i}{\partial y}\tau_{xy} \\ \dfrac{\partial \phi_i}{\partial y}\sigma'_y + \dfrac{\partial \phi_i}{\partial x}\tau_{xy} \\ \theta \dfrac{\Delta t}{\gamma_w}\left[K_x \dfrac{\partial \phi_i}{\partial x} \sum_r \dfrac{\partial \phi_r}{\partial x} u_r^t + K_y \dfrac{\partial \phi_i}{\partial y} \sum_r \dfrac{\partial \phi_r}{\partial y} u_r^t\right] \end{Bmatrix} \tag{8.33}$$

The flowchart for the numerical algorithm of PIM is briefly as follows:

1. Determine all Gauss points (position and weight) over the background mesh.
2. Begin with an initial time.
3. Loop over the time steps.
4. Loop over the Gauss points to assemble the stiffness matrix and load vector.
   a. Determine the support domain for a Gauss point and select neighboring nodes based on a criterion.
   b. Compute a shape function and its derivatives for each Gauss point.
   c. Evaluate the nodal stiffness matrix and load vector.
   d. Assemble the nodal contribution to global matrices/vectors.
5. Solve the system equation to obtain displacement increments and excess pore water pressure at each node.
6. Evaluate strain and effective stress at each Gauss point.
7. Back to 3.
8. End.

8.2.3 Numerical Examples

Example 8.5 One-Dimensional Consolidation Problem

PIM is coded for 2D problems, and applied first to the 1D Terzaghi's foundation consolidation problem schematically described in Figure 8.15. The PIM model with regular nodal distribution for this 1D problem is shown Figure 8.16. In this model, only the upper surface is permeable and the rest is impervious. Two sides and the bottom are all fixed for displacements. Therefore, it simulates a single-sided drainage problem. The thickness of the soil layer is assumed to be $H = 16$ m. Soil parameters are regarded as linearly elastic with $E = 40{,}000$ kPa and $\nu = 0.3$. When a surcharge $\Delta\sigma = 10$ kPa is suddenly applied on the surface of the soil layer, excess pore water pressure will be generated by the program through a short time increment. The excess pore water pressure thereafter dissipates with time.

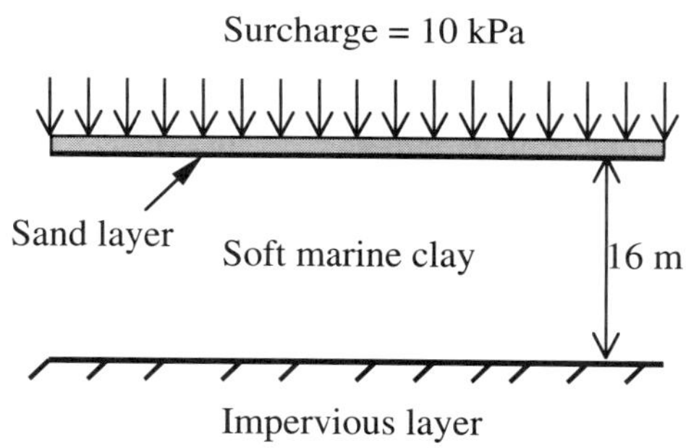

FIGURE 8.15
1D Terzaghi's foundation consolidation problem. (From Wang, J. G. et al., *Comput. Methods Appl. Mech. Eng.*, 190, 5907–5922, 2001. With permission.)

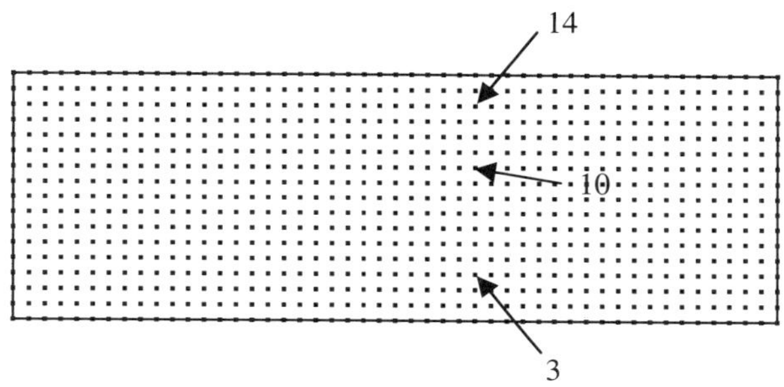

FIGURE 8.16
Nodal distribution in the PIM model for the 1D Terzaghi's foundation consolidation problem. Results at points 3, 10, and 14 are to be plotted. (From Wang, J. G. et al., *Comput. Methods Appl. Mech. Eng.*, 190, 5907–5922, 2001. With permission.)

This 1D Terzaghi's foundation consolidation problem has a closed-form solution (Terzaghi and Peck, 1976) for excess pore water pressure:

$$u = \frac{4}{\pi}\Delta\sigma\sum_{n=1}^{\infty}\frac{1}{2n-1}\sin\left(\frac{(2n-1)\pi y}{2H}\right)e^{-(2n-1)^2\frac{\pi^2}{4}T_V} \tag{8.34}$$

Degree of consolidation U_t is

$$U_t = 1-\frac{8}{\pi^2}\sum_{n=1}^{\infty}\frac{1}{(2n-1)^2}e^{-(2n-1)^2\frac{\pi^2}{4}T_V} \tag{8.35}$$

where the parameters are defined as

$$T_V = \frac{C_V}{H^2}t \qquad C_V = \frac{k}{\gamma_w m_v} \qquad m_v = \frac{(1+\nu)(1-2\nu)}{E(1-\nu)} \tag{8.36}$$

Surface settlement S_t at any time t is given by

$$S_t = U_t m_v \Delta\sigma H \tag{8.37}$$

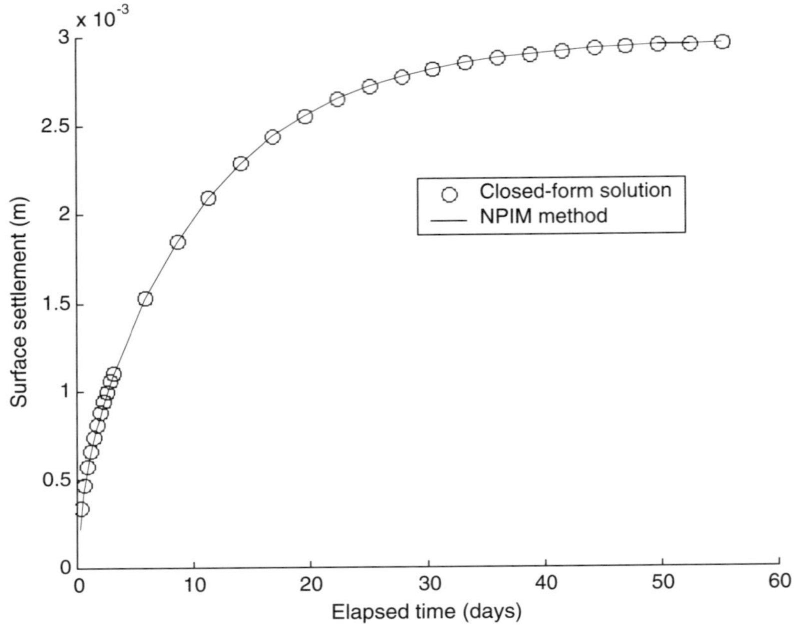

FIGURE 8.17
Surface settlement with time for the 1D Terzaghi's consolidation problem. (From Wang, J. G. et al., *Comput. Methods Appl. Mech. Eng.*, 190, 5907–5922, 2001. With permission.)

In the Crank–Nicholson method, $\theta = 0.5$ is used in the numerical computation. The time step is automatically selected based on the following criteria to maintain stability and freedom from oscillation:

$$\frac{h^2}{6C_v} \leq \Delta t \leq \frac{h^2}{2C_v} \tag{8.38}$$

where h is the characteristic size of the node distance. For example, for the 1D model, h is the nodal spacing.

CASE 1 Constant Permeability along Depth
A constant permeability of $k = 1.728 \times 10^{-3}$ m/day (or 2×10^{-8} m/s) in all directions is first considered. This is a standard Terzaghi problem with the closed-form solution given above. Results obtained using the closed form of equations and the PIM method are plotted in Figures 8.17 through 8.19. Surface settlement is plotted in Figure 8.17 and the history of excess pore water pressure is plotted in Figure 8.18 for three sample points (see Figure 8.16). Figure 8.19 gives the spatial distribution of excess pore water pressure at different times. They are all in very good agreement with the closed-form solutions.

CASE 2 Variable Permeability along Depth
Marine clay is a soft clay and its permeability usually varies with depth. A linearly variable permeability with depth is studied here. The permeability at the top surface is assumed to be 100 times the permeability at the bottom, that is, $k = 1.728 \times 10^{-3}$ m/day. The permeability between the top surface and the bottom is assumed to vary linearly. There is no closed-form solution for this problem. For reference, averaged permeability at the top and bottom is used to obtain a closed-form solution. For quantitative comparison, FEM analysis is performed using four-node isoparametric elements. Figure 8.20 shows

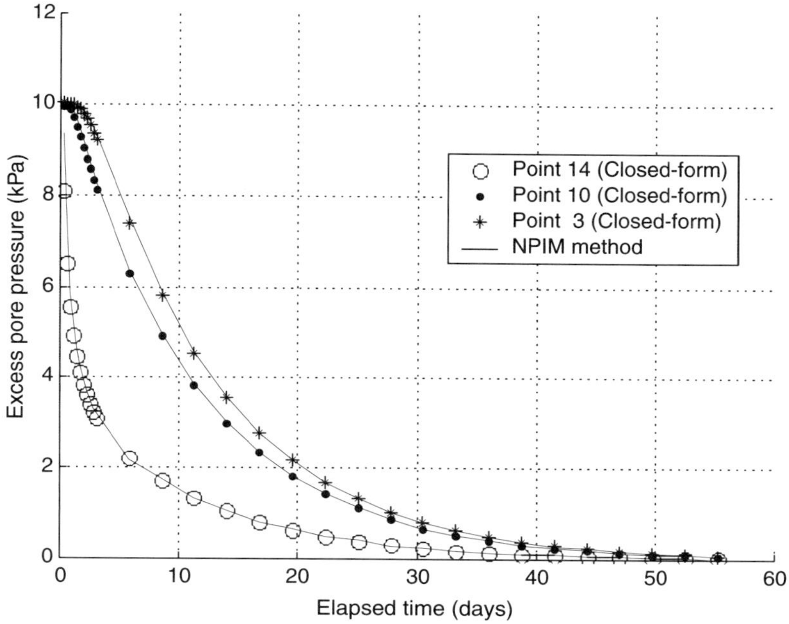

FIGURE 8.18
Dissipation of excess pore water pressure at different points for the 1D Terzaghi's consolidation problem. (From Wang, J. G. et al., *Comput. Methods Appl. Mech. Eng.*, 190, 5907–5922, 2001. With permission.)

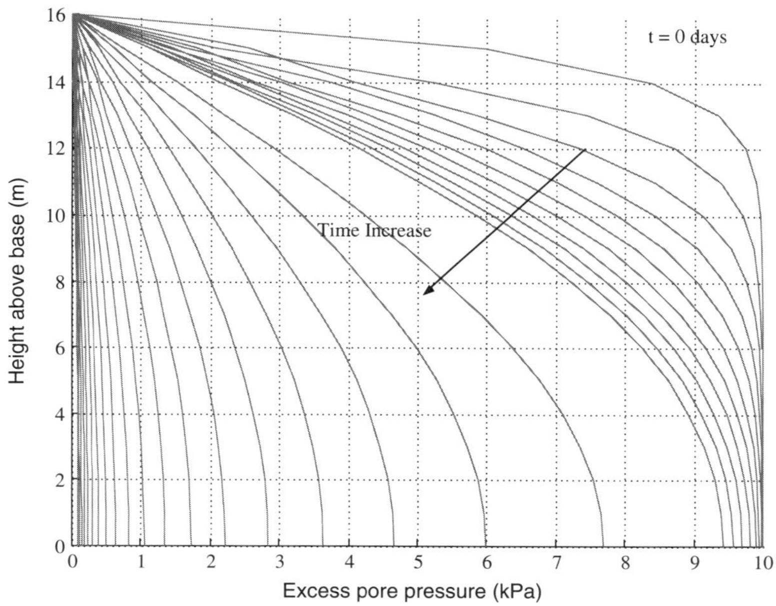

FIGURE 8.19
Distribution of excess pore water pressure at different times for the 1D Terzaghi's consolidation problem. (From Wang, J. G. et al., *Comput. Methods Appl. Mech. Eng.*, 190, 5907–5922, 2001. With permission.)

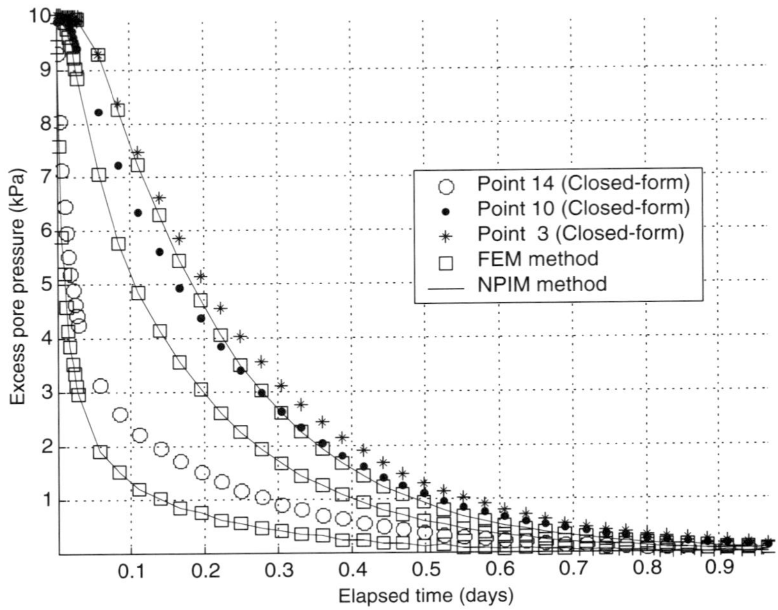

FIGURE 8.20
Distribution of excess pore water pressure for varying permeability (case 2: top permeability is 100 times that of bottom). Note that the closed-form solution shown here is based on average permeability, which serves only as a reference. (From Wang, J. G. et al., *Comput. Methods Appl. Mech. Eng.*, 190, 5907–5922, 2001. With permission.)

the history of excess pore water pressure obtained using NPIM, closed-form solutions, and FEM. The NPIM and FEM results agree very well while the closed-form result is slower than NPIM. Therefore, average permeability can be used only when permeability does not vary very much for 1D consolidation problems. Figure 8.21 is an isotemporal curve computed using the NPIM method. It is confirmed that all these figures are consistent with Schiffman's results (Schiffman and Gibson, 1964).

Example 8.6 Two-Dimensional Consolidation Problem

A 2D plane strain consolidation problem is also studied here using the NPIM code. The foundation is assumed linearly elastic, and is under a strip loading of 10 kPa, simulating a road load. A schematic description of the problem is shown in Figure 8.22. The top surface is fully drained and the rest of the boundaries are all impervious. For displacement boundaries, the horizontal freedom is fixed on the vertical boundaries, and the vertical freedom is fixed on the horizontal boundaries. There is no closed-form solution for this 2D problem; thus FEM is applied for comparison. Figure 8.23 shows the settlement distribution at different consolidation times ($T = 0, 3, 20$ days). The foundation demonstrates an immediate settlement after the action of the load. Its settlement will increase with the dissipation of excess pore water pressure. The dissipation history of excess pore water pressure at two different points is shown in Figure 8.24. After about 20 days, dissipation of excess pore water pressure is almost complete and the settlement, u_y, reaches its stable state, as shown in Figure 8.25. For more details, results on horizontal and vertical sections are sampled to reveal the distributions of excess pore water pressure and effective stress (σ_y'). The horizontal section is 2 m below ground level. Along this section, the excess pore water pressure is shown in Figure 8.26 and the effective stress (σ_y') is in Figure 8.27.

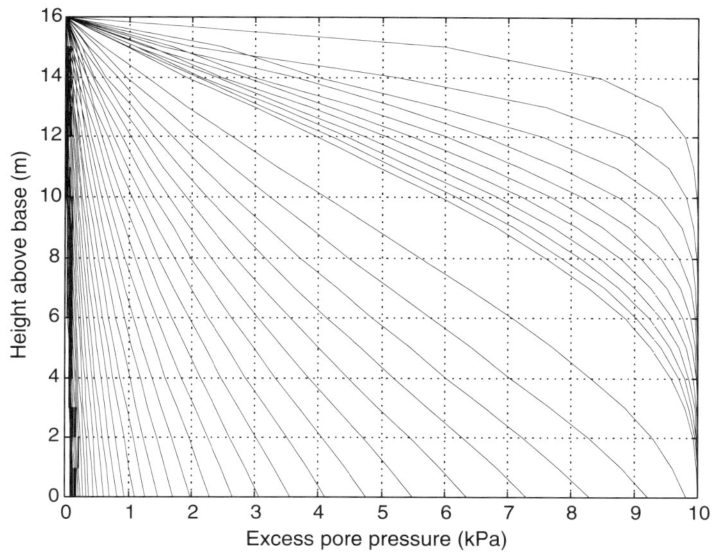

FIGURE 8.21
Isotemporal distribution of excess pore water pressure obtained using NPIM (case 2: top permeability is 100 times that of bottom). (From Wang, J. G. et al., *Comput. Methods Appl. Mech. Eng.*, 190, 5907–5922, 2001. With permission.)

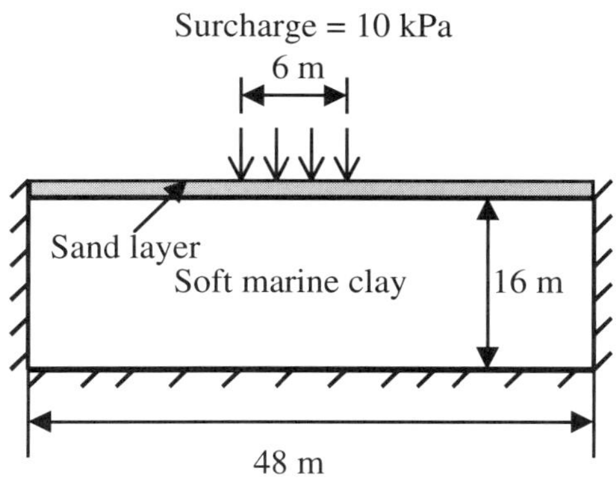

FIGURE 8.22
2D consolidation proble m. The foundation is subjected to a strip load on the surface. (From Wang, J. G. et al., *Comput. Methods Appl. Mech. Eng.*, 190, 5907–5922, 2001. With permission.)

The excess pore water pressure reaches its peak right below the strip load. It approaches zero gradually with the progress of consolidation. The effective stress increases immediately after loading. With consolidation, the excess pore water pressure is transferred to a soil skeleton. This leads to a gradual increase in the effective stress, which reaches its peak at last. The vertical section is located at the middle line of the foundation. Figure 8.28 shows the dissipation of excess pore water pressure, and Figure 8.29 shows its effective stress (σ_y'). All these are in good agreement with the FEM results.

Finally, the effect of the irregularity of nodal distribution is studied for this 2D problem. A typical irregular distribution of nodes is given in Figure 8.30. The average nodal density

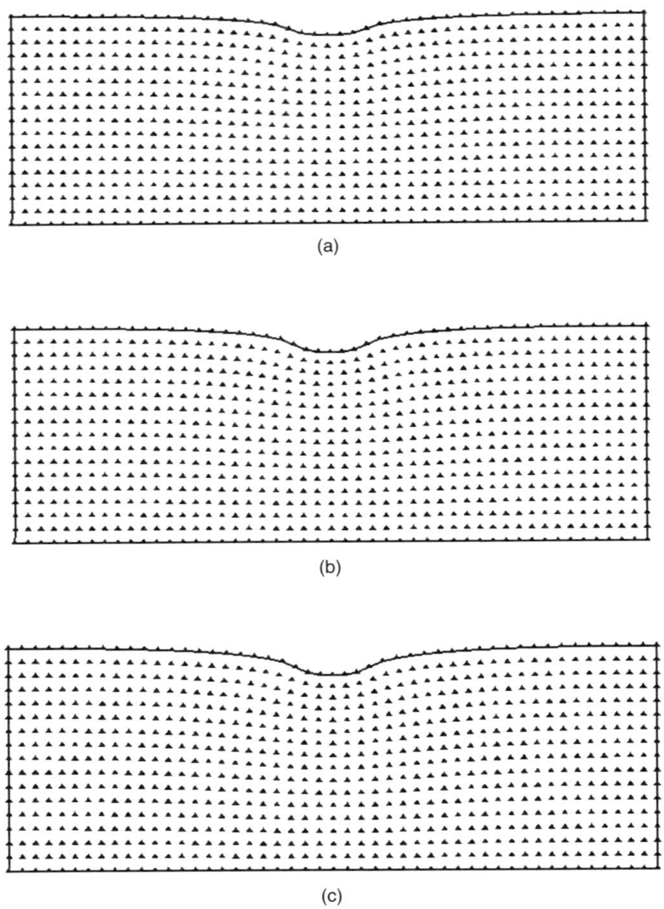

FIGURE 8.23
Settlement distribution at different times for the 2D consolidation problem: (a) $T = 0$ days; (b) $T = 3$ days; (c) $T = 20$ days. (From Wang, J. G. et al., *Comput. Methods Appl. Mech. Eng.*, 190, 5907–5922, 2001. With permission.)

is almost the same as the model of regular nodal distribution shown in Figure 8.16. Figure 8.31 shows little difference in the results obtained using regular and irregular nodes. The greatest difference is observed at the beginning. In conclusion, the irregularity of nodal distribution will affect the accuracy of the NPIM method, but this effect is small.

8.3 Radial Point Interpolation Method

8.3.1 Key Considerations

Chapter 5 demonstrated that the use of radial functions as basis functions can guarantee a nonsingular moment matrix. This section examines an MFree method that uses PIM shape functions constructed using radial basis functions instead of polynomial basis functions, which is called radial PIM or RPIM. Two types of radial functions are utilized. They are Gaussian (EXP) and multiquadric (MQ) radial basis functions (see Table 5.2). We

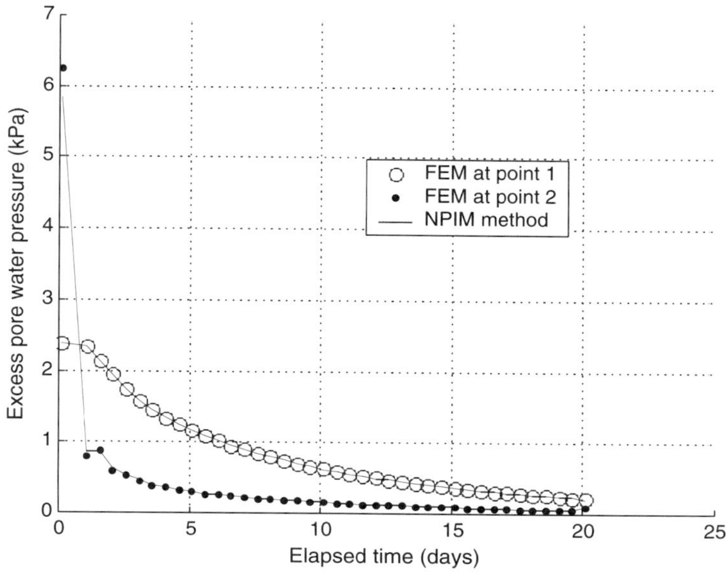

FIGURE 8.24
Excess pore water pressure history at different locations for the 2D consolidation problem. (From Wang, J. G. et al., *Comput. Methods Appl. Mech. Eng.*, 190, 5907–5922, 2001. With permission.)

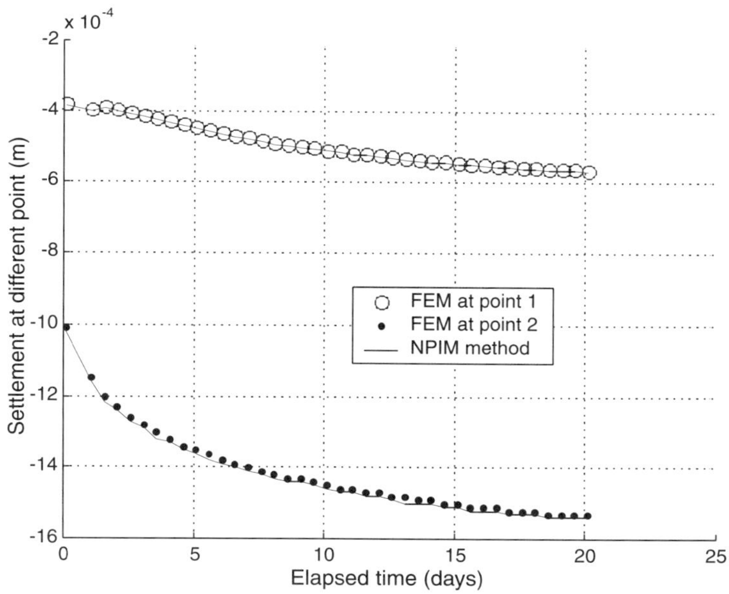

FIGURE 8.25
Settlement history at different points for the 2D consolidation problem. (From Wang, J. G. et al., *Comput. Methods Appl. Mech. Eng.*, 190, 5907–5922, 2001. With permission.)

again use the Galerkin weak form in this section. Similar to the case of polynomial PIM, there are nonconforming RPIM or NRPIM and conforming RPIM or CRPIM. The NRPIM is formulated using the standard Galerkin weak form, and the CRPIM is formulated using the constrained Galerkin weak form.

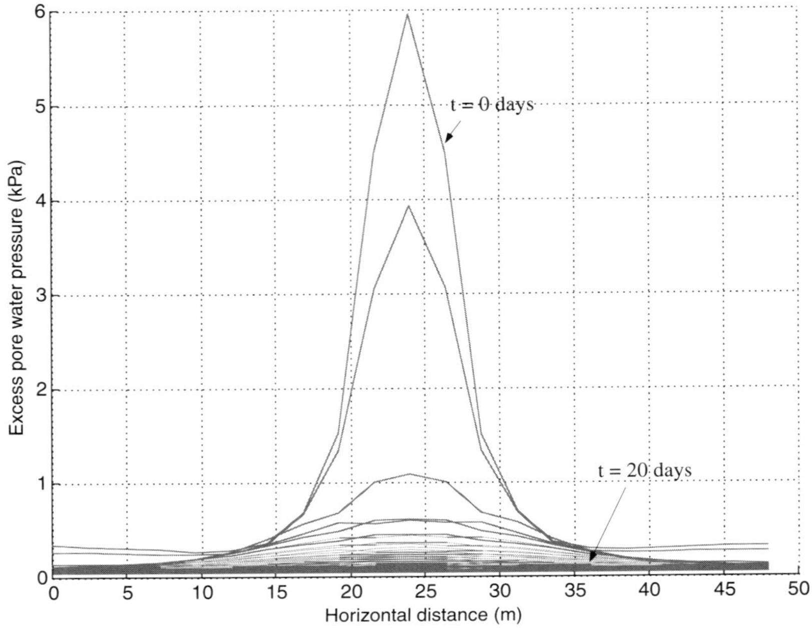

FIGURE 8.26
Dissipation of excess pore water on surface 2 m below ground. (From Wang, J. G. et al., *Comput. Methods Appl. Mech. Eng.*, 190, 5907–5922, 2001. With permission.)

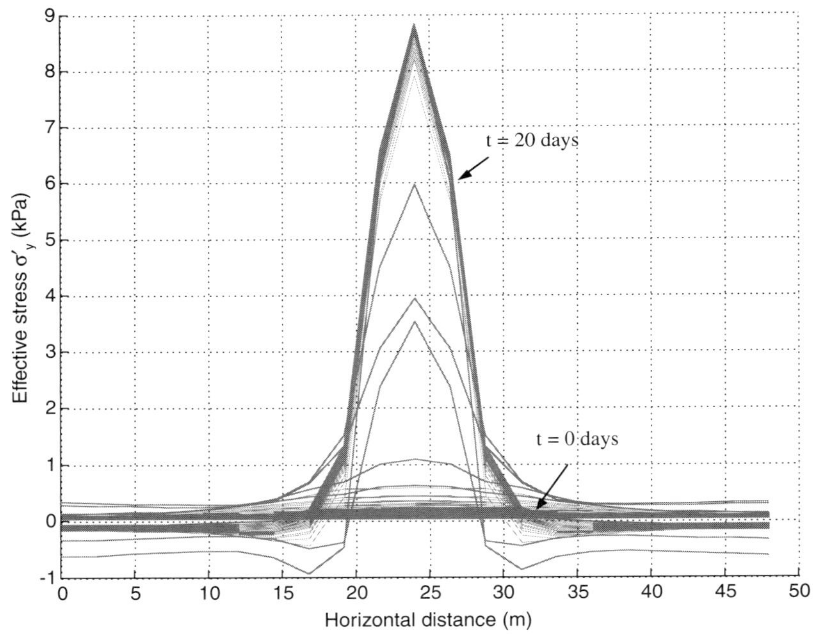

FIGURE 8.27
Effective stress distribution on surface 2 m below ground. (From Wang, J. G. et al., *Comput. Methods Appl. Mech. Eng.*, 190, 5907–5922, 2001. With permission.)

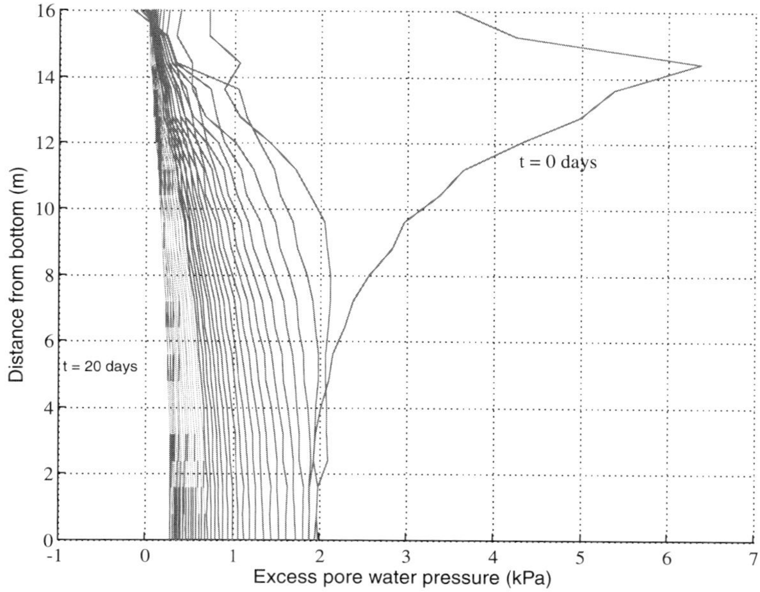

FIGURE 8.28
Excess pore water along vertical middle line of the foundation. (From Wang, J. G. et al., *Comput. Methods Appl. Mech. Eng.*, 190, 5907–5922, 2001. With permission.)

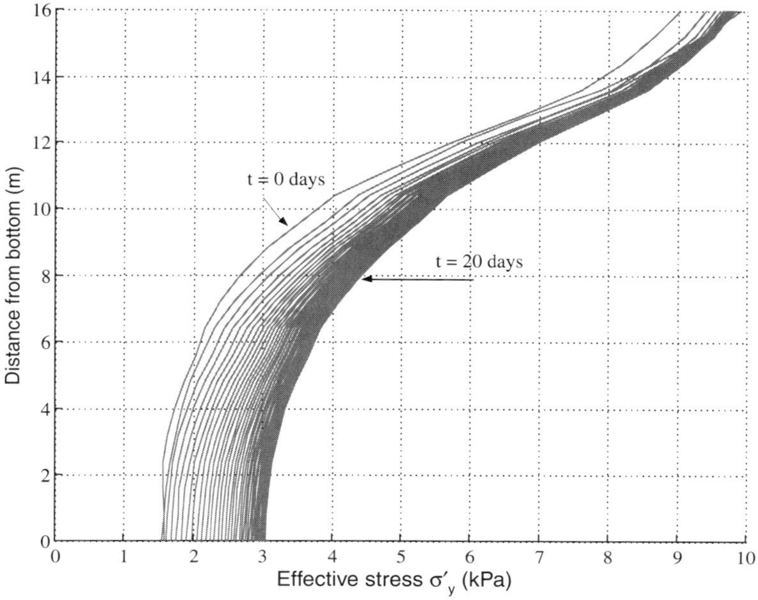

FIGURE 8.29
Effective stress σ_y' along the vertical middle line of the foundation. (From Wang, J. G. et al., *Comput. Methods Appl. Mech. Eng.*, 190, 5907–5922, 2001. With permission.)

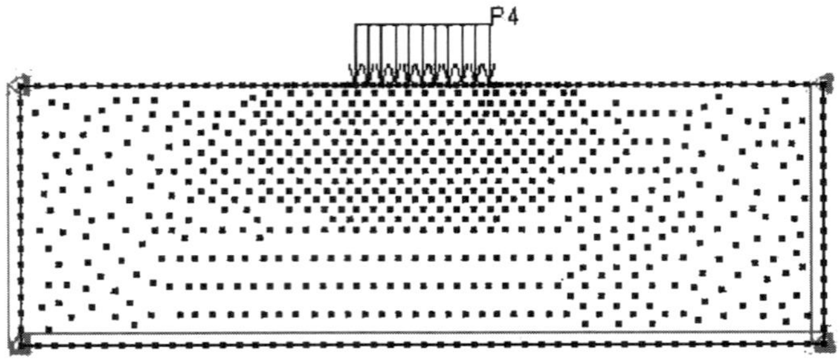

FIGURE 8.30
Irregular distribution of node for the 2D foundation consolidation problem. (From Wang, J. G. et al., *Comput. Methods Appl. Mech. Eng.*, 190, 5907–5922, 2001. With permission.)

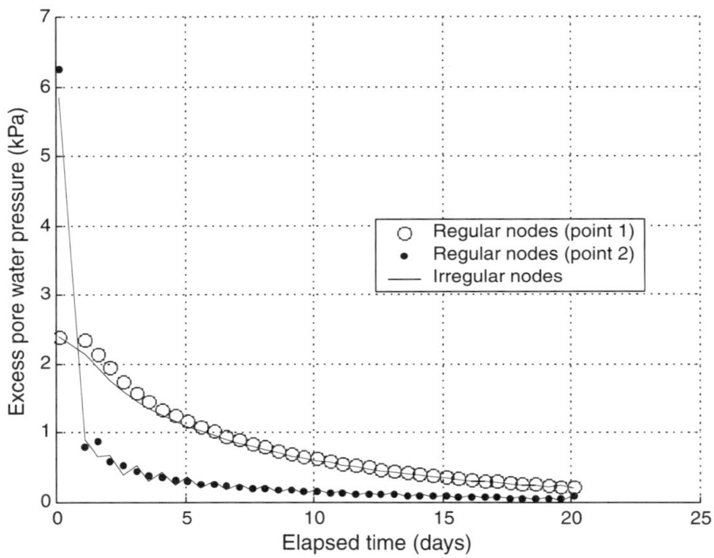

FIGURE 8.31
Effect of irregularity of nodal distribution on excess pore water pressure. (From Wang, J. G. et al., *Comput. Methods Appl. Mech. Eng.*, 190, 5907–5922, 2001. With permission.)

Chapter 5 revealed that RPIM is less accurate compared with polynomial PIM in terms of curve/surface fitting. Using RPIM shape functions for mechanics problems, results error will arise from two sources: radial basis approximation and Galerkin weak form. In meshless methods, the support domain is local instead of global and its property is influenced by shape parameters. This makes the theoretical analysis of error more complicated, but offers an opportunity to fine-tune our numerical algorithms to achieve better performance. The optimal shape parameters in MQ have been a hot topic in approximation theory and solution of partial differential equations (PDEs) (Franke, 1982; Carlson and Foley, 1991; Golberg et al., 1996; Franke and Schaback, 1997; Rippa, 1999). In the approximation theory of data fitting, Franke (1982) compared about 30 interpolation schemes in two dimensions and found that the two most accurate schemes were the methods based on radial basis function interpolation (MQ and TPS thin plate spline). He used

$C = 1.25D/\sqrt{n}$, where D is the diameter of the minimal circle enclosing all data points and n is the number of data points. Hardy (1990) recommended $C = 0.815d$, where

$$d = \frac{1}{n}\sum_{i=1}^{n} d_i$$

and d_i is the distance between the ith data point and its nearest neighbor nodes. Rippa (1999) proposed an algorithm for selecting a good value of shape parameter in multiquadric, inverse multiquadric, and Gaussian interpolants.

In the numerical solutions of PDEs, Golberg et al. (1996, 1999) discussed the error analysis for the dual reciprocity method (DRM) with radial basis function approximation and boundary integral equations. They found that the convergence behavior of the DRM depends on both the interpolation error and the error of the boundary element method (BEM). They used a technique of cross validation to improve the accuracy of MQ. That algorithm is based on statistical cross validation to search for a good shape parameter. Kansa (1990a,b) proposed a scheme that allows the shape parameter to vary with the basis functions. He observed that the more distinct the entries of the MQ coefficient matrix, the lower the MQ coefficient matrix condition number becomes, and the better the accuracy. Carlson and Foley (1991) obtained a new result that the optimal value of shape parameter was most strongly influenced by the magnitude of function values, the number of data points. But their location in the domain has little influence. All of these just optimize a single shape parameter in the radial basis functions.

Chapter 5 also presented studies on the effects on shape parameters of radial functions on curve/surface fitting. What most concerns an analyst of the mechanics of solids and structures is the effects of those parameters on the accuracy of computed displacements, stresses, and strains. This section therefore discusses this aspect further. The formulation of RPIM is exactly the same as the original polynomial PIM discussed in Section 8.1, except that the RPIM shape function is used. Therefore, in this section we emphasize the effects of the shape parameter of radial functions on the accuracy and convergence rate of the results obtained using RPIM via examples of mechanics problems.

8.3.2 Numerical Examples

In the following examples, polynomial terms are not included in the radial basis ($m = 0$), except for the patch test and other specifically mentioned cases. This is to reveal clearly the effects of the shape parameters of the radial functions.

Example 8.7 Patch Test

Node distribution for the patch used for this examination is shown in Figure 8.32. A 2×2 rectangular background mesh of four cells is used for numerical integration and each mesh uses 3×3 Gauss points. The material constants used in this example are $E = 1$ and $v = 0.3$. Two types of essential boundaries are given: constant displacement (rigid motion) and linear displacement are imposed along the boundary of the patch. The CRPIM passes the patch test exactly when three polynomial bases ($m = 3$) are added to both MQ and EXP basis functions. That is, for the rigid displacement case, displacements computed by CRPIM at any point within the patch are the same as that on the boundary, and the stress and strain are all zeros. For the linear displacement case, RPIM reproduced to machine accuracy the linear displacement distribution, and the stress computed is constant within the patch. The NRPIM cannot, in general, pass the patch test even if the linear polynomial terms are used. When the polynomial term is not included ($m = 0$), RPIM failed to pass

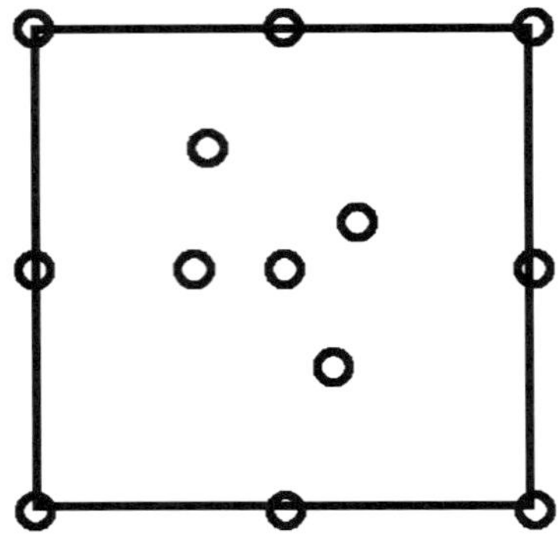

FIGURE 8.32
Patch with arbitrarily distributed internal nodes.

TABLE 8.5
Patch Test Results for Linear Displacement Case (NRPIM, $m = 0$)

Internal Node	Coordinates	EXP ($c = 0.003$)	MQ ($q = 1.03$, $C = 1.0$)
9	(1.0, 1.0)	(0.939, 1.043)	(1.029, 0.9732)
10	(0.65, 1.0)	(0.6548, 1.016)	(0.6496, 1.013)
11	(0.70, 1.5)	(0.817, 1.445)	(0.8607, 1.419)
12	(1.3, 1.2)	(1.246, 1.20)	(1.298, 0.189)
13	(1.2, 0.6)	(1.17, 0.6269)	(1.170, 0.5678)

the linear displacement patch for both radial functions. As an example, the linear displacement case is calculated and a typical distribution of displacement at internal nodes is given in Table 8.5. It is therefore concluded that for an RPIM to pass the patch test of linear displacement, inclusion of a polynomial term is required. The CRPIM will always pass the patch test as long as the polynomial terms are used and the numerical integration is accurate. The NRPIM may or may not pass the patch test even if the linear polynomial terms are included.

Example 8.8 Cantilever Beam

RPIM is benchmarked using Example 6.2 to analyze stresses in a cantilever beam. The beam has length L and height D, as shown in Figure 6.4. The beam is subjected to a parabolic traction at the free end given in Equation 6.27. The beam has a unit thickness and a plane stress problem is considered. The analytical solution is available; it can be found in the textbook by Timoshenko and Goodier (1977) and is listed in Equations 6.48 to 6.53. The parameters are taken as $E = 3.0 \times 10^7$, $\nu = 0.3$, $D = 12$, $L = 48$, and $P = -1000$. Both a regular nodal distribution and an irregular nodal distribution are used.

Effect of Irregular Node Distribution

Two typical node distributions are shown in Figure 8.33. The regular distribution has 637 (49×13) nodes whereas the irregular distribution has 644 nodes. The average node density is almost the same. The background mesh has 576 four-node rectangular cells for Figure 8.33a and 1136 three-node triangular cells for Figure 8.33b. The dimension of support domain is fixed as

$$d_s = \alpha_s \times d_c \tag{8.39}$$

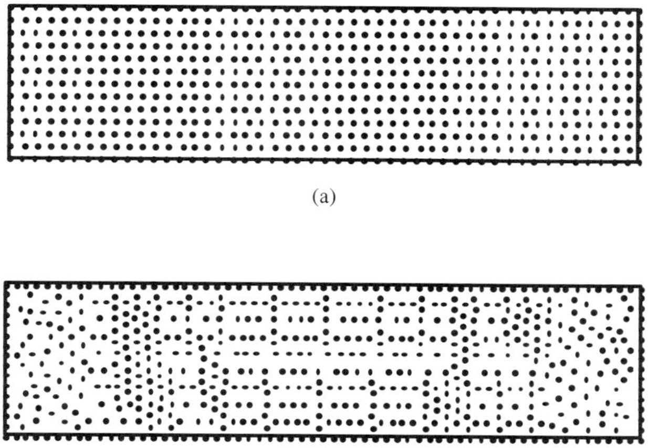

FIGURE 8.33
Nodal distribution in the cantilever beam: (a) regular nodal distribution with 637 = (49 × 13) nodes; (b) irregular nodal distribution with 644 nodes.

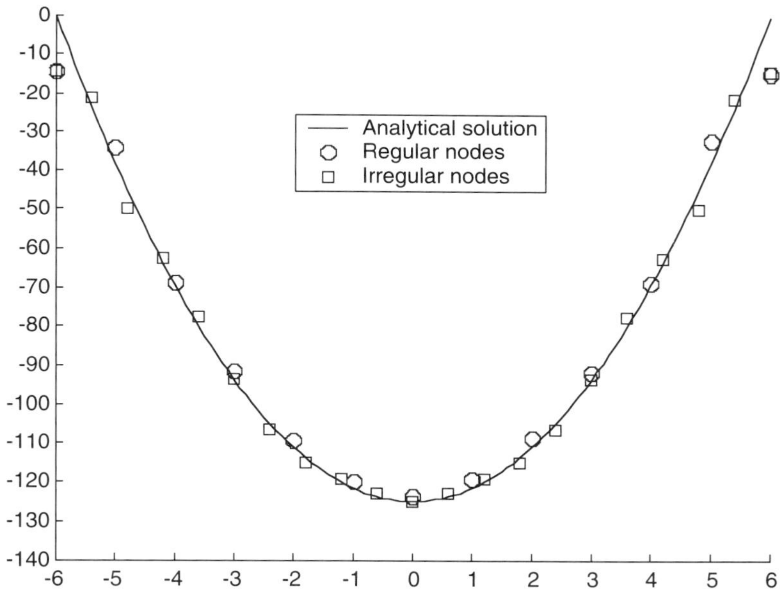

FIGURE 8.34
Effect of irregularity on the nodal arrangement on shear stress distribution on the section of the cantilever beam at $x = 24$ (NRPIM with EXP radial function).

where d_c is the distance between node I and its nearest neighboring node in the support domain (for this example $d_c = 1.0$ m). We use a square support domain to select nodes, and $\alpha_s = 2.0$. Such a parameter selects 9 to 16 nodes for each Gauss point. Figures 8.34 and 8.35 plot the sectional distribution of shear stress at $x = 24$, obtained using NRPIM, and the closed-form solution given by Equation 6.53 is also plotted for comparison. Shape parameters used in RPIM are $c = 0.003$ for EXP and $q = 1.03$ and $C = 1.42$ for MQ. NRPIM results of using both EXP and MQ radial functions are in excellent agreement with the

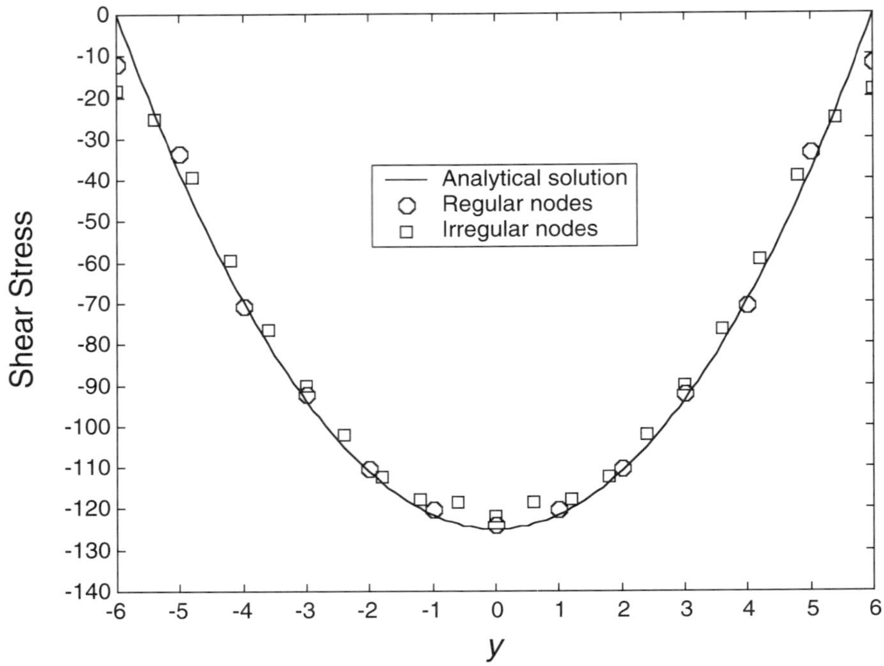

FIGURE 8.35
Effect of irregularity on the nodal arrangement on shear stress distribution on the section of the cantilever beam at $x = 24$ (NRPIM with MQ radial function).

analytic solution for both regular and irregular nodal distributions. Compared with polynomial PIM, RPIM is less sensitive to the irregularity of node distributions. The EXP method seems to give better results than the MQ method in this case. It is also found that the shape parameters using these radial functions affect the numerical results. A detailed study of the effects of the shape parameters is given as follows.

Effect of Shape Parameters

The relative error of displacements is first defined as follows:

$$\delta = \frac{\sum_{i=1}^{n} \left| u_i^{\text{Exact}} - u_i^{\text{RPIM}} \right|}{\sum_{i=1}^{n} \left| u_i^{\text{Exact}} \right|} \times 100\% \tag{8.40}$$

This error indicator will be used for the investigation of the effects of parameters on displacements. Because the stress and strain are much more sensitive to the numerical error, an energy error defined by an energy norm per unit area is also used here as an error indicator.

$$e_e = \frac{1}{2LD} \int_{\Omega} (\varepsilon^{\text{RPIM}} - \varepsilon^{\text{Exact}})^T (\sigma^{\text{RPIM}} - \sigma^{\text{Exact}}) \text{d}\Omega \tag{8.41}$$

In the EXP radial function, there is only one shape parameter c to be studied. In the MQ radial function, there are two parameters: q and C. Our strategy is that for a number of fixed C values, detailed investigation of q is conducted first to lock into a best value of q. For the locked q value, we can then investigate in detail the effects of C.

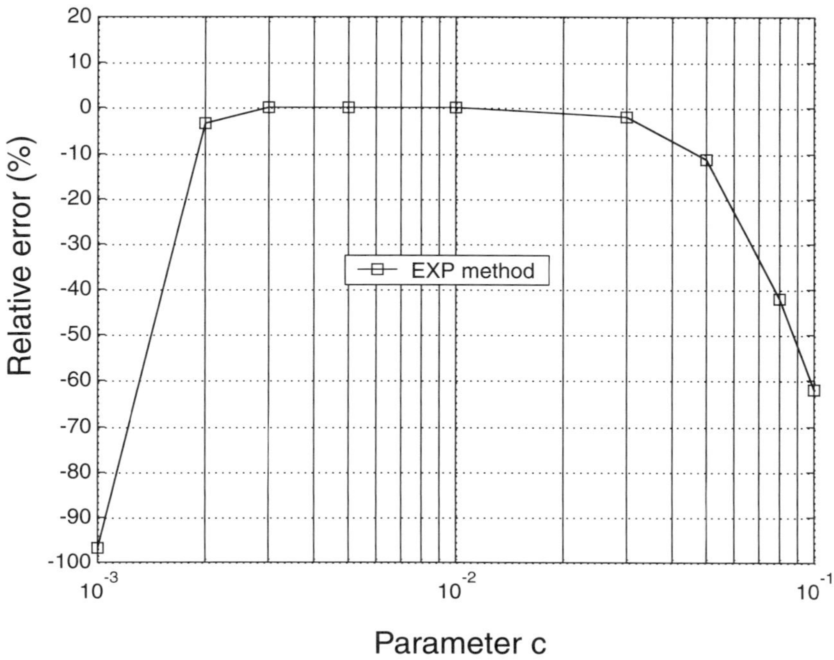

FIGURE 8.36
Effect of shape parameters on relative error of maximum deflection δ defined by Equation 8.40 (NRPIM with EXP radial basis functions; regularly distributed nodes). (From Wang, J. G. and Liu, G. R., *Comput. Methods Appl. Mech. Engrg.*, 191, 2611–2630, 2002. With permission.)

The regular node distribution (637), as shown in Figure 8.33a, is used in the computation. In this study, basis functions do not include polynomials ($m = 0$) so as to reveal the performance of the pure radial functions. Figures 8.36 and 8.37 show the effects of shape parameters on the relative error in the maximum deflection of the beam obtained using NRPIM with EXP radial basis functions. It is found that the results are not sensitive to the shape parameter c in the range of $c = 0.002$ to 0.03. Figure 8.37 shows the effects of shape parameters on the relative error in the maximum deflection of the beam obtained using NRPIM with MQ radial basis functions. This figure indicates that smaller error is observed when q is around 1.03 for all the C values examined. Figure 8.38 shows the effects of shape parameters on the error of the energy norm using the MQ radial basis functions. This figure indicates also a smaller error when q is around 1.03 regardless of C values. Figures 8.39 and 8.40 give a localized view of both the displacement relative error and the energy error in the vicinity of $q = 1.03$ for the MQ radial basis used, respectively. The following findings can be highlighted.

First, when $q = 0.5$, which is the original MQ function widely used in the mathematics community, the relative error is not acceptable for our mechanics problem. In addition, when $q = 0.5$, the accuracy is the most sensitive to the shape parameter C.

Second, the minimum relative error of deflection and error of energy are achieved when $q = 0.98$ to 1.03 (exclusive of 1.0); $q = 1.03$ is found to be the optimal value for this problem. Note that when $q = 1.0$, the radial functions degenerate to polynomial form. Hence, it will be not stable, and the error can be very high depending on the situation. Therefore, $q = 1.0$ should always be avoided for stability reasons.

Third, the shape parameter C has little effect on the accuracy when $q = 1.03$. When q is fixed as 1.03, the effect of shape parameter C on the error of energy is as shown in Figure 8.41, which indicates that an optimal value of C is 1.42.

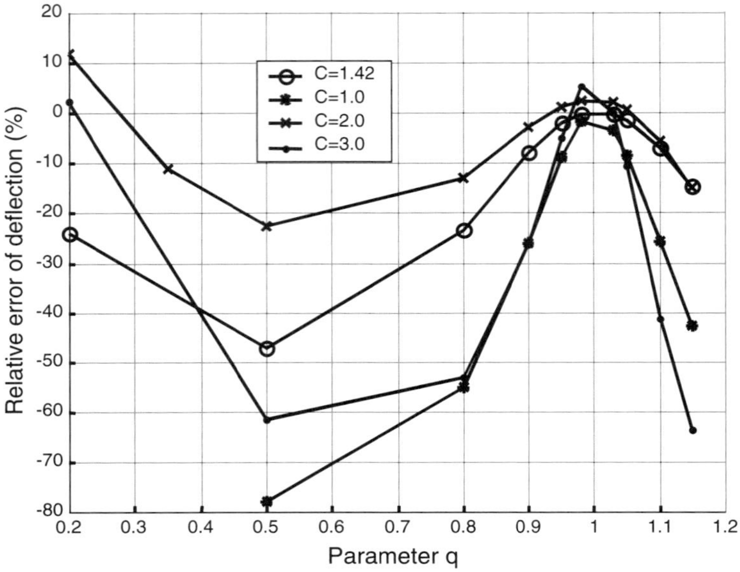

FIGURE 8.37
Effect of shape parameters on relative error of maximum deflection δ defined by Equation 8.40 (NRPIM with MQ radial basis functions; regularly distributed nodes). (From Wang, J. G. and Liu, G. R., *Comput. Methods Appl. Mech. Engrg.*, 191, 2611–2630, 2002. With permission.)

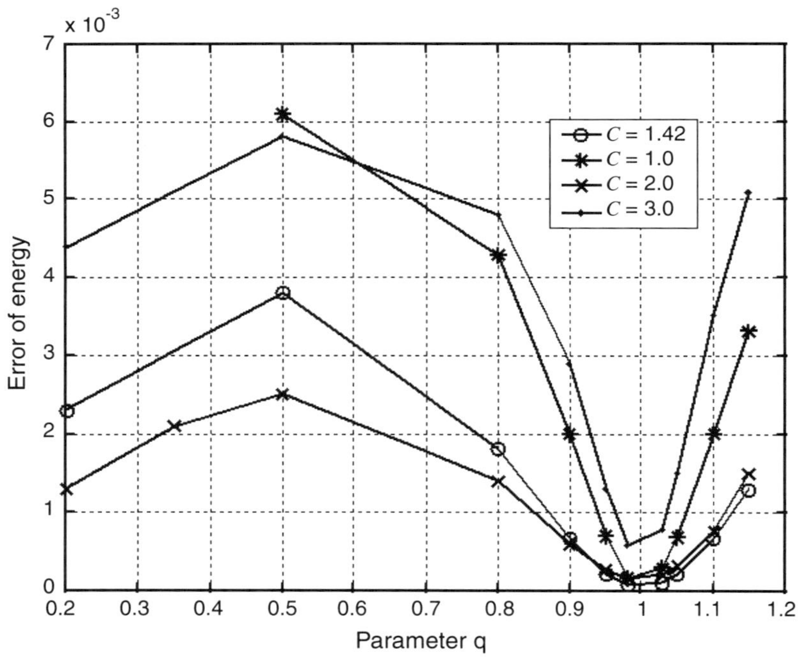

FIGURE 8.38
Effect of shape parameters C and q on the energy norm (NRPIM with MQ radial basis; regularly distributed nodes). (From Wang, J. G. and Liu, G. R., *Comput. Methods Appl. Mech. Engrg.*, 191, 2611–2630, 2002. With permission.)

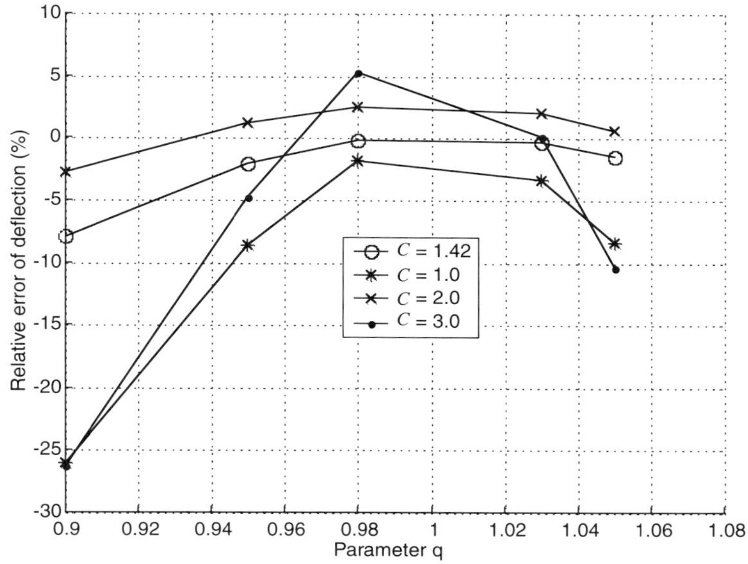

FIGURE 8.39
Effect of shape parameters C and q on the displacement error (NRPIM with MQ radial basis with q in the vicinity 1.03; regularly distributed nodes). (From Wang, J. G. and Liu, G. R., *Comput. Methods Appl. Mech. Engrg.*, 191, 2611–2630, 2002. With permission.)

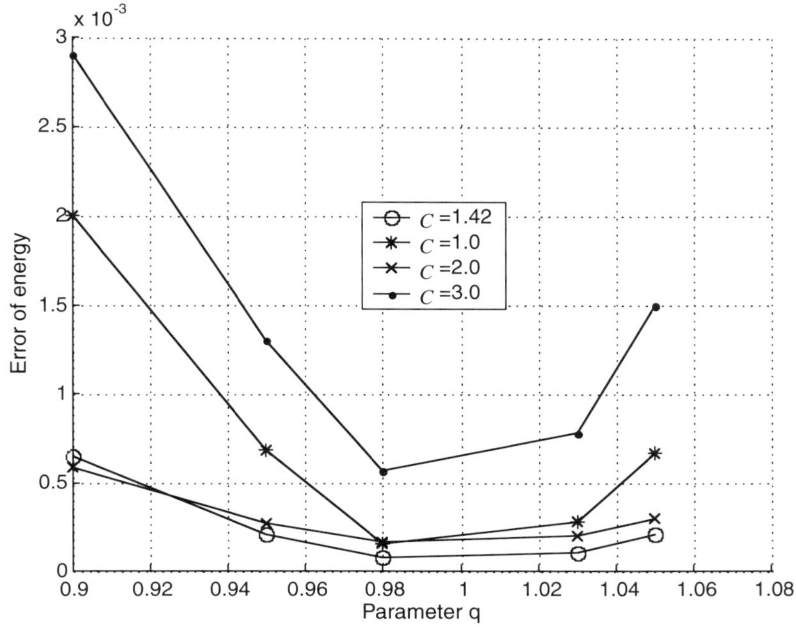

FIGURE 8.40
Effect of shape parameters C and q on the error of energy norm (NRPIM with MQ radial basis with q in the vicinity of 1.03; regularly distributed nodes). (From Wang, J. G. and Liu, G. R., *Comput. Methods Appl. Mech. Engrg.*, 191, 2611–2630, 2002. With permission.)

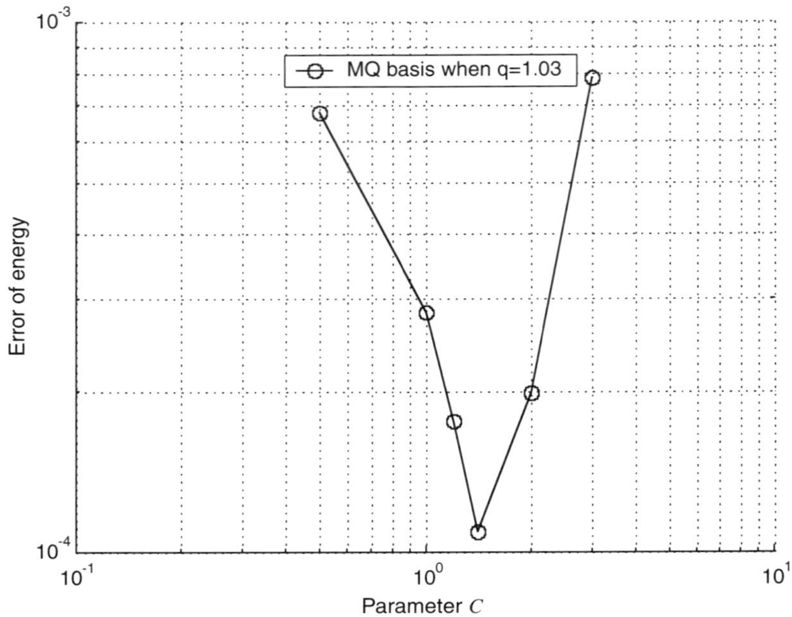

FIGURE 8.41
Optimal value of C for fixed shape parameter (NRPIM with MQ, q = 1.03; regularly distributed nodes). (From Wang, J. G. and Liu, G. R., *Comput. Methods Appl. Mech. Engrg.*, 191, 2611–2630, 2002. With permission.)

TABLE 8.6
Error in the Numerical Results for the Maximum Defection of the Beam Calculated by NRPIM Using Different Radial Functions with Different Shape Parameters

EXP Method			MQ Method (C = 1.42)		
c	$u_{y\,\max}(\times 10^{-3})$	Relative Error (%)	q	$u_{y\,\max}(\times 10^{-3})$	Relative Error (%)
0.001	0.298	−96.7	1.15	7.612	−14.5
0.002	8.59	−3.4	1.10	8.284	−6.9
0.003	8.9	0	1.05	8.767	−1.5
0.005	8.913	0.15	1.03	8.875	−0.28
0.01	8.9	0	0.98	8.883	−0.19
0.03	8.739	−1.8	0.95	8.711	−2.12
0.05	7.901	−11.2	0.8	6.816	−23.4
0.08	5.156	−42.1	0.5	4.716	−47.0
0.1	3.39	−62.0	0.2	6.757	−24.1

Note: Closed-form solution is $u_{y\,\max} = 8.9 \times 10^{-3}$.

Table 8.6 lists the errors in the numerical results for the maximum deflection of the beam calculated by NRPIM using different radial functions with different shape parameters. It clearly indicates that PIM gives the most accurate results when the EXP radial function is used with $c = 0.002$ to 0.03. The minimum error is achieved when $q = 0.98$ and 1.03 and $C = 1.42$ for our mechanics problem.

Further investigation was also conducted for negative q. Figures 8.42 and 8.43 show, respectively, the relative error of deflection and error of energy computed using RPIM with MQ basis functions with the negative q. These two figures do not suggest a better parameter

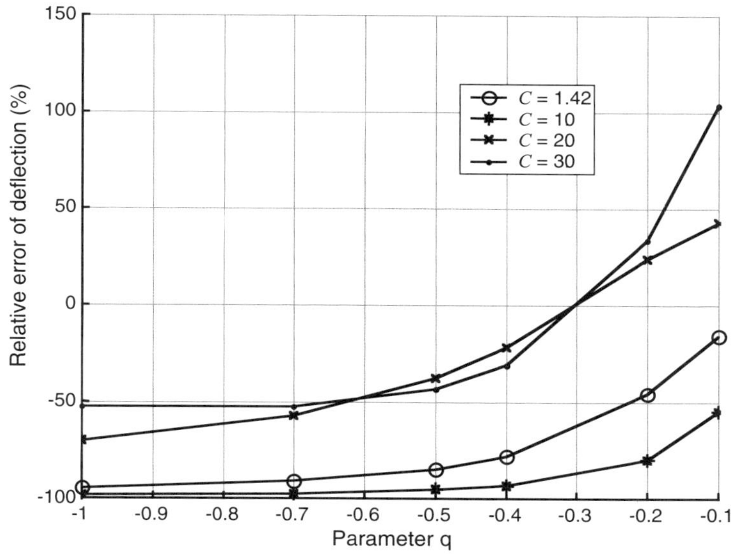

FIGURE 8.42
Relative error of beam deflection at the free end computed using NRPIM with MQ basis functions with negative q (regularly distributed nodes). (From Wang, J. G. and Liu, G. R., *Comput. Methods Appl. Mech. Engrg.*, 191, 2611–2630, 2002. With permission.)

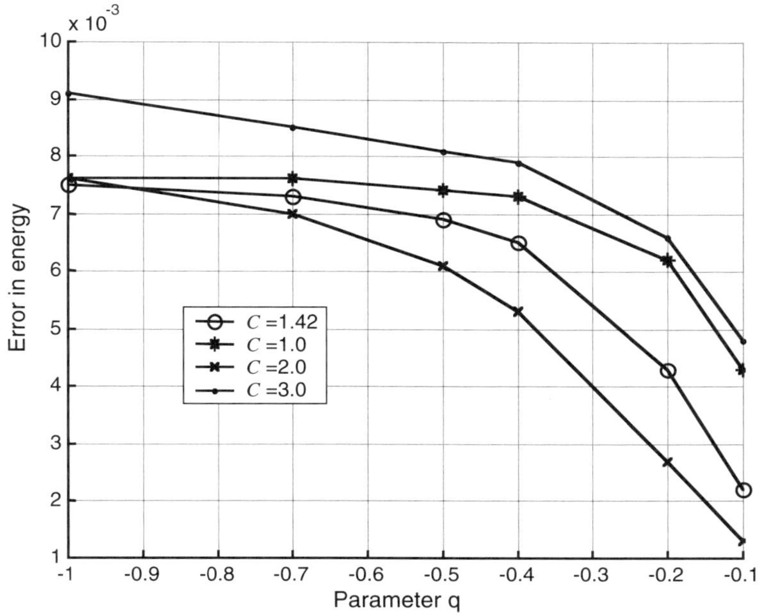

FIGURE 8.43
Energy error of the cantilever beam computed using NRPIM with MQ basis functions with negative q (regularly distributed nodes). (From Wang, J. G. and Liu, G. R., *Comput. Methods Appl. Mech. Engrg.*, 191, 2611–2630, 2002. With permission.)

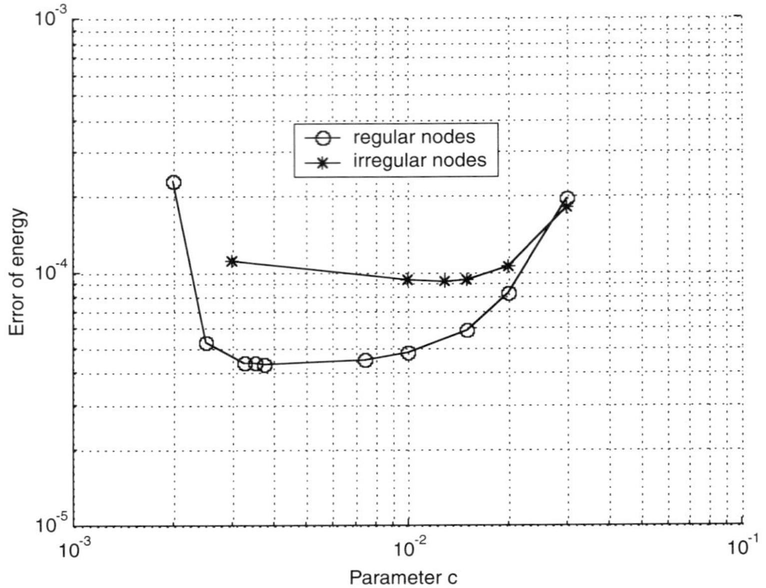

FIGURE 8.44
Effect of node distributions on the error of energy norm (NRPIM with EXP basis). (From Wang, J. G. and Liu, G. R., *Comput. Methods Appl. Mech. Engrg.*, 191, 2611–2630, 2002. With permission.)

of q for the range of values investigated here; $q = -0.5$ is clearly not an optimal shape parameter, although it has been suggested for curve fitting by a number of other researchers.

Effect of Irregularity of Node Distributions

As discussed by Carlson and Foley (1991), the optimal value of shape parameters depends also on the number and the distribution of data points, and on the data vector and the precision of the computation. An investigation is, therefore, also conducted for our mechanics problem. NRPIM with EXP basis function is used in the investigation with both regularly (637 nodes shown in Figure 8.33a) and irregularly distributed nodes (644 nodes shown in Figure 8.33b). The error of energy norm is shown in Figure 8.44 for two types of node distributions. Although node distributions will affect the accuracy of the NRPIM results, the effects are rather small and the optimal range of shape parameter c is almost unaffected. A similar investigation has also been conducted for MQ basis functions. The results are plotted in Figure 8.45. Comparison with Figure 8.38 supports the same conclusion made for the EXP basis; that is, irregularity has little effect on the performance of parameters of both radial basis functions used in RPIM for mechanics problems.

Effect of Nodal Density

The effect of nodal density on the range of the optimal shape parameter is investigated using NRPIM with EXP radial functions. Figure 8.46 plots the error of energy against the shape parameter c used in the EXP basis function for different node densities. Each density has a stable parameter range in which the error of energy is the lowest. This range of shape parameters may slightly vary with node densities. In the figure, the range of the shape parameter c is 0.002 to 0.03 for the cases of 637 and 175 nodes. The case of 795 nodes has a small exception; that is, the range is narrowed to 0.003 to 0.01, as shown in Figure 8.46. Thus, use of denser nodes tends to narrow the range of the parameters.

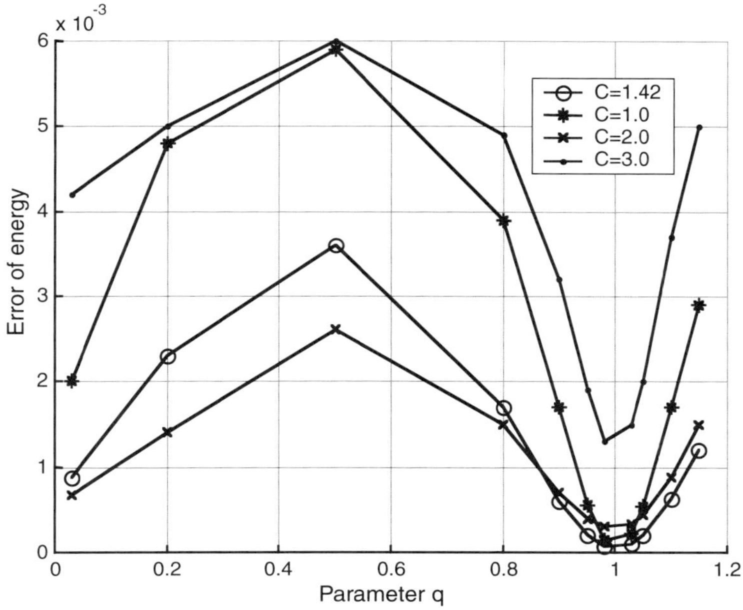

FIGURE 8.45
Effect of node distributions on the error of energy norm computed using NRPIM with MQ basis and irregularly distributed nodes. (From Wang, J. G. and Liu, G. R., *Comput. Methods Appl. Mech. Engrg.*, 191, 2611–2630, 2002. With permission.)

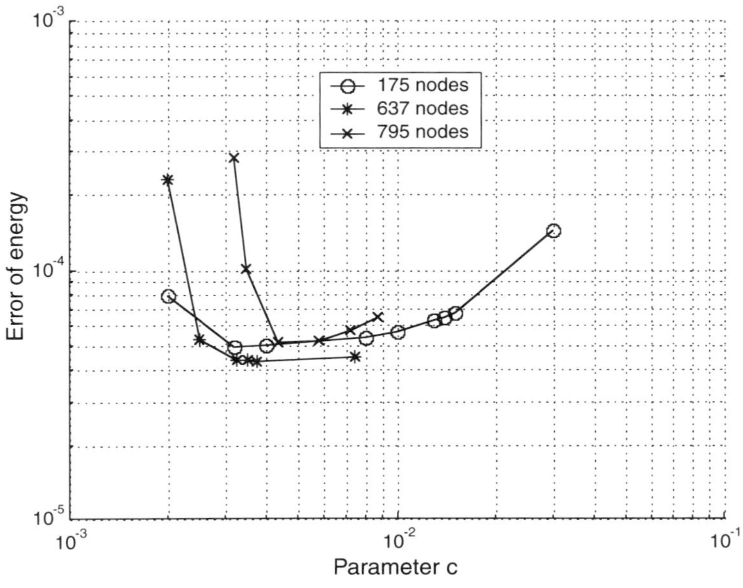

FIGURE 8.46
Error of energy norm for different node densities (NRPIM with EXP basis). (From Wang, J. G. and Liu, G. R., *Comput. Methods Appl. Mech. Engrg.*, 191, 2611–2630, 2002. With permission.)

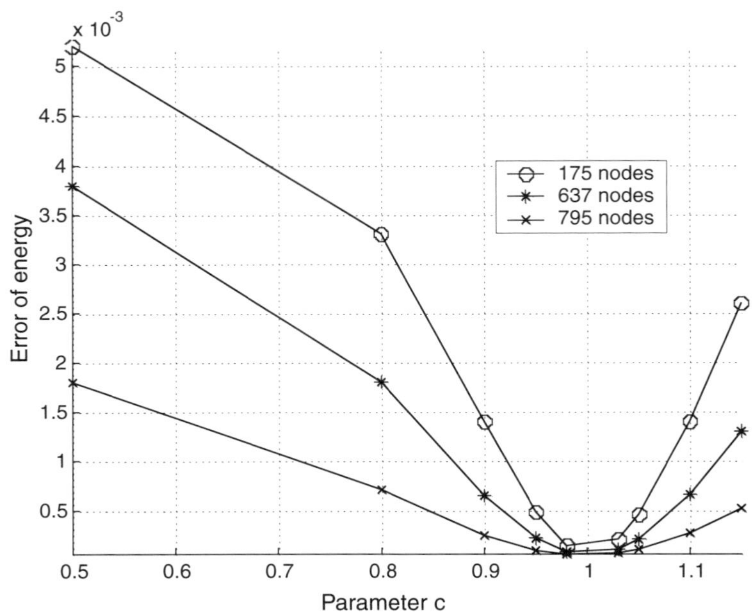

FIGURE 8.47
Error of energy norm for different node densities (NRPIM with MQ basis with $C = 1.42$). (From Wang, J. G. and Liu, G. R., *Comput. Methods Appl. Mech. Engrg.*, 191, 2611–2630, 2002. With permission.)

A similar study for MQ basis functions has also been conducted. The results are shown in Figure 8.47. It is found in this case that denser nodes tend to widen the range of parameter q. This is one of the advantages of MQ radial basis functions. Figure 8.47 also suggests that a q in the vicinity of $q = 1.0$ (but not at 1.0) gives more accurate results.

Convergence Rate

A number of MFree models with regular node distributions of nodes (175, 495, 637, 795, and 2425) are used for this study. Shape parameters are taken as $C = 1.42$ and $q = 1.03$ for the MQ method and $c = 0.003$ for the EXP method. Figure 8.48 plots the convergence rates of the energy norm for NRPIM using EXP radial functions. It is seen that a linear (in logarithm scales) convergence rate of about 2.0 is achieved.

Figure 8.48 has also shown clearly that the original PIM with only polynomial basis is more accurate than that using pure EXP radial functions.

Figure 8.49 shows the convergence rates of the energy norm for NRPIM using MQ radial functions. It is seen that a linear (in logarithm scales) convergence rate is achieved. Figure 8.49 also shows that the original NPIM with an only polynomial basis converges about two times faster than that using MQ radial functions. Comparison of Figure 8.48 with Figure 8.49 shows that the convergence rates of NRPIM using EXP and MQ radial functions are roughly the same.

Example 8.9 Infinite Plate with a Hole

Example 6.11, examined using the EFG method, is now reexamined here using the NRPIM method. The geometry of the plate is plotted in Figure 6.34. Because of the twofold symmetry, only a quarter of the plate shown in Figure 6.35 is modeled with symmetric boundary conditions applied on $x = 0$ and $y = 0$. The parameters and the boundary conditions are exactly the same as those in Example 6.11. The displacement and the stress

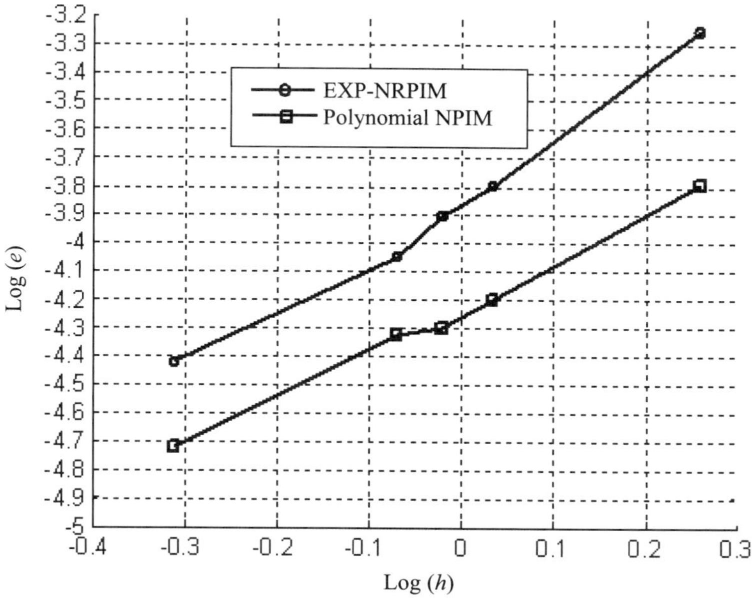

FIGURE 8.48
Convergence of NRPIM using EXP radial basis functions and NPIM with polynomial basis. Convergence rate: about 2.0.

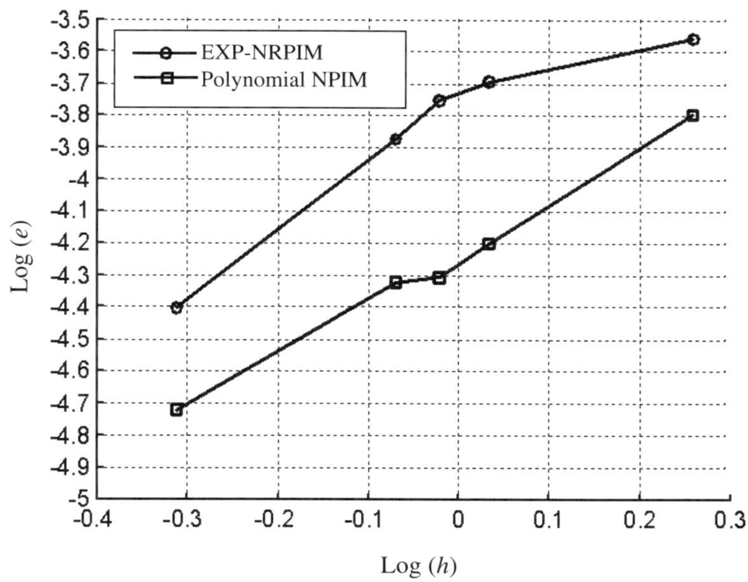

FIGURE 8.49
Convergence of NPRIM using MQ radial basis functions and NPIM with polynomial basis. Convergence rate: about 2.0.

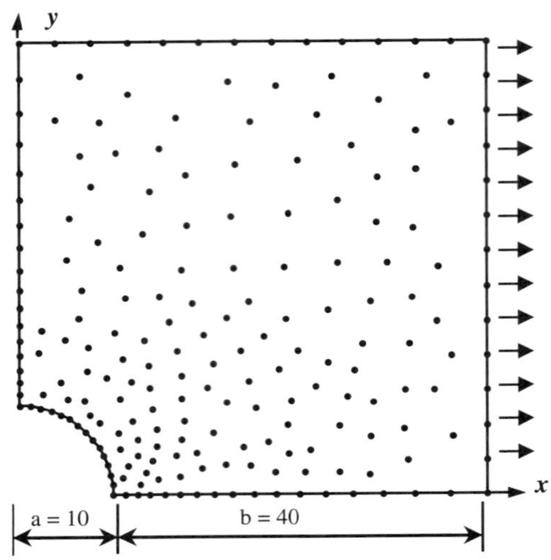

FIGURE 8.50
Distribution of 209 nodes used in an MFree model of a plate with a hole in the center of the plate (a quarter of the plate is modeled). (From Wang, J. G. and Liu, G. R., *Comput. Methods Appl. Mech. Engrg.*, 191, 2611–2630, 2002. With permission.)

field within the plate are provided by Equations 6.121 to 6.126 in the polar coordinates of (r, θ). The parameters are listed as follows:

Loading: $p = 1$ N/m
Young's modulus: $E = 3.0 \times 10^7$ kPa
Poisson's ratio: $\nu = 0.3$
Height of the beam: $a = 10$ m
Length of the beam: $b = 40$ m

The node distribution (209 nodes) is shown in Figure 8.50.

A study is first conducted to examine the effects of parameters used in the MQ basis functions, to determine whether the optimal parameters found in the previous example work for different problems. Figure 8.51 plots the error in energy against the shape parameters C and q. A local view of zooming at $q = 1.0$ is plotted in Figure 8.52. It is found from Figures 8.51 and 8.52 that a $C = 1.42$ and a q near 1.0 (not equal to 1.0) give very good accuracy for this problem, as well from Figure 8.51, it can be seen that $C = 1.42$ and $q = 0.1$ also give very good results. However, we do not have the results for the cantilever beam to support this finding.

In conclusion, for MQ basis functions, one should use $C = 1.42$, $q = 0.1$ or 0.98, or 1.03. In any case, $q = \pm 0.5$ is not a good choice although it has been widely used for curve fitting.

A study has also been carried out for the EXP basis used in RPIM. We state without showing the results that a c value in the range of 0.003 to 0.03 gives the most accurate results for the problem of a plate with a hole at its center. This supports the finding obtained from the cantilever beam problem.

In the following examples, we use $c = 0.003$ when an EXP radial function is used, and $q = 1.03$ and $C = 1.42$ when an MQ radial function is used.

Note that since we used $d_c = 1.0$ in this example, $C = 1.42$ is equivalent to $\alpha_C = 1.42$, and $c = 0.003$ is equivalent to $\alpha_c = 0.003$, if we use the definition of radial functions given in Table 5.3.

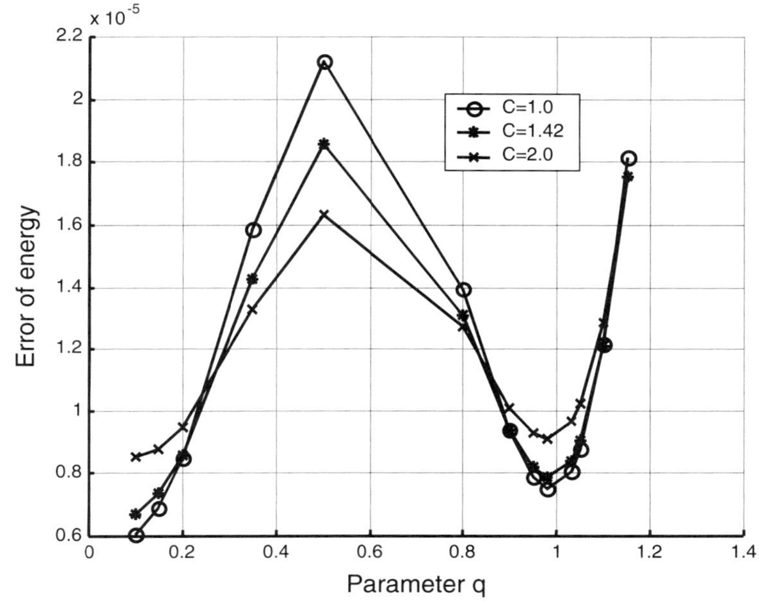

FIGURE 8.51
Effects of shape parameters for the problem of a plate with a hole at its center (MQ basis function). (From Wang, J. G. and Liu, G. R., *Comput. Methods Appl. Mech. Engrg.*, 191, 2611–2630, 2002. With permission.)

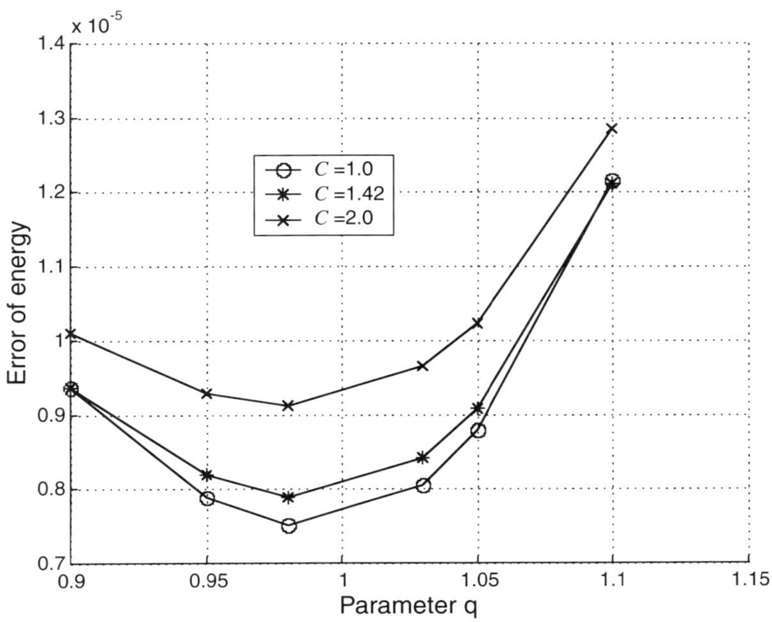

FIGURE 8.52
Effects of shape parameters for the problem of a plate with a hole at its center (MQ basis function). Local view at $q = 1.0$ neighborhood (exclusive of $q = 1.0$). (From Wang, J. G. and Liu, G. R., *Comput. Methods Appl. Mech. Engrg.*, 191, 2611–2630, 2002. With permission.)

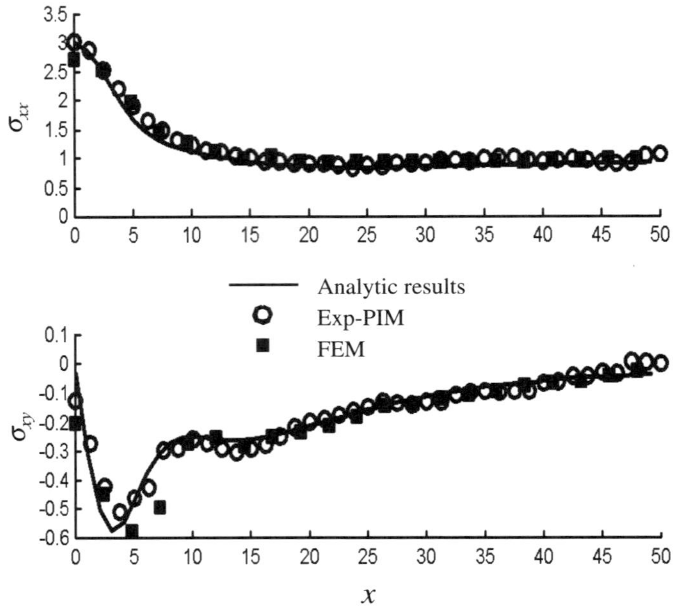

FIGURE 8.53
Comparison of EXP–PIM method, FEM, and analytic solution. Stress distribution along the section of $y = 10$.

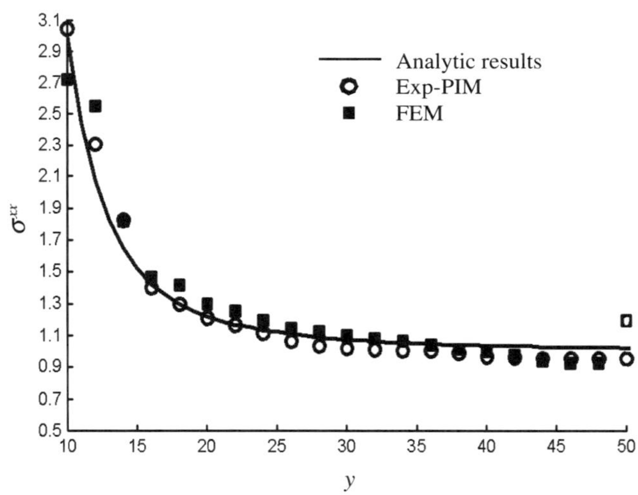

FIGURE 8.54
Comparison of EXP–PIM method, FEM, and analytic solution. Stress distribution along the section of $x = 0$.

Comparison of RPIM with FEM and Analytical Solutions

Figure 8.53 shows the distribution of stress σ_{xy} and σ_{xx} along a $y = 10$ section obtained by NRPIM with EXP radial functions, FEM, and analytical methods. Results from NRPIM and FEM are almost the same. They are all in good agreement with the closed-form solution. Figure 8.54 plots the distribution of stress σ_{xx} at $x = 0$. The closed-form and FEM results

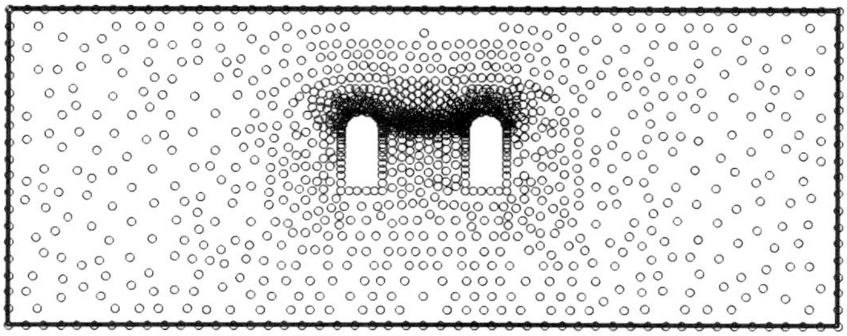

FIGURE 8.55
MFree model for stress analysis of two parallel tunnels.

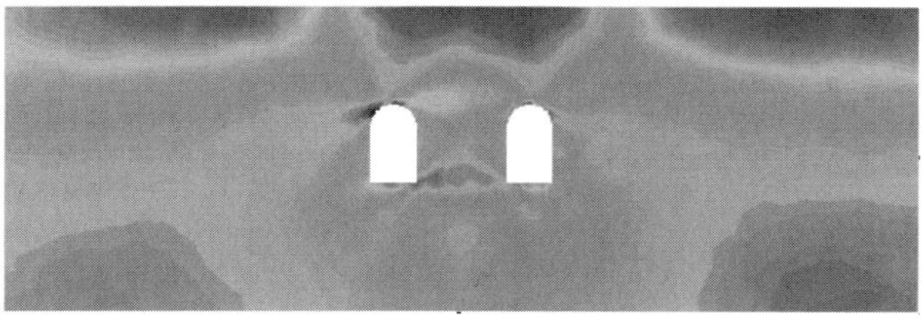

FIGURE 8.56
Stress distribution in the parallel tunnels obtained using NRPIM with EXP radial function. Distribution of stress σ_{xx}.

are also plotted for comparison. It can be observed from these figures that the EXP method gives satisfactory results.

Example 8.10 Parallel Tunnel

RPIM is applied to analyze the stress distribution in a 2D solid with two parallel tunnels. The material is assumed to be linearly elastic.

The parameters are listed as follows:

Loading: $p = 15$ kN/m
Young's modulus: $E = 3.0 \times 10^3$ kPa
Poisson's ratio: $\nu = 0.3$
Soil density = 17 kN/m^3

Figure 8.55 shows our MFree model with a distribution of 1270 nodes. A triangle background mesh (2290 cells whose vertices are the field nodes) is used for numerical integration and one Gauss quadrature point is used for each cell. The same background mesh of triangular elements is used for the FEM analysis. A strip load (15 kN/m^2) is applied on the top ground between two tunnels. Figure 8.56 is the distribution of stress σ_{xx}. The results are almost the same as FEM results.

Note that in using cells of a background mesh, we often found that triangular cells whose vertices are the field nodes give very good results. This is because the division of cells breaks the integrand into smooth pieces allowing the Gauss quadrature scheme to perform efficiently. In MFree2D© to be introduced in Chapter 16, triangular cells of integration are used.

TABLE 8.7

Comparison of CPU Time(s)

Nodes	FEM	NRPIM	Difference (%)
209 (plate with hole)	5 s	5 s	0
1270 (tunnel)	141 s	166 s	18

Note: Tested on a PIII450 PC.

TABLE 8.8

Comparison of Energy Error for NRPIM Using MQ Basis with/without Polynomial Terms Added

q (C = 1.42)	Cantilever Beam 175 Nodes		Cantilever Beam 795 Nodes		Plate with Hole (209 Nodes)	
	$m = 0$	$m = 3$	$m = 0$	$m = 3$	$m = 0$	$m = 3$
1.15	2.60E-3	6.05E-5	5.18E-4	4.49E-5	1.75E-5	7.39E-6
1.10	1.40E-3	6.12E-5	2.65E-4	4.46E-5	1.21E-5	7.31E-6
1.05	4.53E-4	6.18E-5	1.03E-4	4.43E-5	9.07E-6	7.23E-6
1.03	**2.02E-4**	6.21E-5	**6.81E-5**	4.42E-5	**8.42E-6**	7.21E-6
0.98	**1.39E-4**	6.29E-5	**5.69E-5**	4.39E-5	**7.89E-6**	7.14E-6
0.95	4.81E-4	6.33E-5	1.00E-4	4.38E-5	8.18E-6	7.11E-6
0.9	1.40E-3	6.40E-5	2.47E-4	4.35E-5	9.37E-6	7.06E-6
0.8	3.30E-3	6.55E-5	7.10E-4	4.30E-5	1.31E-5	6.96E-6
0.5	5.20E-3	6.88E-5	1.80E-3	4.20E-5	1.86E-5	6.77E-6

Source: Wang, J. G. and Liu, G. R., *Comput. Methods Appl. Mech. Engrg.*, 191, 2611–2630, 2002. With permission.

Efficiency Comparison with FEM

A study of the computational efficiency of NRPIM vs. FEM has found that NRPIM is as efficient as FEM. Table 8.7 shows a comparison of CPU time (seconds) used for the NRPIM and FEM codes for the last two examples presented. It is found that the NRPIM code takes about 18% more CPU time compared with the FEM code for the parallel tunnel problem model with 1270 nodes.

Effect of Polynomial Terms

The above numerical analyses were performed without using polynomial terms ($m = 0$) in the NRPIM interpolation. A study has also been conducted to reveal the effect of polynomial terms. Three terms ($m = 3$) of a linear polynomial were added into the radial basis. Table 8.8 compares the results of different problems under different nodal densities. For the cantilever beam problem, the error of energy can reach 6.0×10^{-5} with a linear polynomial, whereas the energy error without the polynomial term can only reach 1.39×10^{-4}. This shows that adding of linear polynomial terms can double the accuracy at the best-tuned shape parameters, when the nodal density is not too high ($n = 175$). If the shape parameter is not properly chosen, the inclusion of a polynomial basis can improve the accuracy by as much as an order of two. When the nodal density is very high ($n = 795$), the improvement is not as great as that for the low nodal density case, but it is still remarkable. The results for the cantilever beam reveal the following two very important facts:

1. The inclusion of polynomial terms can significantly improve the accuracy of the results, especially for low nodal density cases.

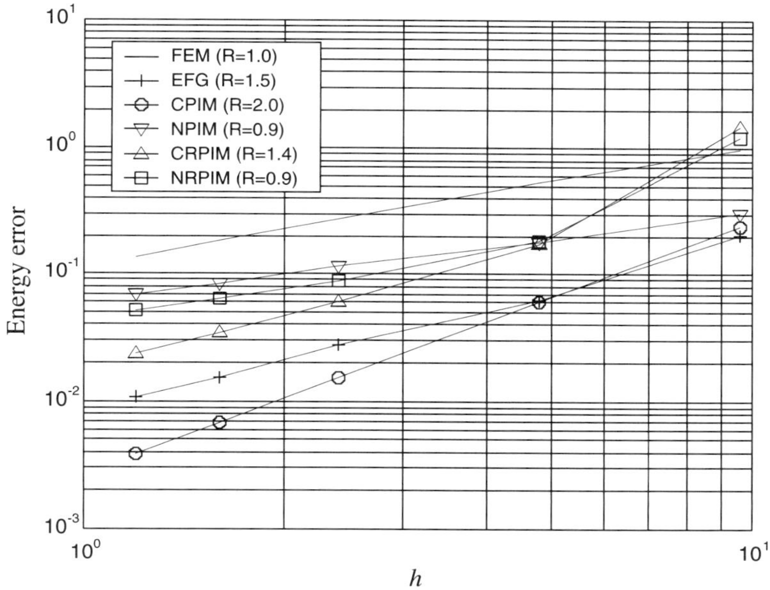

FIGURE 8.57
Comparison of convergence of six different methods. The cantilever beam problem is used for this examination. R is the rate of convergence. In the FEM, 4-node elements are used; in EFG, CRPIM, and NRPIM, the support domain of $\alpha_s = 2.5$ (about 25 nodes) is used for constructing MLS shape functions; in CPIM, one-piece shape functions are constructed using the nearest 9 nodes surrounding the integration cell; in NPIM, the nearest 9 nodes surrounding the Gauss point are used. In CRPIM and NRPIM, the support domain of $\alpha_s = 2.5$ (about 25 nodes) is used for constructing shape functions. NPIM is more accurate than FEM, but converges a little slower. CPIM converges fastest. CRPIM has about the same convergence rate as EFG. When 16 nodes were used in CPIM, exact results were produced for the cantilever beam problem.

2. The inclusion of the polynomial terms will significantly reduce the effects of the shape parameters on the accuracy of the results. When linear polynomial terms are added to the radial function basis, one can practically not worry about choosing of shape parameters.

The findings are also applicable to the problem of the plate with a hole, as shown in Table 8.8.

Figure 8.57 shows a comparison of the rates of convergence of NRPIM and CRPIM with FEM, EFG, NPIM, and CPIM. R denotes the rate of convergence. In FEM, four-node elements are used; in computing the polynomial PIM shape functions, nine nodes are used. In EFG, CRPIM, and NRPIM, the support domain of $\alpha_s = 2.5$ (about 25 nodes) is used for constructing shape functions. In CPIM and CRPIM, one-piece shape functions are used for the integration cell; in NPIM and RPIM, shape functions are constructed for the Gauss points. In addition to Figure 8.9, Figure 8.57 shows that CPIM performs much better than CRPIM, and that CRPIM has about the same convergence rate as that of EFG.

8.3.3 Remarks

PIM with a radial function basis is formulated based on Galerkin formulation. RPIM has been tested and benchmarked for a number of mechanics problems. The features of the RPIM method are summarized as follows:

- RPIM requires a background mesh for integration, as the Galerkin weak form is employed. The standard Galerkian weak form produces NRPIM, and the constrained Galerkin weak form produces CRPIM. CRPIM converges much faster than NRPIM.

- The problem domain can be represented using nodes that are structured or unstructured. The effects of the irregularity of the nodes are small.
- Use of 10 to 17 nodes in a support domain can usually give results of sufficient accuracy for solid mechanics problems.
- No clear preference for either MQ or EXP basis is found.
- Inclusion of polynomial terms helps in two ways. One is the improvement in accuracy; the other is the reduction of the sensitivity of the results to shape parameters of radial basis functions. To pass the patch tests, terms of monomials up to the first order have to be included in the basis.
- For MQ basis, $C = 1.42$ ($\alpha_C = 1.42$), $q = 0.98$ or 1.03 is recommended, if polynomial terms are not used. $q = \pm 0.5$ is not recommended for solid mechanics problems.
- For EXP basis, $c = 0.003$ to 0.03 ($\alpha_c = 0.003$ to 0.03) is recommended, if polynomial terms are not used.
- When linear polynomial terms are added, the results are more accurate and the effects of the shape parameters are significantly reduced. We need not be concerned about what values of parameters to use. The optimal parameters listed above are still the best to use.
- Computational efficiency of RPIM is comparable with that of FEM when the same number of nodes is used.

8.4 Local Point Interpolation Method (LPIM)

Chapter 7 introduced the meshless local Petrov–Galerkin (MLPG) method, and it was demonstrated that MLPG is nearly a truly MFree method. It needs only a local background mesh for constructing the discretized system equations. We found also that MLPG uses MLS for creating the trial function (shape function), and therefore special techniques, such as the penalty method or the direct interpolation method, are required for imposing essential boundary conditions. The use of the penalty method requires a proper choice of the penalty factor. In addition, the use of MLS shape functions requires additional efforts in imposing essential boundary conditions for free vibration problems.

Section 8.1 presented PIM based on the Galerkin formulation using PIM shape functions with the Kronecker delta function property. We demonstrated that the imposition of the essential boundary conditions in PIM can be done in the conventional manner as in FEM. No special treatment is required. However, the penalty method is required to enforce the compatibility of the field variable using PIM shape functions to formulate conformable PIMs.

This section formulates the local point interpolation method (LPIM) based on the ideal of MLPG. LPIM uses the Petrov–Galerkin weak formulation integrated locally, and uses the PIM shape function for field variable interpolation. This work was performed originally by G. R. Liu and Gu (2001a).

Application of the local Petrov–Galerkin weak form requires the trail function to be differentiable up to the kth order (see Section 4.7) in the local quadrature domain. The PIM shape functions can easily satisfy this requirement. The PIM shape functions can be created in two ways: based on either Gauss points or the center of the local quadrature domain (one-piece PIM). LPIM will be reproductive, because PIM approximation is reproductive or consistent (see Section 5.5.2). Examples of patch tests are presented to confirm this feature of LPIM.

8.4.1 LPIM Formulation

We formulate LPIM for a 2D solid mechanics problem that is defined in a domain Ω bounded by Γ. The strong form of system equation is given by Equations 6.1 to 6.3. A generalized local weak form can be written in an integral form over a local quadrature domain Ω_Q bounded by Γ_Q, based on the weighted residual method, i.e.,

$$\int_{\Omega_Q} \widehat{W}(\sigma_{ij,j} + b_i)\,\mathrm{d}\Omega = 0 \tag{8.42}$$

where $\widehat{W}$ is the weight function. The choice of the weight function follows the same principle that is used for choosing the weight function for the MLPG method. The weight function will be nonzero only within the domain Ω_Q. In other words, the weighted domain is chosen to be the same as the local quadrature domain (see Chapter 7 for definition of the types of domains). The local quadrature domain Ω_Q can be arbitrary in theory but such simple shapes as circles and rectangles are often used for convenience of integration. The difference between Equation 8.42 and Equation 7.4 is that there is no term for the essential boundary conditions. This implies that the shape function to be used has to satisfy the Kronecker delta function property.

The first term on the left-hand side of Equation 8.42 can be integrated by parts to become

$$\int_{\Gamma_Q} \widehat{W}\sigma_{ij}n_j\,\mathrm{d}\Gamma - \int_{\Omega_Q} (\widehat{W}_{,j}\sigma_{ij} - \widehat{W}b_i)\,\mathrm{d}\Omega = 0 \tag{8.43}$$

As discussed in Chapter 7, the boundary Γ_Q for the quadrature domain is usually composed of three parts: the internal boundary Γ_{Qi}, and the boundaries Γ_{Qu} and Γ_{Qt}, over which the essential and natural boundary conditions are specified, as shown in Figure 7.1. Imposing the natural boundary condition and noticing that $\sigma_{ij}n_j = \partial u/\partial n \equiv t_i$ in Equation 7.3, Equation 8.43 becomes

$$\int_{\Gamma_{Qi}} \widehat{W}t_i\,\mathrm{d}\Gamma + \int_{\Gamma_{Qu}} \widehat{W}t_i\,\mathrm{d}\Gamma + \int_{\Gamma_{Qt}} \widehat{W}\bar{t}_i\,\mathrm{d}\Gamma - \int_{\Omega_Q} (\widehat{W}_{,j}\sigma_{ij} - \widehat{W}b_i)\,\mathrm{d}\Omega = 0 \tag{8.44}$$

For a support domain located entirely within the global domain, there is no intersection between Γ_Q and the global boundary Γ, $\Gamma_{Qi} = \Gamma_Q$, and the integrals over Γ_{Qu} and Γ_{Qt} vanish.

The weak form of Equation 8.44 works for any node in the problem domain Ω, and therefore it is used to establish discretized system equations for every node that is distributed in the problem domain.

Equation 8.2 is used to interpolate the value of field variables of displacements at a point of interest x_Q in the quadrature domain using the nodes in the support domain of the point. Substituting Equation 8.2 into the local weak form Equation 8.44 for all nodes and following the procedure given in Section 7.1.2 lead to the following global discrete system of equations:

$$\mathbf{KU} = \mathbf{F} \tag{8.45}$$

where **U** is the displacement vector for all the nodes in the entire problem domain and **K** is the global stiffness matrix for the entire problem domain, which is assembled using the nodal stiffness matrix defined by

$$\mathbf{K}_{ij} = \int_{\Omega_Q} \widehat{\mathbf{V}}_i^T \mathbf{c}\mathbf{B}_j \, \mathrm{d}\Omega - \int_{\Gamma_{Qi}} \widehat{\mathbf{W}}_i \mathbf{n}\mathbf{c}\mathbf{B}_j \, \mathrm{d}\Gamma - \int_{\Gamma_{Qu}} \widehat{\mathbf{W}}_i \mathbf{n}\mathbf{c}\mathbf{B}_j \, \mathrm{d}\Gamma \tag{8.46}$$

where **c** is a matrix of material constant given by Equation 3.30 for plane stress problems and by Equation 3.31 for plane strain problems. Matrices $\mathbf{B}_j$, $\widehat{\mathbf{V}}_i$, $\widehat{\mathbf{W}}_i$, and **n** are defined, respectively, by Equations 7.12, 7.14, 7.15, and 7.17.

The force vector **F** in Equation 8.45 is obtained using the nodal force vector consisting of contributions from body forces applied in the problem domain and tractions applied on the natural boundary.

$$\mathbf{f}_I = \int_{\Omega_Q} \widehat{\mathbf{W}}_I \mathbf{b} \, \mathrm{d}\Omega + \int_{\Gamma_{Qt}} \widehat{\mathbf{W}}_I \bar{\mathbf{t}} \, \mathrm{d}\Gamma \tag{8.47}$$

It can be easily seen that the system stiffness matrix **K** in the present method is banded but asymmetric.

8.4.2 Weight Function

As the LPIM is regarded as a weighted residual method, the weight function plays an important role in the performance of the method. Theoretically, as long as the condition of continuity is satisfied, any weight function is acceptable. However, the local weak form is based on the local quadrature domains of a node at the center. It is found that weight functions that decrease in magnitude with increasing distance from the point x_Q to the surrounding nodes yield better results. Therefore, weight functions that depend only on the distance between the two points are used in LPIM. The weight functions listed in Section 5.2.1 can all be used.

Note that weight functions used in Equation 8.44 do not need to satisfy all the conditions listed in Section 5.2, which a weight function has to satisfy to ensure the reproduction property in the integral representation, although any weight function that satisfies these conditions can be used.

It may also be noted here that in constructing MLS shape functions, one also needs weight functions. Such a weight function also does not need to satisfy all the conditions listed in Section 5.2, although any weight function that satisfies these conditions can be used for constructing MLS shape functions. However, the weight functions play an important role in guaranteeing the smoothness of the MLS shape functions, because MLS tends to use lower-order basis functions. Therefore, a quartic spline function that has higher order of continuity is preferred.

The requirement for weight functions used in Equation 8.44 is most relaxed; only the following is necessary:

- It is differentiable up to *k*th order for a partial differential equation of order 2*k*.
- It monotonically decreases with the distance from the center of the quadrature domain Ω_Q.
- It vanishes on the boundary of Ω_Q to simplify the integration on the boundary of Γ_Q.

In this section we test both parabolic and spline weight functions that meet our requirements.

Parabolic weight function:

$$\widehat{W}_i(x) = \begin{cases} 1-\left(\frac{d_i}{d_W}\right)^2 & 0 \le d_i \le d_W \\ 0 & d_i \ge d_W \end{cases} \tag{8.48}$$

Quartic weight function:

$$\widehat{W}_i(x) = \begin{cases} 1-6\left(\frac{d_i}{d_W}\right)^2 + 8\left(\frac{d_i}{d_W}\right)^3 - 3\left(\frac{d_i}{d_W}\right)^4 & 0 \le d_i \le d_W \\ 0 & d_i \ge d_W \end{cases} \tag{8.49}$$

where $d_i = |x_i - x_Q|$ is the distance from node i to a point x_Q in the quadrature domain, and d_W is the size of the weighted domain.

8.4.3 Numerical Examples

LPIM has been coded based on the formulation given above. Cases are run to benchmark and examine LPIM for 2D elastostatics. Because the problem domains in the examples are rectangles, rectangular quadrature domains are used for establishing the weight function. The size of the quadrature domain for node i is defined as

$$d_Q = \alpha_Q d_c \tag{8.50}$$

where α_Q is the dimensionless size of the quadrature domain, which will be varied and examined in the range of 0.5 to 3.0 in this section, and d_c is a characteristic length that relates to the nodal spacing near node i. If the nodes are uniformly distributed, d_c is simply the distance between two neighboring nodes. In the case where the nodes are non-uniformly distributed, d_c can be defined as an "average" nodal spacing in the quadrature domain of node i. The detailed procedure of determining d_c is given in Section 2.10.3.

The local quadrature domain is chosen to be the same as the weighted domain ($d_Q = d_W$). As mentioned in Chapter 7, we often need to divide the quadrature domain further into smaller portions to obtain better accuracy of integration.

Example 8.11 Standard Patch Test (LPIM + MTA)

Numerical examples of standard patch tests using LPIM with the matrix triangulation algorithm (MTA) are presented here. Details on PIM-MTA can be found in Chapter 5. From the discussion given in Example 7.2, we are sure that the LPIM will pass the patch test because PIM shape functions satisfy these two requirements. This example is to confirm this fact numerically and to show that LPIM is indeed reproductive. Three 2.0 × 2.0 square patches are tested. Figure 8.58a shows a patch with 9 nodes of which 1 is an interior node. Figure 8.58b shows a patch of 25 regularly distributed nodes, and Figure 8.58c shows a patch of 25 irregularly distributed nodes. The material constants are taken as $E = 1.0$, and $\nu = 0.3$. The linear displacement functions are $u_x = x + y$ and $u_y = x - y$. Satisfaction of the patch test requires that the displacement of any interior node be given by the same linear function and that the strains and stresses be constant in the patch.

In the 9-node patch test, a square support domain of $\alpha_s = 1.6$ to 2.1 is defined as the support domain for a point x_Q. Therefore, 4 to 9 nodes are usually used in the interpolation.

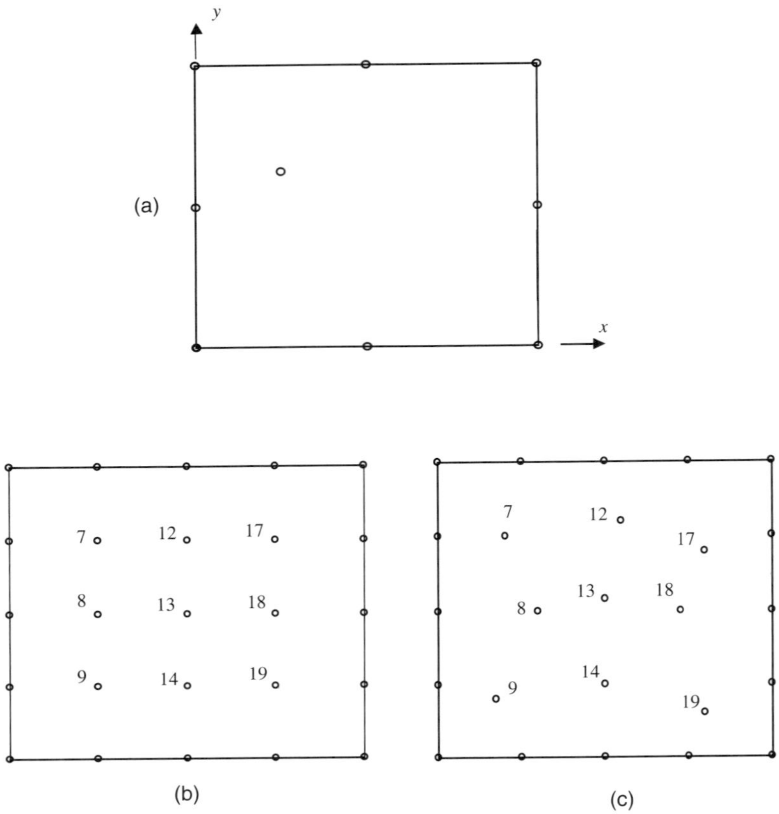

FIGURE 8.58
Three standard square patches of 2.0 × 2.0 for testing LPIM: (a) 9 nodes, (b) regular 25 nodes, (c) irregular 25 nodes.

The dimension of the quadrature domain used is $\alpha_Q = 1.0$. It can be found that LPIM-MTA passes exactly (to machine accuracy) the 9-node patch test, meaning that the LPIM has reproduced the linear field. This is because the LPIM also satisfies the two conditions used on page 222.

In the 25 regularly distributed nodes patch test, a rectangular support domain of $\alpha_s = 1.1$ to 2.1 is defined. The dimension of the quadrature domain used is also $\alpha_Q = 1.0$. The coordinates, displacements, and stresses of the 25-node patch test with regular distribution are obtained and listed in Table 8.9. It is shown that LPIM-MTA passes the patch test exactly. It should be mentioned here that some interpolations in this problem would be singular without the use of MTA. Hence, it demonstrates that MTA efficiently overcomes the interpolation singularity of PIM.

In the 25 irregularly distributed nodes patch test, there are 9 randomly distributed interior nodes. The size of the support domain is $\alpha_s = 1.6$ to 2.1. The coordinates, displacements, and stresses of this 25-node patch test are obtained and listed in Table 8.10. It is shown that LPIM-MTA also passes this patch test exactly. The computational stability and high accuracy for a nonstructured nodal distribution are very significant advantages of LPIM-MTA. These properties are very beneficial for practical applications. Without using MTA, patch tests on PIM (see Example 8.1) are not straightforward, because manually controlled selection of nodes in the support domain must be performed to avoid singularity.

Similar to all the patch tests, an exact numerical integration for the stiffness matrix is also required for LPIM to pass the patch test. Any integration error can lead to a failure in the

TABLE 8.9
LPIM Patch Test Results at Interior Nodes of the Regular Patch of 25 Nodes

(x, y)	For Interior Node	u_x	u_y	σ_x	σ_y	τ_{xy}
7	(0.5, 1.5)	2.000	–1.000	0.76923077	–0.76923077	0.76923077
8	(0.5, 1.0)	1.500	–0.500	0.76923077	–0.76923077	0.76923077
9	(0.5, 0.5)	1.000	0.000	0.76923077	–0.76923077	0.76923077
12	(1.0, 1.5)	2.500	–0.500	0.76923077	–0.76923077	0.76923077
13	(1.0, 1.0)	2.000	0.000	0.76923077	–0.76923077	0.76923077
14	(1.0, 0.5)	1.500	0.500	0.76923077	–0.76923077	0.76923077
17	(1.5, 1.5)	3.000	0.000	0.76923077	–0.76923077	0.76923077
18	(1.5, 1.0)	2.500	0.500	0.76923077	–0.76923077	0.76923077
19	(1.5, 0.5)	2.000	1.000	0.76923077	–0.76923077	0.76923077

TABLE 8.10
LPIM Patch Test Results at Interior Nodes of the Irregular Patch of 25 Nodes

(x, y)	For Interior Node	u_x	u_y	σ_x	σ_y	τ_{xy}
7	(0.4, 1.5)	1.900	–1.100	0.76923077	–0.76923077	0.76923077
8	(0.6, 1.0)	1.600	–0.400	0.76923077	–0.76923077	0.76923077
9	(0.35, 0.4)	0.750	–0.050	0.76923077	–0.76923077	0.76923077
12	(1.1, 1.6)	2.700	–0.500	0.76923077	–0.76923077	0.76923077
13	(1.0, 1.08)	2.080	–0.080	0.76923077	–0.76923077	0.76923077
14	(1.0, 0.5)	1.500	0.500	0.76923077	–0.76923077	0.76923077
17	(1.6, 1.4)	3.000	0.200	0.76923077	–0.76923077	0.76923077
18	(1.45, 1.0)	2.450	0.450	0.76923077	–0.76923077	0.76923077
19	(1.6, 0.3)	1.900	1.300	0.76923077	–0.76923077	0.76923077

patch test. The use of a smaller local rectangular quadrature domain can often help to reduce the difficulty in numerical integration. Subdivision of the quadrature domain and use of more Gauss points are effective means to ensure an accurate numerical integration.

Example 8.12 Higher-Order Patch Test

Example 8.11 confirmed that the LPIM passed the standard patch tests. Here we present a higher-order patch test.

The higher-order patch tests shown in Figure 7.4 are used again to study the effects of the weight functions, the size of the quadrature domain, the number of the subdivision of the weight domain for integration, as well as the number of Gauss points used for numerical integration. Two test cases are conducted. In case 1, a uniform axial stress of unit intensity is applied on the right end. The exact solution for this problem with $E = 1$ and $v = 0.25$ is $u_x = x$ and $u_y = -y/4$. In case 2, a linearly varying normal stress is applied on the right end. The exact solution for this problem with $E = 1$ and $v = 0.25$ is $u_x = 2xy/3$ and $u_y = -(x^2 + y^2/4)/3$. Nine nodes are used in the interpolation. It can be found that case 1 passed the test exactly (to machine accuracy) using both the parabolic and quartic spline weight functions. The computational results for case 2 are shown in Table 8.11. It can be found that case 2 is passed exactly using the parabolic weight function for all support domain sizes. When the quartic weight function is used, case 2 is passed exactly for small support domains of $\alpha_Q = 1.0$. If the size of the support domain grows ($\alpha_Q = 2.0$ and 3.0), the tests fail.

TABLE 8.11

Relative Errors (%) at Point A for Higher-Order Patch Test Case 2 (n_c, number of subdivision; Gauss points in each subdivision: $n_g \times n_g$)

Weight Functions	$\alpha_Q = 1.0$			$\alpha_Q = 1.5$			$\alpha_Q = 2.0$		
	$n_c = 1$, $n_g = 4$	$n_c = 1$, $n_g = 10$	$n_c = 2 \times 2$, $n_g = 4$	$n_c = 1$, $n_g = 4$	$n_c = 1$, $n_g = 10$	$n_c = 2 \times 2$, $n_g = 4$	$n_c = 1$, $n_g = 4$	$n_c = 1$, $n_g = 10$	$n_c = 2 \times 2$, $n_g = 4$
Parabolic	0.0	0.0	0.0	0.0	0.0	0.0	0.0	0.0	0.0
Quartic	0.0	0.0	0.0	0.83	3.3×10^{-3}	2×10^{-4}	1.1	1×10^{-2}	6×10^{-4}

Source: Liu, G. R. and Gu, Y. T., *Struct. Eng. Mech.*, 11(2), 221–236, 2001. With permission.

The reason for the failure is the numerical integration errors caused by the complex integrands. The parabolic weight function is of lower order and therefore is simpler compared with the quartic spline weight function. Hence, exact numerical integration can be achieved more easily using the parabolic weight function. To study the effect of numerical integration, several cases with different Gauss quadrature points and subdivisions of integration (weighted) domain for integration are computed. The results are also shown in Table 8.11. It can be found that the accuracy of the solution improves with the refinement in the numerical integration. With further refinement in the numerical integration, all the tests will eventually be passed.

This example reveals three important facts:

1. Use of a larger quadrature domain does not necessarily always increase the accuracy in results, unless the numerical integration is sufficiently accurate.
2. Use of a higher order of weight function does not necessarily always lead to accurate results, unless the numerical integration is sufficiently accurate.
3. Local numerical integration is crucial in LPIM. If the numerical integration is accurate, LPIM can produce a high-order polynomial field.

However, because the order of consistency of the quartic spline weight function is higher than that of the parabolic weight function, the spline weight function can yield better results for many practical problems with high gradients of stress. One needs to make sure that the integration is sufficiently accurate.

Example 8.13 Cantilever Beam

LPIM is benchmarked using Example 6.2 to analyze stresses in a cantilever beam. The beam has length L and height D as shown in Figure 6.4. The beam is subjected to a parabolic traction at the free end as given in Equation 6.57. The beam has a unit thickness and a plane stress problem is considered. The analytical solution is available; it can be found in the textbook by Timoshenko and Goodier (1977) and is listed in Equations 6.51 to 6.56. The parameters are taken as $E = 3.0 \times 10^7$, $\nu = 0.3$, $D = 12$, $L = 48$, and $P = -1000$. Both a regular nodal distribution and an irregular nodal distribution, as shown in Figure 8.4, are employed.

Figure 8.59 shows a comparison of the analytical solution and the present numerical solution for beam deflection along the x axis. The plot shows excellent agreement between the analytical and numerical results. It is also found that the solutions on the displacement obtained using regular and irregular nodal distribution (see Figure 8.4) are very close. Figure 8.60 shows a comparison of the analytical solution and the present numerical solution for the entire deformation field of the cantilever beam. Three nodal distributions of 55 regular nodes, 189 regular nodes, and 189 irregular nodes are used. The plot shows

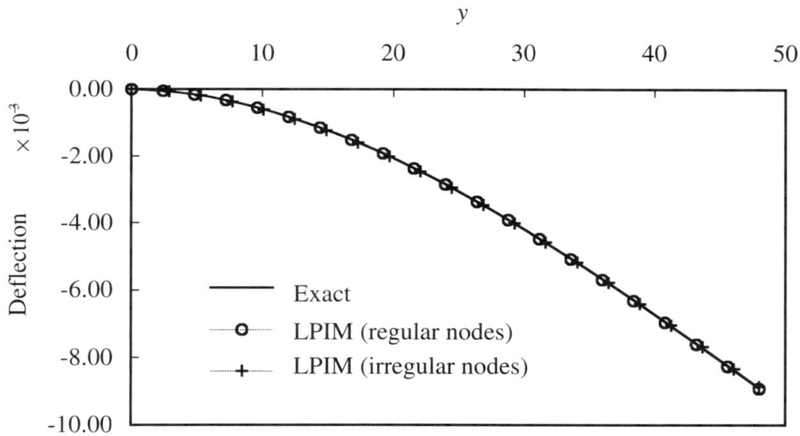

FIGURE 8.59
Deflection of the beam calculated using LPIM (using 9 × 21 = 189 nodes shown in Figure 8.4) and analytical formula. (From Liu, G. R. and Gu, Y. T., *Struct. Eng. Mech.*, 11(2), 221–236, 2001. With permission.)

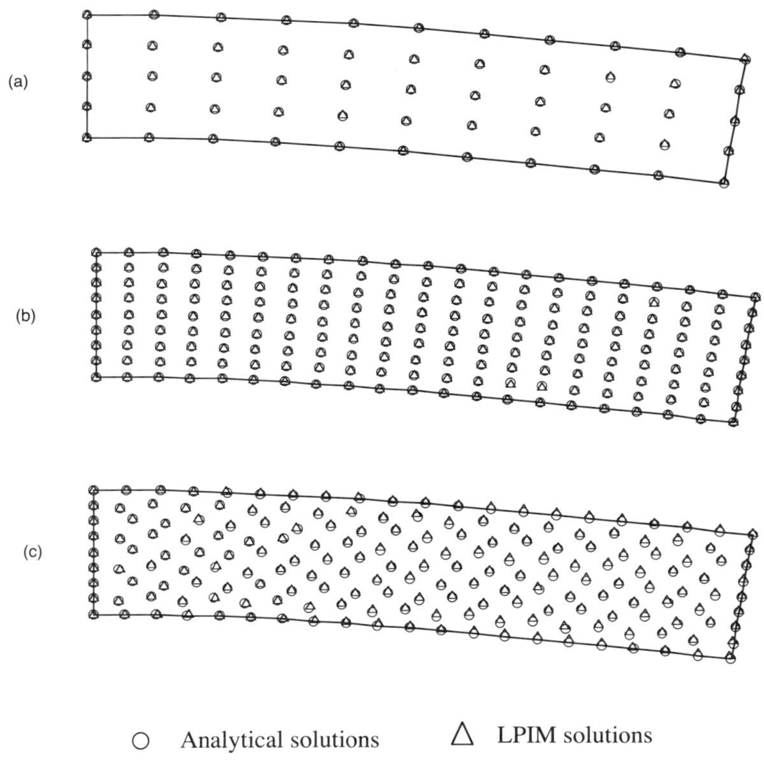

FIGURE 8.60
Deflections of the beam: (a) 55 regular nodes, (b) 189 regular nodes, and (c) 189 irregular nodes.

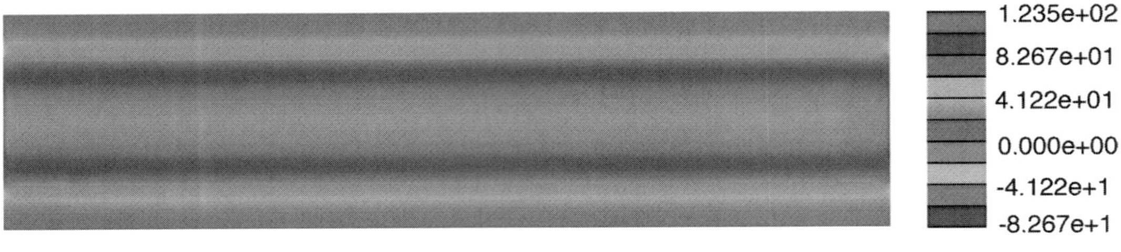

FIGURE 8.61
Distribution of shear stress σ_{xy} in the cantilever beam computed using LPIM-MTA.

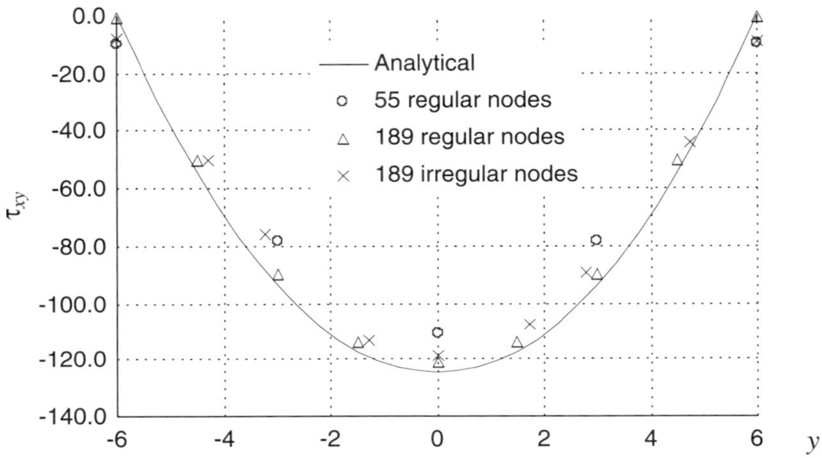

FIGURE 8.62
Shear stress σ_{xy} at section $x = L/2$ of the cantilever beam computed using LPIM-MTA.

excellent agreement between the analytical and numerical results for both regular and irregular nodal distributions.

The stress results are also obtained. As the shear stress is much more critical to the accuracy than the other stress components in the cantilever beam, only shear stress results are presented here. Figure 8.61a shows the shear stress results of the beam using 697 regularly distributed nodes. Figure 8.62 illustrates the comparison between the shear stress at the section $x = L/2$ calculated analytically and using the LPIM-MTA. Again, very good agreement is observed for both regular and irregular distributions of nodes.

For more precise error analysis, the energy norm defined by Equation 6.58 is used as the error indicator, as the accuracy in strain or stress is much more critical than the displacement.

Effects of Weight Functions

The convergence of LPIM is further studied using two different weight functions given in Equations 8.48 and 8.49. The regular distribution of nodes is used for this comparison study. The investigation is done for $\alpha_Q = 1.0$ and 2.0 using both the parabolic weight function and the quartic weight function. The convergence with mesh refinement is shown in Figure 8.63. Here, h is equivalent to the maximum element size in the FEM analysis. It is

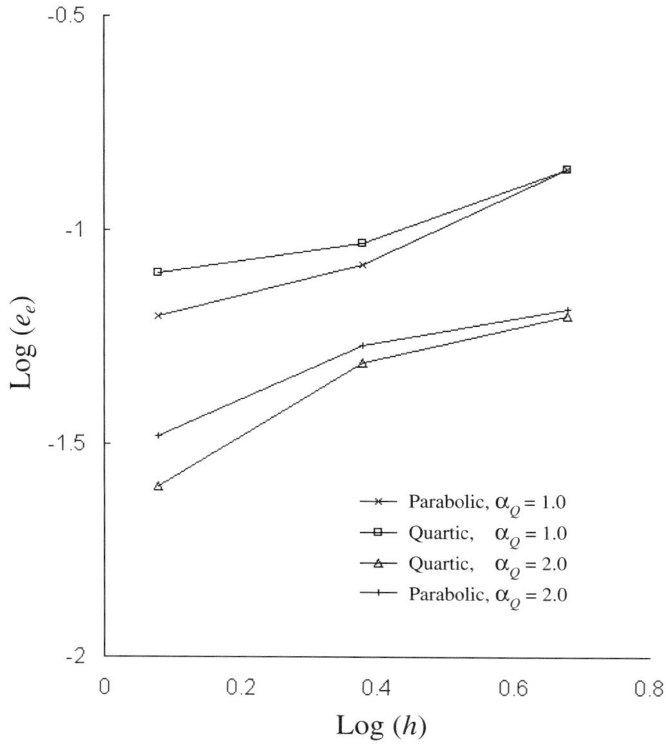

FIGURE 8.63
Convergence in norm of error e_e. Parabolic and quartic weight functions are used. The sizes of the rectangular quadrature domain are $\alpha_Q = 1.0$ and $\alpha_Q = 2.0$. (From Liu, G. R. and Gu, Y. T., *Struct. Eng. Mech.*, 11(2), 221–236, 2001. With permission.)

observed that the convergence of LPIM is very good. Comparing the results obtained using two different sizes of quadrature domain reveals that the use of a larger quadrature domain ($\alpha_Q = 2.0$) gives clearly better accuracy, provided that the numerical integration is accurate. From Figure 8.63, it is also found that using the quartic weight function does not always lead to more accurate results compared with that using a parabolic weight function. For the case of $\alpha_Q = 1.0$, the quartic weight function showed worse performance, but for the case of $\alpha_Q = 2.0$, it showed clearly better performance.

This study shows the choice of weight function does affect the results, but it is not straightforward to determine what kind of weight function performs best. Our recommendation is to use a simple monotonically decaying weight function that satisfies a minimum continuity requirement allowing the integral by part operation in the weighted residual method. Second-order differential equations require the existence of first-order differentiation. The current example is this case, and hence, the parabolic weight function can be used. For fourth-order differential equations that will be dealt with in later chapters for beam, plates, and shells, the requirement is the existence of second-order differentiation on the weight function. In that case, the quartic spine weight function is recommended.

The more important fact that Figure 8.63 reveals is that the dimension of the quadrature domain (weighted domain) plays a more important role affecting the accuracy of the solution, which leads to the following detailed investigation.

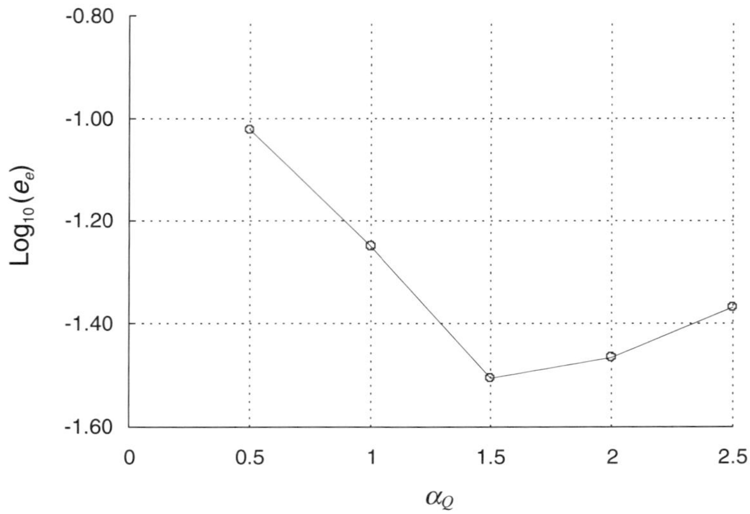

FIGURE 8.64
Influence of dimension parameter α_Q of the local weight (quadrature) domain.

Effects of the Dimension of the Local Quadrature (Weight) Domain

The size of the local quadrature domain affects the accuracy of the LPIM solution, as shown in Figure 8.63. A more detailed investigation is therefore conducted. Several quadrature domains with different sizes are used to compute the energy errors for the cantilever beam using LPIM, and the results are plotted in Figure 8.64. This figure shows that the accuracy of solutions usually increases with the increase of the quadrature domain size for LPIM. When the quadrature domain is too small ($\alpha_Q \leq 1.0$), the results will become unacceptable. This is because a local residual formulation with a very small quadrature domain for the weight function behaves more like a strong form formulation. A strong form formulation is usually less accurate than a weak integral form formulation, in which case the integration smears the error over the integral domain.

When the quadrature domain is large enough ($\alpha_Q \geq 1.5$), results obtained are very good. However, there exist difficulties in performing accurate numerical integrations for a large local quadrature domain. More subdivisions of the quadrature domain and more Gauss quadrature points are needed. The numerical integration for a large quadrature domain becomes computationally expensive and not really necessary. This fact is also clearly evidenced again in Figure 8.64, where the error for $\alpha_Q = 2.5$ is found larger than that for $\alpha_Q = 1.5$ and 2.0. Hence, $\alpha_Q = 1.5$ to 2.0 is a recommended economic choice for LPIM.

Effects of the Size of the Support Domain

The size of the support domain of a quadrature point is defined by the parameter α_s in Equation 7.28. The energy errors obtained using $\alpha_s = 1.1$ to 2.8 are plotted in Figure 8.65. It can be found that results of $\alpha_s = 1.6$ to 2.5 are very good. A too small support domain ($\alpha_s \leq 1.5$) leads to large errors, which occur because there are not enough nodes to perform interpolation for the field variable. On the contrary, a too large support domain ($\alpha_s > 2.5$) also leads to large errors. This is because too large a support domain will increase the error of the numerical integrations because the order of the shape functions is too high.

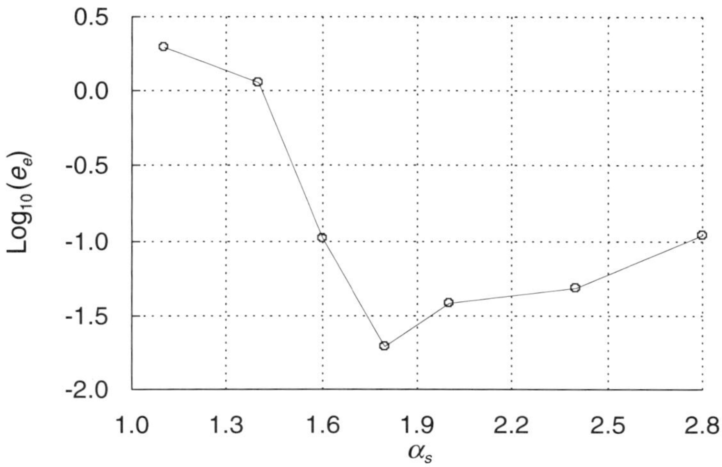

FIGURE 8.65
Influence of parameter α_s of the interpolation domain.

In addition, if too many nodes are chosen, the possibility of the interpolation singularity will increase. To overcome the interpolation singularity, the MTA has to be frequently used to exclude the nodes from the support domains. It might increase the numerical errors and computational cost. Therefore, $\alpha_s = 1.6$ to 2.5 should be chosen for this problem.

Example 8.14 Infinite Plate with a Hole

Example 6.11, examined using the EFG method, is now reexamined here using the RPIM method. The geometry of the plate is plotted in Figure 6.34. Because of the twofold symmetry, only a quarter of the plate is modeled with symmetric boundary conditions applied on $x = 0$ and $y = 0$. The parameters and the boundary conditions are exactly the same as those in Example 6.11. The displacement and the stress field within the plate are provided by Equations 6.121 to 6.126 in the polar coordinates (r, θ). The parameters are listed as follows:

Loading: $p = 1$
Young's modulus: $E = 1.0 \times 10^3$
Poisson's ratio: $\nu = 0.3$
Height of the beam: $a = 1$
Length of the beam: $b = 4$

The results are obtained using two kinds of nodal arrangements: 54 nodes and 165 nodes. The nodal distribution (165 nodes) in the plate is shown in Figure 8.10. It is found that the results for the displacements are identical. As the stresses are more critical, detailed results are presented here. The stress σ_{xx} at $x = 0$ obtained using LPIM is shown in Figure 8.66. It can be observed from Figure 8.66 that LPIM yields very good results for this problem, as well. The convergence of the present method is also demonstrated in this figure. As the number of nodes increases, the results obtained approach the analytically exact solution.

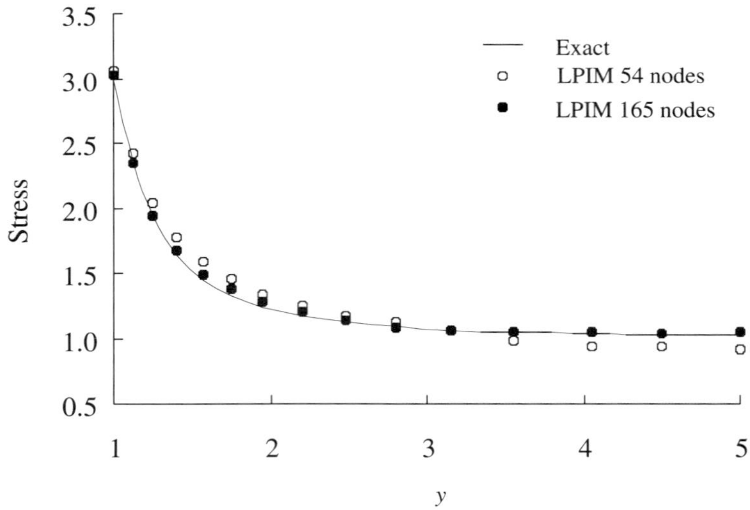

FIGURE 8.66
Stress distribution in plate with a central hole subjected to a unit unidirectional tensile load (σ_{xx} at $x = 0$). (From Liu, G. R. and Gu, Y. T., *Struct. Eng. Mech.*, 11(2), 221–236, 2001. With permission.)

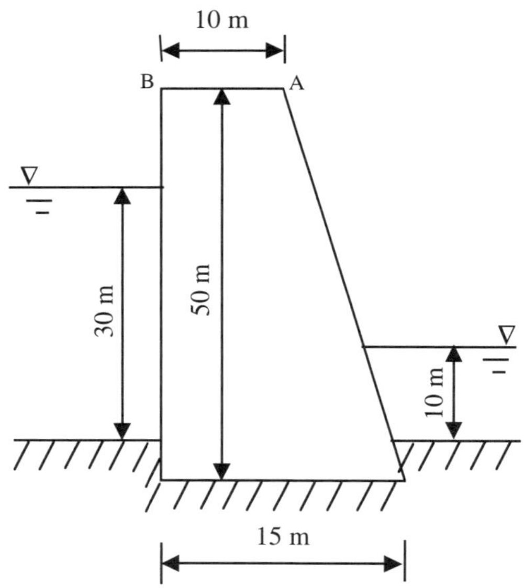

FIGURE 8.67
Water dam subjected to hydrostatic pressure modeled as a 2D plane strain problem. (From Liu, G. R. and Gu, Y. T., *Struct. Eng. Mech.*, 11(2), 221–236, 2001. With permission.)

Example 8.15 Stress Distribution in a Dam

LPIM is applied to the stress analysis of a dam subjected to hydrostatic pressure on both sides of the dam, as shown in Figure 8.67. The problem is solved for the plane strain case with E = 30 GPa and v = 0.15. The nodal arrangement for this LPIM model is shown in

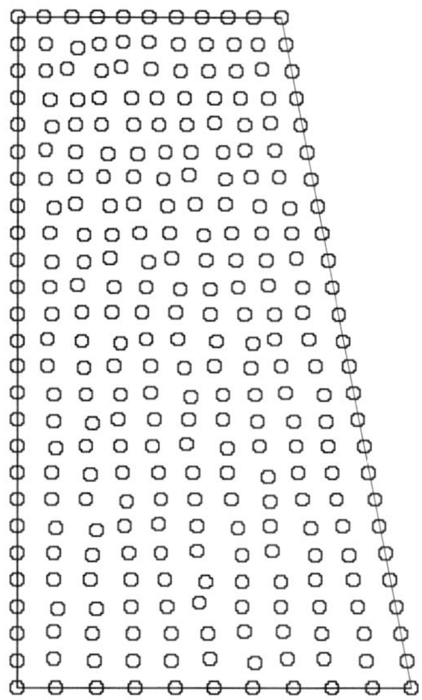

FIGURE 8.68
Arrangement of nodes on the water dam model for LPIM analysis. (From Liu, G. R. and Gu, Y. T., *Struct. Eng. Mech.*, 11(2), 221–236, 2001. With permission.)

TABLE 8.12
Displacements ($\times 10^{-3}$) at Points A and B of the Dam (see Figure 8.67)

			LPIM		FEM	
Nodes	x	y	u_x	u_y	u_x	u_y
A	10.0	50.0	2.468	–0.142	2.445	–0.140
B	0.0	50.0	2.468	0.382	2.445	0.376

Source: Liu, G. R. and Gu, Y. T., *Struct. Eng. Mech.*, 11(2), 221–236, 2001. With permission.

Figure 8.68. The problem is also analyzed using the FEM software ABAQUS using a mesh of quadrilateral elements with the same number of nodes.

The displacements at two corner nodes A and B (see Figure 8.67) are listed in Table 8.12. The results obtained by the present LPIM are in very good agreement with those obtained using FEM. The distribution of the stress σ_{yy} in the domain obtained by FEM and PIM are plotted in Figure 8.69.

8.4.4 Remarks

LPIM has been formulated, coded, and benchmarked. In using LPIM, the following parameters worked well for our tested problems:

- The quadrature domain should be of dimension $\alpha_Q = 1.5$ to 2.0. That is, it should be 1.5 to 2.0 times the local average nodal spacing.
- The support domain should be of dimension $\alpha_s = 1.6$ to 2.5. That is, it should be 1.6 to 2.5 times of the local average nodal spacing.

The important point is that large local domains may not necessarily improve the accuracy of the results, and it could present difficulties in numerical integration over the local quadrature domain.

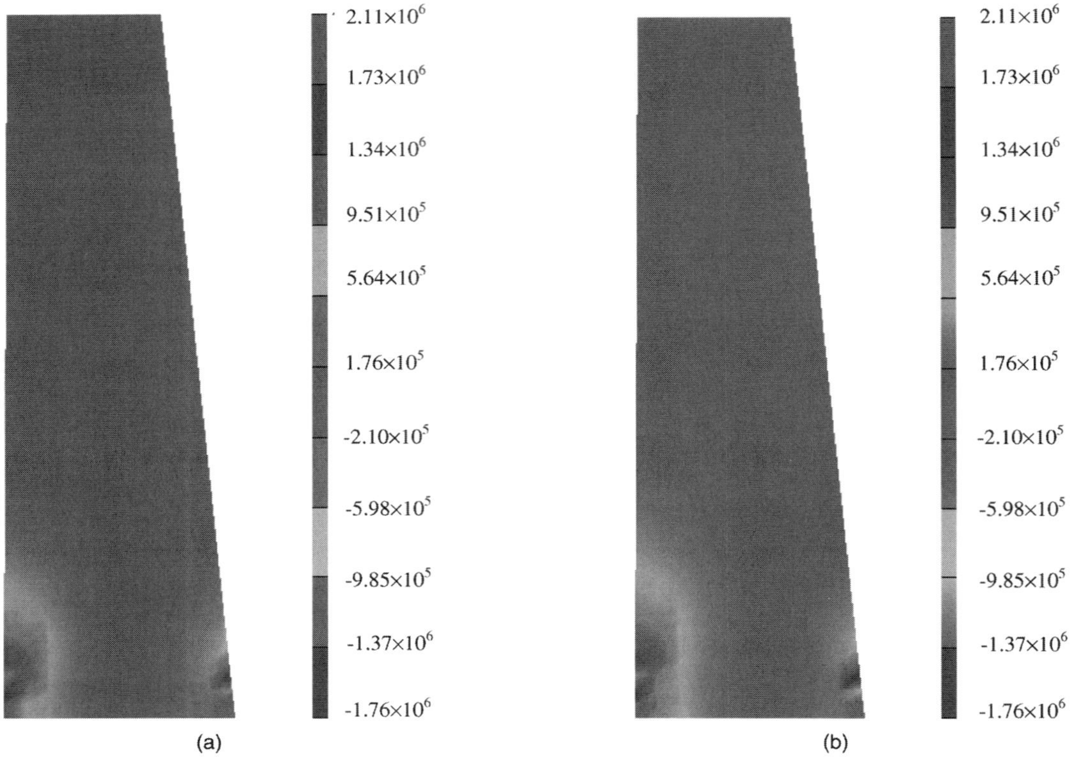

FIGURE 8.69
Distribution of stress σ_{yy}: (a) LPIM; (b) FEM. (From Liu, G. R. and Gu, Y. T., *Struct. Eng. Mech.*, 11(2), 221–236, 2001. With permission.)

8.5 Local Radial Point Interpolation Method

This section presents a local radial point interpolation method (LRPIM) that uses radial basis functions. This study was performed based on MLPG by G. R. Liu et al. (2001) for static problems and G. R. Liu and Gu (2001) for dynamic problems. The LRPIM is briefly introduced here to deal with boundary value problems for both static and dynamic analysis of 2D solids. The difference between this section and the previous section is the use of the RPIM shape function, instead of the polynomial PIM shape function. The formulation of LRPIM in this section will be, therefore, very much similar to that in the previous section for static problems and to Section 7.2 for dynamic problems. This section, therefore, will focus only on investigation of the effects of various parameters. Numerical examples are presented for static stress and free vibration analyses of 2D solids to reveal some of the very important issues in using LRPIM. In the following examples, the quartic spline weight function is used in the local residual formulation.

8.5.1 Examples of Static Problems

The LRPIM method with the first three radial basis functions listed in Table 5.2 is examined. For convenience of presentation, these are noted as MQ-LRPIM, EXP-LRPIM, and TPS-LRPIM for PIMs with a pure radial function basis, and MQ-linear, EXP-linear, and TPS-linear for

PIMs with a radial function basis with linear polynomial basis included. In computation, without special annotation, the shape parameters are chosen as $q = 1.03$ for MQ, $c = 0.003$ for EXP, and $\eta = 4.001$ for TPS. In the local integration, the quadrature domain is divided into $n_c \times n_c$ cells.

Example 8.16 Patch Test

The standard patch test for patches used in Example 6.1 is first performed. For RPIMs that do not use a linear polynomial basis, the interpolation does not reproduce the linear polynomial, and it will not pass the patch tests. Table 8.13 lists the displacement results of the interior node for various radial basis functions. These results confirm that MQ-LRPIM, EXP-LRPIM, and TPS-LRPIM cannot pass the patch test exactly but approximate it with rather high accuracy. It is also confirmed that when the MQ-linear, EXP-linear, and TPS-linear are used, the patch test can be passed exactly to machine accuracy, meaning that these local radial PIMs are linear field reproductive.

Example 8.17 High-Order Patch Test

The high-order patch that was used in Example 6.2 is used for various radial basis functions. It is found that for case 1 of the high-order patch test, similar approximation accuracy is obtained for MQ-LRPIM, EXP-LRPIM, and TPS-LRPIM. Table 8.14 shows the displacements at the right end and their errors. The results show that the accuracy is almost independent of the quadrature (weighted) domain parameter α_Q. For MQ-linear, EXP-linear, and TPS-linear, the exact solution can be obtained for case 1 as they include the complete linear polynomial terms in the basis.

For case 2 of the high-order patch test, the results are shown in Tables 8.15 through 8.17; α_s is 2.5. Results are obtained for a different number of cells in each quadrature domain. It is noted that, in general, the approximation is better with increasing n_c, especially for a large α_Q. In fact, larger α_Q requires finer cells in each subdomain to obtain accurate integration. Moreover, we note no apparent difference among MQ-LRPIM, EXP-LRPIM, and TPS-LRPIM.

For MQ-linear, EXP-linear, and TPS-linear, the results are shown in Table 8.18. There is a slight improvement in the accuracy compared with MQ-LRPIM, EXP-LRPIM, and TPS-LRPIM. All these LRPIMs failed to reproduce the high-order field, because the high-order polynomials are not included in the basis function for constructing these PIM shape function.

TABLE 8.13

Results of u_x and u_y for the Standard Patch Test Shown in Figure 7.3

Patch Model	(a)	(b)	(c)
Nodes	5	5	19
Coordinates (x, y)	(1.0, 0.0)	(1.2, 0.2)	(1.45, 0.35)
MQ-PIM (u_x, u_y)	(0.99997, 0.00000)	(1.20034, 0.20045)	(1.45005, 0.35000)
EXP-PIM (u_x, u_y)	(0.99984, 0.00000)	(1.19366, 0.19409)	(1.45010, 0.35019)
TPS-PIM (u_x, u_y)	(1.00001, 0.00000)	(1.09628, 0.16364)	(1.45004, 0.34986)

TABLE 8.14

Displacement at the Right End ($N = 35$, $n_c = 4$, $\alpha_s = 2.5$) ($u_x^{\text{exact}} = 6.0$)

	MQ-LRPIM		EXP-LRPIM		TPS-LRPIM	
α_Q	u_x	Error	u_x	Error	u_x	Error
1.0	5.99843	–0.026%	6.00603	0.101%	5.99872	–0.021%
2.0	5.99928	–0.012%	6.00233	0.039%	6.00190	0.032%
3.0	5.99791[a]	–0.034%	6.00199[a]	0.033%	6.00032[a]	0.005%

[a] $n_c = 8$.

TABLE 8.15

MQ-LRPIM for High-Order Patch (q = 1.03)

	α_Q = 1.0		α_Q = 2.0		α_Q = 3.0	
n_c	u_x at A	u_y at B	u_x at A	u_y at B	u_x at A	u_y at B
1	–6.169	–12.42	–6.110	–12.32	–5.502	–9.23
2	–6.047	–12.19	–5.986	–11.99	–6.037	–12.19
4	–6.048	–12.01	–6.008	–12.09	–5.998	–12.02

Note: u_x^{exact} = –6.0 at point A and u_y^{exact} = –12.0 at point B.

TABLE 8.16

EXP-LRPIM for High-Order Patch (c = 0.003)

	α_Q = 1.0		α_Q = 2.0		α_Q = 3.0	
n_c	u_x at A	u_y at B	u_x at A	u_y at B	u_x at A	u_y at B
1	–6.111	–12.32	–6.081	–12.17	–5.674	–10.64
2	–6.104	–12.30	–5.934	–11.84	–6.003	–12.05
4	–6.104	–12.30	–5.957	–11.89	–5.953	–11.87

Note: u_x^{exact} = –6.0 at point A and u_y^{exact} = –12.0 at point B.

TABLE 8.17

TPS-PIM for High-Order Patch (η = 4.001)

	α_Q = 1.0		α_Q = 2.0		α_Q = 3.0	
n_c	u_x at A	u_y at B	u_x at A	u_y at B	u_x at A	u_y at B
1	–5.990	–11.97	–6.064	–12.13	–5.929	–10.84
2	–5.986	–11.96	–5.999	–12.00	–6.040	–12.09
4	–5.988	–11.96	–6.007	–12.02	–5.996	–12.00

Note: u_x^{exact} = –6.0 at point A and u_y^{exact} = –12.0 at point B.

TABLE 8.18

RPIM Interpolation with Linear Reproduction for High-Order Patch Test

	MQ-Linear (q = 1.03)		EXP-Linear (b = 0.03)		TPS-Linear (η = 4.001)	
α_Q	u_x at A	u_y at B	u_x at A	u_y at B	u_x at A	u_y at B
1.0	–6.223	–12.60	–6.064	–12.20	–6.000	–12.00
2.0	–5.948	–11.92	–5.995	–12.00	–6.001	–12.00
3.0	–5.962	–11.92	–6.000	–12.01	–5.995	–11.99

Note: u_x^{exact} = –6.0 at point A and u_y^{exact} = –12.0 at point B, n_c = 4.

Example 8.18 Cantilever Beam

The cantilever beam used in Example 7.3 is examined again for benchmarking LRPIM. The model, test function, and parameters are the same as in the previous study. For the convergence investigation, α_s = 2.5, α_Q = 2.0, and n_c = 4 are used. Figure 8.70 plots the convergence in terms of the relative error of energy norm defined by Equation 7.32 for MQ-LRPIM, EXP-LRPIM, and TPS-LRPIM, and Figure 8.71 for MQ-linear, EXP-linear, and TPS-linear. It can be seen that the present LRPIM method based on various radial basis functions has acceptable convergence. Those with polynomial terms in the radial basis have slightly better results.

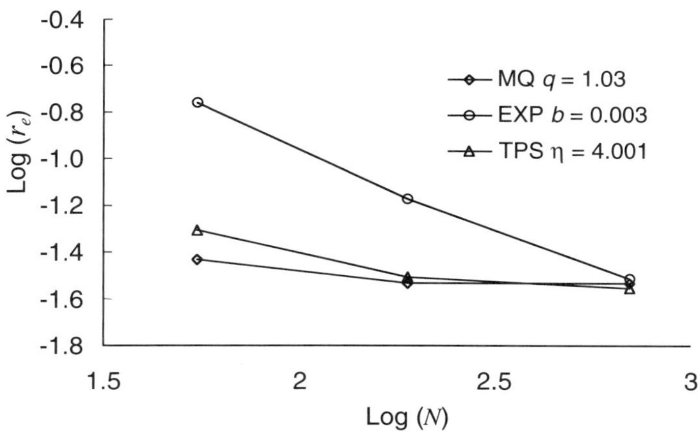

FIGURE 8.70
Convergence of the relative error of energy norm in the problem of a cantilever beam for radial basis function without polynomial terms.

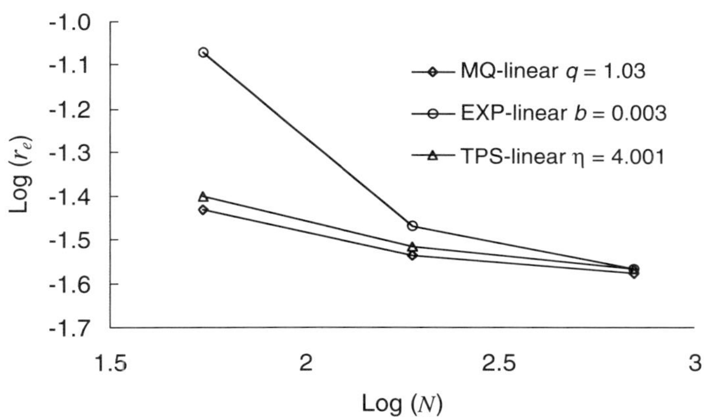

FIGURE 8.71
Convergence of the relative error of energy norm in the problem of a cantilever beam for radial basis function with linear polynomial terms.

The effects of α_s and α_Q on the relative error of energy norm are examined for various radial basis models. Figures 8.72 through 8.74 show, respectively, the results for MQ-LRPIM, EXP-LRPIM, and TPS-LRPIM radial basis without polynomial terms added. Figures 8.75 through 8.77 show results for these radial basis functions with linear polynomial terms added. Good results generally can be obtained using α_s = 2.5 for different α_Q. When a larger support domain is used (α_s = 3.0), the error does not necessarily reduce. This is due to the inaccurate integration resulting from the increased complexity of the integrand as the result of the increase of higher-order RPIM shape functions. Figure 8.78 plots the relative errors obtained using different numbers of cell n_c for integration when MQ-linear, EXP-linear, and TPS-linear are used and α_Q is set at 3.0. It is found that when cells in each support domain are increased while maintaining the same number of the Gauss points in each cell, the accuracy of the integration is improved, which leads to reduction of error, as shown in Figure 8.78.

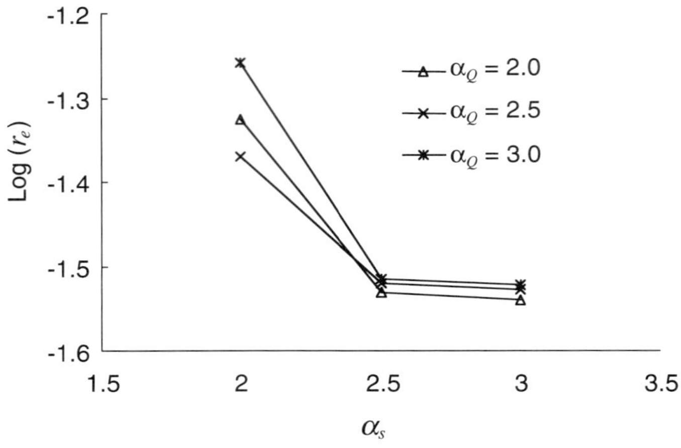

FIGURE 8.72
Relative error of energy norm with different α_s and α_Q (MQ-PIM $q = 1.03$).

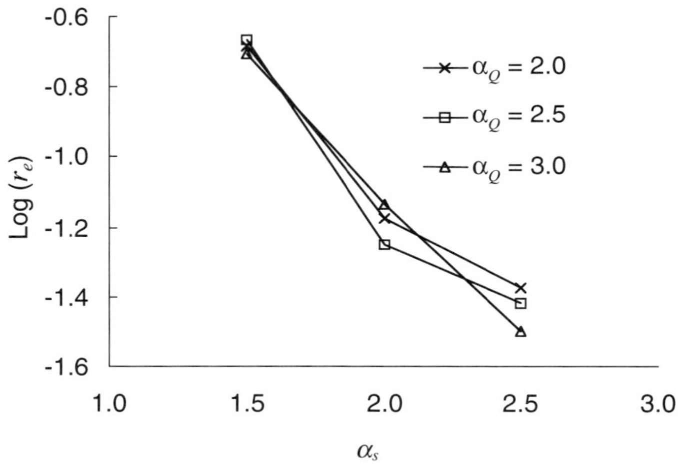

FIGURE 8.73
Relative error of energy norm with different α_s and α_Q (EXP-LRPIM $c = 0.003$).

Irregularity of the nodal arrangement is investigated further. Table 8.19 shows the relative error of the irregular node arrangement for various radial basis functions. In Table 8.19 the coefficient c is defined as the ratio of the (randomly) moved distance of nodes to the interval of the neighboring regular nodes. For example, when $c = 0.1$, a node may be moved by 10% of the interval of the neighboring regular nodes. The following key points are noticed from Table 8.19:

- MQ-LRPIM and TPS-LRPIM are not sensitive to the irregularity of the nodes.
- EXP-LRPIM is sensitive to the irregularity of the nodes, if the linear polynomial terms are not added to the basis. This is one reason that EXP-LRPIM is less favorable than MQ-LRPIM.

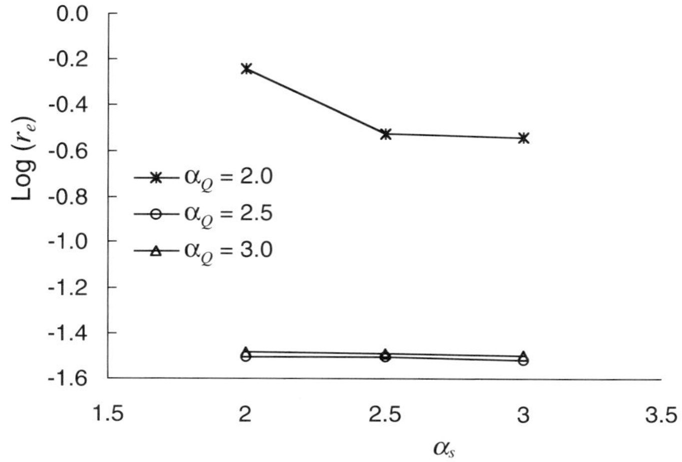

FIGURE 8.74
Relative error of energy norm with different α_s and α_Q (TPS-PIM $\eta = 4.001$).

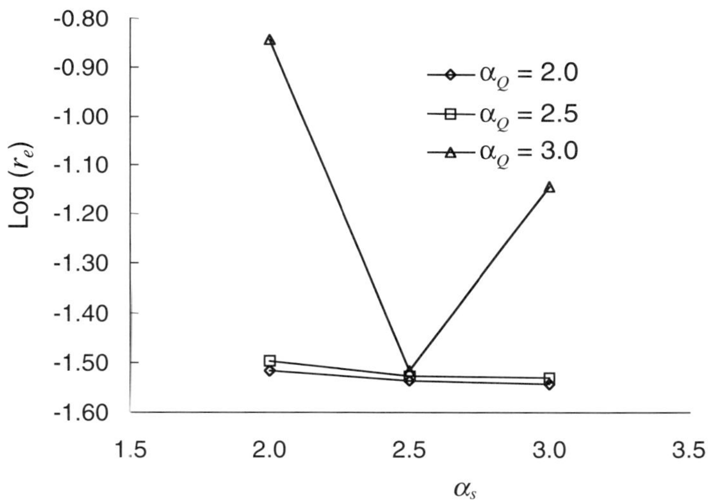

FIGURE 8.75
Relative error of energy norm with different α_s and α_Q (MQ-linear $q = 1.03$).

- With the polynomial terms in the radial basis, the accuracy is improved for all three radial basis functions, especially for the EXP basis. This is another reason that EXP-LRPIM is less favorable than MQ-LRPIM.
- EXP-linear is no longer sensitive to the irregularity of nodes.

In conclusion, LRPIMs are very stable for arbitrary nodal distribution when linear polynomial terms are added to the basis.

Another study has also been carried out to compare LRPIM (MQ) with PIM-MTA using the same benchmark problem of the cantilever beam. The purpose is to examine the difference in the effects of dimensions of the domains and of subdivision of the quadrature domains. Figure 8.79 shows the results of comparison of the effects of the dimension of

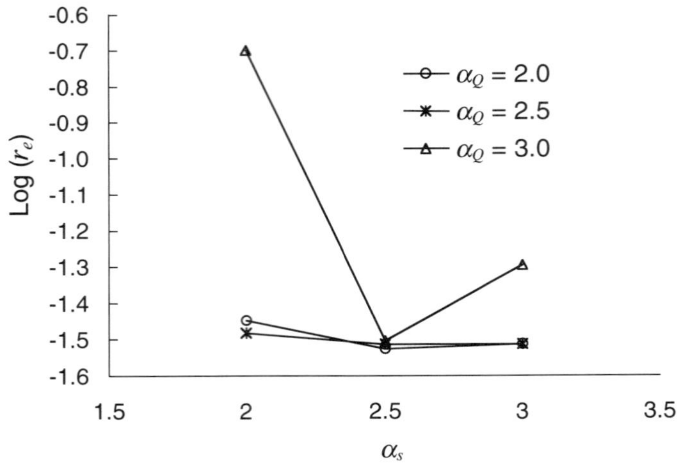

FIGURE 8.76
Relative error of energy norm with different α_s and α_Q (EXP-linear $b = 0.03$).

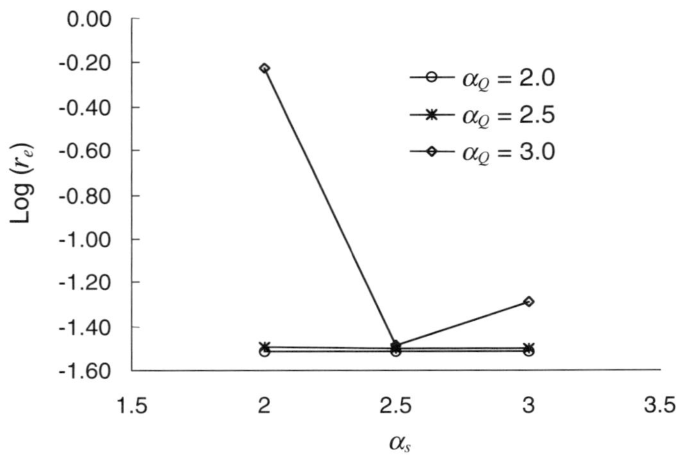

FIGURE 8.77
Relative error of energy norm with different α_s and α_Q (TPS-linear $\eta = 4.001$).

the quadrature (weighted) domain on the results obtained using LPIM (MTA) and LRPIM (MQ, $q = 1.03$, $C = 1.42$). The energy norm defined by Equation 6.58 is plotted. It is shown that a quadrature domain of $\alpha_Q = 1.5$ works for both LPIM and LRPIM. Figure 8.80 shows the same results but for the effects of the dimension of the support (interpolation) domain. It is shown that a support domain of $\alpha_s = 2.5$ works for both LPIM and LRPIM. Figure 8.81 shows the results of comparison of the effects of the number ($n_c \times n_c$) of subdivision of the quadrature domain on the results obtained using LPIM (MTA) and LRPIM. This figure suggests that a subdivision larger than 2×2 works for both LPIM and LRPIM. This finding agrees with what we concluded in Chapter 7.

Figure 8.82 plots the deflection of beam (Figure 8.82a) and shear stress distribution at the central section $x = 24$ (Figure 8.82b) for the irregular node arrangement shown in Figure 7.8 with 189 nodes. Exact solutions are also plotted in the same figure. Very good results have been obtained including the results of shear stress distribution.

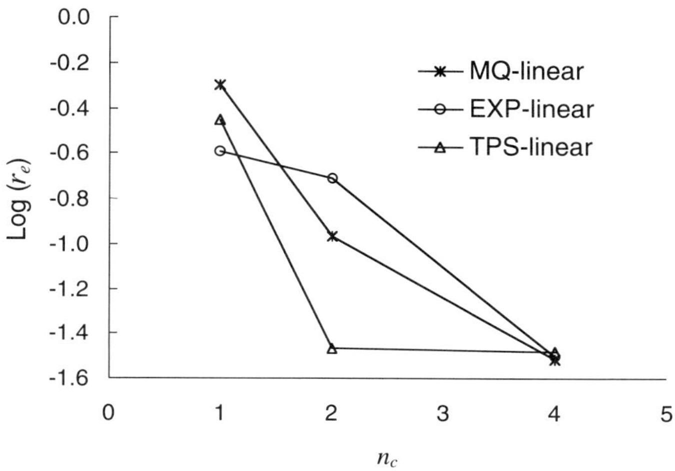

FIGURE 8.78
Relative error of energy norm with respect to the number of cells evenly divided in the quadrature domain (α_Q = 3.0, α_s = 2.5, $N = 9 \times 21 = 189$).

TABLE 8.19
Effects of Irregularity of Nodes on the Relative Error r_e ($\times 10^{-2}$)

c	0.00	0.10	0.20	0.30	0.40	Arbitrary
MQ-LRPIM (q = 1.03)	2.95	3.11	3.06	3.19	3.06	3.11
MQ-linear (q = 1.03)	2.91	3.10	3.05	3.10	3.01	3.08
EXP-LRPIM (b = 0.003)	**4.21**	**19.6**	**25.2**	**29.8**	**32.4**	**28.0**
EXP-linear (b = 0.03)	2.97	3.35	3.10	3.09	2.98	3.22
TPS-LRPIM (η = 4.001)	3.12	3.11	3.11	3.42	3.24	3.09
TPS-linear (η = 4.001)	3.05	3.06	3.09	3.11	3.07	3.08

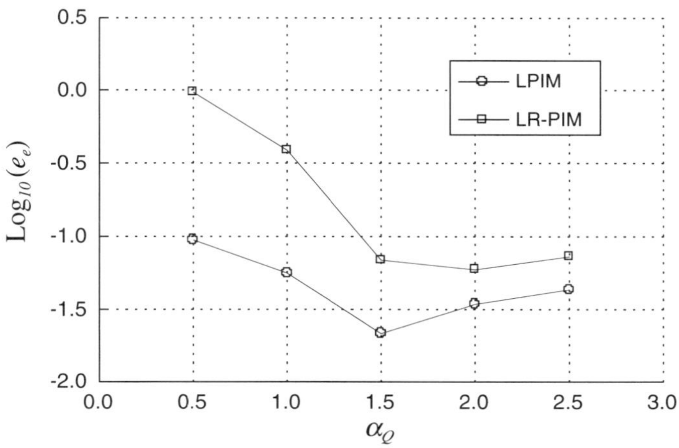

FIGURE 8.79
Effects of the dimension of the quadrature (weighted) domain. Comparison between LPIM (MTA) and LRPIM (MQ, q = 1.03, C = 1.42). The energy norm defined by Equation 6.58 is plotted.

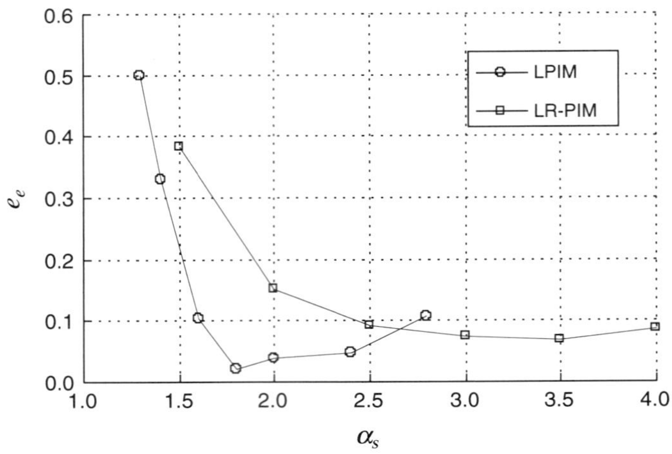

FIGURE 8.80
Effects of the dimension of the support (interpolation) domain. Comparison between LPIM (MTA) and LRPIM (MQ, $q = 1.03$, $C = 1.42$). The energy norm defined by Equation 6.58 is plotted.

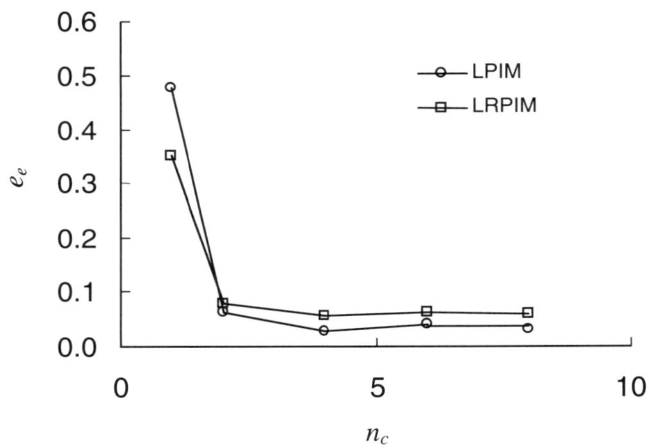

FIGURE 8.81
Effects of the number ($n_c \times n_c$) of subdivision of the quadrature domain. Comparison between LPIM (MTA) and LRPIM (MQ, $q = 1.03$, $C = 2.0$).

Example 8.19 Infinite Plate with a Circular Hole

Example 7.4 is reexamined here. The model, geometry parameters, and node arrangement are kept exactly the same. For the MQ radial basis functions with or without linear polynomial terms, the stress σ_{xx} along $x = 0$ is shown in Figures 8.83 and 8.84, if $\alpha_Q = 2.5$. It is seen that whether they have polynomial terms or not, they all give satisfactory results. However, with linear polynomial terms the results are more accurate because of their linear reproduction property. MQ-linear gives good results even for $\alpha_Q = 2.0$.

For the TPS radial basis functions with or without linear polynomial terms, the stress σ_{xx} along $x = 0$ is shown in Figures 8.85 and 8.86. Similar conclusions can be drawn.

When EXP-LRPIM and EXP-linear are used, an ill-conditioned matrix is observed in this example. Our experience indicates that EXP radial basis is often more likely to become

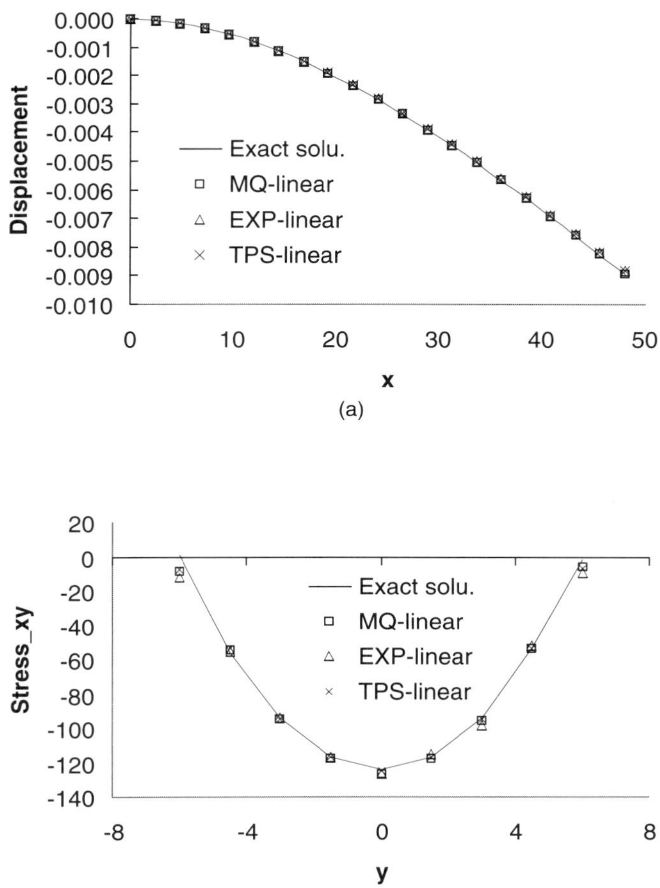

FIGURE 8.82
Numerical results of radial basis for irregular node arrangement of the beam compared with the exact solution. (a) Deflection of the beam; (b) shear stress distribution at central section.

ill-conditioned than the other two radial bases, MQ-LRPIM and TPS-LRPIM. We found that the EXP radial basis function performed the most poorly for this problem.

Example 8.20 Half-Plane Problem

Example 7.5 is reexamined using LRPIMs. The results of normal stress at point A (see Figure 7.15) for half plane problem using various radial basis functions are listed in Table 8.20. The parameters are chosen as $\alpha_s = 2.5$, $\alpha_Q = 2.0$. It is seen that good results can be obtained under various radial basis functions only if the cell number in each quadrature domain is great enough to ensure accurate integration. A typical normal stress and shear stress along the section of *m*–*n* (see Figure 8.15) are shown in Figure 8.87.

8.5.2 Examples of Dynamic Problems

LRPIM is used also for free-vibration analyses of 2D solids. The quartic weight function is used in the following examples.

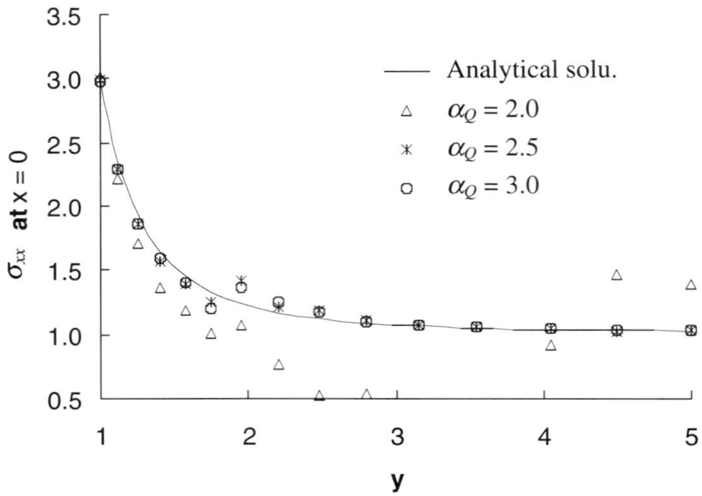

FIGURE 8.83
Comparison between the exact solution and MQ-PIM for σ_{xx} at $x = 0$ ($q = 1.03$).

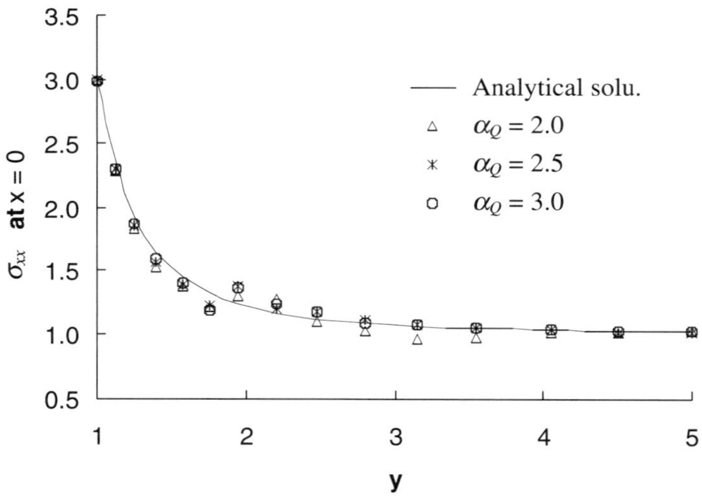

FIGURE 8.84
Comparison between the exact solution and MQ-linear for σ_{xx} at $x = 0$ ($q = 1.03$).

Example 8.21 Cantilever Beam

The LRPIM method is first applied to analyze free vibration of a cantilever beam as shown in Figure 8.88a. The problem was analyzed by Nagashima (1999) using the node-by-node meshless (NBNM) method, which is based on a global weak form and MLS approximation. NBNM is similar to EFG but uses nodal integration by adding stabilization terms.

A slender beam modeled as a plane stress problem is considered here. The parameters used are as follows:

Length: $L = 100$ mm

Height: $D = 10$ mm

TABLE 8.20
Comparison of Normal Stress at Point A with FEM Result (= 35.4)

n_c	MQ-LRPIM (q = 1.03)	MQ-Linear (q = 1.03)	EXP-LRPIM (c = 0.003)	EXP-Linear (c = 0.03)	TPS-LRPIM (η = 4.001)	TPS-Linear (η = 4.001)
1	37.7	21.5	—	—	—	—
2	34.8	35.3	35.8	36.8	35.8	35.8
4	35.2	35.5	35.8	35.5	35.7	35.8

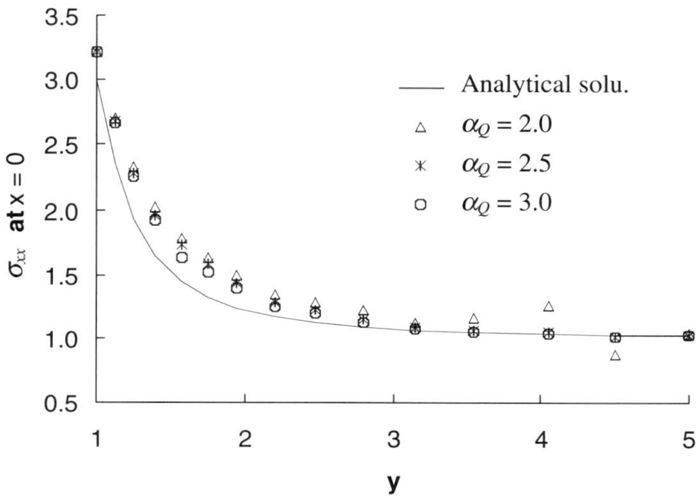

FIGURE 8.85
Comparison between the exact solution and TPS-LRPIM for σ_{xx} at $x = 0$ ($\eta = 4.001$).

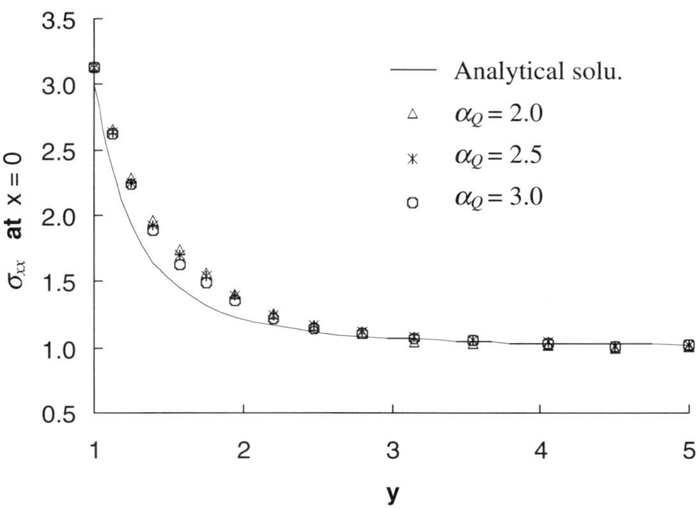

FIGURE 8.86
Comparison between the exact solution and TPS-linear for σ_{xx} at $x = 0$ ($\eta = 4.001$).

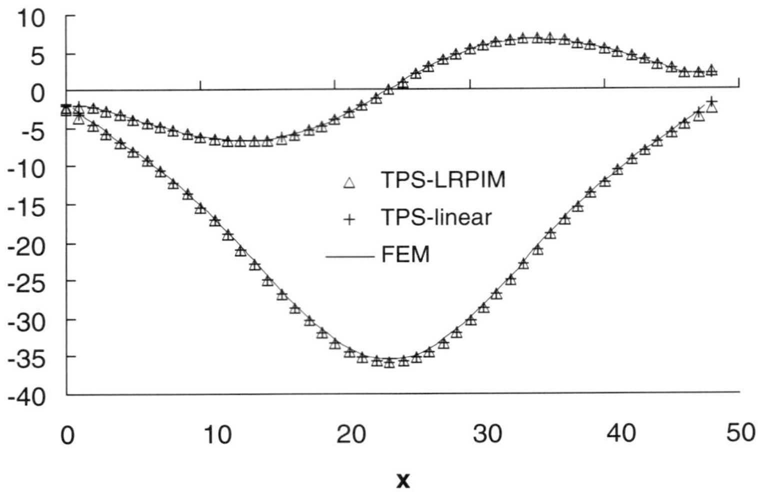

FIGURE 8.87
Stress distribution at section of *m-n* of the half space shown in Figure 8.15. Comparison of results obtained using TPS-LRPIM, TPS-linear, and FEM.

Thickness: $t = 1.0$ mm

Young's modulus: $E = 2.1 \times 10^4$ kgf/mm^2

Poisson's ratio: $\nu = 0.3$

Mass density: $m = 8.0 \times 10^{-10}$ kgfs^2/mm^4

Figure 8.88b, c, and d show three MFree models: a regular coarse model of 63 (21 × 3) nodes ($d_c = 5$), a regular fine model of 306 (51 × 6) nodes ($d_c = 2$), and an irregular fine model of 306 nodes. Parameters on the performance of the LRPIM methods for dynamic problems are investigated first. In the following parameter investigations, regular distributed 306 nodes (Figure 8.88c) are used. As reference, results obtained by FEM software (ABAQUS) using a very fine mesh with 8000 degrees of freedom (DOFs) are used as the base of comparison.

Effects of Radial Function Parameters

Both the multiquadric (MQ) radial function and the Gaussian (EXP) radial function are used as basis functions in this section. As shown in Table 5.2, the shape parameters (C and q for MQ, b for EXP) influence the performance of these radial functions as demonstrated in the previous section of static problems. In this investigation, no polynomial terms are included in the basis.

Effect of C *and* q *for MQ*

In static analyses, it has been found that parameter q influences the performance of MQ more significantly than C. Therefore, q is first investigated for $q = 0.5$, –05, and 1.03, in which $q = 0.5$ and $q = -0.5$ are the traditional parameters in MQ, and $q = 1.03$ was first found to be effective by Wang and Liu (2000). Natural frequencies for $q = -0.5$, 0.5, and

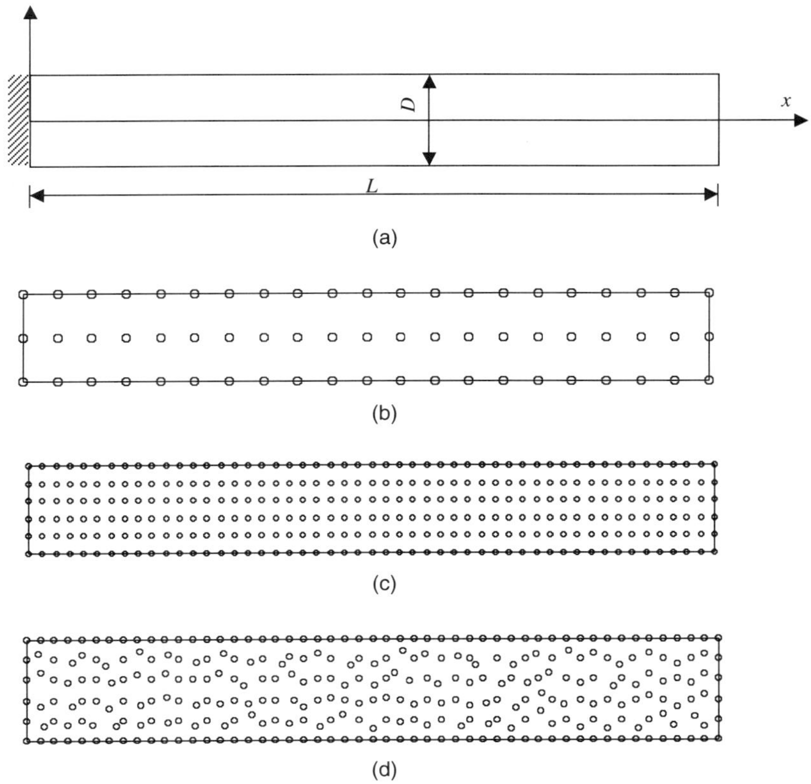

FIGURE 8.88
Slender cantilever beam and MFree models. (a) Beam for vibration analysis; (b) regular coarse nodal distribution with 63 (21 × 3) nodes; (c) regular fine nodal distribution with 306 (51 × 6) nodes; (d) irregular fine nodal distribution with 306 nodes. (From Liu, G. R. and Gu, Y. T., *J. Sound Vib.*, 246(1), 29–46, 2001. With permission.)

1.03 are obtained and compared with the FEM result. Relative error of the numerically computed natural frequency is defined as

$$r_e = \frac{\omega_i^{\text{LRPIM}} - \omega_i^{\text{FEM}}}{\omega_i^{\text{FEM}}} \times 100\% \tag{8.51}$$

where ω_i^{LRPIM} is the ith natural frequency obtained using LRPIM, and ω_i^{FEM} is that obtained by FEM software (ABAQUS) using a very fine mesh of 8000 DOFs.

Relative errors of the frequencies for different q are plotted in Figure 8.89. It can be reconfirmed from this figure that $q = 1.03$ leads to better results in the range of studies. Hence, $q = 1.03$ is used in the following studies.

It has been found that C should not be taken very small or large. Therefore, C is taken as 1.0, 1.42, and 2.0, respectively. Errors in results of natural frequency are plotted in Figure 8.90. It can be found from this figure that the influence of parameter C is smaller than that of q. Any one of these three can be used. For simplicity, $C = 1.0$ is used in the following studies.

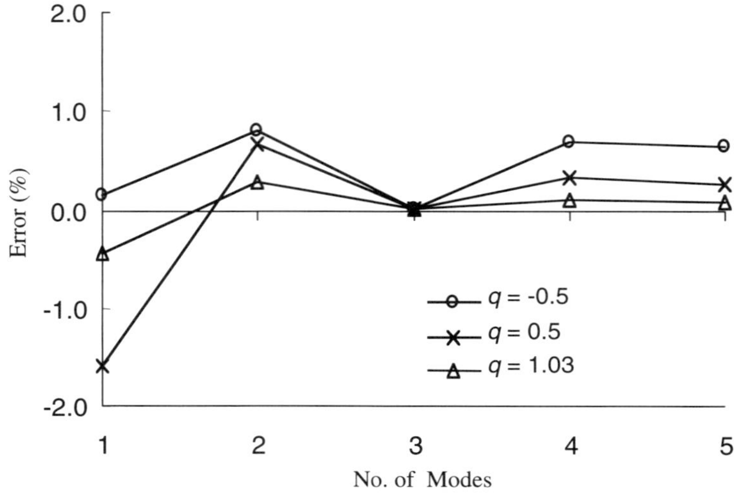

FIGURE 8.89
Influence of parameter q of MQ on the natural frequencies of the cantilever beam (306 regular nodes). (From Liu, G. R. and Gu, Y. T., *J. Sound Vib.*, 246(1), 29–46, 2001. With permission.)

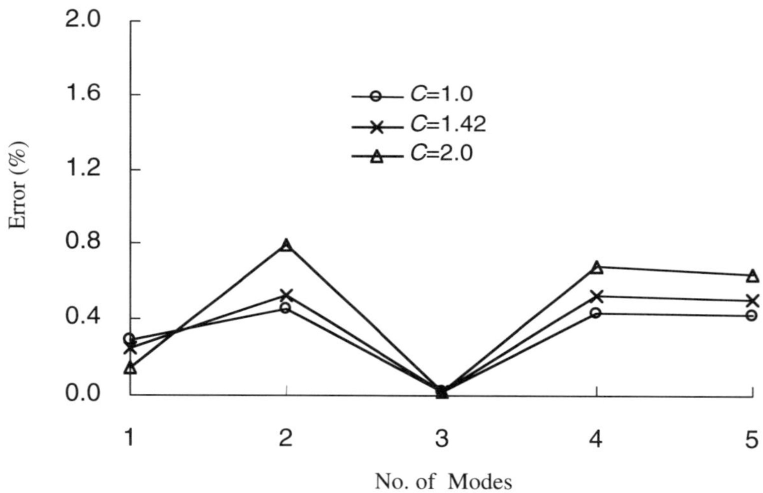

FIGURE 8.90
Influence of parameter c_0 of MQ on the natural frequencies of the cantilever beam (306 regular nodes). (From Liu, G. R. and Gu, Y. T., *J. Sound Vib.*, 246(1), 29–46, 2001. With permission.)

Effect of c *for EXP*

There is only one parameter η in the EXP radial function. A study is performed for $c = 0.3$, 0.03, and 0.003. Errors in results of natural frequency are plotted in Figure 8.91. From this figure, one can observe that a better accuracy can be obtained using a smaller c. This finding agrees with those for static problems. However, a smaller c leads to a larger condition number in the system matrix due to the property of the EXP radial function. In this study, it is found that $c = 0.003$ will lead to an ill-conditioned matrix when a large support domain is used ($\alpha_s \geq 3.5$), and $c = 0.03$ is more robust with acceptable accuracies in the range of cases studied. Therefore, $c = 0.03$ is used in the following studies.

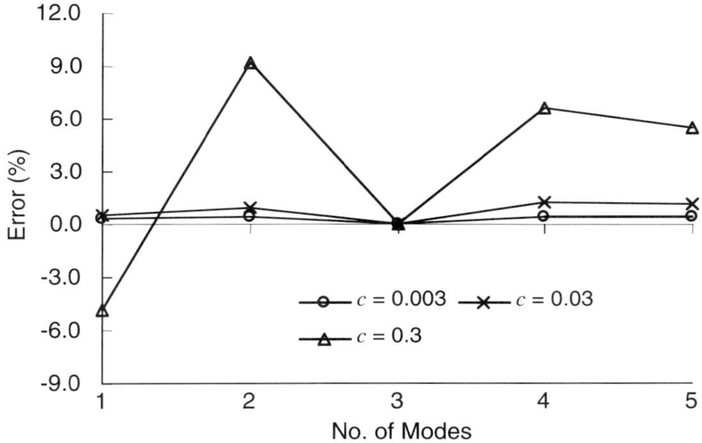

FIGURE 8.91
Influence of parameter η of EXP radial basis function on the natural frequencies of the cantilever beam (306 regular nodes). (From Liu, G. R. and Gu, Y. T., *J. Sound Vib.*, 246(1), 29–46, 2001. With permission.)

Effects of the Dimension of the Quadrature Domain

As LRPIM is a local meshless method, the size of the local quadrature domain (which is the same as the weighted domain) used will affect the accuracy of the solution of dynamic problems. Several quadrature domains with different sizes are therefore investigated for dynamic problems. The errors of the frequencies for the first three modes are plotted in Figure 8.92. From this figure, it can be found that the accuracy for frequencies increases with the increase of the quadrature domain size for both the MQ and EXP bases. When the quadrature domain is too small ($\alpha_Q = 0.5$), the results will become less accurate. This is because a local residual formulation with very small quadrature domain behaves more like a strong form formulation (Xu and Liu, 1999). A strong form formulation is usually less accurate than a weak integral form formulation, because the integration smears the error over the integral domain. For the weak form to work well, the quadrature domain (which is the same as the weighted domain in our LRPIM) should be sufficiently large.

When the quadrature domain is large enough ($\alpha_Q \geq 1.5$), results obtained are very good, as shown in Figure 8.92. However, there exist more difficulties to obtain accurate numerical integrations for a larger integral domain (or weighted domain). As discussed in Chapter 7, subdivision of the integral domain is usually required together with the use of additional Gauss quadrature points. Therefore, the numerical integration for a large quadrature domain becomes computationally expensive. On the other hand, too large a local quadrature domain does not necessarily improve accuracy significantly. This fact, clearly evidenced in Figure 8.92, implies that as long as the integral domain is large enough to "smear" the error, the size of integral domain does not play an important role. From Figure 8.92, we find that $\alpha_Q = 2.0$ can give very good results, and $\alpha_Q = 1.5$ is an economic choice that leads to reasonably accurate results.

The above finding also implies that weak forms based on a global domain of integration are not necessarily required for achieving accuracy for free vibration analyses.

Effects of the Dimension of the Support Domain

The size of the support domain of a quadrature point is determined by the parameter α_s. Because the problem domain of the cantilever beam is rectangular, rectangular support domains are used. Results of $\alpha_s = 1.0$ to 3.5 are obtained and plotted in Figure 8.93. It can

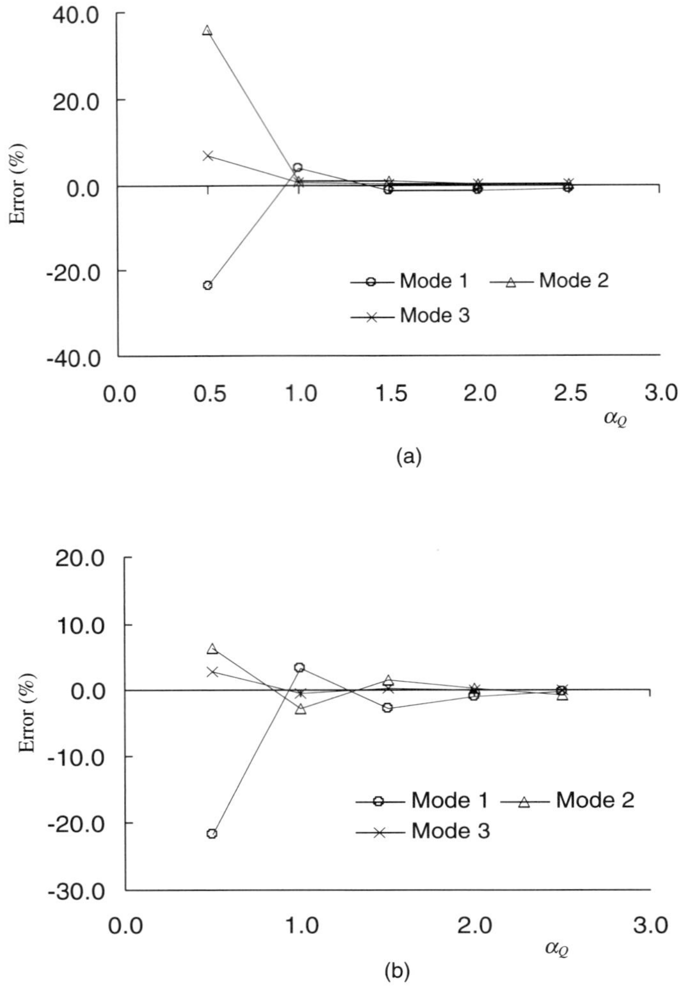

FIGURE 8.92
Influence of parameter α_Q of the quadrature domain on the relative error of natural frequencies: (a) LRPIM(MQ); LRPIM(EXP). (From Liu, G. R. and Gu, Y. T., *J. Sound Vib.*, 246(1), 29–46, 2001. With permission.)

be found that results of α_s = 1.5 to 3.0, which will include about 15 to 40 nodes used in the support domain, are very good. Too small a support domain (α_s = 1.0) and too large a support domain (α_s > 3.5) lead to large errors. Too small a support domain causes poor accuracy because there are not enough nodes (fewer than 8 nodes) to perform interpolation for the field variable. In contrast, too large a support domain increases the numerical error of interpolation because there are too many nodes (more than 70 nodes) to perform interpolation, which results in a very complex shape function and hence integrand. Therefore, α_s = 1.5 to 3.0 is a very good compromise. For both convenience and consistency, α_s = 2.0 is used in the following studies.

Modal Analyses Results of the Beam

By using above-mentioned parameters, frequencies of two regularly distributed nodal arrangements obtained by LRPIM are listed in Table 8.21. The results obtained by NBNM (Nagashima, 1999) and the commercial FEM software ABAQUS, using rectangular elements

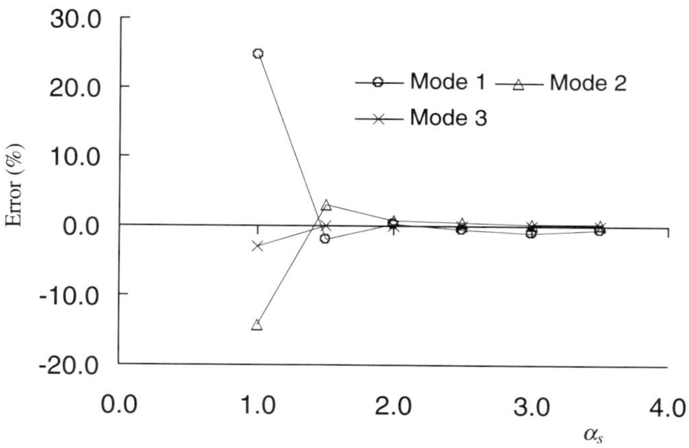

(a)

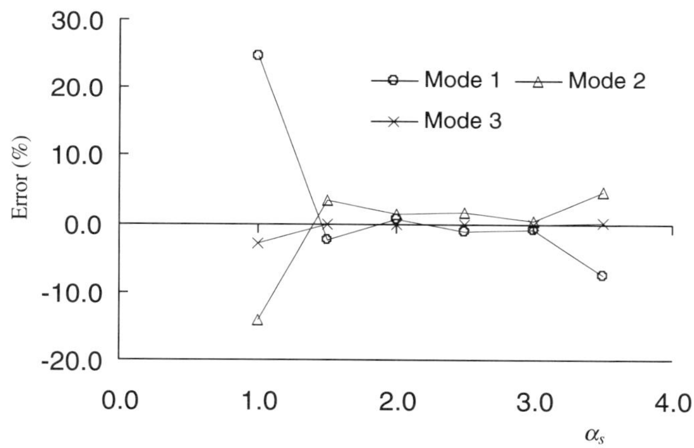

(b)

FIGURE 8.93
Influence of dimension of the support domain on the relative error of natural frequencies: (a) LRPIM(MQ); (b) LRPIM(EXP). (From Liu, G. R. and Gu, Y. T., *J. Sound Vib.*, 246(1), 29–46, 2001. With permission.)

with the same number of nodes, are also listed in the table for comparison. From this table, one can observe that the results by the present LRPIM method are in very good agreement with those obtained using FEM and NBNM. The convergence of the present method is also examined in Table 8.21. As the number of nodes increases, the results obtained by the present LRPIM approach the FEM results with extremely fine mesh, which serves as reference. The results indicate that the LRPIM is more accurate than FEM, which can be considered a whole-domain residual method. We, therefore, note again that the size of the local integral domain in the local residual weak form is not important as long as it is larger than a certain size.

The irregular distribution nodal arrangement, shown in Figure 8.88d, is also used for modal analysis. Resulting frequencies are listed in Table 8.22. From Table 8.22, one can observe that very good results are obtained using the irregular distribution of nodes. This finding supports the same conclusion made from the static problems. The computational

TABLE 8.21

Natural Frequency of a Cantilever Beam with Different Regular Nodal Distributions

	Coarse Node Distribution (63 nodes)				Fine Node Distribution (306 nodes)				
Mode	LRPIM (MQ)[a]	LRPIM (EXP)[b]	Nagashima (1999)	FEM (ABAQUS)	LRPIM (MQ)[a]	LRPIM (EXP)[b]	Nagashima (1999)	FEM (ABAQUS)	FEM (8000 DOF)
1	888.6	1000.3	926.10	870	824.3	825.8	844.19	830	823
2	5309.6	5034.0	5484.11	5199	4976.6	4958.4	5051.21	4979	4937
3	12829.4	12781.1	12831.88	12830	12826.5	12826.0	12827.60	12826	12824
4	13963.9	13643.8	14201.32	13640	13093.5	13058.3	13258.21	13111	13005
5	25311.2	24993.0	25290.04	24685	23781.9	23726.3	23992.82	23818	23632
6	38463.2	38234.5	37350.18	37477	36258.3	36179.8	36432.15	36308	36040
7	38488.0	38324.5	38320.59	38378	38451.6	38450.2	38436.43	38436	38442
8	52832.6	52727.1	50818.64	51322	49910.7	49806.8	49937.19	49958	49616
9	64012.4	63806.1	63283.70	63584	63987.8	63985.6	63901.16	63917	63955
10	67933.3	68087.4	63994.48	65731	64334.8	64202.9	64085.90	64348	63967

Unit: Hz.

[a] MQ $q = 1.03$, $C = 1.0$, $\alpha_Q = 2.0$, $\alpha_s = 2.0$.
[b] EXP $c = 0.03$, $\alpha_Q = 2.0$, $\alpha_s = 2.0$.

Source: Liu, G. R. and Gu, Y. T., *J. Sound Vib.*, 246(1), 29–46, 2001. With permission.

TABLE 8.22

Natural Frequency of a Cantilever Beam with Irregular Nodal Distribution

Mode	LRPIM (MQ)[a]	Error (%)	LRPIM (EXP)[b]	Error (%)	Nagashima (1999)	Error (%)	FEM (ABAQUS)	Error (%)	FEM (8000 DOF)
1	820.2	–0.349	815.0	–0.987	844.19	2.565	830	0.841	823
2	4938.4	0.018	4982.8	0.918	5051.21	2.303	4979	0.841	4937
3	12814.3	–0.076	12815.7	–0.065	12827.60	0.028	12826	0.016	12824
4	13005.0	0.000	13078.3	0.563	13258.21	1.947	13111	0.815	13005
5	23652.2	0.086	23724.0	0.389	23992.82	1.527	23818	0.787	23632
6	36096.7	0.157	36230.0	0.527	36432.15	1.088	36308	0.744	36040
7	38418.8	–0.060	38423.9	–0.047	38436.43	–0.014	38436	–0.016	38442
8	49742.9	0.256	49912.5	0.598	49937.19	0.647	49958	0.689	49616
9	63919.6	–0.055	63938.5	–0.026	63901.16	–0.084	63917	–0.059	63955
10	64198.3	0.362	64419.2	0.707	64085.90	0.186	64348	0.596	63967

Unit: Hz.

[a] MQ $q = 1.03$, $C = 1.0$, $\alpha_Q = 2.0$, $\alpha_s = 2.5$.
[b] EXP $c = 0.03$, $\alpha_Q = 2.0$, $\alpha_s = 2.5$.

Source: Liu, G. R. and Gu, Y. T., *J. Sound Vib.*, 246(1), 29–46, 2001. With permission.

TABLE 8.23

Natural Frequencies of a Shear Wall

	ω (rad/s)			
Mode	**LRPIM (MQ)**[a]	**LRPIM (EXP)**[b]	**FEM (ABAQUS)**	**Brebbia and Georgiou (1979)**
1	2.086	2.090	2.073	2.079
2	7.152	7.133	7.096	7.181
3	7.647	7.645	7.625	7.644
4	12.019	11.987	11.938	11.833
5	15.628	15.617	15.341	15.947
6	18.548	18.508	18.345	18.644
7	20.085	20.087	19.876	20.268
8	22.564	22.518	22.210	22.765

[a] MQ $q = 1.03$, $C = 1.0$, $\alpha_Q = 2.0$, $\alpha_s = 2.0$.
[b] EXP $c = 0.03$, $\alpha_Q = 2.0$, $\alpha_s = 2.0$.

Source: Liu, G. R. and Gu, Y. T., *J. Sound Vib.*, 246(1), 29–46, 2001. With permission.

stability and high accuracy for a nonstructured nodal distribution are very significant advantages of LRPIM. These properties are very beneficial for practical applications of LRPIM.

It may be mentioned again that the natural frequencies obtained using the LRPIM method are not necessarily the upper bounds of the exact solutions of the natural frequencies. This is because LRPIM is a Petrov–Galerkin formulation, not a Galerkin formulation.

Example 8.22 Free Vibration Analysis of a Shear Wall

Example 7.8 of a shear wall shown in Figure 7.18 is reexamined here using LRPIM. The solution of natural frequencies for this problem has been given by Brebbia and Georgiou (1979) using the boundary element method (BEM). The problem is solved for the plane stress case with $E = 1000$, $\nu = 0.2$, $t = 1.0$, and $\rho = 1.0$. A total of 574 uniformed nodes are used to represent the problem domain. The problem is also analyzed using FEM software ABAQUS. Natural frequencies of the first eight modes are calculated and listed in Table 8.23. Results obtained by the present LRPIM are in very good agreement with those obtained using BEM and FEM.

8.5.3 Remarks

LRPIM is examined in the section via a number of example problems including statics and dynamics. The following remarks can be made based on this study.

1. The attractive characteristic of RPIM interpolation is its Kronecker delta property.
2. If polynomial terms are included in the radial basis, the accuracy of the results can be improved and the solution will be less sensitive to the selection of shape parameters. Therefore, adding linear terms of polynomials to the radial basis is recommended. This also helps in passing the standard patch test.
3. For complex field functions rather than the polynomial functions, adding polynomials into the radial basis does not necessarily improve the accuracy of the solution. This also supports the finding from the study using radial functions for curve/surface fitting (see Examples 5.7 and 5.9). However, adding polyno-

mial terms will always improve the results to some extent. In addition, it can surely help in reducing the sensitivity of the shape parameters.

4. Generally, all radial functions tested in this study are very stable and flexible for both regular and irregular nodal distributions. Using an MQ or TPS radial basis is more stable than using an EXP radial basis, although our study indicates that EXP radial basis functions perform well for some problems. Therefore, MQ (q = 1.03, C = 1.0 to 1.42, which translates to α_C = 0.5 to 0.71) and TPS (η = 4.001) are recommended by this study.
5. When the local quadrature domain is large enough ($\alpha_Q \geq 2.0$ for vibration problems, $\alpha_Q \geq 1.5$ for static problems), results obtained are very good. $\alpha_Q = 2.0$ is recommended for consistency. $\alpha_Q = 1.5$ is also an economic choice that leads to reasonably accurate results for vibration problems.
6. The dimension of the support domain should be α_s = 2.0 to 3.0 for static problems and α_s = 1.5 to 3.0 for dynamic problems. This study recommends $\alpha_s = 2.0$.
7. The subdivision of the quadrature should be at least 2×2.

8.6 Application of LRPIM to Diffusion Equations

This section applies the LRPIM method to problems governed by diffusion equations that are time dependent. A foundation consolidation problem is chosen as an example. The problem is modeled based on Terzaghi's consolidation theory, which is different from that dealt with in Section 8.2, where Biot's consolidation theory is used. The reason to choose Terzaghi's consolidation theory is that it leads to a diffusion equation. This work was performed by Yan, Liu, and Wang during 2000, and is reported in Yan's master's thesis. Note that this section will not pay too much attention to the physical problem itself, but rather examines the LRPIM method by solving the problem.

The process of soil consolidation is the dissipation process of excess pore water pressure. We first briefly describe the Terzaghi's consolidation equation, and then its discrete formulation is derived by using a modified MLPG method with radial basis functions. That is, the weak form of Terzaghi's consolidation equation is obtained for each node at each time step. The time domain is discretized by the Crank–Nicholson method. Finally, 1D and 2D examples are used to illustrate the effectiveness of the present method for time domain problems. The results are compared with the closed-form Terzaghi's 1D theoretical solution and 2D FEM results.

8.6.1 Terzaghi's Consolidation Theory

Terzaghi's consolidation theory is established based on the hypothesis of constant total hydrostatic pressure. A general formulation of a 3D consolidation problem is developed and then simplified to a 2D consolidation problem. More details can be obtained in the textbook by Huang (1983).

For saturated soil, Darcy's law can describe the flow. If the deformation of soil particles is ignored, the continuity of water flow is expressed as

$$\frac{k}{\gamma_w}\nabla^2 u = -\frac{\partial \varepsilon_v}{\partial t} \tag{8.52}$$

where u is the excess pore water pressure, k is the permeability of the soil skeleton, γ_w is the density of water, and ε_v is the volumetric strain of the soil skeleton. Three effective principal stresses σ_1', σ_2', σ_3' can be expressed by principal stresses σ_1, σ_2, σ_3 and excess pore water pressure u as

$$\sigma_1' = \sigma_1 - u, \qquad \sigma_2' = \sigma_2 - u, \qquad \sigma_3' = \sigma_3 - u \tag{8.53}$$

The soil skeleton is hereafter considered an elastic body and Hooke's law is applicable. Thus, volumetric strain can be expressed as

$$\varepsilon_v = \frac{1-2\nu}{E}(\sigma_1' + \sigma_2' + \sigma_3') \tag{8.54}$$

Substituting Equation 8.53 into 8.54, one obtains

$$\varepsilon_v = \frac{1-2\nu}{E}(\sigma_1 + \sigma_2 + \sigma_3 - 3u) = \frac{1-2\nu}{E}(\Theta - 3u) \tag{8.55}$$

where $\Theta = \sigma_1 + \sigma_2 + \sigma_3$ is the hydrostatic pressure. According to Terzaghi's assumption, this hydrostatic pressure is constant for constant surcharge; therefore,

$$\frac{\partial \varepsilon_v}{\partial t} = -\frac{3(1-2\nu)}{E}\frac{\partial u}{\partial t} \tag{8.56}$$

By substituting Equation 8.56 into 8.52, a 3D Terzaghi's consolidation equation is obtained as

$$\frac{k}{\gamma_w}\nabla^2 u = \frac{3(1-2\nu)}{E}\frac{\partial u}{\partial t} \tag{8.57}$$

or

$$C_{v3}\nabla^2 u = \frac{\partial u}{\partial t} \tag{8.58}$$

where

$$C_{v3} = \frac{kE}{3\gamma_w(1-2\nu)} \tag{8.59}$$

C_{v3} is called a consolidation parameter. For a 2D consolidation problem

$$C_{v2} = \frac{k}{2\gamma_w}\frac{E}{(1-2\nu)(1+\nu)} \tag{8.60}$$

and a 1D consolidation problem

$$C_{v1} = \frac{k}{\gamma_w} \frac{E(1-\nu)}{(1-2\nu)(1+\nu)} \tag{8.61}$$

The boundary conditions are expressed as

$$\begin{aligned} u &= 0 \quad \text{for permeable boundary } \Gamma_u \\ \frac{\partial u}{\partial n} &= 0 \quad \text{for impermeable boundary } \Gamma_q \end{aligned} \tag{8.62}$$

The initial conditions are expressed as

$$u|_{t=0} = u_0 \tag{8.63}$$

To obtain initial distributions of excess pore water pressure, stress components of soil are first computed. Then, initial distributions of excess pore water pressure are obtained through experiential law for various types of soil.

Equation 8.58 is a typical form of diffusion equation, which governs a large class of engineering problems including heat and mass transfer problems, and so forth. In geomechanics, a diffusion equation governs a simplified formulation of the consolidation process, in which stress components of the soil skeleton are not included. It thus does not reflect some of the factors that may affect the process of consolidation. These factors include surcharge with time elapsing, thickness of soil with time elapsing, soil permeability with time elapsing, and so on. Our purpose here is to use this simplified consolidation problem to examine the applicability of LRPIM for solving diffusion problems. We focus more on the mathematical aspects rather than the engineering aspects. Alternative formulations of the foundation consolidation problem are presented in Chapter 6 using the EFG method and in Section 8.2 using PIM.

8.6.2 Discretized System Equation in the Time Domain

A 2D consolidation problem is considered here. At time t, excess pore water pressure in any point $\mathbf{x}$ can be written as

$$u(\mathbf{x},t) = \sum_{I=1}^{n} \phi_I(\mathbf{x})\omega_I(t) \tag{8.64}$$

where $\omega_I(t)$ is the time function for the excess pore water pressure at node I. The derivatives of the pore water pressure can be obtained by

$$u_{,i}(\mathbf{x},t) = \sum_{I=1}^{n} \phi_{I,i}(\mathbf{x})\omega_I(t) \tag{8.65}$$

$$\frac{\partial u(\mathbf{x},t)}{\partial t} = \sum_{I=1}^{n} \phi_I(\mathbf{x})\frac{\mathrm{d}\omega_I(t)}{\mathrm{d}t} \tag{8.66}$$

where $(\ )_{,i}$ denotes $\partial(\ )/\partial x_i$.

At each node, the weighted residual method is employed for Equation 8.58 within the local quadrature domain Ω_Q

$$\int_{\Omega_Q} \widehat{W}\left(C_{v2}\nabla^2 u - \frac{\partial u}{\partial t}\right) \mathrm{d}\Omega = 0 \tag{8.67}$$

where $\widehat{W}$ is a weight or test function and Ω_Q is a subdomain defined for each node. Similar to the derivation in the Chapter 8 for the MLPG method, the weak form of Terzaghi's consolidation equation can finally be obtained as

$$\int_{\Omega_Q} \widehat{W}_{,i} u_{,i}\, \mathrm{d}\Omega + \int_{\Omega_Q} \widehat{W}\frac{1}{C_{v2}}\frac{\partial u}{\partial t}\, \mathrm{d}\Omega - \int_{\Omega_Q} \widehat{W}\frac{\partial u}{\partial n}\, \mathrm{d}\Gamma = 0 \tag{8.68}$$

Substituting Equations 8.64 through 8.66 into Equation 8.68, one obtains the following set of discretized global system equations:

$$\mathbf{C}\frac{\mathrm{d}\boldsymbol{\omega}(t)}{\mathrm{d}t} + \mathbf{K}\boldsymbol{\omega}(t) = 0 \tag{8.69}$$

where $\boldsymbol{\omega}$ is the vector of all the nodal values of excess pore water pressure, the global $\mathbf{C}$ matrix is assembled using

$$C_{ij} = \frac{1}{C_{v2}}\int_{\Omega_Q} \phi_j(\mathbf{x})\widehat{W}(\mathbf{x}, \mathbf{x}_i)\, \mathrm{d}\Omega \tag{8.70}$$

and the global $\mathbf{K}$ matrix is assembled using

$$K_{ij} = \int_{\Omega_Q} \widehat{W}_{,k}(\mathbf{x}, \mathbf{x}_i)\phi_{j,k}(\mathbf{x})\, \mathrm{d}\Omega - \int_{\Gamma_{Qu}} \widehat{W}(\mathbf{x}, \mathbf{x}_i)\phi_{j,n}(\mathbf{x})\, \mathrm{d}\Gamma \tag{8.71}$$

The Crank–Nicholson method is used to discretize the time domain for Equation 8.69. The general formulation is

$$\int_t^{t+\Delta t} f(x)\, \mathrm{d}x = \Delta t[\theta f(t) + (1-\theta)f(t+\Delta t)] \tag{8.72}$$

Here $0 \le \theta \le 1$. In the thesis, $\theta = 0.5$ is used. The recursive form of Equation 8.69 can thus be written as

$$\left(\frac{\mathbf{C}}{\Delta t} + \mathbf{K}(1-\theta)\right)\boldsymbol{\omega}_{t+\Delta t} + \left(-\frac{\mathbf{C}}{\Delta t} + \mathbf{K}\theta\right)\boldsymbol{\omega}_t = 0 \tag{8.73}$$

which can be solved easily for $\boldsymbol{\omega}$ at time sequences.

8.6.3 Numerical Example

Example 8.23 Two-Dimensional Foundation

A 2D consolidation problem will be investigated in this section. The model parameters are all taken as follows:

Thickness of soil layer: $H = 16$ m

Young's modulus: $E = 4.0 \times 10^7$ Pa

Poisson's ratio: $\nu = 0.3$

Soil permeability: $k = 1.728 \times 10^{-3}$ m/day (or 2×10^{-8} m/second) in all directions

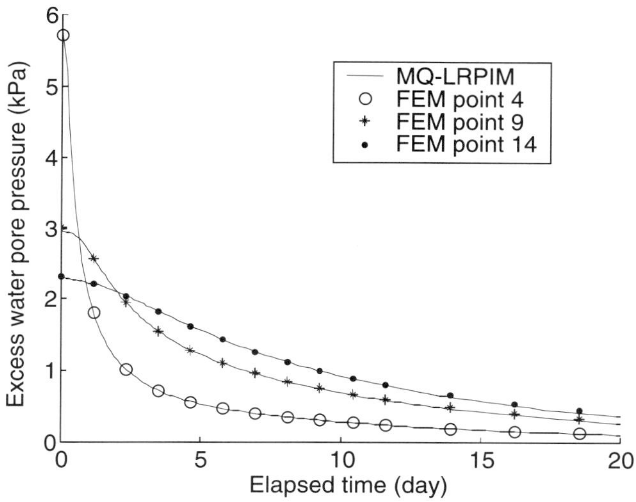

FIGURE 8.94
Excess pore water pressure history at different points for regular arrangement of nodes of 2D consolidation. Comparison between MQ-LRPIMs and FEM ($\nu = 0.3$).

The quadrature domain for a node is chosen as a rectangle of $\alpha_Q \cdot d_{xI} \times \alpha_Q \cdot d_{yI}$, where $\alpha_Q = 2.0$ is used.

The MFree model is schematically shown in Figure 8.22, where a strip load is applied on the central surface. The foundation of saturated clay is assumed to be linearly elastic. Only the upper surface is fully drained and the other three sides are all impermeable.

The initial distribution of excess pore water pressure is obtained through

$$u(\mathbf{x}, 0) = \bar{B}[\sigma_3(\mathbf{x}) + \bar{A}(\sigma_1(\mathbf{x}) - \sigma_3(\mathbf{x}))] \tag{8.74}$$

where $\bar{A}$ and $\bar{B}$ are parameters of soil. For saturated soil, $\bar{B} = 1$. When the skeleton of soil is considered elastic, $\bar{A} = \frac{1}{3}$. $\sigma_1(\mathbf{x})$ and $\sigma_3(\mathbf{x})$ are principal stress components obtained by static computation.

The dissipation history of excess pore water pressure at sample points for regular nodes is compared with FEM results, in which the same regular mesh model (833 nodes as shown in Figure 8.16) is adopted. The results are shown in Figure 8.94 for MQ-LRPIM and in Figure 8.95 for TPS-LRPIM radial basis functions. Good agreement with FEM results is observed. These two radial basis functions give very consistent results.

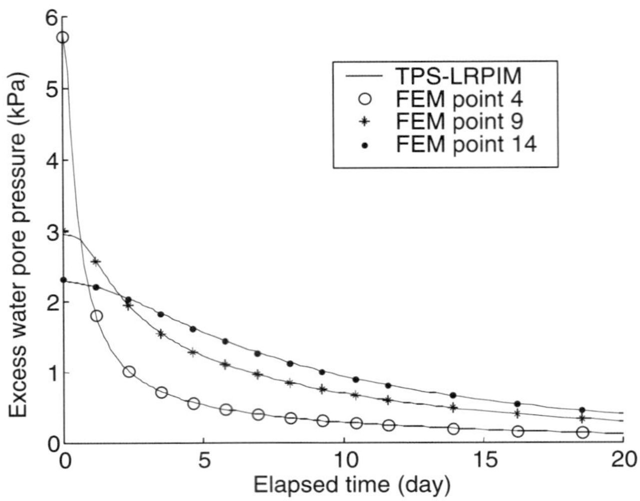

FIGURE 8.95
Excess pore water pressure history at different points for regular arrangement of nodes for 2D consolidation. Comparison between TPS-LRPIMs and FEM ($\nu = 0.3$).

8.7 Comparison Study

8.7.1 Convergence Comparison

Example 8.24 Cantilever Beam (Convergence of LPIM-MTA, MQ-LRPIM, and MLPG)

The convergence of LPIM-MTA, MQ-LRPIM, and MLPG is studied using the benchmark cantilever beam problem (Example 8.2). The study is conducted under exactly the same conditions. Regularly distributed 18 (3×6), 55 (5×11), 112 (7×16), 189 (9×21), 403 (13×31), and 697 (17×41) nodes are used. The convergence with node refinement is shown in Figure 8.96. The energy norm defined by Equation 6.58 is computed for different nodal spacing h. The convergence rates, R, are also given in Figure 8.96. The convergence rate is computed via linear regression. Note that the method of calculating the convergence rate can greatly affect the values of the convergence rate. This difference is caused by the nature of the convergence process. At the early stage, the error reduces much faster than that at later stages, where the results are very close to the exact solution. For example, if only the last two points are used to calculate the convergence rate, the R value can be much higher.

From Figure 8.96, it is observed that convergence rates of these three methods are about double the corresponding convergence rates of the Galerkin FEM, which is 1.0 for linear elements. The following can be observed:

1. Both the convergence rate and the accuracy of LPIM is the highest of these three methods. This is because PIM-MTA exhibits higher interpolation accuracy than RPIM-MLS, as discussed in Examples 5.6 and 5.9.
2. The convergence process of LRPIM is not very smooth. In addition, the parameters chosen in the radial basis functions will also affect the convergence rate and the accuracy.

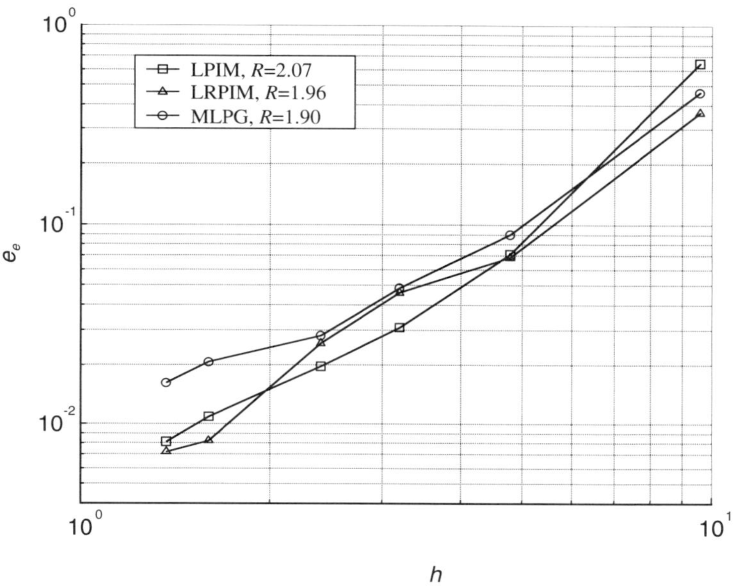

FIGURE 8.96
Convergence in error e_e of energy norm. R is the rate of convergence.

3. Although the convergence rate of the MLPG-MLS is nearly the same as that of LPIM, the accuracy of MLPG is lower.

8.7.2 Efficiency Comparison

Example 8.25 Cantilever Beam (Efficiency of LPIM-MTA, MQ-LRPIM, and MLPG)

Computational cost vs. accuracy is one of the major indicators for evaluating a numerical method. A successful numerical method should obtain high accuracy at a lower computational cost. Regularly distributed 55 and 189 nodes are used to calculate the error–computation time curve of the LPIM-MTA. The size of the interpolation domain with $\alpha_s = 1.6$ is used in the case of 55 nodes, and $\alpha_s = 1.1, 1.6$ are used in the case of 189 nodes. The energy norm defined by Equation 6.58 is computed. The error–computation time curve of the LPIM-MTA is obtained and plotted in Figure 8.97. For comparison, the curves of LRPIM and MLPG are also computed and plotted in the same figure. It is found that the errors of energy norm obtained using LRPIM and MLPG are larger than those using LPIM-MTA, when the same support domain is employed. LRPIM and MLPG must use larger support domains to achieve the same accuracy as LPIM.

It should be noted that the computational cost of an MFree method is mainly the result of two expenses:

1. The first is the cost of interpolation, which mainly comes from computing the inverse of the moment matrix. Therefore, the cost of interpolation is mainly determined by the dimension of the moment matrix. The dimension of the moment matrices of LPIM and LRPIM are $n \times n$ (n is the number of nodes in a support domain), and the dimension of the moment matrices of MLPG is $m \times m$

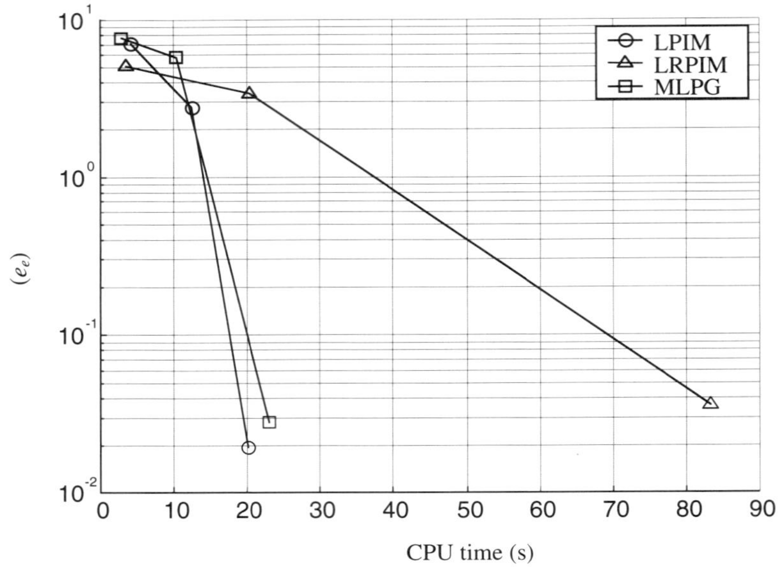

FIGURE 8.97
Computational cost vs. accuracy. Comparison of three methods: LPIM-MAT, MQ-LRPIM, and MLPG.

(m is the number of basis terms). However, the complexity of MLS shape functions and their derivatives will increase the interpolation cost.

2. The second is the cost of solving the final discrete system equation, which depends on the maximum bandwidth of the global stiffness matrix. The maximum bandwidth increases with nodes chosen in the interpolation domains.

From Figure 8.97, the following can be observed:

1. For a desired accuracy (say, an error of 10^{-1}), the cost of LPIM is the lowest, and the cost of LRPIM is the highest. As discussed above, the interpolation domains used in LRPIM and MLPG must be larger than that used in LPIM to obtain the same accuracy as LPIM. It will notably increase the cost of LRPIM and MLPG both in the interpolation cost and in the cost for solving the final system equation.
2. For a given cost (say, 20 s of CPU time), the accuracy of LPIM is the best. If the cost is given, it means that the numbers of nodes in the interpolation domain for these three methods are nearly same. Hence, LPIM-MTA obtains better accuracy due to the accurate PIM interpolation formulation.
3. It can be seen that when the support domains are the same, the computational costs of these three methods are very close. However, the accuracy of LPIM is usually higher than that of LRPIM and MLPG.

Summarizing the above discussion, one can conclude that the efficiency of LPIM-MTA is the best of all these three methods. LPIM-MTA can obtain high accuracy at lower computational cost. The efficiency of LRPIM is the worst.

8.8 Summary

A research group led by G. R. Liu originated the PIM, and has been working on its improvement over the past years. The advantages and disadvantages are summarized as follows:

- The ideal PIM produces shape functions that possess the Kronecker delta function property, and hence solves all the complex problems associated with the use of MLS approximation.
- PIM is applicable to any type of formulation including Galerkin and Petrov–Galerkin. When PIM is used with Galerkin weak form, nonconforming PIM is formulated; when the constrained Galerkin weak form is used instead, the PIM can be conforming. Using the local Petrov–Galerkin weak form, PIM can always be reproductive.
- The battle so far in the development of PIM has been concentrating on the removal of the singularity in the moment matrix. A number of methods have been suggested, including moving nodes, coordinate transformation, use of radial basis function, and MTA. Among those, MTA works best and is most convenient, as everything is automatically coded.
- The MFree procedure has made MTA possible, because it gives MTA more than enough freedom in excluding nodes.
- MTA currently works well if the nodes chosen initially are in the range of 7 to 16, which is robust enough for most of the problems because PIM needs only about 7 to 9 nodes to achieve best performance and accuracy. MTA may have a problem if the initially chosen number of nodes is too great, say, 40; then MTA may need a few iterations to find a nonsingular moment matrix.
- There is large room for improvement on MTA to achieve better efficiency and robustness. This is a continuing direction of research in PIM.
- RPIM, used either in Galerkin or Petrov–Galerkin, is very stable. Our experience led us to favor (not exclusively) MQ radial basis functions.
- The problem with RPIM is its computational efficiency. There is large room for improvement. Its current poor performance is because too little effort has been made in the improvement of computational efficiency. The effort has been focused on merely making it work. This is another direction of research. If the efficiency of RPIM or LRPIM can be improved, they will become very popular MFree methods.

9

Mesh Free Methods for Fluid Dynamics Problems

9.1 Introduction

Previous chapters have discussed a number of MFree methods to solve problems of solid mechanics. These methods can also be applied to problems of fluid mechanics, because they basically provide a means of discretizing partial differential equations in the spatial domain. This chapter discusses some of the methods that have been applied to solve or to attempt to solve computational fluid mechanics problems.

Simulation and analysis of problems of fluid dynamics have been generally performed using traditional methods, such as the finite difference method (FDM), the finite volume method (FVM), and the finite element method (FEM). These traditional numerical methods have been widely applied to practical problems and have dominated the subject of computational fluid dynamics. An important feature of these methods is that a corresponding Eulerian (for FDM and FVM) grid or a Lagrangian (for FEM) mesh or both are required as a computational frame to solve the governing equations. When simulating some special problems with large distortions, moving material interfaces, deformable boundaries, and free surfaces, these methods encounter many difficulties. FEM cannot resolve the problems of extreme distortion, and Eulerian-based methods have difficulty treating the moving material interfaces, deformable boundaries, free surfaces, and so on. Although a large amount of work on numerical schemes for the solution of fluid dynamics problems has emerged, special difficulties still exist for problems with the above-mentioned features. Attempts have also been made to combine the best features of FDM, FVM, and FEM by using such two-grid systems as arbitrary Lagrange–Eulerian (ALE) coupling, where one is a Lagrangian grid and the other is a Eulerian grid (Shin and Chisum, 1997). Computational information is exchanged either by mapping or by special interface treatment between these two grids. This approach is rather complicated and also can cause problems related to stability and accuracy. The search for better methods and techniques is still ongoing.

MFree methods offer very promising alternatives for solving problems of computational fluid dynamics (CFD). The most attractive feature of the MFree methods is that there is no need for a mesh to solve the problem. This opens a new opportunity to conduct adaptive analyses for CFD problems. There are basically three types of methods that have been explored for CFD problems:

1. Finite integral representation methods including the smoothed particle hydrodynamics (SPH) method (Lucy, 1977; Gingold and Monaghan, 1977, 1982; Monaghan, 1987, 1988) and the reproducing kernel particle method (RKPM) (Liu, W. K. et al., 1995)
2. Finite series representation methods including the meshless Petrov–Galerkin (MLPG) method (Lin and Atluri, 2001; Liu, G. R. and Yan, 2001), and the local radial point interpolation method (LRPIM) (Liu, G. R. and Yan, 2001)

3. Finite differential representation methods including the finite point method (Onate et al., 1996) and the finite difference method with arbitrary irregular grids (Liszka and Orkisz, 1980; Jensen, 1980).

This chapter details three of these methods—SPH, MLPG, and LRPIM—that fall into the first two categories listed above. For the finite point method in category 3, the reader is referred to the literature given above. It might be wise to start with an excellent summary by Orkisz (1998).

9.2 Smoothed Particle Hydrodynamics Method

SPH was developed and advanced by Lucy (1977) and Gingold and Monaghan (1977) to solve astrophysical problems in three-dimensional (3D) open space. Since its invention, SPH has been heavily studied and extended to dynamic response with material strength by Libersky and Petscheck (1991, 1993), and Johnson et al. (1996), fracture simulation (Benz and Asphaug, 1993), impact simulation (Benz and Asphaug, 1994; Randles et al., 1995), brittle solids (Benz and Asphaug, 1995), and metal forming simulation (Bonet and Kulasegaram, 2000). SPH has also been explored for simulating dynamic fluid flows with large distortions (Swegle and Attaway, 1995), explosion processes (Liu, G. R. et al., 2001a,b; Liu, M. B. et al., 2002), underwater shock (Lam et al., 2000; Liu, M. B. et al., 2000a,b), as well as other CFD problems (Campbell, 1989, Liu, W. K. et al., 1997a,b; Cleary, 1998; Chen, J. K. et al., 1999a,b; Liu, M. B. et al., 2000a,b, 2001a,b, 2002; Campbell et al., 2000). Many issues related to SPH can be found in publications by Monaghan and co-workers (Monaghan, 1982, 1987, 1989, 1992, 1994, 1998; Monaghan and Poinracic 1985; Monaghan and Gingold, 1993; Monaghan and Kocharyan, 1995). As a mesh free, particle method of pure Lagrangian nature, SPH uses smoothed particles as interpolation points to represent materials at discrete locations, so it can easily trace material interfaces, free surfaces, and moving boundaries. The MFree nature of SPH overcomes the difficulties due to large deformations because SPH uses particles or points rather than mesh as a computational frame to interpolate. These nice features of SPH make it fairly attractive, as can be seen from the large literature that has emerged during the last decade.

Efforts to improve some of the important features in SPH have also been made by W. K. Liu et al. (1995); these efforts have led to the reproducing kernel particle method (RKPM) discussed in Chapter 5 especially with regard to issues related to consistency.

This section presents a systematic implementation (Lam et al., 2000; Liu, M. B. 2000a,b; Liu, G. R. et al., 2001a,b) of the SPH to solve the Navier–Stokes equations for computational fluid dynamic applications. As the presented SPH formulations are based on the Navier–Stokes equation, physical viscosity can be modeled. Some modifications and improvements in numerical techniques such as the smoothing kernel function, smoothing length, nearest neighboring particle searching, treatment of solid boundaries, and artificial compressibility are made to suit the needs of simulating dynamic fluid flows. The presented SPH method includes general numerical aspects and various techniques for implementation. Therefore, it can simulate different flow scenarios such as inviscid or viscous flows, compressible or incompressible flows. The present SPH method is applied to different flow scenarios, which include incompressible flows with solid boundaries, free surface flows, and complex compressible flow in explosion. Numerical examples show that the presented SPH method can simulate these problems fairly well at reasonable accuracy with less computational effort, and it is an effective alternative to traditional numerical methods.

9.2.1 SPH Basics

Fundamental to SPH is the theory of integral representation of functions, which is discussed in the first two sections of Chapter 5. In SPH convention, the integral representation of function is termed *kernel approximation*.

In SPH implementation, the state of a system can be represented by a collection of arbitrarily distributed particles while forces are calculated through interparticle interactions in a smoothed fashion. These particles move freely in space, carry all the computational information, and thus can be regarded as interpolation points or field nodes, which form the computational frame for spatial discretization in numerically solving the partial differential equations describing the conservation law of the continuum fluid dynamics. The partial differential equations can be transformed into corresponding integral forms using a specially selected smoothing kernel function. There are basically two steps in the SPH procedure:

1. *Kernel approximation*. Integration of the multiplication of the field variable function and a smoothing kernel function gives the kernel approximation of the function of the field variable.
2. *Particle approximation*. Summing over the nearest neighboring particles yields the particle approximation of the function at a certain discrete point or particle.

In the kernel approximation, a function f is approximated by multiplying f with a smoothing kernel function, and then integrating over the computational domain. We have detailed the theoretical background of this type of function representation in Chapter 5 in a more general approach. Here we follow the conventional SPH notation to derive the system questions for CFD problems. The kernel approximation of f is denoted as $\langle f\rangle$ and written in the form:

$$\langle f(\mathbf{x})\rangle = \int f(\boldsymbol{\xi})\widehat{W}(\mathbf{x}-\boldsymbol{\xi}, h)\,\mathrm{d}\boldsymbol{\xi} \tag{9.1}$$

where $\mathbf{x}$ and $\boldsymbol{\xi}$ are the position vectors at different points, $\widehat{W}(\mathbf{x} - \boldsymbol{\xi}, h)$ is termed the smoothing function or smoothing kernel function in SPH or the weight function in general. The conditions that a smoothing function has to satisfy are listed in Equations 5.3 through 5.7. In Equation 9.1, h is the smoothing length representing the effective width of the smoothing kernel function, which is equivalent to the dimension of the support domain in the EFG (Chapter 6), MLPG (Chapter 7), and PIM (Chapter 8). The determination and updating of h during the computation will be discussed in detail later.

If the smoothing kernel function is assumed to be an even function, which is often but not always true (see Chapter 5), and if r_{ij} is the distance between particles i and j, we then have

$$\widehat{W}_{ij} = \widehat{W}(\mathbf{x}_i-\mathbf{x}_j, h) = \widehat{W}(|\mathbf{x}_i-\mathbf{x}_j|, h) \tag{9.2}$$

$$\nabla_i\widehat{W}_{ij} = \frac{\mathbf{x}_i-\mathbf{x}_j}{r_{ij}}\frac{\partial\widehat{W}_{ij}}{\partial r_{ij}} = \frac{\mathbf{x}_{ij}}{r_{ij}}\frac{\partial\widehat{W}_{ij}}{\partial r_{ij}} \tag{9.3}$$

Equation 9.1 is discretized into a form of summation over all the nearest neighboring particles that are within the region controlled by the smoothing length for a given particle i at a certain instant of time.

In the particle approximation, we write for any function of field variable

$$\langle f_i \rangle = \sum_{j=1}^{n} \left(\frac{m_j}{\rho_j} \right) \times f_j \times \widehat{W}_{ij} \tag{9.4}$$

where $f_i = f(\mathbf{x}_i)$, m_j and ρ_j are the mass and density of particle $j, m_j/\rho_j$ is the volume element associated with particle j, and n is the total number of particles within the smoothing length that affects particle i.

The approximation of spatial derivatives of the function of field variable can also be obtained in the same way in terms of the function values at particles, and is derived simply through integration by parts to transform the differential operation on function f into an operation on the smoothing kernel function that is known. To this end, we have

$$\langle \nabla f_i \rangle = \sum_{j=1}^{n} \left(\frac{m_j}{\rho_j} \right) f_j \nabla_i \widehat{W}_{ij} \tag{9.5}$$

From Equations 9.1, 9.4, and 9.5, the numerical value of a function f and its spatial derivatives can be obtained by SPH kernel and particle approximation over a collection of smoothing particles rather than over a mesh. This is the essence of the SPH method, and the differences between the SPH method and the traditional numerical methods of FDM, FVM, and FEM.

The above has shown that the SPH is a very simple and straightforward numerical procedure. The following will introduce its applications to fluid dynamics problems that are governed by the Navier–Stokes equation.

9.2.2 SPH Formulations for Navier–Stokes Equation

The standard SPH method is generally used to solve the Euler equation, which is limited to inviscid flows since it is not easy to obtain the SPH expressions of the second derivatives in the physical viscous term for the more general Navier–Stokes equation. Many solutions have been proposed in recent years to treat the physical viscosity. Monaghan (1995) employed an SPH approximation of the viscous term to model heat conduction, which seems to be more acceptable in low velocity flows. Takeda et al. (1994) directly used the second-order derivative of the smoothing kernel, which is analytical, rather than the second derivative of the physical variable. However, this approach is limited to constant viscosity and is very susceptible to error at low resolution. Flebbe et al. (1994) proposed using a nested sum over concerned particles to obtain an SPH expression for the physical viscosity. Unfortunately, the density evolution is regarded as incorrect (Whitworth et al., 1995), and so the whole approach is also unconvincing. Nevertheless, the approach in treating the physical viscosity is straightforward and should be fairly attractive if the whole SPH algorithm is properly arranged to minimize the computational effort arising from the nested summations. In the present SPH implementation, we retain this merit in treating physical viscosity while overcoming the disadvantage by modifying the density evolution.

Navier–Stokes Equation

We start our SPH formulation from the Navier–Stokes equation in Lagrangian form for general-purpose dynamic flow simulations. In the following expressions, the Greek superscripts α, β, and γ are used to denote the coordinate directions.

The continuity equation is

$$\frac{D\rho}{Dt} = -\rho\nabla \cdot \mathbf{v} \tag{9.6}$$

where $\mathbf{v}$ is the velocity.

The momentum equation is

$$\frac{D\mathbf{v}^{\alpha}}{Dt} = \frac{1}{\rho}\frac{\partial \sigma^{\alpha\beta}}{\partial \mathbf{x}^{\beta}} + \boldsymbol{F}^{\alpha} \tag{9.7}$$

where F is the external body force per unit mass and σ is the total internal stress. It is made up of two parts, one part comes from the isotropic pressure p and the other part is the viscous shear stress τ; i.e.,

$$\sigma^{\alpha\beta} = -p\delta^{\alpha\beta} + \tau^{\alpha\beta} \tag{9.8}$$

The different definitions of the viscous shear stress lead to different SPH applications. For Newtonian fluids, the viscous shear stress should be proportional to the shear denoted by ε through the dynamic viscosity μ; i.e.,

$$\tau^{\alpha\beta} = \mu\varepsilon^{\alpha\beta} \tag{9.9}$$

where

$$\varepsilon^{\alpha\beta} = \frac{\partial \mathbf{v}^{\beta}}{\partial \mathbf{x}^{\alpha}} + \frac{\partial \mathbf{v}^{\alpha}}{\partial \mathbf{x}^{\beta}} - \frac{2}{3}(\nabla \cdot \mathbf{v})\delta^{\alpha\beta} \tag{9.10}$$

The time rate of change of the internal energy e comes from two parts, the work done by isotropic pressure multiplying the volumetric strain and the energy dissipation due to viscous shear forces:

$$\frac{De}{Dt} = -\frac{p}{\rho}\nabla \cdot \mathbf{v} + \frac{\mu}{2\rho}\varepsilon^{\alpha\beta}\varepsilon^{\alpha\beta} \tag{9.11}$$

Density Evolution

There are two methods to evolve density in the standard SPH. The first one is the *density summation* method, which directly approximates the density of a given particle by simply substituting f in Equation 9.4 with ρ, and then summing over the neighboring particles within the effective width of the smoothing kernel; i.e.,

$$\langle \rho \rangle_i = \sum_{j=1}^{n} m_j \widehat{W}_{ij} \tag{9.12}$$

This implies that the density at particle i is approximated by a weighted average of those of all the neighboring particles within the effective width of the smoothing kernel.

The second approach is to evolve the density from the continuity Equation 9.6, and after some simple transformation, we arrive at the following *continuity density* form:

$$\left\langle \frac{D\rho}{Dt} \right\rangle_i = \sum_{j=1}^{n} m_j(\mathbf{v}_i - \mathbf{v}_j) \cdot \nabla_i \widehat{W}_{ij} \tag{9.13}$$

There are advantages and disadvantages for both approaches. The density summation approach conserves the mass exactly, while the continuity density approach does not. However, the density summation approach has an edge effect when applied to particles at the edge of the fluid, and will smooth out the density of the concerned particles, thus leading to spurious results. Another disadvantage of the density summation approach is that it requires more computational effort because the density must be evaluated before other parameters can be interpolated (approximated to be exact) and because it requires the calculation of the smoothing kernel itself.

In our formulation, these two approaches are both implemented. For problems in which mass conservation plays a significant role, the density summation approach is used. For flows with free surfaces, a compromise is made to employ the continuity density approach for particles within the smoothing area of λh inside from the free surfaces (λ is related to the actual dimension of the smoothing kernel and is described later), while other particles still use the density summation approach. This minimizes the edge effect due to the density summation approach and the mass nonconservation due to the continuity density approach.

Momentum Evolution

From Equation 9.7, the SPH approximation of the momentum equation for particle i can be written as

$$\begin{aligned} \left\langle \frac{D\mathbf{v}^{\alpha}}{Dt} \right\rangle_i &= \left\langle \frac{1}{\rho}\frac{\partial \sigma^{\alpha\beta}}{\partial \mathbf{x}^{\beta}} \right\rangle_i + \boldsymbol{F}_i^{\alpha} \\ &= \frac{1}{\rho_i}\left\langle \frac{\partial \sigma^{\alpha\beta}}{\partial \mathbf{x}^{\beta}} \right\rangle_i + \boldsymbol{F}_i^{\alpha} \\ &= \frac{1}{\rho_i}\sum_{j=1}^{n} m_j \frac{\sigma_j^{\alpha\beta}}{\rho_j}\frac{\partial \widehat{W}_{ij}}{\partial \mathbf{x}_i^{\beta}} + \boldsymbol{F}_i^{\alpha} \end{aligned} \tag{9.14}$$

In order to anti-symmetrize Equation 9.14, the following identity Equation 9.15 is used to obtain Equation 9.16:

$$\sum_{j=1}^{n} m_j \frac{\sigma_i^{\alpha\beta}}{\rho_i\rho_j}\frac{\partial \widehat{W}_{ij}}{\partial \mathbf{x}_i^{\beta}} = \frac{\sigma_i^{\alpha\beta}}{\rho_i}\sum_{j=1}^{N}\frac{m_j}{\rho_j}\frac{\partial \widehat{W}_{ij}}{\partial \mathbf{x}_i^{\beta}} = \frac{\sigma_i^{\alpha\beta}}{\rho_i}\left\langle \frac{\partial 1}{\partial \mathbf{x}^{\beta}} \right\rangle_i = 0 \tag{9.15}$$

is used to obtain

$$\left\langle \frac{D\mathbf{v}^{\alpha}}{Dt} \right\rangle_i = \sum_{j=1}^{n} m_j \frac{\sigma_i^{\alpha\beta} + \sigma_j^{\alpha\beta}}{\rho_i\rho_j}\frac{\partial \widehat{W}_{ij}}{\partial \mathbf{x}_i^{\beta}} + \boldsymbol{F}_i^{\alpha} \tag{9.16}$$

By substituting Equation 9.8 into Equation 9.16, the discretized moment equation can be written as

$$\left\langle \frac{D\mathbf{v}^{\alpha}}{Dt} \right\rangle_i = -\sum_{j=1}^{N} m_j \frac{p_i + p_j}{\rho_i \rho_j} \frac{\partial \widehat{W}_{ij}}{\partial \mathbf{x}_i^{\alpha}} + \sum_{j=1}^{N} m_j \frac{\mu_i \varepsilon_i^{\alpha\beta} + \mu_j \varepsilon_j^{\alpha\beta}}{\rho_i \rho_j} \frac{\partial \widehat{W}_{ij}}{\partial \mathbf{x}_i^{\beta}} + F_i^{\alpha} \tag{9.17}$$

The first part of the right-hand side of Equation 9.17 is the standard SPH expression for pressure force. It is the second part that concerns the physical viscosity. By using Equation 9.10, the SPH approximation of $\varepsilon^{\alpha\beta}$ for particle i can be approximated as

$$\langle \varepsilon^{\alpha\beta} \rangle_i = \sum_{j=1}^{N} \frac{m_j}{\rho_j} \mathbf{v}_j^{\beta} \frac{\partial \widehat{W}_{ij}}{\partial \mathbf{x}_i^{\alpha}} + \sum_{j=1}^{N} \frac{m_j}{\rho_j} \mathbf{v}_j^{\alpha} \frac{\partial \widehat{W}_{ij}}{\partial \mathbf{x}_i^{\beta}} - \left(\frac{2}{3} \sum_{j=1}^{N} \frac{m_j}{\rho_j} \mathbf{v}_j \cdot \nabla_i \widehat{W}_{ij} \right) \delta^{\alpha\beta} \tag{9.18}$$

The following identities are subtracted from Equation 9.18:

$$\sum_{j=1}^{n} \frac{m_j}{\rho_j} \mathbf{v}_i^{\beta} \frac{\partial \widehat{W}_{ij}}{\partial \mathbf{x}_i^{\alpha}} = \mathbf{v}_i^{\beta} \left\langle \frac{\partial 1}{\partial \mathbf{x}^{\alpha}} \right\rangle_i = 0 \tag{9.19}$$

$$\sum_{j=1}^{n} \frac{m_j}{\rho_j} \mathbf{v}_i^{\alpha} \frac{\partial \widehat{W}_{ij}}{\partial \mathbf{x}_i^{\beta}} = \mathbf{v}_i^{\alpha} \left\langle \frac{\partial 1}{\partial \mathbf{x}^{\beta}} \right\rangle_i = 0 \tag{9.20}$$

$$\sum_{j=1}^{n} \frac{m_j}{\rho_j} \mathbf{v}_i \cdot \nabla_i \widehat{W}_{ij} = \mathbf{v}_i \cdot \langle \nabla 1 \rangle_i = 0 \tag{9.21}$$

We obtain the final SPH approximation of $\varepsilon^{\alpha\beta}$ for particle i:

$$\langle \varepsilon^{\alpha\beta} \rangle_i = \sum_{j=1}^{n} \frac{m_j}{\rho_j} \mathbf{v}_{ji}^{\beta} \frac{\partial \widehat{W}_{ij}}{\partial \mathbf{x}_i^{\alpha}} + \sum_{j=1}^{n} \frac{m_j}{\rho_j} \mathbf{v}_{ji}^{\alpha} \frac{\partial \widehat{W}_{ij}}{\partial \mathbf{x}_i^{\beta}} - \left(\frac{2}{3} \sum_{j=1}^{n} \frac{m_j}{\rho_j} \mathbf{v}_{ji} \cdot \nabla_i \widehat{W}_{ij} \right) \delta^{\alpha\beta} \tag{9.22}$$

which relates the velocity differences with the viscous shear and shear stress.

The SPH approximation of $\varepsilon^{\alpha\beta}$ for particle j can be obtained in a similar way by summing over the nearest neighboring particles of j. After $\varepsilon^{\alpha\beta}$ for particles i and j have been calculated, the acceleration can by calculated be Equation 9.17. This approach is straightforward and can model variable viscosity and viscosities for different fluids.

Energy Equation

The SPH formulation for the discretized energy equation can be obtained by following a procedure similar to the momentum equation. The time rate of change of the internal energy e for a particle i can be calculated once ε has been calculated:

$$\left\langle \frac{De}{Dt} \right\rangle_i = \frac{1}{2} \sum_{j=1}^{n} m_j \frac{p_i + p_j}{\rho_i \rho_j} \mathbf{v}_{ij} \cdot \nabla_i \widehat{W}_{ij} + \frac{\mu_i}{2\rho_i} \varepsilon_i^{\alpha\beta} \varepsilon_i^{\alpha\beta} \tag{9.23}$$

The procedure for solving this set of discretized system equations of SPH is rather standard. However, the following implementation issues need to be considered.

9.2.3 Major Numerical Implementation Issues

Smoothing Kernel

The smoothing kernel function is important in the SPH method because it determines the pattern to approximate all the field variables, and the dimension of the influence area if the smoothing kernel has compact support. The kernel function generally needs to satisfy conditions listed in Equations 5.3 through 5.7. In conventional SPH, however, Equation 5.4 is often written is the following form:

$$\widehat{W}(\mathbf{x}-\boldsymbol{\xi}) = 0 \quad \text{for } |\mathbf{x}-\boldsymbol{\xi}| > \lambda h \tag{9.24}$$

where h is the smoothing length and λ is a constant controlling the actual dimension of the smoothing domain.

The most widely used smoothing kernel functions are the cubic and quartic spline functions listed in Section 5.2. In SPH implementations, they are often written in the following forms. For example, the cubic spline is given by

$$\widehat{W}_{ij}(S,h) = \alpha_D \times \begin{cases} \frac{2}{3} - S^2 + \frac{1}{2}S^3, & 0 \le S \le 1 \\ \frac{1}{6}(2-S)^3, & 1 \le S \le 2 \\ 0, & S \ge 2 \end{cases} \tag{9.25}$$

where $S = r_{ij}/h$, with r_{ij} the distance between two particles i and j, and α_D is a factor depending on the dimension of the problem. For 2D problems, $\alpha_D = 15/7\pi h^2$. It is clear that the $\lambda = 2.0$ is used in the cubic spline kernel given in Equation 9.25.

The derivative of the cubic spline kernel can be easily obtained for $\lambda = 2.0$ as follows:

$$\widehat{W}'_{ij} = \alpha_D \times \begin{cases} \frac{1}{h}\left(-2S + \frac{3}{2}S^2\right), & 0 \le S \le 1 \\ -\frac{1}{2h}(2-S)^2, & 1 \le S \le 2 \\ 0, & S \ge 2 \end{cases} \tag{9.26}$$

The shape of this kernel function and its derivative is plotted in Figures 9.1 and 9.2. Figure 9.1 shows that the value of this cubic spline function increases as the two particles approach each other, and it makes sense naturally because our instinct based on the nature of the physics tells us that the closer the two neighboring particles are, the greater influence they have on each other. However, its first derivative of the cubic spline function has its maximum value at the point $S = \frac{2}{3}$, as shown in Figure 9.2. The first derivative decreases for $S < \frac{2}{3}$ as the distance between two particles decreases. This seems unnatural. The violation of the physical nature of the cubic spline function sometimes can cause instability (Swegle and Attaway, 1995).

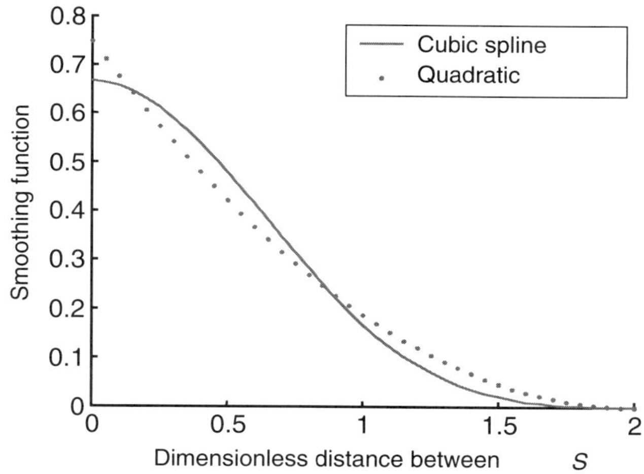

FIGURE 9.1
SPH smoothing functions ($\widehat{W}_{ij}/\alpha_D$).

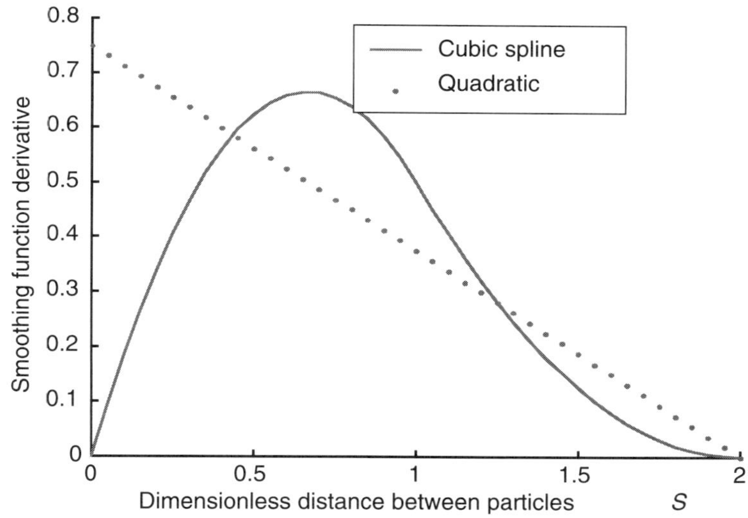

FIGURE 9.2
Derivatives of SPH smoothing functions ($\widehat{W}'_{ij}h/\alpha_D$).

The SPH quadratic spline kernel function formulated for $\lambda = 2.0$ can be written in the form of (Johnson and Beissel, 1996)

$$\widehat{W}_{ij} = \alpha_D\left[\frac{3}{16}S^2 - \frac{3}{4}S + \frac{3}{4}\right] \quad 0 \le S \le 2 \tag{9.27}$$

$$\widehat{W}'_{ij} = \alpha_D\left[\frac{3}{8}S - \frac{3}{4}\right] \quad 0 \le S \le 2 \tag{9.28}$$

where the dimension-dependent factor $\alpha_D = 2/\pi h^2$ for 2D cases. The quadratic smoothing function and its derivative are also shown in Figures 9.1 and 9.2. These figures show that a larger distance between two neighboring particles always leads to a smaller function value, as well as a smaller first derivative of the function value, and thus gives rise to less mutual influence. The problem with the quadratic smooth function is that it has a lower order of reproduction of functions. This can be easily examined using the procedure described in Chapter 5.

The cubic spline smoothing functions and the quadratic smoothing functions in one or three dimensions behave in the same manner as their counterparts in two dimensions except for the difference in the dimension-dependent factor α_D, which can be determined by imposing the zero-order reproduction condition (or normalization condition) (Equation 5.5).

In our implementation of SPH, the cubic spline function is originally used. It works well for common problems. For flows that may involve tensile instability, the quadratic smoothing function is employed.

Artificial Viscosity

In the standard SPH expressions, the artificial viscosity is used to resolve shocks numerically and to prevent nonphysical particle penetration. The artificial viscosity term is added to the pressure term in the momentum and energy equation. The most commonly used artificial viscosity is

$$\Pi_{ij} = \begin{cases} \dfrac{-\alpha \bar{c}_{ij}\theta_{ij} + \beta\theta_{ij}^2}{\bar{\rho}_{ij}} & \mathbf{v}_{ij}\cdot\mathbf{x}_{ij} < 0 \\ 0 & \mathbf{v}_{ij}\cdot\mathbf{x}_{ij} \geq 0 \end{cases} \tag{9.29}$$

where the parameters in the above equation are given by

$$\theta_{ij} = \frac{h_{ij}\mathbf{v}_{ij}\cdot\mathbf{x}_{ij}}{r_{ij}^2 + \eta^2} \tag{9.30}$$

$$\bar{c}_{ij} = \frac{1}{2}(c_i + c_j) \tag{9.31}$$

$$\bar{\rho}_{ij} = \frac{1}{2}(\rho_i + \rho_j) \tag{9.32}$$

$$h_{ij} = \frac{1}{2}(h_i + h_j) \tag{9.33}$$

$$\mathbf{v}_{ij} = \mathbf{v}_i - \mathbf{v}_j \tag{9.34}$$

In Equations 9.29 and 9.30, α, β, and η are constants that are typically set around 1, 1, and $0.1h_{ij}$, and c_i and c_j represent the speed of sound for particle i and j, respectively. The first term in Π_{ij} is similar to the Navier–Stokes shear and bulk viscosity, while the second term is similar to the Von Neumann–Richtmyer viscosity in FEM. The second term is very important in preventing nonphysical particle penetration, especially for particles that are approaching each other at high speed and almost head-on.

In our implementation, since the Navier–Stokes-based SPH formulation can resolve the general physical shear and bulk viscosity, it is not necessary to have the first term in Π_{ij}. However, the second term must be retained to prevent nonphysical particle penetration. This is different from the approach in Libersky and Petscheck (1993), where the whole artificial viscosity is added to the pressure term in the corresponding SPH equations.

Artificial Compressibility

In standard SPH for solving compressible flows, the particle motion is driven by the pressure gradient, while the particle pressure is calculated by the local particle density and internal energy through the equation of state. However, for incompressible flows, there is no equation of state for pressure. Moreover, the actual equation of state of the fluid will lead to prohibitive time steps. Although it is possible to include the constraint of the constant density into the SPH formulations, the resultant equations are too cumbersome to be solved.

In this implementation, the artificial compressibility technique is used. It is based on the fact that every incompressible fluid is theoretically compressible and, therefore, it is feasible to use a quasi-incompressible equation of state to model the incompressible flow. The purpose of introducing the artificial compressibility is to produce a time derivative of pressure. Monaghan (1994) applied the following equation of state for water to model free surface flow:

$$p = B\left(\left(\frac{\rho}{\rho_0}\right)^{\gamma} - 1\right) \tag{9.35}$$

where the constant $\gamma = 7$ is used in most circumstances, ρ_0 is the reference density, and B is a problem dependent parameter that exerts a limit for the maximum change of the density. In the artificial compressibility technique, the sound speed should be much larger than the maximum speed of the bulk flow. The subtraction of 1 in Equation 9.35 can remove the boundary effect for free surface flows. It can be seen that small oscillation in density may result in large variation in pressure. Therefore, by employing the proper equation of state for the concerned fluid, this artificial compressibility technique models an incompressible flow as a slightly compressible fluid.

The other reason to prefer this artificial compressibility approach is that in reality all fluids are compressible. The difference is in the amount of compressibility.

Smoothing Length

The smoothing length h is significant in SPH, and has a direct influence on the efficiency and accuracy. If h is too small, there may not be enough particles in the designated smoothing range of λh to exert forces on the particle concerned. This lack of influence will result in low accuracy of the numerical solution. If the smoothing length is too large, all details of the particle or local properties may be smoothed out. This oversmoothing will also affect the accuracy. Various forms of smoothing length (Hernquist and Katz, 1989; Steinmetz and Muller, 1993; Nelson and Papaloizou, 1994) have been suggested. They are generally problem dependent, and are not suitable for general CFD problems.

The smoothing length is directly related to fluid density. In different flow problems, the fluid density differs dramatically. To make SPH more adaptive to various flows, specifically, more accurate and more applicable to flows with large density inhomogeneities, as well as to those with uniformly distributed density or slightly changing density, an effective and robust smoothing length model is necessary. The following presents an optimal, adaptive smoothing length calculation procedure.

This procedure consists of three steps: the first is to initialize the smoothing length of each particle spatially, the second is to calculate the time rate of change of the smoothing length of each particle, and the third is a smoothing length optimization and relaxation process. When the smoothing length is calculated in the present procedure, only a minimum and necessary number of neighboring particles will be determined and involved in the discrete summations.

Initial Distribution of Smoothing Length

Suppose h_i^0 is the initial smoothing length of particle i, while N_i^0 is the number of the initial nearest neighboring particles within the smoothing range of λh_i^0 for particle i. In one, two, or three dimensions, N_i^0 is taken as 5, 21, and 57, respectively. This N_i^0 corresponds to the smoothing kernel extending to $2h$. The mass of the particle is m_i, which is determined at the beginning and remains unchanged during the computation. If the initial density of particle i is ρ_i^0, then for 3D cases (assuming the particles to have taken the spherical shape):

$$\sum_{j=1}^{N_i^0} m_j = \frac{4}{3}\pi(2h_i^0)^3 \rho_i^0 \tag{9.36}$$

We can obtain the initial smoothing length simply, using

$$h_i^0 = c_3 \times \left\{\frac{\sum_{j=1}^{N_i^0} m_j}{\rho_i^0}\right\}^{1/3}, \quad c_3 = \left(\frac{3}{32\pi}\right)^{1/3} \tag{9.37}$$

Similar results can also be derived for 2D and 1D cases. For 2D cases, we have

$$h_i^0 = c_2 \times \left\{\frac{\sum_{j=1}^{N_i^0} m_j}{\rho_i^0}\right\}^{1/2}, \quad c_2 = \left(\frac{1}{4\pi}\right)^{1/2} \tag{9.38}$$

and for 1D cases, we have

$$h_i^0 = c_1 \times \left\{\frac{\sum_{j=1}^{N_i^0} m_j}{\rho_i^0}\right\}, \quad c_1 = \frac{1}{4} \tag{9.39}$$

In general, h_i^0 can be written as

$$h_i^0 = c_{n_D} \times \left\{\frac{\sum_{j=1}^{N_i^0} m_j}{\rho_i^0}\right\}^{1/n_D} \tag{9.40}$$

where n_D is the number of dimensions.

Updating of Smoothing Length

The smoothing length of each particle is then treated as a variable and updated during the computation by the following most commonly used expression.

If h_i^n, ρ_i^n, $\mathbf{v}_i^n$, $\nabla_i^n W$, N_i^n are denoted as the smoothing length, density, velocity, smoothing kernel derivative, and nearest neighboring particles of particle i at time step n, we have

$$\frac{Dh_i^n}{Dt} = -\frac{h_i^n}{\rho_i^n n_D}\frac{D\rho_i^n}{Dt} \tag{9.41}$$

By using Equation 9.13, Equation 9.41 can be written in the following SPH form:

$$\frac{Dh_i^n}{Dt} = -\frac{h_i^n}{\rho_i^n n_D}\sum_{j=1}^{N_i^n} m_j(\mathbf{v}_i^n - \mathbf{v}_j^n)\cdot\nabla_i^n \widehat{W}_{ij} \tag{9.42}$$

Equation 9.42 works well for flow problems with homogeneous density or slightly changing density, such as slowly expanding or contracting gases. However, for flows with large deformation, or those with large density inhomogeneities, it may not be very accurate or stable. Hence, an optimization and relaxation procedure is suggested in the following to improve the updating of the smoothing length.

Optimal Procedure

The procedure is used to optimize the updating of h_i^{n+1}, so that each particle interacts with a roughly constant number of nearest neighboring particles. It consists of two steps: the prediction step and the correction step.

PREDICTION STEP Update h_i^{n+1} with expression, $h_i^{n+1} = h_i^n + \alpha_h \times \mathrm{d}h_i^n/\mathrm{d}t$. Here, α_h is a relaxation factor. α_h is taken as 1.0 at first and then adjusted slightly around 1.0 in the later optimal and relaxation step.

CORRECTION STEP Once h_i^{n+1} is obtained, the current number of neighbors N_i^{n+1} can be determined. If N_i^{n+1} is found to be roughly the same as N_i^0, it is desirable, whereas if it is different from N_i^0 by more than a small, prescribed tolerance of a few percent, then adjust the relaxation factor α_h around 1.0 to obtain a new h_i^{n+1} until N_i^{n+1} is roughly the same as N_i^0.

Nearest Neighboring Particle Searching

For SPH implementations, the nearest neighboring particle search algorithm is another vital numerical aspect because the repeated calculation of interactions between the neighbors drastically affects the efficiency of the whole simulation, especially in simulations with a large number of particles. Direct summation by simply comparing the interparticle distance with λh for all particles from the given particle is of order $O(N^2)$, thus would entail an intolerable amount of computational time for large number of particles. The grid-based searching algorithm is efficient for constant smoothing length, but is not suitable for variable smoothing lengths. The hierarchy tree (Hernquist and Katz, 1989) method is very attractive for problems with variable smoothing length and self-gravity. This tree method recursively splits the maximum computational region into octants that contain particles, until the leaves on the tree are individual particles.

Past literature has focused on how to find the neighboring particles and then on calculating the interactions between each pair of neighboring particles. In such implementations, for a given particle, an array recording the total number of nearest neighboring particles and another large array relating to the serial number of the neighbors or something of the sort are necessary. Then, during the particle approximation process, the kernel or its derivatives are repeatedly calculated for later use in interaction summation.

This SPH implementation uses a technique that calculates the interactions between each pair of neighboring particles. In this technique, at first, the hierarchy tree method is used to construct the tree structure. For a given particle i, a cube of $2\lambda h_i$ is used to enclose the particle. The procedure is as follows. When the tree structure is constructed, check the volume of the search cube with that represented by a node or tree cell to see if they overlap. If they overlap, subdivide the cell and proceed to the next level to check repeatedly, until the node or leaf represents a particle. Then, check if the particle is within smoothing area λh_i of the given particle i. If yes, it shows that interaction between these two particles exists. Record the serial number of the pair of particles for this pairwise interaction, increase the number of nearest neighboring particles of each particle, calculate and store the kernel or its derivatives for this pairwise interaction.

The technique of calculating the pairwise interactions between nearest neighboring particles is carried out once for each time step, and can improve the efficiency of the simulation dramatically as there is no need to calculate the kernel or its derivative recursively for every particle in every summation process. The concept of pairwise interaction also saves storage and computational effort. Moreover, this treatment can greatly simplify the parallel computation of the entire SPH implementation.

Solid Boundary Treatment

There is no direct boundary condition in SPH simulations. For particles near the solid boundary, only particles inside the boundary contribute to the summation of the particle interaction, and no contribution comes from outside since there are no particles. This one-sided contribution does not lead to a correct solution, because on the solid surface, although the velocity is zero, other physical quantities such as density are not necessarily zero.

In our work, virtual (or ghost) particles are used to treat the solid boundary condition. Two types of virtual particles are used. One type fills in the boundary region and is similar to that used by Randles and Libersky (1996). The other type is located right on the solid boundary and is similar to that used by Monaghan (1994).

The Libersky type of particle is constructed in the following way. For a certain field particle i within the smoothing area λh_i inside the boundary, a virtual particle is placed symmetrically on the outside of the boundary. These virtual particles have the same density and pressure as the mirror counterpart of field particles but opposite velocity. However, the virtual particles of Libersky type are not enough to prevent the field particles from penetrating the solid boundary. This leads to the application of the virtual particles of the Monaghan type, which are used to produce a sufficient repulsive boundary force when a particle approaches the solid boundary.

These two types of virtual particles are specially marked for contribution in the later summation on the field particles. The virtual particle does not evolve its parameters, because in the next time step, another set of virtual particles of Libersky type will be produced in a similar way, while the virtual particles of Monaghan type always stay at the original location. Our numerical tests have shown that the combination of these two types of virtual particles not only improves the accuracy of the SPH approximation, but also prevents the nonphysical particle penetration into the boundary.

9.2.4 SPH Code Structure

Based on the basic SPH methodology and different algorithms for various numerical aspects of SPH, an explicit code has been implemented. Except for some high-level computing techniques such as parallel computing and message passing interface (MPI), the structured code is written from top down with different modules for various purposes, and is programmed for software reusability, extendibility, and maintainability. The computational procedure is as follows:

1. Initialization module, which includes the input of the initial configuration of the computing geometry (dimensions and boundaries), discretization of the initial geometry with smoothing particles, material properties, and time step. This initial setup containing the internal particle generation module can also be replaced with external input files. It should be noted that energy and momentum should be conserved in particle generation to prevent unnecessary noise in the simulation results.
2. Main SPH processor, which contains the major modules in the SPH simulation, and is represented in the time integration module. As far as the time integration technique alone is concerned, standard methods such as Leap-Frog, predictor-corrector, and Runge-Kutta can be employed, which may have advantages and disadvantages. The following modules are included in the time integration process:
   - 2.1 Generating boundary (virtual or ghost) particles.
   - 2.2 Nearest neighbor particle searching (NNPS)—three NNPS techniques are provided in the code for different uses: direct search, linked-list search, and tree search methods.
   - 2.3 Calculating smoothing function (for the summation density approach) and its derivatives from the generated message of interaction particle pairs.
   - 2.4 Updating density if using the summation density approach (Equation 9.12).
   - 2.5 Calculating the artificial viscous force.
   - 2.6 Calculating the internal forces due to the particle interactions according to Equations 9.17 and 9.22, respectively. It should be noted that the particle pressure is derived from the density and energy from the equation of state.
   - 2.7 Calculating the external forces if necessary. The boundary and interface forces can also be regarded as external forces.
   - 2.8 Calculating the change of momentum, energy, and density (if using the continuity density approach).
   - 2.9 Updating the smoothing length for the next time step.
   - 2.10 Updating particle momentum, energy, and density; updating particle position and velocity; checking the conservation of the energy and momentum.
3. Output module. When the time step reaches a prescribed one or at some interval, the resultant information is output and recorded for later analysis or postprocessing.

The flowchart of the present SPH code is schematically shown in Figure 9.3.

9.2.5 Applications

A series of numerical tests have been carried out to test the ability and efficiency of the presented SPH method and the code implemented in simulating fluid dynamic problems. As can be seen from the previous discussions, the presented SPH method is able to simulate

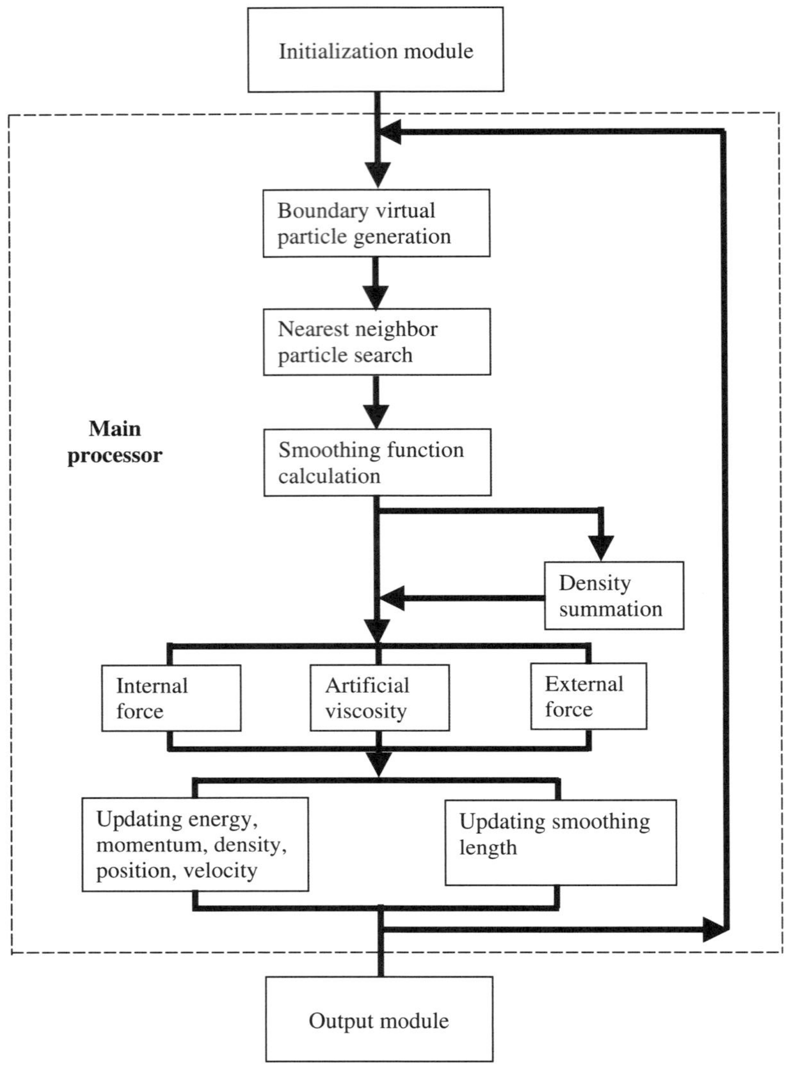

FIGURE 9.3
Flowchart of the present SPH code.

different flow scenarios, such as inviscid or viscous flows, compressible or incompressible flows. Following are some example applications to incompressible flows, free surface flows, and explosion simulation.

Applications to Incompressible Flows

In FDM, incompressible flows have been widely studied because of their special features. In our SPH simulation of incompressible flows, the above-described artificial compressibility technique is employed to model the incompressible fluid as slightly compressible by selecting a proper equation of state. Three simple simulation cases, Poiseuille flow, Couette flow, and shear driven cavity flow, are simulated using our SPH code. The corresponding numerical results are presented below. In these numerical examples, Equation 9.35 is used to model the compressibility of water.

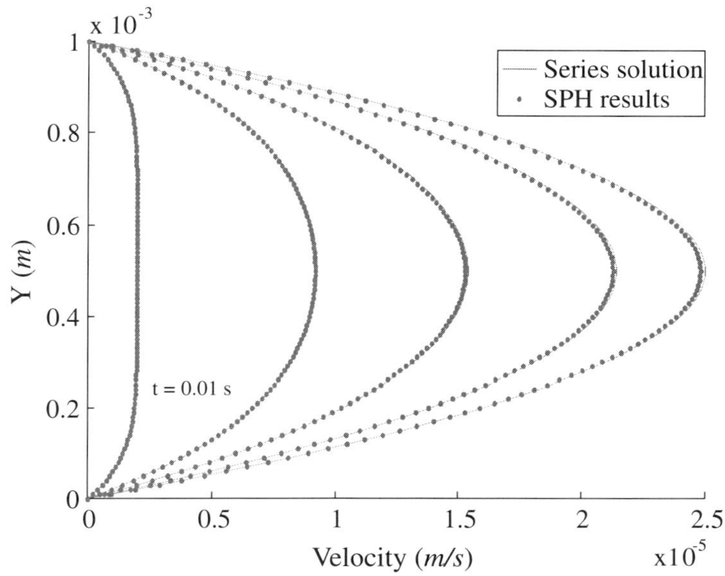

FIGURE 9.4
Velocity profiles for Poiseuille flow.

Example 9.1 Poiseuille Flow

Poiseuille flow is an often-used benchmarking CFD problem. It is a steady flow between two stationary infinite plates placed at $y = 0$ and $y = l$. The originally stationary fluid is driven by some body force F, gradually flows between the two plates, and finally arrives at an equilibrium steady flow state. In our simulation, the parameters are as follows:

Spacing of the plates: $l = 10^{-3}$ m
Kinetic viscosity of the fluid: $\nu = 10^{-6}$ m^2/s
Density of the fluid: $\rho = 10^3$ kg/m^3
Driven body force: $F = 2 \times 10^{-4}$ m/s^2
Peak fluid velocity: $\mathbf{v}_0 = 2.5 \times 10^{-5}$ m/s, which corresponds to a Reynolds number of Re $= 2.5 \times 10^{-2}$

In our SPH simulation, a total of 101 particles are placed in the y direction. Figure 9.4 shows the comparison between the velocity profiles obtained by the SPH method and those by series solution (Morris et al., 1997; Morris and Monaghan, 1997) at $t = 0.01$ s, 0.05 s, 0.1 s, 0.2 s, and the final steady state $t = \infty$. It is found that they are in good agreement; the difference is within 0.5%. Our SPH results are more accurate than series solution has obtained. This results from the different physical viscosity model, smoothing length, and solid boundary treatment.

Example 9.2 Couette Flow

Couette flow is another often-used benchmarking CFD problem. It is a flow between two initially stationary infinite plates placed at $y = 0$ and $y = l$ when the upper plate moves at a certain constant velocity $\mathbf{v}_0$. In our computation, we use $l = 10^{-3}$ m, $\nu = 10^{-6}$ m^2/s, $\rho = 10^3$ kg/m^3, $\mathbf{v}_0 = 2.5 \times 10^{-5}$ m/s, and the corresponding Reynolds number is Re $= 2.5 \times 10^{-2}$. Again, 101 particles are placed in the span direction. Comparison between the

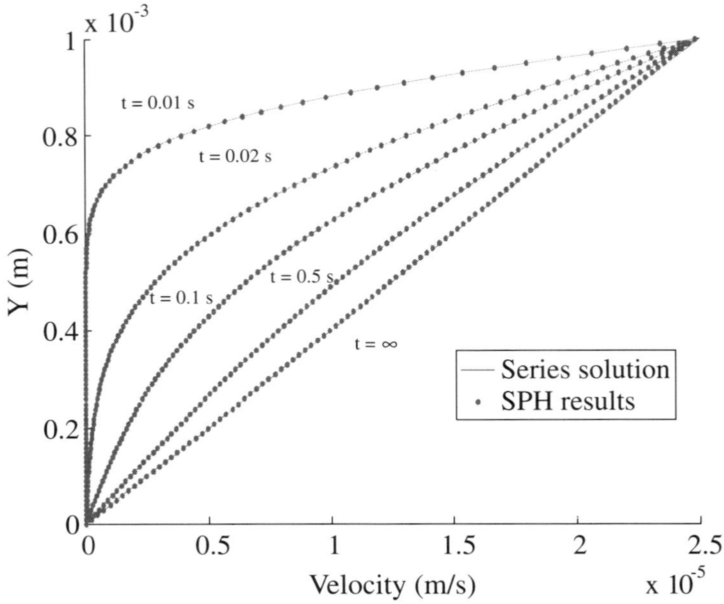

FIGURE 9.5
Velocity profiles for Couette flow.

velocity profiles obtained by the SPH method and those by series solution (Morris et al., 1997; Morris and Monaghan, 1997) at t = 0.01 s, 0.05 s, 0.1 s, 0.2 s, and the steady state $t = \infty$ is shown in Figure 9.5. It is found that they are in good agreement, and the difference is within 0.8%.

Example 9.3 Shear-Driven Cavity Problem

The classic shear-driven cavity problem is a flow within a closed square with the topside moving at a constant velocity V_{top} while the other three sides stay stationary. The flow will reach an equilibrium state, which behaves in a recirculation pattern. This is also a popular and critical benchmarking problem.

In our SPH simulation, a flow of a Reynolds number of 10 is considered. A total of $41 \times 41 = 1681$ field particles are initially placed in the square region. SPH results are compared with those by FDM with the same density of grids. Figure 9.6 shows the dimensionless vertical velocity profile along the horizontal centerline. Figure 9.7 shows the dimensionless horizontal velocity profile along the vertical centerline. It can be seen from Figures 9.6 and 9.7 that the results from the present method and those from FDM are comparable, while the SPH method slightly underpredicts the values compared to FDM.

Figure 9.8 shows the velocity distribution in the entire cavity computed using the present SPH code.

Example 9.4 Free Surface Flows

The study of free surface flows is very important in many industrial applications. Special treatment is necessary to deal with the arbitrary free surface. We have simulated a water discharge problem with the gate partly or fully opened. The viscous effect of the water is neglected here.

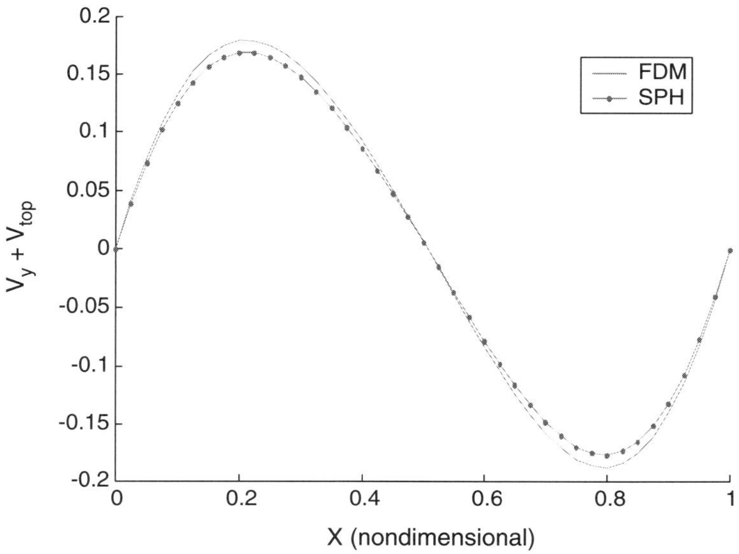

FIGURE 9.6
Dimensionless vertical velocity along the horizontal centerline.

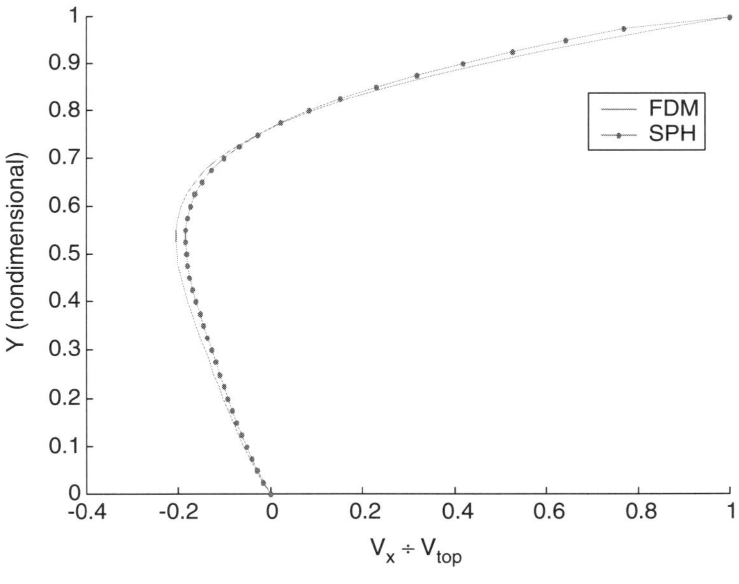

FIGURE 9.7
Dimensionless horizontal velocity along the vertical centerline.

Figure 9.9 shows the particle distribution of water discharge at three representative instants, after the gate is opened by 12% at the bottom of the gate. It can be seen that the particles distribute in an orderly fashion with the flow of water before the gate. The streamline is obvious. Water particles eject off the gate bottom due to the pressure force, splash high outside due the momentum, and finally fall to the ground due to gravity. Near the region of the gate bottom, the water flows rather evenly with potential energy

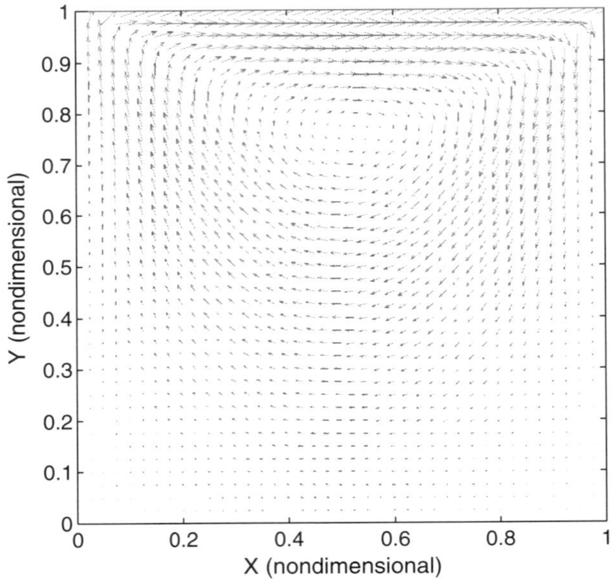

FIGURE 9.8
Velocity distributions for the shear-driven cavity problem.

TABLE 9.1
Water Level H and Surge Front S Computed Using SPH Methods

Time	H_{exp}	H_m	H_p	S_{exp}	S_m	S_p
0.71	0.90	0.90	0.90	1.33	1.56	1.45
1.39	0.76	0.75	0.75	2.25	2.50	2.38
2.10	0.57	0.56	0.56	3.22	3.75	3.50
3.20	0.32	0.37	0.35	4.80	5.00	4.88

Note: Subscript exp: experimental results; subscript p: present SPH code; subscript m: results by Monaghan (1994).

transformed into kinetic energy. Because the water splashes high outside the gate and then falls to the ground, a cavity occurs during the course of the flow.

The case of the water discharge with a fully opened gate is similar to the problem of a dam collapsing suddenly. In this case, the initial water level H_0 and the initial surge front S_0 are 25 m. The water particles flow in an orderly fashion forward with increasing surge front S and decreasing water level H. The numerical results from the present work (denoted by the subscript p), the experimental data (denoted by the subscript exp), and the results by Monaghan (denoted by the subscript m) are compared and shown in Table 9.1. The surge front S and the water level H at different instants are nondimensionalized by the initial water level H_0 while the time is nondimensionalized by $\sqrt{H_0/g}$, where g is the gravity constant. Our results, especially the surge fronts, are more accurate than the results Monaghan obtained. This is due to the contribution of the virtual particles of Libersky type in the calculation process, which increases the drag force for the particles near the bottom.

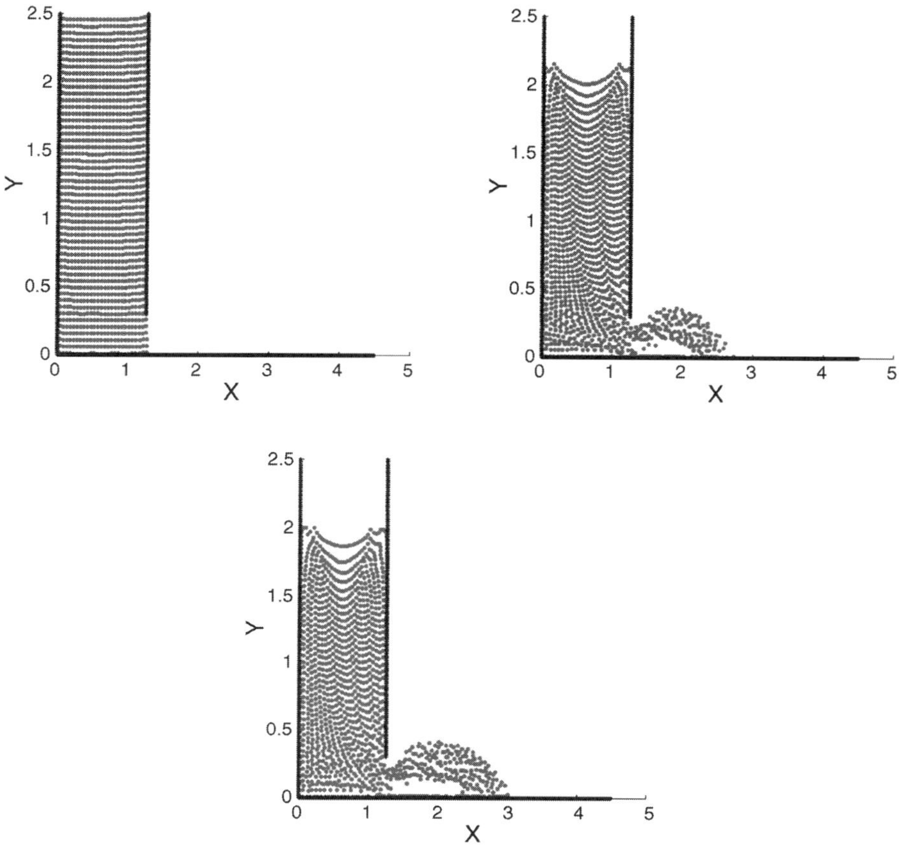

FIGURE 9.9
Particle distribution at three representative instants. Boundary virtual particle is not shown.

Applications to Explosion Simulations

An explosion consists of a complicated sequence of energy-releasing processes and is difficult to simulate. In an explosion, especially an underwater explosion, there exist large deformations, large inhomogeneities, and moving material interfaces. In light of such factors, some numerical simulations use two grids, one Eulerian grid for treating large deformations and inhomogeneities, the other Lagrangian grid for tracking different material interfaces. In this section, the SPH simulation of explosion is performed, and two numerical cases are presented. The first case is an explosion problem in a vacuum; the second is a water mitigation problem with moving gas/water/air interfaces.

Example 9.5 Explosion in Vacuum

In the first example, a clump of cylindrical explosive with high energy explodes in a vacuum. The initial total energy and density of explosive are 4.29×10^6 J/kg and 1630 kg/m^3, respectively, while the initial radius is 0.1 m. Figure 9.10 shows the density and pressure profiles in the radial direction at four different, randomly selected instances using the SPH method. The results are compared with those computed using the commercial software MSC/DYTRAN (1997), which is coded using FVM for fluid flow with explicit time marching.

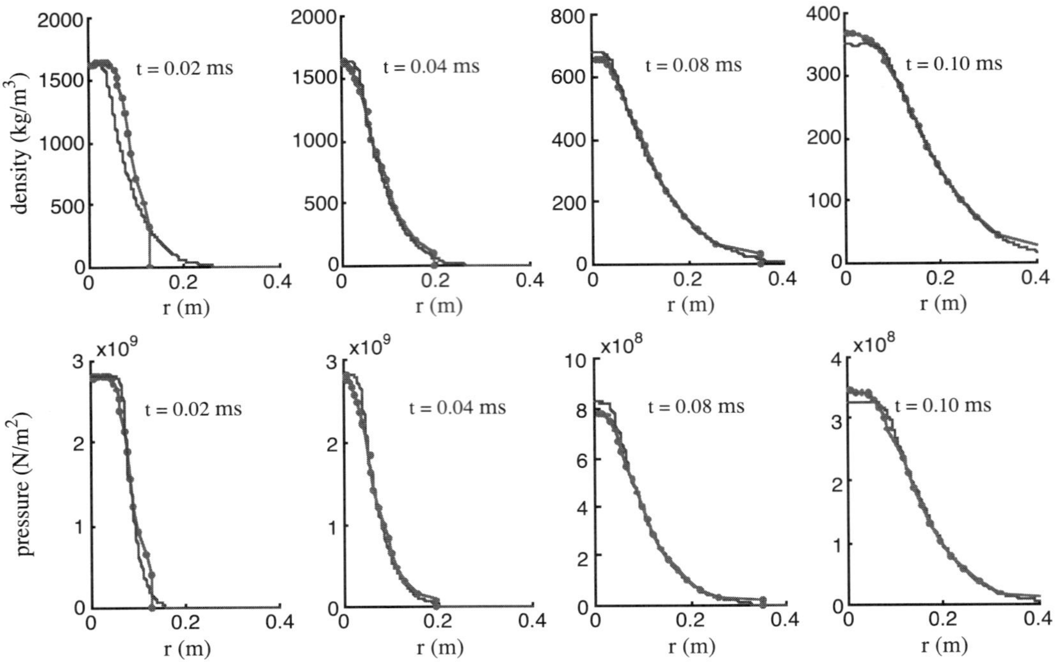

FIGURE 9.10
Density and pressure profiles for the 2D explosion problem at t = 0.02, 0.04, 0.08, and 0.10 ms. Lines with dots represent the results by SPH; other lines represent the results by DYTRAN.

The good agreement between the results by the two approaches suggests that the presented SPH method can simulate the explosion process well.

We have also checked on the computational efficiency of our SPH code using this example in comparison with MSC/DYTRAN. We were not be able to have our SPH code and MSC/DYTRAN run on a common computer for some trivial reasons, and therefore we cannot provide quantitative results for this comparison study. Nevertheless, we report here some indicative findings. The machine running MSC/DYTRAN was an SGI Origin 2000, and the machine running our SPH code was an HP workstation. The clock of these two machines is roughly the same. To run Example 9.5 to 0.1 ms, the MSC/DYTRAN code took about three times more wall-time compared with our SPH code.

Example 9.6 Simulation of Explosion Mitigated by Water

In the example, a more complicated case involving underwater shock is considered. To reduce potential damage resulting from an accidental explosion, high-energy explosives are sometimes stored with a layer of water cover, as shown in Figure 9.11. Outside the water is air. In the case of an accidental explosion, the water layer covering the explosive may mitigate shock pressure greatly. At the beginning of the simulation, the water and the air are all at atmospheric condition. The simulation starts at the initial stage when the gas globe is surrounded by water. It is assumed that the energy contained in the gas globe is equal to that in the explosive charge. As the explosive charge detonates in water, it will drive a shock wave into the surrounding water. Figure 9.12 shows the density and pressure contours. With the progress of the explosion, the produced high-pressure gas pushes water to the outside and tends to occupy more space, while the water layer becomes thinner.

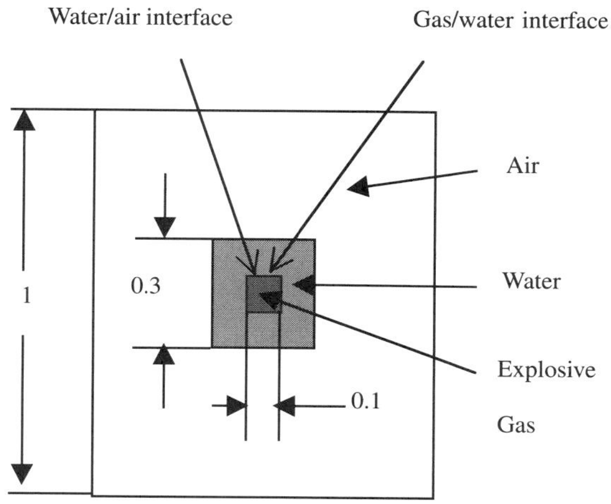

FIGURE 9.11
Initial explosive/water/air configuration of materials. The explosive is wrapped by water in a confined square space of air.

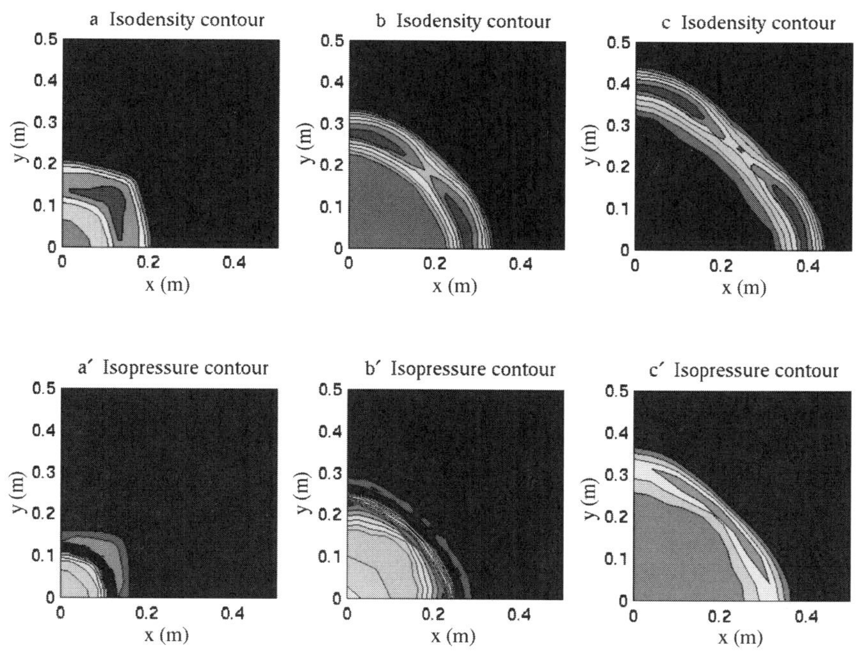

FIGURE 9.12
Density and pressure contour at t = 0.09, 0.21, and 0.30 ms in a quarter of the computational domain for the water mitigation problem, after the explosive detonated in water in a confined square space of air.

The gas/water/air interfaces and the latter reflection waves can be seen clearly either from the density or pressure contours. The density around the four corners gradually becomes sparser. The final penetration of particles of different materials will first occur here, which clearly shows that particles of different materials should first mix near the place where the interactions between the same kinds of particles are weakest.

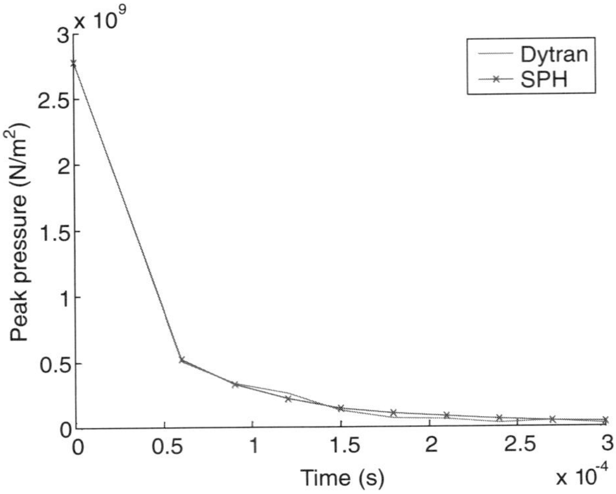

FIGURE 9.13
Time history of peak shock pressure after the detonation of the explosive wrapped by water in a confined square space of air.

Figure 9.13 gives the peak shock pressure curve vs. time obtained by the SPH method. It nearly coincides with the peak shock pressure curve obtained by MSC/DYTRAN. As the simulation by the SPH method and that by MSC/DYTRAN start at the same initial condition, the original pressures for the two simulations should be the same. Later, as time elapses, the shock wave moves farther from the center of the original explosive charge, and the peak shock pressure gradually decreases in an exponential fashion. The peak shock pressure obtained by MSC/DYTRAN seems to change more slowly than the peak shock pressure obtained by the SPH method, while as time goes on, the two curves gradually become closer. This simulation shows that the presented SPH method can provide a good prediction of the peak shock pressure for an underwater explosion both qualitatively and quantitatively.

9.2.6 Remarks

SPH has a number of advantages over traditional numerical methods, and is an effective technique for solving fluid dynamic problems. It is very effective, especially for problems that are difficult to simulate by traditional numerical methods. This section presents an SPH method that is extended to general dynamic fluid flows. This modified SPH method can simulate different flow scenarios, such as inviscid or viscous flows, compressible or incompressible flows. While the traditional artificial viscosity is used to model the inviscid flow problems in resolving shock waves, an SPH expression of physical viscosity constructed from the viscous shear force is employed for viscous flow problems. For incompressible flow problems, the concept of artificial compressibility is applied by selecting a special equation of state to model the incompressible fluid as quasi-incompressible. An adaptive, optimal smoothing length model is proposed to suit the needs of flows with large inhomogeneities, which can also be used for general flow problems. A technique is introduced in this section that directly calculates the interactions between every pair of neighboring particles, and thus improves the efficiency significantly. Extra virtual particles are used to treat the solid boundary condition. This treatment employs virtual particles of Monaghan type and Libersky type, which can prevent field particles from penetrating outside the boundary and can also improve the accuracy.

A series of numerical tests have been carried out for different dynamic flow problems. For incompressible flows such as Poiseuille flow, Couette flow, and the shear driven cavity flow, the presented SPH method can obtain satisfactory results. The advantages of the SPH method in treating free surface flows can be clearly seen in the simulation of the water discharge problem with a partly opened gate and the dam collapse. To test its ability to simulate problems with large deformations and large inhomogeneities, a 2D explosion problem in vacuum and a water mitigation problem are simulated. Compared with a grid-based method, this modified SPH method can successfully simulate such problems at reasonable accuracy with less computational effort. It can be concluded that SPH with proper modifications is an effective alternative to traditional numerical methods for dynamic flow applications.

9.3 Local Petrov–Galerkin Method

In Chapter 7 we introduced the MLPG method originally proposed by Atluri and Zhu (1998) for 2D solid mechanics problems, and demonstrated the advantages of this method. MLPG has also been implemented by Lin and Atluri (2001) and Liu, Wu, and Gu (2001) for solving incompressible Navier–Stokes equations of different forms. The MLPG is based on a local weak form integrated over a local quadrature domain, which can be of simple geometry, such as a sphere, cube, or ellipsoid in 3D (or circles, rectangles, ellipses in 2D). The field variables at any point in the problem domain are approximated using MLS approximation with the field nodes in the support domain of the point. This chapter introduces an MLPG formulation and its application to CFD problems. The formulation introduced in this chapter is based on the work done by Liu, Wu, and Gu (2001).

9.3.1 MLPG Formulation

The MLPG procedure used in this section is as follows.

1. A local weak form of integrated weighted residual is used over a local quadrature domain Ω_Q of a node. The local quadrature domain Ω_Q is conveniently taken to be a simple shape, such as a cube or sphere (for 3D cases) or a square or circle (for 2D cases), centered at the node. We have found in Chapter 7 that the dimension of the quadrature domain should be 1.5 to 2.0 times the average nodal spacing.
2. The weak form is then discretized using MLS approximation (see Section 5.4). This is performed for each and every node in the entire problem domain, except those on the essential boundary. In computing the MLS shape functions, we use a quartic spline function as the weight function. In computing the derivatives of the approximated field variables, we use a "diffuse derivative" instead of the "full derivative," which is much more expensive to compute. The derivative of the MLS shape function is approximated by

$$\Phi'^{T}(\mathbf{x}) = \mathbf{p}'(\mathbf{x})\mathbf{A}^{-1}(\mathbf{x})\mathbf{B}(\mathbf{x}) \tag{9.43}$$

As shown by Breitkopf et al. (1998) the diffuse derivative often provides a very good approximation of the derivative and it converges toward the full derivative of the approximation. Moreover, the diffuse derivative does not require the weight functions to be differentiable. Hence, we will adopt the diffuse derivative in our following calculations.

3. Because the MLPG method establishes equations node by node, we can deal with interior nodes and essential boundary nodes separately. In this section, we use different equations obtained directly using MLS approximation for essential boundary nodes.
4. The discretized equations are then solved numerically for field variables.

In applying the above procedure to CFD problems, one needs, of course, an iteration procedure to solve the equations, as they are basically nonlinear. The procedure can be the same as that used in FDM and FEM, where a large number of methods have been developed. However, we will show later that some of the existing iterative schemes do not work well with the present MLPG formulation. Modifications to these existing methods or even new methods that suit different MFree methods are required.

9.3.2 Numerical Integration in MLPG

In the MLPG method, there exist difficulties obtaining the exact numerical integration as mentioned in Chapters 7 and 8. The method of subdivision of quadrature domains suggested by Atluri et al. (1999b) works well when the number of subdivisions is sufficiently large, as investigated in Chapter 7. In this chapter, the method proposed by Liu, Wu, and Gu (2001) for complex domains is used. This method is based on the coordinate transform to transfer annular sectors into standard squares for the Gauss quadrature scheme. The advantages of this method are as follows:

1. The use of only four subdivisions can obtain very accurate results.
2. The method guarantees the exactitude of the double integral over an annular circle or sector.
3. The method is very easy to extend to domains of irregular shape, such as polygonal shapes produced by the intersection of the circle and the global boundaries.
4. The transformation is simple, cheap, and effective.

The following provides a detailed transformation procedure for the numerical integration in MFree methods based on nodal formulation.

If a nonrectangular quadrature domain is used for a node, we often use circles centered by the node as our integration cell because (1) a circle has no bias on directions and (2) it is very easy to use the formulation of the weight functions. The following proposes an effective procedure for accurate integration over circular quadrature domains.

Division of Quadrature Domain into Four Sectors

As demonstrated by examples in Chapters 6 and 7, numerical integration errors arise from complexities of the integrand. Further division of the quadrature domain into many small partitions yields better solutions, because it breaks the complex integrand into pieces of simpler ones. As a result, the computing effort will increase when the number of small partitions is very large. We should also divide the domain in a proper manner to achieve best accuracy with minimum divisions. This can only be done after a careful examination of the characteristics of the integrand.

The question now is why we need to divide the quadrature domain into four sections. Why not two or three sectors?

The reason lies in the weight function. From Figures 5.2 through 5.4, we know that all these weight functions continue at least up to the second derivatives at the center point ($x = 0$). However, the accurate (to machine accuracy) numerical integration of the weight function itself (forget about the derivatives of the shape functions) over any domain that contains $x = 0$ can never be done, no matter how small the domain that is used in the Gauss integration scheme.

To further emphasize this point, we take a 1D case as an example, where we try to integrate numerically the quartic spline weight function given by Equation 5.12. In this example, we assume $d_W = 0.5$, and the weight function is used for a node at $x_I = 2.0$. This weight function is plotted in Figure 9.14 from which we see a well-behaved function, and

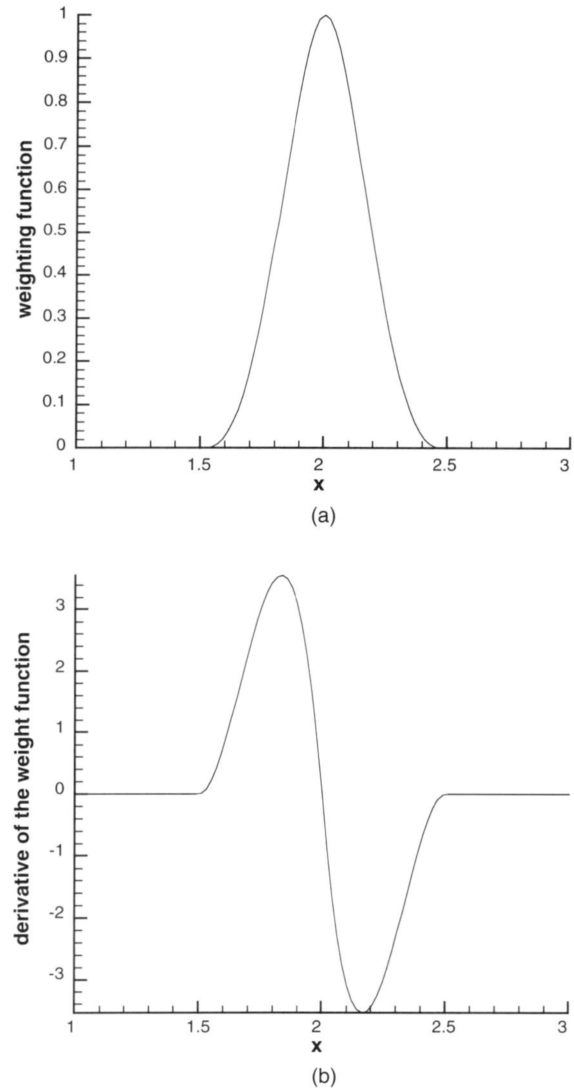

FIGURE 9.14
Quartic weight function for a node at $x_I = 2.0$ and its first derivative. (a) Weight function; (b) the first derivative of the weight function.

have no doubt about obtaining an accurate integration for this kind of function. However, from Equations 5.12 and 5.14, we know that $\bar{d} = |x_I - x|$, that is, the distance from node x_I to point x, where the absolute operation does not behave like a polynomial of any order in the domain that contains x_I as an interior point. Therefore, it is impossible to obtain the exact results no matter how many Gauss points have been used, if it is integrated numerically in one piece. This is because the Gauss quadrature can only give the exact result when applied to a polynomial. If, however, the integration is performed in two pieces: one integration over the right-hand side (RHS) of x_I and the another over the left-hand side (LHS), we can then easily obtain the exact numerical integration for this weight function.

To confirm and further enforce our point, we used the following three methods to perform the integration:

METHOD I Gauss quadrature is carried out over the entire weighted function domain.

METHOD II The weighted function domain is divided into two equal portions, and three Gauss points are used, respectively, on right- and left-hand side of point x_I.

METHOD III The weighted domain is divided evenly into 3, 4, 5, and 6 portions and three Gauss points are used in each portion.

The testing is done for the integration of the quartic spline weight function $\int_{\Omega_Q} \widehat{W}_I d\Omega$, where the dimension of the Ω_Q is 0.1. A total of 21 nodes are uniformly distributed in the weighted domain of [1.5. 2.5]. The dimension of the integration domain is two times the distance between two adjacent nodes. The maximum relative errors of integration at all the interior nodes are computed and listed in Table 9.2. This table shows that when x_I is located in the interior of a subdomain of integration (which happens when dividing the entire weighted domain into 1, 3, 5 partitions), the integration will never obtain accurate results no matter how many Gauss points are used. When x_I is one of the dividing points, the results are exact (to machine accuracy) and it requires only three Gauss quadrature points, as highlighted in Table 9.2. In such a case, two divisions are enough, and the total quadrature points are six.

In summary, to integrate a function that contains the quartic weight function in 1D space, two subdivisions divided by the point where the weight function peaks are required to obtain exact results of integration. Naturally, for 2D cases, four subdivisions corresponding to the four quadrants are required the integration is carried out in the Cartesian coordinate system.

TABLE 9.2
Errors of Integration Obtained Using Three Methods

Method	Gauss Points Used	Error of Integration
I	3	1.640E-01
	6	8.942E-03
	9	2.787E-02
	12	6.001E-04
II	**2 × 3**	**4.078E-08**
III	3 × 3	2.024E-03
	4 × 3	**3.049E-08**
	5 × 3	2.624E-04
	6 × 3	**2.994E-08**

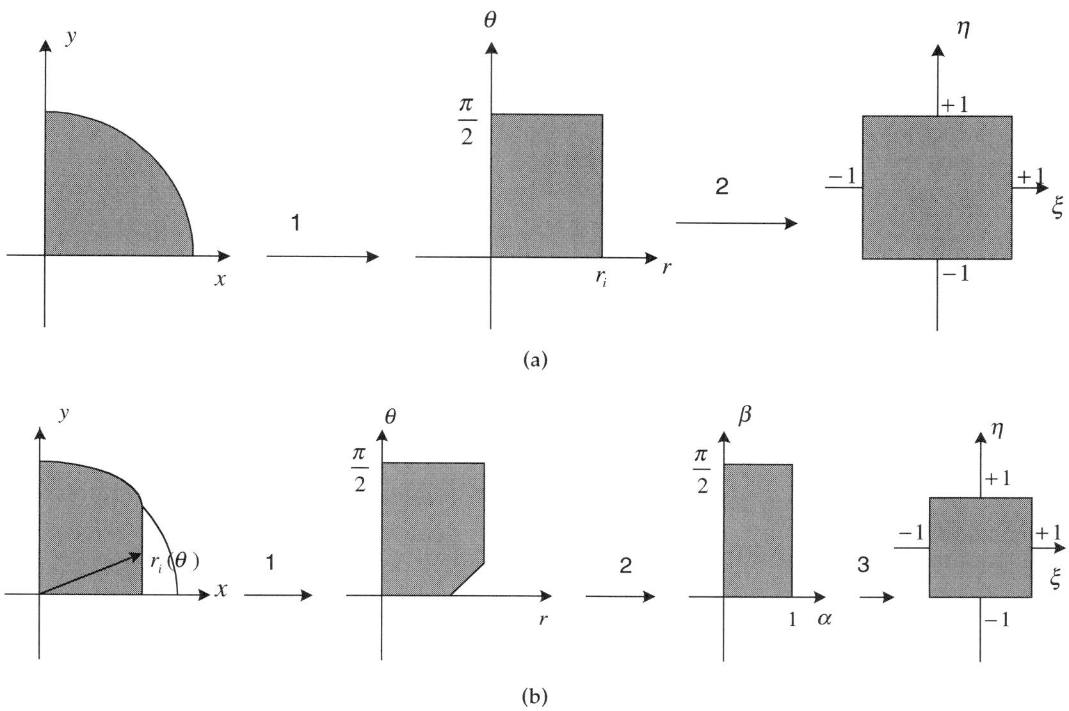

FIGURE 9.15
(a) Transformation of first quarter of a circular quadrature domain into a standard square. (b) Transformation of first quarter of a circular quadrature domain that intersects with the global boundaries.

Note that when the cubic spline weight function is used, the problem will be more complex and require more divisions, as the weight function itself is made of pieces. If there is a reason to require the use of the cubic spline function, because of the particular shape of the cubic function, we recommend use of the quartic weight function defined by Equation 5.15, which behaves very similarly to the cubic spline function but is one piece, as we can see from Figures 5.2 to 5.4.

If the polar coordinate system is used for the integration of 2D cases, subdivision is not required to obtain the exact solution for quartic weight functions. However, the Cartesian coordinate system is used far more often in practice, as trial functions are usually defined in the Cartesian coordinate system.

Interior Circle

If a circular quadrature domain Ω_Q is located entirely within the global domain Ω, and there is no intersection between the boundary of the quadrature domain Γ_Q and the global boundary Γ, the transformation for the first quarter of the circle consists of the following two steps, as shown in Figure 9.15.

STEP 1 Map the given annulus in the xy plane into a rectangle region in $r\theta$ plane using

$$\begin{cases} x = r\cos\theta \\ y = r\sin\theta \end{cases} \tag{9.44}$$

STEP 2 Maps the rectangular region into the standard square of $-1 \le \xi,\ \eta \le +1$.

$$\begin{cases} r = \frac{r_Q}{2}\xi + \frac{r_Q}{2} \\ \theta = \frac{\pi}{4}\eta + \frac{\pi}{4} \end{cases} \tag{9.45}$$

where r_Q is the radius of the quadrature domain. Combining these two changes of variables, it follows that

$$\begin{cases} x = \frac{r_Q}{2}(\xi+1)\cos\left[\frac{\pi}{4}(\eta+1)\right] \\ y = \frac{r_Q}{2}(\xi+1)\sin\left[\frac{\pi}{4}(\eta+1)\right] \end{cases} \tag{9.46}$$

This simple equation will transform the annulus directly into the standard square in a single operation.

The Jacobian resulting from this transformation can be expressed as a simple form:

$$|J| = \begin{vmatrix} \frac{\partial x}{\partial \xi} & \frac{\partial y}{\partial \xi} \\ \frac{\partial x}{\partial \eta} & \frac{\partial y}{\partial \eta} \end{vmatrix} = \frac{\pi}{4}\left(\frac{r_Q}{2}\right)^2(\xi+1) \tag{9.47}$$

Similarly, simple formulations for the second, third, and fourth quarter of the circle can be obtained. The general formulation for all four quadrants of the subdomain can be written concisely as

$$\begin{cases} x = \frac{r_Q}{2}(\xi+1)\cos\left[\frac{\pi}{4}(\eta+2\cdot k-1)\right] \\ y = \frac{r_Q}{2}(\xi+1)\sin\left[\frac{\pi}{4}(\eta+2\cdot k-1)\right] \end{cases} \tag{9.48}$$

where $k = 1, 2, 3$, and 4 represents the first, second, third, and fourth quadrant, respectively.

The corresponding Jacobian of the transformation for all four quarters are the same as Equation 9.47.

The double integral of a function can then be performed using the Gauss quadrature scheme over the squared domain.

$$\iint_{\Omega_Q} f(x,y)\,\mathrm{d}x\mathrm{d}y = \int_{-1}^{+1}\int_{-1}^{+1} g(\xi,\eta)\,\mathrm{d}\xi\mathrm{d}\eta \approx \sum_{1}^{n_g}\sum_{1}^{n_g} w_i w_j g(\xi_i,\eta_j) \tag{9.49}$$

where w_i ($i = 1, 2,\ldots$) are the Gauss weights at n_g Gauss points, and

$$g(\xi,\eta) = f(x(\xi,\eta),y(\xi,\eta))|J| \tag{9.50}$$

Note that the transformation requires very little additional computation because of the simplicity of these equations.

Intersection with the Global Boundaries

In MFree methods that are based on the local weak form, the other challenge is integration over the regions intersecting with the global boundaries, which produces quadrature domains of complex shape. This problem can be solved using coordinate transformation. We again take the first quarter, for example, to explain the transformation process. Figure 9.15 schematically shows the process of transformation. The mathematic operation of the transformation consists of the following three steps.

STEP 1 Map the given annulus in the xy plane into a rectangular region in $r\theta$ plane using Equation 9.44.

STEP 2 Map the polygonal shapes into a rectangle in the $\alpha\beta$ plane.

$$\begin{cases} r = r_Q(\beta)\alpha \\ \theta = \beta \end{cases} \tag{9.51}$$

where $0 \leq \alpha \leq 1$, $0 \leq \beta \leq \pi/2$. Note here that the radius of the quadrature domain r_Q is not constant any longer; it is a function of the variable β.

STEP 3 Map the rectangular region into the standard square of $-1 \leq \xi$, $\eta \leq +1$.

$$\begin{cases} \alpha = \frac{1}{2}\xi + \frac{1}{2} \\ \beta = \frac{\pi}{4}\eta + \frac{\pi}{4} \end{cases} \tag{9.52}$$

Combining the three steps, we can write

$$\begin{cases} x = \dfrac{r_Q\left[\frac{\pi}{4}(\eta+1)\right]}{2}(\xi+1)\cos\left[\frac{\pi}{4}(\eta+1)\right] \\ y = \dfrac{r_Q\left[\frac{\pi}{4}(\eta+1)\right]}{2}(\xi+1)\sin\left[\frac{\pi}{4}(\eta+1)\right] \end{cases} \tag{9.53}$$

The corresponding Jacobian in this case becomes

$$|J| = \begin{vmatrix} \dfrac{\partial x}{\partial \xi} & \dfrac{\partial y}{\partial \xi} \\ \dfrac{\partial x}{\partial \eta} & \dfrac{\partial y}{\partial \eta} \end{vmatrix} = \frac{\pi}{4}\left(\frac{r_Q\left[\frac{\pi}{4}(\eta+1)\right]}{2}\right)^2(\xi+1) \tag{9.54}$$

Similarly, formulation for the intersection region of the second, third, and fourth quarter of the circle with the global boundaries can be obtained in a similar manner. The double integral can be obtained using Equation 9.49, where the locations of the sampling points are calculated using

$$\begin{cases} x = \dfrac{r_Q\left[\frac{\pi}{4}(\eta+1)\right]}{2}(\xi+1)\cos\left[\frac{\pi}{4}(\eta+2\cdot k-1)\right] \\ y = \dfrac{r_Q\left[\frac{\pi}{4}(\eta+1)\right]}{2}(\xi+1)\sin\left[\frac{\pi}{4}(\eta+2\cdot k-1)\right] \end{cases} \tag{9.55}$$

where $k = 1, 2, 3$, and 4 represents the first, second, third, and fourth quadrant, respectively.

It is very clear here that the key is to calculate $r_Q[\pi/4(\eta + 1)]$, and it is easy to calculate as long as the intersection line of the subdomain and the global boundaries are known.

9.3.3 Governing Equations and Their Discretized Form

Using the MLPG method with the integration scheme described above, we are now set to solve incompressible Navier–Stokes equations. The 2D governing equations in terms of vorticity-stream function formulation are as follows:

$$\frac{\partial^2 \psi}{\partial x^2} + \frac{\partial^2 \psi}{\partial y^2} = \omega \tag{9.56}$$

$$u\frac{\partial \omega}{\partial x} + v\frac{\partial \omega}{\partial y} = \mathrm{Pr}\left(\frac{\partial^2 \omega}{\partial x^2} + \frac{\partial^2 \omega}{\partial y^2}\right) - \mathrm{Pr} \cdot \mathrm{Ra} \cdot \frac{\partial T}{\partial x} \tag{9.57}$$

$$u\frac{\partial T}{\partial x} + v\frac{\partial T}{\partial y} = \frac{\partial^2 T}{\partial x^2} + \frac{\partial^2 T}{\partial y^2} \tag{9.58}$$

where ω, ψ, T, Pr, and Ra are the vorticity, stream function, temperature, Prandtl number, and Rayleigh number; u, v are the components of velocity in the x and y directions, which can be calculated using

$$u = \frac{\partial \psi}{\partial y}, \qquad v = -\frac{\partial \psi}{\partial x} \tag{9.59}$$

It is clear that this problem is a nonlinear fluid mechanics problem with both Dirichlet (essential) and Neumann (natural) boundary conditions.

The boundary conditions can be written as

1. $x = 0, 0 \le y \le 1$:

$$\psi = 0, \qquad \frac{\partial \psi}{\partial x} = 0, \qquad T = 0$$

2. $x = 1, 0 \le y \le 1$:

$$\psi = 0, \qquad \frac{\partial \psi}{\partial x} = 0, \qquad T = 1$$

3. $y = 0, 0 \le x \le 1$:

$$\psi = 0, \qquad \frac{\partial \psi}{\partial y} = 0, \qquad \frac{\partial T}{\partial y} = 0$$

4. $y = 1, 0 \le x \le 1$:

$$\psi = 0, \qquad \frac{\partial \psi}{\partial y} = 0, \qquad \frac{\partial T}{\partial y} = 0$$

It is clear that, at each boundary, there are two conditions for ψ: one is of Dirichlet type ($\psi = 0$), the other is of Neumann type.

Following the MLPG procedure outlined in Section 9.3.1, consider a node, say, node I. A quadrature domain Ω_Q centered by node I is first defined. The field variables ψ, ω, T at any point $\mathbf{x}$ are approximated as

$$\psi(\mathbf{x}) = \sum_{i=1}^{n} \phi_i(\mathbf{x})\psi_i \quad (9.60)$$

$$\omega(\mathbf{x}) = \sum_{i=1}^{n} \phi_i(\mathbf{x})\omega_i \quad (9.61)$$

$$T(\mathbf{x}) = \sum_{i=1}^{n} \phi_i(\mathbf{x})T_i \quad (9.62)$$

where $\phi_i(\mathbf{x})$ is the MLS shape function for node i located in the support domain Ω_s centered by the point of interest $\mathbf{x}$, n is the number of the nodes in the support domain, and ψ_i, ω_i, T_i are the values of stream function, vorticity, and temperature at node i.

Using the local residual formulation and the strong form of system Equations 9.56 through 9.58, we have

$$\int_{\Omega_Q} \widehat{W}_I\left(\frac{\partial^2\psi}{\partial x^2} + \frac{\partial^2\psi}{\partial y^2} - \omega\right)\mathrm{d}\Omega = 0 \quad (9.63)$$

$$\int_{\Omega_Q} \widehat{W}_I\left(u\frac{\partial\omega}{\partial x} + v\frac{\partial\omega}{\partial y} - \Pr\left(\frac{\partial^2\omega}{\partial x^2} + \frac{\partial^2\omega}{\partial y^2}\right) + \Pr \cdot \mathrm{Ra} \cdot \frac{\partial T}{\partial x}\right)\mathrm{d}\Omega = 0 \quad (9.64)$$

$$\int_{\Omega_Q} \widehat{W}_I\left(u\frac{\partial T}{\partial x} + v\frac{\partial T}{\partial y} - \frac{\partial^2 T}{\partial x^2} - \frac{\partial^2 T}{\partial y^2}\right)\mathrm{d}\Omega = 0 \quad (9.65)$$

where $\widehat{W}_I$ is the weight function centered by node I. Any weight function listed in Equations 5.11 through 5.13 and 5.15 can be used. Substituting Equations 9.60 through 9.62 into Equations 9.63 through 9.65, we obtain

$$C_{ij}\psi_j - P_{ij}\psi_j = -A_{ij}\omega_j \quad (9.66)$$

$$B_{ijk}\psi_j\omega_k + \Pr \cdot C_{ij}\omega_j - \Pr \cdot P_{ij}\omega_j = -\Pr \cdot \mathrm{Ra} \cdot G_{ij}T_j \quad (9.67)$$

$$B_{ijk}\psi_j T_k + C_{ij}T_j - P_{ij}T_j = 0 \quad (9.68)$$

where

$$A_{ij} = \int_{\Omega_Q} \phi_j \widehat{W}_i \,\mathrm{d}\Omega \quad (9.69)$$

$$B_{ijk} = \int_{\Omega_Q} \left(\frac{\partial\phi_j}{\partial y}\frac{\partial\phi_k}{\partial x} - \frac{\partial\phi_j}{\partial x}\frac{\partial\phi_k}{\partial y}\right)\widehat{W}_i \,\mathrm{d}\Omega \quad (9.70)$$

$$C_{ij} = \int_{\Omega_Q} \left(\frac{\partial \phi_j}{\partial x} \frac{\partial \widehat{W}_i}{\partial x} + \frac{\partial \phi_j}{\partial y} \frac{\partial \widehat{W}_i}{\partial y} \right) \mathrm{d}\Omega \tag{9.71}$$

$$G_{ij} = \int_{\Omega_Q} \frac{\partial \phi_j}{\partial x} \widehat{W}_i \, \mathrm{d}\Omega \tag{9.72}$$

$$P_{ij} = \int_{\Gamma_Q} \frac{\partial \phi_j}{\partial n} \widehat{W}_i \, \mathrm{d}\Gamma \tag{9.73}$$

Note that, for different variables with different boundary conditions, P_{ij} will take different forms. Note also that for essential boundary conditions, we use Equations 9.60 through 9.62 directly to obtain the relationship of the field variables at the essential boundary nodes with those at the nodes that fall into the support domains of these boundary nodes.

9.3.4 Boundary Condition for Vorticity

It can be seen from the stream function equations that the implementation of vorticity condition is equivalent to the approximation of the second-order derivatives of the stream function on the boundary. There are two methods to give the discrete boundary condition for vorticity.

METHOD I This method is based on the Taylor series expansion. Suppose w is the node on the wall, i is the interior node nearest to the w, and the distance between w and i is l. Taylor series expansion gives

$$\psi_i = \psi_w + l\left(\frac{\partial \psi}{\partial l}\right)_w + \frac{1}{2} l^2 \left(\frac{\partial^2 \psi}{\partial l^2}\right)_w + o(l^3) \tag{9.74}$$

Neglecting the terms higher than second order, and introducing the Neumann boundary condition for ψ, we have

$$\omega_w = \left(\frac{\partial^2 \psi}{\partial l^2}\right)_w = \frac{2}{l^2}(\psi_i - \psi_w) \tag{9.75}$$

METHOD II The vorticity condition can be derived from the stream function equation.

$$\frac{\partial^2 \psi}{\partial x^2} + \frac{\partial^2 \psi}{\partial y^2} = \omega \tag{9.76}$$

Suppose that w is the point at the wall, i is the interior point nearest to the w, and the distance between w and i is l. Equation 9.76 becomes

$$\omega_w = \left(\frac{\partial^2 \psi}{\partial l^2}\right)_w \tag{9.77}$$

The derivative of ω at the wall is

$$\left(\frac{\partial \omega}{\partial l}\right)_w = \left(\frac{\partial^3 \psi}{\partial l^3}\right)_w \tag{9.78}$$

where

$$\left(\frac{\partial \omega}{\partial l}\right)_w = \frac{\omega_i - \omega_w}{l}, \quad \text{and} \quad \left(\frac{\partial^3 \psi}{\partial l^3}\right)_w$$

can be given by Taylor series expansion. Neglecting the terms higher than second order introducing the Neumann boundary condition for ψ, and using Equation 9.76, we have

$$\left(\frac{\partial^3 \psi}{\partial l^3}\right)_w = \frac{6}{l^3}\left(\psi_i - \psi_w + \frac{l^2}{2}\omega_w\right) \tag{9.79}$$

Therefore, the boundary condition for ω_w is actually an essential condition, and

$$\omega_w = \frac{3}{l^2}(\psi_i - \psi_w) - \frac{1}{2}\omega_i \tag{9.80}$$

Method II has been found to provide more accurate results. Therefore, in the following two sections, we use method II in the implementation of MLPG and LRPIM.

9.3.5 Numerical Results and Discussion

Because Equations 9.66 through 9.68 are three sets of nonlinear equations coupled together, they can be solved using alternate iteration method. At each iteration, the resultant linear algebraic equations are solved by an amended Gaussian elimination procedure to obtain a better approximation for the next iteration. The iteration stops when the steady-state resolution is reached and the results are converged, at which the differences of field function values between the nth and the $(n + 1)$th iterations, $\Delta\psi$, $\Delta\omega$, and ΔT are smaller than a predefined threshold. In our later examples, when the maximum relative values of these three field variables $\frac{\|\Delta\psi\|}{\|\psi^{n+1}\|}$, $\frac{\|\Delta\omega\|}{\|\omega^{n+1}\|}$, $\frac{\|\Delta T\|}{\|T^{n+1}\|}$ are less than 10^{-3}, the results are considered convergent.

Example 9.7 Natural Convection in a Square Cavity Problem

We consider now a natural convection in a square cavity. The cavity flow is a standard test case in CFD for validating numerical methods. In the case here, the cavity flow is wall driven as well as temperature driven.

The geometry and boundary condition of the problem are given in Figure 9.16. The domain of the cavity is 1×1. Flows with a Prandtl number of 0.71 and Rayleigh numbers of 10^3, 10^4, and 10^5 are investigated. To compare the performance of the present method with the "true" solution and that of FDM, the following quantities are calculated:

$|\psi_{max}|$, the maximum absolute value of the stream function
u_{max}, the maximum horizontal velocity on the vertical midplane of the cavity
v_{max}, the maximum vertical velocity on the horizontal midplane of the cavity
Nu_{max}, the maximum value of the local Nusselt number on the boundary at $x = 0$
Nu_{min}, the minimum value of the local Nusselt number on the boundary at $x = 0$

Because there is no analytical solution for the problem, the benchmark numerical solution of Davis (1983) is adopted as the "true" solution to this cavity problem. The Davis solution

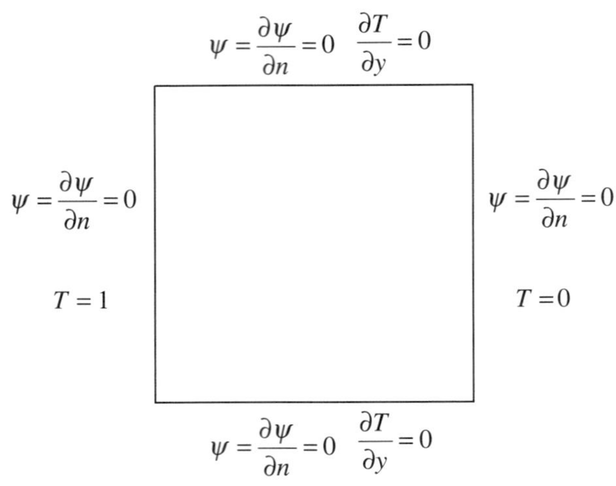

FIGURE 9.16
Schematic of natural convection in square cavity flow problem.

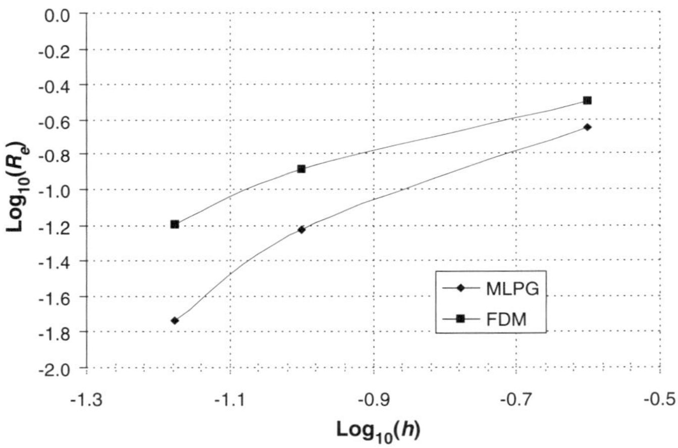

FIGURE 9.17
Comparison of convergence rate of MLPG and FDM. Shear- and temperature-driven cavity flow problem.

is obtained by using an extrapolation strategy from finite difference results of the different mesh sizes. Thus, it can be expected that the solution is more accurate than the single-mesh finite difference results.

We compare the rates of convergence of the present method and FDM in terms of an "energy norm" r_e, which is defined as the relative error norm of $|\psi_{max}|$ for the case of Ra = 10^3. The nodal distributions used in the compilation are type I: uniformly distributed node set, and type II: arbitrarily scattered nodes set, as shown in Figure 9.18. Figure 9.17 shows the results of convergence rates of both the present MLPG and FDM, where h is the nodal distance (both methods use the uniform nodal distribution). It is very clear that MLPG method is much more accurate than FDM if h is at same level. Table 9.3 lists the numerical results for two different sets of nodes for Rayleigh numbers of 10^3 and 10^4, and 10^5, respectively. It can be observed that for all the sets of nodes, the results of the MLPG

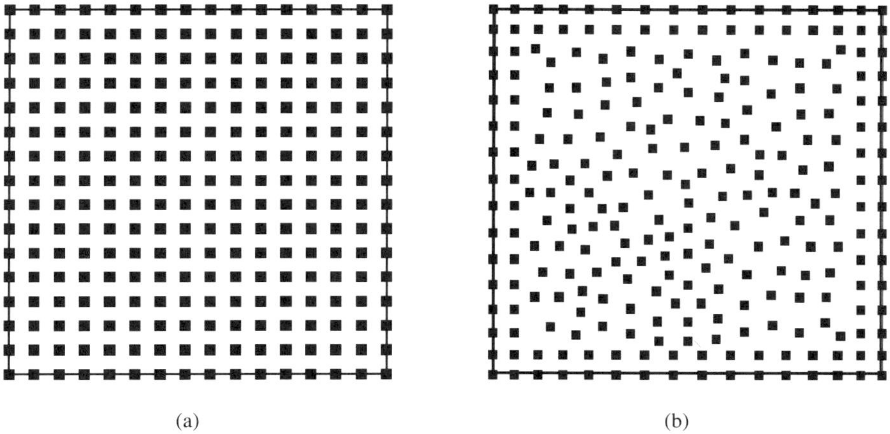

FIGURE 9.18
Two types of nodal distributions. (a) Type I: uniformly distributed nodes set (256 nodes). (b) Type II: arbitrarily scattered nodes set (257 nodes).

TABLE 9.3
Comparison of Numerical Results Obtained by MLPG for the Natural Convection in a Square Cavity Problem for Different Rayleigh Numbers

Ra		$\lvert\psi_{max}\rvert$	u_{max}	v_{max}	Nu_{max}	Nu_{min}
10^3	TYPE I (256 nodes)	1.161	3.566	3.593	1.521	0.684
	TYPE II (257 nodes)	1.124	3.525	3.580	1.527	0.691
	Davis	1.174	3.649	3.697	1.505	0.692
10^4	TYPE I (256 nodes)	5.391	16.726	18.244	3.479	0.582
	TYPE II (257 nodes)	5.248	16.681	18.960	3.608	0.621
	Davis	5.071	16.178	19.617	3.528	0.586
10^5	TYPE I (256 nodes)	10.317	36.455	58.218	8.036	0.647
	TYPE II (257 nodes)	10.818	41.658	58.256	7.976	0.840
	Davis	9.612	34.730	68.590	7.717	0.729

method agree well with the benchmark solution using very few nodes. The streamlines and isotherms of Ra = 10^5 are shown in Figure 9.19.

9.3.6 Remarks

In this section, the MLPG method is adopted and formulated with some modifications to simulate CFD problems. The simulation of natural convection in a square cavity is used as an example to test the present formulation of the MLPG method, and it is found that the method works well with some distributions of nodes. The accuracy achieved for the converged results is much higher than that obtained by FDM. One of the major advantages of the MLPG method is that the nodal distribution in the problem domain can be arbitrary.

We found, however, some problems in implementing the present MLPG. These problems are listed as follows:

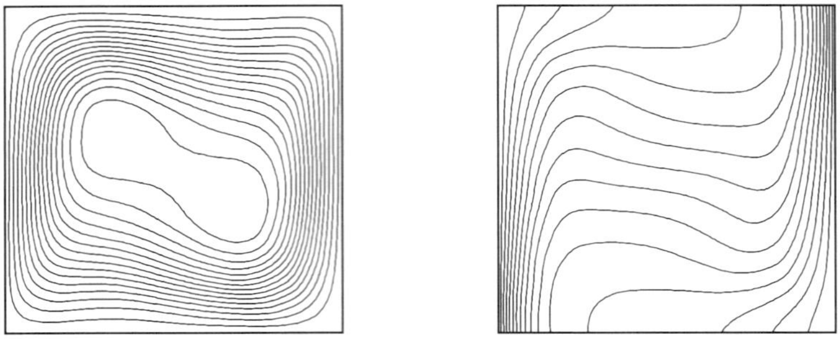

FIGURE 9.19
Streamlines and isotherms for Ra = 10^5. Shear- and temperature-driven cavity flow problem. (a) Streamlines; (b) isotherms.

1. We encountered problems if we used a support domain that is too large and that contains too many nodes. For regular nodal distributions, the size of the support domain can be as large as 6.5 times the nodal spacing, and the solution still converges. For arbitrary nodal distributions, however, the size of the support domain has to be less than 3.0 times the nodal spacing. Otherwise, the iteration may diverge. This finding may imply that there may be some kind of overdiffusing when too many nodes are used in MLS approximation.
2. High computational cost is a noticeable problem for the present MLPG, as the iteration must be used to obtain the final solution. Because the local weak form is used, the resulting linear system matrix may not have the diagonal domination feature. Therefore, many iteration schemes that are often used in FDM (such as the Jacobi and Gauss-Seidel iteration schemes) may not guarantee convergence of the solution. One has to seek a direct solver, which is very expensive if many iteration steps are needed. A more efficient algorithm is necessary.

Lin and Atluri (2001) introduced upwinding schemes into the MLPG method. In their work, formulation based on the primitive variables is used. They have applied the Newtonian iteration method, a more expensive iteration scheme, in implementing the MLPG. However, the initial solution guess can be a problem as a good initial estimate is very important for the Newtonian iteration to achieve the convergent solution. In addition, for high Reynolds number flows, Newtonian iteration seems not to work.

The next section examines the local radial PIM (LRPIM) approach for CFD problems.

9.4 Local Radial Point Interpolation Method

In the MLPG method described in the previous section, the MLS approximation is used to create the shape functions that lack the delta function property. This makes it more complex to implement essential boundary conditions. The natural progression of the research at our group is then to ask what if we use PIM. The first stage of our next study phase is to use the local radial point interpolation method (LRPIM) for CFD problems. Below is the report on this venture.

In Chapter 8, we demonstrated that the LRPIM method works well for solid mechanics problems. The advantages of this method include (1) the shape function has the delta function property, so that the essential boundary conditions can be implemented as easily as in FEM, and (2) stability for randomly scattered nodal distribution, which demonstrates that the RBF interpolation scheme is effective for interpolation of arbitrarily scattered data points.

In this section, the LRPIM is adopted to simulate the impressible fluid problems. Natural convection in closed domains, a typical benchmarking problem in fluid mechanics, is selected as a sample problem. To show the LRPIM MFree method is applicable to irregular geometry, the computation is carried for problems of different geometries that coincide as well as do not coincide with the Cartesian coordinate axis. Uniformly distributed nodes and randomly scattered nodes are also used in the computation.

In this study, we decided to use MQ radial basis functions, as given in Table 5.3, and our study started with the investigation of the effects of the shape parameters. We found that α_C work well in a very wide range from 0.01 to 13, and $q = 1.03$ is one of the good choices. We use $\alpha_C = 7.75$ and $q = 1.03$. Support domains are chosen so that each contains about 20 to 40 field nodes.

9.4.1 LRPIM Formulation

The formulation of LRPIM is very much the same as in Chapter 7 and Section 9.3. Most of the equations in Section 9.3 are valid, except that RPIM shape functions (see Chapter 5) are used. In this study, we made further effort to improve the efficiency of the LRPIM, as the efficiency is now a crucial issue for solving nonlinear problems that require many iterations. As LRPIM requires performing a large number of numerical integrations that consume a large amount of CPU time, we suggest improving computational efficiency in the following ways.

Numerical Integration in LRPIM

The integration procedures discussed in Section 9.3.2 are used here to divide the quadrature domain into four portions, and to transform them into standard squared domains. The subdivision of the quadrature domain into four sectors solves the problem caused by the weight functions. To further improve the efficiency of the numerical integrations, we suggest the following additional modification.

A careful examination reveals that the complexity of the integrand is also because the shape functions are constructed in a piece-wise fashion for each and every Gauss quadrature point. Note that each of these pieces is basically a polynomial type of function and behaves well. The problem results from the union of too many pieces in too complex a manner. For example, if 6×6 Gauss points were used (which is often true), we would have an integrand that is formed by 36 pieces. To obtain an accurate numerical integration, one needs a large number of subdivisions.

The modification here is, therefore, that for all the Gauss quadrature points in one sector, we use a common set of shape functions that are created using the support domain of the center point of the sector. The use of one set of shape functions for one quadrant not only can reduce the cost in integration, but also can improve the accuracy, as the integrand becomes one piece and hence behaves much better in a quadrant.

9.4.2 Implementation Issue in LRPIM for CFD Problems

To further save the repetitive computational work in the iteration process of solving CFD problems, one should try to store intermediate results that are to be used and unchanged in the iteration process. When the Eulerian formulation is used, the nodes are not changed in the iteration process. Therefore, information required for computing the shape functions needs be done only once and stored for use in the rest of the iterations.

In performing integration in the quadrature domain, we have suggested in previous subsections that four support (interpolation) domains can be used for constructing shape functions for each of the four sectors. An alternative, and even simpler approach, is that we use the same support (interpolation) domain for all the quadrature points in the quadrature domain of a node. In this way, the support domain for each node can be determined first and stored for use throughout the whole iteration process. The computation of the inverse of the moment matrix can also be precomputed for each node before iteration. The computation can then be saved. Note that the division of the quadrature domain is still required because of the nature of the weight function. In the following examples, we continue to divide the quadrature domain into four pieces.

It should be noted that using the same support (interpolation) domain, and hence the same nodes, for all the quadrature points in the quadrature domain should not affect the accuracy of interpolation, as long as the support domain is sufficiently large with sufficient overlap to ensure the minimum consistency between support domains. It, however, will affect the complexity of the shape functions constructed, as it will result in using more nodes for interpolation.

9.4.3 Numerical Results and Discussion

We now present the results of testing the LRPIM method for CFD problems of incompressible Navier–Stokes equations. Since the natural convection problem can serve as a good numerical experiment for testing and validating numerical approaches, two such problems are chosen as our test cases. In this study, pure radial functions are used, and no polynomial terms are added ($m = 0$). In the following examples, the boundary condition for vorticity is given by method II, described in Section 9.3.4.

Example 9.8 Natural Convection in a Square Cavity

Example 9.7 is used again to test the LRPIM. The configuration of the flow domain is shown in Figure 9.14. Two kinds of nodal arrangements, regular and irregular, are used for this study, as shown in Figure 9.20.

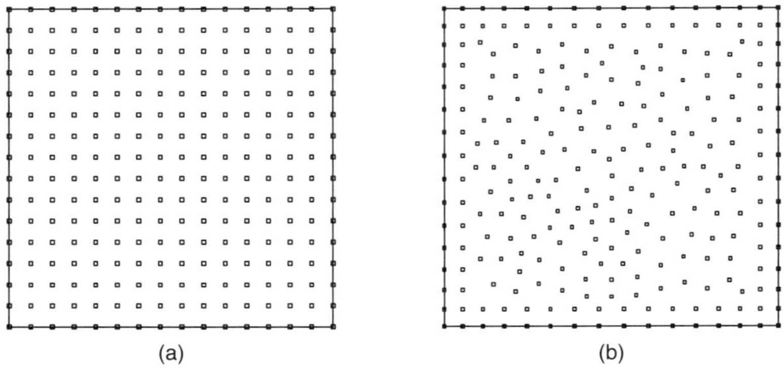

FIGURE 9.20
Two sets of nodal distribution for the square cavity flow problem. (a) Uniformly distributed node set (256 nodes); (b) irregularly scattered node set (257 nodes).

The control of the iteration is the same as in Example 9.7, and the comparison is also performed with the benchmark numerical solution of Davis (1983), which is regarded as a good reference solution. Table 9.4 lists the numerical results for different sets of nodes for Rayleigh numbers of 10^3, 10^4, and 10^5, respectively. It can be observed that for all sets of nodes, the results of the LRPIM method agree very well with the benchmark solution. From Table 9.4, we found a very important feature of LRPIM: the results obtained using the randomly scattered node set are as good as those obtained using the uniformly distributed node set, which is very difficult to achieve using MLPG. This fact shows again that RBF is particularly well suited to irregularly scattered nodes.

The rates of convergence of the present method and FDM are computed in terms of the "energy norm" r_e, which is defined as the mean square of the relative error norm of the above five quantities for the case of Ra = 10^3. Figure 9.21 shows the results of the energy norm against h, which is the nodal distance (both methods use the uniform nodal distribution). It is very clear that the LRPIM method is much superior in both accuracy as well as in rate of convergence compared with FDM.

Figure 9.22 shows the contour for the streamlines and isotherms for the cavity flow problem with Rayleigh numbers of Ra = 10^3, 10^4, and 10^5.

TABLE 9.4

Comparison of Numerical Results Obtained by LRPIM for the Natural Convection Square Cavity Flow Problem for Different Rayleigh Numbers

Ra		$\lvert\psi_{max}\rvert$	u_{max}	v_{max}	Nu_{max}	Nu_{min}
10^3	TYPE I	1.173	3.632	3.680	1.507	0.692
	TYPE II	1.175	3.649	3.694	1.505	0.691
	Davis	1.174	3.649	3.697	1.505	0.692
10^4	TYPE I	5.063	16.104	19.641	3.585	0.588
	TYPE II	5.066	16.119	20.206	3.540	0.658
	Davis	5.071	16.178	19.617	3.528	0.586
10^5	TYPE I	9.842	35.494	69.537	9.642	0.729
	TYPE II	9.824	35.537	69.329	9.594	0.729
	Davis	9.612	34.730	68.590	7.717	0.729

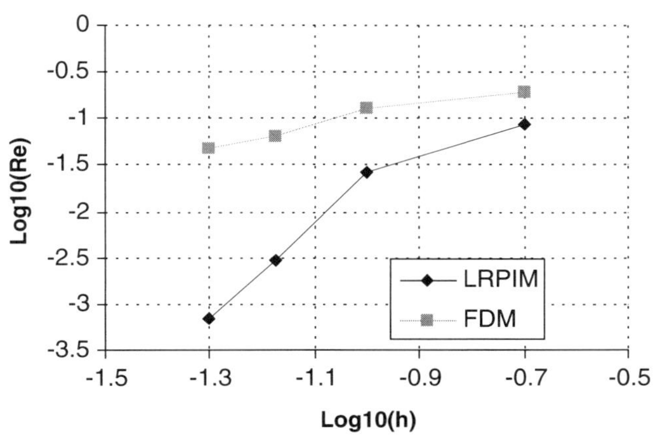

FIGURE 9.21
Comparison of the rate of convergence between LRPIM and FDM for the square cavity flow problem.

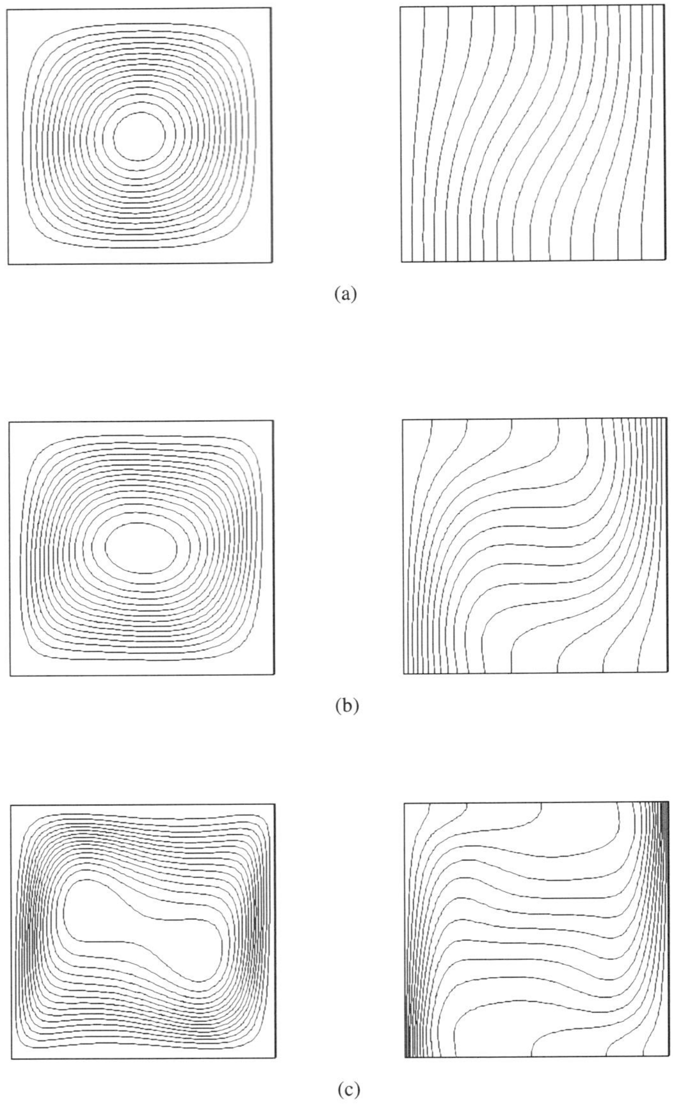

FIGURE 9.22
Streamlines (left images) and isotherms (right images) for cavity flow for different Rayleigh numbers. (a) Ra = 10^3; (b) Ra = 10^4; (c) Ra = 10^5.

Example 9.9 Natural Convection in a Concentric Annulus

Because concentric annulus can be regarded as irregular in geometry in the Cartesian coordinate system, and there are many results available in the literature, it can serve as a good benchmark example for our LRPIM. A concentric annulus with an inner radius of 0.625 and an outer radius of 1.625 is considered. Arbitrarily scattered nodes shown in Figure 9.23 are used in LRPIM. The number of nodes used is 967. The governing equations are the same as in Example 9.8. The boundary conditions on two impermeable isothermal walls are given by

$$\psi = u = v = 0, \qquad \omega = \frac{\partial^2 \psi}{\partial r^2}, \qquad T = 1 \tag{9.81}$$

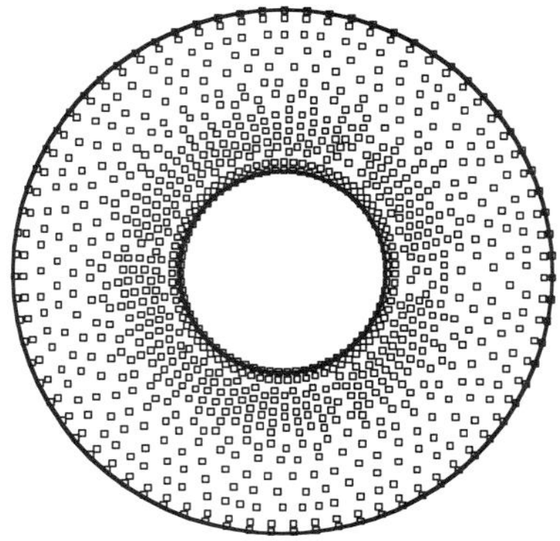

FIGURE 9.23
Nodal distribution for concentric annuli (967 nodes). The inner radius is 0.625, and the outer radius is 1.625.

on the inner cylinder, and

$$\psi = u = v = 0, \qquad \omega = \frac{\partial^2 \psi}{\partial r^2}, \qquad T = 0 \tag{9.82}$$

on the outer cylinder.

For the natural convection between concentric annuli, if the radius ratio is defined as $r_r = \bar{R}_o/\bar{R}_i$, then the dimensionless radii of the inner and outer cylinders are $R_i = 1/(r_r - 1)$ and $R_o = r_r/(r_r - 1)$. The average equivalent conductivity, defined in Shu (1999)

$$\bar{k}_{eqi} = -\frac{\ln(r_r)}{2\pi(r_r - 1)} \int_0^{2\pi} \frac{\partial T}{\partial r} \cdot d\theta \tag{9.83}$$

for the inner cylinder, and

$$\bar{k}_{eqo} = -\frac{r_r \cdot \ln(r_r)}{2\pi(r_r - 1)} \int_0^{2\pi} \frac{\partial T}{\partial r} \cdot d\theta \tag{9.84}$$

for the outer cylinder, will be computed to compare their performances. Table 9.5 compares the results of $\bar{k}_{\text{eqi}}$ and $\bar{k}_{\text{eqo}}$ computed using the LRPIM method with a general scattered distributed node set for the case of Pr = 0.71 and $r_r = 2.6$ and Rayleigh numbers of 10^2, 10^3, 10^4, 5×10^4. The results obtained by Shu (1999) and Kuehn and Goldstein (1976) are also included in the table for comparison. The results of Shu (1999) were obtained by using the PDQ and FDQ methods. They are from the grid-independent study, and can be considered as the benchmark solution. The results of Kuehn and Goldstein (1976) were obtained from the second-order finite difference scheme. It can be obviously observed from Table 9.5 that the present results of LRPIM agree very well with the benchmark solution of Shu (1999). Figure 9.24 shows the streamlines and the isotherms of LRPIM results for Ra = 10^4 and 10^5. The separation of the inner and outer cylinder thermal boundary layer and the symmetry of flow pattern can be seen clearly.

TABLE 9.5

Comparison of Average Equivalent Heat Conductivity for Concentric Annuli

	$\bar{k}_{eqi}$ (inner cylinder)			$\bar{k}_{eqo}$ (outer cylinder)		
Ra	**Present**	**Kuehn**	**Shu**	**Present**	**Kuehn**	**Shu**
10^2	1.000	1.000	1.001	1.001	1.002	1.001
10^3	1.082	1.081	1.082	1.081	1.084	1.082
10^4	1.969	2.010	1.979	1.964	2.005	1.979
5×10^4	2.964	3.024	2.958	2.914	2.973	2.958

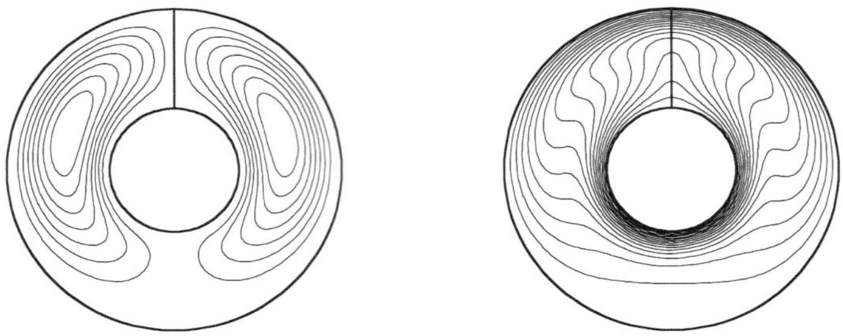

(a)

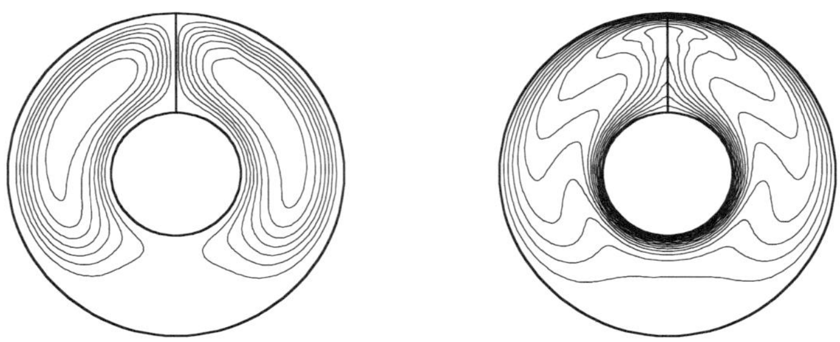

(b)

FIGURE 9.24
Streamlines (left images) and isotherms (right images) for natural convection flow in a concentric annuli. (a) Ra = 10^4; (b) Ra = 5×10^4.

9.4.4 Remarks

In this section, the LRPIM method is adopted with some modifications to simulate CFD problems. The simulation of natural convections in closed domains of different geometries is performed to test LRPIM for CFD problems. The following remarks can be drawn from this research:

1. The LRPIM method works very stably in using different distributions of nodes. The more nodes used, the greater the accuracy.
2. The LRPIM converges for most of the cases we tested.
3. The accuracy achieved is much higher than that obtained by FDM.

The only drawback of LRPIM compared with FDM is that it is far slower than our FDM code, which is well tuned. If we factor in the accuracy, the performance of LRPIM is still worse than FDM in terms of accuracy per CPU time. This is because of the high computational cost needed for numerical integration and solving the resultant linear system by a direct method. The unchallengeable advantage of LRPIM is that it works on irregular nodes without having *any* preknowledge of the nodal arrangement. This opens an excellent opportunity for developing adaptive analysis algorithms for CFD problems. The challenge that we currently face is to make the LRPIM more computationally efficient for CFD problems.

The ongoing effort at our research group is to implement polynomial PIM and LPIM for CFD problems. We hope to achieve better performance in terms of both accuracy and efficiency. We have reason to believe that the performance of LPIM can be much more comparable with FDM even in terms of efficiency, as we have demonstrated in Chapter 8 that LPIM performs far better than LRPIM (see Figure 8.97) for solid mechanics problems.

We also need effective iteration schemes that suit the MFree methods. In using FDM, one can use many iteration schemes, such as Jacobi, Gauss–Seidel, SOR, ADI, Block iteration, etc. These schemes allow one to avoid solving the linear algebraic system equations by a direct method at each iteration, and allow the FDM to work much more efficiently. Unfortunately, our study revealed that these stationary iteration schemes do not work in either MLPG or LRPIM. In all the examples analyzed using MLPG and LRPIM, and benchmarked with the FDMs, a direct solver of linear algebraic equation systems is employed. Beyond that is the iteration. This is the reason the code is slow, and this is the major drawback of these two MFree methods tested for CFD problems. More work is needed to overcome these drawbacks.

10

Mesh Free Methods for Beams

A beam is a simple but a very common and important structural component. A huge volume of earlier research work has focused on analysis of beams. The finite element method (FEM) is now the mainstream method for analysis of all kinds of problems involving beams. In recent years, MFree methods have also been applied to analyzing beams, such as EFG method for modal analyses of Euler–Bernoulli beams and Kirchhoff plates (Ouatouati and Johnson, 1999), MLPG for thin beams (Atluri et al., 1999a), and LPIM and LRPIM for both thin and thick beams (Gu and Liu, G. R., 2001d,e).

In Chapter 8, we demonstrated that LPIM is an efficient method for dealing with mechanics problems of two-dimensional (2D) solids. LPIM combines the ideal of MLPG, which leads to a nearly truly MFree method, with the PIM shape function, which possesses the Kronecker delta function property. In this chapter, an LPIM formulation is introduced to deal with fourth-order boundary value and initial value problems for analyzing static stability, and free and forced vibration of beams. This work was initially performed by Gu and G. R. Liu (2001d) and G. R. Liu and Gu (2001e).

As a beam is geometrically one dimensional, PIM works perfectly well, and there is no singular issue in the moment matrix. Therefore, there is no need at all to use elements for field variable (deflection and rotation/slope) interpolation. Moreover, PIM shape functions provide a higher order of consistency for the field variables. The procedure of formulating LPIM for beams is as follows.

The beam is first represented using a set of nodes scattered properly in the span of the beam. Local Petrov–Galerkin weak forms are then developed using the weighted residual method based on the local concept of MLPG. PIM shape functions are constructed based only on a group of arbitrarily distributed points in the support domain. Discretized system equations are then produced by substituting the 1D PIM shape functions into the local residual weak form. Because the PIM shape functions possess the delta function property, the essential boundary conditions can be implemented with ease, as in conventional FEM. The validity and efficiency of the present LPIM formulation are demonstrated through numerical examples of static, stability, and dynamic analyses of beams under various loads and boundary conditions.

Note that as discussed in Section 5.11, the PIM approximation could be incompatible. Similar to the PIM methods for 2D solids, we can also formulate nonconforming and conforming PIMs for beams, if the Galerkin formulation is followed.

In using nonconforming PIM, the support domain is determined for each Gauss point. In nonconforming PIM, however, we use so-called one-piece shape functions for the entire integration cell to ensure the compatibility of the field function approximation within the cells. Therefore, the support domain is defined for the cell, meaning all the Gauss points in the cell share the same support domain and hence the same PIM shape functions. The support domain of a cell is determined in a manner similar to that discussed in the previous paragraph. The only difference is that the domain is centered at the geometrical center of the cell and not at the Gauss point.

Note also that in formulating PIMs for beams, since the integration cells are connected at points (not lines), compatibility is automatically enforced, as long as the one-piece PIM shape functions are used. There is no need to use the penalty method to enforce compatibility as we have done in Chapter 8 for two-dimensional solids.

This chapter deals with only straight and thin beams that are governed by Euler–Bernoulli beam theory. Only bending deformation is considered, and it is also assumed that the beam is planar, meaning it deforms only within a plane that is defined as an x–y plane. Although this chapter discusses only LPIMs, PIMs for beams can also be formulated in a similar manner based on a global integration mesh and Galerkian formulation.

10.1 PIM Shape Function for Thin Beams

10.1.1 Formulation

The procedure for constructing the PIM shape function is largely the same as that detailed in Chapter 5. The major point to note is that one needs to use higher-order terms of polynomial basis, because the governing equations for thin beams are of fourth order, in contrast to second order for 2D solids.

The thin beam is represented by a line of its neutral axis that is defined in a 1D domain Ω, as shown in Figure 10.1. This neutral axis line is then represented by a set of nodes properly distributed on the line of beam. This set of nodes is often termed *field nodes*, as it is used to register the values of field variables. For the field variables for our Euler–Bernoulli beam, deflection is denoted by $v(x)$ and rotation of the cross section of the beam (or slope of the deflection) is denoted by $\theta(x)$. The deflection $v(x)$ is approximated using polynomial PIM shape functions that are to be constructed using a set of nodes included in the 1D support domain Ω_s of a point of interest x_Q, as shown in Figure 10.1.

$$v(x, x_Q) = \sum_{i=1}^{2n} p_i(x)a_i(x_Q) = \mathbf{p}^{\mathrm{T}}(x)\mathbf{a}(x_Q) \tag{10.1}$$

where $p_i(x)$ are monomials of x, n is the number of nodes in the support domain of x_Q, and $a_i(x_Q)$ is the coefficient for $p_i(x)$, corresponding to the given point x_Q. Matrix $\mathbf{p}^T(x)$ in Equation 10.1 consists of monomials in 1D space in the form:

$$\mathbf{p}^{\mathrm{T}}(x) = \{1, x, x^2, x^3, x^4, \ldots, x^{2n-1}\} \tag{10.2}$$

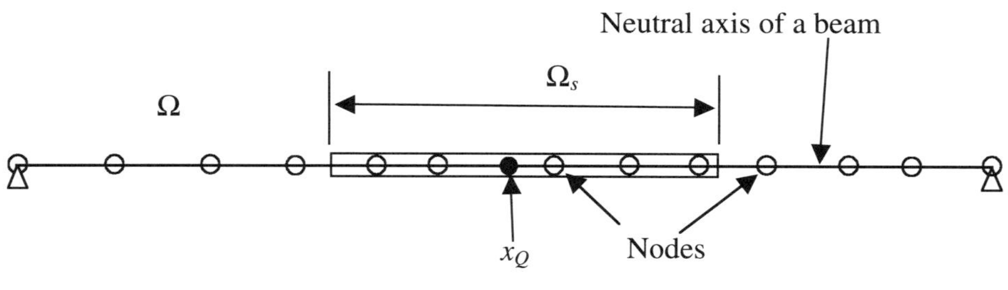

FIGURE 10.1
A straight beam represented by its neutral axis, which is represented by a set of field nodes. PIM shape functions are constructed using a subset of the nodes included in the support domain Ω_s of a point of interest x_Q that can be a quadrature point or the center of an integration cell.

The number of terms in matrix $\mathbf{p}^{\mathrm{T}}(x)$ depends on the number of nodes included in the support domain. In our Euler–Bernoulli beam, we choose to have two variables, deflection and slope, for each node. Therefore, if there are n nodes in the support domain, we should have $2n$ terms in matrix $\mathbf{p}^{\mathrm{T}}(x)$. In practical implementation, one can also choose nodes based on the requirement of the number of terms in $\mathbf{p}^{\mathrm{T}}(x)$.

In our thin beam formulation, the derivatives of the field variable of the deflection are required. We assume that the derivatives up to an order l of the field variable are required. By using Equation 10.1, these derivatives of deflection can be obtained simply, as follows:

$$v^{(l)}(x, x_Q) = \sum_{i=1}^{2n} p_i^{(l)}(x) a_i(x_Q) = \{\mathbf{p}^{(l)}(x)\}^{\mathrm{T}} \mathbf{a}(x_Q) \tag{10.3}$$

where $v^{(l)}(x, x_Q)$ and $\{\mathbf{p}^{(l)}(x)\}^{\mathrm{T}}$ are the lth order derivatives of the field variable $v(x, x_Q)$ and the basis function $\mathbf{p}^{\mathrm{T}}(x)$.

According to the Euler–Bernoulli beam theory, the slope θ of the beam can be obtained by the first derivative of the deflection, i.e.,

$$\theta(x) = -\frac{dv}{dx} = -\sum_{i=1}^{2n} p_i^{(1)}(x) a_i(x_Q) = -\{\mathbf{p}^{(1)}(x)\}^{\mathrm{T}} \mathbf{a}(x_Q) \tag{10.4}$$

where

$$\begin{aligned} \mathbf{p}^{(1)} &= \left\{ \frac{dp_1(x)}{dx} \;\; \frac{dp_2(x)}{dx} \;\; \frac{dp_2(x)}{dx} \;\; \cdots \;\; \frac{dp_{2n}(x)}{dx} \right\}^{\mathrm{T}} \\ &= \{0 \;\; 1 \;\; 2x \;\; \ldots \;\; (2n-1)x^{2n-2}\}^{\mathrm{T}} \end{aligned} \tag{10.5}$$

The coefficients a_i in Equations 10.1 and 10.4 can be determined by enforcing Equations 10.1 and 10.4 to be satisfied at the n nodes surrounding point x_Q. At node i we have equation

$$\mathbf{v}_i = \begin{Bmatrix} v_i \\ \theta_i \end{Bmatrix} = \begin{Bmatrix} \sum_{i=1}^{2n} p_i(x_i) a_i \\ -\sum_{i=1}^{2n} \frac{dp_i(x_i)}{dx} a_i \end{Bmatrix} = \begin{Bmatrix} \mathbf{p}^{\mathrm{T}}(x_i)\mathbf{a} \\ -[\mathbf{p}^{(1)}(x_i)]^{\mathrm{T}}\mathbf{a} \end{Bmatrix} \tag{10.6}$$

where v_i and θ_i are the nodal value of v and θ at $x = x_i$. Equation 10.6 can be written in the following matrix form:

$$\mathbf{v}_s = \mathbf{P}_Q \mathbf{a} \tag{10.7}$$

where $\mathbf{v}_s$ is the (generalized) displacement vector of the node in the support domain arranged in the form:

$$\mathbf{v}_s = [v_1, \theta_1, v_2, \theta_2, \ldots, v_n, \theta_n]^{\mathrm{T}} \tag{10.8}$$

$\mathbf{P}_Q$ is the (generalized) moment matrix formed using alternately vectors $\mathbf{p}$ and $\mathbf{p}^{(1)}$ evaluated at n nodes at x_i ($i = 1,\ldots, n$) in the support domain.

$$\mathbf{P}_Q^{\mathrm{T}} = [\mathbf{p}(x_1), -\mathbf{p}^{(1)}(x_1), \mathbf{p}(x_2), -\mathbf{p}^{(1)}(x_2), \ldots, \mathbf{p}(x_n), -\mathbf{p}^{(1)}(x_n)] \tag{10.9}$$

Solving Equation 10.7 for $\mathbf{a}$, we have

$$\mathbf{a} = \mathbf{P}_Q^{-1}\, \mathbf{v}_s \tag{10.10}$$

Note that for 1D point interpolations, $\mathbf{P}_Q^{-1}$ always exists unless there are duplicated nodes in the support domain, which will not happen in practical implementation, because there is no reason to create a mode that has duplicated nodes. Even where there are duplicated nodes, they can be easily detected and eliminated.

Substituting Equation 10.10 into Equation 10.1, we should have

$$v(x) = \mathbf{\Phi}^{\mathrm{T}}(x)\mathbf{v}_s \tag{10.11}$$

where $\mathbf{\Phi}(x)$ is the matrix of shape functions arranged in the form:

$$\begin{aligned}\mathbf{\Phi}^{\mathrm{T}}(x) &= [\phi_1(x)\ \ \phi_2(x)\ \ \phi_3(x)\ \ \phi_4(x)\ \ \ldots\ \ \phi_{2n-1}(x)\ \ \phi_{2n}(x)] \\ &= [\phi_{v1}(x)\ \ \phi_{\theta 1}(x)\ \ \phi_{v2}(x)\ \ \phi_{\theta 2}(x)\ \ \ldots\ \ \phi_{vn}(x)\ \ \phi_{\theta n}(x)]\end{aligned} \tag{10.12}$$

The above equation can be rewritten in the following form similar to that used in FEM (Petyt, 1990):

$$v(x) = \mathbf{\Phi}_v^{\mathrm{T}}(x)\hat{\mathbf{v}}^s + \mathbf{\Phi}_\theta^{\mathrm{T}}(x)\hat{\boldsymbol{\theta}}^s \tag{10.13}$$

where $\hat{\mathbf{v}}^s$ collects only the deflections at the nodes, and $\hat{\boldsymbol{\theta}}^s$ collects only the slopes at the nodes in the support domain. The shape matrix corresponding to the deflection is then given by

$$\mathbf{\Phi}_v^{\mathrm{T}}(x) = \mathbf{p}^{\mathrm{T}}(x)\mathbf{P}_Q^{-1} = [\phi_{v1}(x), \phi_{v2}(x),\ldots, \phi_{vn}(x)] \tag{10.14}$$

The shape matrix corresponding to the slope is given by

$$\mathbf{\Phi}_\theta^{\mathrm{T}}(x) = -\mathbf{p}_x^{\mathrm{T}}(x)\mathbf{P}_Q^{-1} = [\phi_{\theta 1}(x), \phi_{\theta 2}(x),\ldots, \phi_{\theta n}(x)] \tag{10.15}$$

10.1.2 Example

Example 10.1 PIM Shape Functions for Thin Beams

The computation for shape functions in $\mathbf{\Phi}(x)$, or in $\mathbf{\Phi}_v(x)$ and $\mathbf{\Phi}_\theta(x)$ are straightforward. It requires a numerical inversion of the moment matrix $\mathbf{P}_Q$. Typical shape functions of $\mathbf{\Phi}_v(x)$, $\mathbf{\Phi}_\theta(x)$ and its derivatives obtained using evenly distributed nodes and $n = 3$ are shown in Figure 10.2. The first and second derivatives of the shape functions are shown in Figures 10.3 and 10.4, respectively. It can be found that the shape function $\mathbf{\Phi}_w(x)$ and $\mathbf{\Phi}_\theta(x)$ obtained through the above procedure satisfies

$$\phi_{vi}(x_j) = \delta_{ij} \tag{10.16}$$

$$\sum_{i=1} \phi_{vi} = 1 \tag{10.17}$$

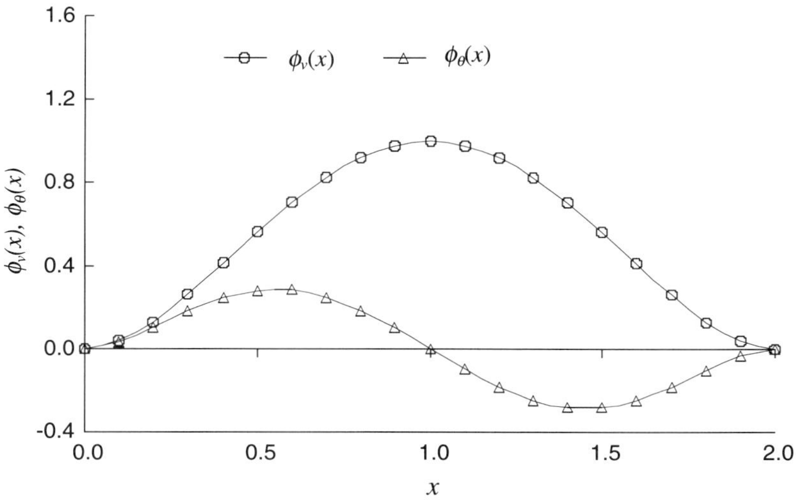

FIGURE 10.2
PIM shape functions for field variables of displacement and rotation (slope). Computed using three nodes at $x = 0.0$, 1.0, 2.0, based on the deflection at node at $x = 1.0$. (From Gu, Y. T. and Liu, G. R., *Comput. Methods Appl. Mech. Eng.*, 190, 5515–5528, 2001. With permission.)

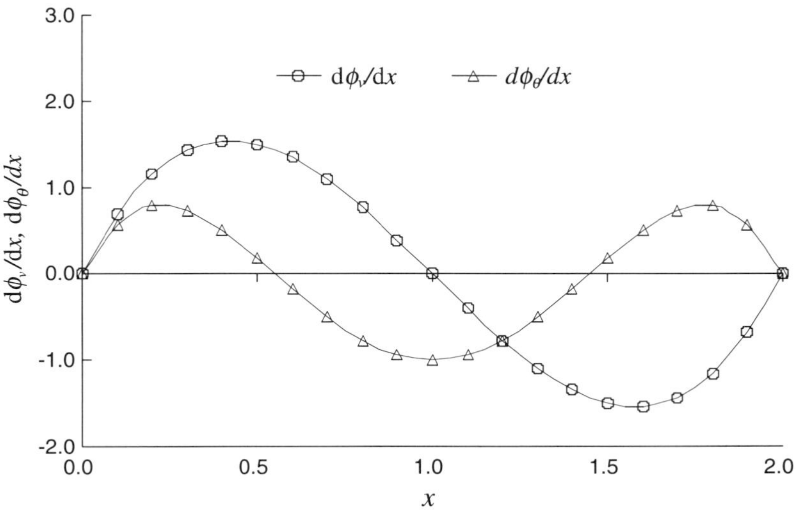

FIGURE 10.3
The first derivatives of PIM shape functions for field variables of displacement and rotation (slope). Obtained using evenly distributed nodes and $n = 3$ for the deflection at node $x = 1.0$. (From Gu, Y. T. and Liu, G. R., *Comput. Methods Appl. Mech. Eng.*, 190, 5515–5528, 2001. With permission.)

where δ_{ij} is the Kronecker delta function given by

$$\delta_{ij} = \begin{cases} 1 & i = j \\ 0 & i \neq j \end{cases} \tag{10.18}$$

Therefore, the shape functions constructed possess the delta function property, and the essential boundary conditions can be easily imposed in the present LPIM.

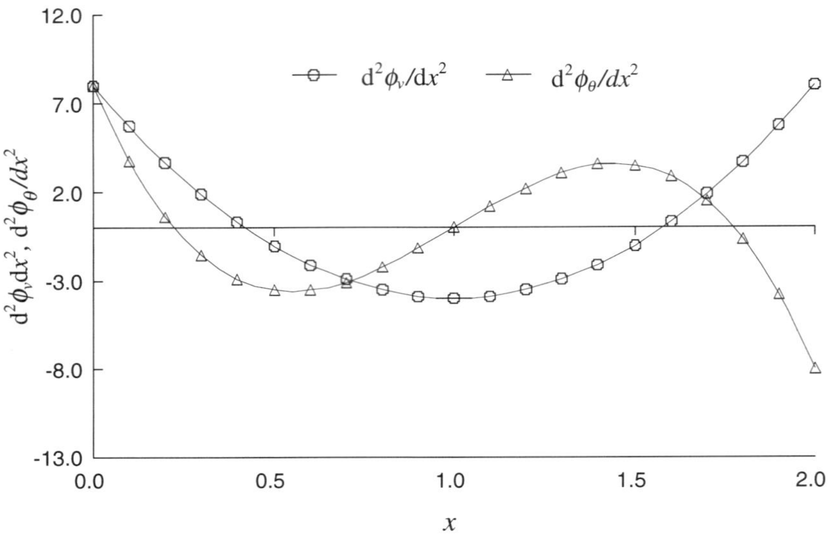

FIGURE 10.4
The second derivatives of PIM shape functions for field variables of displacement and rotation (slope). Obtained using evenly distributed nodes and $n = 3$ for the deflection at node at $x = 1.0$. (From Gu, Y. T. and Liu, G. R., *Comput. Methods Appl. Mech. Eng.*, 190, 5515–5528, 2001. With permission.)

10.2 Elastostatic Analysis of Thin Beams

10.2.1 Local Weighted Residual Weak Form

The weighted residual method formulates the weak form of system equations using the strong form of governing equations. The strong form of governing equation is derived in Chapter 3 for thin beams based on Euler–Bernoulli thin beam theory, which leads to a fourth-order differential equation. For beams of the constant bending stiffness, *EI*, we have (see Equation 3.55)

$$EI\frac{\mathrm{d}^4 v}{\mathrm{d}x^4} = b_y \quad \text{in domain } \Omega \tag{10.19}$$

where v is transverse deflection and b_y is distributed load over the beam. A single-span Euler–Bernoulli beam has four boundary conditions, two at each end. The boundary conditions are combinations of the following:

$$v(x_0) = \bar{v}, \quad \text{on } \Gamma_v \tag{10.20}$$

$$-\frac{\mathrm{d}v(x_0)}{\mathrm{d}x} = \bar{\theta}, \quad \text{on } \Gamma_\theta \tag{10.21}$$

$$M(x_0) = EI\frac{\mathrm{d}^2 v}{\mathrm{d}x^2} = \overline{M}, \quad \text{on } \Gamma_M \tag{10.22}$$

$$V(x_0) = -EI\frac{\mathrm{d}^3 v}{\mathrm{d}x^3} = \overline{V}, \quad \text{on } \Gamma_V \tag{10.23}$$

where M and V denote the moment and the shear force, respectively. Γ_w, Γ_θ, Γ_M, and Γ_V denote, respectively, the boundary locations where deflection, slope, moment, and shear force are specified. Note that boundary conditions Equations 10.20 and 10.21 are essential (displacement type) conditions, and Equations 10.22 and 10.23 are natural (force) boundary conditions. If the MLS shape function is employed, care must be taken in handling the essential boundary conditions. In this chapter, we use PIM shape functions, and therefore the handling of the essential boundary conditions is the same as the practice in conventional FEM.

A local weak form for the differential Equation 10.19, over a local quadrature domain Ω_Q bounded by Γ_Q, can be obtained using the weighted residual method

$$\int_{\Omega_Q} \widehat{W}(EIv'''' - b_y)\mathrm{d}x = 0 \tag{10.24}$$

where $\widehat{W}$ is the weight function. The first term on the left-hand side of Equation 10.24 can be integrated by parts to become

$$\int_{\Omega_Q} (EI\widehat{W}''v'' - \widehat{W}b_y)\mathrm{d}x - [\bar{n}EI\widehat{W}'v'']_{\Gamma_Q} + [\bar{n}EI\widehat{W}v''']_{\Gamma_Q} = 0 \tag{10.25}$$

where $\bar{n}$ is the unit outward normal to domain Ω_Q, and it is either +1 or −1 for straight beams. The quadrature domain Ω_Q is usually centered by a node. Figure 10.5 shows a schematic drawing of the quadrature domain together with some other domains that are involved in the LPIM formulation. One is the weighted domain Ω_W of a node i at x_i, which is a domain where $\widehat{W}(x) \neq 0$. The weighted domain can be theoretically independent of other domains, but it is often taken as the same as the quadrature domain Ω_Q. This book uses the same domain for both the quadrature domain Ω_Q and the weighted domain Ω_W. The other domain is the support domain Ω_s, which is usually centered by the point of interest (usually the Gauss quadrature point). The support domain Ω_s determines the nodes used for the field variable interpolation, and therefore can also be termed the *interpolation domain*.

For beam problems, the boundary Γ_Q for the quadrature domain is usually composed of five parts: the internal boundary Γ_{Qi}, and the boundaries Γ_{Qv}, $\Gamma_{Q\theta}$, Γ_{QM}, and Γ_{QV}, at which the essential boundary conditions v, θ and the natural boundary conditions M, V are specified. The boundaries Γ_{Qv} with Γ_{QV} and $\Gamma_{Q\theta}$ with Γ_{QM} are mutually exclusive.

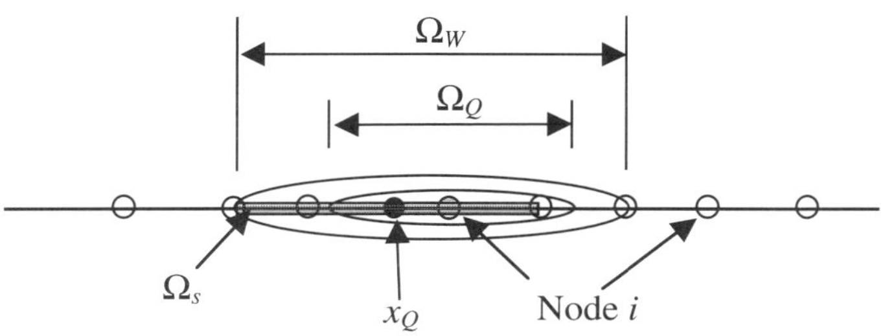

FIGURE 10.5
Weight function domain Ω_W and quadrature domain Ω_Q for node i; support domain Ω_s for Gauss integration point x_Q. Note that the weighted function domain is usually chosen to be the same as the quadrature domain ($\Omega_W = \Omega_Q$). (From Gu, Y. T. and Liu, G. R., *Comput. Methods Appl. Mech. Eng.*, 190, 5515–5528, 2001. With permission.)

Imposing the natural boundary condition given in Equation 10.25, we obtain

$$\int_{\Omega_Q}(EI\widehat{W}''v''-\widehat{W}b_y)\mathrm{d}x-[\bar{n}\overline{M}\widehat{W}']_{\Gamma_{QM}}-[\bar{n}\overline{V}\widehat{W}]_{\Gamma_{QV}}-[\bar{n}EI\widehat{W}'v'']_{\Gamma_{Q\theta}}$$
$$+[\bar{n}\widehat{W}EIv''']_{\Gamma_{Qv}}-[\bar{n}\widehat{W}'EIv'']_{\Gamma_{Qi}}+[\bar{n}\widehat{W}EIv''']_{\Gamma_{Qi}}=0 \tag{10.26}$$

If the weight function is so chosen that both itself and its first derivatives vanish at Γ_{Qi}, the last two terms in Equation 10.26 also vanish, and the integration can be greatly simplified as

$$\int_{\Omega_Q}(EI\widehat{W}''v''-\widehat{W}b_y)\mathrm{d}x-[\bar{n}\overline{M}\widehat{W}']_{\Gamma_{QM}}-[\bar{n}\overline{V}\widehat{W}]_{\Gamma_{QV}}-[\bar{n}EI\widehat{W}'v'']_{\Gamma_{Q\theta}}+[\bar{n}\widehat{W}EIv''']_{\Gamma_{Qv}}=0 \tag{10.27}$$

In LPIM, Equation 10.26 is to be satisfied for all the local quadrature domains for each and every node in the entire problem domain including the boundaries. This implies that the equilibrium equation and the boundary conditions are satisfied node by node in a weak sense of local weighted residual. This node-by-node local formulation provides freedom in the solving system equation numerically and the potential to remove the mesh.

10.2.2 Discretized System Equations

As LPIM is basically regarded as a weighted residual method, the weight function plays an important role in the performance of the method. Theoretically, as long as the condition of continuity is satisfied, any weight function is acceptable. However, the local weak form is based on the local subdomains centered by nodes. It can be found that the weight function with the following properties performs well:

1. It decreases in magnitude as the distance from a point x_Q to the node at x_i increases.
2. We prefer also that the weight function and its first derivative vanish at the boundary, so that integration in Equation 10.27 can be used.

Any weight function listed in Equations 5.11 through 5.13 and 5.15 satisfies these two preferences, and can be used. Here we introduce an alternative way of constructing a weight function. From Figures 10.2 and 10.3, it can be found that the shape functions constructed using Equations 10.14 and 10.15 satisfy the above two preferences. Therefore, we can construct weight functions using the same procedure that we use to construct shape functions. This will lead to a *form* very similar to that of the Galerkin formulation. The weight functions so constructed should have the following form:

$$\widehat{W}(x_Q)=\mathbf{\Psi}_v^{\mathrm{T}}(x)\boldsymbol{\alpha}+\mathbf{\Psi}_\theta^{\mathrm{T}}(x)\boldsymbol{\beta} \tag{10.28}$$

where $\boldsymbol{\alpha}$ and $\boldsymbol{\beta}$ denote the fictitious nodal coefficients and $\mathbf{\Psi}_v$ and $\mathbf{\Psi}_\theta$ are constructed using Equations 10.14 and 10.15, respectively. It should be noted that the support domain used to construct $\mathbf{\Psi}_v$ and $\mathbf{\Psi}_\theta$ can be independent of the support domain used to construct $\mathbf{\Phi}_v$ and $\mathbf{\Phi}_\theta$. Because of this difference, we are not able to obtain a symmetric stiffness matrix.

Substituting Equations 10.13 and 10.28 into the local weak form Equation 10.26 for all nodes leads to the following global discrete equation for the entire beam:

$$\mathbf{KU}=\mathbf{F} \tag{10.29}$$

where **U** is the vector collecting all unknown nodal deflections and rotations for all the nodes in the entire problem domain.

The global stiffness matrix **K** is formed by assembling the nodal stiffness matrix defined by

$$\mathbf{k}_{ij} = EI\begin{bmatrix} \int_{\Omega_Q}\psi''_{vi}\phi''_{vj}\mathrm{d}x & \int_{\Omega_Q}\psi''_{vi}\phi''_{\theta j}\mathrm{d}x \\ \int_{\Omega_Q}\psi''_{\theta i}\phi''_{vj}\mathrm{d}x & \int_{\Omega_Q}\psi''_{\theta i}\phi''_{\theta j}\mathrm{d}x \end{bmatrix} + \bar{n}EI\begin{bmatrix} \psi_{vi}\phi'''_{vj} & \psi_{vi}\phi'''_{\theta j} \\ \psi_{\theta i}\phi'''_{vj} & \psi_{\theta i}\phi'''_{\theta j} \end{bmatrix}\Bigg|_{\Gamma_{Qv}} - \bar{n}EI\begin{bmatrix} \psi'_{vi}\phi''_{vj} & \psi'_{vi}\phi''_{\theta j} \\ \psi'_{\theta i}\phi''_{vj} & \psi'_{\theta i}\phi''_{\theta j} \end{bmatrix}\Bigg|_{\Gamma_{Q\theta}}$$

$$+ \bar{n}EI\begin{bmatrix} \psi_{vi}\phi'''_{vj} & \psi_{vi}\phi'''_{\theta j} \\ \psi_{\theta i}\phi'''_{vj} & \psi_{\theta i}\phi'''_{\theta j} \end{bmatrix}\Bigg|_{\Gamma_{Qi}} - \bar{n}EI\begin{bmatrix} \psi'_{vi}\phi''_{vj} & \psi'_{vi}\phi''_{\theta j} \\ \psi'_{\theta i}\phi''_{vj} & \psi'_{\theta i}\phi''_{\theta j} \end{bmatrix}\Bigg|_{\Gamma_{Qi}} \tag{10.30}$$

If the weight function is so chosen that both itself and its first derivatives vanish at Γ_{Qi}, the last two terms in Equation 10.30 vanish. It can be easily seen that the matrix **K** is usually banded but asymmetric.

The global force vector **F** is formed by assembling the nodal force vector defined by

$$\mathbf{f}_i = \begin{Bmatrix} \int_{\Omega_Q}\psi_{vi}\,f dx \\ \int_{\Omega_Q}\psi_{\theta i}\,f dx \end{Bmatrix} + \bar{n}\overline{M}\begin{Bmatrix} \psi'_{vi} \\ \psi'_{\theta i} \end{Bmatrix}_{\Gamma_{QM}} + \bar{n}\overline{V}\begin{Bmatrix} \psi_{vi} \\ \psi_{\theta i} \end{Bmatrix}_{\Gamma_{Qv}} \tag{10.31}$$

Note from Equations 10.30 and 10.31 that the integrands are basically polynomials. Therefore, these integrals can be carried out analytically. A numerical integration using the Gauss quadrature scheme is performed here to evaluate the integration in Equations 10.30 and 10.31. For node i at x_i, a local integration cell is needed to employ Gauss quadrature. For each Gauss quadrature point x_Q, the point interpolation is performed to obtain the integrand. The quadrature domain Ω_Q is shown in Figure 10.5.

In LPIM, because the shape functions possess the delta function property, the essential boundary conditions can be implemented in the conventional way as in FEM. Note also that because the system equations of LPIM are assembled based on nodes as in the finite difference method (FDM), the items of the row in the matrix **K** for the nodes on the essential boundary need not even be computed. This reduces the computational cost.

10.2.3 Numerical Example for Static Problems

A number of numerical examples of static deflection analyses of beams will be presented in this section. For simplification, the units are omitted, and any consistent set of units can be assigned.

Example 10.2 Simply-Simply Supported Beams under Various Loads

Simply-simply supported thin beams under uniformly distributed load, concentrated load, and linearly distributed load are analyzed. A total of 21 irregularly distributed nodes are used to represent the beam. In this example, the parameters for the beam are taken as $EI = 1.0$, the length of beam $l = 1.0$, and $q_0 = 1.0$. Figure 10.6 plots the distribution of the deflection and slope results for the simply-simply supported beams subjected to uniformly distributed load. Both numerical results of LPIM and analytical results are plotted together for comparison. The marks indicate the locations of the nodes. From Figure 10.6 very good

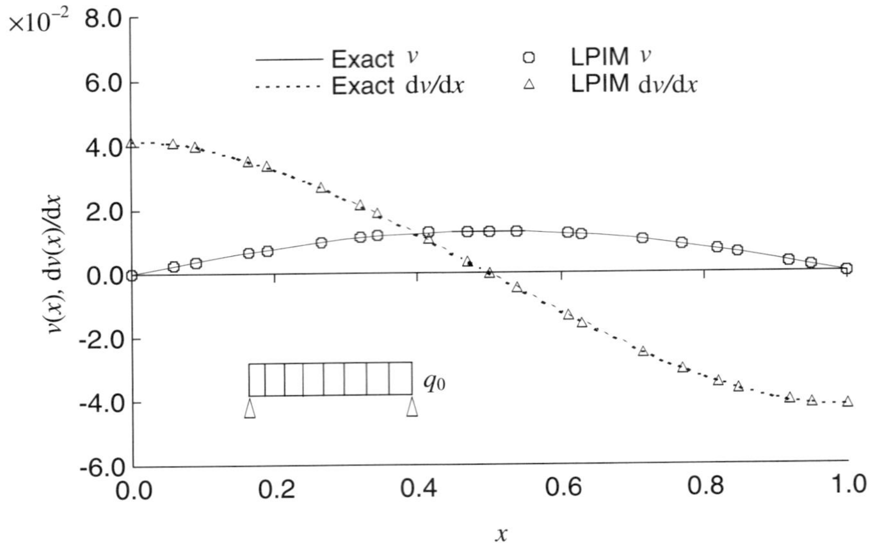

FIGURE 10.6
Distribution of deflection and slope of a simply-simply supported thin beam subjected to a uniformly distributed load. (From Gu, Y. T. and Liu, G. R., *Comput. Methods Appl. Mech. Eng.*, 190, 5515–5528, 2001. With permission.)

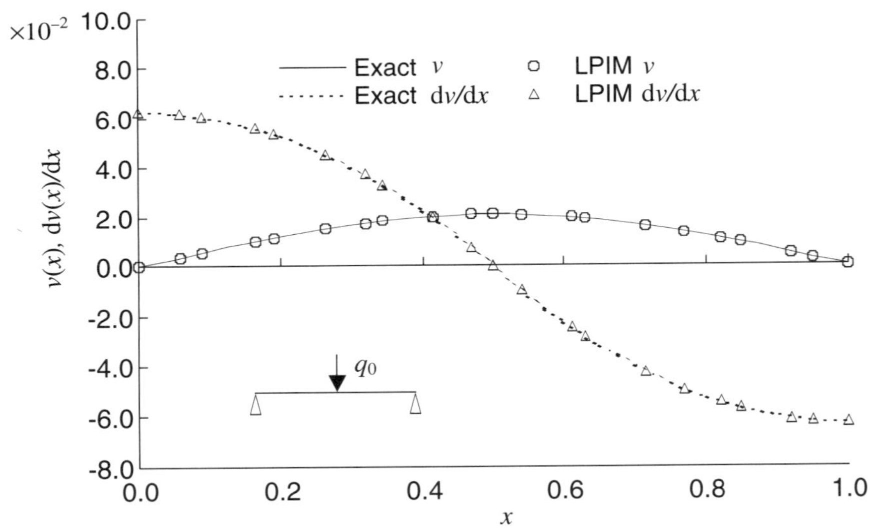

FIGURE 10.7
Distribution of deflection and slope of a simply-simply supported thin beam subjected to a concentrated load. (From Gu, Y. T. and Liu, G. R., *Comput. Methods Appl. Mech. Eng.*, 190, 5515–5528, 2001. With permission.)

results are obtained in comparison with the analytical solution. Figure 10.7 plots the distribution of the deflection and slope results for the simply-simply supported beams subjected to a concentrated load. Figure 10.8 shows the same results but the beam is subjected to a triangularly distributed load. All these results confirm that the LPIM solution agrees very well with the analytical results. This validates the LPIM method.

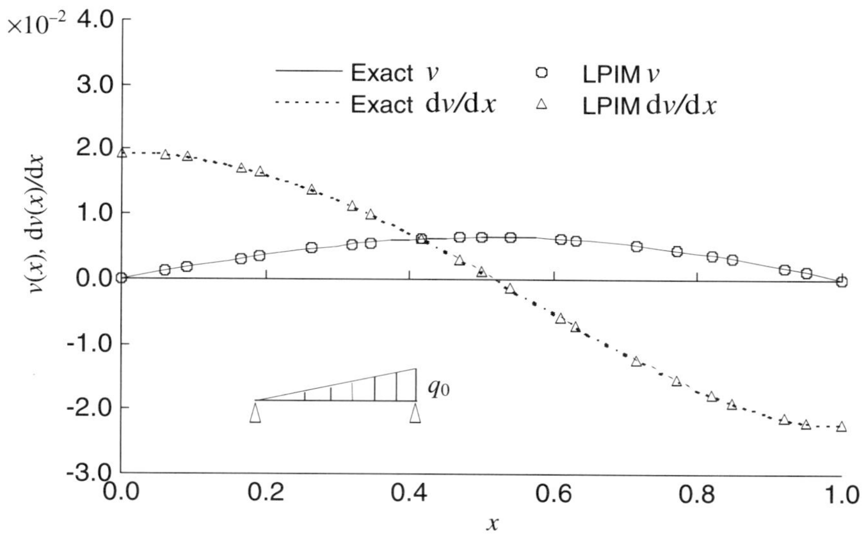

FIGURE 10.8
Distribution of deflection and slope of a simply-simply supported thin beam subjected to a triangularly distributed load. (From Gu, Y. T. and Liu, G. R., *Comput. Methods Appl. Mech. Eng.*, 190, 5515–5528, 2001. With permission.)

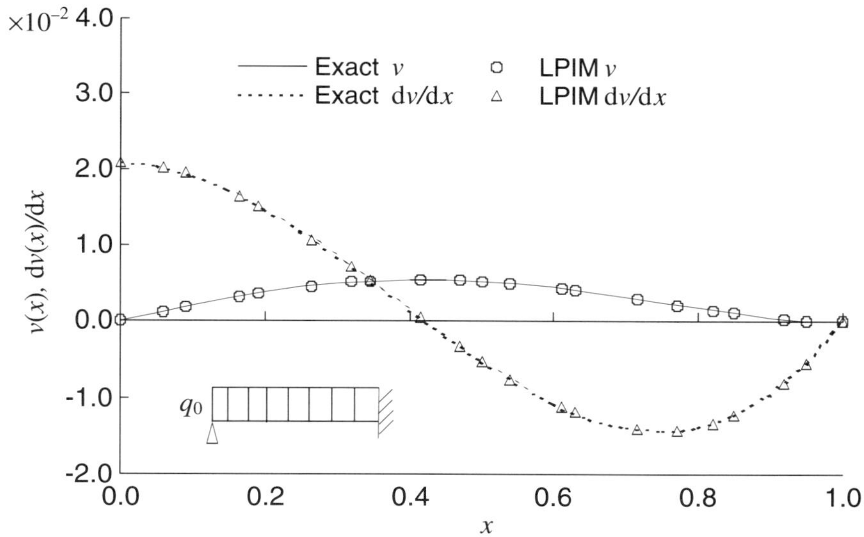

FIGURE 10.9
Distribution of the deflection and slope of a thin beam under uniformly distributed load. The boundary condition for the beam is pinned-fixed. (From Gu, Y. T. and Liu, G. R., *Comput. Methods Appl. Mech. Eng.*, 190, 5515–5528, 2001. With permission.)

Example 10.3 Beams under Uniformly Distributed Load with Different Boundary Conditions

To further validate LPIM for different boundary conditions, thin beams under uniformly distributed load with different boundary conditions are analyzed. The results are shown in Figure 10.9 for pinned-fixed, Figure 10.10 for cantilever, and Figure 10.11 for fixed-fixed

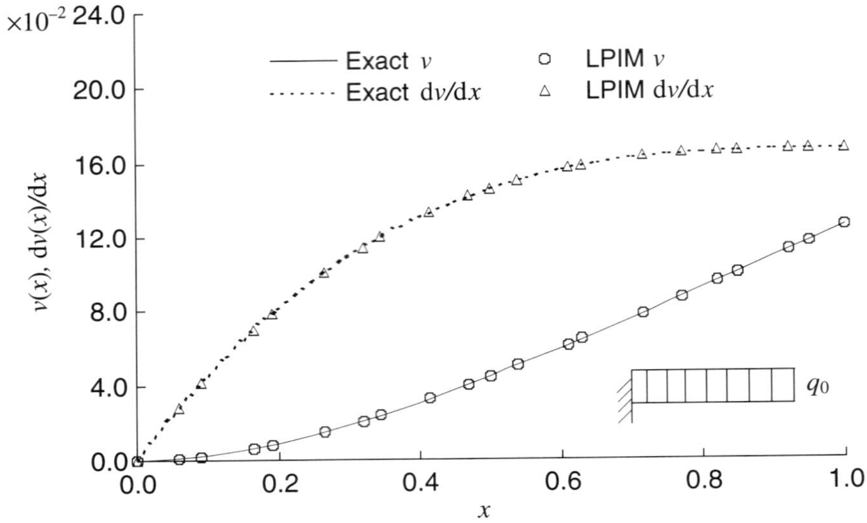

FIGURE 10.10
Distribution of the deflection and slope of a cantilever thin beam under uniformly distributed load. (From Gu, Y. T. and Liu, G. R., *Comput. Methods Appl. Mech. Eng.*, 190, 5515–5528, 2001. With permission.)

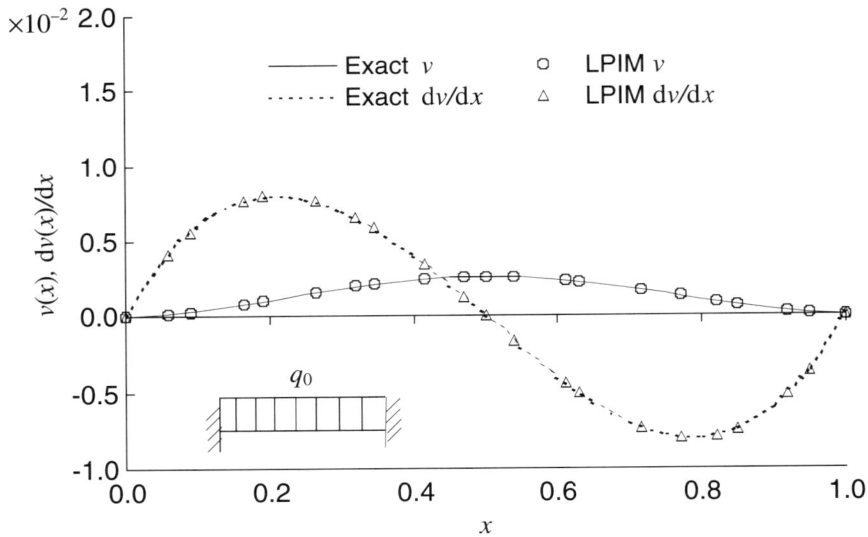

FIGURE 10.11
Distribution of the deflection and slope of a thin beam under uniformly distributed load. The boundary condition for the beam is fixed-fixed. (From Gu, Y. T. and Liu, G. R., *Comput. Methods Appl. Mech. Eng.*, 190, 5515–5528, 2001. With permission.)

supported beam. Both numerical results of LPIM and analytical results are plotted together for comparison. Again, excellent agreements between the analytical and numerical results are observed.

10.3 Buckling Analysis of Thin Beams (Eigenvalue Problem)

10.3.1 Local Weak Form

The equation that governs the buckling process of a Euler–Bernoulli beam can be given by

$$EI\frac{d^4v}{dx^4} - p\frac{d^2v}{dx^2} = 0 \quad \text{in domain } \Omega \tag{10.32}$$

where p is the compressive axial load applied on two ends of the beam. The boundary conditions for buckling analysis are

$$v(x_0) = \bar{v}, \quad \text{on } \Gamma_v \tag{10.33}$$

$$-\frac{dv(x_0)}{dx} = \bar{\theta}, \quad \text{on } \Gamma_\theta \tag{10.34}$$

$$M(x_0) = EI\frac{d^2v}{dx^2} = 0, \quad \text{on } \Gamma_M \tag{10.35}$$

$$V(x_0) = -EI\frac{d^3v}{dx^3} = 0, \quad \text{on } \Gamma_V \tag{10.36}$$

Similar to Section 10.2, a local weak form of the system equation can be formulated in an integral form over a local domain Ω_s bounded by Γ_s using the weighted residual method

$$\int_{\Omega_Q} \widehat{W}(EIv'''' - pv'')dx = 0 \tag{10.37}$$

where $\widehat{W}$ is the weight function. Integrating by parts for the first term on the left-hand side of Equation 10.37, and imposing the natural boundary condition (Equations 10.35 and 10.36), we obtain

$$\int_{\Omega_Q} (EI\widehat{W}''v'' - \widehat{W}pv'')dx - [\bar{n}EI\widehat{W}'v'']_{\Gamma_{Q\theta}} + [\bar{n}\widehat{W}EIv''']_{\Gamma_{Qv}}$$
$$- [\bar{n}\widehat{W}'EIv'']_{\Gamma_{Qi}} + [\bar{n}\widehat{W}EIv''']_{\Gamma_{Qi}} = 0 \tag{10.38}$$

Note that if the weight function is so chosen that both itself and its first derivatives vanish at Γ_{Qi}, the last two terms in Equation 10.38 vanish.

10.3.2 Discretized System Equations

Substituting Equations 10.13 and 10.28 into the local weak form Equation 10.38 for all nodes in the support domain leads to the following discrete equations:

$$\mathbf{KU} - p\mathbf{BU} = 0 \tag{10.39}$$

where the stiffness matrix **K** is given by Equation 10.30 and **B** is defined by

$$\mathbf{B}_{ij} = \begin{bmatrix} \int_{\Omega_Q} \psi_{vi}\phi''_{vj}\,\mathrm{d}x & \int_{\Omega_Q} \psi_{vi}\phi''_{\theta j}\,\mathrm{d}x \\ \int_{\Omega_Q} \psi_{\theta i}\phi''_{vj}\,\mathrm{d}x & \int_{\Omega_Q} \psi_{\theta i}\phi''_{\theta j}\,\mathrm{d}x \end{bmatrix} \tag{10.40}$$

To obtain a nontrivial solution of Equation 10.39, the determinant of the coefficient matrices in Equation 10.39 should be zero:

$$|\mathbf{K} - p\mathbf{B}| = 0 \tag{10.41}$$

which is the characteristic equation that the critical buckling load, p, has to satisfy. Note that Equation 10.39 can be written in the following general eigenvalue equation form:

$$[\mathbf{K} - p\mathbf{B}]\mathbf{U} = 0 \tag{10.42}$$

Solving this eigenvalue equation using standard routines, one should obtain a set of eigenvalues and a corresponding set of eigenvectors. The eigenvalues give the buckling loads, and the eigenvectors give the buckling modes corresponding to the buckling loads. Note that matrices **K** and **B** are usually asymmetric. Therefore, routines for asymmetric matrices have to be used to solve the eigenvalue equation.

10.3.3 Numerical Example

Example 10.4 Buckling Analysis of Thin Beams

LPIM is used for buckling analyses of thin beams with different boundary conditions. The results of buckling forces are computed using three kinds of nodal arrangement: 21 nodes, 41 nodes, and 61 nodes. Table 10.1 shows the comparison between the critical buckling loads calculated analytically and numerically using LPIM. The dimensionless parameter $\lambda = p_{cr}l^2/EI$ is used in Table 10.1, where l is the length of the beam. It can be observed that the results obtained by LPIM are in very good agreement with the analytical results. The convergence of the present method can also be examined from the data shown in this table. As the number of nodes increases, the numerical results of LPIM approach the analytical solution.

TABLE 10.1

Normalized Critical Buckling Load ($\lambda = p_{cr}l^2/EI$) for Beams with Different Boundary Conditions (Computed Using LPIM and Analytically)

Nodes No.	Pinned-Pinned	Pinned-Fixed	Fixed-Free	Fixed-Fixed
Analytical	9.86960	20.19073	2.46740	39.47842
21	9.87019	20.19330	2.47014	39.48875
41	9.86912	20.19022	2.46721	39.47992
61	9.86954	20.19096	2.46745	39.47878

Source: Gu, Y. T. and Liu, G. R., *Comput. Methods Appl. Mech. Eng.*, 190, 5515–5528, 2001. With permission.

10.4 Free-Vibration Analysis of Thin Beams (Eigenvalue Problem)

10.4.1 Local Weak Form

The dynamic equation that governs vibration of Euler–Bernoulli beams can be written as

$$m_b \frac{d^2 v(x,t)}{dt^2} + \eta_c \frac{dv(x,t)}{dt} + EI \frac{d^4 v(x,t)}{dx^4} = b_y(x,t) \quad \text{in domain } \Omega \tag{10.43}$$

where $v(x, t)$ is the deflection of the beam, which is a function of both coordinate and time, m_b is the mass, and η_c is the damping coefficient. The mass m and damping η_c are assumed constant for simplification.

The governing equation for free vibration of the thin beam is given by

$$m_b \frac{d^2 v(x,t)}{dt^2} + EI \frac{d^4 v(x,t)}{dx^4} = 0 \quad \text{in domain } \Omega \tag{10.44}$$

The boundary conditions are usually of the same form as Equations 10.33 through 10.36. In the free vibration analysis, $v(x, t)$ should be harmonic and, therefore, have the form:

$$v(x,t) = Y(x)\sin(\omega t + \varphi) \tag{10.45}$$

where ω is the frequency. Substituting Equation 10.45 into 10.44 leads to the following differential equation of the deflection amplitude Y with respect to coordinate x:

$$EI \frac{d^4 Y(x)}{dx^4} - \omega^2 m_b Y(x) = 0 \tag{10.46}$$

The local weak form of the differential Equation 10.46 can be written in an integral form over a local domain Ω_Q bounded by Γ_Q using the weighted residual method.

$$\int_{\Omega_Q} \widehat{W}(EIY'''' - \omega^2 m_b Y)dx = 0 \tag{10.47}$$

where $\widehat{W}$ is the weight function. The first term on the left-hand side of Equation 10.47 can be integrated by parts. In doing so, and imposing the natural boundary conditions, we obtain

$$\int_{\Omega_Q} (EI\widehat{W}''Y'' - \widehat{W}\omega^2 m_b Y)dx - [\bar{n}EI\widehat{W}'Y'']_{\Gamma_{Q\theta}} + [\bar{n}\widehat{W}EIY''']_{\Gamma_{Qw}}$$
$$- [\bar{n}\widehat{W}'EIY'']_{\Gamma_{Qi}} + [\bar{n}\widehat{W}EIY''']_{\Gamma_{Qi}} = 0 \tag{10.48}$$

10.4.2 Discretized System Equations

Substituting Equations 10.13 and 10.28 into the local weak form (Equation 10.48) for all nodes leads to the following discrete equations:

$$\mathbf{KY} - \omega^2 \mathbf{MY} = 0 \tag{10.49}$$

where the global stiffness matrix **K** has the same form of Equation 10.30 and **M** is the global mass matrix, which can be assembled using the nodal mass matrix computed using

$$\mathbf{M}_{ij} = m_b \begin{bmatrix} \int_{\Omega_Q} \psi_{vi}\phi_{vj}\,\mathrm{d}x & \int_{\Omega_Q} \psi_{vi}\phi_{\theta j}\,\mathrm{d}x \\ \int_{\Omega_Q} \psi_{\theta i}\phi_{vj}\,\mathrm{d}x & \int_{\Omega_Q} \psi_{\theta i}\phi_{\theta j}\,\mathrm{d}x \end{bmatrix} \tag{10.50}$$

To obtain a nontrivial solution of Equation 10.49, the determinant of the coefficient matrix should be zero:

$$|\mathbf{K} - \omega^2\mathbf{M}| = 0 \tag{10.51}$$

which is the characteristic equation that the natural frequency, ω, has to satisfy. Note that Equation 10.51 can be written in the following general eigenvalue equation form:

$$[\mathbf{K} - \omega^2\mathbf{M}]\mathbf{Y} = 0 \tag{10.52}$$

Solving this eigenvalue equation using standard routines, one should obtain a set of eigenvalues and a corresponding set of eigenvectors. The eigenvalues give the natural frequencies of the free vibration, and the eigenvectors give the free vibration modes corresponding to the natural frequencies. Note that matrices **K** and **M** are usually asymmetric. Therefore, routines for asymmetric matrices have to be used to solve the eigenvalue equation.

10.4.3 Numerical Results

Example 10.5 Free-Vibration Analysis of Thin Beams

LPIM as formulated above is used for free vibration analysis of thin beams with different boundary conditions. A total of 21 uniformly distributed nodes are used. Table 10.2 shows the natural frequencies of the first eight modes calculated analytically and numerically using LPIM. It can be observed that the results obtained by LPIM are in very good agreement with the analytical results. The first three free vibration modes obtained using LPIM and the analytic solution are shown in Figure 10.12. Again, very good results are obtained using LPIM.

TABLE 10.2

Natural Frequency Parameters $\beta_i l$ Computed Using LPIM and Analytical Formulae for Thin Beams of Various Boundary Conditions

	Pinned-Pinned		Pinned-Fixed		Fixed-Free		Fixed-Fixed	
Modes	Analytical	LPIM	Analytical	LPIM	Analytical	LPIM	Analytical	LPIM
1	3.14159	3.14164	3.92699	3.92667	1.87510	1.87510	4.73004	4.73014
2	6.28318	6.28357	7.06858	7.06907	4.69409	4.69419	7.85398	7.85382
3	9.42477	9.42612	10.21018	10.21179	7.853982	7.855357	10.99557	10.99753
4	12.56637	12.56965	13.35177	13.35555	10.99557	10.9974	14.13717	14.14151
5	15.70796	15.71444	16.49336	16.50057	14.13717	14.14134	17.27876	17.28670
6	18.84956	18.86057	19.63495	19.64675	17.27876	17.28635	20.42035	20.43282
7	21.99115	22.00764	22.77655	22.79353	20.42035	20.43221	23.56194	23.57913
8	25.13274	25.15466	25.91814	25.93971	23.56194	23.57813	26.70354	26.72425

Note: Natural frequency $\omega_i = (\beta_i l)^2\sqrt{EI/m_b l^4}$.

Source: Gu, Y. T. and Liu, G. R., *Comput. Methods Appl. Mech. Eng.*, 190, 5515–5528, 2001. With permission.

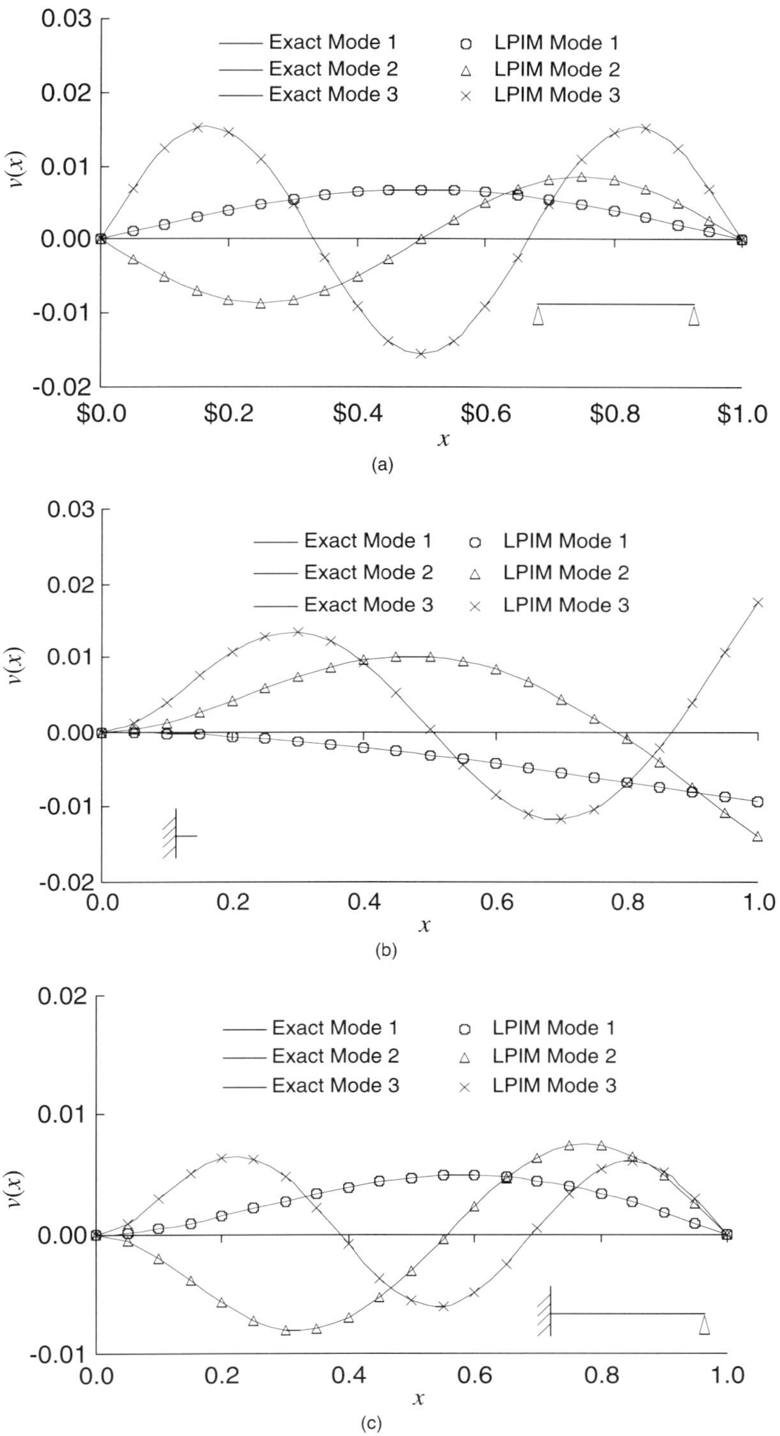

FIGURE 10.12
Free vibration modes of thin beams with different boundary conditions. (a) Pinned-pinned; (b) fixed-pinned; (c) fixed-free; (d) fixed-fixed. (From Gu, Y. T. and Liu, G. R., *Comput. Methods Appl. Mech. Eng.*, 190, 5515–5528, 2001. With permission.) (Continued)

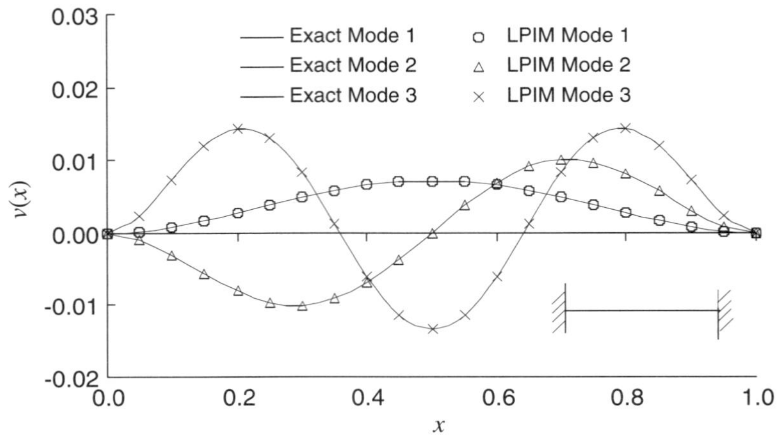

(d)

FIGURE 10.12
(Continued)

10.5 Forced Vibration Analysis of Thin Beams (Time-Dependent Problem)

10.5.1 Local Weak Form

If the external force is not zero, the beam will undergo a forced vibration, which is governed by Equation 10.43. In this case, the boundary conditions can be generally given in Equations 10.20 through 10.23, and the initial conditions are given as

$$v(x, t)\big|_{t=t_0} = \bar{v}(x, t_0) \tag{10.53}$$

$$v'(x, t)\big|_{t=t_0} = \bar{\theta}(x, t_0) \tag{10.54}$$

$$\dot{v}(x, t)\big|_{t=t_0} = \dot{\bar{v}}(x, t_0) \tag{10.55}$$

$$\dot{v}'(x, t)\big|_{t=t_0} = \dot{\bar{\theta}}(x, t_0) \tag{10.56}$$

where the dot denotes the differentiation with respect to time, for example, $\dot{v} = \mathrm{d}v/\mathrm{d}t$ is the velocity, and $\ddot{v} = \mathrm{d}^2v/\mathrm{d}t^2$ is acceleration of the vibrating beam.

The strategy for solving the forced vibration problem is similar to that used in conventional FEM. We use LPIM to discretize in space, and use FDM to integrate over time.

The local weak form of the system equation of forced vibration can be obtained in an integral form over a local domain Ω_Q bounded by Γ_Q using the strong form of Equation 10.43 and the weighted residual method.

$$\int_{\Omega_Q} \widehat{W}(m_b\ddot{v}(x, t) + \eta_c\dot{v}(x, t) + EIv(x, t)'''' - b_y(x, t))\mathrm{d}x = 0 \tag{10.57}$$

The third term on the left-hand side of Equation 10.57 can be integrated by parts. Using the natural boundary conditions, Equations 10.22 through 10.23, we obtain

$$\int_{\Omega_Q}(\widehat{W}m_b\ddot{v}+\widehat{W}\eta_c\dot{v}+EI\widehat{W}''v''-\widehat{W}b_y(x,t))\mathrm{d}x-[\bar{n}\bar{M}\widehat{W}']_{\Gamma_{QM}}-[\bar{n}\bar{V}\widehat{W}]_{\Gamma_{QV}}$$
$$-[\bar{n}EI\widehat{W}'v'']_{\Gamma_{Q\theta}}+[\bar{n}\widehat{W}EIv''']_{\Gamma_{Qv}}-[\bar{n}\widehat{W}'EIv'']_{\Gamma_{Qi}}+[\bar{n}\widehat{W}EIv''']_{\Gamma_{Qi}}=0 \tag{10.58}$$

10.5.2 Discretized System Equations

The field variables are now discretized in space using PIM shape functions in the form of (see Equations 10.13 and 10.28)

$$v(x,t) = \mathbf{\Phi}_v^{\mathrm{T}}(x)\mathbf{v}(t)+\mathbf{\Phi}_\theta^{\mathrm{T}}(x)\mathbf{\theta}(t) \tag{10.59}$$

$$\widehat{W}(x,t) = \mathbf{\Psi}_v^{\mathrm{T}}(x)\boldsymbol{\alpha}(t)+\mathbf{\Psi}_\theta^{\mathrm{T}}(x)\boldsymbol{\beta}(t) \tag{10.60}$$

Substituting Equations 10.59 and 10.60 into the local weak form Equation 10.58 and performing the integration for all the nodes in the problem domain lead to the following discrete equations:

$$\mathbf{M}\ddot{\mathbf{U}}(t)+\mathbf{C}\dot{\mathbf{U}}(t)+\mathbf{K}\mathbf{U}(t) = \mathbf{F}(t) \tag{10.61}$$

where the global stiffness matrix **K** has the same form of Equation 10.30, the global mass matrix **M** can be formed using Equation 10.50, and the global dumping matrix **C** is formed by assembling the nodal matrix given by

$$\mathbf{C}_{ij} = \eta_c\begin{bmatrix}\int_{\Omega_Q}\psi_{vi}\phi_{vj}\,\mathrm{d}x & \int_{\Omega_Q}\psi_{vi}\phi_{\theta j}\,\mathrm{d}x\\ \int_{\Omega_Q}\psi_{\theta i}\phi_{vj}\,\mathrm{d}x & \int_{\Omega_Q}\psi_{\theta i}\,\phi_{\theta j}\,\mathrm{d}x\end{bmatrix} \tag{10.62}$$

The global force vector **F** is formed by assembling the nodal force vector defined by

$$\mathbf{f}_i(t) = \begin{Bmatrix}\int_{\Omega_Q}\psi_{vi}f(x,t)\mathrm{d}x\\ \int_{\Omega_Q}\psi_{\theta i}f(x,t)\mathrm{d}x\end{Bmatrix}+\bar{n}\bar{M}(t)\begin{Bmatrix}\psi'_{vi}\\ \psi'_{\theta i}\end{Bmatrix}_{\Gamma_{QM}}+\bar{n}\bar{V}(t)\begin{Bmatrix}\psi_{vi}\\ \psi_{\theta i}\end{Bmatrix}_{\Gamma_{QV}} \tag{10.63}$$

Equation 10.61 is the discretized dynamic system equation for the entire beam, which has a form similar to FEM. All the techniques developed for solving the finite element systems can be utilized to solve Equation 10.61. There are basically two types of methods. One is the modal superimposition method, and the other is the direct integration method. Both are applicable in the LPIM formulation. Note that matrices **K**, **M**, and **C** produced by the LPIM formulation are asymmetric, and hence routines for asymmetric matrices have to be used.

In using direct integration techniques to solve Equation 10.61, several methods have been developed to solve the dynamic systems like Equation 10.61, such as the central difference method (CDM) and the Newmark method (see, e.g., Petyt, 1990). The central difference method and the Newmark method are used in this section. The central difference method expresses the velocity and acceleration at time t in terms of a displacement, say, v, at time $t - \Delta t$, t, and $t + \Delta t$ in the form:

$$v(t) = \frac{1}{\Delta t^2}(v(t-\Delta t) - 2v(t) + v(t+\Delta t)) \tag{10.64}$$

$$\dot{v}(t) = \frac{1}{2\Delta t}(-v(t-\Delta t) + v(t+\Delta t)) \tag{10.65}$$

where Δt is the time step.

10.5.3 Numerical Results

Example 10.6 Vibration of a Pinned-Pinned Thin Uniform Beam Subject to Harmonic Loading

The forced vibration of a pinned-pinned thin uniform beam is analyzed. In this numerical example of forced vibration analysis, we assume $f(t) = q_0 \sin(gt)$, $q_0 = 1.0$, $g = 20.0$, $m_b = 1.0$, $EI = 1.0$, $l = 1.0$, and $\omega_i = (\beta_i l)^2 \sqrt{EI/m_b l^4} = 0.0$. For comparison, an analytical solution for this problem is obtained with the mode-superposition method using the exact formulation of natural modes of the beam. This comparison can therefore serve as validation of LPIM.

The analytic solutions of (angular) natural frequencies for the pinned-pinned uniform beam are

$$\omega_i = (i\pi)^2 \sqrt{\frac{EI}{m_b l^4}} \tag{10.66}$$

The analytic shapes of normal modes for the pinned-pinned uniform beam are

$$\phi_i(x) = \sqrt{2}\sin\frac{i\pi x}{l} \tag{10.67}$$

By using the modal superimposition technique, the deflection of the beam is expressed by the summation

$$w(x, t) = \sqrt{2}\sum_{i=1}^{\infty} q_i(t)\sin\frac{i\pi x}{l} \tag{10.68}$$

where $q_i(t)$ is defined as

$$q_i(t) = A_i\left(\frac{\sin(gt)}{\omega_i^2 - g^2} - \frac{g\sin(\omega_i t)}{\omega_i(\omega_i^2 - g^2)}\right) \tag{10.69}$$

$$A_i = \sqrt{2}\frac{1}{i\pi}[(-1)^{i+1} + 1] \tag{10.70}$$

LPIM is used for forced vibration analysis of the pinned-pinned thin beam using 21 uniformly distributed nodes. The time step is set to be $\Delta t = T_0/\pi$, where T_0 is the free vibration period of one sub-beam of length l_s, where l_s is the shortest distance between two nodes.

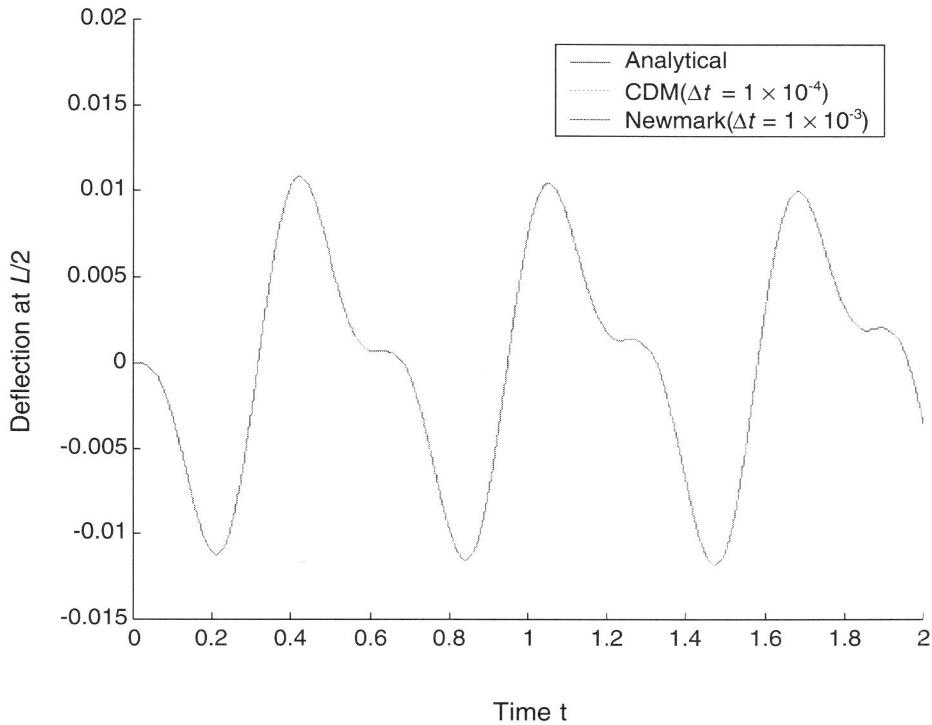

FIGURE 10.13
Deflections at $x = L/2$ of a pinned-pinned beam subjected to simple harmonic force excitation. Results obtained by three approaches coincide.

CDM and the Newmark method are used to obtain the deflection results. Results of different time steps are obtained and plotted in Figure 10.13. Results obtained by both methods are in very agreement with the analytical result when $\Delta t = 1.0 \times 10^{-4}$ is used for CDM and $\Delta t = 1.0 \times 10^{-3}$ is used for the Newmark method. When $\Delta t > \Delta t_{crit}$ is used, e.g., $\Delta t = 1.0 \times 10^{-3}$, which works for the Newmark method, CDM becomes unstable. This demonstrates that CDM is a conditionally stable method and the Newmark method is an unconditionally stable method. Note that we already have a very good understanding of these issues in FEM. Here we are simply revisiting one of them using an MFree method.

Many time steps have been tried and tested to check the stability of the presented LPIM. The calculated deflection results of the beam by LPIM are listed in Table 10.3, together with the analytical results of modal superimposition. Table 10.3 shows that the LPIM results are in very good agreement with the analytical results, and the stability of LPIM is also very good, as long as $\Delta t < \Delta t_{crit}$ is satisfied.

Example 10.7 Vibration of a Pinned-Pinned Thin Uniform Beam Subject to Transient Loading

A simply supported uniform beam, as shown in Figure 10.14, is loaded by a transient loading of triangular distribution defined by

$$b_y(x, t) = \frac{f_0 x}{2} g(t) \tag{10.71}$$

where the time history of the force is given in Figure 10.15.

TABLE 10.3

Comparison of Deflections between the LPIM Results and Analytical Results of Mode Superposition for the Forced Vibration Problem of the Pinned-Pinned Thin Beam

Time Step	Node No.	LPIM Solution	Analytical Solution	Error (%)
100000	3	2.28378E-03	2.28245E-03	–5.82E-02
	5	4.37707E-03	4.37468E-03	–5.47E-02
	7	6.06858E-03	6.06544E-03	–5.17E-02
	9	7.16980E-03	7.16631E-03	–4.88E-02
	11	7.55206E-03	7.54853E-03	–4.68E-02
200000	3	–1.10688E-03	–1.10557E-03	–1.19E-01
	5	–2.08017E-03	–2.07788E-03	–1.10E-01
	7	–2.83107E-03	–2.82823E-03	–1.00E-01
	9	–3.30369E-03	–3.30034E-03	–1.02E-01
	11	–3.46505E-03	–3.46137E-03	–1.06E-01
300000	3	2.18601E-03	2.18557E-03	–2.00E-02
	5	4.14165E-03	4.14071E-03	–2.27E-02
	7	5.67674E-03	5.67464E-03	–3.70E-02
	9	6.65263E-03	6.64922E-03	–5.13E-02
	11	6.98690E-03	6.98293E-03	–5.69E-02
400000	3	–3.82316E-03	–3.82220E-03	–2.51E-02
	5	–7.31284E-03	–7.31128E-03	–2.14E-02
	7	–1.01196E-02	–1.01185E-02	–1.02E-02
	9	–1.19398E-02	–1.19399E-02	1.08E-03
	11	–1.25703E-02	–1.25710E-02	5.41E-03
500000	3	1.44739E-03	1.45008E-03	1.85E-01
	5	2.73840E-03	2.74272E-03	1.58E-01
	7	3.75132E-03	3.75637E-03	1.34E-01
	9	4.39674E-03	4.40251E-03	1.31E-01
	11	4.61835E-03	4.62459E-03	1.35E-01

Source: Gu, Y. T. and Liu, G. R., *Comput. Methods Appl. Mech. Eng.*, 190, 5515–5528, 2001. With permission.

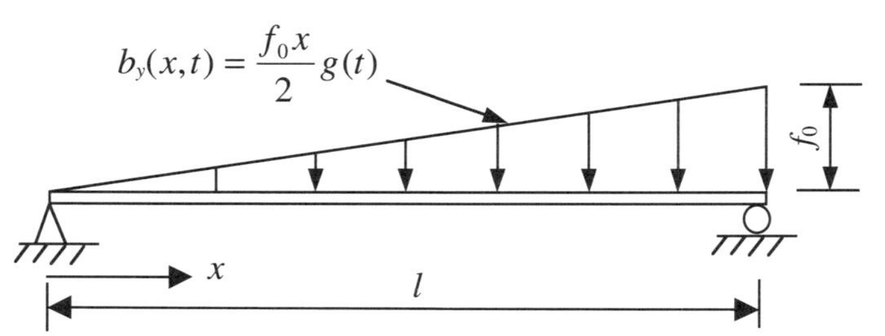

FIGURE 10.14
Simply supported beam subjected to the transient loading of triangular distribution.

The parameter of the beam is considered as $m = 1.0$, $EI = 1.0$, $l = 1.0$, and $c = 0.0$. The analytical solution for this problem is obtained using the mode-superposition method as follows:

$$v(x,\, t) \;=\; \sqrt{2}\sum_{i=1}^{\infty} q_i(t)\sin\frac{i\pi x}{l} \tag{10.72}$$

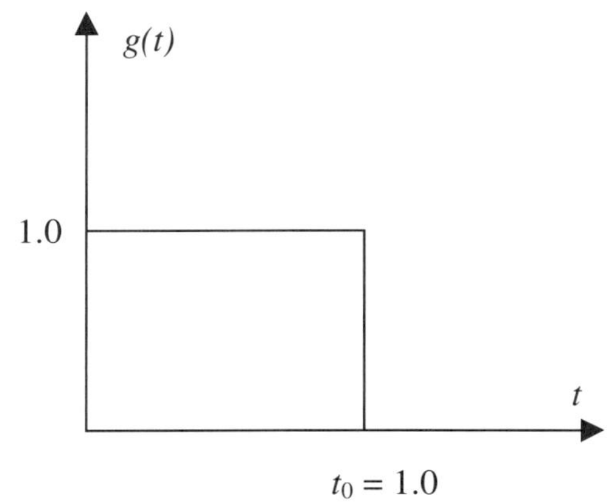

FIGURE 10.15
The history of the excitation force $g(t)$.

where $q_i(t)$ is found to be

$$q_i(t) = \frac{-\sqrt{2l}f_0}{n\pi}\frac{(-1)^n}{\omega_n^2}(1 - \cos\omega_n t) \quad \text{when } 0 \le t \le t_0 \tag{10.73}$$

$$q_i(t) = \frac{-\sqrt{2l}f_0}{n\pi}\frac{(-1)^n}{\omega_n^2}(1 - \cos\omega_n t) + \frac{\sqrt{2l}f_0}{n\pi}\frac{(-1)^n}{\omega_n^2}[1 - \cos\omega_n(t - t_0)] \quad \text{when } t_0 \le t \le \infty \tag{10.74}$$

The Newmark method is used to obtain the deflection. Results obtained using $\Delta t = 1.0 \times 10^{-3}$ and $\Delta t = 1.0 \times 10^{-2}$ are obtained and plotted in Figure 10.16. This figure shows that the results of $\Delta t = 1.0 \times 10^{-3}$ coincide with the analytical results. When the time step is increased to $\Delta t = 1.0 \times 10^{-2}$, the numerical error will increase, but the results are still stable. This demonstrates the feature of unconditional stability of the Newmark method.

10.6 Timoshenko Beams

In previous sections, LPIM has been formulated for thin beams based on the Euler–Bernoulli beam theory. The thin beam theory works for beams whose height is less than about one tenth of its span. For thick (or deep) beams, one needs to use theories for thick beams. This section formulates LPIM and LRPIM equations based on the Timoshenko beam theory, which work for relatively thicker beams. In the LPIM formulation, the polynomial PIM shape functions are used, and in the LRPIM formulation, RPIM shape functions are used (see Chapter 5 for PIM shape functions). This work was initially performed by G. R. Liu and Gu in 2001. Because the procedure of creating the PIM or RPIM shape functions are very much the same as described in Section 10.1 and Chapter 5, we skip the formulation of shape functions for Timoshenko beams. The only difference in creating the PIM shape functions for Timoshenko beams compared to that for thin Euler–Bernoulli beams is that the deflection and rotation should be interpolated separately.

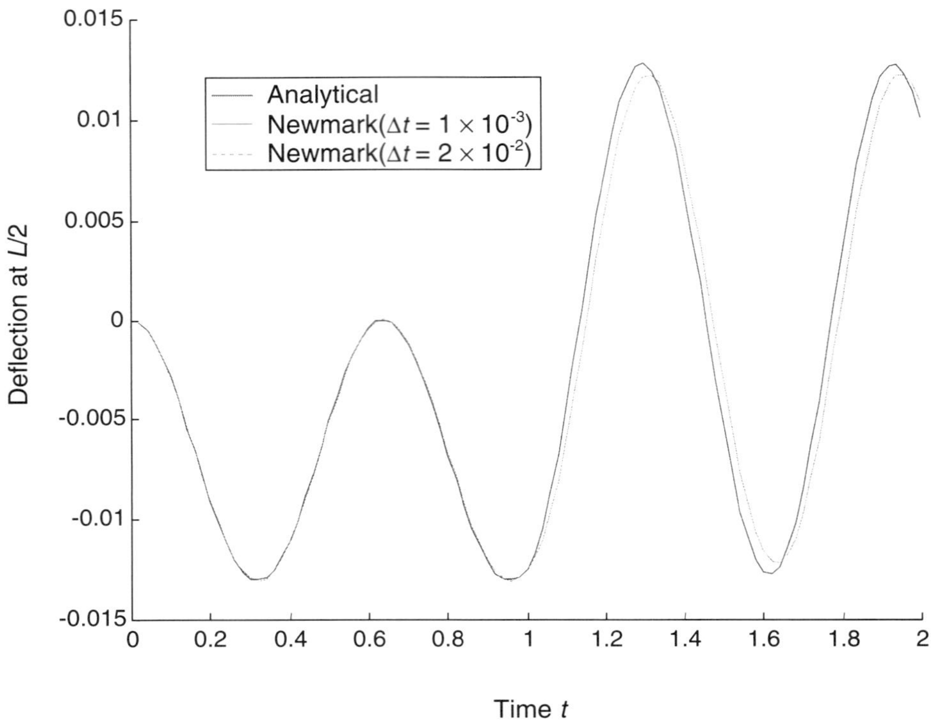

FIGURE 10.16
Deflections at $x = L/2$ of a simply supported beam subjected to transient loading.

10.6.1 Local Weak Form

The Timoshenko beam theory accounts for transverse shear deformation, which is neglected in Euler–Bernoulli beam theory, because the shear deformation effect is not negligible if the thickness of a beam becomes larger. The general equation of static deformation of Timoshenko beam theory is written as (for undamped systems, Equation 3.56)

$$\begin{aligned} -\frac{\partial}{\partial x}\left[GA\kappa\left(\frac{\partial v}{\partial x}+\theta\right)\right]-b_y &= 0 \\ -\frac{\partial}{\partial x}\left(EI\frac{\partial\theta}{\partial x}\right)+GA\kappa\left(\frac{\partial v}{\partial x}+\theta\right) &= 0 \quad \text{in domain } \Omega \end{aligned} \tag{10.75}$$

where v is the deflections of the beam, A is the cross-sectional area, G is the shear modulus, and K is the shear correction coefficient.

The boundary conditions are given as

$$v(x_0) = \bar{v}, \quad \text{on } \Gamma_v \tag{10.76}$$

$$\theta(x_0) = \bar{\theta}, \quad \text{on } \Gamma_\theta \tag{10.77}$$

$$M(x_0) = EI\frac{\mathrm{d}^2 v}{\mathrm{d}x^2} = \overline{M}, \quad \text{on } \Gamma_M \tag{10.78}$$

$$V(x_0) = -EI\frac{\mathrm{d}^3 v}{\mathrm{d}x^3} = \overline{V}, \quad \text{on } \Gamma_V \tag{10.79}$$

The local weak form of the differential Equation 10.75 can be written in an integral form over a local quadrature domain Ω_Q bounded by Γ_Q using the weighted residual method:

$$\begin{cases} \int_{\Omega_Q} \widehat{W}_1 \left\{ -\frac{\partial}{\partial x}\left[GA\kappa\left(\frac{\partial v}{\partial x} + \theta \right) \right] - b_y \right\} \mathrm{d}\Omega = 0 \\ \int_{\Omega_Q} \widehat{W}_2 \left[-\frac{\partial}{\partial x}\left(EI\frac{\partial \theta}{\partial x} \right) + GA\kappa\left(\frac{\partial v}{\partial x} + \theta \right) \right] \mathrm{d}\Omega = 0 \end{cases} \tag{10.80}$$

where $\widehat{W}_1$ and $\widehat{W}_2$ are two weight functions that can be constructed using the point interpolation method. They can be the same or independent. In this chapter they are the same:

$$\widehat{W}_2 = \widehat{W}_1 \tag{10.81}$$

10.6.2 Discretized System Equations

In using Timoshenko beam theory, the deflection, v, and the rotation, θ, are treated as independent variables. They are, hence, interpolated separately using PIM shape functions:

$$v = \mathbf{\Phi}_v \mathbf{v}_s \tag{10.82}$$

$$\theta = \mathbf{\Phi}_\theta \boldsymbol{\theta}_s \tag{10.83}$$

where $\mathbf{\Phi}_v$ and $\mathbf{\Phi}_\theta$ are matrices of PIM shape functions constructed separately for the deflection and rotation using the exact procedure described in Chapter 5. These shape functions can be constructed using either polynomial basis functions (for LPIM) or radial basis functions (for LRPIM).

Performing integration by parts to the left-hand side of Equation 10.80, imposing the natural boundary condition (Equations 10.78 and 10.79), and substituting Equations 10.82 and 10.83 into the resultant two equations, we can obtain the discretized global system equations of LPIM or LRPIM for beams. The global system equations are assembled using the nodal system equation for node i given as follows:

$$\begin{bmatrix} K_{ij}^{11} & K_{ij}^{12} \\ K_{ij}^{21} & K_{ij}^{22} \end{bmatrix} \begin{Bmatrix} v_i \\ \theta_i \end{Bmatrix} = \begin{Bmatrix} f_i^v \\ f_i^\theta \end{Bmatrix} \tag{10.84}$$

where

$$K_{ij}^{11} = \int_{\Omega_Q} GA\kappa \frac{\mathrm{d}\widehat{W}_{1i}}{\mathrm{d}x} \frac{\mathrm{d}\Phi_j^v}{\mathrm{d}x} \mathrm{d}\Omega + \left[GA\kappa \widehat{W}_{1i} \frac{\mathrm{d}\Phi_j^v}{\mathrm{d}x} \right] \Bigg|_{\Gamma_{Qi}+\Gamma_{Qv}+\Gamma_{Q\theta}+\Gamma_{QM}} \tag{10.85}$$

$$K_{ij}^{12} = \int_{\Omega_Q} GA\kappa \frac{\mathrm{d}\widehat{W}_{1i}}{\mathrm{d}x} \Phi_j^\theta \mathrm{d}\Omega + [GA\kappa \widehat{W}_{1i} \Phi_j^\theta] \Big|_{\Gamma_{Qi}+\Gamma_{Qv}+\Gamma_{Q\theta}+\Gamma_{QM}} \tag{10.86}$$

$$K_{ij}^{21} = \int_{\Omega_s} GA\kappa \widehat{W}_{2i} \frac{\mathrm{d}\Phi_j^v}{\mathrm{d}x} \mathrm{d}\Omega \tag{10.87}$$

$$K_{ij}^{22} = \int_{\Omega_Q} \left(EI \frac{\mathrm{d}\widehat{W}_{2i}}{\mathrm{d}x} \frac{\mathrm{d}\Phi_j^{\theta}}{\mathrm{d}x} + GA\kappa \widehat{W}_{2i} \Phi_j^{\theta} \right) \mathrm{d}\Omega + \left[EI \widehat{W}_{2i} \frac{\mathrm{d}\Phi_j^{\theta}}{\mathrm{d}x} \right] \Bigg|_{\Gamma_{Qi}+\Gamma_{Qv}+\Gamma_{Q\theta}+\Gamma_{QV}} \tag{10.88}$$

$$f_i^{v} = \int_{\Omega_Q} \widehat{W}_{1i} b_y \,\mathrm{d}\Omega + \widehat{W}_{1i} \bar{V} \Big|_{\Gamma_{QV}} \tag{10.89}$$

$$f_i^{\theta} = \widehat{W}_{2i} \overline{M} \Big|_{\Gamma_{QM}} \tag{10.90}$$

In the above equations, $\widehat{W}_{1i}\widehat{W}_{2i}$ denote, respectively, the weight functions $\widehat{W}_1$ and $\widehat{W}_2$ for node i.

10.6.3 Numerical Example

Example 10.8 Static Deflection of Timoshenko Beams

LPIM has been used for static deflection analysis for Timoshenko beams under different loads and boundary conditions. The detailed results for a pinned-pinned beam are presented here. The following error indicator is defined to examine the LPIM methods. Both polynomial LPIM and MQ-LRPIM are coded and used in this study.

For precise investigation, we define the following error indicator using the deflections of the beam computed numerically, using LPIM or LRPIM, and analytically.

$$e = \frac{\sqrt{\int_{\Omega} (v_{\mathrm{NUM}} - v_{\mathrm{EXACT}})^2 \mathrm{d}\Omega}}{\sqrt{\int_{\Omega} v_{\mathrm{EXACT}}^2 \,\mathrm{d}\Omega}} \tag{10.91}$$

where v_{NUM} and v_{EXACT} are deflections of the beam obtained by numerical methods and the analytical method, respectively. The integration is performed over the entire span of the beam.

Shear Locking

As well known, there is a shear-locking phenomenon in analyses of thick beams if the length and thickness ratio (i.e., L/D) of a beam becomes much larger (see Reddy, 1997). Figure 10.17 shows displacements of different L/D ratios obtained by local PIM using lower-order point interpolation. This figure shows that the shear locking occurs when the beam becomes too thin ($L/D > 50$). Both LPIM and MQ-LRPIM failed to predict the thin beam behavior. In FEM, a number of methods have been developed to deal with this kind of shear-locking problem. Details can be found in the textbook by Reddy (1993). The simple and straightforward method is to use a shape function for the rotation that is one order lower than that for the deflection. This idea has been implemented by Kanok–Nukulchai et al. (2001) in their EFG formulation for beams, and achieved good results. We will also use this method to handle the shear locking in Chapter 11, when we discuss thick plates.

Another method is to use equal-order interpolation for both deflection and rotation, but to evaluate the energy in a proper manner. A similar approach was implemented in MLPG by Cho and co-workers (2000). They added the transverse shear strain as another variable to proposed a "locking-free" formulation. It is reported that the method is very effective in analyses of thick beams. These ideas can also be used in LPIM and LRPIM to avoid shear locking. However, we introduce here alternative and simpler ways to deal with the shear-locking problem in our LPIM methods.

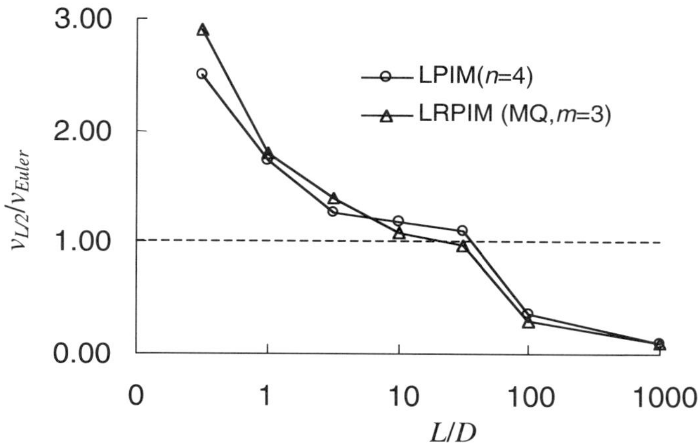

FIGURE 10.17
Observation of shear locking; $v_{L/2}$: middle deflection obtained by numerical methods; v_{Euler}: middle deflection obtained by Euler–Bernoulli beam theory.

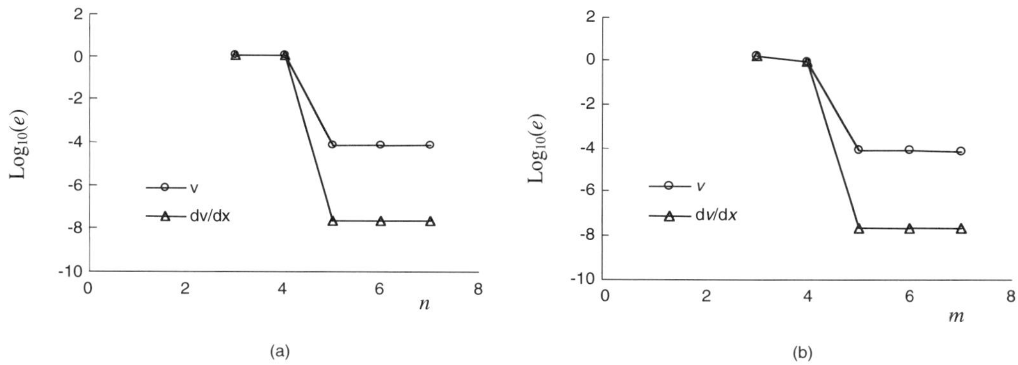

FIGURE 10.18
Errors computed using LPIM with different order of point interpolation. (a) LPIM; (b) MQ-LRPIM.

In FEM analysis, it has also been found that there is no shear locking if high-order elements are used. Because there is no difficulty in using point interpolations to construct high-order shape functions, we therefore studied the possibility of using high-order point interpolation to eliminate shear locking when the deflection and rotation use the same interpolation formulation. In the study of the shear locking, 20 regularly distributed nodes are used in LPIM to discretize the beam.

In LPIM, the order of shape functions is determined by the number of nodes, n, chosen in the support domains. The errors e defined in Equation 10.91 are computed using LPIM and MQ-PIM for different orders of polynomial basis and plotted in Figure 10.18a. This figure shows that there is shear locking when the order polynomial basis is lower than five. When the order of point interpolation is larger than five, or when quartic polynomial basis functions are used, there is no shear locking at all. Usually, LPIM can easily satisfy the requirement of higher-order shape functions. Therefore, the shear-locking phenomenon can be easily avoided by simply using high-order shape functions in LPIM.

In MQ-LRPIM, the radial basis point interpolation includes two parts: the radial basis functions and the polynomial basis functions (see Chapter 5). The order of the radial part has been predetermined, and cannot change freely, because one of the nodes is fixed. The order of the polynomial part in LRPIM can be changed freely by changing m. Our studies found that the radial part does not contribute to overcoming the shear locking. Therefore, errors for different m are obtained and plotted in Figure 10.18b. This figure shows that there is shear locking when fewer ($m \leq 4$) polynomial terms are used in the radial basis point interpolation, even though more nodes are used in creating the RPIM shape function. When there are more polynomial terms ($m \geq 5$), or quartic polynomial basis functions are included, there is no shear locking. The result is the same as the result in LPIM. Therefore, we can conclude that only the order of the polynomial basis in LRPIM affects the shear-locking phenomenon, and a sufficient number of polynomial terms are necessary in LRPIM to avoid shear locking.

It may also be mentioned here that using shape functions of too high an order does not necessarily improve the accuracy significantly. This fact, clearly evidenced in Figure 10.18, implies that as long as the order of shape functions is high enough to avoid shear locking, the order of shape functions does not play an important role. This reason is that, for our beam problem, the true solution does not vary very drastically. There is no need to use a polynomial basis of too high order.

It may also be mentioned here that, because polynomial terms have to be included in the radial basis to avoid shear locking, and also because PIM has no problem in the moment matrix, the use of LRPIM gives no advantage for analyzing beams.

The Study of Convergence

The convergences of LPIM and LRPIM for Timoshenko beams are studied. The beam is of length $L = 1.0$ and height $D = 0.01$. Evenly distributed nodes are used. The convergence of the norm e defined in Equation 10.91 is computed for different node refinements and is plotted in Figure 10.19. The h is the distance between two neighboring nodes. It is observed that the convergences of LPIM and MQ-LRPIM are very good, and that LPIM performs better than MQ-LRPIM.

LPIM and MQ-LRPIM have also been used for buckling analysis, free vibration analysis, and forced vibration analysis of Timoshenko beams. The treatments are very much the same as those discussed in the previous sections of this chapter. We state without further citing the numerical results that very accurate results have been obtained.

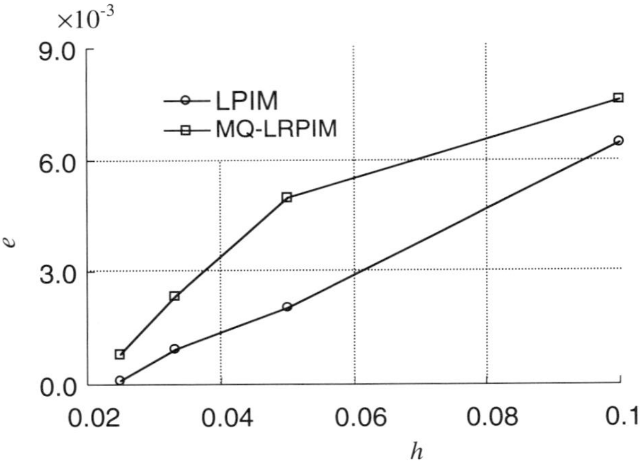

FIGURE 10.19
Convergence of numerical results obtained by LPIM and MQ-LPIM.

10.7 Remarks

LPIM have been presented to solve fourth-order boundary and initial problems of both the thin beam based on Euler–Bernoulli beam theory and the thick beam based on Timoshenko beam theory. The following may be mentioned:

1. For geometrically 1D problems, there is no need at all to use elements.
2. For geometrically 1D problems, there is no need at all to use MLS approximation. PIM works much better without any difficulty.
3. PIM methods have a clear advantage in problems governed by high-order differential equations, as the order of interpolation can be as high as desired.
4. Shear locking can be easily resolved simply by using more nodes in creating PIM shape functions.

In analyzing beams, creating cells of background mesh for integration is not a problem. Therefore, using PIM based on the constrained Galerkin formulation can be a good (even better) choice, because it leads to a set of symmetric system equations that can be solved in a much more efficient manner.

11

Mesh Free Methods for Plates

With the wide application of plate structures of complex geometry, static and dynamic analyses of plates of complicated shape become very important. However, exact analyses of such plates are usually very difficult. Therefore, numerical techniques with different discretization schemes such as finite element method (FEM) have been developed. FEM has achieved remarkable success in static and dynamic analyses of plates.

MFree methods have also been developed for analysis of plates. Krysl and Belytschko (1996a,b) have extended the element free Galerkin (EFG) method to static analysis of thin plates and shells. In their work, the essential boundary conditions are enforced by a method of Lagrange multipliers. An EFG method has also been formulated for modal analyses of Euler-Bernoulli beams and Kirchhoff plates (Ouatouati and Johnson, 1999). Analysis of thin plates with straight lines has been performed, where the essential boundary conditions are enforced directly at each constraint boundary point. EFG has also been formed for dynamic problems (Liu, G. R. and Chen, 2001), buckling problems of thin plates (Liu, G. R. and Chen, 2000; 2002a), as well as composite laminates (Chen, X. L. et al., 2001; 2002b). In their work, the essential boundary conditions are imposed using orthogonal transform techniques.

EFG formulations for thick plates were presented by G. R. Liu and Chen in 2001 for both static and dynamic problems based on first- and third-order shear deformation theories. When using higher-order plate theories, the well-known issue of shear locking arises. Therefore, different methods for eliminating shear locking have also been proposed.

All the EFG formulations require a background mesh of cells for integration. The MLPG method, which needs only a local mesh of cells for integration, has also been formulated by Gu and G. R. Liu (2001f) for thin plates for both static and dynamic problems. A detailed investigation of the effects of various parameters has been conducted.

PIM methods that produce shape functions with the Kronecker delta function property have also been formulated by G. R. Liu and Chen since early 2001. The work in this direction is still in progress, and some of the current work will be presented in this chapter.

This chapter is dedicated to MFree methods for analysis of plates. Both formulations and applications of EFG, MLPG, as well as PIM will be presented in detail. Problems considered include static deformation, buckling, and dynamic response of thin and thick plates, including composite laminated plates. Issues related to shear locking when using theories of thick plates are also addressed.

11.1 EFG Method for Thin Plates

This section presents an EFG method for static and free-vibration analyses of plates following the work done by G. R. Liu and Chen (2001). Formulations of discretized system equations based on Kirchhoff's thin plate theory (or classic plate theory) are provided. Methods are introduced for the imposition of essential boundary conditions. For static deflection of thin plates, a penalty method is formulated. For analysis of free vibration of thin plates, the essential boundary conditions are formulated via a weak form, which is distinct from the weak form of the system equation. The boundary conditions are then imposed using orthogonal transform techniques. The eigenvalue equation derived using the present approach is of smaller dimension than that in FEM.

A number of numerical examples are presented. Static deflections of thin rectangular plates with fully clamped and simply supported boundaries are computed using the present approach. Natural frequencies of thin square, elliptical, hexagonal, and complicated plates with various boundaries such as free, simply supported, and fully clamped are also calculated. Both regularly and irregularly distributed nodes are used in the computation to reveal the sensitivity of the results to the irregularity of the nodes. Examples are presented to demonstrate the convergence and validation of the EFG formulation compared with analytical solutions.

It is shown that the present EFG formulation has a clear advantage over element-based formulations, as the nodal variable in the EFG methods is only one (deflection) compared to three in the element-based formulation (one deflection and two rotations). The dimension of the discretized system equations generated using MFree methods is therefore one third of that generated using FEM. In addition, in EFG there is no "conformability" issue, which exists on the interface between the finite elements, as there is no element boundary in the MFree methods, and higher-order consistency can be achieved in MFree methods easily by including higher-order terms in the polynomial basis.

11.1.1 Approximation of Deflection

Shape Function

Consider a plate of Ω shown in Figure 3.12. A Cartesian coordinate system is used to establish the system equations. The plate is represented by its neutral plane of the plate. The deflections of the plate in the x, y, z directions are denoted as u, v, w, respectively.

Based on Kirchhoff's thin plate assumption, the deflection $w(\mathbf{x})$ of its neutral plane at $\mathbf{x} = \{x, y\}^T$ can be taken as an independent variable, and the other two displacement components $u(\mathbf{x})$ and $v(\mathbf{x})$ can be expressed in terms of $w(\mathbf{x})$.

The plate is represented by a set of field nodes scattered in the domain of the plate. The field variables should be the deflections and the two rotations with respect to the x and y axis at all the nodes. Moving least square (MLS) approximation is employed to approximate $w(\mathbf{x})$ using nodes included in the support domain of $\mathbf{x}$, and hence the two rotations are also approximated in relation to the deflection. The process of approximating the deflection is the same as that discussed in Section 5.4. The only differences are as follows:

1. For thin plates, we should use more terms of polynomial basis functions. This is because the rotations that relate to the derivative of deflection are also the field variables, and they depend on the deflection function. Therefore, a higher order of consistency is required. In the examples given in this section, quadratic (complete second order, $m = 6$) of polynomial basis functions are used.
2. Derivatives of higher orders of shape functions are required in deriving system equations.

Using MLS approximation, the deflection of the plates can be approximated using parameters of nodal deflection, w_I, in the following form:

$$w^h(\mathbf{x}) = \sum_{I=1}^{n} \phi_I(\mathbf{x}) w_I \tag{11.1}$$

where n is the number of nodes in the support domain of a point of interest $\mathbf{x}$, and ϕ_I is the MLS shape function obtained in exactly the same way described in Section 5.6, but using quadratic polynomial basis.

11.1.2 Variational Forms

The strong form of system equations that govern thin plates is given in Chapter 3. The boundary conditions are given at global boundary, Γ, as follows. The essential boundaries Γ_u consist of

$$w = \bar{w}, \quad \text{on essential boundary } \Gamma_w \tag{11.2}$$

$$\theta_n = \frac{\partial w}{\partial n} = \bar{\theta}, \quad \text{on essential boundary } \Gamma_\theta \tag{11.3}$$

where θ_n is the rotation on the boundary about the boundary line, $\bar{w}$ and $\bar{\theta}$ are the prescribed deflection and rotation on the essential boundaries, and n denotes the normal of the boundary of the problem domain Ω. On the natural boundaries

$$M_n = \bar{M}_n, \quad \text{on natural boundary } \Gamma_M \tag{11.4}$$

$$M_{ns} = \bar{M}_{ns}, \quad \text{on natural boundary } \Gamma_M \tag{11.5}$$

$$V_n = \bar{V}, \quad \text{on natural boundary } \Gamma_V \tag{11.6}$$

where M_n, M_{ns}, and V_n denote the moment, torsional moment, and shear force on the boundary of the plate, which are defined by

$$M_n = n_x^2 M_{xx} + 2n_x n_y M_{xy} + n_y^2 M_{yy} \tag{11.7}$$

$$M_{ns} = -n_y n_x M_{xx} + n_x n_y M_{yy} \tag{11.8}$$

in which (n_x, n_y) is the unit outward normal vector on the boundary, M_{xx}, M_{yy}, and M_{xy} are obtained using Equations 3.76 through 3.78, 3.81, and 3.82. $\bar{M}_n$, $\bar{M}_{ns}$, and $\bar{V}_n$ are, respectively, the prescribed moments and shear force on the plate edges of natural boundary.

The essential boundary condition can be written in a concise form of

$$\tilde{\mathbf{u}} = \bar{\mathbf{u}} \quad \text{on essential boundary } \Gamma_u = \Gamma_w \cup \Gamma_\theta \tag{11.9}$$

where $\bar{\mathbf{u}}$ is a vector consisting of the prescribed deflection and rotation at the essential boundary of the plate. Vector $\tilde{\mathbf{u}}$ is given by

$$\tilde{\mathbf{u}} = \mathbf{L}_b w \tag{11.10}$$

where $\mathbf{L}_b$ is a vector of differential operators given by

$$\mathbf{L}_b = \begin{Bmatrix} 1 \\ \frac{\partial}{\partial n} \end{Bmatrix} \quad \text{for clamped boundary} \tag{11.11}$$

and

$$\mathbf{L}_b = \begin{Bmatrix} 1 \\ 0 \end{Bmatrix} \quad \text{for simply supported boundary} \tag{11.12}$$

Note in Equation 11.12 that the zero entry in $\mathbf{L}_b$ ensures that the deflection is constrained for simply supported boundaries.

Because the Kronecker delta condition is not satisfied by the MLS shape function, the constraints of essential boundary conditions (Equation 11.9) need to be imposed in a proper manner. For static deflection analysis of thin plates, we use the penalty method to enforce essential boundary conditions. The constrained Lagrangian weak form can be written as

$$\int_A \delta\boldsymbol{\varepsilon}_p^T \boldsymbol{\sigma}_p \, dA - \int_\Omega \delta\mathbf{u}^T \mathbf{b} \, d\Omega - \int_{S_t} \delta\mathbf{u}^T \bar{\mathbf{t}} \, dS - \frac{1}{2}\int_{\Gamma_u} \delta(\tilde{\mathbf{u}} - \bar{\mathbf{u}})^T \boldsymbol{\alpha} (\tilde{\mathbf{u}} - \bar{\mathbf{u}}) \, d\Gamma = 0 \tag{11.13}$$

where A stands for the area of the plate, S_t is the surface of the plate edge of natural boundary, $\boldsymbol{\varepsilon}_p$ is the pseudo-strain defined by Equation 3.68, $\boldsymbol{\sigma}_p$ is the pseudo-stress defined by Equation 3.74, $\mathbf{b}$ is a body force vector, and $\boldsymbol{\alpha}$ is a diagonal matrix of penalty factors, which are usually very large numbers. In Equation 11.13, the first term relates to the virtual work done by the internal stress in the thin plate, and the second term relates to that by the body force, which may be distributed over the entire volume of the plates. The third term relates to the virtual work done by the forces applied on the natural boundary, and the last term counts for that on the essential boundaries. This term is needed only when MLS shape functions are used. If a shape function with the Kronecker delta function property is used, the last term will vanish and can hence be removed.

For free-vibration analyses of thin plates, the Galerkin weak form of the elastodynamic undamped equilibrium equations can be written as follows:

$$\int_A \delta\boldsymbol{\varepsilon}_p^T \boldsymbol{\sigma}_p \, dA + \int_\Omega \delta\mathbf{u}^T \rho \ddot{\mathbf{u}} \, d\Omega = 0 \tag{11.14}$$

where ρ is the mass density. Because the plate is free of external forces, it is easy to understand that the second and third terms in Equation 11.13 vanish. However, we need to justify why we left out the last term in Equation 11.13.

For free-vibration analysis, one accurate method is to formulate the boundary condition equation separately from the system equation (see note 2 in Section 4.9). The weak form of the essential boundary conditions with Lagrange multipliers is employed to produce the discretized essential boundary conditions as given below:

$$\int_{\Gamma_u} \delta\boldsymbol{\lambda}^T (\tilde{\mathbf{u}} - \bar{\mathbf{u}}) \, d\Gamma = 0 \tag{11.15}$$

where $\boldsymbol{\lambda}$ is a vector of Lagrange multipliers each of which, λ, can be interpolated as follows:

$$\lambda(\mathbf{x}) = N_I(s)\lambda_I, \quad \mathbf{x} \in \Gamma_u \tag{11.16}$$

where s and $N_I(s)$ are, respectively, the arc-length and Lagrange interpolant along the boundary, which we have detailed in Chapter 6. The variation of the Lagrange multiplier can be written as

$$\delta\lambda(\mathbf{x}) = N_I(s)\delta\lambda_I, \quad \mathbf{x} \in \Gamma_u \tag{11.17}$$

The displacement fields in a Kirchhoff plate have simple relations of (see Equation 3.63)

$$\mathbf{u} = \begin{Bmatrix} u \\ v \\ w \end{Bmatrix} = \underbrace{\begin{Bmatrix} -z\dfrac{\partial}{\partial x} \\ -z\dfrac{\partial}{\partial y} \\ 1 \end{Bmatrix}}_{\mathbf{L}_u} w = \mathbf{L}_u w \tag{11.18}$$

In Equations 11.13 and 11.14, the pseudo-strains of the plate are defined by Equations 3.69 and 3.74. The relationship between the pseudo-strain and the stress is given by Equation 3.75. The constrained Galerkin weak form for static problems of thin plates can be rewritten as

$$\begin{aligned} &\int_A \delta(\mathbf{L}w)^T\mathbf{D}(\mathbf{L}w)\,\mathrm{d}A - \int_\Omega \delta(\mathbf{L}_u w)^T\mathbf{b}\,\mathrm{d}\Omega - \int_{S_t} \delta(\mathbf{L}_u w)^T\bar{\mathbf{t}}\,\mathrm{d}S \\ &-\frac{1}{2}\int_{\Gamma_u} \delta(\mathbf{L}_b w - \bar{\mathbf{u}})^T\boldsymbol{\alpha}(\mathbf{L}_b w - \bar{\mathbf{u}})\mathrm{d}\Gamma = 0 \end{aligned} \tag{11.19}$$

where the differential operator $\mathbf{L}$ is given by Equation 3.68. The Galerkin weak form for free vibration of Equation 11.14 can be rewritten as

$$\int_A \delta(\mathbf{L}w)^T\mathbf{D}(\mathbf{L}w)\mathrm{d}A + \int_\Omega \rho\delta(\mathbf{L}_u w)^T\mathbf{L}_u\ddot{w}\,\mathrm{d}\Omega = 0 \tag{11.20}$$

with the weak form for essential boundary condition of Equation 11.15 rewritten as

$$\int_{\Gamma_u} \delta\boldsymbol{\lambda}^T(\mathbf{L}_b w - \bar{\mathbf{u}})\mathrm{d}\Gamma = 0 \tag{11.21}$$

11.1.3 Discrete Equations

By substituting the displacement field (Equation 11.1) into the variational form (Equation 11.19), the final static discrete equation is as follows:

$$(\mathbf{K} + \tilde{\mathbf{K}})\mathbf{U} = \mathbf{F} \tag{11.22}$$

where the global stiffness matrix **K** is assembled using the nodal stiffness (a scalar) defined by

$$\mathbf{K}_{IJ} = \int_A \mathbf{B}_I^T \mathbf{D} \mathbf{B}_J \mathrm{d}A \tag{11.23}$$

in which

$$\mathbf{B}_I = \mathbf{L}\phi_I = \begin{Bmatrix} -\phi_{I,xx} \\ -\phi_{I,yy} \\ -2\phi_{I,xy} \end{Bmatrix} \tag{11.24}$$

Matrix $\tilde{\mathbf{K}}$ in Equation 11.22 is obtained using

$$\tilde{\mathbf{K}}_{IJ} = \int_{\Gamma_u} \mathbf{\Psi}_I^T \boldsymbol{\alpha} \mathbf{\Psi}_J \mathrm{d}\Gamma \tag{11.25}$$

where

$$\mathbf{\Psi}_I = \begin{Bmatrix} \phi_I \\ \phi_{I,n} \end{Bmatrix} \quad \text{for clamped boundary} \tag{11.26}$$

and

$$\mathbf{\Psi}_I = \begin{Bmatrix} \phi_I \\ 0 \end{Bmatrix} \quad \text{for simply supported boundary} \tag{11.27}$$

in which n is the unit normal on the essential boundary surface Γ_u.

The force vector **F** is assembled using the nodal force given by

$$f_I = \int_\Omega \underbrace{(\mathbf{L}_u\phi_I)}_{\mathbf{B}_u}{}^T \mathbf{b}\mathrm{d}\Omega + \int_{S_t} \underbrace{(\mathbf{L}_u\phi_I)}_{\mathbf{B}_u}{}^T \bar{\mathbf{t}}\mathrm{d}S = \int_\Omega \mathbf{B}_u^T \mathbf{b}\mathrm{d}\Omega + \int_{S_t} \mathbf{B}_u^T \bar{\mathbf{t}}\mathrm{d}S \tag{11.28}$$

where

$$\mathbf{B}_u = \begin{Bmatrix} -z\phi_{I,x} \\ -z\phi_{I,y} \\ \phi_I \end{Bmatrix} \tag{11.29}$$

Because the plate considered here is subject only to transverse (in the z direction) load, we have

$$\mathbf{b} = \begin{Bmatrix} 0 \\ 0 \\ b_z \end{Bmatrix} \tag{11.30}$$

Also the external traction on the edge of the plate $\bar{\mathbf{t}}$ can be given by stresses on the surface of the edge:

$$\bar{\mathbf{t}} = \begin{bmatrix} n_x & 0 & 0 & n_y & 0 \\ 0 & n_y & 0 & 0 & n_x \\ 0 & 0 & n_y & n_x & 0 \end{bmatrix} \begin{Bmatrix} \sigma_{xx} \\ \sigma_{yy} \\ \sigma_{yz} \\ \sigma_{xz} \\ \sigma_{xy} \end{Bmatrix} = \begin{Bmatrix} \sigma_{xx}n_x + \sigma_{xz}n_y \\ \sigma_{xy}n_x + \sigma_{yy}n_y \\ \sigma_{xz}n_x + \sigma_{yz}n_y \end{Bmatrix} \tag{11.31}$$

where (n_x, n_y) is the unit outward normal vector on the boundary.

$$\begin{aligned} f_I &= \int_\Omega \phi_I b_z \,\mathrm{d}\Omega + \int_{S_t} \{-z\phi_{I,x} \quad -z\phi_{I,y} \quad \phi_I\} \begin{Bmatrix} \sigma_{xx}n_x + \sigma_{xy}n_y \\ \sigma_{xy}n_x + \sigma_{yy}n_y \\ \sigma_{xz}n_x + \sigma_{yz}n_y \end{Bmatrix} \mathrm{d}S \\ &= \int_\Omega \phi_I b_z \,\mathrm{d}\Omega + \int_{\Gamma_t} \Bigg[-\phi_{I,x}\Bigg(\underbrace{n_x \int_{-h/2}^{h/2} z\sigma_{xx}\,\mathrm{d}z}_{M_{xx}} + \underbrace{n_y \int_{-h/2}^{h/2} z\sigma_{xy}\,\mathrm{d}z}_{M_{xy}} \Bigg) \\ &\quad -\phi_{I,y}\Bigg(\underbrace{n_x \int_{-h/2}^{h/2} z\sigma_{xy}\,\mathrm{d}z}_{M_{xy}} + \underbrace{n_y \int_{-h/2}^{h/2} z\sigma_{yy}\,\mathrm{d}z}_{M_{yy}} \Bigg) + \phi_I \Bigg(n_x \underbrace{\int_{-h/2}^{h/2} \sigma_{xz}\,\mathrm{d}z}_{V_{xz}} + n_y \underbrace{\int_{-h/2}^{h/2} \sigma_{xz}\,\mathrm{d}z}_{V_{yz}} \Bigg) \Bigg] \mathrm{d}\Gamma \\ &= h\int_A \phi_I p_z \mathrm{d}A + \int_{\Gamma_t} [-\phi_{I,x}\, \underbrace{(n_x M_{xx} + n_y M_{xy})}_{\bar{M}_{xn}} - \phi_{I,y}\, \underbrace{(n_x M_{xy} + n_y M_{yy})}_{\bar{M}_{yn}} + \phi_I\, \underbrace{(n_x V_{xz} + n_y V_{yz})}_{\bar{V}_n}]\mathrm{d}\Gamma \end{aligned} \tag{11.32}$$

where p_z is the pressure (force per unit area) applied on the neutral plane of the plate. Note in Equation 11.32 that the integration over the surface of the plate edge has been changed to a curve integration on the natural boundary (on the neutral plane). We now examine the first two terms in the integrand of the last integral in Equation 11.32. Note that on the boundary of the plate, we should have

$$\begin{aligned} \phi_{I,x} &= \phi_{I,n}n_x - \phi_{I,s}n_y \\ \phi_{I,y} &= \phi_{I,n}n_y + \phi_{I,s}n_x \end{aligned} \tag{11.33}$$

Therefore,

$$\begin{aligned} &\phi_{I,x}(n_x M_{xx} + n_y M_{xy}) + \phi_{I,y}(n_x M_{xy} + n_y M_{yy}) \\ &\quad = (\phi_{I,n}n_x - \phi_{I,s}n_y)(n_x M_{xx} + n_y M_{xy}) + (\phi_{I,n}n_y + \phi_{I,s}n_x)(n_x M_{xy} + n_y M_{yy}) \\ &\quad = \phi_{I,n}\, \underbrace{[n_x^2 M_{xx} + 2n_x n_y M_{xy} + n_y^2 M_{yy}]}_{\bar{M}_n} + \phi_{I,s}\, \underbrace{[-n_y n_x M_{xx} + n_x n_y M_{yy} + (n_x^2 - n_y^2) M_{xy}]}_{\bar{M}_{ns}} \\ &\quad = \phi_{I,n}\bar{M}_n + \phi_{I,s}\bar{M}_{ns} \end{aligned} \tag{11.34}$$

where $\overline{M}_n$ and $\overline{M}_{ns}$ are, respectively, the specified moment and torsional moment on the natural boundary (see Equations 11.4 and 11.5).

Substituting Equation 11.34 into Equation 11.32 leads to

$$f_I = h\int_A \phi_I p_z \,\mathrm{d}A + \int_{\Gamma_t} [-\phi_{I,n}\overline{M}_n - \phi_{I,s}\overline{M}_{ns} + \phi_I \overline{V}]\mathrm{d}\Gamma \tag{11.35}$$

The vector **U** in Equation 11.22 has the form:

$$\mathbf{U} = \{w_1, \ldots, w_{n_t}\}^T \tag{11.36}$$

where n_t is the total number of nodes in the entire domain of the plate. Note that the total degrees of freedom of the thin plates is the number of nodes, and rotations are not involved as an unknown in the global system equation.

Similarly, by substituting the displacement field (Equation 11.1) into the variational form (Equation 11.20), the final dynamic discrete equation for free vibration of thin plates is obtained as follows:

$$\mathbf{M}\ddot{\mathbf{U}} + \mathbf{K}\mathbf{U} = 0 \tag{11.37}$$

where **M** is the global mass matrix that is assembled using the nodal mass given by the following integral form over the entire area of the plate:

$$M_{IJ} = \int_A (\rho\phi_I\phi_J h + \phi_{I,x}\phi_{J,x} I + \phi_{I,y}\phi_{J,y} I)\mathrm{d}A \tag{11.38}$$

where I is the mass moment of inertial for the plate computed using

$$I = \int_{-h/2}^{h/2} \rho z^2 \mathrm{d}z \tag{11.39}$$

For homogeneous plates, $I = \rho(h^3/12)$.

The first integral in Equation 11.38 is the mass inertial corresponding to the vertical translational vibration of the plate, and the second and third terms are rotational inertia corresponding the rotational vibrations of the cross section of the thin plate. For *very* thin plates, these two terms for rotational vibrations can be neglected (see, e.g., G. R. Liu and Xi, 2001).

Equations for Essential Boundary Conditions

Substituting the displacement expression of Equation 11.1 into the boundary condition weak form of Equation 11.21 yields a set of linear algebraic constraint equations:

$$\mathbf{H}_{(2n_b \times n)}\mathbf{U}_{(n\times 1)} = \mathbf{q}_{(2n_b\times 1)} \tag{11.40}$$

where n_b is the number of the nodes that are included in the union of all the support domains of all the quadrature points on the essential boundary for the integration in Equation 11.43. Nonzero entries in matrix **H** are computed using

$$\mathbf{H}_{KI} = \int_{\Gamma_u} \mathbf{N}_K \mathbf{\Psi}_I \,\mathrm{d}\Gamma \tag{11.41}$$

in which $\mathbf{\Psi}_I$ is the matrix of MLS shape functions given by Equation 11.26 or 11.27, depending on the type of boundary condition, and $\mathbf{N}_k$ has the form:

$$\mathbf{N}_K = \begin{bmatrix} N_K & 0 \\ 0 & N_K \end{bmatrix} \tag{11.42}$$

where N_K is the Lagrange interpolation function for node K on the essential boundary. A Lagrange interpolant formed using Equation 6.11 can be used. Vector $\mathbf{q}$ in Equation 11.40 is defined by

$$\mathbf{q}_K = \int_{\Gamma_u} \mathbf{N}_K \bar{\mathbf{u}} \mathrm{d}S \tag{11.43}$$

Note that matrix $\mathbf{H}$ in Equation 11.40 has n columns and $2n_b$ rows, where n_b is usually a very small number compared to n. All the entries without contribution from Equation 11.41 will be zero. Thus, $\mathbf{H}$ is in general a very "short-fat" sparse matrix.

For free-vibration analysis, the essential boundary conditions have to be homogeneous, that is, $\bar{\mathbf{u}} = \mathbf{0}$ on the essential boundary. Using Equation 11.43, we have $\mathbf{q} = 0$ and, hence,

$$\mathbf{HU} = \mathbf{0} \tag{11.44}$$

This is the discretized essential boundary condition for free-vibration analysis.

11.1.4 Eigenvalue Problem

Eigenvalue Equation

Consider now that the plate is undergoing a harmonic vibration. The deflection $\mathbf{U}$ can be expressed in the form

$$\mathbf{U} = \mathbf{W}\exp(i\omega t) \tag{11.45}$$

where i is the imaginary unit, ω is the angular frequency, and $\mathbf{W}$ is the amplitude of the vibration. Substitution of the foregoing equation into Equation 11.37 leads to the following eigenvalue equation:

$$(\mathbf{K} - \omega^2\mathbf{M})\mathbf{W} = 0 \tag{11.46}$$

where $\mathbf{W}$ is an eigenvector that has the form of

$$\mathbf{W} = \{W_1 \ \ W_2 \ \ \ldots \ \ W_n\}^T \tag{11.47}$$

For a system with n_t nodes, the dimension of matrices $\mathbf{K}$ and $\mathbf{M}$ should be $n_t \times n_t$, and there should be n_t eigenvalues ω_i $(i = 1, 2, \ldots, n_t)$, which correspond to n_t natural frequencies, and n_t corresponding eigenvectors.

Note, however, that Equation 11.46 has to be solved subject to the constraints of Equation 11.44 that can be restated for free vibration as

$$\mathbf{HW} = \mathbf{0} \tag{11.48}$$

Using singular value decomposition (Griffel, 1989), **H** can be decomposed as

$$\mathbf{H}_{2n_b\times n} = \mathbf{R}_{2n_b\times 2n_b}\begin{bmatrix}\Sigma_{r\times r} & \mathbf{0}\\ \mathbf{0} & \mathbf{0}\end{bmatrix}_{2n_b\times n}\mathbf{V}^T_{n\times n} \tag{11.49}$$

where **R** and **V** are orthogonal matrices, $\Sigma_{r\times r}$ has diagonal form, in which diagonal elements are equal to singular values of **H**, and r is the rank of **H**, which is the same as the number of independent constraints.

The orthogonal matrix **V** can be partitioned as follows:

$$\mathbf{V}^T = \{\mathbf{V}_{n\times r}, \mathbf{V}_{n\times(n-r)}\}^T \tag{11.50}$$

where $\mathbf{V}_{n\times r}$ corresponds to the portion of $\Sigma_{r\times r}$. We now examine

$$\begin{aligned}\mathbf{HV} &= \mathbf{H}[\mathbf{V}_{n\times r} \quad \mathbf{V}_{n\times(n-r)}] = [\mathbf{HV}_{n\times r} \quad \mathbf{HV}_{n\times(n-r)}]\\ &= \mathbf{R}_{2n_b\times 2n_b}\begin{bmatrix}\Sigma_{r\times r} & \mathbf{0}\\ \mathbf{0} & \mathbf{0}\end{bmatrix}_{2n_b\times n}\mathbf{V}^T_{n\times n}\mathbf{V}_{n\times n}\end{aligned} \tag{11.51}$$

Invoking the orthogonal condition of **V**, we have $\mathbf{V}^T_{n\times n}\mathbf{V}_{n\times n} = \mathbf{I}$. The foregoing equation becomes

$$\begin{aligned}\mathbf{HV} &= \mathbf{H}[\mathbf{V}_{n\times r} \quad \mathbf{V}_{n\times(n-r)}] = [\mathbf{HV}_{n\times r} \quad \mathbf{HV}_{n\times(n-r)}]\\ &= \mathbf{R}_{2n_b\times 2n_b}\begin{bmatrix}\Sigma_{r\times r} & \mathbf{0}\\ \mathbf{0} & \underbrace{\mathbf{0}}_{\mathbf{HV}_{n\times(n-r)}}\end{bmatrix}\end{aligned} \tag{11.52}$$

This implies

$$\mathbf{HV}_{n\times(n-r)} = \mathbf{0}_{n\times(n-r)} \tag{11.53}$$

or a null transformation, or $\mathbf{V}_{n\times(n-r)}$ is a basis of the null space of the linear transformation **H**. Therefore, the following orthogonal matrix transformation:

$$\mathbf{W} = \mathbf{V}_{n\times(n-r)}\tilde{\mathbf{W}} \tag{11.54}$$

satisfies Equation 11.48. Substituting Equation 11.54 into Equation 11.46 and premultiplying $\mathbf{V}^T_{(n-r)\times n}$ to the resultant equation lead to

$$[\tilde{\mathbf{K}} - \omega^2\tilde{\mathbf{M}}]\tilde{\mathbf{W}} = 0 \tag{11.55}$$

which is the condensed eigenvalue equations, where

$$\tilde{\mathbf{K}} = \mathbf{V}^T_{(n-r)\times n}\mathbf{K}\mathbf{V}_{n\times(n-r)} \tag{11.56}$$

is the condensed stiffness matrix and

$$\tilde{\mathbf{M}} = \mathbf{V}^T_{(n-r)\times n}\mathbf{M}\mathbf{V}_{n\times(n-r)} \tag{11.57}$$

is the condensed mass matrix.

Note that these condensed matrices are, in general, non-negative definite. Solving Equation 11.55 using standard routines of eigenvalue equation solvers gives natural frequencies of the free vibration of thin plates. This orthogonal transformation technique was used by Ouatouati and Johnson (1999) for imposing constraints for eigenvalue equations formulated using the EFG method.

Note that matrix **H** in Equation 11.48 is formed using the weak form of the constraint equation, which requires integration on the boundary and ensures the satisfaction of the essential boundary conditions on the entire essential boundary. We can obtain the discrete constraint equations directly using MLS approximation and obtain the **H** matrix, as shown by Ouatouati and Johnson (1999). Such direct formulation ensures the constraints at the nodes on the essential boundary.

Note that the penalty method can also be applied for imposing essential boundary conditions for both free vibration and forced vibration. In such cases, orthogonal transformation is not needed.

11.1.5 Numerical Examples

In all the examples given in this section, complete quadratic polynomial basis functions are used ($m = 6$) for constructing MLS shape functions. Cells of background mesh are used for integrating the system matrices and the Gauss quadrature scheme is utilized. The dimension of support domain is chosen as 3.5 to 3.9 times the nodal distance for the nodes in the support domain of the Gauss quadrature point.

Example 11.1 Static Deflection of Rectangular Thin Plates

Consider now a benchmarking problem of a rectangular plate, as shown in Figure 11.1. Analytical solutions are available for this problem (Timoshenko and Woinowsky-Krieger, 1995). The following parameters are used:

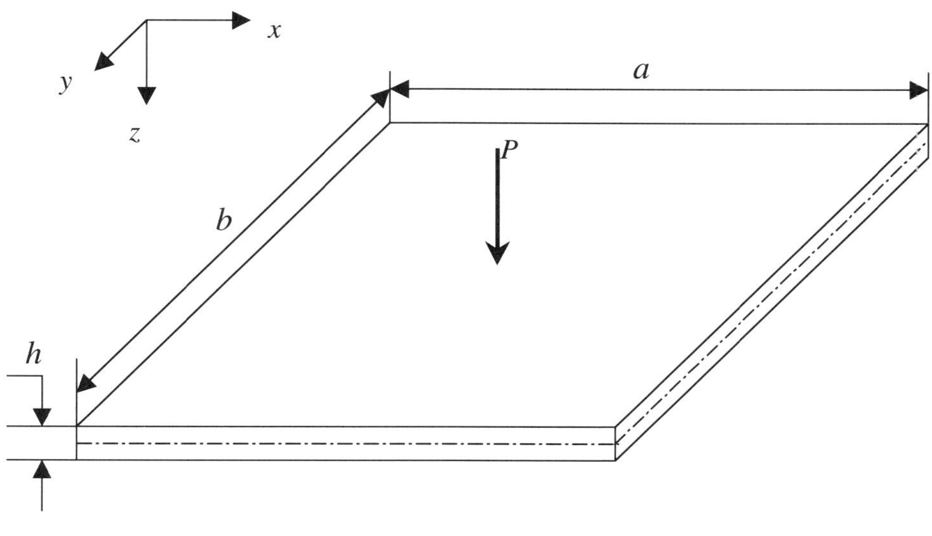

FIGURE 11.1
Thin rectangular plate of uniform thickness h. (From Liu, G. R. and Chen, X. L., *J. Sound Vib.*, 241(5), 839–855, 2001. With permission.)

TABLE 11.1

Deflection of a Square Plate $\beta = w_{max}D/Pa^2$

	Present Method (EFG)					Timoshenko and Woinowsky-Krieger (1995)
Nodes	6 × 6	9 × 9	12 × 12	15 × 15	18 × 18	
SSSS	0.01032	0.01141	0.01145	0.01155	0.01157	0.01160
CCCC	0.00452	0.00538	0.00546	0.00552	0.00554	0.00560

Source: Liu, G. R. and Chen, X. L., *J. Sound Vib.*, 241(5), 839–855, 2001. With permission.

TABLE 11.2

Deflection of Simply Supported Rectangular Plates $\beta = w_{max}D/Pa^2$

b/a	1.0	1.2	1.4	1.6	1.8	2.0
Present method (EFG)	0.01157	0.01344	0.01476	0.01556	0.01603	0.01632
Timoshenko and Woinowsky-Krieger (1995)	0.01160	0.01353	0.01484	0.01570	0.01620	0.01651

Source: Liu, G. R. and Chen, X. L., *J. Sound Vib.*, 241(5), 839–855, 2001. With permission.

TABLE 11.3

Deflection of Fully Clamped Rectangular Plates

b/a	1.0	1.2	1.4	1.6	1.8	2.0
β (Present method)	0.00552	0.00637	0.00680	0.00698	0.00703	0.00704
β (Timoshenko and Woinowsky-Krieger, 1995)	0.00560	0.00647	0.00691	0.00712	0.00720	0.00722

Source: Liu, G. R. and Chen, X. L., *J. Sound Vib.*, 241(5), 839–855, 2001. With permission.

Length in x direction: $a = 0.6$ m for rectangular plates with various widths

Length: $a = b = 0.6$ m for square plate

Thickness: $h = 0.001$ m

Young's modulus: $E = 1.0 \times 10^9$ N/m^2

Poisson's ratio: $\nu = 0.3$

A concentrated force of $P = 100.0$ N is applied at the center of the plates.

A dimensionless deflection coefficient β of the center of a thin rectangular plate is defined as $\beta = w_{max}D/Pa^2$, where w_{max} is the deflection at center of the plates and $D = Eh^3/[12(1-\nu^2)]$ is the flexural rigidity of the plates.

To analyze the convergence of the present method, we calculate deflections of a square plate using different densities of nodes. Two kinds of boundary conditions are imposed: simply supported and fully clamped. In the notation of the boundary conditions, S denotes simply supported and C means clamped. The edges of the plate are denoted clockwise using S or C depending on the type of boundary on the edge. For example, notation SCSC means that the left and right edges of the plate are simply supported while the upper and lower edges are clamped, and notation SCCS means that the left and lower edges are simply supported while the upper and right edges are clamped.

The results of deflection of the square plate are shown in Table 11.1 together with the analytical results. Good convergence has been achieved.

Further examinations are performed for plates with different width/length ratios. The deflections are calculated using 16 × 16 nodes in the present method. The results are shown in Tables 11.2 and 11.3. Compared with the analytical results (Timoshenko and Woinowsky-Krieger, 1995), good agreement has been achieved for all tested cases.

Example 11.2 Natural Frequency Analysis of Thin Square Plates

Consider now a square plate with the following parameters:

Length: $a = b = 10.0$ m
Thickness: $h = 0.05$ m
Young's modulus: $E = 200 \times 10^9$ N/m^2
Poisson's ratio: $\nu = 0.3$
Mass density: $\rho = 8000$ kg/m^3

CASE 1 Free thin square plate

We calculate the natural frequencies of free vibration of a free thin square plate. The frequency coefficients are computed using regular nodes of different density, and the results are shown in Table 11.4 together with the FEM results. The frequency coefficients in the tables are defined as

$$\varpi_1 = \left(\frac{\omega^2 \rho h a^4}{D}\right)^{1/4}$$

In the FEM results, HOE denotes an eight-noded semiloof thin shell element (4 × 4 mesh); LOE denotes a four-noded isoparametric shell element (8 × 8 mesh). The first three frequencies corresponding to the rigid displacements are zero, and therefore are not listed in the table. The results obtained using the present method are between those of FEMs using HOE and LOE. The present results show good convergence and good agreement with the analytical solution. When a 9 × 9 nodal density is used, the present results are more accurate than either FEM result.

CASE 2 Simply supported and fully clamped thin square plate

Natural frequencies of lateral free vibration of a simply supported and fully clamped thin square plate are computed using the present method. To analyze the effectiveness of the present method using irregular nodes, we calculate frequencies using 13 × 13 regular nodes shown in Figure 11.2 and 169 irregular nodes shown in Figure 11.3. The results are shown in Tables 11.5 and 11.6. It is found that the results of using both regular and irregular nodes show good agreement with each other and with the analytical solutions.

TABLE 11.4

Natural Frequency Coefficients $\varpi_1 = (\omega^2 \rho h a^4/D)^{1/4}$ of Lateral Free Vibration of a Free Square Plate

Mode	Analytical Solution (Abbassian et al., 1987)	Present Method (EFG)				FEM (Abbassian et al., 1987)	
		5 × 5	9 × 9	13 × 13	17 × 17	HOE	LOE
4	3.670	3.700	3.670	3.670	3.670	3.567	3.682
5	4.427	4.468	4.434	4.430	4.429	4.423	4.466
6	4.926	5.000	4.939	4.933	4.930	4.875	4.997
7	5.929	6.010	5.907	5.903	5.901	5.851	5.942
8	5.929	6.010	5.907	5.903	5.901	5.851	5.942
9	7.848	8.189	7.855	7.840	7.832	7.820	8.079

HOE denotes an eight-noded semiloof thin shell element (4 × 4 mesh); LOE denotes a four-noded isoparametric shell element (8 × 8 mesh).

Source: Liu, G. R. and Chen, X. L., *J. Sound Vib.*, 241(5), 839–855, 2001. With permission.

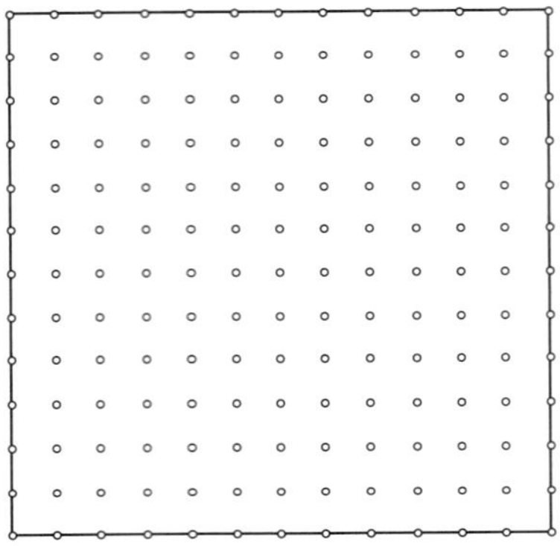

FIGURE 11.2
A square plate with 13 × 13 = 169 regular nodes. (From Liu, G. R. and Chen, X. L., *J. Sound Vib.*, 241(5), 839–855, 2001. With permission.)

FIGURE 11.3
Square plate with 169 irregular nodes. (From Liu, G. R. and Chen, X. L., *J. Sound Vib.*, 241(5), 839–855, 2001. With permission.)

TABLE 11.5

Natural Frequency Coefficients $\varpi_1 = (\omega^2 \rho h a^4/D)^{1/4}$ of Lateral Free Vibration of a Simply Supported Square Plate

Mode	Analytical Solution (Abbassian et al., 1987)	Present Method (EFG) Regular Nodes 13 × 13	Present Method (EFG) Irregular Nodes 169
1	4.443	4.443	4.453
2	7.025	7.031	7.033
3	7.025	7.036	7.120
4	8.886	8.892	8.912
5	9.935	9.959	9.966
6	9.935	9.966	10.010
7	11.327	11.341	11.345
8	11.327	11.347	11.540
9	—	13.032	12.994
10	—	13.036	13.064

Source: Liu, G. R. and Chen, X. L., *J. Sound Vib.*, 241(5), 839–855, 2001. With permission.

TABLE 11.6

Natural Frequency Coefficients $\varpi_1 = (\omega^2 \rho h a^4/D)^{1/4}$ of Lateral Free Vibration of a Fully Clamped Square Plate

		Present Method	
Mode	**Analytical Solution (Robert, 1979)**	**Regular Nodes 13 × 13**	**Irregular Nodes 169**
1	5.999	6.017	5.999
2	8.568	8.606	8.596
3	8.568	8.606	8.602
4	10.407	10.439	10.421
5	11.472	11.533	11.507
6	11.498	11.562	11.528
7	—	12.893	12.925
8	—	12.896	12.986
9	—	14.605	14.570
10	—	14.606	14.604

Source: Liu, G. R. and Chen, X. L., *J. Sound Vib.*, 241(5), 839–855, 2001. With permission.

TABLE 11.7

Natural Frequency Coefficients $\varpi_1 = 2a(\omega^2 \rho h/D)^{1/4}$ of Lateral Free Vibration of a Free Elliptical Plate

	Present Method		
Mode	**97 Nodes**	**241 Nodes**	**289 Nodes**
4	5.197	5.176	5.173
5	6.533	6.509	6.505
6	8.288	8.244	8.234
7	9.451	9.405	9.397
8	10.602	10.559	10.547
9	11.333	11.256	11.249
10	12.223	12.168	12.160

Source: Liu, G. R. and Chen, X. L., *J. Sound Vib.*, 241(5), 839–855, 2001. With permission.

Example 11.3 Natural Frequency Analysis of Elliptical Plates

An elliptical plate with radii $a = 5.0$ m and $b = 2.5$ m are investigated. Other parameters are the same as the square plate examined in Example 11.2.

Frequency coefficients of free vibration are computed. Table 11.7 shows frequencies of a free thin elliptical plate using regular nodes. The first three frequencies corresponding to rigid displacements are zero, and are not listed in the table. Good convergence of the results has been achieved. Table 11.8 shows the frequencies of a fully clamped thin elliptical plate. The frequencies are calculated using regular nodes (Figure 11.4) and irregular nodes (Figure 11.5). Good agreement between the results using regular and irregular nodes has been observed.

Example 11.4 Natural Frequency Analysis of Polygonal Plates

Free-vibration analysis of a square and hexagonal plate is performed. The length of each side is $a = 10.0$ m. Other parameters are the same as the square plate examined in Example 11.2.

TABLE 11.8

Natural Frequency Coefficients $\varpi_1 = 2a(\omega^2 \rho h/D)^{1/4}$ of Lateral Free Vibration of a Fully Clamped Elliptical Plate

	Present Method	
Mode	**Regular Nodes 201**	**Irregular Nodes 201**
1	10.467	10.454
2	12.619	12.621
3	15.009	14.992
4	16.726	16.716
5	17.629	17.658
6	18.838	18.840
7	20.604	20.508
8	21.081	21.060
9	22.913	22.890
10	23.610	23.591

Source: Liu, G. R. and Chen, X. L., *J. Sound Vib.*, 241(5), 839–855, 2001. With permission.

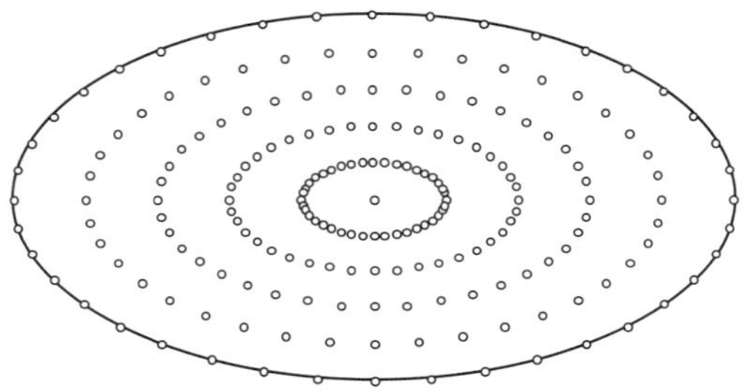

FIGURE 11.4
An elliptical plate with 201 regular nodes. (From Liu, G. R. and Chen, X. L., *J. Sound Vib.*, 241(5), 839–855, 2001. With permission.)

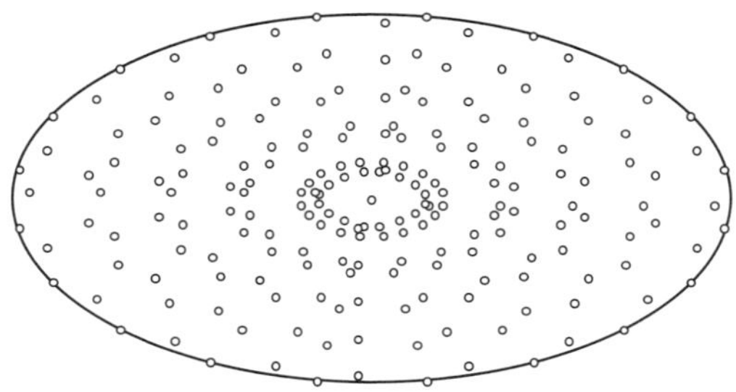

FIGURE 11.5
An elliptical plate with 201 irregular nodes. (From Liu, G. R. and Chen, X. L., *J. Sound Vib.*, 241(5), 839–855, 2001. With permission.)

The natural frequencies of a square plate with fully clamped boundaries are first calculated. A total of 524 irregular nodes, as shown in Figure 11.6, are used. The frequency coefficients are defined as

$$\varpi_2 = \left(\frac{\omega^2 \rho h a_p^4}{D}\right)^{1/4}$$

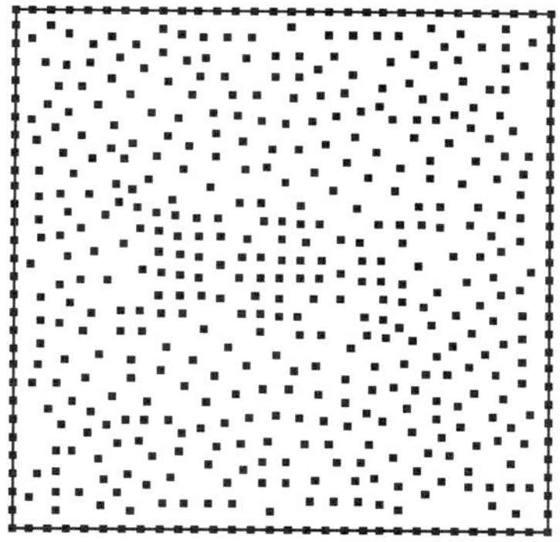

FIGURE 11.6
A square plate with 524 irregular nodes. (From Liu, G. R. and Chen, X. L., *J. Sound Vib.*, 241(5), 839–855, 2001. With permission.)

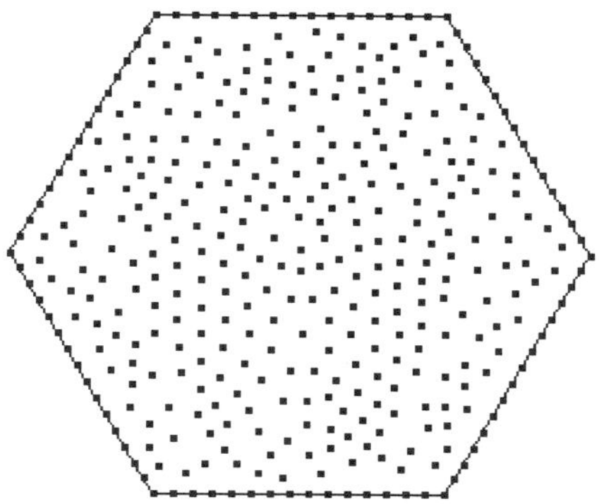

FIGURE 11.7
A hexagonal plate with 380 irregular nodes. (From Liu, G. R. and Chen, X. L., *J. Sound Vib.*, 241(5), 839–855, 2001. With permission.)

where a_p is the radius of the inscribing circle for regular polygonal plates. The natural frequencies of a hexagonal plate with fully clamped boundaries are also calculated, where 380 irregular nodes, as shown in Figure 11.7, are used. Table 11.9 lists the natural frequency coefficients of the lowest 10 modes for these two plates. Table 11.10 shows a comparison between the frequency coefficients using the EFG method and those given by Schinzinger and Laura (1991). For the square plate, the natural frequency coefficient of the first mode using the EFG method agrees very well with the exact solution and the numerical result in the textbook by Schinzinger and Laura (1991). For the hexagonal plate, the natural frequency coefficient of the first mode using the EFG method is slightly smaller than the numerical result in the textbook by Schinzinger and Laura (1991).

Example 11.5 Natural Frequency Analysis of a Plate of Complex Shape

A plate of very complicated shape is also studied. The geometric parameters are shown in Figure 11.8. The unit is meters. Other parameters are the same as the square plate examined in Example 11.2. The plate is chosen hypothetically, but it serves the purpose of demonstrating the applicability of the present EFG method to plates of complicated shape.

Different boundary conditions are considered to examine the present method in imposing boundary conditions.

TABLE 11.9

Natural Frequency Coefficients $\varpi_2 = (\omega^2 \rho h a_p^4 / D)^{1/4}$ of Lateral Free Vibration of Fully Clamped Regular Polygonal Plate

Mode	Present Method	
	Square	Hexagon
1	9.089	9.042
2	18.700	17.805
3	18.829	20.586
4	28.121	29.802
5	33.515	34.740
6	33.649	37.368
7	42.550	46.622
8	43.529	51.846
9	53.896	55.799
10	54.156	59.138

Source: Liu, G. R. and Chen, X. L., *J. Sound Vib.*, 241(5), 839–855, 2001. With permission.

TABLE 11.10

Comparison of Natural Frequency Coefficients $\varpi_2 = (\omega^2 \rho h a_p^4 / D)^{1/4}$ of First Mode of Fully Clamped Regular Polygonal Plates

	Present Method	Exact (Schinzinger and Laura, 1991)	Schinzinger and Laura (1991)
Square	9.089	8.997	9.122
Hexagon	9.042	—	9.638

Source: Liu, G. R. and Chen, X. L., *J. Sound Vib.*, 241(5), 839–855, 2001. With permission.

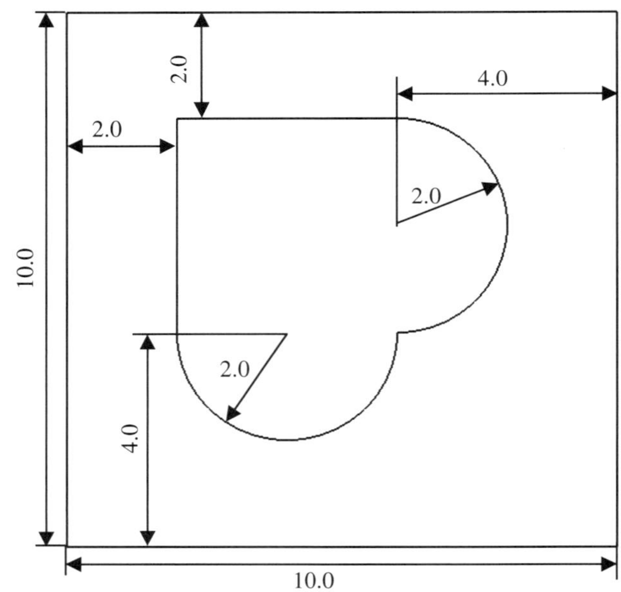

FIGURE 11.8
Plate with a hole of complicated shape. The unit is meters. (From Liu, G. R. and Chen, X. L., *J. Sound Vib.*, 241(5), 839–855, 2001. With permission.)

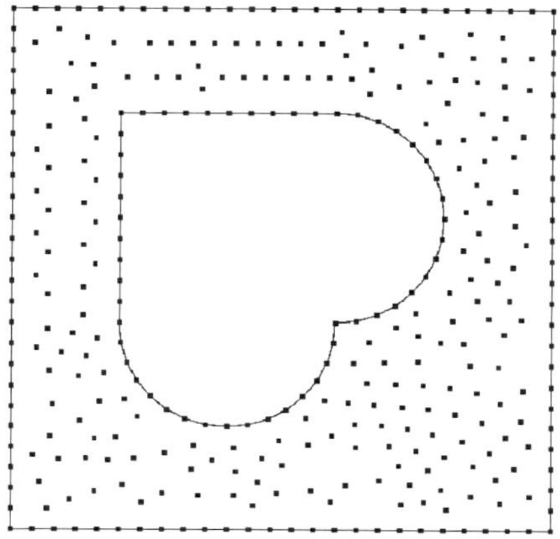

FIGURE 11.9
Nodal distribution in a plate with a hole of complicated shape. (From Liu, G. R. and Chen, X. L., *J. Sound Vib.*, 241(5), 839–855, 2001. With permission.)

TABLE 11.11
Natural Frequency Coefficients $\varpi_1 = (\omega^2 \rho h a^4/D)^{1/4}$ of Lateral Free Vibration of a Plate with a Hole of Complicated Shape

	Present Method			
Mode	**SSSS**	**CCCC**	**SCSC**	**SCCS**
1	5.453	7.548	7.170	6.079
2	8.069	10.764	10.343	9.204
3	9.554	11.113	11.415	10.837
4	10.099	11.328	12.572	11.273
5	11.328	12.862	12.811	12.278
6	12.765	13.300	13.272	13.322
7	13.685	14.168	13.997	14.308
8	14.305	15.369	14.627	14.900
9	15.721	16.205	15.743	15.170
10	17.079	17.137	16.391	16.302

Source: Liu, G. R. and Chen, X. L., *J. Sound Vib.*, 241(5), 839–855, 2001. With permission.

The nodal distribution is plotted in Figure 11.9. Table 11.11 lists the frequencies obtained for the plate with different boundary conditions. As expected, the natural frequencies of the plate with clamped boundaries are generally higher than those with simply supported boundaries.

11.2 EFG Method for Thin Composite Laminates

Composite laminates are widely used in modern structures due to their advantages of high strength, high stiffness, and low weight. Therefore, buckling analysis of laminates becomes very important in the process of designing such composite structures. Exact buckling solutions of laminates for arbitrary geometries and lamination schemes are usually very difficult. Therefore, numerical methods such as FEM have been used for analyzing laminated plate problems. However, it is not easy to construct conventionally conformable plate elements of high order as required for thin plates (C^1 consistency), and requires element connectivity to form the finite element equations, whose generation is often a time-consuming procedure.

In the previous section, an EFG method has been formulated for analyzing static deflection and free vibration of thin plates. This section introduces an EFG formulation to solve the static buckling problems of thin plates and symmetrically laminated composite plates. The eigenvalue equations of the static buckling of the plates are established by applying energy principles and classical (Kirchhoff) plate theory. Similar to the formulation in the previous section, the deflection of plates is the only unknown at a node; therefore, the dimension of the discrete eigenvalue equations obtained by the present formulation is only one third of that in FEM. Thus, solving the eigenvalue equation is more computationally efficient compared with FEM. To demonstrate the efficiency of the present method, static buckling loads for thin plates have been calculated and examined in detail by comparison with other available solutions. The convergence of static buckling loads of thin plates is analyzed. The static buckling loads for thin plates of complicated shape and symmetrically laminated composite plates with different boundaries have also been calculated using the present EFG method.

11.2.1 Governing Equation for Buckling

A symmetrically laminated composite plate with thickness h in the z direction is schematically drawn in Figure 11.10. The laminated plate may consist of n_L layers of plies. The reference plane $z = 0$ is located at the undeformed neutral plane of the laminated plate. The fiber orientation of a layer is indicated by α, as shown in Figure 11.11. The laminate plate is subjected to in-plane forces within the plane of symmetry of the plate on its edges. It is assumed that the applied edge forces are independent of each other. We can then write

$$N_x = -N_0, \qquad N_y = -\mu_1 N_0, \qquad N_{xy} = -\mu_2 N_0 \tag{11.58}$$

where N_0 is a constant and μ_1 and μ_2 are possibly functions of coordinates.

The strain energy of bending of the laminates can be obtained using

$$\Pi_b = \frac{1}{2}\int_A \boldsymbol{\varepsilon}_p^T \boldsymbol{\sigma}_p \, dA \tag{11.59}$$

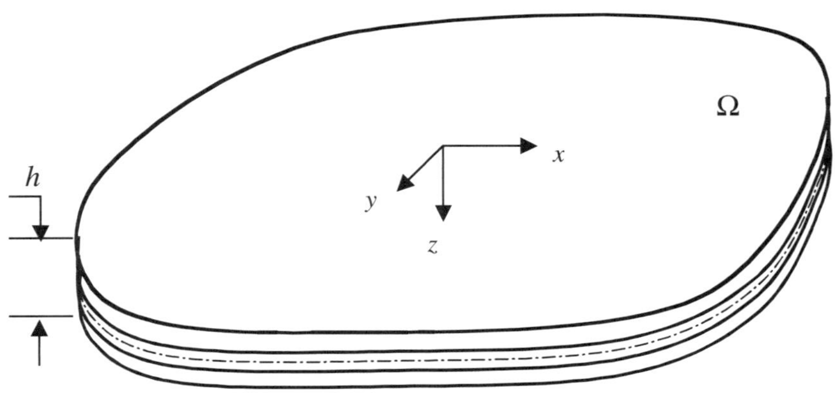

FIGURE 11.10
Thin laminate with symmetrically stacked layers of composites and its coordinate system.

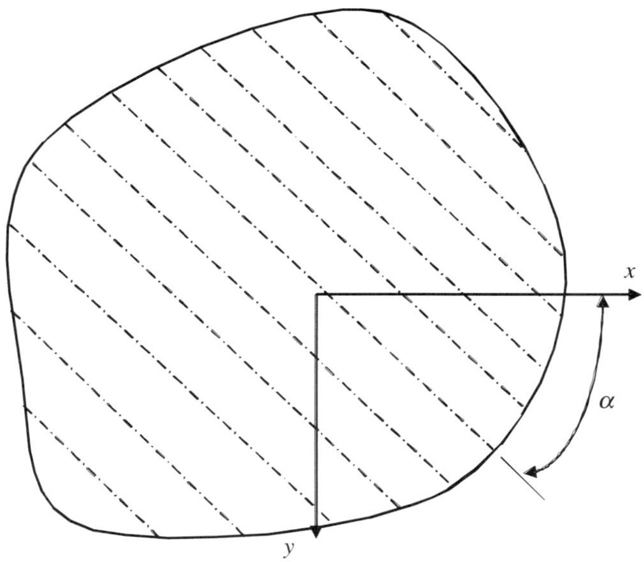

FIGURE 11.11
Ply with a fiber orientation of α in a laminated plate.

where A stands for the area of the plate, $\boldsymbol{\varepsilon}_p$ is the pseudo-strain defined by Equation 3.68, and $\boldsymbol{\sigma}_p$ is the pseudo-stress defined by Equation 3.74. Substituting Equations 3.68 and 3.74 into Equation 11.59, we obtain

$$\Pi_b = \frac{1}{2}\int_A \left[D_{11}\left(\frac{\partial^2 w}{\partial x^2}\right)^2 + 2D_{12}\frac{\partial^2 w}{\partial x^2}\frac{\partial^2 w}{\partial y^2} + D_{22}\left(\frac{\partial^2 w}{\partial y^2}\right)^2 + 4D_{66}\left(\frac{\partial^2 w}{\partial x \partial y}\right)^2 + 4\left(D_{16}\frac{\partial^2 w}{\partial x^2} + D_{26}\frac{\partial^2 w}{\partial y^2}\right)\frac{\partial^2 w}{\partial x \partial y}\right] \mathrm{d}A \tag{11.60}$$

The strain energy caused by in-plane forces can be expressed by (Reddy, 1997)

$$\Pi_i = \frac{1}{2}\int_A \left[N_x\left(\frac{\partial w}{\partial x}\right)^2 + N_y\left(\frac{\partial w}{\partial y}\right)^2 + 2N_{xy}\frac{\partial w}{\partial x}\frac{\partial w}{\partial y}\right] \mathrm{d}A \tag{11.61}$$

The matrices of elastic constants of the laminates $\mathbf{D}$ can be obtained using Equation 3.72, and are given as follows (Reddy, 1997):

$$D_{IJ} = \frac{1}{3}\sum_{K=1}^{n_L} (\bar{Q}_{IJ})_K (z_K^3 - z_{K-1}^3), \quad I, J = 1, 2, 6 \tag{11.62}$$

$$\bar{Q}_{11} = Q_{11}\cos^4\alpha + 2(Q_{12} + 2Q_{66})\sin^2\alpha\cos^2\alpha + Q_{22}\sin^4\alpha \tag{11.63}$$

$$\bar{Q}_{12} = (Q_{11} + Q_{22} - 4Q_{66})\sin^2\alpha\cos^2\alpha + Q_{12}(\sin^4\alpha + \cos^4\alpha) \tag{11.64}$$

$$\bar{Q}_{16} = (Q_{11} - Q_{12} - 2Q_{66})\sin\alpha\cos^3\alpha + (Q_{12} - Q_{22} + 2Q_{66})\sin^3\alpha\cos\alpha \tag{11.65}$$

$$\bar{Q}_{22} = Q_{11}\sin^4\alpha + 2(Q_{12} + 2Q_{66})\sin^2\alpha\cos^2\alpha + Q_{22}\cos^4\alpha \tag{11.66}$$

$$\bar{Q}_{26} = (Q_{11} - Q_{12} - 2Q_{66})\sin^3\alpha\cos\alpha + (Q_{12} - Q_{22} + 2Q_{66})\sin\alpha\cos^3\alpha \tag{11.67}$$

$$\bar{Q}_{66} = (Q_{11} + Q_{22} - 2Q_{12} - 2Q_{66})\sin^2\alpha\cos^2\alpha + Q_{66}(\sin^4\alpha + \cos^4\alpha) \tag{11.68}$$

$$Q_{11} = \frac{E_1}{1 - \nu_{12}\nu_{21}} \tag{11.69}$$

$$Q_{12} = \frac{\nu_{12}E_2}{1 - \nu_{12}\nu_{21}} \tag{11.70}$$

$$Q_{22} = \frac{E_2}{1 - \nu_{12}\nu_{21}} \tag{11.71}$$

$$Q_{66} = G_{12} \tag{11.72}$$

$$\nu_{21}E_1 = \nu_{12}E_2 \tag{11.73}$$

where E_1 and E_2 are the Young's moduli parallel to and perpendicular to the orientation of fibers, ν_{12} and ν_{21} are the corresponding Poisson's ratios, and α is the angle of fiber orientation of the ply.

The total potential energy of the laminated composite, both strain energies of bending and in-plane forces, is

$$\Pi = \Pi_b + \Pi_i \tag{11.74}$$

11.2.2 Discretized Equation for Buckling Analysis

The laminated plate is represented using a set of nodes on the symmetric plane ($z = 0$). Using MLS approximation described in Section 5.3, the deflection at any point in the plate can then be approximated using Equation 11.1. The expression of the total energy can be obtained by substituting the deflection w of Equation 11.1 into Equations 11.60 and 11.61 and then Equation 11.74. The stationary condition for the potential energy gives (see Section 4.6)

$$\frac{\partial \Pi}{\partial w_i} = 0 \tag{11.75}$$

which leads to a set of discrete eigenvalue equations for the laminated plate:

$$[\mathbf{K} - N_0\mathbf{B}]\mathbf{W} = 0 \tag{11.76}$$

where

$$\mathbf{W} = \{w_1, w_2, \ldots, w_{n_t}\}^T \tag{11.77}$$

is the deflection vector consisting of deflections at all the n_t nodes in the plate domain and **K** is the global stiffness matrix for the entire plate, which is assembled using the nodal stiffness (a scalar) defined by

$$K_{ij} = \frac{1}{2}\int_A \left[2D_{11}\frac{\partial^2\phi_i}{\partial x^2}\frac{\partial^2\phi_j}{\partial x^2} + 2D_{12}\left(\frac{\partial^2\phi_i}{\partial x^2}\frac{\partial^2\phi_j}{\partial y^2} + \frac{\partial^2\phi_j}{\partial x^2}\frac{\partial^2\phi_i}{\partial y^2}\right) + 2D_{22}\frac{\partial^2\phi_i}{\partial y^2}\frac{\partial^2\phi_j}{\partial y^2} + 8D_{66}\frac{\partial^2\phi_i}{\partial x\partial y}\frac{\partial^2\phi_j}{\partial x\partial y} \right.$$
$$\left. + 4D_{16}\left(\frac{\partial^2\phi_i}{\partial x^2}\frac{\partial^2\phi_j}{\partial x\partial y} + \frac{\partial^2\phi_j}{\partial x^2}\frac{\partial^2\phi_i}{\partial x\partial y}\right) + 4D_{26}\left(\frac{\partial^2\phi_i}{\partial y^2}\frac{\partial^2\phi_j}{\partial x\partial y} + \frac{\partial^2\phi_j}{\partial y^2}\frac{\partial^2\phi_i}{\partial x\partial y}\right)\right] \mathrm{d}A \tag{11.78}$$

$$B_{ij} = \frac{1}{2}\int_A \left[2\frac{\partial\phi_i}{\partial x}\frac{\partial\phi_j}{\partial x} + 2\mu_1\frac{\partial\phi_i}{\partial y}\frac{\partial\phi_j}{\partial y} + 2\mu_2\left(\frac{\partial\phi_i}{\partial x}\frac{\partial\phi_j}{\partial y} + \frac{\partial\phi_j}{\partial x}\frac{\partial\phi_i}{\partial y}\right)\right] \mathrm{d}A \tag{11.79}$$

For isotropic thin plates, the expression of potential energy of bending can be reduced as

$$\Pi_b = \frac{D}{2}\int_A \left[\left(\frac{\partial^2 w}{\partial x^2} + \frac{\partial^2 w}{\partial y^2}\right)^2 - 2(1-\nu)\left\{\frac{\partial^2 w}{\partial x^2}\frac{\partial^2 w}{\partial y^2} - \left(\frac{\partial^2 w}{\partial x\partial y}\right)^2\right\}\right] \mathrm{d}A \tag{11.80}$$

The strain energy caused by in-plane forces is the same as Equation 11.61. Minimizing the total potential energy yields the same form of discrete eigenvalue equations as Equation 11.76, where matrix **B** is defined by Equation 11.79 and the elements in matrix **K** have a much simpler form:

$$K_{ij} = \frac{D}{2}\int_A \left[2\left(\frac{\partial^2\phi_i}{\partial x^2} + \frac{\partial^2\phi_i}{\partial y^2}\right)\left(\frac{\partial^2\phi_j}{\partial x^2} + \frac{\partial^2\phi_j}{\partial y^2}\right) - 2(1-\nu)\left\{\left(\frac{\partial^2\phi_i}{\partial x^2}\frac{\partial^2\phi_j}{\partial y^2} + \frac{\partial^2\phi_j}{\partial x^2}\frac{\partial^2\phi_i}{\partial y^2}\right) - 2\frac{\partial^2\phi_i}{\partial x\partial y}\frac{\partial^2\phi_j}{\partial x\partial y}\right\}\right] \mathrm{d}A \tag{11.81}$$

where D is the bending stiffness of the plate defined by Equation 3.85. It can be easily confirmed that the foregoing equation is an alternative form of Equation 11.23.

The treatment on essential boundary conditions can be performed exactly as in Section 11.1.4.

Eigenvalue Equations of the Buckling

By performing orthogonal matrix transformation

$$\mathbf{W} = \mathbf{V}_{n_t\times(n_t-r)}\tilde{\mathbf{W}} \tag{11.82}$$

to Equation 11.76, the condensed eigenvalue equation of the static buckling can be rewritten as

$$[\tilde{\mathbf{K}} - N_0\tilde{\mathbf{B}}]\tilde{\mathbf{W}} = 0 \tag{11.83}$$

where $\tilde{\mathbf{W}}$ is an eigenvector,

$$\tilde{\mathbf{K}} = \mathbf{V}^T_{(n_t-r)\times n_t}\mathbf{K}\mathbf{V}_{n_t\times(n_t-r)} \tag{11.84}$$

is the dimension stiffness matrix, and

$$\tilde{\mathbf{B}} = \mathbf{V}^T_{(n_t-r)\times n_t}\mathbf{B}\mathbf{V}_{n_t\times(n_t-r)} \tag{11.85}$$

Solving Equation 11.83 using standard routines of eigenvalue solvers gives the static buckling values of thin laminated plates including thin plates of isotropic homogeneous materials.

11.2.3 Discretized Equation for Free-Vibration Analysis

The discretized dynamic equation has the same form as Equation 11.22. The only difference is that the stiffness matrix **K** needs to be replaced with that in Equation 11.76 for composite laminates. The treatment of essential boundary conditions is also exactly the same as in Section 11.1.4, and the final eigenvalue equation is given by Equation 11.55, where the stiffness matrix **K** needs to be replaced with that in Equation 11.76 for composite laminated plates. The mass matrix formulated using Equation 11.38 is still valid, but the mass moment of inertial I needs to be computed using Equation 11.39 with the consideration that ρ could be a function of z for composite laminated plates.

11.2.4 Numerical Examples for Buckling Analysis

In all the examples given in this section, complete second-order polynomial basis functions are used (m = 6) for constructing MLS shape functions. The dimension of the support domain is chosen as 3.5 to 3.9 times the averaged nodal distance.

Example 11.6 Static Buckling of Rectangular Plates (Validation)

The convergence of static buckling loads of square plates obtained by the present method is studied. The dimensions of the rectangular plates are noted by a and b, respectively, for the dimensions in the x and y directions. An in-plane compressive load is applied in the x direction. The static buckling loads of thin plates with different aspect ratios and boundaries are calculated and compared with analytical results (Timoshenko and James, 1985). The effects of different boundary conditions are also investigated. In the notation of the boundary conditions, the same convention used in the previous section is adopted. The buckling loads are represented via the factor of buckling load defined as $k = N_0b^2/\pi^2D$, where $D = Eh^3/[12(1-\nu^2)]$.

The geometry of the plates is shown in Figure 11.12. The geometric parameters and material properties of the thin rectangular plates are as follows: length $b = 10.0$ m, thickness $h = 0.06$ m, and the width of the plate can be determined by the aspect ratio, which changes from case to case. The material constants are elastic constant $E = 2.45\times10^6$ N/m^2, Poisson's ratio $\nu = 0.23$.

The buckling factors for square plates with three kinds of aspect ratios and simply supported boundaries are calculated using different densities of field nodes. The results are listed in Table 11.12. It is seen that highly accurate results can be obtained using a small number ($6\times6 = 36$) of field nodes. The convergence is very rapid.

Table 11.13 shows the buckling factors of rectangular plates of different aspect ratios with simply supported boundaries. Field nodes (13×13) regularly distributed in the plates are used. The buckling factors obtained using the present EFG method agrees very well with Timoshenko's analytical solutions. The buckling factor decreases as the aspect ratio increases from 0.2 to 1.0, while the buckling factor increases as the aspect ratio increases from 1.0 to 1.41. The minimum buckling factor occurs when the aspect ratio is 1.0, namely, when the plate is squared.

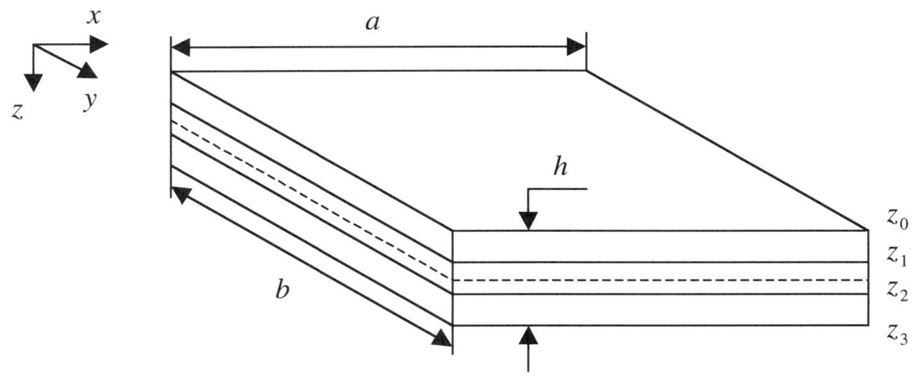

FIGURE 11.12
Symmetrically laminated composite plate and its coordinate system.

TABLE 11.12

Convergence of Buckling Factor $k = N_0 b^2/\pi^2 D$ for Rectangular Plates (b = 10.0 m, h = 0.06 m; BC: SSSS)

	Nodes							Timoshenko and James (1985)
	5 × 5	6 × 6	7 × 7	8 × 8	9 × 9	10 × 10	11 × 11	
k a/b = 0.8	4.60	4.20	4.17	4.20	4.19	4.20	4.20	4.20
a/b = 1.0	4.25	3.99	3.97	4.01	4.03	4.02	4.00	4.00
a/b = 1.2	4.03	4.13	4.14	4.15	4.12	4.14	4.14	4.13

TABLE 11.13

Buckling Factor $k = N_0 b^2/\pi^2 D$ for Rectangular Plates (b = 10.0 m, h = 0.06 m; nodes: 13 × 13, BC: SSSS)

	a/b													
	0.2	0.3	0.4	0.5	0.6	0.7	0.8	0.9	1.0	1.1	1.2	1.3	1.4	1.41
k EFG	27.09	13.22	8.42	6.25	5.13	4.53	4.20	4.04	4.04	4.05	4.13	4.28	4.48	4.48
k Timoshenko and James (1985)	27.0	13.2	8.41	6.25	5.14	4.53	4.20	4.04	4.00	4.04	4.13	4.28	4.47	4.49
Error (%)	0.33	0.15	0.12	0.00	−0.19	0.00	0.00	0.00	1.00	0.25	0.00	0.00	0.22	−0.22

Table 11.14 shows the buckling factors of plates with clamped boundaries. A total of 169 (= 13 × 13) regularly distributed nodes are used in the computation. Very good agreement between the buckling factors in the EFG method and Timoshenko's analytical solutions has also been achieved. The buckling factor obtained using the EFG method is slightly larger than the analytical solution and decreases as the aspect ratio increases from 0.75 to 4.0. The minimum buckling factor occurs when the aspect ratio is 4.0. The error between the EFG result and Timoshenko's solution for plates with clamped boundaries is larger than that for plates with simply supported boundaries.

Table 11.15 shows the buckling factors for SCSC boundaries. The nodal distribution is also 13 × 13 (regular). The errors between the EFG results and analytic solutions are very small. The minimum buckling factor is observed at the aspect ratio of 0.7.

TABLE 11.14

Buckling Factor $k = N_0 b^2/\pi^2 D$ for Rectangular Plates ($b = 10.0$ m, $h = 0.06$ m; nodes: 13 × 13, BC: CCCC)

	a/b													
	0.75	**1.00**	**1.25**	**1.50**	**1.75**	**2.00**	**2.25**	**2.50**	**2.75**	**3.00**	**3.25**	**3.50**	**3.75**	**4.00**
k EFG	11.79	10.18	9.42	8.47	8.22	8.00	7.75	7.70	7.57	7.49	7.48	7.41	7.39	7.38
k Timoshenko and James (1985)	11.69	10.07	9.25	8.33	8.11	7.88	7.63	7.57	7.44	7.37	7.35	7.27	7.24	7.23
Error (%)	0.86	1.09	1.84	1.68	1.36	1.52	1.57	1.72	1.75	1.63	1.77	1.93	2.07	2.07

TABLE 11.15

Buckling Factor $k = N_0 b^2/\pi^2 D$ for Rectangular Plates ($b = 10.0$ m, $h = 0.06$ m; nodes: 13 × 13, BC: SCSC)

	a/b						
	0.4	**0.5**	**0.6**	**0.7**	**0.8**	**0.9**	**1.0**
k EFG	9.48	7.72	7.11	7.05	7.38	7.96	7.77
k Timoshenko and James (1985)	9.44	7.69	7.05	7.00	7.29	7.83	7.69
Error (%)	0.42	0.39	0.85	0.71	1.23	1.66	1.04

TABLE 11.16

Buckling Factor $k = N_0 b^2/\pi^2 D$ for Rectangular Plates (BC: SCCS)

	a/b					
	1.0	**1.2**	**1.4**	**1.6**	**1.8**	**2.0**
k EFG	6.28	6.26	6.07	5.98	5.77	5.92

Table 11.16 shows the buckling factors for SCCS boundaries. Nodes of 13 × 13 that are regularly distributed in the plates are used. The buckling factors from aspect ratio 1.0 to 2.0 are calculated using the EFG method. The minimum buckling factor occurs at the aspect ratio of 1.8.

Example 11.7 Static Buckling of a Square Plate (Efficiency)

To study the efficiency of the EFG method, the buckling of a thin square plate with clamped boundaries is studied using both EFG and an FEM that was also developed in-house. In FEM, four-node isoparametric elements are used. The results obtained using EFG and FEM for different densities of nodes are shown in Table 11.17. The dimensions of the plate are length $a = b = 10.0$ m, thickness $h = 0.7$ m. The material properties are elastic constant $E = 1.0 \times 10^9$ N/m^2, Poisson's ratio $\nu = 0.3$.

Note that the dimension of the eigenvalue equation produced by EFG is only one third of that produced by FEM. Because of the smaller dimension of the eigenvalue equation in the EFG formulation, the CPU time for solving the eigenvalue equation in EFG is much less than that in FEM, as shown in Table 11.17, especially when the node number is large. Note that the total CPU time for solving the problem is affected by many factors, such as the construction of shape functions, imposition of boundary conditions, and the number of eigenvalues required. In EFG, selecting the nodes in the support domain of a quadrature point and constructing the shape functions for the point are much more expensive procedures as compared with FEM. The total CPU time of EFG is therefore greater than that of

TABLE 11.17

Comparison of Results of $k = N_0 b^2/\pi^2 D$ for a Square Plate Obtained by EFG and FEM ($a = b = 10.0$ m, $h = 0.7$ m, BC: CCCC)

		Nodes					
		9 × 9	**11 × 11**	**13 × 13**	**15 × 15**	**17 × 17**	**19 × 19**
Dimension of equation	EFG	81	121	169	225	289	361
	FEM	243	363	507	675	867	1083
CPU time (s) for eigenvalue solver	EFG	0.02	0.04	0.1	0.3	0.6	1.2
	FEM	0.0	0.7	1.4	5.2	16.6	63.8
Total CPU time (s)	EFG	**14.3**	47.2	155.7	531.5	1302.3	3789.8
	FEM	8.5	33.2	76.0	217.5	554.6	**1530.1**
Buckling factor k	EFG	**10.41**	10.23	10.18	10.16	10.14	10.13
	FEM	16.23	13.60	12.20	11.37	10.84	**10.47**

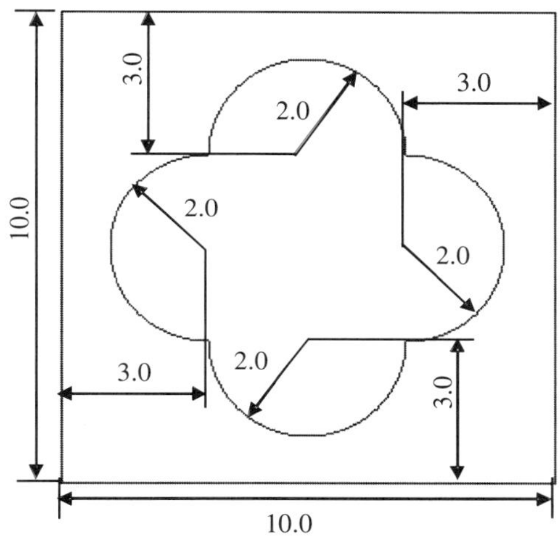

FIGURE 11.13
Plate with a hole of complicated shape.

FEM for the same density of nodes. This is the price EFG has to pay for not using an element mesh. However, the results obtained by EFG are much more accurate than those of FEM. For this square plate the analytical result of buckling factor is $k = 10.07$. EFG produces a result of $k = 10.41$ using only 81 nodes, but FEM requires 361 nodes to obtain a result of the same accuracy. Therefore, to obtain a result of the same accuracy, the EFG method is still much more computationally efficient than FEM for solving static buckling problems. For this particular case, EFG is about 100 times faster than FEM in terms of total CPU time to obtain results of the same accuracy.

Example 11.8 Static Buckling of a Plate with Complicated Shape (Application)

To demonstrate the applicability of the present method to plates of complicated shape, the static buckling loads of a plate with a hole of complicated shape are also calculated. The geometries of the plates shown in Figure 11.13 are calculated. The field nodes irregularly distributed in the plate are shown in Figure 11.14. The number of the nodes is 339. Triangular background meshes with 493 nodes are used for integration and all the field nodes are in the vertex of the integration cells. It can be found from Table 11.18 that the

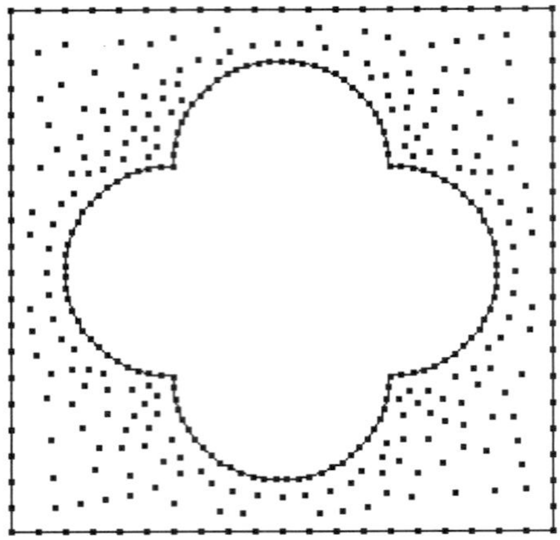

FIGURE 11.14
Nodal distribution in a plate with a hole of complicated shape.

TABLE 11.18
Buckling Factor $k = N_0b^2/\pi^2D$ for the Plate with a Hole of Complicated Shape

	Boundaries			
	SSSS	CCCC	SCSC	SCCS
k	2.64	16.66	5.43	3.97

TABLE 11.19
Buckling Factor $k = N_0b^2/\pi^2D_0$ for Symmetrically Laminated Composite Square Plates with Different Angle Plies and Boundaries ($a = b = 10.0$ m, $h = 0.06$ m, nodes: 13 × 13)

k \ BC / Angle Ply	SSSS	CCCC	SCSC	SCCS
(0°, 0°, 0°)	2.39	6.78	4.34	3.97
(10°, −10°, 10°)	2.42	6.72	4.39	3.97
(15°, −15°, 15°)	2.45	6.64	4.46	3.96
(20°, −20°, 20°)	2.49	6.55	4.56	3.96
(30°, −30°, 30°)	2.57	6.36	4.84	3.96
(40°, −40°, 40°)	2.63	6.21	4.91	3.94
(45°, −45°, 45°)	2.64	6.16	4.79	3.93
(0°, 90°, 0°)	2.39	6.78	4.43	3.97

plate with SSSS boundaries has the lowest buckling factor, and the plate with CCCC boundaries has the highest buckling factor. The static buckling factors of the plates with a hole are smaller compared with those of the corresponding plates without a hole.

Example 11.9 Static Buckling of a Laminated Plate (Application)

The static buckling loads of three layers of symmetrically laminated composite plates of E-glass/epoxy materials for four cases of boundaries are calculated, as shown in Table 11.19. The in-plane compressive loads are applied in the x direction. The factor of the buckling load is defined by $k = N_0b^2/\pi^2D_0$, where $D_0 = E_1h^3/[12(1-\nu_{12}\nu_{21})]$. The geometric parameters and material properties of the laminates are length $a = b = 10.0$ m, thickness $h = 0.06$ m, ratio of elastic constant $E_1/E_2 = 2.45$, ratio of elastic constant $G_{12}/E_2 = 0.48$, Poisson's ratio $\nu_{12} = 0.23$. The buckling load factors of the laminates in the EFG method are calculated using 13×13 nodes regularly distributed in the plate domain. For simply supported boundaries, the buckling value increases as the ply angle increases. In contrast, for clamped boundaries, the buckling value decreases as the ply angle increases. For SSSS, CCCC, and SCCS boundaries, the buckling value for the plies of (0°, 0°, 0°) is the same as that for the plies of (0°, 90°, 0°). For SCCS boundaries, the buckling factor is little affected by the angle ply. As expected, the buckling values of the laminates with clamped boundaries are generally larger than those with simply supported boundaries.

11.2.5 Numerical Examples for Free-Vibration Analysis

Example 11.10 Frequency Analysis of Free Vibration of Orthotropic Square Plates

Consider a square plate of a single layer of orthotropic material. The following parameters are used in the computation:

Length: $a = b = 10.0$ m

Thickness: $h = 0.05$ m

Mass density: $\rho = 8000$ kg/m^3

To compare the results obtained by the present method with those given by Y. Z. Chen (1998), we use the same elastic constants as Chen for the following two cases: (1) If $E_1 \geq E_2$, $\nu_{12} = 0.3$; (2) If $E_1 < E_2$, $\nu_{21} = 0.3$.

The dimensionless fundamental frequency is defined as $\xi = (\omega^2 a^4 \rho h / D_3)^{1/4}$, where $D_3 = D_{12} + 2D_{66}$. Regular rectangular background meshes (12 × 12) and regularly distributed field nodes (13 × 13) are used. The vertices of the background mesh are used as the field nodes. The dimensionless fundamental frequencies are listed in Tables 11.20 through 11.23 for four cases of different boundary conditions. In these tables, the subscript of ξ indicates the number of mode. In each table, frequencies for different ratios of elastic constants are listed. Table 11.20 shows the results for a square plate simply supported at all edges. Good agreement between the frequencies of the first mode obtained using the present method and those given by Y. Z. Chen (1998) has been observed. Table 11.21 shows the results for a square plate clamped at all edges. The frequencies of the first mode obtained by the present method fall in between those of Y. Z. Chen (1998) and those of Gontkewitz (1964).

TABLE 11.20

The Dimensionless Fundamental Natural Frequencies ξ for Orthotropic Square Plates (BC: SSSS)

D_{22}/D_3		0.5			1.0			2.0		
D_{11}/D_3		0.5	1.0	2.0	0.5	1.0	2.0	0.5	1.0	2.0
EFG	ξ_1	4.130	4.295	4.576	4.295	4.443	4.700	4.576	4.700	4.921
	ξ_2	6.333	6.387	6.479	6.387	7.031	7.104	6.479	7.104	8.008
	ξ_3	6.341	6.996	7.936	6.996	7.036	7.961	7.936	7.961	8.011
	ξ_4	8.273	8.600	8.781	8.600	8.892	9.404	8.781	9.404	9.843
	ξ_5	8.714	8.743	9.159	8.743	9.959	9.988	9.159	9.988	11.575
	ξ_6	8.732	9.949	10.903	9.949	9.966	11.560	10.903	11.560	11.578
	ξ_7	10.411	10.587	11.277	10.587	11.341	11.604	11.277	11.604	12.713
	ξ_8	10.422	11.205	11.552	11.205	11.347	12.518	11.552	12.518	12.714
	ξ_9	11.242	11.259	12.416	11.259	13.032	13.048	12.416	13.048	14.775
	ξ_{10}	11.249	12.891	13.515	12.891	13.036	14.110	13.515	14.110	15.285
ξ_1 Chen (1998)		4.118	4.279	4.557	4.279	4.425	4.678	4.557	4.678	4.897

TABLE 11.21

The Dimensionless Fundamental Natural Frequencies ξ for Orthotropic Square Plates (BC: CCCC)

D_{22}/D_3		0.5			1.0			2.0		
D_{11}/D_3		0.5	1.0	2.0	0.5	1.0	2.0	0.5	1.0	2.0
EFG	ξ_1	5.312	5.697	6.288	5.697	6.017	6.532	6.288	6.532	6.948
	ξ_2	7.554	7.710	7.982	7.710	8.605	8.805	7.982	8.805	9.976
	ξ_3	7.554	8.495	9.767	8.495	8.606	9.841	9.767	9.841	9.976
	ξ_4	9.430	9.973	10.203	9.973	10.439	11.209	10.203	11.209	11.847
	ξ_5	9.978	10.069	10.840	10.069	11.533	11.637	10.840	11.637	13.500
	ξ_6	10.006	11.497	12.459	11.497	11.562	13.334	12.459	13.334	13.525
	ξ_7	11.592	11.911	12.644	11.911	12.893	13.455	12.644	13.455	14.690
	ξ_8	11.597	12.565	13.423	12.565	12.896	14.367	13.423	14.367	14.692
	ξ_9	12.524	12.651	14.192	12.651	14.605	14.655	14.192	14.655	16.761
	ξ_{10}	12.525	14.202	14.584	14.202	14.606	15.902	14.584	15.902	17.185
ξ_1 Gontkewitz (1964)		5.324	5.712	6.307	5.712	6.034	6.553	6.307	6.553	6.972
ξ_1 Chen (1998)		5.171	5.551	6.131	5.551	5.866	6.371	6.131	6.371	6.779

TABLE 11.22

The Dimensionless Fundamental Natural Frequencies ξ for Orthotropic Square Plates (BC: SCSC)

D_{22}/D_3		0.5			1.0			2.0		
D_{11}/D_3		0.5	1.0	2.0	0.5	1.0	2.0	0.5	1.0	2.0
EFG	ξ_1	4.829	4.936	5.130	5.319	5.400	5.551	6.017	6.073	6.181
	ξ_2	6.655	7.239	7.468	6.876	7.413	8.231	7.246	7.715	8.456
	ξ_3	7.360	7.397	8.106	8.362	8.388	8.439	9.206	9.701	9.735
	ξ_4	8.901	9.215	9.690	9.020	9.799	10.205	9.685	10.287	11.032
	ξ_5	8.954	9.909	9.952	9.577	10.156	11.469	10.545	10.717	11.776
	ξ_6	9.891	10.081	11.628	11.262	11.445	11.689	11.505	12.474	13.376
	ξ_7	10.881	11.415	11.703	11.405	11.919	12.720	11.913	13.197	13.444
	ξ_8	11.285	11.609	12.466	11.439	12.525	12.982	13.387	13.396	14.259
	ξ_9	11.347	12.453	12.727	12.414	13.131	14.541	13.810	14.109	15.380
	ξ_{10}	12.446	13.093	13.908	13.384	14.214	14.879	13.993	14.915	15.921
ξ_1 Chen (1998)		4.729	4.838	5.037	5.205	5.288	5.443	5.883	5.941	6.052

TABLE 11.23

The Dimensionless Fundamental Natural Frequencies ξ for Orthotropic Square Plates (BC: SCCS)

D_{22}/D_3		0.5			1.0			2.0		
D_{11}/D_3		0.5	1.0	2.0	0.5	1.0	2.0	0.5	1.0	2.0
EFG	ξ_1	4.726	4.997	5.428	4.997	5.228	5.611	5.428	5.611	5.927
	ξ_2	6.935	7.037	7.213	7.037	7.807	7.941	7.213	7.941	8.980
	ξ_3	6.953	7.750	8.855	7.750	7.827	8.905	8.855	8.905	8.997
	ξ_4	8.844	9.277	9.489	9.277	9.656	10.296	9.489	10.296	10.836
	ξ_5	9.356	9.407	9.988	9.407	10.752	10.811	9.988	10.811	12.542
	ξ_6	9.376	10.735	11.679	10.735	10.770	12.469	11.679	12.469	12.556
	ξ_7	11.004	11.250	11.957	11.250	12.120	12.518	11.957	12.518	13.703
	ξ_8	11.011	11.909	12.498	11.909	12.121	13.445	12.498	13.445	13.705
	ξ_9	11.885	11.930	13.305	11.930	13.821	13.850	13.305	13.850	15.771
	ξ_{10}	11.895	13.458	13.725	13.458	13.828	15.011	13.725	15.011	16.237
ξ_1 Chen (1998)		4.637	4.905	5.332	4.905	5.136	5.515	5.332	5.515	5.829

TABLE 11.24

Convergence of Frequency of Laminated Composite Square Plates (BC: SSSS)

	Field Nodes				
	9 × 9	11 × 11	13 × 13	15 × 15	17 × 17
β	20.02	16.05	15.88	15.89	15.86

The results shown in Tables 11.22, and 11.23 also indicate that the frequencies of the first mode obtained by the present method agree well with those of Y. Z. Chen (1998). In all four cases, the frequencies of the first mode obtained by the present method are slightly larger than those given by Y. Z. Chen (1998).

Example 11.11 Natural Frequency Analysis of Composite Laminated Plates

Symmetric laminated composite plates with three layers of E-glass/epoxy materials are considered. Natural frequencies of square, elliptical, and complicated shaped plates are calculated using the present EFG method. The dimensionless frequency parameters are $\beta = (\rho h\omega^2 a^4/D_0)^{1/2}$, where $D_0 = E_1h^3/[12(1 - \nu_{12}\nu_{21})]$.

The geometric parameters and material properties of the square plates are as follows:

Length: $a = b = 10.0$ m
Thickness: $h = 0.06$ m
Mass density: $\rho = 8000$ kg/m^3
Ratio of elastic constant: $E_1/E_2 = 2.45$
Ratio of elastic constant: $G_{12}/E_2 = 0.48$
Poisson's ratio: $\nu_{12} = 0.23$

The radii of the elliptical plates are $a = 5.0$ m and $b = 2.5$ m, respectively. Other geometric parameters and material properties are the same as the square plates.

Convergence of Natural Frequency

We use different densities of regularly distributed field nodes in the present method to calculate natural frequencies of the laminated composite square plates with simply supported boundaries. Regular rectangular meshes are used and all the field nodes are in the vertexes of the meshes. The angle ply is arranged as (30°, –30°, 30°). Table 11.24 shows the frequency parameters. Good convergence can be observed from the table for the present method.

Effectiveness of Irregularly Distributed Nodes

To analyze the effectiveness of the present method using irregularly distributed nodes, we calculate frequencies of square laminates using $17 \times 17 = 289$ regularly distributed nodes and 289 irregularly distributed nodes (Figure 11.15). A uniform mesh of 16 × 16 mesh is used for integration for the square laminates. We also calculate frequencies of elliptical laminates using 201 regularly distributed nodes (Figure 11.4) and 201 irregularly distributed nodes (Figure 11.5). A regular mesh of 180 cells is used for integration for the elliptical laminates. The angle ply of both the laminates is arranged as (30°,–30°,30°). The results are shown in Tables 11.25 and 11.26. It is found from the two tables that the frequency parameters of using both regularly and irregularly distributed nodes show very good agreement with each other. This confirms that node irregularity does not affect the results significantly, provided there is no significant variation on nodal density across the domain.

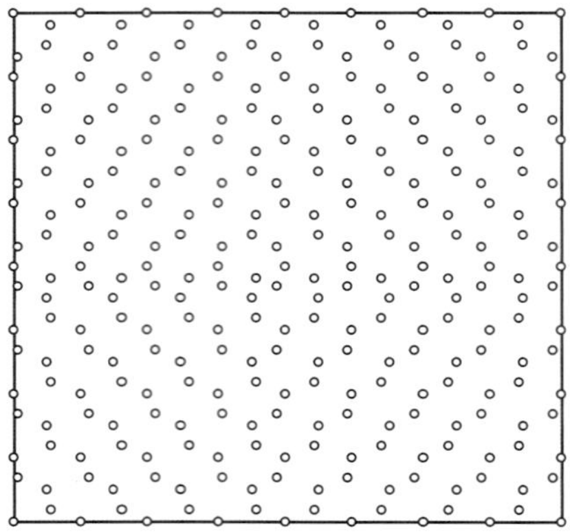

FIGURE 11.15
Square plate with 289 irregularly distributed nodes.

TABLE 11.25
Natural Frequency Parameters β of Laminated Composite Square Plates (BC: SSSS)

	Present Method	
Modes	Regular Nodes 289	Irregular Nodes 289
1	15.86	15.92
2	35.86	35.99
3	42.58	42.66
4	61.47	61.66
5	71.83	72.35
6	85.97	86.45
7	93.95	94.19
8	109.00	109.06
9	119.71	120.45
10	133.43	133.58

TABLE 11.26
Natural Frequency Parameters β of Laminated Composite Elliptical Plates (BC: CCCC)

	Present Method	
Modes	Regular Nodes 201	Irregular Nodes 201
1	19.89	19.87
2	30.44	30.44
3	44.34	44.25
4	49.95	49.90
5	61.94	61.98
6	65.77	65.76
7	84.63	84.19
8	84.81	84.75
9	93.36	93.12
10	107.33	107.06

TABLE 11.27

Natural Frequency Parameters β of Laminated Composite Square Plates (BC: SSSS)

Angle Ply	Source of Result	Modes 1	2	3	4	5	6	7	8	9	10
(0°, 0°, 0°)	Present	15.18	33.34	44.51	60.79	64.80	90.39	94.23	109.4	109.9	136.8
	Chow et al. (1992)	15.19	33.31	44.52	60.78	64.55	90.31	93.69	108.7	—	—
	Leissa and Narita (1989)	15.19	33.30	44.42	60.77	64.53	90.29	93.66	108.6	—	—
(15°, –15°, 15°)	Present	15.41	34.15	43.93	60.91	66.94	91.74	92.01	109.3	112.4	135.0
	Chow et al. (1992)	15.37	34.03	43.80	60.80	66.56	91.40	91.51	108.9	—	—
	Leissa and Narita (1989)	15.43	34.09	43.87	60.85	66.67	91.40	91.56	108.9	—	—
(30°, –30°, 30°)	Present	15.88	35.95	42.63	61.54	72.12	86.32	94.08	109.2	120.6	134.3
	Chow et al. (1992)	15.86	35.77	42.48	61.27	71.41	85.67	93.60	108.9	—	—
	Leissa and Narita (1989)	15.90	35.86	42.62	61.45	71.71	85.72	93.74	108.9	—	—
(45°, –45°, 45°)	Present	16.11	37.04	41.80	61.94	78.03	80.11	95.07	109.3	132.3	134.1
	Chow et al. (1992)	16.08	36.83	41.67	61.65	76.76	79.74	94.40	109.0	—	—
	Leissa and Narita (1989)	16.14	36.93	41.81	61.85	77.04	80.00	94.68	109.0	—	—
(0°, 90°, 0°)	Present	15.18	33.82	44.14	60.79	66.12	91.16	93.31	108.8	112.4	136.8

TABLE 11.28

Natural Frequency Parameters β of Laminated Composite Square Plates (BC: CCCC)

Angle Ply	Source of Result	Modes 1	2	3	4	5	6	7	8	9	10
(0°, 0°, 0°)	Present	29.27	51.21	67.94	86.25	87.97	119.3	127.6	138.5	144.0	169.8
	Chow et al. (1992)	29.13	50.82	67.29	85.67	87.14	118.6	126.2	137.5	—	—
(15°, –15°, 15°)	Present	29.07	51.82	66.54	85.17	90.56	120.0	124.1	140.8	143.2	167.6
	Chow et al. (1992)	28.92	51.43	65.92	84.55	89.76	119.3	122.7	139.9	—	—
(30°, –30°, 30°)	Present	28.69	53.57	63.24	84.43	96.13	115.4	121.5	139.6	150.8	166.3
	Chow et al. (1992)	28.55	53.15	62.71	83.83	95.21	114.1	120.7	138.6	—	—
(45°, –45°, 45°)	Present	28.50	55.11	60.91	84.25	103.2	106.7	122.3	138.3	165.2	166.1
	Chow et al. (1992)	28.38	54.65	60.45	83.65	102.0	105.6	121.4	137.3	—	—
(0°, 90°, 0°)	Present	29.27	51.93	67.40	86.25	89.76	120.3	126.4	141.6	143.2	172.2

Square Plate

Four cases of different boundary conditions together with five cases of different angle plies are considered. Regular rectangular background meshes (12 × 12) and regularly distributed field nodes (13 × 13) are used. Table 11.27 shows the results for laminates simply supported at all edges. The present frequencies agree very well with those given by both Chow et al. (1992) and Leissa and Narita (1989). Table 11.28 shows the results for laminates clamped at all edges. The present frequencies are very close to and slightly

TABLE 11.29

Natural Frequency Parameters β of Laminated Composite Square Plates (BC: SCSC)

	Modes									
Angle Ply	1	2	3	4	5	6	7	8	9	10
(0°, 0°, 0°)	20.55	46.41	47.18	70.52	85.02	96.06	107.5	116.7	136.0	150.6
(15°, −15°, 15°)	20.89	45.70	48.28	71.23	86.87	94.03	108.7	117.1	138.7	149.7
(30°, −30°, 30°)	21.85	44.45	51.19	72.63	88.65	93.23	109.0	121.1	147.7	150.0
(45°, −45°, 45°)	23.16	43.08	55.56	73.61	83.50	102.6	108.4	126.3	136.4	151.0
(0°, 90°, 0°)	20.77	46.92	47.32	70.90	86.92	95.22	108.8	116.3	139.3	151.2

TABLE 11.30

Natural Frequency Parameters β of Laminated Composite Square Plates (BC: SCCS)

	Modes									
Angle Ply	1	2	3	4	5	6	7	8	9	10
(0°, 0°, 0°)	21.62	41.64	55.66	72.81	75.92	104.3	110.5	123.7	126.2	150.3
(15°, −15°, 15°)	21.69	42.40	54.72	72.48	78.18	105.5	107.7	125.3	126.9	149.5
(30°, −30°, 30°)	21.81	44.27	52.49	72.48	83.53	100.6	107.4	124.4	135.3	149.7
(45°, −45°, 45°)	21.88	45.70	50.89	72.62	89.90	93.18	108.3	123.9	148.4	150.0
(0°, 90°, 0°)	21.62	42.24	55.21	72.82	77.47	105.2	109.4	125.4	126.6	152.5

TABLE 11.31

Natural Frequency Parameters β of Laminated Composite Elliptical Plates (BC: CCCC)

	Modes									
Angle Ply	1	2	3	4	5	6	7	8	9	10
(0°, 0°, 0°)	18.48	29.38	44.97	45.72	60.44	65.33	79.24	85.31	91.50	102.8
(15°, −15°, 15°)	18.83	29.70	44.73	46.72	62.06	64.07	81.09	87.14	88.90	104.5
(30°, −30°, 30°)	19.89	30.44	44.34	49.95	61.94	65.77	84.63	84.81	93.36	107.3
(45°, −45°, 45°)	21.60	31.38	44.11	55.17	60.19	70.21	81.64	88.25	103.7	110.2
(0°, 90°, 0°)	18.81	29.58	44.99	46.72	61.34	65.14	79.99	87.23	91.16	103.4

larger than those given by Chow et al. (1992). Table 11.29 shows the effect of fiber orientation on the frequency parameters of laminated plates with SCSC boundary conditions. Maximum natural frequency parameters of the fifth, seventh, and ninth modes are observed at the plate with $\alpha = 30°$, the first, third, fourth, sixth, and eighth modes occur at the plate with $\alpha = 45°$, and the second and tenth modes occur at the plate of (0°, 90°, 0°). Table 11.30 shows the effect of fiber orientation on the natural frequency parameters of laminated plates with SCCS boundary conditions. Maximum natural frequency parameters of the third and seventh modes are observed at $\alpha = 0°$, the sixth mode at $\alpha = 15°$, the first, second, fifth, and ninth modes occur at the plate with $\alpha = 45°$, and the fourth, eighth, and tenth modes are registered at the plate with (0°, 90°, 0°).

Elliptical Plate

The natural frequencies are computed for elliptical laminates using the present EFG code. Regularly distributed nodes (201) and regular meshes (180) are used. Table 11.31 shows the effect of fiber orientation on the natural frequency parameters of laminated plates clamped at all edges. Maximum frequency parameters of the fifth mode are registered at

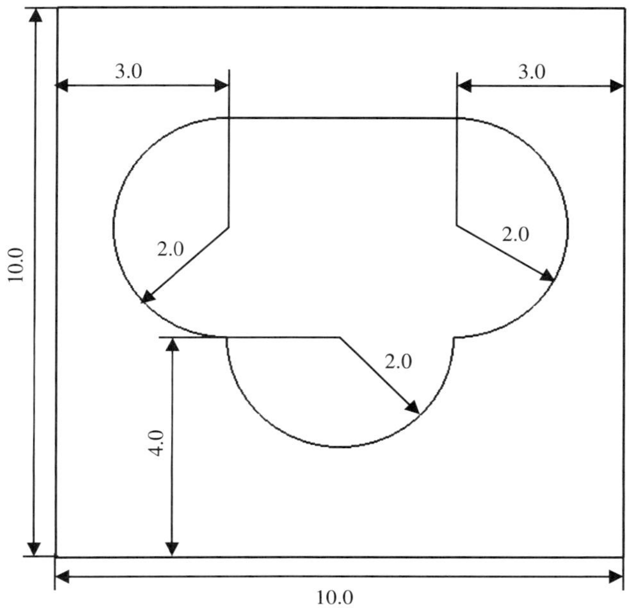

FIGURE 11.16
Square plate with a hole of complicated shape.

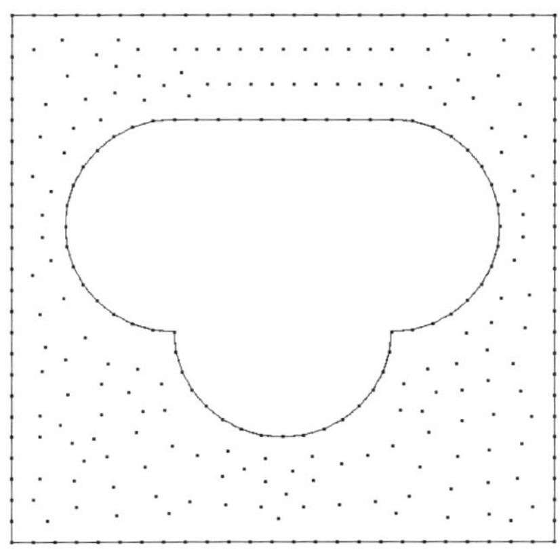

FIGURE 11.17
Nodal distribution of the square plate with a hole of complicated shape.

the plate with $\alpha = 15°$, the seventh mode at the plate with $\alpha = 30°$, the first, second, fourth, sixth, eighth, ninth, and tenth modes at the plate with $\alpha = 45°$, and the third mode at the plate with (0°, 90°, 0°).

Plate of Complicated Shape

The natural frequencies are calculated for laminates with a hole of complicated shape, as shown in Figure 11.16. The 310 nodes and 462 triangular background meshes are plotted in Figure 11.17. Four cases of different boundary conditions together with five cases of different angle ply were considered. Table 11.32 shows the effect of fiber orientation on

TABLE 11.32

Frequency Parameters β of Laminated Composite Plates of Complicated Shape (BC: SSSS)

	Modes									
Angle Ply	1	2	3	4	5	6	7	8	9	10
(0°, 0°, 0°)	25.64	79.94	88.32	118.2	169.2	179.3	205.0	241.0	259.1	288.1
(15°, –15°, 15°)	26.51	79.31	88.37	118.3	168.0	175.4	210.8	240.1	250.6	291.0
(30°, –30°, 30°)	28.33	79.48	88.18	118.7	165.7	172.1	213.0	238.4	247.0	294.1
(45°, –45°, 45°)	29.53	79.28	86.85	118.6	161.4	171.5	210.9	237.5	245.2	295.1
(0°, 90°, 0°)	25.70	79.95	88.30	118.4	169.0	180.1	206.1	241.7	259.9	288.7

TABLE 11.33

Frequency Parameters β of Laminated Composite Plates of Complicated Shape (BC: CCCC)

	Modes									
Angle Ply	1	2	3	4	5	6	7	8	9	10
(0°, 0°, 0°)	88.48	95.16	107.1	157.0	166.9	185.2	250.1	291.0	299.3	325.3
(15°, –15°, 15°)	88.86	95.35	107.8	156.4	166.5	186.8	244.8	290.4	297.8	323.5
(30°, –30°, 30°)	88.62	94.62	110.0	155.8	166.8	191.1	240.0	284.8	294.2	320.1
(45°, –45°, 45°)	87.74	93.19	113.3	156.1	167.4	194.7	235.4	281.4	290.4	316.9
(0°, 90°, 0°)	88.67	95.15	107.9	157.8	167.5	186.4	250.3	291.3	299.8	325.4

TABLE 11.34

Frequency Parameters β of Laminated Composite Plates of Complicated Shape (BC: SCSC)

	Modes									
Angle Ply	1	2	3	4	5	6	7	8	9	10
(0°, 0°, 0°)	64.07	72.44	98.25	139.3	160.0	170.8	230.7	254.6	266.0	273.0
(15°, –15°, 15°)	64.73	72.00	100.0	138.6	163.3	168.7	229.3	250.4	264.0	274.1
(30°, –30°, 30°)	66.36	72.99	103.5	138.4	161.5	171.1	228.0	247.4	259.9	276.5
(45°, –45°, 45°)	67.69	74.39	108.5	139.7	158.3	172.8	230.6	244.8	254.7	277.3
(0°, 90°, 0°)	64.37	72.86	99.32	140.1	160.5	171.2	231.9	255.0	266.3	273.4

the frequency parameters of laminates simply supported at all edges. Maximum frequency parameters of the fifth mode occur at $\alpha = 0°$, the third mode occurs at $\alpha = 15°$, the fourth and seventh modes occur at $\alpha = 30°$, the first and tenth modes occur at $\alpha = 45°$, and the second, sixth, eighth, and ninth modes occur at (0°, 90°, 0°). Table 11.33 shows the effect of fiber orientation on the frequency parameters of laminates clamped at all edges. Maximum frequency parameters of the first and second modes occur at the plate with $\alpha = 15°$, the third and sixth modes occur at the plate with $\alpha = 45°$, and the fourth, fifth, seventh, eighth, ninth, and tenth modes occur at the plate of (0°, 90°, 0°). Table 11.34 shows the effect of fiber orientation on the frequency parameters of laminates for SCSC boundary conditions. Maximum frequency parameters of the fifth mode occur at the plate with $\alpha =$ 15°, the first, second, third, sixth, and tenth modes occur at the plate with $\alpha = 45°$, and the fourth, seventh, eighth, and ninth modes occur at the plate of (0°, 90°, 0°). Table 11.35 shows the effect of fiber orientation on the frequency parameters of laminates for SCCS boundary conditions. Maximum frequency parameters of the second, seventh, and tenth modes occur at the plate with $\alpha = 15°$, the third, fourth, and ninth modes occur at the plate with $\alpha = 45°$, and the first, fifth, sixth, and eighth modes occur at the plate with (0°, 90°, 0°).

TABLE 11.35

Frequency Parameters β of Laminated Composite Plates of Complicated Shape (BC: SCCS)

	Modes									
Angle Ply	**1**	**2**	**3**	**4**	**5**	**6**	**7**	**8**	**9**	**10**
(0°, 0°, 0°)	68.95	93.87	114.2	140.6	170.2	191.5	221.9	251.5	264.3	301.7
(15°, −15°, 15°)	68.43	94.46	115.2	141.0	166.8	190.6	223.2	244.8	265.9	304.1
(30°, −30°, 30°)	68.64	94.08	117.1	141.9	164.8	189.8	223.1	238.9	270.8	301.6
(45°, −45°, 45°)	68.95	93.10	119.8	142.6	164.4	188.4	221.1	236.6	273.6	292.3
(0°, 90°, 0°)	69.17	94.02	114.9	141.1	170.9	191.8	222.5	251.9	265.2	302.5

It also can be seen that the natural frequencies of the laminates with clamped boundaries are generally higher than those with simply supported boundaries.

11.3 EFG Method for Thick Plates

In the previous two sections, we discussed EFG formulations for problems of static, buckling, and vibration of thin plates that are governed by the simplest displacement-based theory: classical plate theory (CPT). In CPT, the transverse shear strain is neglected. For analyses of thin plates with a thickness-to-width ratio of up to about 1:10, CPT can give good results with sufficient accuracy for engineering application, because the effect of the transverse shear deformation is very small. For thick plates, however, transverse shear strains cannot be neglected. The CPT underpredicts deflections and overpredicts natural frequencies and buckling loads for thick plates.

Thick plates are very important structural elements and have wide applications. There are a few displacement-based theories for the analyses of plates. A slightly more complex displacement-based theory, often termed the first-order shear deformation theory (FSDT) or Mindlin plate theory, must be used to take into account transverse shear strain. In FSDT, in-plane displacements through the thickness are assumed to vary linearly. FSDT gives a state of constant shear strain through the thickness, which cannot satisfy the natural condition that the shear strains in the top and bottom surfaces of plates should vanish. In FSDT, a shear correction factor that depends on material property, geometry, and boundary conditions is introduced to correct for the discrepancy in the shear forces. However, FSDT still does not reflect the high-order variation of shear strain through the thickness. To ensure the vanishing of the transverse shear strains on the top and bottom surfaces of the plates and more accurately to reflect the high-order variation of shear strain through the thickness, the third-order shear deformation theory (TSDT) was developed. In the formulations of finite plate element based on TSDT, C^1 displacement interpolants must be constructed, which can be difficult. Moreover, application of high-order displacement-based elements makes mesh generation algorithms much more complicated, leading to an increment in computational costs.

MFree methods have been applied for shear deformable beams and plates (Donning and Liu, W. K., 1998). This section formulates the EFG method to solve free-vibration and static buckling problems for shear deformable plates. The material presented in this section is based on the work done by G. R. Liu and Chen during 2000–2001. Based on both FSDT and TSDT of plates, weak forms of governing equation for free-vibration and static buckling are established. The essential boundary conditions are formulated through a weak form separately from the system equation, and are imposed using the orthogonal transform techniques used in Section 11.1. To examine the validity of the EFG method in the application of eigenvalue problems for shear deformable plates, natural frequencies of square thick plates with different hard-type boundary conditions are calculated. Buckling loads of square thick plates with different boundary conditions and a square thick plate with a circular hole are also investigated.

11.3.1 Field Variables for Thick Plates

The basic equations for thick plates based on both FSDT and TSDT are provided in Section 3.6. There is, however, a small complication in the boundary conditions. For shear deformable plates, there are two kinds of essential boundary conditions: soft-type and hard-type (Haggblad and Bathe, 1990). In this section, only hard-type conditions are considered.

1. At simply supported edges:

   For plates based on FSDT and TSDT,

$$w|_{\text{at edges}} = 0 \tag{11.86}$$

   where w is the deflection on the neutral plane of the plate.

$$\varphi_s|_{\text{at edges}} = 0 \tag{11.87}$$

   where φ_s is the rotation with respect to the axis perpendicular to the normal direction of the boundary.

2. At clamped edges:

   For plates based on FSDT,

$$w|_{\text{at edges}} = 0 \tag{11.88}$$

$$\varphi_n|_{\text{at edges}} = 0 \tag{11.89}$$

$$\varphi_s|_{\text{at edges}} = 0 \tag{11.90}$$

   where φ_n is the rotation with respect to the axis parallel to the normal direction of the boundary.

   For plates based on TSDT,

$$w|_{\text{at edges}} = 0 \tag{11.91}$$

$$\varphi_n|_{\text{at edges}} = 0 \tag{11.92}$$

$$\varphi_s|_{\text{at edges}} = 0 \tag{11.93}$$

$$\left.\frac{\partial w}{\partial n}\right|_{\text{at edges}} = 0 \tag{11.94}$$

11.3.2 Approximation of Field Variables

Using the EFG method, the plate is represented by a set of nodes scattered in the domain of the plate. The field variables should be the deflections and the rotations at all the nodes. MLS approximation is employed to approximate all these field variables w, φ_x, and φ_y, i.e.,

$$w = \sum_{I=1}^{n} \phi_I w_I \tag{11.95}$$

$$\varphi_x = \sum_{I=1}^{n} \phi_{xI} \varphi_{xI} \tag{11.96}$$

$$\varphi_y = \sum_{I=1}^{n} \phi_{yI} \varphi_{yI} \tag{11.97}$$

where ϕ_I, ϕ_{xI}, and ϕ_{yI} are the shape functions for the three field variables w, φ_x, and φ_y, respectively. These shape functions do not have to be the same. Formulations that use

$$\phi_{xI}(x, y) = \frac{\partial \phi_I(x, y)}{\partial x} \tag{11.98}$$

$$\phi_{yI}(x, y) = \frac{\partial \phi_I(x, y)}{\partial y} \tag{11.99}$$

can avoid shear locking (Kanok-Nukulchai et al., 2001). In our formulation, we will assume that ϕ_I, ϕ_{xI}, and ϕ_{yI} are different shape functions and are independent of each other. Equations 11.95 through 11.97 can be written in matrix form:

$$\mathbf{u} = \begin{Bmatrix} w \\ \varphi_x \\ \varphi_y \end{Bmatrix} = \sum_{I=1}^{n} \underbrace{\begin{bmatrix} \phi_I & 0 & 0 \\ 0 & \phi_{xI} & 0 \\ 0 & 0 & \phi_{yI} \end{bmatrix}}_{\mathbf{\Phi}_I} \underbrace{\begin{Bmatrix} w_I \\ \varphi_{xI} \\ \varphi_{yI} \end{Bmatrix}}_{\mathbf{u}_I} \tag{11.100}$$

where $\mathbf{\Phi}_I$ is a matrix of shape functions given by

$$\mathbf{\Phi}_I = \begin{bmatrix} \phi_I & 0 & 0 \\ 0 & \phi_{xI} & 0 \\ 0 & 0 & \phi_{yI} \end{bmatrix} \tag{11.101}$$

and $\mathbf{u}_I$ is the vector of nodal variables for node I.

$$\mathbf{u}_I = \begin{Bmatrix} w_I \\ \varphi_{xI} \\ \varphi_{yI} \end{Bmatrix} \tag{11.102}$$

11.3.3 Variational Forms of System Equations

For free-vibration analysis of shear deformable plates, we can start with the Galerkin weak form of elastic solids:

$$\int_{\Omega} \delta\boldsymbol{\varepsilon}^T \boldsymbol{\sigma}\, d\Omega + \int_{\Omega} \delta\mathbf{u}^T \rho \ddot{\mathbf{u}}\, d\Omega = 0 \tag{11.103}$$

where ρ is the mass density. Note that the terms related to external forces have been removed, because we are interested only in free vibration, and hence there should be no external forces. The formulation can then be based on the third-order shear deformation theory, as it contains the first-order shear deformation theory as its special case. Using the expressions for displacement (Equation 3.104), strains (Equation 3.105), stress (Equation 3.93), for thick plates given in Section 3.6, we can easily arrive at the following equations for thick plate structures:

$$\int_{\Omega} \delta(\mathbf{L}\mathbf{u})^T \mathbf{D}\mathbf{L}\mathbf{u}\, d\Omega + \int_{\Omega} \rho\delta(\mathbf{L}_u\mathbf{u})^T \mathbf{L}_u \ddot{\mathbf{u}}\, d\Omega = 0 \tag{11.104}$$

where $\mathbf{u}$, $\mathbf{L}_u$, $\mathbf{L}$, and $\mathbf{D}$ are defined, respectively, in Equations 3.96 through 3.98 and 3.106.

For static buckling analysis of shear deformable plate, the total potential energy of the plate is (Wang, C. M. et al., 1993)

$$\Pi_N = \frac{1}{2}\int_{\Omega} \boldsymbol{\varepsilon}^T \mathbf{D}\boldsymbol{\varepsilon}\, d\Omega + \int_{\Omega} \boldsymbol{\varepsilon}_N^T \boldsymbol{\tau}_N\, d\Omega \tag{11.105}$$

where the first term is same as that in Equation 11.103 and $\boldsymbol{\varepsilon}_N$ denotes the nonlinear strains. The higher-order terms associated with the in-plane displacements in $\boldsymbol{\varepsilon}_N$ are neglected. $\boldsymbol{\varepsilon}_N$ has the following form:

$$\boldsymbol{\varepsilon}_N^T = \left\{ \frac{1}{2}\left(\frac{\partial w}{\partial x}\right)^2 \quad \frac{1}{2}\left(\frac{\partial w}{\partial y}\right)^2 \quad \frac{\partial w}{\partial x}\frac{\partial w}{\partial y} \quad 0 \quad 0 \right\} \tag{11.106}$$

The stress tensor $\boldsymbol{\tau}_N$ is

$$\boldsymbol{\tau}_N = \left\{ \begin{array}{c} -\frac{N_x}{h} \\ -\frac{N_y}{h} \\ -\frac{N_{xy}}{h} \\ 0 \\ 0 \end{array} \right\} = -\frac{1}{h}\mathbf{P} \tag{11.107}$$

The variational form of the total potential energy is

$$\delta\Pi = \delta(\Pi_s + \Pi_N) = \int_{\Omega} \delta\boldsymbol{\varepsilon}^T \boldsymbol{\sigma}\, d\Omega + \int_{\Omega} \delta\boldsymbol{\varepsilon}_N^T \boldsymbol{\tau}_N\, d\Omega = 0 \tag{11.108}$$

Using Equation 11.108 and the expressions for strains (Equation 3.105) and stress (Equation 3.93), for thick plate given in Section 3.6, as well as Equations 11.106 and 11.107, we can easily arrive at the following weak form for the static buckling problem of thick plates.

$$\int_{\Omega} \delta(\mathbf{L}\mathbf{u})^T \mathbf{D}\mathbf{L}\mathbf{u}\, d\Omega - \int_{A} \delta(\mathbf{L}_{N1} w)^T (\mathbf{L}_{N2}^T w)\mathbf{P}\, dA = 0 \tag{11.109}$$

where **L** and **D** are defined, respectively, in Equations 3.98 and 3.105, and

$$\mathbf{L}_{N1}^{T} = \left\{ \frac{\partial}{\partial x}, \frac{\partial}{\partial y} \right\} \tag{11.110}$$

$$\mathbf{L}_{N2}^{T} = \begin{bmatrix} \frac{\partial}{\partial x} & 0 & \frac{\partial}{\partial y} & 0 & 0 \\ 0 & \frac{\partial}{\partial y} & \frac{\partial}{\partial x} & 0 & 0 \end{bmatrix} \tag{11.111}$$

$$\mathbf{P} = \begin{Bmatrix} N_x \\ N_y \\ N_{xy} \\ 0 \\ 0 \end{Bmatrix} \tag{11.112}$$

In deriving Equation 11.109, the integration in the z direction has been carried out for the second term.

11.3.4 Discrete System Equations

Substituting the displacement interpolants Equations 11.95 through 11.97 into Equation 11.104 or 11.109, we obtain the eigenvalue equation for both free vibration and static buckling of plate structures as follows:

$$[\mathbf{K} - \eta \mathbf{G}]\mathbf{q} = 0 \tag{11.113}$$

where η is the eigenvalues. For free-vibration problems, $\eta = \omega^2$, where ω represents the angular frequencies, and static buckling problem, $\eta = N_0$, where N_0 is the buckling load. Vector **q** is the eigenvector.

In Equation 11.113, **K** is the stiffness matrix for both free-vibration and static buckling problems, which is assembled using a nodal stiffness matrix of 3 × 3 given by

$$\mathbf{K}_{IJ} = \int_{\Omega} \mathbf{B}_I^T \mathbf{D} \mathbf{B}_J \, d\Omega \tag{11.114}$$

where the strain matrix **B** is given by

$$\mathbf{B}_I = \mathbf{L}\mathbf{\Phi}_I = \begin{bmatrix} -\alpha z^3 \frac{\partial^2}{\partial x^2} & (z - \alpha z^3)\frac{\partial}{\partial x} & 0 \\ -\alpha z^3 \frac{\partial^2}{\partial y^2} & 0 & (z - \alpha z^3)\frac{\partial}{\partial y} \\ -2\alpha z^3 \frac{\partial^2}{\partial x \partial y} & (z - \alpha z^3)\frac{\partial}{\partial y} & (z - \alpha z^3)\frac{\partial}{\partial x} \\ (1 - \beta z^2)\frac{\partial}{\partial x} & (1 - \beta z^2) & 0 \\ (1 - \beta z^2)\frac{\partial}{\partial y} & 0 & (1 - \beta z^2) \end{bmatrix} \begin{bmatrix} \phi_I & 0 & 0 \\ 0 & \phi_{xI} & 0 \\ 0 & 0 & \phi_{yI} \end{bmatrix} \tag{11.115}$$

In Equation 11.113, **G** is the mass matrix for the free-vibration problem,

$$\mathbf{G}_{IJ} = \int_\Omega \rho \mathbf{B}_{uI}^T \mathbf{B}_{uJ} \, d\Omega \tag{11.116}$$

where

$$\mathbf{B}_{uI} = \mathbf{L}_u \mathbf{\Phi}_I \begin{bmatrix} -\alpha z^3 \frac{\partial}{\partial x} & z - \alpha z^3 & 0 \\ -\alpha z^3 \frac{\partial}{\partial y} & 0 & z - \alpha z^3 \\ 1 & 0 & 0 \end{bmatrix} \begin{bmatrix} \phi_I & 0 & 0 \\ 0 & \phi_{xI} & 0 \\ 0 & 0 & \phi_{yI} \end{bmatrix} \tag{11.117}$$

For the static buckling problem, **G** is assembled using a nodal matrix of

$$\mathbf{G}_{IJ} = \begin{bmatrix} G_{IJ}^w & 0 & 0 \\ 0 & 0 & 0 \\ 0 & 0 & 0 \end{bmatrix} \tag{11.118}$$

where

$$G_{IJ}^w = \frac{1}{2} \int_A \left[2 \frac{\partial \phi_I}{\partial x} \frac{\partial \phi_J}{\partial x} + 2\mu_1 \frac{\partial \phi_I}{\partial y} \frac{\partial \phi_J}{\partial y} + 2\mu_2 \left(\frac{\partial \phi_I}{\partial x} \frac{\partial \phi_J}{\partial y} + \frac{\partial \phi_J}{\partial x} \frac{\partial \phi_I}{\partial y} \right) \right] dA \tag{11.119}$$

where μ_1 and μ_2 are defined in Equation 11.58.

Equation 11.113 is the eigenvalue equation for thick plates without essential boundary conditions imposed.

11.3.5 Discrete Form of Essential Boundary Conditions

Because the Kronecker delta condition $\phi_I(x_J) = \delta_{IJ}$ at each node is not satisfied by the MLS shape function, the above essential boundary conditions are imposed in a manner similar to that described in Section 11.1. The essential boundary conditions are represented by a weak form with Lagrange multipliers to produce the discretized essential boundary conditions as given below:

$$\int_{\Gamma_u} \delta \boldsymbol{\lambda}^T (\tilde{\mathbf{u}} - \bar{\mathbf{u}}) \, d\Gamma = 0 \tag{11.120}$$

where $\boldsymbol{\lambda}$ is a vector of Lagrange multipliers and $\bar{\mathbf{u}}$ is the prescribed essential boundary conditions. For free-vibration and static buckling analyses, we should have $\bar{\mathbf{u}} = 0$, because the problem is homogeneous. Vector **u** is the displacement approximated on the essential boundary conditions. For the essential boundary of plates based on FSDT,

$$\tilde{\mathbf{u}} = \begin{Bmatrix} w \\ \varphi_n \\ \varphi_s \end{Bmatrix} = \underbrace{\begin{bmatrix} 1 & 0 & 0 \\ 0 & n_x & n_y \\ 0 & -n_y & n_x \end{bmatrix}}_{\mathbf{L}_b} \underbrace{\begin{Bmatrix} w \\ \varphi_x \\ \varphi_y \end{Bmatrix}}_{\mathbf{u}} = \mathbf{L}_b \mathbf{u} \tag{11.121}$$

where n_x and n_y are direction cosines of the outward normal on the boundary.

For the essential boundary of plates based on TSDT,

$$\tilde{\mathbf{u}} = \begin{Bmatrix} w \\ \varphi_n \\ \varphi_s \\ \dfrac{\partial w}{\partial n} \end{Bmatrix} = \underbrace{\begin{bmatrix} 1 & 0 & 0 \\ 0 & n_x & n_y \\ 0 & -n_y & n_x \\ \dfrac{\partial}{\partial n} & 0 & 0 \end{bmatrix}}_{\mathbf{L}_b} \underbrace{\begin{Bmatrix} w \\ \varphi_x \\ \varphi_y \end{Bmatrix}}_{\mathbf{u}} = \mathbf{L}_b \mathbf{u} \tag{11.122}$$

The discrete essential boundary conditions derived from Equation 11.120 can be written in the form of (for cases of TSDT)

$$\mathbf{H}_{4n_b \times 3n} \mathbf{Q}_{3n \times 1} = 0 \tag{11.123}$$

where n_b is the number of constraint points on the supported boundaries. In computing **H**, one-point Gauss quadrature is used along each span between the constraint points. For example, for a clamped edge of plates based on TSDT and any one span, **H** is assembled from the nodal contributions defined as

$$\mathbf{H}_{KI} = \int_{\Gamma_u} \begin{bmatrix} N_k \phi_I & 0 & 0 \\ 0 & n_x N_K \phi_{xI} & n_y N_K \phi_{yI} \\ 0 & -n_y N_K \phi_{xI} & n_x N_K \phi_{yI} \\ \phi_{I,n} & 0 & 0 \end{bmatrix} \mathrm{d}\Gamma \tag{11.124}$$

where N_K ($K = 1, 2$) are Lagrange linear interpolations given in Equation 6.10 for the span between two constraint points on the essential boundary.

Generally, the matrix **H** is sparse and singular. Using singular-value decomposition technique, it can be decomposed as

$$\mathbf{H}_{4n_b \times 3n} = \mathbf{U}_{4n_b \times 4n_b} \begin{bmatrix} \Sigma_{r \times r} & 0 \\ 0 & 0 \end{bmatrix}_{4n_b \times 3n} \mathbf{V}^T_{3n \times 3n} \tag{11.125}$$

where **R** and **V** are the orthogonal matrices, Σ is the singular value of **H**, and r is the rank of **H**, which represents the number of independent constraints.

The matrix **V** can be written as

$$\mathbf{V}^T_{3n \times 3n} = \{\mathbf{V}_{3n \times r}, \mathbf{V}_{3n \times (3n-r)}\}^T \tag{11.126}$$

By performing orthogonal transformation in Equation 11.113,

$$\mathbf{q} = \mathbf{V}_{3n \times (3n-r)} \tilde{\mathbf{q}} \tag{11.127}$$

The condensed eigenvalue equation of both free vibration and static buckling can be expressed as

$$(\tilde{\mathbf{K}} - \eta \tilde{\mathbf{G}}) \tilde{\mathbf{q}} = 0 \tag{11.128}$$

where

$$\tilde{\mathbf{K}} = \mathbf{V}^T_{(3n-r) \times 3n} \mathbf{K}_{3n \times 3n} \mathbf{V}_{3n \times (3n-r)} \tag{11.129}$$

and

$$\tilde{\mathbf{G}} = \mathbf{V}^T_{(3n-r)\times 3n}\mathbf{G}_{3n\times 3n}\mathbf{V}_{3n\times(3n-r)} \tag{11.130}$$

For plates of FSDT, the above procedure is valid, except $4n_b$ should be changed to $3n_b$, and $\mathbf{H}_{KI}$ is given by

$$\mathbf{H}_{KI} = \int_{\Gamma_u} \begin{bmatrix} N_K\phi_I & 0 & 0 \\ 0 & n_x N_K \phi_{xI} & n_y N_K \phi_{yI} \\ 0 & -n_y N_K \phi_{xI} & n_x N_K \phi_{yI} \end{bmatrix} \mathrm{d}\Gamma \tag{11.131}$$

Solving Equation 11.128 for eigenvalues yields the square of the circle frequencies for the free-vibration problem and the buckling loads for the static buckling problem. The eigenvectors, after being transformed back using Equation 11.127, give the vibration modes or buckling modes.

11.3.6 Equations for Static Deformation Analysis

For static deflection analysis of thick plates, the penalty method can be employed to enforce the essential boundary conditions by adding an additional essential boundary condition term in the Galerkin weak form of the static elastic equilibrium. The constrained Galerkin weak form for thick plates can be written as

$$\int_\Omega \delta(\mathbf{Lu})^T\mathbf{DLu}\,\mathrm{d}\Omega - \int_\Omega \delta(\mathbf{L}_u\mathbf{u})^T\mathbf{b}\,\mathrm{d}\Omega - \int_{S_t}\delta(\mathbf{L}_u\mathbf{u})^T\bar{\mathbf{t}}\,\mathrm{d}S + \delta\int_{\Gamma_u}\frac{1}{2}(\tilde{\mathbf{u}}-\bar{\mathbf{u}})^T\boldsymbol{\alpha}(\tilde{\mathbf{u}}-\bar{\mathbf{u}})\mathrm{d}\Gamma = 0 \tag{11.132}$$

where S_t is the edge surface of the plate where the natural boundary condition is specified, $\boldsymbol{\alpha}$ is a diagonal matrix of the penalty factors, and vector $\tilde{\mathbf{u}}$ is given by either Equation 11.121 or 11.122, depending on the plate theory used. The dimension of $\boldsymbol{\alpha}$ is 3×3 or 4×4 depending also on the plate theory used.

The discretized system equation of static deflections of the entire plate can be expressed as

$$[\mathbf{K} + \tilde{\mathbf{K}}]\mathbf{U} = \mathbf{F} \tag{11.133}$$

where global stiffness matrix $\mathbf{K}$ is the same as that in Equation 11.113. The additional stiffness matrix $\tilde{\mathbf{K}}$ is formed using

$$\tilde{\mathbf{K}}_{IJ} = \int_{\Gamma_u} \boldsymbol{\Phi}^T_{bI}\boldsymbol{\alpha}\boldsymbol{\Phi}_{bJ}\,\mathrm{d}\Gamma \tag{11.134}$$

and $\boldsymbol{\Phi}_{bI}$ is the matrix of shape functions for node I. For FSDT, they can be written as

$$\boldsymbol{\Phi}_{bI} = \begin{bmatrix} \phi_I & 0 & 0 \\ 0 & n_x\phi_{xI} & n_y\phi_{yI} \\ 0 & -n_y\phi_{xI} & n_x\phi_{yI} \end{bmatrix} \tag{11.135}$$

For TSDT,

$$\boldsymbol{\Phi}_{bI} = \begin{bmatrix} \phi_I & 0 & 0 \\ 0 & n_x\phi_{xI} & n_y\phi_{yI} \\ 0 & -n_y\phi_{xI} & n_x\phi_{yI} \\ \dfrac{\partial \phi_I}{\partial n} & 0 & 0 \end{bmatrix} \tag{11.136}$$

The force vector **F** in Equation 11.133 is the global force vector assembled using the nodal force vector of

$$\mathbf{f}_I = \int_{\Omega} (\mathbf{L}_u \mathbf{\Phi}_I)^T \mathbf{b}\, d\Omega + \int_{S_t} (\mathbf{L}_u \mathbf{\Phi}_I)^T \bar{\mathbf{t}}\, dS + \int_{\Gamma_u} \mathbf{\Phi}_{bI}^T \boldsymbol{\alpha} \bar{\mathbf{u}}\, d\Gamma \quad (11.137)$$

Further derivation on the first two terms in Equation 11.137 can be performed following the procedure given in Section 11.1.3.

Equation 11.133 is an algebraic equation, and can be solved using standard routines for the deflections and rotations for all the nodes in the plate.

11.3.7 Numerical Examples of Static Deflection Analyses

To examine the efficiency of the present formulation of the EFG method for static deflection analyses of shear deformable thick plates, Examples 11.12 to 11.16 are studied. The common geometric and material property parameters used in these five examples are as follows:

Length: $a = b = 10.0$ m
Young's modulus: $E = 1.0 \times 10^9$ N/m^2
Poisson's ratio: $\nu = 0.3$
Mass density: $\rho = 8000$ kg/m^3

The deflection coefficient $\xi = w_{max}D/Pb^2$ is defined for concentrated load P, and $\xi = w_{max}D/qb^4$ for uniform load q, where w_{max} is the maximum deflection at the center of the plates. Elastic rigidity of the plate is $D = Eh^3/[12(1 - \nu^2)]$.

In all the following examples, the size of the support domain is chosen to be 3.9 times the average nodal distance, except for special illustration. The nodes used in the following three examples are regularly distributed nodes. Uniform rectangular cells of background mesh are used for the integration, and the vertices of the background cells coincide with the field nodes.

In the following examples, the three shape functions used are the same, i.e.,

$$\phi_I(x, y) = \phi_{xI}(x, y) = \phi_{yI}(x, y) \quad (11.138)$$

Therefore, only ϕ_I needs to be constructed. The issue of shear locking is addressed using Example 11.16.

Example 11.12 Comparison of Deflection of Thin and Thick Square Plates with Different Types of Boundary Conditions

Using 21 × 21 nodes in the EFG method, two types of simply supported boundary conditions—soft-type and hard-type—are considered, and the maximum deflections of thin plate ($h/a = 0.01$) and thick plate ($h/a = 0.1$) are computed. The results for the thin plate are shown in Table 11.36. The deflections calculated based on both FSDT and TSDT are very close to Timoshenko's solution (Timoshenko and Woinowsky-Krieger, 1995) for concentrated load and uniform load. It is also seen that the deflections of thin plate calculated using the soft-type simply supported boundary condition are very close to those obtained using the hard-type simply supported boundary condition.

The maximum deflections of a thick plate are shown in Table 11.37. The deflections calculated using FSDT are very close to those obtained using TSDT for concentrated and

TABLE 11.36

Maximum Deflection Coefficients ξ of Simply Supported Thin Plate ($h/a = 0.01$)

Load	Boundary Condition	FSDT	TSDT	Timoshenko and Woinowsky-Krieger (1995)
Concentrated	$w = 0$ (soft)	0.01161	0.01162	0.01160
	$w = 0,\ \varphi_t = 0$ (hard)	0.01156	0.01157	
Uniform	$w = 0$ (soft)	0.004084	0.004085	0.004062
	$w = 0,\ \varphi_t = 0$ (hard)	0.004062	0.004063	

TABLE 11.37

Maximum Deflection Coefficients ξ of Simply Supported Thick Plate ($h/a = 0.1$)

Load	Boundary Condition	FSDT	TSDT
Concentrated	$w = 0$ (soft)	0.01398	0.01391
	$w = 0,\ \phi_s = 0$ (hard)	0.01323	0.01318
Uniform	$w = 0$ (soft)	0.004619	0.004615
	$w = 0,\ \phi_s = 0$ (hard)	0.004273	0.004275

TABLE 11.38

Maximum Deflection Coefficients ξ of Soft-Type Simply Supported Thin Plate ($h/a = 0.01$) Subjected to a Concentrated Load at the Center of the Plate

	Nodes						Timoshenko and Woinowsky-Krieger (1995)
	6 × 6	9 × 9	12 × 12	15 × 15	18 × 18	21 × 21	
FSDT	0.009835	0.01085	0.01139	0.01151	0.01158	0.01161	0.01160
TSDT	0.01001	0.01102	0.01144	0.01154	0.01159	0.01162	
CPT	0.01032	0.01141	0.01145	0.01155	0.01157	0.01157	

uniform loads. However, the deflections of thick plate calculated using the soft-type simply supported boundary condition are significantly larger than those obtained using the hard-type simply supported boundary condition, especially for uniform load.

Example 11.13 Convergence of Deflection of a Thin Square Plate

The convergence of the maximum deflections at the center of a square thin plate applied with concentrated load at the center of the plate is analyzed. The plate is simply supported with soft-type. The thickness of $h = 0.1$ m and the aspect ratio of $h/a = 0.01$ are considered. Various nodal densities in the plate domain are used to calculate the deflections based on FSDT, TSDT, and CPT in the present EFG method. The maximum deflections are shown in Table 11.38. The present EFG deflections based on both FSDT and TSDT rapidly converge to Timoshenko's solution. However, the rate of convergence of deflection based on CPT is higher than that based on FSDT and TSDT. All the deflections have a monotonous convergence.

Example 11.14 Convergence of Deflection of a Thick Square Plate

The convergence on maximum deflections of the center of a hard-type simply supported square thick plate subjected to a uniform load is analyzed. The thickness of $h = 1.0$ m and an aspect ratio of $h/b = 0.1$ are used. Based on FSDT and TSDT in the present EFG method,

TABLE 11.39

Maximum Deflection Coefficients ξ of Hard-Type Simply Supported Thick Plate ($h/a = 0.1$) Subjected to a Uniform Load

	Nodes						Senthilnathan (1989)
	4 × 4	6 × 6	8 × 8	10 × 10	12 × 12	14 × 14	
FSDT	0.003948	0.004306	0.004281	0.004272	0.004275	0.004273	0.004249
TSDT	0.003945	0.004311	0.004303	0.004278	0.004279	0.004276	

TABLE 11.40

Maximum Deflection Coefficients ξ of Thick Plate ($h/a = 0.1$)

Load	Boundaries	CPT	FSDT	TSDT	Senthilnathan (1989)
Concentrated	SSSS	0.01159	0.01323	0.01318	
	CCCC	0.005566	0.007322	0.007248	
	SCSC	0.006990	0.008845	0.008768	
	SCCS	0.007379	0.009190	0.009115	
Uniform	SSSS	0.004066	0.004273	0.004275	0.004249
	CCCC	0.001259	0.001505	0.001494	0.001496
	SCSC	0.001910	0.002209	0.002198	0.002191
	SCCS	0.002098	0.002375	0.002366	

the various nodal densities in the plate domain are used to calculate the deflections, as shown in Table 11.39. The present EFG deflections based on both FSDT and TSDT rapidly converge to the analytical solution (Senthilnathan, 1989).

Example 11.15 Maximum Deflections of Thick Plates under Several Kinds of Boundaries

Thick plates with thickness $h = 1.0$ m and aspect ratio $h/b = 0.1$ are considered. Based on CPT, FSDT, and TSDT in the present EFG method, 21 × 21 nodes are used to calculate the maximum deflections of the thick plates for four kinds of hard-type boundary conditions and two types of loads. The maximum deflections are listed in Table 11.40. The present EFG results of deflections calculated based on FSDT and TSDT are very close to each other and larger than those obtained based on CPT. The present deflections based on FSDT and TSDT agree well with the available analytical solutions (Senthilnathan, 1989). This demonstrates that the present EFG formulation based on shear deformation plate theory is accurate in calculating deflections of thick plates.

Example 11.16 Elimination of Shear Locking

Similar to the phenomenon discussed in Chapter 10 for beams, there is also a shear-locking issue in using thick plate theory to analyze thin plates. Techniques to avoid shear locking have been well developed in FEM. An excellent discussion of this issue can be found in the textbook by Reddy (1993). All these techniques, listed as follows, can be adopted here for our MFree methods:

1. Use sufficiently high-order elements.
2. Construct shape functions for rotations φ_x and φ_y from the first-order derivative of the shape function used for transverse displacement w to make rotation field shape functions exactly match the approximate deflection field.
3. Evaluate the shear energy in a proper manner.

The second method has also recently been proved effective for EFG methods (Kanok-Nukulchai et al., 2001). We study here the first two methods in detail.

Use of High-Order Basis

A soft-type simply supported Mindlin plate with uniform load of $p = 1.0$ N/m^2 is considered to analyze the shear-locking problem. To visualize the shear-locking phenomenon clearly, an extremely small aspect ratio of $h/a = 1.0 \times 10^{-4}$ is utilized. The other parameters are as follows:

Lengths of two sides: $a = b = 10.0$ m

Young's modulus: $E = 1.0 \times 10^9$ N/m^2

Poisson's ratio: $\nu = 0.3$

Regularly distributed 21×21 nodes are used. The dimensionless support domain is $\alpha_s = 3.9$ to 4.1. Three polynomial terms $(1, x, y)$ are used for both methods. The deflection coefficient is defined as $\xi = w_{\max} D/b_z b^4$, where $w_{\max}$ is the maximum deflection of the center of the plates. The analytical solution for the maximum deflection at the center of the plates is $\xi = 0.004062$ (Timoshenko's solution).

The same polynomial bases are used for the approximations of displacements w_0, φ_x, and φ_y, but we use different orders of polynomial basis functions in constructing MLS shape functions, which will be then used to compute the deflection of the beam based on FSDT. Different orders of polynomial bases to be used are listed in Table 11.41. The deflections of the plates computed using EFG based on FSDT are shown in Figure 11.18. The shear-locking phenomenon is observed very clearly when only three polynomial bases are used in MLS approximation. When the polynomial terms up to 18 are used, shear-locking is eliminated, and the deflections of the thin plate calculated based on FSDT are very close to and slightly larger than those obtained using EFG based on CPT. As the polynomial terms increase, more nodes need to be included in the influence domain of a selected point to approximate displacements. In this study, the dimension of the support domain is chosen to be 4.1 times the average nodal distance for 18 polynomial terms. Note that a *plate* of an aspect ratio of $h/a = 10 \times 10^{-4}$ is very extreme and will never exist in reality.

Comparison Study

Using the present EFG method for Mindlin plates (FSDT), we perform a comparison study to further analyze the shear-locking issue. The following two schemes are employed.

TABLE 11.41

Basis Functions for MLS Displacement Approximation

m	$\mathbf{p}^T$
3	1, x, y
6	1, x, y, x^2, xy, y^2
10	1, x, y, x^2, xy, y^2, x^3, x^2y, xy^2, y^3
15	1, x, y, x^2, xy, y^2, x^3, x^2y, xy^2, y^3, x^3y, x^2y^2, xy^3, x^3y^2, x^2y^3
18	1, x, y, x^2, xy, y^2, x^3, x^2y, xy^2, y^3, x^3y, x^2y^2, xy^3, x^3y^2, x^2y^3, x^3y^3, x^4, y^4

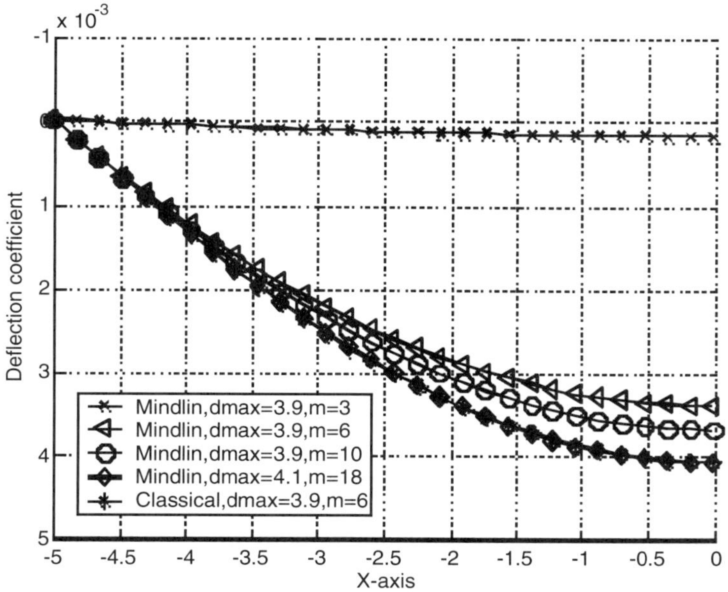

FIGURE 11.18
Shear locking in simply supported thin plate of a thickness-to-width ratio of $h/a = 1.0 \times 10^{-4}$. The plate is subjected to a uniform load. Results are computed using EFG based on FSDT.

SCHEME 1 Use of high-order basis functions but the same set of shape functions for three field variables, as we have done above.

SCHEME 2 The shape functions for rotations are the first-order derivatives of that for deflection.

$$w = \sum_{I=1}^{n} \phi_I w_I, \qquad \varphi_x = \sum_{I=1}^{n} \phi_{I,x} \varphi_{xI}, \qquad \varphi_y = \sum_{I=1}^{n} \phi_{I,y} \varphi_{yI} \tag{11.139}$$

This approach has been used by Kanok-Nukulchai et al. (2001) in the EFG method. A simply supported square plate under uniform load is considered for this analysis of the shear-locking problem, and all the parameters are exactly the same as the previous case.

The results are listed in Table 11.42 and plotted in Figure 11.19. We make the following points.

1. There is no shear locking, if $h/a > 0.01$, even if $m = 3$ is used. Usually, we use at least $m = 6$ for plate problems.
2. The higher the order of polynomial used, the less the shear locking. When $m = 18$, no shear locking is observed until $h/a = 1.0 \times 10^{-5}$, which is not actually a plate.
3. When Scheme 2 is used, no shear locking is observed.

Although Scheme 1 can practically solve the shear-locking problem, Scheme 2 is the sure way to eliminate this issue completely.

TABLE 11.42

Maximum Deflections $\xi = w_{max}D/b_z b^4$ of a Square Plate with Different Aspect Ratio

		h/a				
		1.0 × 10^{-1}	**1.0 × 10^{-2}**	**1.0 × 10^{-3}**	**1.0 × 10^{-4}**	**1.0 × 10^{-5}**
Scheme 1	$m = 3$	0.4618×10^{-2}	0.4068×10^{-2}	**0.3307×10^{-2}**	**0.1715×10^{-3}**	**0.1791×10^{-5}**
	$m = 6$	0.4618×10^{-2}	0.4084×10^{-2}	**0.3928×10^{-2}**	**0.3401×10^{-2}**	**0.5461×10^{-3}**
	$m = 10$	0.4620×10^{-2}	0.4088×10^{-2}	0.4065×10^{-2}	**0.3691×10^{-2}**	**0.2639×10^{-2}**
	$m = 15$	0.4621×10^{-2}	0.4090×10^{-2}	0.4069×10^{-2}	**0.3810×10^{-2}**	**0.3556×10^{-2}**
	$m = 18$	0.4621×10^{-2}	0.4088×10^{-2}	0.4064×10^{-2}	0.4089×10^{-2}	0.4009×10^{-2}
Scheme 2	$m = 3$	0.004612	0.004073	0.004059	0.004059	0.004025
	$m = 6$	0.004617	0.004085	0.004066	0.004066	0.004065
	$m = 10$	0.004620	0.004088	0.004066	0.004066	0.004066
	$m = 15$	0.004621	0.004090	0.004068	0.004067	0.004067
	$m = 18$	0.004621	0.004089	0.004068	0.004068	0.004068
Timoshenko				0.004062		

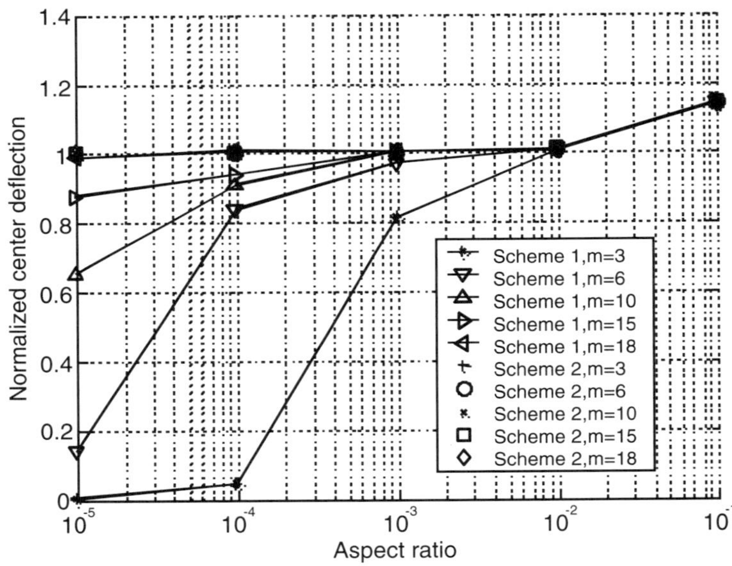

FIGURE 11.19
Normalized maximum deflections of a square plate with different aspect ratios.

11.3.8 Numerical Examples of Vibration Analyses

To examine the efficiency of the EFG method for eigenvalue analyses of shear deformable thick plates, some numerical examples on free vibration and static buckling of plates are presented. The common geometric and material property parameters are as follows:

Length: $a = b = 10.0$ m
Thickness: $h = 1.0$ m
Young's modulus: $E = 200 \times 10^9$ N/m^2
Poisson's ratio: $\nu = 0.3$
Mass density: $\rho = 8000$ kg/m^3

TABLE 11.43

Natural Dimensionless Frequencies $\varpi = (\omega^2 \rho h a^4/D)^{1/4}$ of Free Vibration of a Simply Supported Square Thick Plate Based on FSDT

Modes	Analytical Solution (Abbassian et al., 1987)	Present Method 5 × 5	7 × 7	9 × 9	11 × 11	FEM (Abbassian et al., 1987) HOE[a]	LOE[b]
1	4.37	4.43	4.37	4.37	4.37	4.37	4.40
2	6.74	7.33	6.75	6.75	6.75	6.77	6.94
3	6.74	7.53	6.77	6.75	6.75	6.77	6.94
4	8.35	9.79	8.38	8.36	8.36	8.41	8.59
5	9.22	10.91	9.31	9.24	9.23	9.40	9.84
6	9.22	11.14	9.33	9.24	9.23	9.40	9.84
7	10.32	15.07	10.41	10.35	10.33	10.59	10.85
8	10.32	15.99	10.48	10.35	10.33	10.59	10.85

[a] HOE denotes an eight-noded isoparametric thick shell element (16 elements, 65 nodes).

[b] LOE denotes a four-noded isoparametric shell element (64 elements, 81 nodes).

In all the tables of numerical examples, the dimensionless frequency coefficient is

$$\varpi = \left(\frac{\omega^2 \rho h a^4}{D}\right)^{1/4}$$

the factor of buckling load is $k = N_0 b^2/\pi^2 D$, where the flexural rigidity is $D = Eh^3/[12(1-\nu^2)]$.

Example 11.17 Frequency Analysis of Thick Plates (FSDT)

Based on FSDT, the natural dimensionless frequencies of a simply supported square thick plate are calculated using different densities of nodes regularly distributed in the plate. Quadrilateral background meshes are applied in the plate for the Gauss integration. In each background mesh, 4 × 4 Gauss points are used. All the nodes are in the vertices of the background meshes. The natural dimensionless frequencies are shown in Table 11.43 together with analytical solutions and FEM results. In the FEM results, HOE denotes an eight-noded isoparametric thick shell element (4 × 4 elements, 65 nodes); LOE denotes a four-noded isoparametric shell element (8 × 8 elements, 9 × 9 nodes). The frequencies obtained using the present EFG method rapidly converge and are in very good agreement with analytical solutions for all eight modes. Given the same density of nodes 9 × 9 in the plate, the present EFG results are better than the FEM results using LOE when compared with the analytical solution. The present EFG method results using 7 × 7 nodes are also closer to the analytical solutions than the FEM results using HOE with 65 nodes.

Example 11.18 Frequency Analysis of Thick Plates (FSDT and TSDT)

Based on FSDT and TSDT, the natural dimensionless frequencies of a square thick plate with different boundaries are calculated using 10 × 10 nodes regularly distributed in the plate. Quadrilateral background meshes are also applied for the Gauss integration and 4 × 4 Gauss points are used in each background mesh. All the nodes are in the vertices of the background meshes. The frequencies are shown in Table 11.44. In the notation of boundary conditions, FFFF denotes fully free at all edges. For all the five cases of boundaries, the frequencies of the square plate based on FSDT agree well with those based on TSDT

TABLE 11.44

Natural Dimensionless Frequencies $\varpi = (\omega^2 \rho h a^4/D)^{1/4}$ of Free Vibration of a Square Thick Plate with Different Boundaries Based on FSDT and TSDT

		Modes							
Boundary	**Theory**	**1**	**2**	**3**	**4**	**5**	**6**	**7**	**8**
FFFF	FSDT	0	0	0	3.57	4.35	4.83	5.65	5.65
	TSDT	0	0	0	3.57	4.35	4.83	5.66	5.66
SSSS	FSDT	4.37	6.75	6.75	8.36	9.23	9.23	10.34	10.34
	TSDT	4.37	6.75	6.75	8.36	9.23	9.23	10.34	10.34
CCCC	FSDT	5.71	7.88	7.88	9.33	10.13	10.18	11.14	11.14
	TSDT	5.72	7.92	7.92	9.39	10.19	10.23	11.21	11.22
SCSC	FSDT	5.17	7.01	7.70	8.88	9.33	10.08	10.60	10.92
	TSDT	5.17	7.02	7.73	8.92	9.33	10.13	10.63	10.96
SCCS	FSDT	5.03	7.31	7.33	8.85	9.70	9.71	10.74	10.76
	TSDT	5.04	7.33	7.35	8.87	9.72	9.74	10.77	10.79

TABLE 11.45

Convergence of Axial Buckling Factor $k = N_0 b^2/\pi^2 D$ along the x Axis for a Square Thick Plate Based on FSDT and TSDT (BC: SSSS)

	Nodes						Srinivas et al.	Senthilnathan
	4 × 4	**5 × 5**	**6 × 6**	**7 × 7**	**8 × 8**	**9 × 9**	**(1969)**	**(1989)**
FSDT	4.060	3.803	3.783	3.790	3.787	3.788	3.741	3.787
TSDT	4.095	3.797	3.780	3.785	3.785	3.785		

for all modes. The frequencies for FFFF boundaries are lowest. The frequencies for the plate with SSSS boundaries are lower than those with CCCC boundaries. Both frequencies for the plate with SCSC and SCCS boundaries are between those with SSSS and CCCC boundaries.

11.3.9 Numerical Examples of Buckling Analyses

Example 11.19 Buckling Analysis of Thick Plates (FSDT and TSDT)

Based on FSDT and TSDT, the factors $k = N_0 b^2/\pi^2 D$ of axial buckling loads along the x axis of a simply supported square plate are calculated using different densities of nodes regularly distributed in the plate. Quadrilateral background meshes and 4 × 4 Gauss points in each background mesh are chosen in the plate for the Gauss integration as for the frequency analyses. All the nodes are located at the vertices of the background meshes. The results of buckling factors are shown in Table 11.45. The factors based on both FSDT and TSDT have good convergences, agree very well with each other and with analytical results by Senthilnathan (1989), and are a little larger than the analytical results given by Srinivas et al. (1969).

Example 11.20 Buckling Loads of a Square Plate Based on FSDT and TSDT with Different Loads and Boundaries

Based on FSDT and TSDT, the factors of buckling loads of a square plate with different loads and boundaries are calculated using 9 × 9 nodes regularly distributed in the plate. The choices of background meshes, Gauss points, and nodal position are the same as in Example 11.17. The factors of axial buckling loads along the x axis and shear buckling loads and biaxial buckling loads along both the x and y axes for different boundaries are listed in Tables 11.46 through 11.48. The factors for SSSS boundaries are smaller than those

TABLE 11.46

Axial Buckling Factor $k = N_0 b^2/\pi^2 D$ along the x Axis for a Square Thick Plate Based on FSDT and TSDT with Different Boundaries

	Boundaries			
Theory	**SSSS**	**CCCC**	**SCSC**	**SCCS**
FSDT	3.788	8.327	6.398	5.546
TSDT	3.785	8.471	6.451	5.569

TABLE 11.47

Shear Buckling Factor $k = N_0 b^2/\pi^2 D$ for a Square Thick Plate Based on FSDT and TSDT with Different Boundaries

	Boundaries			
Theory	**SSSS**	**CCCC**	**SCSC**	**SCCS**
FSDT	7.859	10.757	9.693	9.226
TSDT	7.853	11.669	10.542	9.349

TABLE 11.48

Biaxial Buckling Factor $k = N_0 b^2/\pi^2 D$ along the x and y Axes for a Square Thick Plate Based on FSDT and TSDT with Different Boundaries ($N_x = N_y = N$)

	Boundaries			
Theory	**SSSS**	**CCCC**	**SCSC**	**SCCS**
FSDT	1.894	4.563	3.384	2.941
TSDT	1.892	4.657	3.396	2.944

for CCCC boundaries. The factors for both SCSC and SCCS boundaries are between those for simply supported and clamped boundaries. Except for CCCC and SCSC boundaries, the factors of shear buckling loads based on TSDT are slightly larger than those based on FSDT; the factors based on TSDT are very close to those based on FSDT in Tables 11.46 through 11.48.

Example 11.21 Buckling Loads of a Square Plate with a Circular Hole Based on FSDT and TSDT

The factors of buckling loads are computed for a square plate with a circular hole, as shown in Figure 11.20. The plate is chosen to demonstrate the applicability of the present EFG method to plates of complicated shape. Triangular background meshes (196) for Gauss integration are applied in the plate as shown in Figure 11.21. Three Gauss points are used for each background mesh. The 144 nodes are located at the vertices of the background meshes. The influence domain d_{max} is chosen to be 3.5 times the average nodal distance. The factors of axial buckling loads along the x axis and shear buckling loads and biaxial buckling loads along both the x and y axes for different boundaries are listed in Tables 11.49 through 11.51. Except for CCCC boundaries, the factor of shear buckling load

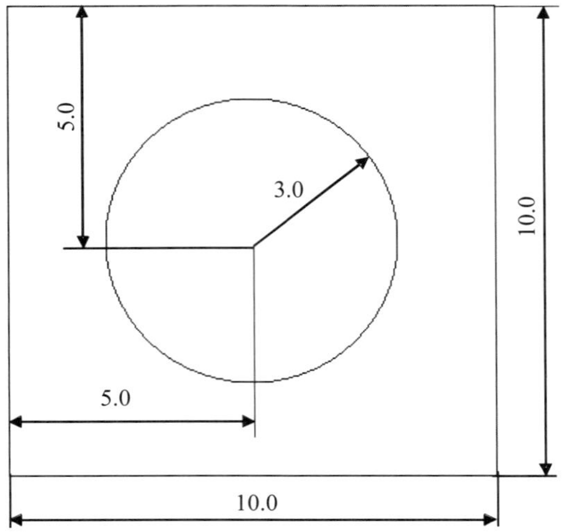

FIGURE 11.20
Square plate with a circular hole.

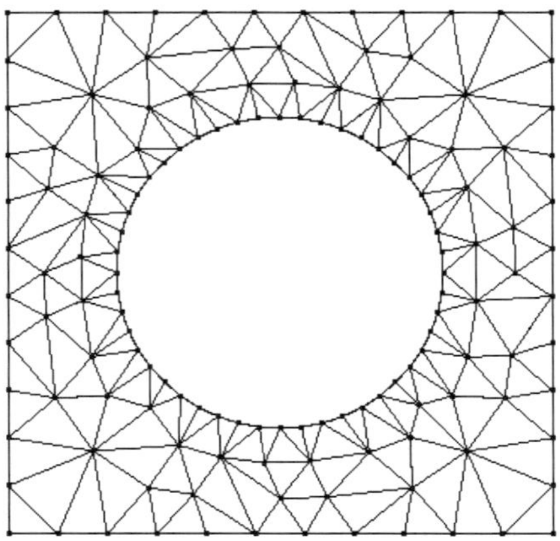

FIGURE 11.21
Distributions of 144 nodes and 196 background meshes in a plate with a circular hole.

TABLE 11.49

Axial Buckling Factor $k = N_0 b^2/\pi^2 D$ along the x Axis for a Square Thick Plate with a Circular Hole Based on FSDT and TSDT with Different Boundaries

	Boundaries			
Theory	**SSSS**	**CCCC**	**SCSC**	**SCCS**
FSDT	1.986	7.995	4.096	3.221
TSDT	1.969	8.097	4.130	3.226

based on TSDT is slightly larger than that based on FSDT; the factors based on TSDT are very close to those based on FSDT in Tables 11.49 through 11.51. Factors of shear buckling loads are larger than those of biaxial buckling loads along both the x and y axes. Factors of axial buckling loads along the x axis are between them.

TABLE 11.50

Shear Buckling Factor $k = N_0 b^2/\pi^2 D$ for a Square Thick Plate with a Circular Hole Based on FSDT and TSDT with Different Boundaries

	Boundaries			
Theory	**SSSS**	**CCCC**	**SCSC**	**SCCS**
FSDT	7.867	12.669	10.596	8.968
TSDT	7.873	13.357	10.861	9.043

TABLE 11.51

Biaxial Buckling Factor $k = N_0 b^2/\pi^2 D$ along the x and y Axes for a Square Thick Plate with a Circular Hole Based on FSDT and TSDT with Different Boundaries ($N_x = N_y = N$)

	Boundaries			
Theory	**SSSS**	**CCCC**	**SCSC**	**SCCS**
FSDT	1.032	4.781	2.559	1.763
TSDT	1.021	4.860	2.556	1.774

11.4 RPIM for Thick Plates

11.4.1 Formulation

In the previous two sections, we presented EFG formulations for both thin and thick plates using MLS approximations. We have shown that special treatment is needed to deal with the essential boundary conditions, because of the lack of Kronecker delta function property in the MLS shape functions. The natural progress of our research is then to use point interpolation method (PIM) shape functions, as we have already seen the advantages of PIM shape functions in Chapters 8 to 10 for solids, fluids, and beams. This section therefore presents radial PIM (RPIM) methods that use Galerkin formulation and RPIM shape functions for thick (Mindlin) plates. The RPIM is then tested, validated, and used for the static analysis of several benchmark problems.

The formulation of RPIM was presented in detail in Chapter 5, and the formulation of plates has also been detailed in the previous three sections of this chapter. In formulating RPIM for plates, all we need to do is replace MLS shape functions with RPIM shape functions, and ignore all the terms related to essential boundary conditions. Therefore, we need not repeat the process, and proceed straight to examining the performance of the RPIM code that we was recently developed by G. R. Liu and Chen for thick plates. Note that the RPIM used in the following examples are nonconforming.

11.4.2 Numerical Examples

Terms including the conditions, parameters, and notation conventions for examples presented in this section are exactly the same as those stated in Section 11.3.7 for all the common terms.

An intensive study on the effects of shape parameters of both EXP and MQ radial functions has been conducted. Those parameters are very important and greatly affect the performance of RPIM, as we have seen in Chapters 5 through 10 for curve/surface fitting and mechanics problems of 2D solids and beams, as well as fluid flows. We have decided to revisit the problem for the following reasons:

- We have included quadratic polynomials in the basis, and we need to know the effects.
- The findings for 2D solids and fluid flows may be applicable for plate structures.
- We need to confirm the effect on boundary and loading conditions.

Therefore, the shape parameter c of the EXP radial basis is investigated again for plates, through the calculation of maximum deflections of Mindlin plates with two extreme boundaries: fully clamped and simply supported at all edges. The plate is subjected to two extreme loading situations: concentrated and uniform loads.

Example 11.22 Deflection of a Thick Square Plate (Effects of the EXP Shape Parameters)

This example is used to examine the effects of the shape parameters of EXP radial basis functions used in the RPIM code. In this section, we use the radial functions defined in Table 5.3. The EXP radial function has the form:

$$R_i(r) = \exp\left[-\alpha_c\left(\frac{r_i}{d_c}\right)^2\right] \tag{11.140}$$

where d_c is the characteristic length that relates the nodal spacing and α_c is the dimensionless shape parameter defined by

$$\alpha_c = c \times d_c \tag{11.141}$$

where c is the shape parameter with a dimension of reciprocal length for the original form of EXP radial function defined in Table 5.2. The characteristic length r_c is defined in this section by

$$d_c = \Delta r = \sqrt{\Delta x^2 + \Delta y^2} \tag{11.142}$$

where Δx and Δy are the average nodal space along the x and y directions, respectively.

Six polynomial terms (complete quadratic, $m = 6$) and 21×21 regularly distributed nodes are used. Because the plate is 10×10 m, the nodal spacings in x and y directions are $\Delta x = 0.5$ m and $\Delta y = 0.5$ m, respectively. The characteristic length d_c is then chosen to be $d_c = \sqrt{\Delta x^2 + \Delta y^2} = 0.707$ m. The effects of the dimensionless shape parameter α_c on maximum deflections of Mindlin plates with different boundary and loading conditions are computed using RPIM and summarized in Table 11.52. The relative errors of maximum deflections calculated in reference to the EFG results computed using EFG code for thick plates introduced in Section 11.3 are drawn in Figure 11.22. From Table 11.52 and Figure 11.22, the following conclusions can be drawn:

- The parameter α_c has a very wide range, due to the inclusion of the polynomial basis. We have also examined the effects of α_c without the use of the polynomial basis, and have found that the results are very sensitive to the shape parameters, as shown in cases in previous chapters.

TABLE 11.52

Effects of Parameter α_c of EXP Radial Basis on Maximum Deflections ξ of Mindlin Plates

Parameter	Concentrated Load				Uniform Load			
	SSSS		CCCC		SSSS		CCCC	
α_c ($c \times r_c$)	ξ	Error (%)	ξ	Error (%)	ξ	Error (%)	ξ	Error (%)
0.1	0.01399	0.07	0.007345	0.31	0.004616	−0.06	0.001509	0.27
0.3	0.01403	0.36	0.007389	0.92	0.004609	−0.22	0.001504	−0.07
0.5	0.01398	0.00	0.007361	0.53	0.004596	−0.50	0.001500	−0.33
0.75	0.01392	−0.43	0.007330	0.11	0.004577	−0.91	0.001495	−0.66
1.0	0.01388	−0.72	0.007315	−0.10	0.004562	−1.23	0.001492	−0.86
3.0	0.01403	0.36	0.007446	1.69	0.004585	−0.74	0.001506	0.07
5.0	0.01424	1.86	0.007607	3.89	0.004635	0.35	0.001527	1.46
10.0	0.01403	0.36	0.007500	2.43	0.004601	−0.39	0.001546	2.72
15.0	0.01431	2.36	0.007778	6.23	0.004653	0.74	0.001601	6.38
20.0	0.01504	7.58	0.008386	14.53	0.004810	4.14	0.001693	12.49
50.0	0.01977	41.42	0.01251	70.85	0.005705	23.51	0.002223	47.71
EFG[a]	0.01398		0.007322		0.004619		0.001505	

[a] EFG: see Section 11.3.

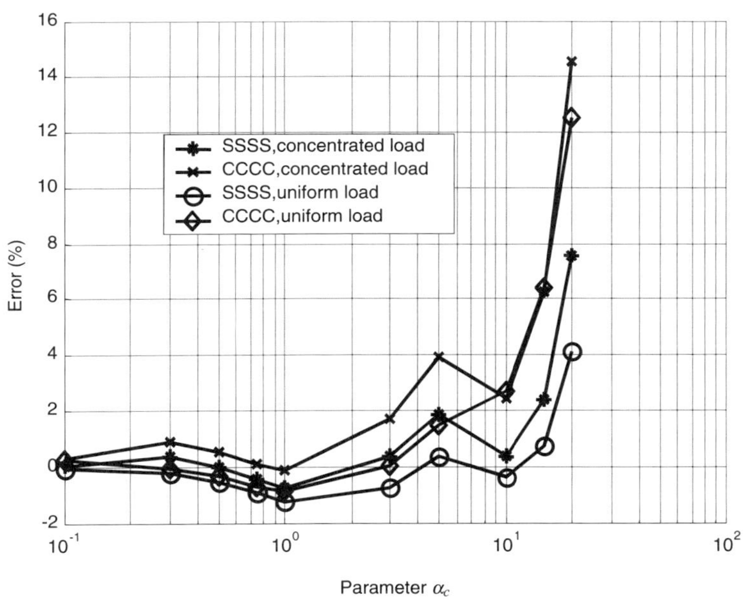

FIGURE 11.22
Effects of the dimensionless shape parameter α_c of EXP radial basis on the maximum deflections of Mindlin plates.

- When $0.1 \leq \alpha_c \leq 1.0$, good results can be obtained for all kinds of boundary and loading conditions. We observed that the results for plates with simply supported boundary conditions are more accurate than for those with fully clamped boundary conditions, and that the results for uniform loading are more accurate than for concentrated loading. However, most of the errors are all within 1%, when $0.1 \leq \alpha_c \leq 1.0$.
- Figure 11.23 shows the condition numbers of moment matrices $\mathbf{R}_Q$ and $\mathbf{G}$, which decrease as the parameter increases. The trends of the condition numbers of both

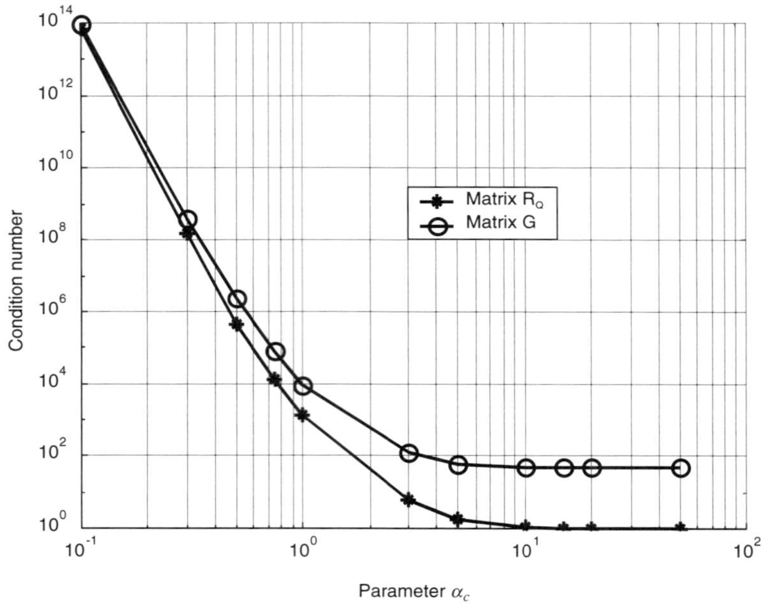

FIGURE 11.23
Condition numbers of matrices **G** and $\mathbf{R}_Q$ for different parameter α_c of EXP radial basis.

matrices $\mathbf{R}_Q$ and **G** are similar. We observe again the phenomenon that good parameters that give good results for RPIM are those that produce large condition numbers for the moment matrices.

- We give α_c a lower bound of 0.1 because beyond which the condition numbers of matrices $\mathbf{R}_Q$ and **G** are too large, and it has a higher chance to run into a numerical problem. We, however, use a much smaller value in Chapters 5 and 9 in pursuing good accuracy. The upper bound is given simply based on the accuracy of the results.

Example 11.23 Deflection of a Thick Square Plate (Effects of the MQ Shape Parameters)

This example is used to examine the effects of the shape parameters of MQ radial basis functions used in the RPIM code. The MQ radial function has the form of (see Table 5.3)

$$R_i(r) = [r_i^2 + (\alpha_C d_c)^2]^q \tag{11.143}$$

where d_c is the characteristic length that relates the nodal spacing defined by Equation 11.142 and α_C is the dimensionless shape parameter defined by

$$\alpha_C = \frac{C}{d_c} \tag{11.144}$$

where C is the shape parameter with a dimension of length for the original form of MQ radial function defined in Table 5.2.

Similar to the previous study, six polynomial terms (complete quadratic, $m = 6$) and 21×21 regularly distributed nodes are used, which gives $d_c = 0.707$.

TABLE 11.53

Effects of Parameter q of MQ Radial Basis on Maximum Deflections ξ of Mindlin Thick Plates ($\alpha_C = 2.0$)

Parameter	Concentrated Load				Uniform Load			
	SSSS		CCCC		SSSS		CCCC	
q	ξ	Error (%)	ξ	Error (%)	ξ	Error (%)	ξ	Error (%)
−15.5	0.01417	1.36	0.007561	3.26	0.004613	−0.13	0.001517	0.80
−10.5	0.01401	0.21	0.007432	1.50	0.004582	−0.80	0.001504	−0.07
−5.5	0.01392	−0.43	0.007350	0.38	0.004569	−1.08	0.001497	−0.53
−3.5	0.01392	−0.43	0.007340	0.25	0.004576	−0.93	0.001497	−0.53
−1.5	0.01395	−0.21	0.007343	0.29	0.004592	−0.58	0.001500	−0.33
−0.5	0.01397	−0.07	0.007349	0.37	0.004601	−0.39	0.001502	−0.20
−0.05	0.01398	0.00	0.007352	0.41	0.004605	−0.30	0.001503	−0.13
0.05	0.01398	0.00	0.007352	0.41	0.004606	−0.28	0.001503	−0.13
0.5	0.01399	0.07	0.007355	0.45	0.004609	−0.22	0.001504	−0.07
1.03	0.01400	0.14	0.007359	0.51	0.004611	−0.17	0.001505	0.00
1.5	0.01400	0.14	0.007363	0.56	0.004613	−0.13	0.001506	0.07
3.5	0.01403	0.36	0.007386	0.87	0.004615	−0.09	0.001507	0.13
5.5	0.01410	0.86	0.007429	1.46	0.004651	0.69	0.001526	1.40
10.5	0.01412	1.00	0.007558	3.22	0.004594	−0.54	0.001526	1.40
15.5	0.01447	3.51	0.007944	8.49	0.004637	0.39	0.001594	5.91
EFG[a]	0.01398		0.007322		0.004619		0.001505	

[a] EFG: see Section 11.3.

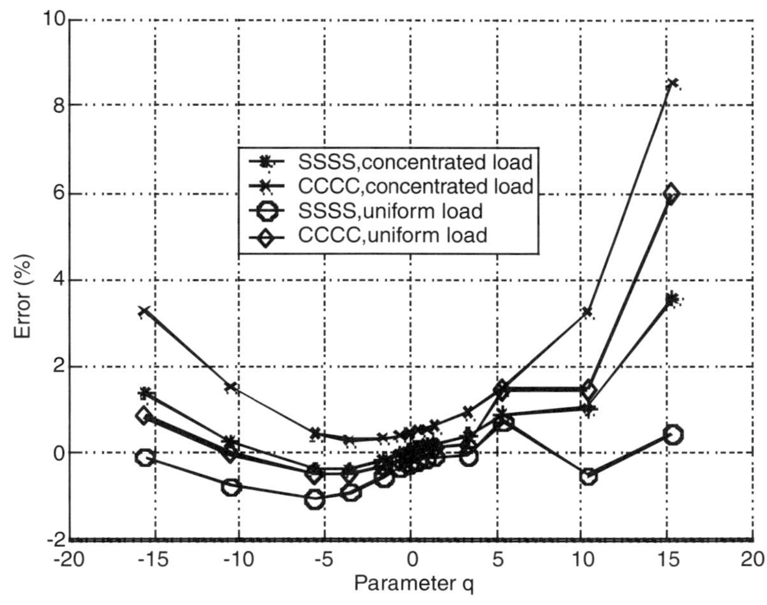

FIGURE 11.24
Effects of parameter q of MQ radial basis on maximum deflections of Mindlin plates ($\alpha_C = 2.0$).

First, we fix $\alpha_C = 2.0$, and vary the parameter q. The effects of parameter q on maximum deflections of Mindlin plates are computed and listed in Table 11.53. The relative errors of maximum deflections calculated in reference to the EFG results (Section 11.3) are summarized in Figure 11.24. From Table 11.53 and Figure 11.24, the following conclusions can be drawn:

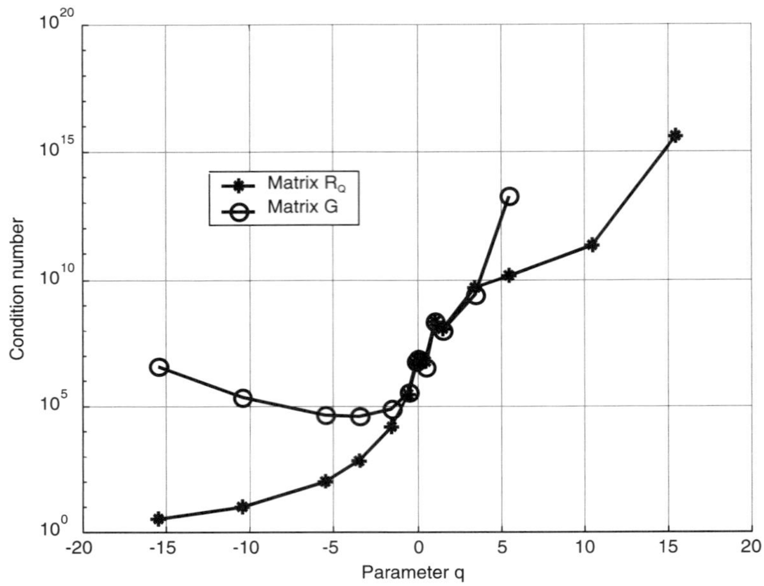

FIGURE 11.25
Condition numbers of matrices **G** and $\mathbf{R}_Q$ for different parameter q of MQ radial basis ($\alpha_C = 2.0$).

- The parameter q can be chosen from a wide range for reasonable results. A q in the range of $-3.5 \leq q \leq 3.5$ (exclusive of integers and zero) can lead to good results; $q = 1.03$, which we often used in previous chapters, is one of the best.
- When q is set to zero, the property of interpolation function will be determined only by polynomial terms. Because the inverse of matrix $\mathbf{R}_Q$ does not exist, when $q = 0$, the formulation has to follow that of polynomial PIM.
- The inverse of matrix $\mathbf{R}_Q$ does not exist if q is chosen to be a positive integer. Figure 11.25 shows the condition numbers of matrices $\mathbf{R}_Q$ and **G** computed using $\alpha_C = 2.0$ and various q. When $q \geq 5.5$, both condition numbers are too large and the condition number of matrix **G** rises much faster than that of matrix $\mathbf{R}_Q$, in this case. When $q \leq -3.5$, the condition number of matrix **G** will also increase as q decreases. The condition number of matrix $\mathbf{R}_Q$ increases as q increases in the entire range of our investigation.
- The limits imposed on the q are based on the accuracy of the RPIM results of deflection of plates. The condition numbers are under control for a q in the range of $-3.5 \leq q \leq 3.5$.

Because good results can be obtained using $q = 1.03$, not only for this study but also for studies discussed in previous chapters, we will now fix $q = 1.03$ and investigate the effects of parameter α_C on the maximum deflections of a Mindlin plate. The results of maximum deflections of Mindlin plates are computed using different α_C and are listed in Table 11.54. The relative errors of maximum deflections are computed in reference to the results obtained using EFG (Section 11.3) and plotted in Figure 11.26. Table 11.54 and Figure 11.26, show that $1.0 \leq \alpha_C \leq 3.0$ can lead to good results. The following examples use $\alpha_C = 2.0$, which corresponds to $C \approx 1.42$, which we used very often in previous chapters. It is observed from Figure 11.27 that the condition numbers of matrices $\mathbf{R}_Q$ and **G** are almost the same and increase as parameter α_C increases, and both become too large when $\alpha_C \geq 3.8$.

TABLE 11.54

Effects of Parameter α_C of MQ Radial Basis on Maximum Deflections ξ of Mindlin Plates (q = 1.03)

Parameter	Concentrated Load				Uniform Load			
	SSSS		CCCC		SSSS		CCCC	
r_0	ξ	Error (%)	ξ	Error (%)	ξ	Error (%)	ξ	Error (%)
0.02	0.01413	1.07	0.007511	2.58	0.004601	−0.39	0.001504	−0.07
0.2	0.01407	0.64	0.007449	1.73	0.004600	−0.41	0.001504	−0.07
1.0	0.01399	0.07	0.007360	0.52	0.004605	−0.30	0.001503	−0.13
2.0	0.01400	0.14	0.007359	0.51	0.004611	−0.17	0.001505	0.00
3.0	0.01402	0.29	0.007377	0.75	0.004614	−0.11	0.001506	0.07
3.4	0.01403	0.36	0.007386	0.87	0.004615	−0.09	0.001507	0.13
3.8	0.01408	0.72	0.007413	1.24	0.004624	0.11	0.001510	0.33
EFG[a]	0.01398		0.007322		0.004619		0.001505	

[a] EFG: see Section 11.3.

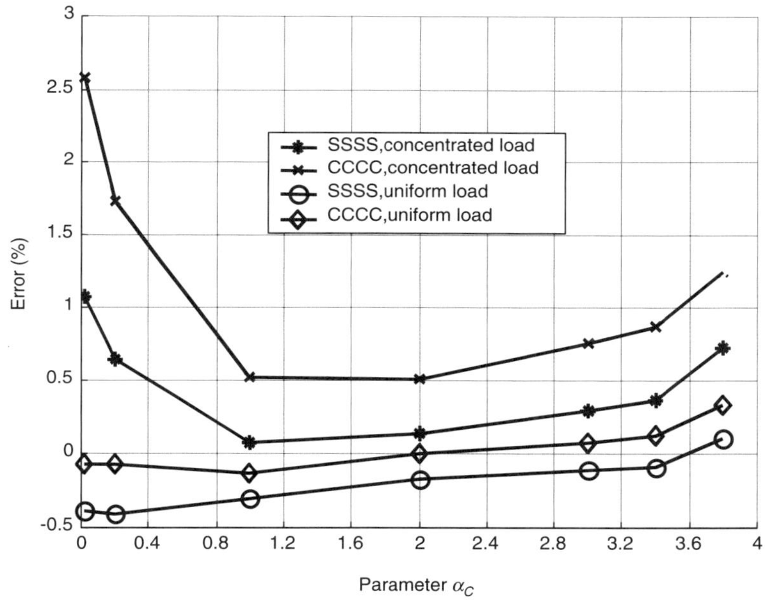

FIGURE 11.26
Effects of parameter α_C of MQ radial basis on maximum deflections of Mindlin plates (q = 1.03).

It may be noted that the inclusion of polynomial terms helps not only to improve the accuracy of the results but also to reduce the reliance on the choice of shape parameters. Our study on problems for plates has reached the same conclusion as that in Chapter 8 for 2D solids. That is, we need not worry too much about choosing the shape parameters, if polynomial terms are included in the basis.

There could, however, be problems related to the singularity of the moment matrix if we include too many polynomial terms. More detailed study is conducted to address this concern.

Example 11.24 Deflection of a Thick Square Plate (Effects of Polynomial Terms)

Further study is conducted to investigate the effects of inclusion of polynomial basis. Table 11.55 shows the effects of the inclusion of polynomial terms in MQ RPIM on efficiency in computing the maximum deflections of the square Mindlin plate. Regularly distributed

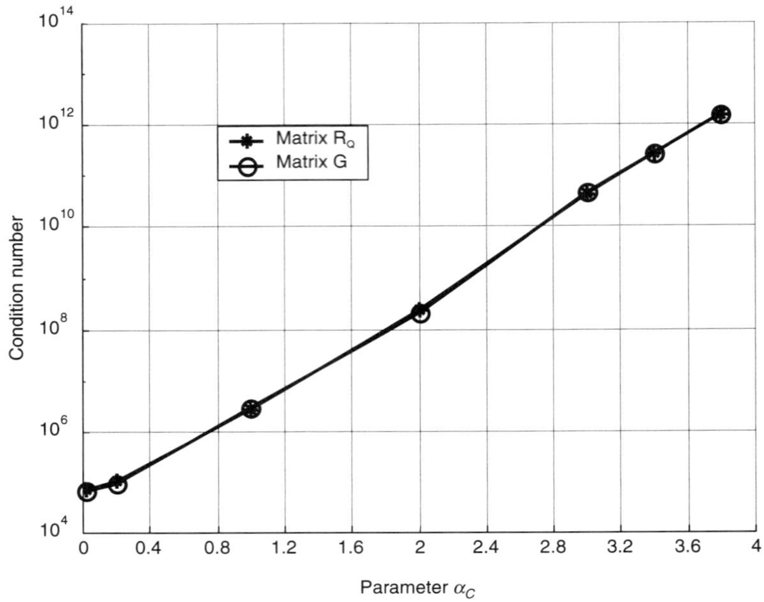

FIGURE 11.27
Condition numbers of matrices **G** and $\mathbf{R}_Q$ for different parameter α_C of MQ radial basis ($q = 1.03$).

TABLE 11.55

Effects of the Inclusion of Polynomial Terms in MQ RPIM on Computational Efficiency in Computing the Maximum Deflections of Mindlin Plates ($\alpha_C = 2.0$, $q = 1.03$)

Polynomial Terms	Concentrated Load				Uniform Load			
	SSSS		CCCC		SSSS		CCCC	
	ξ	CPU (s)	ξ	CPU (s)	ξ	CPU (s)	ξ	CPU (s)
0	0.01399	502	0.007355	501	0.004607	1148	0.001504	1148
3	0.01399	528	0.007356	526	0.004608	1218	0.001504	1218
6	0.01400	557	0.007359	558	0.004611	1294	0.001505	1294
EFG[a]	0.01398		0.007322		0.004619		0.001505	

[a] EFG: see Section 11.3.

TABLE 11.56

Effects of Polynomial Terms in MQ RPIM on Condition Number of Matrix $\mathbf{G}$($\alpha_C = 2.0$, $q = 1.03$)

	Polynomial Terms m					
	0	3	6	10	15	17
Condition number	0.2277e+9	0.3727e+9	0.2060e+9	0.3719e+9	0.3899e+9	0.4070e+9

nodes (21 × 21) are again used in the investigation, and shape parameters used are $\alpha_C = 2.0$ and $q = 1.03$. It is observed from Table 11.55 that the inclusion of polynomial terms increases the computational time, but the effect is very small and will not be a concern.

Table 11.56 shows the effects of different numbers of polynomial terms in MQ RPIM on the condition number of matrix **G**. Polynomial terms included are 3, 6, 10, 15, and 17. Table 11.56 shows that the number of polynomial terms also has little effect on the condition number of matrix **G**.

Example 11.25 Deflection of a Thick Square Plate (Convergence of Maximum Deflections)

The maximum deflections of Mindlin plates with simply supported and clamped boundaries under concentrated and uniform loads are calculated using different densities of regularly distributed nodes in the MQ RPIM. The parameters $\alpha_C = 2.0$ and $q = 1.03$ are used and six polynomial terms are included. The maximum deflections are listed in Table 11.57 for different nodal densities. The relative errors of maximum deflections compare with EFG results (Section 11.3) and are shown in Figure 11.28. Rapid convergences of all the maximum deflections can be seen in Table 11.57 and Figure 11.28, and the maximum deflections of plates under uniform load converge much faster compared with those of concentrated load. We observe, again, a nonsmooth convergence process for cases of concentrated load.

TABLE 11.57

Convergence of Maximum Deflections of Mindlin Plates Obtained Using MQ RPIM ($\alpha_C = 2.0$, $q = 1.03$)

	Concentrated Load				Uniform Load			
	SSSS		CCCC		SSSS		CCCC	
Nodes	ξ	Error (%)	ξ	Error (%)	ξ	Error (%)	ξ	Error (%)
11 × 11	0.01354	−3.15	0.006999	−4.41	0.004576	−0.93	0.001507	0.13
13 × 13	0.01369	−2.07	0.007106	−2.95	0.004590	−0.63	0.001507	0.13
16 × 16	0.01368	−2.15	0.007063	−3.54	0.004603	−0.35	0.001505	0.00
19 × 19	0.01394	−0.29	0.007307	−0.20	0.004609	−0.22	0.001505	0.00
21 × 21	0.01400	0.14	0.007359	0.51	0.004611	−0.17	0.001505	0.00
EFG[a]	0.01398		0.007322		0.004619		0.001505	

[a] EFG: see Section 11.3.

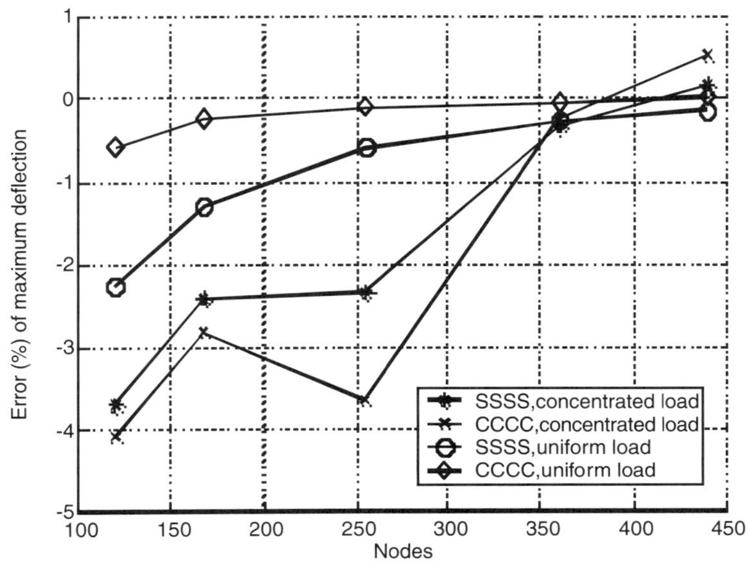

FIGURE 11.28
Convergence of maximum deflections of Mindlin plates for different densities of nodes ($\alpha_C = 2.0$, $q = 1.03$).

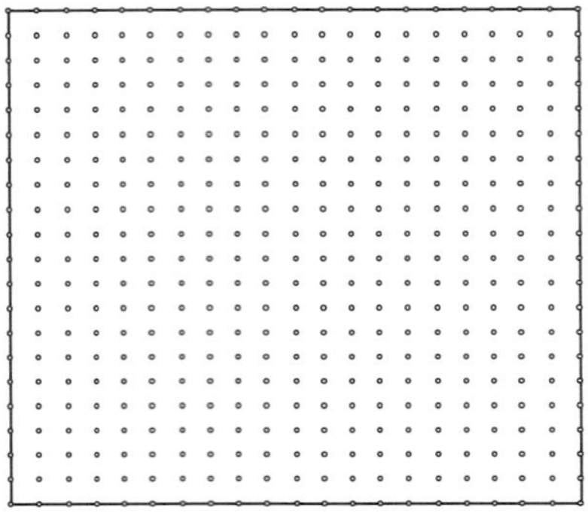

FIGURE 11.29
441 regularly distributed nodes in a square plate.

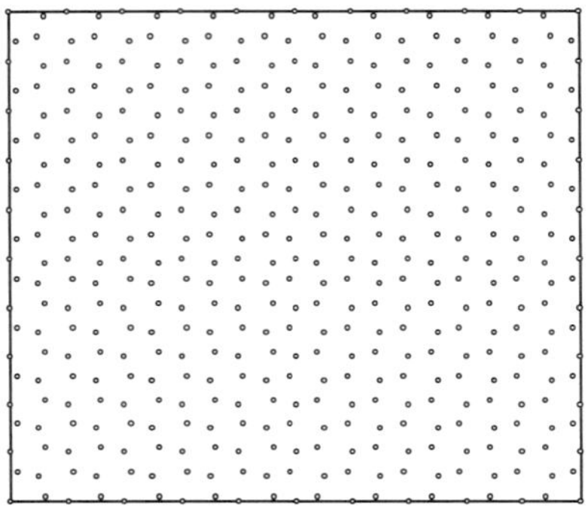

FIGURE 11.30
441 irregularly distributed nodes in a square plate.

Example 11.26 Deflection of a Thick Square Plate (Effects of Irregularly Distributed Nodes)

Effects of irregularly distributed nodes on the results of maximum deflections of square Mindlin plates with simply supported and clamped boundaries under uniform load are analyzed using RPIM with MQ radial basis functions. The parameters used are $\alpha_C = 2.0$ and $q = 1.03$. Regularly and irregularly distributed 21×21 nodes, as shown in Figures 11.29 and 11.30, respectively, are used. The same mesh of 20×20 regular quadrilateral cells is used for the background integration for both regularly and irregularly distributed nodes. It is observed from Figure 11.31 that the deflections of Mindlin plates are little affected by irregularity of nodes. This confirms again that the RPIM is robust when using irregular nodes.

Example 11.27 Deflection of a Thick Square Plate (Effects of Shear Locking)

The issue of shear locking in plates has been discussed in Section 11.3. Higher-order polynomial basis functions are suggested to eliminate shear locking. This example examines the same method for RPIM using MQ basis functions.

In this example, a simply supported Mindlin thin plate with an extremely small thickness-to-width ratio ($h/a = 1.0 \times 10^{-4}$) subjected to a uniform load of $p = 1.0$ N/m^2 is considered. The same shape functions are used for the approximations of displacements w and rotations Q_x and Q_y. The shape parameters used in the MQ basis function are $\alpha_C = 2.0$ and

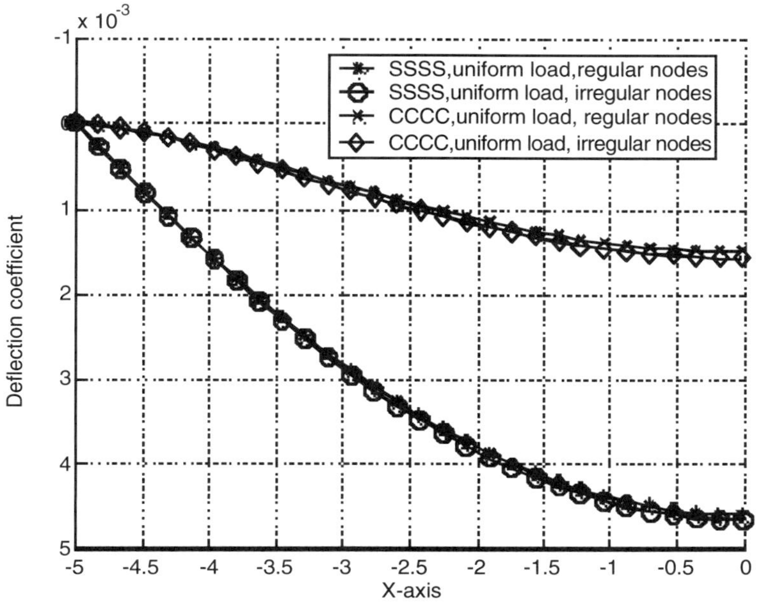

FIGURE 11.31
Deflections of Mindlin thick plates for regularly and irregularly distributed nodes ($\alpha_C = 2.0$, $q = 1.03$).

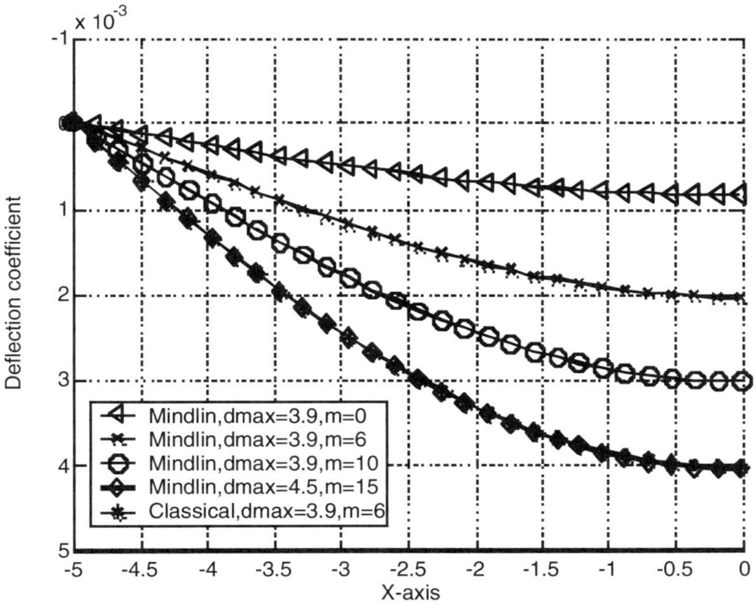

FIGURE 11.32
Shear locking in a simply supported thin square plate ($h/a = 1.0 \times 10^{-4}$) subjected to uniform load ($\alpha_C = 2.0$, $q = 1.03$).

$q = 1.03$. The plate is represented using 21×21 regularly distributed nodes. The deflections of the Mindlin plates are calculated using the present RPIM-MQ using different numbers of polynomial terms. The results are shown in Figure 11.32. It should be noted that the use of a greater number of polynomial terms requires the use of a larger support domain, which gives more nodes for constructing RPIM shape functions. In this study, the size of the support domain d_{max} is chosen to be 3.9 to 4.1 times the average nodal distance. The deflections of the thin plate based on classical plate theory are also computed using RPIM-MQ and plotted in Figure 11.32. Figure 11.32 shows that shear locking appears very clearly in the results obtained using pure radial basis functions without adding any polynomial terms. The results obtained based on Mindlin plate theory are much smaller than those based on classical plate theory meant for this thin plate. With the increase of the polynomial terms, shear locking is gradually relieved, and the results approach those based on classical plate theory. Shear locking disappears when 15 polynomial terms (complete quartic) are included in the MQ basis, and the maximum deflection of the thin plate based on Mindlin plate theory is slightly larger than that based on classical plate theory, which is exactly what we expect.

11.5 MLPG for Thin Plates

In the previous sections, we presented EFG formulations for static, free-vibration, as well as buckling analysis for both thin and thick plates, including composite laminated plates. As we have seen, EFG requires a background mesh for the integration required to compute the system matrices. In Chapter 7, we introduced the MLPG method, which has the advantage that it requires no global domain mesh for integration. The MLPG is an almost truly MFree method.

In this section, an MLPG formulation is proposed for static and free-vibration analyses of thin plates governed by the Kirchhoff plate theory. The formulation follows that presented by Gu and G. R. Liu (2001f). The local weak form is developed using the weighted residual method locally from the fourth-order partial differential equation of Kirchhoff plates. MLS approximation is used to construct shape functions with the desired high-order consistency. Several examples of static and free-vibration analyses of thin plates under various loads and boundary conditions are presented to demonstrate the convergence, validity, and efficiency of the present method. Some important parameters on the performance of the present method are also investigated, and reported here in detail.

11.5.1 Governing Equations

In MLPG, it is convenient to use the weighted residual method. In using the weighted residual method, the strong form based on the Kirchhoff plate is required. The partial differential equation governing the deflection w for static and free-vibration analyses of thin plates is (see Equation 3.86)

$$\frac{\partial^2}{\partial x^2}\left(D_{11}\frac{\partial^2 w}{\partial x^2}+D_{12}\frac{\partial^2 w}{\partial y^2}\right)+\frac{\partial^2}{\partial y^2}\left(D_{12}\frac{\partial^2 w}{\partial x^2}+D_{22}\frac{\partial^2 w}{\partial y^2}\right)+2\frac{\partial^2}{\partial x\partial y}\left(2D_{66}\frac{\partial^2 w}{\partial x\partial y}\right)$$
$$+I_0\frac{\partial^2 w}{\partial t^2}-I_2\frac{\partial^2}{\partial t^2}\left(\frac{\partial^2 w}{\partial x^2}+\frac{\partial^2 w}{\partial y^2}\right)=b_z(x,y,t),\quad \text{in } \Omega \tag{11.145}$$

where D_{ij} are the plate rigidities given by Equation 3.74 for plates of isotropic materials, and I_0 and I_2 are the mass per unit area and mass moments of inertia given by Equations 3.88 and 3.90.

11.5.2 Local Weak Form of MLPG

In MLPG, the problem domain is represented by a set of nodes scattered in the problem domain. Local weighted residual formulation is established for a node over a local quadrature domain of the node. The definition of the domains used in MLPG was explained in detail in Chapter 7 and shown in Figure 7.2. The penalty method is used here to enforce the essential boundary conditions. The constrained Petrov–Galerkin statement of Equation 11.145 can be given in an integral form over a local quadrature domain Ω_Q bounded by Γ_Q:

$$\int_{\Omega_Q} \widehat{W}\left\{\frac{\partial^2}{\partial x^2}\left(D_{11}\frac{\partial^2 w}{\partial x^2}+D_{12}\frac{\partial^2 w}{\partial y^2}\right)+\frac{\partial^2}{\partial y^2}\left(D_{12}\frac{\partial^2 w}{\partial x^2}+D_{22}\frac{\partial^2 w}{\partial y^2}\right)+2\frac{\partial^2}{\partial x\partial y}\left(2D_{66}\frac{\partial^2 w}{\partial x\partial y}\right)-b_z(x,y)\right.$$
$$\left.+I_0\frac{\partial^2 w}{\partial t^2}-I_2\frac{\partial^2}{\partial t^2}\left(\frac{\partial^2 w}{\partial x^2}+\frac{\partial^2 w}{\partial y^2}\right)\right\}\mathrm{d}\Omega-\int_{\Gamma_{Qw}}\alpha\widehat{W}(w-\bar{w})\mathrm{d}\Gamma-\int_{\Gamma_{Q\theta}}\beta\frac{\partial\widehat{W}}{\partial n}(\theta_n-\bar{\theta})\,\mathrm{d}\Gamma=0 \tag{11.146}$$

where $\widehat{W}$ is the weight function and α and β are the penalty constants for deflection and slope, respectively.

Equation 11.146 can be integrated by parts to become

$$\int_{\Omega_Q}\left\{\frac{\partial^2\widehat{W}}{\partial x^2}\left(D_{11}\frac{\partial^2 w}{\partial x^2}+D_{12}\frac{\partial^2 w}{\partial y^2}\right)+\frac{\partial^2\widehat{W}}{\partial y^2}\left(D_{12}\frac{\partial^2 w}{\partial x^2}+D_{22}\frac{\partial^2 w}{\partial y^2}\right)+4D_{66}\frac{\partial^2\widehat{W}}{\partial x\partial y}\frac{\partial^2 w}{\partial x\partial y}-\widehat{W}b_z\right\}\mathrm{d}\Omega$$
$$+\int_{\Gamma_Q}A\,\mathrm{d}\Gamma-\int_{\Gamma_{Qw}}\alpha\widehat{W}(w-\bar{w})\,\mathrm{d}\Gamma-\int_{\Gamma_{Q\theta}}\beta\frac{\partial\widehat{W}}{\partial n}(\theta_n-\bar{\theta})\,\mathrm{d}\Gamma=0 \tag{11.147}$$

where

$$A=-\widehat{W}\left[\left(\frac{\partial M_x}{\partial x}+\frac{\partial M_{xy}}{\partial y}\right)n_x+\left(\frac{\partial M_{xy}}{\partial x}+\frac{\partial M_y}{\partial x^2}\right)n_y+I_2\frac{\partial^2}{\partial t^2}\left(\frac{\partial w}{\partial x}n_x+\frac{\partial w}{\partial y}n_y\right)\right]$$
$$+\frac{\partial\widehat{W}}{\partial x}(M_x n_x+M_{xy}n_y)+\frac{\partial\widehat{W}}{\partial y}(M_{xy}n_x+M_y n_y) \tag{11.148}$$

where M_x, M_y, and M_{xy} are bending moments and (n_x, n_y) is the unit vector outward normal to domain Ω_Q. The moments M_x, M_y, and M_{xy} can be expressed by the deflection, w, as shown in Equations 3.77 through 3.79.

As shown in Figure 7.2, the local quadrature domain Ω_Q of a node is a domain in which $\widehat{W}(\mathbf{x}) \neq 0$, since we choose to use the same domain for the quadrature and weighted domains. As an arbitrary shape support domain can be used, we choose to use simple rectangular and circular support domains for our thin plate problems. It can be found that the boundary Γ_Q for the local quadrature domain Ω_Q is usually composed by five parts: the internal boundary Γ_{Qi} and the boundaries Γ_{Qw}, $\Gamma_{Q\theta}$, Γ_{QM}, Γ_{QV} over which the essential boundary conditions and natural boundary conditions are specified. The boundaries Γ_{Qw} with Γ_{QV} and $\Gamma_{Q\theta}$ with Γ_{QM} are mutually disjoint. Because the boundary conditions are often given locally in normal and tangential coordinates (n, s), the derivatives in the integrations on the Γ_{QM} and Γ_{QV} are converted to the coordinates (n, s) from the global coordinates (x, y). Imposing the natural boundary condition given in Equations 11.4 and 11.6, we obtain

$$\int_{\Omega_Q}\left\{\frac{\partial^2\widehat{W}}{\partial x^2}\left(D_{11}\frac{\partial^2 w}{\partial x^2}+D_{12}\frac{\partial^2 w}{\partial y^2}\right)+\frac{\partial^2\widehat{W}}{\partial y^2}\left(D_{12}\frac{\partial^2 w}{\partial x^2}+D_{22}\frac{\partial^2 w}{\partial y^2}\right)+4D_{66}\frac{\partial^2\widehat{W}}{\partial x\partial y}\frac{\partial^2 w}{\partial x\partial y}-\widehat{W}b_z\right\}\mathrm{d}\Omega$$

$$+\int_{\Omega_Q}\left\{I_0\widehat{W}\frac{\partial^2 w}{\partial t^2}+I_2\frac{\partial^2}{\partial t^2}\left(\frac{\partial\widehat{W}}{\partial x}\frac{\partial w}{\partial x}+\frac{\partial\widehat{W}}{\partial y}\frac{\partial w}{\partial y}\right)\right\}\mathrm{d}\Omega$$

$$+\int_{\Gamma_{Qi}}A\,\mathrm{d}\Gamma+\int_{\Gamma_{Qw}}[A-\alpha\widehat{W}(w-\bar{w})]\,\mathrm{d}\Gamma+\int_{\Gamma_{Q\theta}}\left[A-\beta\frac{\partial\widehat{W}}{\partial n}(\theta_n-\bar{\theta}_n)\right]\mathrm{d}\Gamma$$

$$-\int_{\Gamma_{QM}}\left[\widehat{W}\left(\hat{Q}_n+\frac{\partial M_{ns}}{\partial s}\right)-\frac{\partial\widehat{W}}{\partial n}\bar{M}_n\right]\mathrm{d}\Gamma-\int_{\Gamma_{QV}}\left[\widehat{W}\bar{V}_n-\frac{\partial\widehat{W}}{\partial n}M_n\right]\mathrm{d}\Gamma=0 \tag{11.149}$$

where

$$M_n = M_x n_x^2 + M_y n_y^2 + M_{xy} n_x n_y \tag{11.150}$$

$$M_{ns} = (M_y - M_x)n_x n_y + M_{xy}\left(n_x^2 - n_y^2\right) \tag{11.151}$$

$$\hat{Q}_n = \left(\frac{\partial M_x}{\partial x}+\frac{\partial M_{xy}}{\partial y}\right)n_x+\left(\frac{\partial M_{xy}}{\partial x}+\frac{\partial M_y}{\partial y}\right)n_y+I_2\frac{\partial^2}{\partial t^2}\left(\frac{\partial w}{\partial x}n_x+\frac{\partial w}{\partial y}n_y\right) \tag{11.152}$$

$$V_n = \hat{Q}_n + \frac{\partial M_{ns}}{\partial s} \tag{11.153}$$

$$\frac{\partial}{\partial n} = n_x\frac{\partial}{\partial x}+n_y\frac{\partial}{\partial y} \tag{11.154}$$

$$\frac{\partial}{\partial s} = n_x\frac{\partial}{\partial y}-n_y\frac{\partial}{\partial x} \tag{11.155}$$

For quadrature domains located entirely within the global domain, there is no intersection between Γ_Q and the global boundary Γ, then $\Gamma_{Qi} = \Gamma_Q$, and the integrals over Γ_{Qw}, $\Gamma_{Q\theta}$, Γ_{QM}, and Γ_{QV} vanish.

11.5.3 Discretized System Equations

For a node at $\mathbf{x}_i$, the quadrature domain is first determined. The deflection at any point $\mathbf{x}_Q$ in the quadrature domain (see Figure 7.1) can be obtained using Equation 11.1, where the shape function is constructed using nodes that are included in the support domain of point $\mathbf{x}_Q$. The approximated deflection is then entered into Equation 11.149 to form an algebraic equation for the node at $\mathbf{x}_i$ that gives the relationship of all the nodes near the support domain. The following global system equation is then assembled using all the equations for all the nodes in the entire problem domain.

$$\mathbf{M}\ddot{\mathbf{W}} + \mathbf{K}\mathbf{W} = \mathbf{F} \tag{11.156}$$

where the global "mass" matrix **M** is assembled using entries of

$$M_{ij} = \int_{\Omega_Q} [I_0 \widehat{W}_i \phi_j + I_2(\widehat{W}_{i,x}\phi_{j,x} + \widehat{W}_{i,y}\phi_{j,y})]\, d\Omega - \int_{\Gamma_Q - \Gamma_{QV}} I_2 \widehat{W}_i(\phi_{j,x}n_x + \phi_{j,y}n_y)\, d\Gamma \quad (11.157)$$

The global stiffness matrix **K** is assembled using entries of nodal stiffness given by

$$K_{ij} = k_{ij}^{\Omega_Q} + k_{ij}^{\Gamma_{Qi}} + k_{ij}^{\Gamma_{Qw}} + k_{ij}^{\Gamma_{Q\theta}} + k_{ij}^{\Gamma_{QM}} + k_{ij}^{\Gamma_{QV}} \quad (11.158)$$

where

$$k_{ij}^{\Omega_Q} = \int_{\Omega_Q} \left\{ \widehat{W}_{i,xx}(D_{11}\phi_{j,xx} + D_{12}\phi_{j,yy}) + \widehat{W}_{i,yy}(D_{12}\phi_{j,xx} + D_{22}\phi_{j,yy}) + 4D_{66}\widehat{W}_{i,xy}\phi_{j,xy} \right\} d\Omega \quad (11.159)$$

$$k_{ij}^{\Gamma_{Qi}} = \int_{\Gamma_{Qi}} B_{ij}\, d\Gamma \quad (11.160)$$

where

$$\begin{aligned} B_{ij} = {} & \widehat{W}_i[(D_{11}\phi_{j,xxx} + (D_{12} + 2D_{66})\phi_{j,xyy})n_x + (D_{22}\phi_{j,yyy} + (D_{12} + 2D_{66})\phi_{j,xxy})n_y] \\ & - \widehat{W}_{i,x}[(D_{11}\phi_{j,xx} + D_{12}\phi_{j,yy})n_x + 2D_{66}\phi_{j,xy}n_y] \\ & - \widehat{W}_{i,y}[2D_{66}\phi_{j,xy}n_x + (D_{12}\phi_{j,xx} + D_{22}\phi_{j,yy})n_y] \end{aligned} \quad (11.161)$$

In Equation 11.158,

$$k_{ij}^{\Gamma_{Qw}} = \int_{\Gamma_{Qw}} (B_{ij} - \alpha \widehat{W}_i \phi_j)\, d\Gamma \quad (11.162)$$

$$k_{ij}^{\Gamma_{Q\theta}} = \int_{\Gamma_{Q\theta}} [B_{ij} - \beta \widehat{W}_{i,n}(\phi_{j,x}n_x + \phi_{j,y}n_y)]\, d\Gamma \quad (11.163)$$

$$k_{ij}^{\Gamma_{QM}} = \int_{\Gamma_{QM}} C_{ij}\, d\Gamma \quad (11.164)$$

where

$$\begin{aligned} C_{ij} = {} & \widehat{W}_i\{[(D_{11}\phi_{j,xxx} + (D_{12} + 2D_{66})\phi_{j,xyy})n_x + (D_{22}\phi_{j,yyy} + (D_{12} + 2D_{66})\phi_{j,xxy})n_y] \\ & + n_x[(D_{12}\phi_{j,xxy} + D_{22}\phi_{j,yyy} - D_{11}\phi_{j,xxy} - D_{12}\phi_{j,yyy})n_x n_y + 2D_{66}\phi_{j,xyy}(n_x^2 - n_y^2)] \\ & + n_y[(D_{12}\phi_{j,xxx} + D_{22}\phi_{j,xyy} - D_{11}\phi_{j,xxx} - D_{12}\phi_{j,xyy})n_x n_y + 2D_{66}\phi_{j,xxy}(n_x^2 - n_y^2)]\} \end{aligned} \quad (11.165)$$

$$\widehat{W}_{i,n} = \widehat{W}_{i,x}n_x + \widehat{W}_{i,y}n_y \quad (11.166)$$

In Equation 11.158,

$$k_{ij}^{\Gamma_{QV}} = -\int_{\Gamma_{QV}} [\widehat{W}_{i,n}((D_{11}\phi_{j,xx} + D_{12}\phi_{j,yy})n_x^2 + (D_{12}\phi_{j,xx} + D_{22}\phi_{j,yy})n_y^2 + 4D_{66}\phi_{j,xy}n_x n_y)]\,\mathrm{d}\Gamma \tag{11.167}$$

The load vector **f** in Equation 11.156 is defined by

$$f_i = \int_{\Omega_Q} \widehat{W}_i b_z\,\mathrm{d}\Omega - \int_{\Gamma_{Qw}} \alpha\widehat{W}_i\bar{w}\,\mathrm{d}\Gamma - \int_{\Gamma_{Q\theta}} \beta\widehat{W}_{i,n}\bar{\theta}_n\,\mathrm{d}\Gamma - \int_{\Gamma_{QM}} \widehat{W}_{i,n}\overline{M}_n\,\mathrm{d}\Gamma + \int_{\Gamma_{QV}} \widehat{W}_i\overline{V}_n\,\mathrm{d}\Gamma \tag{11.168}$$

The nodal deflection in Equation 11.156 is defined by

$$\mathbf{W} = \{w_1, w_2, w_3, \ldots, w_n\}^T \tag{11.169}$$

It can be easily seen that the system stiffness matrix **K** in the present method is banded but asymmetric. For static analysis, the first term of Equation 11.156, related time variable, t, vanishes. For free vibration, the loading, **F**, vanishes and a linear eigenvalue equation must be solved.

11.5.4 Weight Function

As MLPG is regarded as a weighted residual method, the weight function plays an important role in the performance of the method. Theoretically, as long as the condition of continuity is satisfied, any weight function is acceptable. For plate problems, the order of the governing equation is high; hence, shape functions with higher order of continuity are preferred. In Equation 11.147 it is seen that second partial derivatives of weight functions in the domain of Ω_Q are required, and a first derivative is required on the boundary. In addition, we also require that the weight function and its first derivative vanish on the boundary of Ω_Q for all internal quadrature domains, so that many integrals along the boundary vanish, which leads to a savings in numerical integration. Therefore, for plates, all the weight functions listed in Equations 5.11 through 5.13 and Equation 5.15 can be used, as all satisfy these requirements, as shown in Figures 5.2 through 5.4. The quadratic spline function (Equation 5.13) and quadratic smooth function (Equation 5.15) are preferred for their higher-order continuity.

Note also that our weak form is based on a local subdomain of a node with the node at the center. It can be shown that weight functions that decrease in magnitude with increasing distance from the node yield better results. In the examples presented in the following section we use the quartic spline function defined by Equation 5.12.

11.5.5 Numerical Integration

Numerical integration is needed to evaluate the integration in computing the system matrices. Gauss quadrature is used in the present work. For the quadrature domain of a node at $\mathbf{x}_i$, a local integration cell is needed to employ Gauss quadrature. For each Gauss quadrature point $\mathbf{x}_Q$, MLS approximation is performed to compute the integrand. Therefore, as shown in Figure 7.1, for a node at $\mathbf{x}_i$, there exist three local domains: local quadrature domain Ω_Q, weight function domain Ω_w (same as Ω_Q) for $\widehat{W} \neq 0$, and support domain Ω_s for a quadrature point $\mathbf{x}_Q$ in which the nodes are used for the approximation of the field variable. Because the problem domains in the following examples are rectangles, rectangular quadrature domains are used to establish the weight functions. The size of the quadrature (weighted) domain for node *i* is defined as

$$r_Q = \alpha_Q d_c \tag{11.170}$$

where α_Q is the dimensionless parameter controlling the dimension, which is to be determined by the analyst; d_c is the distance between the node *i* and its nearest neighbor node. The size of the support domain for a quadrature point $\mathbf{x}_Q$ is defined as

$$r_s = \alpha_s d_c \tag{11.171}$$

where α_s is the dimensionless parameter controlling the dimension, which is to be determined by the analyst. The effects of α_Q and α_S are investigated in the following numerical examples.

There exist difficulties in obtaining exact numerical integration in meshless methods. These have been discussed in detail for 2D solids by Atluri et al. (1999b) and G. R. Liu and Yan (1999). The numerical integration errors result from the complexities of the integrand. Because the integrand of the plate analysis is more complex than 2D plane problem, additional attention should be paid to the numerical integration. To guarantee the accuracy of the numerical integration, the Ω_Q should be divided into smaller subpartitions. In each subpartition, more Gauss quadrature points should be used. In previous chapters, we recommended a minimum of four subdivisions.

11.5.6 Numerical Examples

Example 11.28 Static Analysis of Thin Square Plates

The MLPG method is used for static bending analysis of thin plates. The geometry of the plates is shown in Figure 11.1. Except as specified, in the following examples for static analyses, the units can be any set of consistent units, and the parameters are taken as thickness $h = 1.0 \times 10^{-2}$, Young's modulus $E = 1.0 \times 10^9$, and Poisson ratio $\nu = 0.3$. For the boundary conditions in the rectangular plates, SSSS denotes fully simple supported, CCCC denotes fully clamped, and SCSC denotes two opposite edges simply supported and the other edges ($y = 0$, b) clamped.

Parameters on the performance of the present method are investigated first. The square plate under various loads is a well-known benchmark with a large number of numerical and analytical solutions available for comparison. The analytical results given by Timoshenko and Woinowsky-Krieger (1995) are used for comparison with the present results. In the following parameter investigations, the fully simply supported and clamped square plates are used. A square plate with $a = b = 4$ is considered. Figure 11.33 shows two kinds of nodal arrangements with regularly distributed 81 nodes and irregularly distributed 81 nodes, respectively. Regularly distributed 81 and 289 nodes are utilized in the following parameter investigations. The plate is subject to a uniformly distributed load of $b_z = 100$. The analytical solution is available and can be found in the textbook by Timoshenko and Woinowsky-Krieger (1995). The maximum deflection, $w_{\max}$, is located in the center of the plate:

$$w_{\max} = C_0 \frac{a^4 b_z}{D_{11}} \tag{11.172}$$

The analytical results of C_0 for this case are 0.004062 for fully simply supported plates and 0.00126 for fully clamped plates.

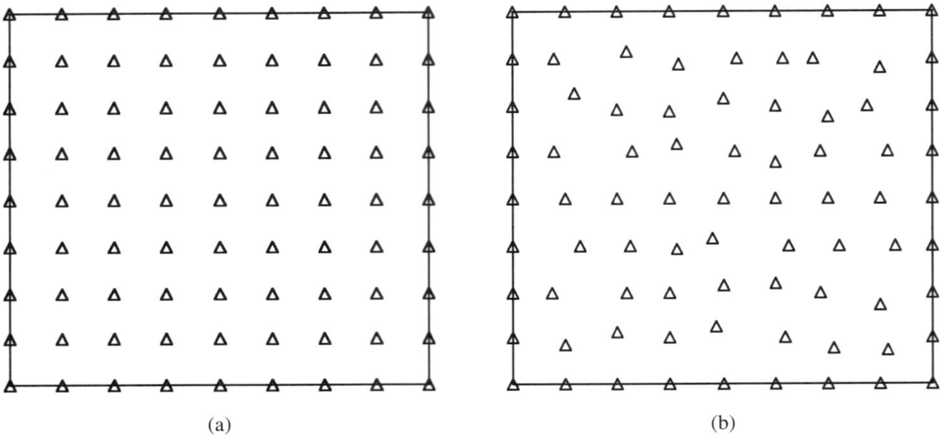

FIGURE 11.33
Nodal arrangement for a square thin plate: (a) regular distribution, (b) irregular distribution.

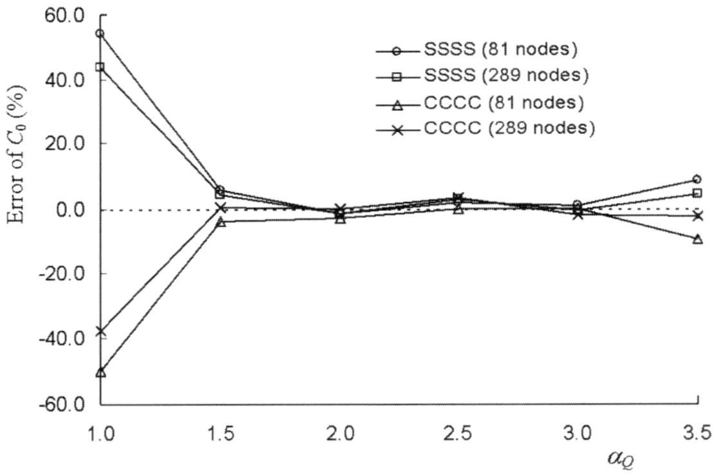

FIGURE 11.34
Influence of the dimension of the quadrature domain on the MLPG results of deflection of a thin square plate subjected to a uniformly distributed load.

Effects of Local Support Domain

As MLPG is a local meshless method, the size of the local quadrature domain used will affect the accuracy of the solution. Quadrature domains with different sizes of r_Q, which are determined by the parameter α_Q in Equation 11.170, are therefore investigated. The errors of the deflections (C_0) for two nodal arrangements are plotted in Figure 11.34. This figure shows that the accuracy for deflections increases with the increase of the dimension of the quadrature domain.

When the quadrature domain is too small ($\alpha_Q = 1.0$), the results become unacceptably erroneous. This is because a local residual formulation with a very small quadrature domain for the weight function behaves more like a strong form formulation. A strong form formulation is usually less accurate than a weak integral form formulation (Lu et al., 1994), which integration smears the error over the integral domain.

When the quadrature domain is large enough ($\alpha_Q \geq 1.5$), results obtained are very good. However, there exist difficulties in obtaining accurate numerical integrations for a large quadrature domain. When the quadrature domain becomes too large ($\alpha_Q \geq 3.0$), the errors will increase a little due to an increase in integration errors. For a large quadrature domain, more regular small partitions and Gauss quadrature points are needed to obtain accurate integrations. The numerical integration will become computationally expensive and is not really necessary. Hence, $\alpha_Q = 2.0$ is an economic choice for our thin plate problem. Note that for 2D solids we use 1.5 to 2.0.

It may also be mentioned here that too large a local quadrature domain does not necessarily improve accuracy significantly. This fact, clearly evidenced in Figure 11.34, implies that as long as the integral domain is large enough to "smear" the error, the size of integral domain does not play an important role.

Effects of Support Domain

The size of the support domain of a quadrature point is determined by the parameter α_s in Equation 11.171. Because the problem domain is a rectangle, rectangular influence domains are used. The errors of the deflections (C_0) of $\alpha_s = 2.0$ to 4.5 are obtained and plotted in Figure 11.35. It can be found that the accuracy for deflections increases with the increase of the support domain size. Results of $\alpha_s = 3.0$ to 4.5 (which includes 40 to 70 nodes for MLS approximation) are very good. Too small an influence domain ($\alpha_s \leq 2.5$) leads to large errors. The inaccuracy of too small an influence domain is because there are not enough nodes to perform the MLS approximation for the field variable that is the deflection for thin plates.

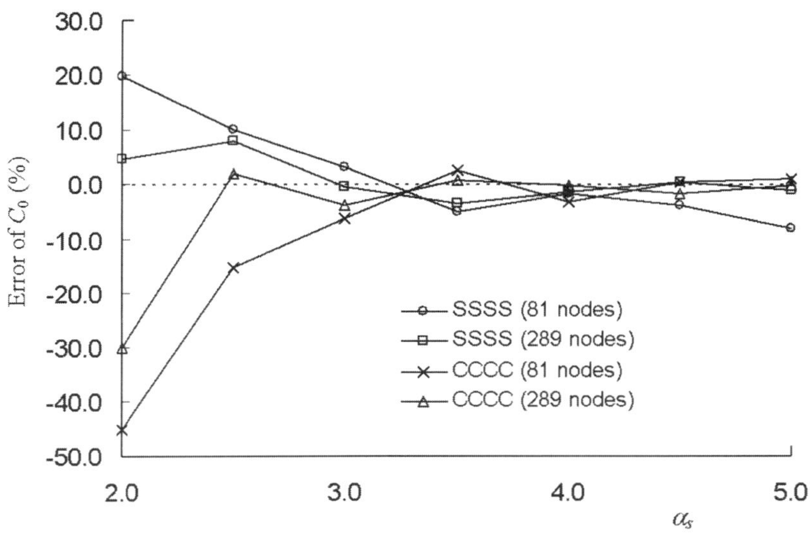

FIGURE 11.35
Influence of the dimension of the support domain on the MLPG results of deflection of a thin square plate subjected to a uniformly distributed load.

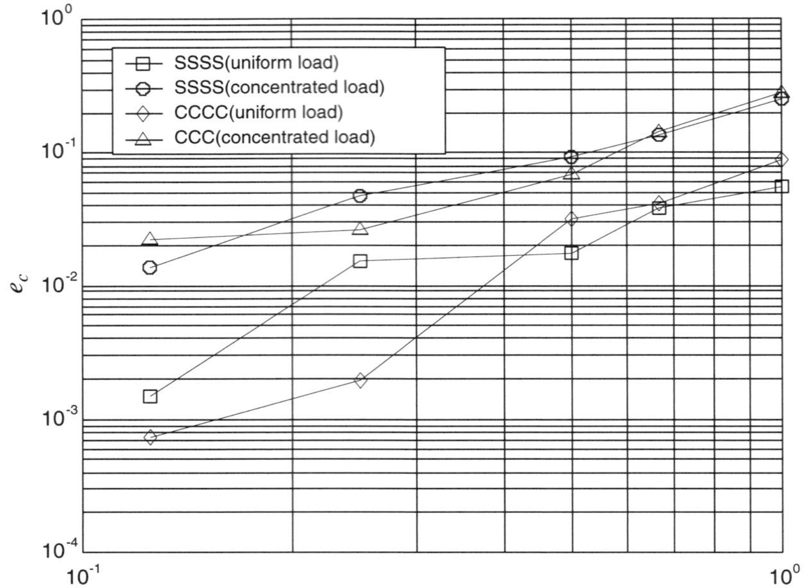

FIGURE 11.36
Convergence of results in relative deflection in a thin square plate subjected to a uniformly distributed and concentrated load.

Using too large a support domain will increase the computational cost in interpolation. Therefore, $\alpha_s = 3.0$ to 4.5 can obtain an acceptable result. For convenience and consistency, $\alpha_s = 4.0$ is used in the following studies. Note that for 2D solids we recommend using 2.5 to 3.0.

Note that analyzing problems of thin plates requires both larger quadrature and larger support domains compared with 2D solids. This is because the order of the governing equation for thin plates is higher than that for 2D solids.

Convergence Study

The convergence of the MLPG method is also investigated using regularly distributed nodes. The following error indicator is defined using C_0

$$e_C = \frac{|C_0 - \bar{C}_0|}{\bar{C}_0} \tag{11.173}$$

where C_0 is the dimensionless deflection coefficient obtained by numerical method, and $\bar{C}_0$ is the analytical result.

The e_C computed using various nodal densities is shown in Figure 11.36. Here, h is the shortest nodal distance. From Figure 11.36, it can be observed that the convergence of the MLPG method is very good.

Example 11.29 Square Plate under Different Load with Different Support

By using the above-mentioned parameters, deflections of an SSSS square 4 × 4 thin plate using regular nodal arrangement (81 nodes) obtained by MLPG is plotted in Figure 11.37. The results are also plotted in Figure 11.38 together with the analytical solution. From this figure, one can observe that the result by the present MLPG method is in very good agreement with the analytical solution.

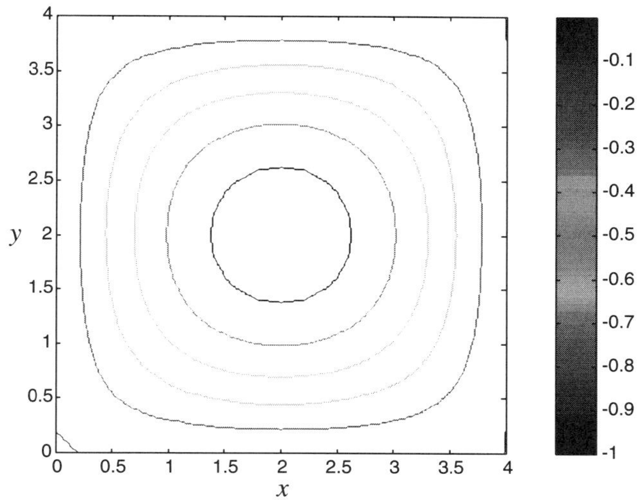

FIGURE 11.37
Deflection of the square plate. The plate is subjected to a uniform pressure.

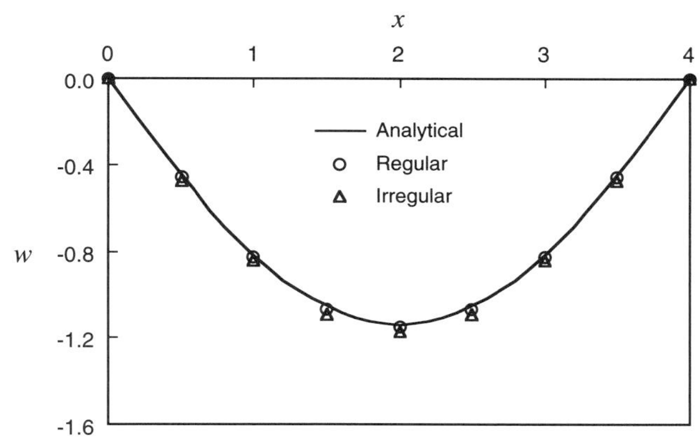

FIGURE 11.38
Deflection of a square plate at the middle line ($x = 2$). The plate is subjected to a uniform pressure.

The irregular distribution nodal arrangement shown in Figure 11.33b is also used for static analysis. Deflection results computed using both regular and irregular nodal arrangements are plotted in Figure 11.38. The figure shows that very good results are obtained using the irregular distribution nodal arrangement. The computational stability and high accuracy for nonstructured nodal distribution are very significant advantages of MLPG. These properties are very beneficial for practical applications.

A square thin plate subjected to a concentrated force P at the center of the plate is examined, and the results are compared with those by Krysl and Belytschko (1995) and Liu and Chen (2001), as well as the FEM solution using software package ABAQUS, as listed in Table 11.58a. From this table, we find that these three methods can obtain very good results for this problem. The accuracy of the results of MLPG is nearly the same as that of EFG, and slightly lower than that of FEM. The reason seems to be the effect of integration error. Although the interpolation accuracy in MLPG and EFG is higher than

TABLE 11.58a

Comparison of Results of Deflection of a Square Plate Computed Using MLPG, EFG, and FEM

	MLPG		EFG (Liu and Chen, 2001)		FEM (ABAQUS)		
	C_0	Error (%)	C_0	Error (%)	C_0	Error (%)	Analytical
SSSS	0.0111	−4.31	0.0115	−0.86	0.0115	−0.86	0.0116
CCCC	0.0058	3.57	0.0055	−1.78	0.0055	−1.78	0.0056

Note: $C_0 = w_{max}D/Pa^2$.

TABLE 11.58b

Deflections of a Square Plate of (4 × 4) Computed Using MLPG and the Analytic Method

		$h^a = 0.4$	$h^a = 0.25$	Analytical
Uniform Load[b]				
SSSS	C_0	0.00399	0.00400	0.00406
	Error (%)	−1.724	−1.477	0
CCCC	C_0	0.00122	0.00126	0.00126
	Error (%)	−3.174	0.000	0
SCSC	C_0	0.00188	0.00196	0.00192
	Error (%)	−2.083	2.083	0
Concentrated Load[c]				
SSSS	C_0	0.01091	0.01214	0.01160
	Error (%)	−5.948	4.655	0
CCCC	C_0	0.00598	0.00575	0.00560
	Error (%)	6.785	2.678	0
SCSC	C_0	0.00737	0.00717	0.00704
	Error (%)	4.687	1.846	0

[a] h = nodal spacing.
[b] $C_0 = w_{max}D/qa^4$ for uniform load.
[c] $C_0 = w_{max}D/Pa^2$ for concentrated load.

that in FEM, larger integration errors will affect the results negatively. Note that FEM is more sensitive to mesh irregularity. For heavily distorted meshes, FEM tends to perform much worse, and can even fail. MLPG and EFG can work normally. Krysl and Belytschko (1995) and Liu and Chen (2001) obtained the same conclusions in their papers. The robustness for irregular nodal arrangements is a very important advantage of meshfree methods.

The performance of the MLPG method is also examined and compared with the EFG method and FEM. In solving the thin plate problem, the computational efficiency of MLPG is lower than that of EFG (with penalty method). This is because the system matrix created in MLPG is asymmetrical. FEM needs a much shorter computational time than the MLPG and EFG methods. However, the computational time of FEM does not include the time for meshing, which can be very time-consuming for analysts. As for the EFG method, the requirement of a background integration mesh is also a practical drawback. MLPG eases the task of background mesh creation for the global problem domain. Background meshes for the local quadrature domains are still required in MLPG.

Thin square plates under different distributed loads with different boundary conditions are also analyzed. Two loading cases (uniform load and a concentrated load on the plate

TABLE 11.59

Dimensionless Deflection Coefficients C_0 of a Rectangular Plate

	b/a	1.2	1.4	1.6	1.8	2.0	5.0
*Uniform Load**							
SSSS	$h = 0.4$	0.00557	0.00701	0.00825	0.00925	0.01008	0.01310
	$h = 0.25$	0.00558	0.00701	0.00823	0.00928	0.01013	—
	Analytical	0.00564	0.00705	0.00830	0.00931	0.01013	0.01297
CCCC	$h = 0.4$	0.00170	0.00203	0.00226	0.00240	0.00248	0.00254
	$h = 0.25$	0.00172	0.00206	0.00229	0.00243	0.00252	—
	Analytical	0.00172	0.00207	0.00230	0.00245	0.00254	0.00260(∞)
SCSC	$h = 0.4$	0.00315	0.00458	0.00598	0.00729	0.00845	0.01315
	$h = 0.25$	0.00317	0.00458	0.00599	0.00728	0.00841	—
	Analytical	0.00319	0.00460	0.00603	0.00732	0.00840	0.01302(∞)
*Concentrated Load***							
SSSS	$h = 0.4$	0.01405	0.01571	0.01659	0.01710	0.01721	0.01912
	$h = 0.25$	0.01410	0.01536	0.01637	0.01687	0.01701	—
	Analytical	0.01353	0.01484	0.01570	0.01620	0.01651	0.01695(∞)
CCCC	$h = 0.4$	0.00664	0.00711	0.00730	0.00721	0.00723	0.00724
	$h = 0.25$	0.00649	0.00694	0.00712	0.00739	0.00742	—
	Analytical	0.00647	0.00691	0.00712	0.00720	0.00722	0.00725(∞)
SCSC	$h = 0.4$	0.00986	0.01207	0.01401	0.01518	0.01616	0.01914
	$h = 0.25$	0.00979	0.01203	0.01379	0.01502	0.01580	—
	Analytical	0.00945	0.01157	0.01326	0.01450	0.01537	0.01694

* $C_0 = w_{\max}D/qa^4$ for uniform load.
** $C_0 = w_{\max}D/Pa^2$ for concentrated load.

center with $P = 100$) and three support cases, SSSS, CCCC, and SCSC, are considered. The comparisons between the deflection results calculated analytically and using MLPG are listed in Table 11.58. Excellent agreement between the analytical and numerical results is obtained.

Example 11.30 Static Analysis of Thin Rectangular Plates

The MLPG method is used to analyze thin plates of different width/length ratios of length, a, and width, b, as shown in Figure 11.1. In our investigation, we keep a unchanged at $a = 4$, and vary b. When changing b, the nodal density is maintained by adding or removing nodes according to the width. Two nodal densities of $h = 0.4$ and 0.25 are used. Dimensionless deflection coefficients of rectangular plates under different distributed loads with different boundary conditions are listed in Table 11.59. For comparison, analytical results are also listed in the same table. This table shows that very good results can be obtained using MLPG for rectangular plates.

Example 11.31 Static Deflection Analysis of a Circular Plate

MLPG is also used to analyze a thin circular plate with two different boundary conditions: simply supported and clamped on the entire edge. The plate is subjected to uniform transverse pressure b_z. The analytical solution of deflection along the radial section is given for the thin circular plate as follows.

For plates simply supported on the edge

$$w = \frac{b_z(a^2 - r^2)}{64D}\left(\frac{5 + v}{1 + v}a^2 - r^2\right) \tag{11.174}$$

TABLE 11.60

Convergence of MLPG Solution ($C_0 = w_{max}D/qa^4$) for a Circular Plate Subjected to a Uniformly Distributed Pressure q

		18 Nodes	193 Nodes	321 Nodes	Analytical
Simply supported	C_0	0.06712	0.06513	0.06391	0.06370
	Error (%)	5.369	2.245	0.430	
Clamped	C_0	0.01478	0.01527	0.01556	0.01562
	Error (%)	–5.377	–3.241	–0.384	

TABLE 11.61

Natural Frequencies (Hz) of Lateral Free Vibration of a Free Square Plate

		MLPG							
	Analytical	81 Nodes		289 Nodes		441 Nodes		FEM[a]	
Mode	Solutions[a]	C_0	Error (%)	C_0	Error (%)	C_0	Error (%)	HOE[b]	LOE[c]
4	1.622	1.655	2.03	1.631	0.55	1.623	0.06	1.532	1.632
5	2.360	2.467	4.53	2.340	–0.85	2.351	–0.38	2.356	2.402
6	2.922	3.040	4.04	2.960	1.30	2.944	0.75	2.861	3.006
7	4.233	4.329	2.27	4.281	1.13	4.273	0.94	4.122	4.251
8	4.233	4.329	2.27	4.281	1.13	4.273	0.94	4.122	4.251
9	7.416	7.814	5.37	7.592	2.37	7.464	0.65	7.363	7.859
10	7.416	7.814	5.37	7.592	2.37	7.464	0.65	7.363	7.859

[a] Results from Abbassian et al. (1987).
[b] HOE: eight-noded semiloof thin shell element (4 × 4 mesh).
[c] LOE: four-noded isoparametric shell element (8 × 8 mesh).

For plates clamped at the edge

$$w = \frac{b_z}{64D}(a^2 - r^2)^2 \tag{11.175}$$

where a is the radius of the plate.

The maximum deflection, w_{max}, located at the center of the plate is $w_{max} = C_0qa^4/D$. For $\nu = 0.3$, the analytical results of C_0 are 0.06370 and 0.01562 for simply supported and clamped circular plates, respectively. The data used for the present analysis are $a = 2.0$, $b_z = 100$. The comparisons between the deflection results calculated analytically and using the MLPG are listed in Table 11.60. A good result is obtained by the MLPG method for circular thin plates.

Example 11.32 Free-Vibration Analysis of Thin Plates

Consider now a square plate with the following parameters: $a = b = 10$ m, $h = 0.05$ m, $E = 2 \times 10^{11}$ N/m^2, $\nu = 0.3$, and $\rho = 8000$ kg/m^3.

A fully free square plate is analyzed first. Three regular nodal arrangements are used. Frequencies obtained by the MLPG method are listed in Table 11.61. The analytical results and FEM results are also listed in the same table. The first three frequencies corresponding to the rigid movement are zero. The table shows that the present results are in good agreement with analytical results and FEM results. The convergence is also demonstrated in this table. As the number of nodes increases, the result obtained approaches the analytical solution.

TABLE 11.62

Natural Frequencies (Hz) of Lateral Free Vibration of SSSS and CCCC Square Plates

	Mode	Analytical Solutions	MLPG (Regular Nodes)			MLPG (Irregular Nodes)
			81 Nodes	169 Nodes	441 Nodes	81 Nodes
SSSS Plate[a]	1	2.377	2.3938	2.3935	2.3815	2.3645
	2	5.942	6.2042	6.0659	5.9629	6.1336
	3	5.942	6.2042	6.0659	5.9629	6.4473
	4	9.507	9.7235	9.6760	9.5564	10.0679
	5	11.884	12.7908	12.2820	12.0049	12.5034
	6	11.884	12.8753	12.2823	12.0064	12.8266
	7	15.449	15.8641	15.7687	15.5824	15.6306
	8	15.449	15.8641	15.7687	15.5824	16.1682
	9		19.2520	21.2358	20.5548	19.5129
	10		19.2520	21.2358	20.5548	19.9391
CCCC Plate[b]	1	4.333	4.4235	4.3577	4.3426	4.3797
	2	8.839	8.7535	8.8883	8.8669	8.4847
	3	8.839	8.7535	8.8883	8.8669	8.8120
	4	13.040	13.0986	13.1092	13.0668	12.9692
	5	15.845	15.1527	15.7818	15.9034	14.6923
	6	15.918	15.1883	15.8529	15.9789	15.3032
	7		19.9624	19.9194	19.9285	19.7194
	8		19.9624	19.9194	19.9285	20.0986
	9		24.7665	24.6326	25.3951	24.3576
	10		24.7665	24.6326	25.3951	24.3576

[a] The analytical results from Abbassian et al. (1987).
[b] The analytical results from Robert (1979).

Natural frequencies of fully simply supported and fully clamped plates are obtained and listed in Table 11.62. Three regularly distributed arrangements and one irregularly distributed nodal arrangement are used. It can be found that a good result can be obtained by the MLPG method for free-vibration analysis of these square plates. One can also observe that very good results are also obtained by the MLPG method using the irregular distribution nodal arrangement.

11.6 Remarks

Three formulations of the MFree methods—EFG, MLPG, and RPIM—are presented in this chapter for static, buckling, and vibration problems of thin and thick plates including composite laminated plates. Detailed studies on validation of these formulations of codes are reported. The following remarks may be made in summarizing all the materials presented in this chapter:

1. All these methods are very stable for irregular nodal distributions.
2. For thin plates, the degree of freedom (DOF) at a node is only one, compared to three in the conventional FEM formulation. The dimension of the system matrix in MFree formulation is only one third of that in FEM. This is a clear advantage of an MFree method. The reason that an MFree method has such a formulation is that it can use nodes beyond any frame like the "element."

3. In the formulations for thick plates, the DOFs at a node is three for all the methods: EFG, RPIM, MLPG, and FEM.
4. In using EFG, the dimension of the support domain should be around 3.0 to 4.5 times the average nodal distance for the nodes in the support domain of the quadrature point. Such a support domain will include about 40 to 70 nodes for MLS approximation.
5. In using RPIM for bending analyses of Mindlin plates, the formulation is the simplest as the RPIM shape functions possess the Kronecker delta function property. It is recommended that polynomial terms be added in the radial basis for both accuracy and freedom of choice of shape parameters.
6. When quadratic polynomial terms are added, the parameters of the EXP and MQ radial basis functions in the RPIM can be chosen from a wide range. For EXP radial basis function, $0.1 \leq \alpha_c \leq 1.0$; for MQ radial basis function, $-3.5 \leq q \leq 3.5$ and $1.0 \leq \alpha_c \leq 3.0$ can lead to good results. For EXP radial basis function, $\alpha_c = 0.1$; for MQ radial basis function, $\alpha_C = 2.0$ and $q = 1.03$ were used in this section.
7. In using MLPG, the support domain size can be the same as above. The local quadrature domain of a node should be about two times the distance of the node to its nearest neighboring node.
8. In using MFree methods for thick plates governed by higher-order shear deformation theories, shear locking is observed at small thickness/width ratios of about 1.0^{-2}. Methods used in FEM can be applied to avoid shear locking. The use of high-order basis functions can solve this problem, but the most effective way is to use the derivatives of the deflection shape functions as the shape functions for the rotations.

Challenging issues in MFree methods for plates exist:

- Connection of plates of different orientations
- Connection of plates with solids

12

Mesh Free Methods for Shells

Previous chapters have introduced a number of MFree methods for solids, fluids, beams, and plates. This chapter formulates MFree methods for shell structures.

Spatial thin shell structures are used very extensively in many engineering structures, including aircraft, pressure vessels, storage tanks, and so on, due to their outstanding efficiency in utilizing materials. Many shells are designed using advanced computer-aided design (CAD) technology to be the main load-carrying structure. Advanced CAD design of shell structures must consider both static and dynamic loads. Hence, both static and frequency analysis are very important and are integral parts of the shell design. Because of the complex nature, both structurally and mechanically, numerical means have to be utilized for analyses of shells. Advances in computing technology have made this possible and have led to the widespread use of numerical analysis of complex shells. The finite element method (FEM) remains the most popular numerical technique as evident from numerous publications since the 1970s (Yang et al., 1990). However, FEM requires the design of meshes, which is an extremely tedious, costly, and time-consuming process. In addition, meshing a shell structure into elements carries a number of risks that are often hidden by colorful pictures, as the curved surface of the shell is usually modeled by shell elements that are flat. This kind of geometric simplification in FEM can lead to severely erroneous results, due to the mechanics coupling effects of bending forces and membrane forces.

MFree methods present a promising alternative to FEM, as they offer opportunities to relieve the manual meshing process in modeling a structure. This is particularly important for shells, as shell structures are very complex both in field variable variation and in geometric configuration. MFree methods can also offer a very important additional capability in representing the complex curved geometry of shell structures using MFree approximation techniques. These MFree approximations both in field variables and in the structure itself provide much more accurate results compared with conventional FEM.

Because MFree methods are still very young, very few works have been reported in the development of MFree methods for shell structures. The first contribution in this regard was made by Krysl and Belytschko (1996b) based on the thin shell theory using moving least square (MLS) approximation with Lagrange multipliers for essential boundary conditions. Noguchi et al. (2000) developed a formulation for thick shell using MLS approximation with penalty method for handling essential boundary conditions. Li et al. (2000) formulated an MFree method based on the reproducing kernel particle method for thin shells with large deformation. In this work, the essential boundary condition is imposed by modifying shape functions for nodes near and on the essential boundaries. These three papers discuss only static deformation problems.

This chapter formulates MFree methods for shell structures based on the element free Galerkin (EFG) method introduced in Chapter 6, and the point interpolation method (PIM) discussed in Chapter 8 for two-dimensional (2D) solids. The work has been performed by L. Liu, G. R. Liu, and V. B. C. Tan during the past 2 years. This chapter covers the following topics:

- MFree formulations for both thin and thick shells based on Galerkin formulation
- Use of both MLS and RPIM shape functions
- Both static and dynamic problems
- Approximation of both field variables and the geometry of the shell using MFree shape functions

12.1 EFG Method for Spatial Thin Shells

This section formulates an EFG method for thin shells governed by Kirchhoff–Love shell theory. In the EFG method, the generalized displacement (deflections and rotations) at an arbitrary point is approximated from nodal displacements using MLS approximation. A compact support domain is used to determine field nodes to be used for constructing MLS shape functions. As discussed in Chapter 6, use of MLS approximation requires special treatment for essential boundary conditions for shell structures because EFG interpolations lack the Kronecker delta property. These techniques include the penalty method (Noguchi et al., 2000), Lagrange multipliers (Krysl and Belytschko, 1996b), and a method that modifies shape functions for nodes near and on the essential boundaries (Li et al., 2000). This chapter discusses both the penalty method and Lagrange multipliers method for analyzing static problems. For dynamic analysis, the Lagrange multipliers method is used for transient analyses and the matrix transformation method is used for free-vibration analyses, as discussed in Chapters 10 and 11 for beams and plates.

The formulation presented in this section is based on the work by L. Liu et al. (2001). It begins with a brief discussion of MLS approximation, which is basically the same as that provided in Chapter 5, except that a higher order of polynomial basis needs to be included. The governing equations for the analysis of general shells and membrane structures are then introduced. Numerical formulation based on a geometrically exact theory accounting for the Kirchhoff hypothesis is presented. This is followed by the definition of curved surfaces, kinematics of shells, stress and strain measures, and the constitutive relations adopted in the formulation. The final discrete equations for the entire shell structure for static, free vibration, and transient vibration then are obtained. For free vibration, the essential boundary conditions are imposed using orthogonal transform techniques to solve the eigenvalue equation (Ouatouati and Johnson, 1999; G. R. Liu and Chen, 2001). For static problems, essential boundary conditions are imposed through the Lagrange multipliers method and the penalty method. Finally, the method is applied to several numerical examples of shells of different geometries to illustrate the efficiency and accuracy of the present EFG method.

12.1.1 Moving Least Squares Approximation

The derivation of shape functions from the MLS approximation method is the same as that provided in Chapter 5, except that a higher order of polynomial basis needs to be included. A 2D field approximation is needed for modeling thin shells. A component of the displacement vector is approximated by a polynomial function as follows:

$$u^h(\mathbf{x}) = \sum_{I=1}^{n} \phi_I(\mathbf{x}) u_I \tag{12.1}$$

where shape function $\phi_I(\mathbf{x})$ is the MLS shape function obtained following the procedure described in Section 5.4 using polynomial basis. The formulations for computing the derivatives of the shape functions are given in Equations 5.62 through 5.67. Note that for shell structures the order of the polynomial basis should be higher than that for 2D solids. In this chapter, the two different orders of polynomial basis are primarily used. The following six terms of basis functions up to quadratic terms are used for shells where shear effect is significant.

$$\mathbf{P}^T(\mathbf{x}) = \{1, x, y, x^2, xy, y^2\} \tag{12.2}$$

The following 15 terms of basis functions up to quartic terms are used to ameliorate membrane locking in bending-dominated cases:

$$\mathbf{P}^T(\mathbf{x}) = \{1, x, y, x^2, xy, y^2, x^3, x^2y, xy^2, y^3, x^4, x^3y, x^2y^2, xy^3, y^4\}^T \tag{12.3}$$

In computing the MLS shape functions, the quartic spline weight function given in Equation 5.12 is used in this section because of requirements on the continuity of the MLS shape functions and their derivatives.

12.1.2 Governing Equation for Thin Shell

The shells considered in this section are assumed to be thin with arbitrarily deep and Gaussian curvature governed by Kirchhoff–Love theory. The governing equations used in this section are based on geometrically exact theory of shells formulated by Simo and Fox (1989) with some modifications to account for the Kirchhoff hypothesis. Here, we outline only the basic concepts of the formulation. Details can be found in the paper by Simo and Fox (1989).

Kinematic Description of Shell

The reference frame coordinates are illustrated in Figure 12.1. The shell in 3D space is described in a global Cartesian coordinate system, $\mathbf{x}$. The Gauss intrinsic coordinates defined locally are used to describe the configuration of the shell. $\boldsymbol{\varphi}(\xi^1, \xi^2)$ gives the position of the point on the shell neutral surface, and $\mathbf{t}(\xi^1, \xi^2)$ is a direction unit vector

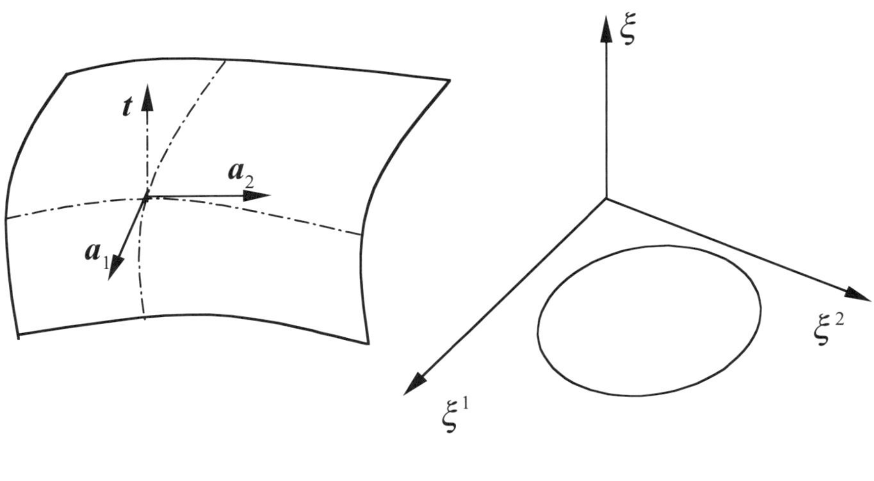

FIGURE 12.1
Reference frames of coordinates on the neutral surface of a thin shell.

normal to the shell neutral surface both in the unformed reference and deformed states according to the Kirchhoff–Love hypothesis. The pair ($\boldsymbol{\varphi}$, $\mathbf{t}$) defines the position of an arbitrary point in the shell. The configuration of the shell can be expressed mathematically as

$$\psi = \{\mathbf{x} \in \mathbb{R}^3 | \mathbf{x} = \boldsymbol{\varphi}(\xi^1, \xi^2) + \xi \mathbf{t}(\xi^1, \xi^2) \quad \text{with} \quad \xi^1, \xi^2 \in \mathbb{A} \quad \text{and} \quad \xi \in \langle h^-, h^+ \rangle\} \tag{12.4}$$

Here $\mathbb{A}$ denotes the parametric space for the shell; (h^-, h^+) are the distances of the "lower" and "upper" surfaces of the shell measured from the shell neutral surface.

The convective basis vectors g_I are defined as

$$\nabla \mathbf{x} = \frac{\partial \mathbf{x}}{\partial \xi^I} \otimes \mathbf{E}^I \equiv \mathbf{g}_I \otimes \mathbf{E}^I \tag{12.5}$$

where ∇ denotes the gradient operator, $\otimes$ denotes tensor product, and $\mathbf{E}$ is the unit basis vector in the global Cartesian basis. A contravariant basis $\mathbf{g}^I$ can be obtained from the standard relation of

$$\mathbf{g}^J \cdot \mathbf{g}_I = \delta_I^J \tag{12.6}$$

The unit area in the neutral surface is defined by the differential two-form:

$$\mathrm{d}\mathbb{A} = \boldsymbol{\varphi}_{,1} \times \boldsymbol{\varphi}_{,2} \mathrm{d}\xi^1 \mathrm{d}\xi^2 \tag{12.7}$$

where "×" denotes vector cross product. The determinants of the tangent maps in the deformed and reference configuration will be denoted subsequently as j and j^0, respectively, with $\bar{j}$ and $\bar{j}^0$ denoting the Jacobians on the reference surface

$$j = \det[\nabla \mathbf{x}], \; j^0 = \det[\nabla \mathbf{x}^0], \; \bar{j} = j|_{\xi=0}, \; \bar{j}^0 = j^0|_{\xi=0} \tag{12.8}$$

where the superscript "0" is used to denote quantities in the reference configuration. The surface convected frame, which spans the tangent space to the neutral surface, is defined as $a_\alpha = \boldsymbol{\varphi}_{,\alpha}$ ($\alpha = 1, 2$). Hence, the first fundamental form on the reference surface is

$$\mathbf{a} = a_{\alpha\beta} \mathbf{a}^\alpha \otimes \mathbf{a}^\beta, \quad a_{\alpha\beta} = \boldsymbol{\varphi}_{,\alpha} \cdot \boldsymbol{\varphi}_{,\beta} \tag{12.9}$$

where a^I ($I = 1, 2$) denotes the dual surface convected basis through the standard relation $\mathbf{a}^I \cdot \mathbf{a}_J = \delta_J^I$ and $\cdot$ denotes dot product. The second fundamental form is defined as

$$\kappa_{\alpha\beta} = \boldsymbol{\varphi}_{,\alpha} \cdot \mathbf{t}_{,\beta} \tag{12.10}$$

Strain Measures

The linear membrane and bending strain measures can be derived from the kinematic variables in Equations 12.9 and 12.10 as

$$\varepsilon_{\alpha\beta} = \frac{1}{2}(\boldsymbol{\varphi}^0_{,\alpha} \cdot \mathbf{u}_{,\beta} + \boldsymbol{\varphi}^0_{,\beta} \cdot \mathbf{u}_{,\alpha}) \tag{12.11}$$

and

$$\rho_{\alpha\beta} = \frac{1}{2}(\boldsymbol{\varphi}^0_{,\alpha} \cdot \Delta\mathbf{t}_{,\beta} + \boldsymbol{\varphi}^0_{,\beta} \cdot \Delta\mathbf{t}_{,\alpha} + \mathbf{u}_{,\alpha} \cdot \mathbf{t}^0_{,\beta} + \mathbf{u}_{,\beta} \cdot \mathbf{t}^0_{,\alpha}) \tag{12.12}$$

where only the symmetric part of the bending strain measure is considered.

The Kirchhoff–Love hypothesis needs to be introduced explicitly to obtain the definite forms for the strain measures. The mathematical form of this hypothesis is expressed as

$$\mathbf{t} = (\bar{j})^{-1}(\boldsymbol{\varphi}_{,1} \times \boldsymbol{\varphi}_{,2}), \quad \|\mathbf{t}\| = 1 \tag{12.13}$$

where $\| \ \|$ denotes the norm of a vector. Hence, the derivatives of the normal vector in the reference configuration $\mathbf{t}^0$ and partial derivatives of the increment $\Delta\mathbf{t}$ can be

$$\mathbf{t}^0_{,\alpha} = (\bar{j}^0)^{-1} (\boldsymbol{\varphi}^0_{,1\alpha} \times \boldsymbol{\varphi}^0_{,2} + \boldsymbol{\varphi}^0_{,1} \times \boldsymbol{\varphi}^0_{,2\alpha}) \tag{12.14}$$

$$\Delta\mathbf{t}_{,\alpha} = (\bar{j}^0)^{-1}(\mathbf{u}_{,1\alpha} \times \boldsymbol{\varphi}^0_{,2} + \mathbf{u}^0_{,1} \times \boldsymbol{\varphi}^0_{,2\alpha} + \boldsymbol{\varphi}^0_{,1\alpha} \times \mathbf{u}_{,2} + \boldsymbol{\varphi}^0_{,1} \times \mathbf{u}_{,2\alpha}) \tag{12.15}$$

Note that the membrane strain measures of Equation 12.11 are not affected by the introduction of the Kirchhoff–Love hypothesis. Considering the symmetry with respect to partial differentiation $\boldsymbol{\varphi}^0_{,12} = \boldsymbol{\varphi}^0_{,21}$ and $\mathbf{u}_{,12} = \mathbf{u}_{,21}$, the bending strain measures can be rewritten as

$$\rho_{11} = -\mathbf{u}_{,11} \cdot \mathbf{t}^0 + (\bar{j}^0)^{-1}[\mathbf{u}_{,1} \cdot (\boldsymbol{\varphi}^0_{,11} \times \boldsymbol{\varphi}^0_{,2}) + \mathbf{u}_{,2} \cdot (\boldsymbol{\varphi}^0_{,1} \times \boldsymbol{\varphi}^0_{,11})] \tag{12.16}$$

$$\rho_{22} = -\mathbf{u}_{,22} \cdot \mathbf{t}^0 + (\bar{j}^0)^{-1}[\mathbf{u}_{,1} \cdot (\boldsymbol{\varphi}^0_{,22} \times \boldsymbol{\varphi}^0_{,2}) + \mathbf{u}_{,2} \cdot (\boldsymbol{\varphi}^0_{,1} \times \boldsymbol{\varphi}^0_{,22})] \tag{12.17}$$

$$\begin{aligned} \rho_{12} &= -\frac{1}{2}(\mathbf{u}_{,12} + \mathbf{u}_{,21}) \cdot \mathbf{t}^0 + \frac{1}{2}(\bar{j}^0)^{-1}[\mathbf{u}_{,1} \cdot ((\boldsymbol{\varphi}^0_{,12} + \boldsymbol{\varphi}^0_{,21}) \times \boldsymbol{\varphi}^0_{,2}) + \mathbf{u}_{,2} \cdot (\boldsymbol{\varphi}^0_{,1} \times (\boldsymbol{\varphi}^0_{,12} + \boldsymbol{\varphi}^0_{,21}))] \\ &= -\mathbf{u}_{,12} \cdot \mathbf{t}^0 + (\bar{j}^0)^{-1}[\mathbf{u}_{,1} \cdot (\boldsymbol{\varphi}^0_{,12} \times \boldsymbol{\varphi}^0_{,2}) + \mathbf{u}_{,2} \cdot (\boldsymbol{\varphi}^0_{,1} \times \boldsymbol{\varphi}^0_{,12})] \end{aligned} \tag{12.18}$$

Stress Resultants and Stress Couples

A section in the current configuration is described by

$$\psi^\alpha = \left\{\mathbf{x} \in \mathbb{R}^3 | \mathbf{x} = \mathbf{x}|_{\xi^\alpha = \text{const}}\right\}, \quad \alpha = 1, 2 \tag{12.19}$$

The stress resultants and resultant couples are defined by normalizing the force and torque with the surface Jacobian $j = \|\boldsymbol{\varphi}_{,1} \times \boldsymbol{\varphi}_{,2}\|$ as follows

$$\mathbf{n}^\alpha = (\bar{j})^{-1}\int_{h^-}^{h^+} \sigma \mathbf{g}^\alpha \, d\xi, \quad \alpha = 1, 2 \tag{12.20}$$

$$\mathbf{m}^\alpha = (\bar{j})^{-1}\int_{h^-}^{h^+} (\mathbf{x} - \boldsymbol{\varphi}) \times \sigma \mathbf{g}^\alpha \, j d\xi, \quad \alpha = 1, 2 \tag{12.21}$$

The director stress couple $\tilde{\mathbf{m}}^{\alpha}$ can also be defined through the expression

$$\mathbf{m}^{\alpha} = \mathbf{t} \times \tilde{\mathbf{m}}^{\alpha} \Rightarrow \tilde{\mathbf{m}}^{\alpha} = (\bar{j})^{-1} \int_{h^-}^{h^+} \xi \sigma \mathbf{g}^{\alpha} j \, d\xi \tag{12.22}$$

The through-thickness stress resultant has been omitted because it does not play a role in Kirchhoff–Love theory.

Constitutive Equations

For the isotropic elastic shell structures, the effective membrane and stress couple resultant for the isotropic hyperelastic material can be written as

$$\begin{Bmatrix} \tilde{n}^{11} \\ \tilde{n}^{22} \\ \tilde{n}^{12} \end{Bmatrix} = \frac{Eh}{1 - \nu^2} \mathbf{C} \begin{Bmatrix} \varepsilon_{11} \\ \varepsilon_{22} \\ 2\varepsilon_{12} \end{Bmatrix} \tag{12.23}$$

$$\begin{Bmatrix} \tilde{m}^{11} \\ \tilde{m}^{22} \\ \tilde{m}^{12} \end{Bmatrix} = \frac{Eh^3}{12(1 - \nu^2)} \mathbf{C} \begin{Bmatrix} \rho_{11} \\ \rho_{22} \\ 2\rho_{12} \end{Bmatrix} \tag{12.24}$$

where the matrix $\mathbf{C}$ is given by

$$\mathbf{C} = \begin{bmatrix} (a^{011})^2 & (\nu a^{011} a^{022} + (1 - \nu)(a^{012})^2) & a^{011} a^{012} \\ & (a^{022})^2 & a^{022} a^{012} \\ \text{symm} & & \frac{1}{2}((1 + \nu)(a^{012})^2 + (1 - \nu) a^{011} a^{022}) \end{bmatrix} \tag{12.25}$$

Here $a^{0\alpha\beta}$ are the components of the first fundamental form in the dual basis.

12.1.3 Strain–Displacement Relations

The displacement vector can be expressed in the global Cartesian basis $\mathbf{E}_K$ as

$$\mathbf{u}(\zeta) = \sum_{I=1}^{n} \phi_I(\zeta) [u_I \mathbf{E}_1 + v_I \mathbf{E}_2 + w_I \mathbf{E}_3] \tag{12.26}$$

where n is the number of nodes in the neighborhood (support domain) of ζ; u_I, v_I, and w_I are the components of the displacement vector at the Ith node in $\mathbf{E}_1$, $\mathbf{E}_2$, and $\mathbf{E}_3$ directions, respectively. The membrane strain–displacement relation for the Ith node is obtained by substituting the displacement approximation Equation 12.26 into Equation 12.11 to give

$$\begin{Bmatrix} \varepsilon_{11} \\ \varepsilon_{22} \\ 2\varepsilon_{12} \end{Bmatrix} = \begin{bmatrix} \phi_{I,1} \boldsymbol{\varphi}_{,1}^0 \cdot \mathbf{E}_1 & \phi_{I,1} \boldsymbol{\varphi}_{,1}^0 \cdot \mathbf{E}_2 & \phi_{I,1} \boldsymbol{\varphi}_{,1}^0 \cdot \mathbf{E}_3 \\ \phi_{I,2} \boldsymbol{\varphi}_{,2}^0 \cdot \mathbf{E}_1 & \phi_{I,2} \boldsymbol{\varphi}_{,2}^0 \cdot \mathbf{E}_2 & \phi_{I,2} \boldsymbol{\varphi}_{,2}^0 \cdot \mathbf{E}_3 \\ (\phi_{I,1} \boldsymbol{\varphi}_{,2}^0 + \phi_{I,2} \boldsymbol{\varphi}_{,1}^0) \cdot \mathbf{E}_1 & (\phi_{I,1} \boldsymbol{\varphi}_{,2}^0 + \phi_{I,2} \boldsymbol{\varphi}_{,1}^0) \cdot \mathbf{E}_2 & (\phi_{I,1} \boldsymbol{\varphi}_{,2}^0 + \phi_{I,2} \boldsymbol{\varphi}_{,1}^0) \cdot \mathbf{E}_3 \end{bmatrix} \begin{Bmatrix} u_I \\ v_I \\ w_I \end{Bmatrix} \tag{12.27}$$

The bending strain–displacement matrix $[\mathbf{B}_{(b)I}]$ for the Ith node is obtained by substituting the displacement approximation Equation 12.26 into Equation 12.12

$$\begin{Bmatrix} \rho_{11} \\ \rho_{22} \\ 2\rho_{12} \end{Bmatrix} = [\mathbf{B}_{(b)I}] \begin{Bmatrix} u_I \\ v_I \\ w_I \end{Bmatrix} \tag{12.28}$$

where the elements of the strain–displacement matrix $[\mathbf{B}_{(b)I}]$ are given by

$$\begin{aligned} [\mathbf{B}_{(b)I}]_{1n} &= (\bar{j}^0)^{-1}[-\bar{j}^0\phi_{I,11}\mathbf{t}^0 + \phi_{I,1}(\boldsymbol{\varphi}^0_{,11} \times \boldsymbol{\varphi}^0_{,2}) + \phi_{I,2}(\boldsymbol{\varphi}^0_{,1} \times \boldsymbol{\varphi}^0_{,11})] \cdot \mathbf{E}_m \\ [\mathbf{B}_{(b)I}]_{2n} &= (\bar{j}^0)^{-1}[-\bar{j}^0\phi_{I,22}\mathbf{t}^0 + \phi_{I,1}(\boldsymbol{\varphi}^0_{,22} \times \boldsymbol{\varphi}^0_{,2}) + \phi_{I,2}(\boldsymbol{\varphi}^0_{,1} \times \boldsymbol{\varphi}^0_{,22})] \cdot \mathbf{E}_m \\ [\mathbf{B}_{(b)I}]_{3n} &= 2(\bar{j}^0)^{-1}[-\bar{j}^0\phi_{I,12}\mathbf{t}^0 + \phi_{I,1}(\boldsymbol{\varphi}^0_{,12} \times \boldsymbol{\varphi}^0_{,2}) + \phi_{I,2}(\boldsymbol{\varphi}^0_{,1} \times \boldsymbol{\varphi}^0_{,12})] \cdot \mathbf{E}_m \end{aligned} \tag{12.29}$$

12.1.4 Principle of Virtual Work

The effective membrane and bending forces are defined to describe the weak formulation of shells

$$\tilde{\mathbf{n}} = \tilde{n}^{\beta\alpha} \boldsymbol{a}_\beta \otimes \boldsymbol{a}_\alpha \tag{12.30}$$

$$\tilde{\mathbf{m}} = \tilde{m}^{\beta\alpha} \boldsymbol{a}_\beta \otimes \boldsymbol{a}_\alpha \tag{12.31}$$

By making use of the basic kinematic assumption (Equation 12.4), the weak form of the governing equation for thin shells under static load can be written as

$$W_{\text{Sta}}(\delta\mathbf{x}) = \int_{\mathbb{A}} [\tilde{n}^{\beta\alpha} \cdot \delta\varepsilon_{\beta\alpha} + \tilde{m}^{\beta\alpha} \cdot \delta\rho_{\beta\alpha}] \mathrm{d}\mathbb{A} - W_{\text{ext}}(\delta\mathbf{x}) \tag{12.32}$$

Here W_{ext} is the virtual work of the external loading given by

$$W_{\text{ext}} = \int_{\mathbb{A}} [\bar{\mathbf{n}} \cdot \delta\boldsymbol{\varphi} + \bar{\bar{\mathbf{m}}} \cdot \delta\bar{\mathbf{t}}] \mathrm{d}\mathbb{A} + \int_{\Gamma_n} \bar{\bar{\mathbf{n}}} \cdot \delta\boldsymbol{\varphi}\bar{j}\,\mathrm{d}\Gamma + \int_{\Gamma_m} \bar{\bar{\mathbf{m}}}\delta\mathbf{t}\bar{j}\,\mathrm{d}\Gamma \tag{12.33}$$

where $\bar{\mathbf{n}}$ is the applied resultant force per unit length and $\bar{\bar{\mathbf{m}}}$ is the applied direct couple per unit length. $\bar{\bar{\mathbf{n}}}$ and $\bar{\bar{\mathbf{m}}}$ are the prescribed resultant force and the prescribed director couple on the boundaries Γ_n and Γ_m, respectively. For static analysis of thin shells, the penalty method is used to enforce essential boundary conditions by adding an additional boundary condition term to Equation 12.33 to obtain

$$W_{\text{Sta}}(\delta\mathbf{x}) - \int_{\Gamma_u} \boldsymbol{\alpha} \cdot (\mathbf{u} - \bar{\mathbf{u}})\delta\mathbf{u}\,\mathrm{d}\Gamma = 0 \tag{12.34}$$

Here $\mathbf{u}, \bar{\mathbf{u}}$ are the nodal displacement vector and the prescribed displacement vector on all the essential boundary Γ_u, and $\boldsymbol{\alpha}$ is a diagonal matrix of penalty factors, which are usually very large numbers (see Chapter 4).

For free-vibration analyses of the thin shells, the discrete system equation and the boundary conditions are formulated separately. The variational form of the elastic dynamic undamped equilibrium equation can be written as follows:

$$W_{\mathrm{Dya}}(\delta \mathbf{x}) = \int_{\mathbb{A}} [\tilde{n}^{\beta\alpha} \cdot \delta\varepsilon_{\beta\alpha} + \tilde{m}^{\beta\alpha} \cdot \delta\rho_{\beta\alpha}] \mathrm{d}\mathbb{A} + \int_{\mathbb{R}} \delta\mathbf{u} \cdot \rho\ddot{\mathbf{u}} \mathrm{d}\mathbb{R} - W_{\mathrm{ext}}(\delta\mathbf{x}) \tag{12.35}$$

The weak form of the essential boundary conditions with Lagrange multipliers is employed to obtain the discretized essential boundary conditions

$$\int_{\Gamma_u} \delta\boldsymbol{\lambda}^T \cdot (\mathbf{u} - \bar{\mathbf{u}}) \mathrm{d}\Gamma = 0 \tag{12.36}$$

where $\boldsymbol{\lambda}$ is a matrix of Lagrange multipliers, each of them can be interpolated on the essential boundary using its nodal values.

$$\lambda(\mathbf{x}) = \sum_I N_I(s)\lambda_I, \quad \mathbf{x} \in \Gamma_u \tag{12.37}$$

and

$$\delta\lambda(\mathbf{x}) = \sum_I N_I(s)\delta\lambda_I, \quad \mathbf{x} \in \Gamma_u \tag{12.38}$$

where s is the arc-length along the essential boundaries, and $N_I(s)$ are the Lagrange interpolates discussed in Equation 6.10.

12.1.5 Surface Approximation

The (neutral) surface of the shell is also approximated using MLS shape functions. The procedure is exactly the same as for approximating the displacement field variables. The approximated surface can be described by

$$\boldsymbol{\varphi}(\boldsymbol{\zeta}) = \phi_I(\boldsymbol{\zeta})\mathbf{x}_I \tag{12.39}$$

where $\boldsymbol{\zeta}$ is the coordinate in the parameter space for the neutral surface of the shell. A deficiency of this approximation is that the constructed surface does not pass through the prescribed points, unlike that in finite element meshes. This is due to the use of MLS shape functions.

12.1.6 Discretized Equations

For static analyses of shells, the penalty term to impose essential boundary conditions in Equation 12.34 can be discretized as follows:

$$\boldsymbol{\alpha}\int_{\Gamma_u} \mathbf{u} \cdot \delta\mathbf{u} \mathrm{d}\Gamma = \sum_{I=1}^{n} \delta\tilde{\mathbf{u}}^{\mathrm{T}} \boldsymbol{\alpha} \boldsymbol{\Phi}_I \boldsymbol{\Phi}_I^{\mathrm{T}} \tilde{\mathbf{u}} \tag{12.40}$$

and

$$\boldsymbol{\alpha}\int_{\Gamma_u} \bar{\mathbf{u}} \cdot \delta\mathbf{u}\,\mathrm{d}\Gamma = \sum_{I=1}^{n} \delta\tilde{\mathbf{u}}^{\mathrm{T}}\bar{\mathbf{u}}_I\boldsymbol{\alpha}\boldsymbol{\Phi}_I \tag{12.41}$$

where n is the number of sampling points for integration on surface Γ_u; $\tilde{\mathbf{u}}$ is the nodal vector corresponding to a degree of freedom in the translation and rotation fields. The penalty matrix in Equation 12.40 and the penalty vector in Equation 12.41 are assembled into the global stiffness matrix and the global external force vector, respectively.

For dynamic analysis of shells, Equations 12.1, 12.23, 12.24, 12.27, and 12.28 are substituted into the variational weak form (Equation 12.35). This gives the dynamic discrete equation of

$$\mathbf{M}\ddot{\mathbf{U}} + \mathbf{C}\dot{\mathbf{U}} + \mathbf{K}\mathbf{U} = \mathbf{F} \tag{12.42}$$

where **K**, **M**, and **F** are, respectively, the global stiffness, global mass matrices, and global force vector, which are assembled using the corresponding nodal matrices and vectors formed in the similar manner as those for plates (see Chapter 10).

12.1.7 Static Analysis

For static analysis of shells, all the terms in Equation 12.42 related to dynamic effects should vanish, and the equation system is simplified as

$$\mathbf{K}\mathbf{U} = \mathbf{F} \tag{12.43}$$

which is a set of linear algebraic equations that can be solved for the deflection using standard routines of equation solvers.

12.1.8 Free Vibration

For free-vibration analysis, the external force vector **F** should vanish, and the damping should not be considered; we then have

$$\mathbf{M}\ddot{\mathbf{U}} + \mathbf{K}\mathbf{U} = \mathbf{0} \tag{12.44}$$

Considering harmonic vibration, the eigenvalue equations for shells derived from Equation 12.44 is of the form

$$(\mathbf{K} - \omega^2\mathbf{M})\mathbf{Q} = 0 \tag{12.45}$$

where ω is the natural frequency and **Q** is a vector that collects the nodal values corresponding the amplitudes of the displacements given by

$$\mathbf{Q} = \{Q_1, \ldots, Q_n\}^{\mathrm{T}} \tag{12.46}$$

We now derive the discretized form of essential boundary conditions. Substituting the displacement field **u** of Equation 12.1 into the weak form (Equation 12.36) yields a set of linear algebraic constraint equations:

$$\mathbf{B}\mathbf{u} = \bar{\mathbf{B}} \tag{12.47}$$

where

$$\mathbf{B}_{IJ} = \phi_J(x_I), \qquad \bar{\mathbf{B}}_I = \bar{\mathbf{u}}(x_I), \quad I \in \Gamma_u, \; J \in \mathbb{A} \tag{12.48}$$

In general, **B** is a very sparse and singular matrix. For eigenvalue analysis, the essential boundary conditions are homogeneous, i.e., $\bar{\mathbf{B}} = 0$.

In treating the essential boundary conditions, we follow the procedure described in Section 11.1. An orthogonal transformation is performed to produce a positive-definite stiffness matrix for the eigenvalue equation in computing natural frequencies. Using singular-value decomposition, **B** can be decomposed as

$$\mathbf{B} = \mathbf{U}\mathbf{S}\mathbf{V}^{\mathrm{T}} \tag{12.49}$$

where **U** and **V** are orthogonal matrices. **S** has a diagonal form whose diagonal elements are equal to singular values of **B**. The matrix **V** can be written as

$$\mathbf{V}^{\mathrm{T}} = (\mathbf{V}_{n\times r}, \mathbf{V}_{n\times(n-r)})^{\mathrm{T}} \tag{12.50}$$

where the rank r of **B** is equal to the number of independent constraints, and the others are redundant. The following change of coordinates satisfied the constraint Equation 12.47:

$$\mathbf{Q} = \mathbf{V}_{n\times(n-r)}\tilde{\mathbf{Q}} \tag{12.51}$$

By substituting Equation 12.51 into Equation 12.45 and left-multiplying the result by the transpose of $\mathbf{V}_{n\times(n-r)}$, the reduced order eigenvalue problem for the structure is obtained

$$(\tilde{\mathbf{K}} - \omega^2\tilde{\mathbf{M}})\tilde{\mathbf{Q}} = 0 \tag{12.52}$$

where $\tilde{\mathbf{K}} = \mathbf{V}^{\mathrm{T}}_{(n-r)\times n}\mathbf{K}\mathbf{V}_{n\times(n-r)}$ and $\tilde{\mathbf{M}} = \mathbf{M}^{\mathrm{T}}_{(n-r)\times n}\mathbf{K}\mathbf{M}_{n\times(n-r)}$ are the dimension reduced stiffness and mass matrices. Equation 12.52 is now non-negative definite and can be solved using standard eigenvalue solvers for eigenvalues that relate to the natural frequencies and eigenvectors that relate to the vibration modes of the thin shells.

12.1.9 Forced (Transient) Vibration

For transient response analysis, Equation 12.42 can be solved using conventional direct integration techniques based on finite difference approaches, where the time space is divided into small time steps. The Newmark method is applied in the section for time integration to obtain the time history of the displacement response of the shell. The procedure is the same as that described in Chapter 7 for 2D solids.

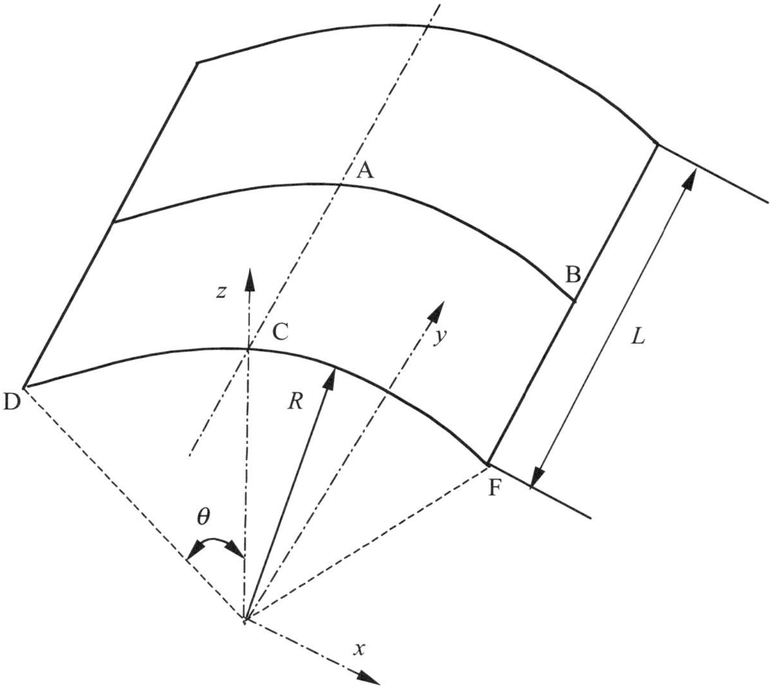

FIGURE 12.2
Barrel vault roof and the coordinate system.

12.1.10 Numerical Example for Static Problems

Example 12.1 Static Deflection of a Barrel Vault Roof under Gravity Force
The present EFG method is used to investigate the response of a barrel vault roof under self-weight that is calculated simply using the density of gravity. The problem has been analyzed by several researchers using the EFG method (Krysl and Belytschko, 1996b; Noguchi et al., 2000) and FEM (Simo et al., 1989). The barrel vault roof is a standard benchmarking test because it undergoes complex membrane and inextensional bending states of stress. The example is used to evaluate the convergence and accuracy of the present EFG method for the static analysis of shells. Figure 12.2 shows the barrel roof and defines the parameters used in the description of its geometry. The two curved edges of the roof are diaphragm supported, which allows displacement in the axial direction and rotation about the tangent to shell boundary. The following parameters are used in the computation.

Length	Radius	Thickness	Semi-Span Angle	Young's Modulus	Poisson's Ratio	Mass Density
$L = 600$	$R = 300$	$h = 3.0$	$\theta = 40°$	$E = 3.0 \times 10^6$	$\nu = 0$	$\rho = 0.20833$

Due to symmetry, only a quarter of the vault roof is modeled, and symmetric boundary conditions are introduced along the planes of symmetry. To evaluate the effectiveness of the present method, both regular and irregular nodal arrangements in the parametric space shown in Figure 12.3 are used in the analyses. Both quadratic ($m = 6$) and quartic

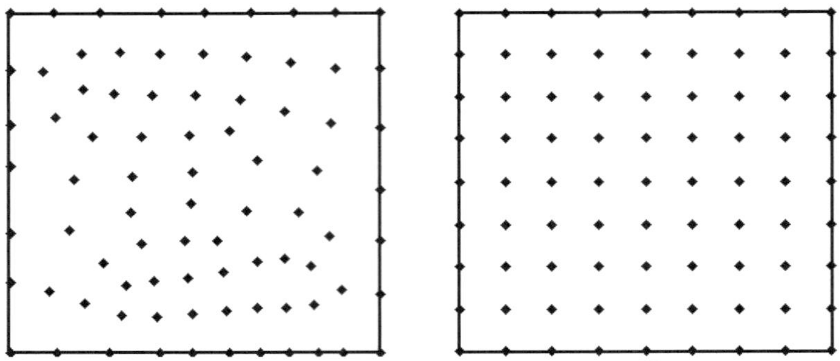

FIGURE 12.3
Nodal arrangement in parameter space.

(m = 15) basis functions are used in the analysis. Figures 12.4 and 12.5 show, respectively, the distributions of the vertical deflection along edges AB and AC, where the FEM solution is obtained using a general-purpose program, i.e., ANSYS©. The EFG results using 81 nodes agree well with the finite element results of ANSYS using 2520 nodes. The reason for the demand by FEM for a very fine element mesh may be the simplification of the geometry, because the field variable (deflection) do not vary drastically and, as shown in Figures 12.4 and 12.5, a coarse mesh should produce a good approximation. However, in ANSYS, the shell is modeled using flat shell elements. This simplification cannot model the mechanics coupling effects of bending forces and membrane forces, which is very significant for shell structures, unless a very fine mesh is used. In our EFG code, however, the geometry of the shell is modeled using MLS approximation, which very accurately represents the curvature of the geometry of the shell, and hence produces very accurate results using very coarse nodes. This example clearly demonstrates the advantage of the MFree method in modeling shell structures. Note also that under the action of gravity force, which is downward, the central point on the roof moves upward against the direction of body force. This fact provides very clear evidence of the importance of the coupling of bending force and membrane force. The existence of the membrane force causes the rise of the rooftop.

The convergence of the EFG results is also investigated in detail, and the results are summarized in Figure 12.6, where the curves are normalized using the converged value of the vertical deflection at B, that is, –3.618. It can be seen that the performance of the EFG method compares well with the finite element solution of Belytschko et al. (1985) and Simo et al. (1989). The results given by quartic polynomial basis are usually better than that by quadratic basis when nodal points are sufficiently dense. However, worse results are obtained when low nodal densities are employed. The reason could be again the approximation of the geometry of the shell. When the nodes are too coarse, even EFG will fail to approximate the geometry well.

Lagrange Multiplier Method vs. Penalty Method

The penalty method and Lagrange multipliers can both be used to enforce the essential boundary conditions. Here we discuss the advantages and disadvantages of these two methods using the example of barrel vault roof. Figure 12.5 shows the distribution of the vertical deflection along edge DF, where deflection is supposed to be zero because a zero essential boundary condition was imposed in the EFG analysis. It can be seen from Figure 12.5

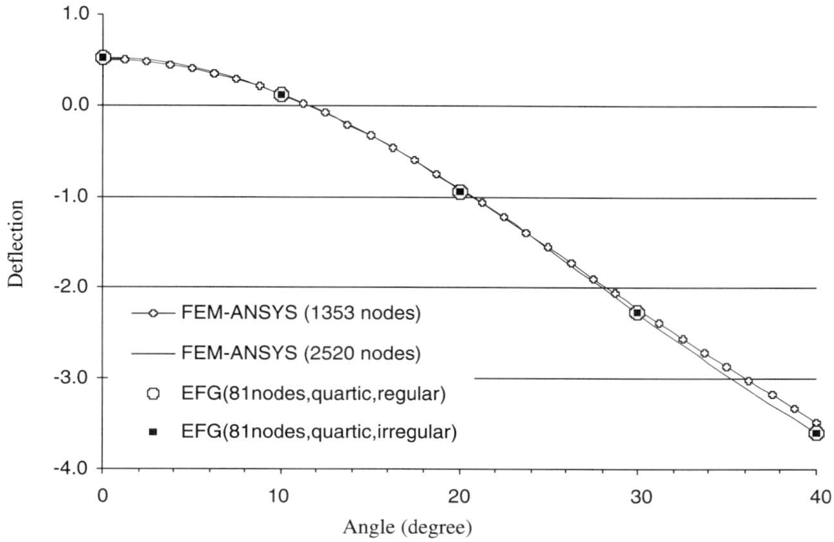

(a)

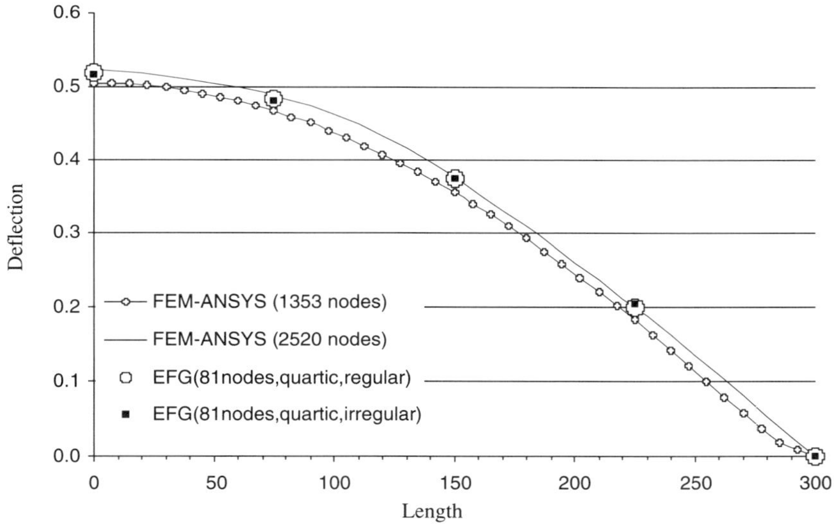

(b)

FIGURE 12.4
(a) Vertical deflection of the barrel vault roof along section AB (see Figure 12.2); (b) vertical deflection along section AC.

that both methods could not give exact results for the zero displacement. Very small errors are observed especially at the two corners. The Lagrange multipliers method is more accurate than that of the penalty method. In the penalty method, a penalty factor of 5000 times the value of the elastic modulus is used. This confirms the discussion given in Chapter 4 on the comparison of the Lagrange multipliers method and penalty method. Note again that the penalty method is much more computationally efficient.

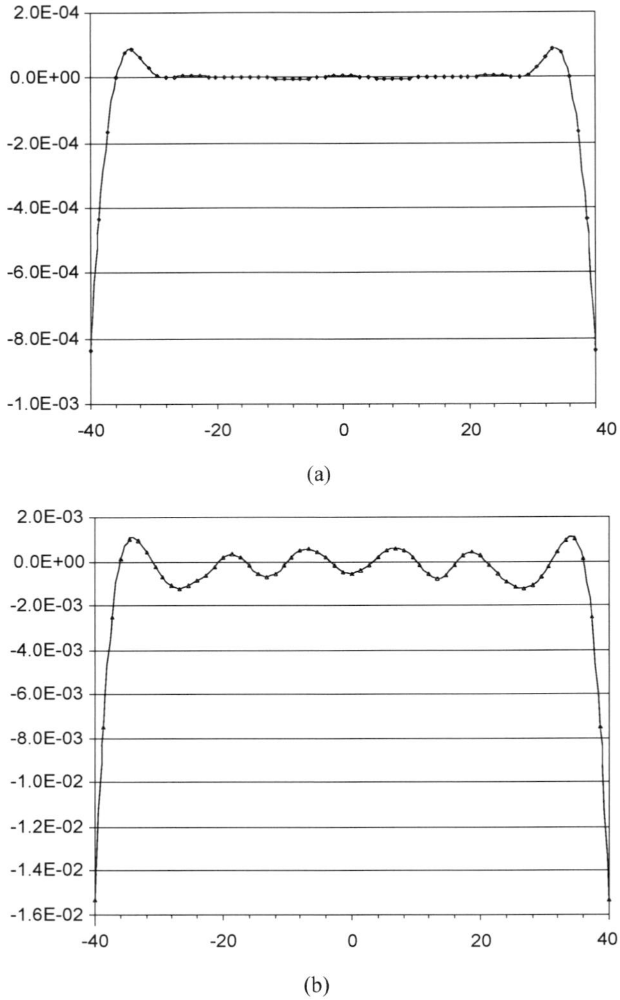

FIGURE 12.5
EFG solution of vertical deflection of the barrel vault roof along section DF. The vertical deflections are supposed to be zero. The numerical solution gives very small values. The largest errors were observed at corners. The Lagrange multipliers method is more accurate than the penalty method. (a) Lagrange multipliers method; (b) penalty method.

12.1.11 Numerical Examples for Free Vibration of Thin Shells

The performance of the EFG method for free-vibration analysis is also evaluated. The results of several examples are presented in comparison with analytical solutions and other results.

Example 12.2 Free Vibration of a Clamped Cylindrical Shell Panel

A panel of cylindrical thin shallow shell, which has been investigated by Petyt (1971), is also examined here to benchmark our EFG code for free-vibration analysis. The thin shell panel is clamped at all edges, and the natural frequencies are computed using the present EFG. The geometry and boundary conditions of the shell panel are shown in Figure 12.7 and the following parameters are used in the analysis.

Length L (mm)	Radius R (mm)	Thickness h (mm)	Span Angle θ	Young's Modulus E (N/m^2)	Poisson's Ratio ν	Mass Density ρ (kg/m^3)
76.2	762	0.33	7.64°	6.8948×10^{10}	0.33	2657.3

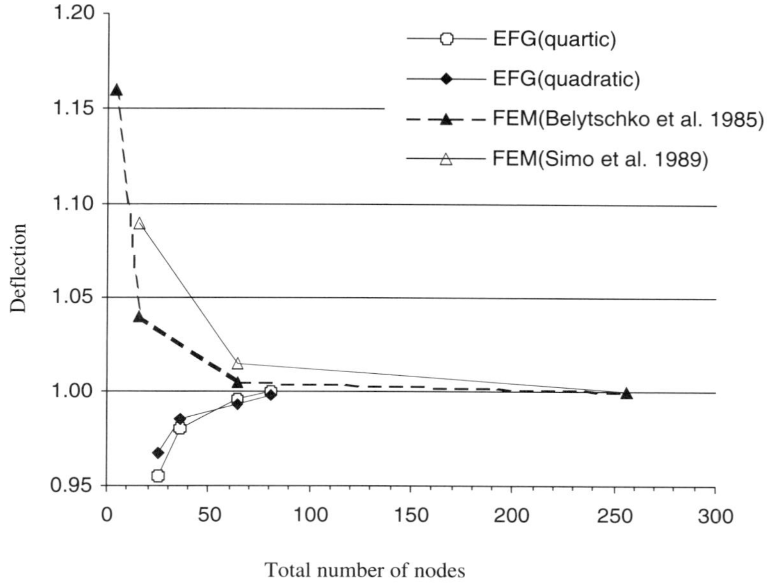

FIGURE 12.6
Convergence of results of vertical deflection at B.

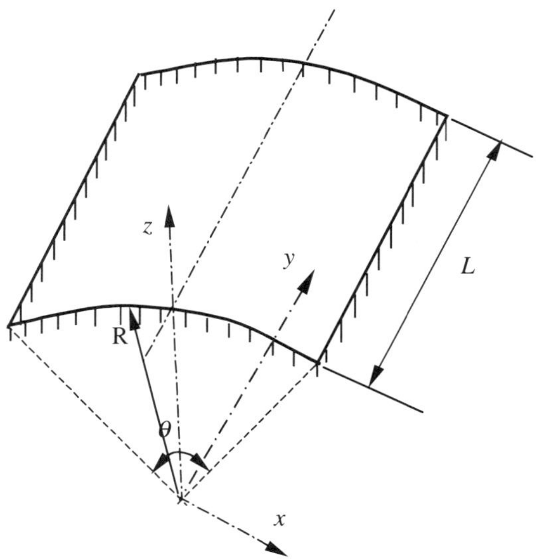

FIGURE 12.7
Clamped panel of cylindrical thin shell.

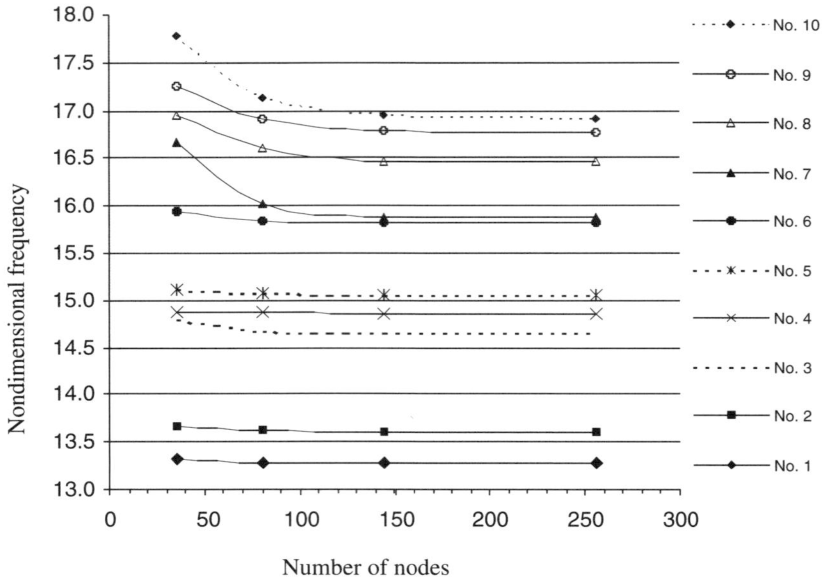

FIGURE 12.8
Convergence of results for natural frequency parameter λ of a clamped panel of a cylindrical shallow thin shell.

TABLE 12.1
Natural Frequency Parameter λ of a Clamped Panel of Cylindrical Thin Shell

Mode	Experimental Results	ERR	FET	FER	Present Method (EFG) 6 × 6	9 × 9	12 × 12	16 × 16
1	13.06	13.28	13.28	13.35	13.32	13.28	13.28	13.28
2	13.54	13.60	13.60	13.65	13.67	13.62	13.60	13.60
3	14.57	14.65	14.65	14.71	14.79	14.68	14.65	14.65
4	14.70	14.86	14.85	14.88	14.88	14.87	14.86	14.85
5	15.09	15.06	15.06	17.21	15.11	15.08	15.06	15.06
6	15.93	15.82	15.83	15.87	15.95	15.85	15.83	15.82
7	15.78/15.86	15.91	15.88	15.96	16.67	16.03	15.89	15.88
8	16.55	16.46	16.46	16.49	16.95	16.61	16.47	16.46
9	16.79	16.78	16.78	16.81	17.25	16.91	16.80	16.78
10	16.89	16.93	16.92	16.96	17.77	17.14	16.95	16.92

ERR: extended Raleigh–Ritz method; FET: triangular finite element method;
FER: rectangular finite element method.

Regular nodes of different densities are used to investigate the convergence characteristics of natural frequencies. Results are given in terms of a frequency parameter defined as $\lambda = \sqrt{\Omega} = (\omega^2 \rho h R^4/D)^{1/8}$, where $D = (Eh^3/(12(1 - \nu^2))$, and are shown in Figure 12.8. As can be seen from Figure 12.8, very high convergence rates for the various vibration modes are obtained. The values of λ are also tabulated in Table 12.1 together with experimental results and results obtained by other, different numerical methods (Petyt, 1971). In this table, ERR denotes the extended Raleigh–Ritz method, FET stands for the triangular finite element method, and FER denotes rectangular finite element method. It is found that the EFG results show good convergence and good agreement with other methods.

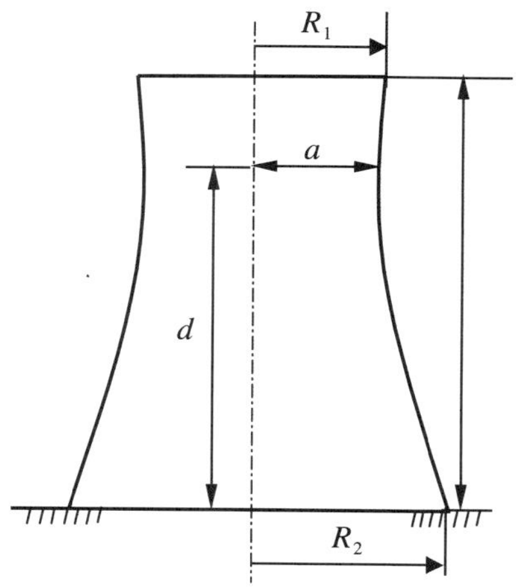

FIGURE 12.9
Geometry of a hyperbolical shell that is clamped at the bottom circular edge and free at the top circular edge.

Example 12.3 Free Vibration of a Hyperbolical Shell

Free-vibration analysis of the clamped-free hyperbolical shell shown in Figure 12.9 is performed. The shell geometry is defined by the following equation of its meridian:

$$\left(\frac{R}{a}\right)^2 - \left(\frac{L-d}{b}\right)^2 = 1 \tag{12.53}$$

where b is a characteristic dimension of the shell, which can be calculated as $b = ad/\sqrt{R_2^2 - a^2}$. The following geometric and material properties are used in the analysis:

Length L (m)	Height h (m)	Height d (m)	Radius a (m)	Radius b (m)	Young's Modulus E (N/m^2)	Poisson's Ratio ν	Mass Density ρ (kg/m^3)
100.787	0.127	82.194	25.603	63.906	2.069×10^{10}	0.15	2405

A regular nodal arrangement in the axial and circumferential direction is used in the analysis. The results of the present study using different numbers of nodes are given in Table 12.2. Results obtained by Carter et al. (1969) using a numerical integration technique and Ozakca and Hinton (1994) using an FEM are also listed in the same table for an easy comparison. Close agreements are again observed.

Example 12.4 Free Vibration of a Cylindrical Shell

The natural frequencies of a cylindrical shell are examined using the present EFG code. The geometry of the shell is schematically drawn in Figure 12.10. The shell is clamped at one edge and free at the other. The following geometry and material properties are used:

Length L (mm)	Thickness h (mm)	Radius R (mm)	Young's Modulus E (N/m^2)	Poisson's Ratio ν	Mass Density ρ (kg/m^3)
226.8	1.021	106.1	2.069×10^{11}	0.3	7868

TABLE 12.2
Natural Frequencies (Hz) of a Clamped-Free Hyperbolical Shell

Mode	Ozakca and Hinton (1994)	Carter et al. (1969)	Present Method (EFG)	
			12× 16	18× 24
1	1.0354	1.0348	1.0351	1.0325
2	1.1508	1.1467	1.1486	1.1450
3	1.1826	1.1808	1.1809	1.1780
4	1.3061	1.3015	1.3043	1.2998
5	1.3293	1.3231	1.3254	1.3223
6	1.3758	1.3749	1.3799	1.3753
7	1.4329	1.4293	1.4284	1.4259
8	1.4488	1.4475	1.4497	1.4470

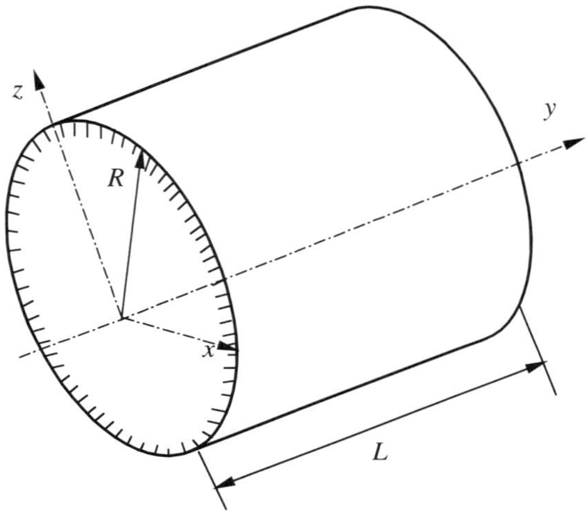

FIGURE 12.10
Geometry of a cylindrical shell.

TABLE 12.3
Natural Frequencies (Hz) of the Clamped-Free Cylinder

Mode	FEM (Subir and Gould, 1974)	HPFEM (Tou and Wong, 1987)	Present Method (FEG)	
			8 × 16	12 × 24
1	487	482	490	483
2	565	561	564	562
3	621	616	629	624
4	NR	NR	875	869
5	982	981	979	980

NR = no results were given.

The nodes used in our EFG are regularly arranged in the axial and circumferential directions. The EFG results are tabulated in Table 12.3. The results obtained by Tou and Wong (1987) and Subir and Gould (1974) using high-precision finite element method (HPFEM) and standard FEM, respectively, are also listed in the same table for comparison. Again, good agreement is evident.

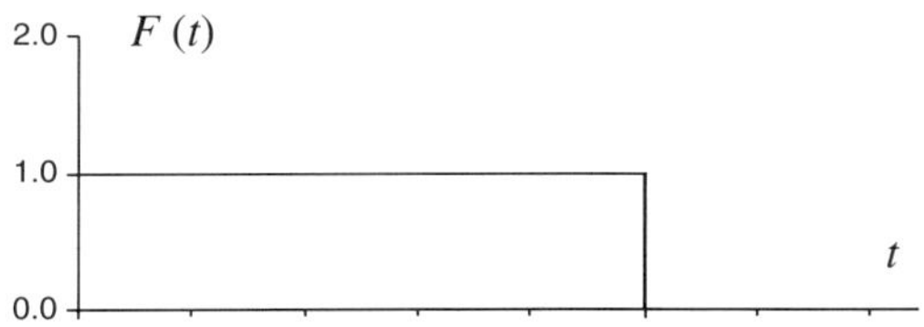

FIGURE 12.11
Transient force of a rectangular pulse applied to the circular plate.

12.1.12 Numerical Examples for Forced Vibration of Thin Shells

To investigate the accuracy as well as the capability of the EFG method for forced vibration problems for thin shells, numerical examples for thin shells subjected to different transient excitations are presented, and the results obtained are compared with those of ordinary FEM and other numerical methods.

Example 12.5 Clamped Circular Plate Subject to an Impulsive Load

A code developed for shells should also be able to work for plates. The first test case is a circular plate subjected to rectangular impulsive force. The parameters used in this calculation are given as follows:

Thickness h (mm)	Radius R (mm)	Young's Modulus E (N/m^2)	Poisson's Ratio ν	Mass Density ρ (kg/m^3)
0.05	1.1	1000.0	0.3	0.229

The natural frequencies are first computed by the EFG method, and the results are compared with those obtained using the boundary element method (BEM) and FEM. We state without showing the results that they are in very good agreement.

The circular plate is loaded by a concentrated vertical impulsive force $F(t)$ at its center. The magnitude of the force is 1 N, and the duration of pulse is $t_0 = 0.121$ s as shown in Figure 12.11. In time stepping, the time increment used is $\Delta t = 0.001$ s. Figure 12.12 shows the history of the vertical deflection at the center of the plate obtained by the presented method, the analytical solution of Sneddon (1945), SAP IV (Bathe et al., 1973), and the domain-boundary element method by Beskos (1990). It is observed that the results obtained using the present EFG code agree well with all these methods including the analytical results by Sneddon (1945).

Example 12.6 Clamped Cylindrical Shell Subject to a Sine Load

The transient response of a clamped cylindrical shell as shown in Figure 12.10 is investigated. The following geometry and material properties are used:

Length L (mm)	Thickness h (mm)	Radius R (mm)	Young's Modulus E (N/m^2)	Poisson's Ratio ν	Mass Density ρ (kg/m^3)
600	3.0	300.0	2.1×10^{11}	0.3	7868

Figure 12.13 shows the history of the transient excitation of half a sine function of time. The excitation is at the center of the meridian ($y = L/2$, $z = R$). The force is expressed as

$$F = F_0 \sin(1000t) \tag{12.54}$$

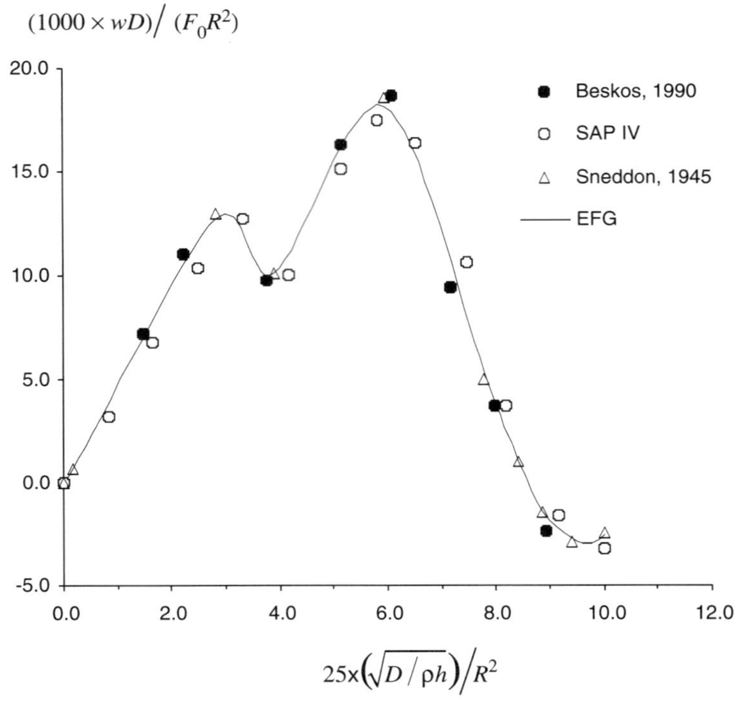

FIGURE 12.12
The time history of the deflection at the center of a clamped circular plate subjected to a rectangular impulsive excitation.

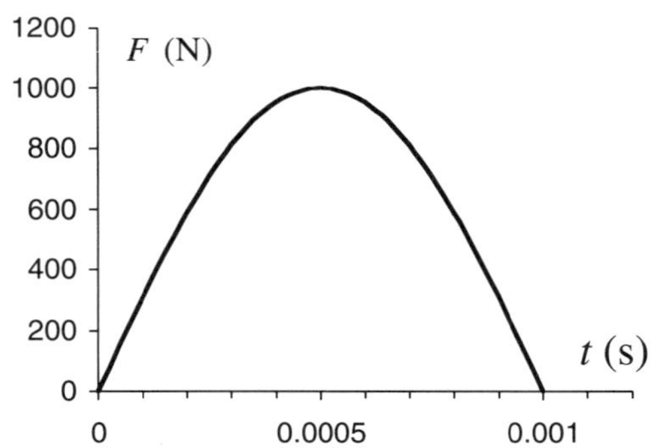

FIGURE 12.13
Time history of the external force excitation of half a sine function of time.

where $F_0 = 1000.0$ N. Good agreement is observed. In performing the time marching, the time step $\Delta t = 2.5e - 5s$ is used, which is almost the 1/35 of the fundamental period of the cylinder. Regular nodes (12×16) are arranged in the axial and circumferential directions. Figure 12.14 shows the transient response of a clamped cylinder subjected to a transient excitation of half a sine function of time. The results are obtained using both the present EFG code and FEM.

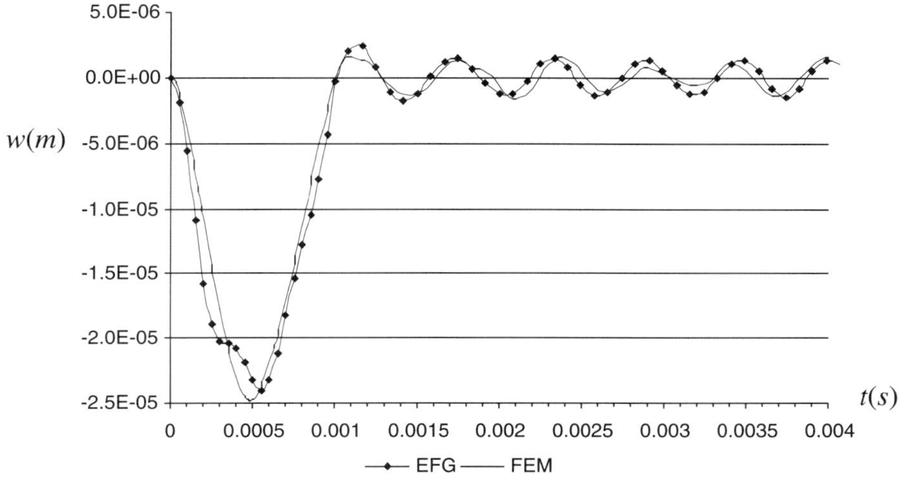

FIGURE 12.14
Transient response of the vertical displacement at the central point in the meridian of the cylindrical shell ($y = L/2$, $z = R$).

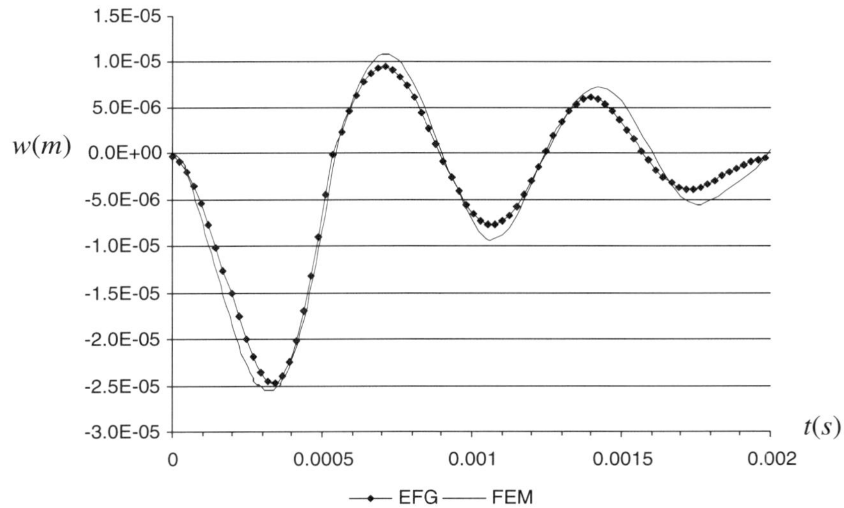

FIGURE 12.15
Same as Figure 12.14, but $F = F_0 \sin(2000t)$.

Figure 12.15 shows the transient response for the cylindrical shell, but the external excitation force becomes $F = F_0 \sin(2000t)$. Again, very good agreement between the results of EFG and FEM is obtained.

Example 12.7 Clamped Spherical Shell Subject to a Sine Curve Load

A spherical cap of thin shell with a central angle of $\theta = 120°$ shown in Figure 12.16 are investigated. The shell is subjected to an excitation of a force of half a circle sine function of time at the apex. The time history of the loading is given by Equation 12.54, but $F_0 = 200.0$ N.

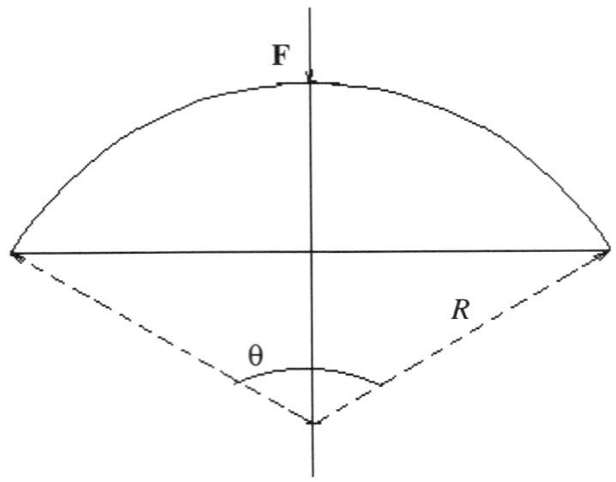

FIGURE 12.16
Cross section of the spherical cap of a thin shell.

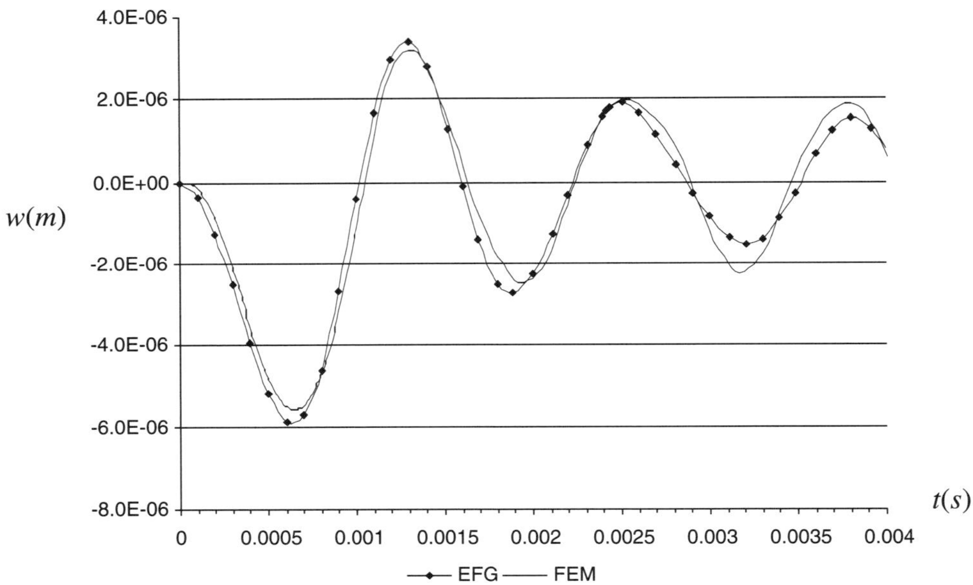

FIGURE 12.17
Transient response of the vertical displacement at the apex of the spherical cap.

This spherical cap is clamped on the circular boundary at the bottom. The following geometry and material properties are used in the computation:

Thickness h (mm)	Radius R (mm)	Young's Modulus E (N/m^2)	Poisson's Ratio ν	Mass Density ρ (kg/m^3)
10	900.0	$2.0e^{11}$	0.3	7800

The time step used is $\Delta t = 5.0e - 5s$, which is almost the 1/25 of the fundamental period of the spherical cap; 185 nodes are used in the calculation. Figure 12.17 shows the history of the vertical displacement response at the apex of the spherical cap obtained by both the present EFG method and FEM. Very good agreement is obtained.

12.1.13 Remarks

The EFG method has been formulated for static, free-vibration, and forced vibration analysis of spatial thin shell structures. In the EFG method, the MLS technique is used for approximations of both the surface geometry of the shell and the field variables.

The present EFG results are benchmarked with a number of examples by comparison with the results obtained by other methods. We have more than enough evidence to show that EFG works for shells.

The present EFG method offers distinct computational advantages over classical FEM: (1) mesh free, (2) accurate geometry representation, (3) fast convergence.

12.2 EFG Method for Thick Shells

The Kirchhoff–Love shell theory works only for very thin shells. The reason is very much similar to the reason we argued for the theories for plates (see Chapter 11). For thick shells, formulations need to be based on thick shell theories. The formulation for thick shells in this section is based on the geometrically exact theory of flexible shells proposed by Simo et al. (1989). We collect the necessary equations in the following, but refer the reader to the articles by Simo et al. (1989) for details.

12.2.1 Fundamental Relations

The Gauss intrinsic coordinates are used to describe the configuration of the shell as described in previous sections. Making use of the definition of spatial tensors, the corresponding linearized strain measures are defined relative to the dual spatial surface basis as

$$\varepsilon_{\alpha\beta} = \frac{1}{2}(\boldsymbol{\varphi}^0_{,\alpha} \cdot \mathbf{u}_{,\beta} + \boldsymbol{\varphi}^0_{,\beta} \cdot \mathbf{u}_{,\alpha}) \tag{12.55}$$

$$\gamma_\alpha = (\boldsymbol{\varphi}^0_{,\alpha} \cdot \Delta\mathbf{t} + \mathbf{u}_{,\alpha} \cdot \mathbf{t}^0) \tag{12.56}$$

$$\rho_{\alpha\beta} = \frac{1}{2}(\boldsymbol{\varphi}^0_{,\alpha} \cdot \Delta\mathbf{t}_{,\beta} + \boldsymbol{\varphi}^0_{,\beta} \cdot \Delta\mathbf{t}_{,\alpha} + \mathbf{u}_{,\alpha} \cdot \mathbf{t}^0_{,\beta} + \mathbf{u}_{,\beta} \cdot \mathbf{t}^0_{,\alpha}) \tag{12.57}$$

Here, $\Delta\mathbf{t}$ is the incremental spatial rotation defined by

$$\Delta\mathbf{t} = \bar{\boldsymbol{\Lambda}} \cdot \Delta T \tag{12.58}$$

where

$$\Delta T = \Delta T^1 \mathbf{E}_1 + \Delta T^2 \mathbf{E}_2 \tag{12.59}$$

and $\bar{\boldsymbol{\Lambda}}$ is the (3×2) matrix given by

$$\bar{\boldsymbol{\Lambda}} = [t_1 \ t_2] = \begin{bmatrix} \Lambda_{11} & \Lambda_{12} \\ \Lambda_{21} & \Lambda_{22} \\ \Lambda_{31} & \Lambda_{32} \end{bmatrix} \tag{12.60}$$

which can be obtained by deleting the third column of $\mathbf{\Lambda}$, where $\mathbf{\Lambda}: \mathbb{A} \subset \mathbb{R}^2 \rightarrow \mathrm{S}_E^2$ is the orthogonal transformation, and can be expressed as

$$\mathbf{\Lambda} = (\mathbf{t} \cdot \mathbf{E})\mathbf{1} + [\widehat{\mathbf{E} \times \mathbf{t}}] + \frac{1}{1 + \mathbf{t} \cdot \mathbf{E}}(\mathbf{E} \times \mathbf{t}) \otimes (\mathbf{E} \times \mathbf{t}) \tag{12.61}$$

The constitutive relations for the effective membrane $\tilde{\mathbf{n}}$ and for the stress couple resultant $\tilde{\mathbf{m}}$ have the form of Equations 12.23 and 12.24. The constitutive relations for the effective shear stress resultants $\tilde{\mathbf{q}}$ can be written as

$$\begin{Bmatrix} \tilde{q}^1 \\ \tilde{q}^2 \end{Bmatrix} = \kappa G h \begin{Bmatrix} \gamma^1 \\ \gamma^2 \end{Bmatrix} \tag{12.62}$$

Here κ is the shear reduction coefficient and G is the shear modulus.

In calculating the effective stress couple resultant $\tilde{\mathbf{m}}$, Equations 12.24 and 12.31 should be used. In calculating the effective membrane $\tilde{\mathbf{n}}$, Equation 12.30 and the following equation should be used.

$$\tilde{n}^{\beta\alpha} = n^{\beta\alpha} - \lambda_\mu^\beta \tilde{m}^{\alpha\mu} \tag{12.63}$$

In calculating the effective shear resultant forces, we use

$$\tilde{\mathbf{q}} = \tilde{q}^\alpha \varphi_{,\alpha} \tag{12.64}$$

where $\tilde{q}^\alpha$ is defined as

$$\tilde{q}^\alpha = q^\alpha - \lambda_\mu^3 \tilde{m}^{\alpha\mu} \tag{12.65}$$

In calculating $\mathbf{t}_{,\alpha}$, we use

$$\mathbf{t}_{,\alpha} = \lambda_\alpha^\beta \varphi_{,\beta} + \lambda_\alpha^3 \mathbf{t} \tag{12.66}$$

The above equation shows that the difference between thick shell theory and thin shell theory is that the shear effects have been taken into account in the thick shell theory.

12.2.2 Principle of Virtual Work

The weak form of the governing equation for the thick shells under static load can be written as

$$W_{\mathrm{Sta}}(\delta\mathbf{x}) = \int_{\mathbb{A}} [\tilde{n}^{\beta\alpha} \cdot \delta\varepsilon_{\beta\alpha} + \tilde{m}^{\beta\alpha} \cdot \delta\kappa_{\beta\alpha} + \tilde{q}^\alpha \delta\gamma_\alpha] d\mathbb{A} - W_{\mathrm{ext}}(\delta\mathbf{x}) \tag{12.67}$$

Here $d\mathbb{A} = \bar{j}\, d\xi^1 d\xi^2$ is the current surface measure. W_{ext} is the virtual work of the external loading given by

$$W_{\mathrm{ext}} = \int_{\mathbb{A}} [\bar{\mathbf{n}} \cdot \delta\boldsymbol{\varphi} + \bar{\mathbf{m}} \cdot \delta\mathbf{t}] d\mathbb{A} + \int_{\Gamma_n} \bar{\bar{\mathbf{n}}} \cdot \delta\boldsymbol{\varphi}\, \bar{j}\, d\Gamma + \int_{\Gamma_m} \bar{\bar{\mathbf{m}}} \delta\mathbf{t}\, \bar{j}\, d\Gamma \tag{12.68}$$

where $\bar{\mathbf{n}}$ is the applied resultant force per unit length and $\bar{\mathbf{m}}$ is the applied direct couple per unit length. $\bar{\bar{\mathbf{n}}}$ and $\bar{\bar{\mathbf{m}}}$ are the prescribed resultant force and the prescribed director couple on their corresponding boundaries Γ_n and Γ_m, respectively.

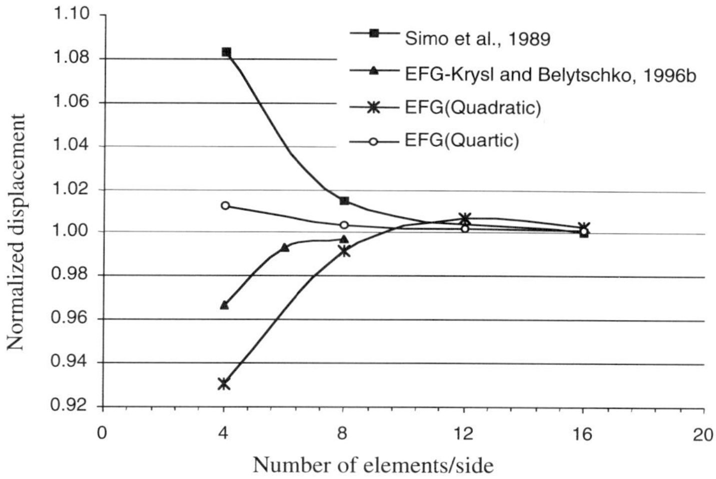

FIGURE 12.18
Convergence of vertical displacement at point B in the barrel vault roof.

The MLS shape functions are then used to approximate the displacement fields. Substituting Equation 12.1 into the weak forms leads to a set of global system equations. The procedure is very much the same as that in the previous section.

The treatment of essential boundaries is also very much the same as that in the previous section. We need to use different ways to handle the essential boundary condition for different problems, whenever MLS approximation is employed.

12.2.3 Numerical Examples

Example 12.8 Static Deflection of a Barrel Vault Roof under Gravity Force

Example 12.1 is reexamined using the EFG formulation for thick shells. The barrel vault roof is shown in Figure 12.2. All the parameters and conditions are exactly the same as those in Example 12.1. In constructing the MLS shape function, both quadratic and quartic basis functions are used. Figure 12.18 shows the results of convergence of vertical displacement at point B in the barrel vault roof. Figure 12.18 shows that the convergence by the present analysis is excellent in comparison with the results of using high-performance FEM by Simo et al. (1989) and EFG for thin shells by Krysl and Belytschko (1996b). It can also be seen that the convergent rate for quartic basis always exceeds that for quadratic basis. This was not very clear from the results using thin shell theory (see Figure 12.6). The result for quartic basis functions approaches the exact result from below with the increase of EFG nodes, while that for quadratic basis functions oscillates along the exact result with the increase of EFG nodes.

Figure 12.19 shows the variation of fractions of shear, membrane, and bend energies in total energy, with respect to the size of the support domain. The computation is performed using quartic basis function. It can be seen that the membrane and bending energies are stable with the increase of the support domain; the difference between the membrane and bending energies are no more than 4.0%. The shear energy is very small and approaches zero with the increase of EFG nodes, which means that the shear stress plays a very minor role in deformation of the vault roof.

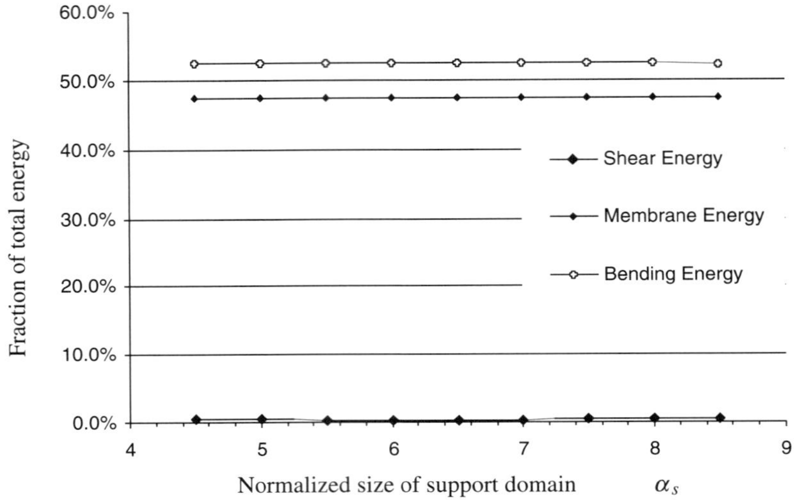

FIGURE 12.19
Variation of membrane, shear, and bending energy with respect to size of the support domain.

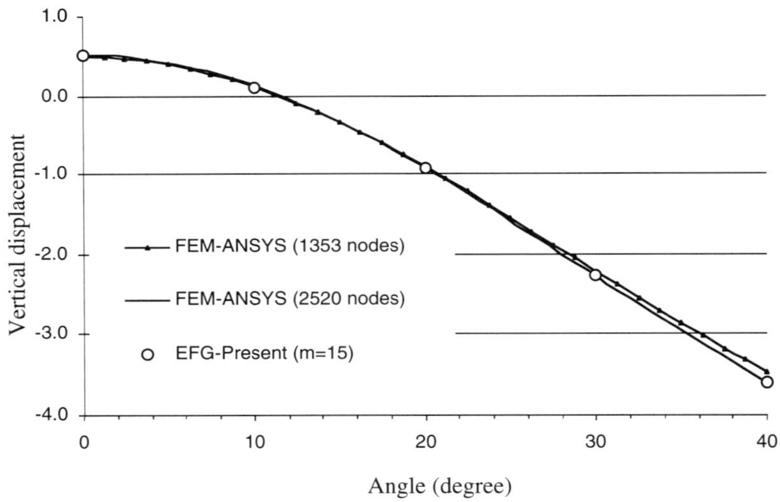

FIGURE 12.20
Vertical displacement in the barrel vault roof along AB.

Figures 12.20 and 12.21 show the distribution of the vertical deflection in the roof along the edges AB and AC, respectively. The central point A moves upward, that is, in the reverse direction of the body force, for the effects of the membrane forces. It can be seen that thousands of nodes are needed to obtain the converged solution using a general-purpose program, e.g., ANSYS, for the reason mentioned in Example 12.1. Many fewer nodes are needed for the present EFG method.

Example 12.9 Pinched Cylindrical Shell

The deformation of the cylindrical shell shown in Figure 12.22 is analyzed using the present EFG method. The shell is loaded by a pair of centrally located and diametrically opposed concentrated forces. The cylindrical boundaries are supported by a rigid diaphragm

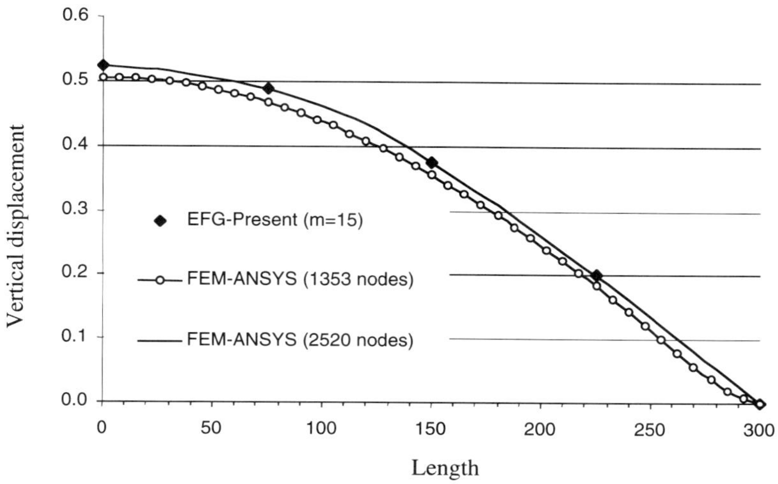

FIGURE 12.21
Vertical displacement in the barrel vault roof along AC.

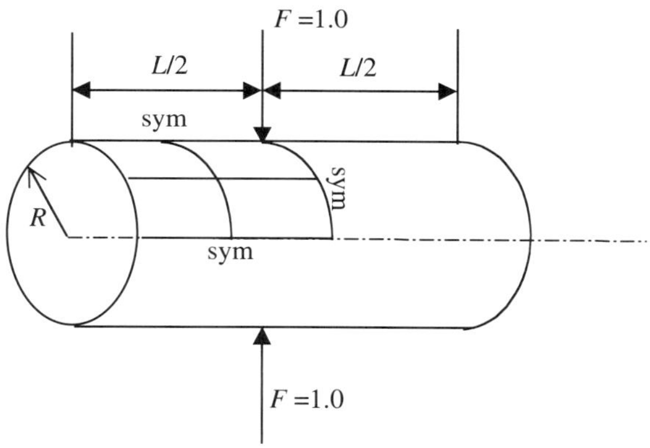

FIGURE 12.22
Pinched cylindrical shell.

that allows displacement in the axial direction and rotation about the tangent to the shell boundary. The parameters for this problem are given in the following table:

Length L (mm)	Thickness h (mm)	Radius R (mm)	Young's Modulus E (N/m^2)	Poisson's Ratio ν
600	3.0	300.0	3.0×10^6	0.3

This problem is one of the most critical tests for numerical methods for both inextensional bending and complex membrane states of stress. Because of its double symmetry, only 1/8 of the cylinder has been modeled. The results are shown in Figure 12.23, which are normalized by the value of 1.8248e-5, which is the convergent numerical solution of magnitude of the radial displacement at loaded points. The convergence by the present method is excellent in comparison with the results by FEM (Simo et al., 1989) and by EFG for thin

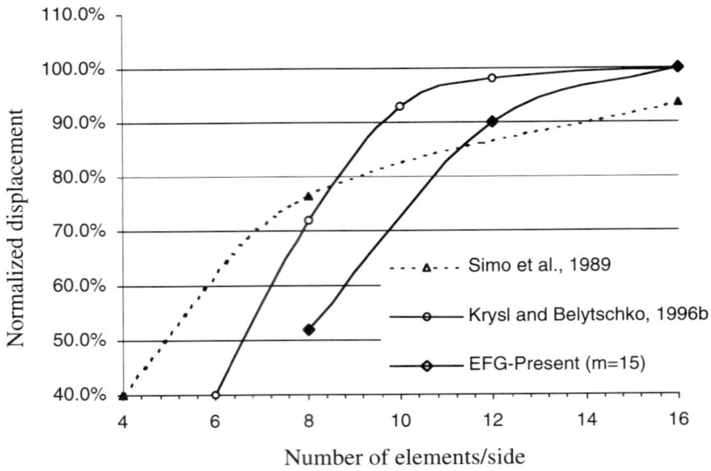

FIGURE 12.23
Convergence of the results of the radial displacement at the center of the cylindrical shell.

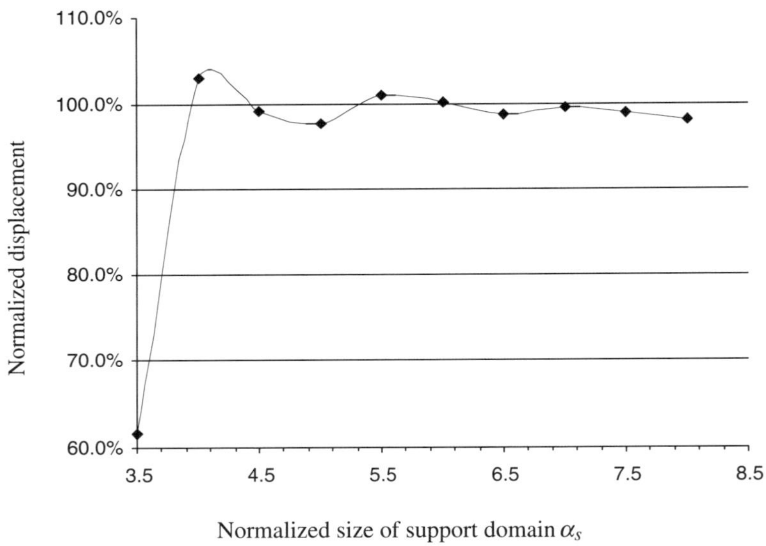

FIGURE 12.24
Effects of the dimension of the support domain on the radial displacement at the loading point.

shells (Krysl and Belytschko 1996b). It is seen that the EFG results reported by Krysl and Belytschko (1996b) converge faster than the present EFG code.

The effects of the dimension of the support domain on the radial displacement at the loading point are also investigated, and the results are shown in Figure 12.24. The shortest nodal distance between nodes is used to normalize the size of the support domain. The fluctuation of the displacement is visible but it is less than 3% when the support domain is larger than 3.8. Effects of the dimension of the support domain on the energy components (bending, membrane, and shear energies) are also investigated, and the results are summarized in Figure 12.25. The energies show a trend of fluctuation similar to that of the displacement. Figure 12.26 shows the convergence of membrane (M), shear (S), and

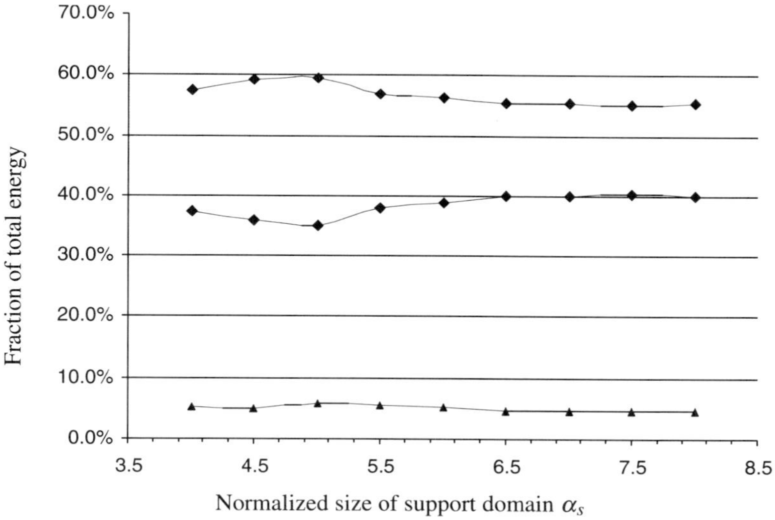

FIGURE 12.25
Effects of the dimension of the support domain on the energy components (from top to bottom: bending, membrane, and shear energy).

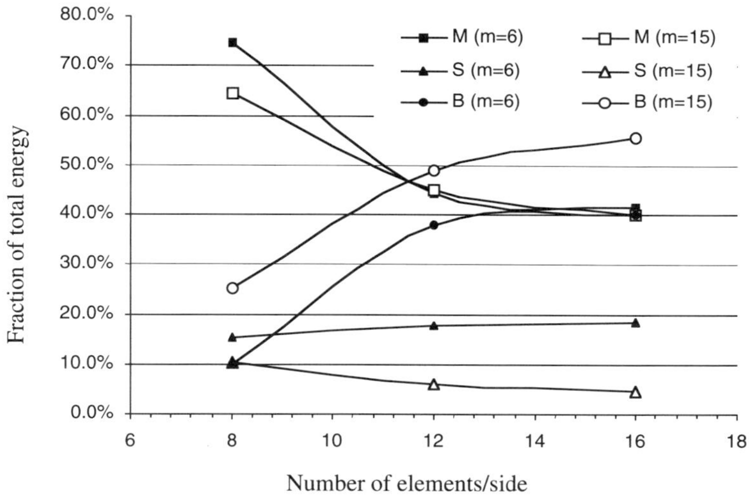

FIGURE 12.26
Convergence of membrane (M), shear (S), and bending (B) energies.

bending (B) energy fractions in the total energy. The results are calculated using both quartic and quadratic polynomial basis functions. As can be seen, for quartic polynomial basis, the shear energy converges to a value of less than 4.0% as the field nodes increase, which means shear stress plays but a small role in the deformation. The membrane energy fraction approaches approximately 40.0% which gives very satisfactory results in comparison with the results using nine-noded gamma FEM by Belytschko et al. (1985). For the quadratic polynomial basis, the membrane fraction approaches 40.0%, same as that for the quartic polynomial. However, the shear energy increases with the increase of field nodes. It can also be observed that for coarse field nodes, membrane locking is dominant while for the fine field nodes, shear locking becomes more important.

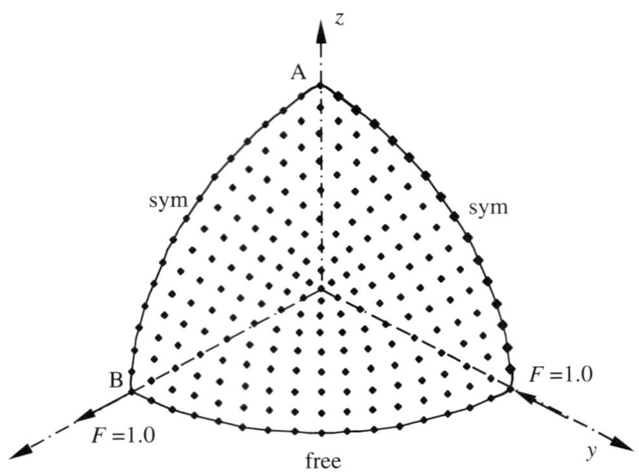

FIGURE 12.27
Pinched hemisphere shell. One quarter of the shell is shown.

Example 12.10 Pinched Hemispherical Shell

In this example, a pinched hemispherical shell shown in Figure 12.27 is analyzed using the EFG method. The shell is pinched by two pairs of opposed radial point loads of magnitude $P = 2.0$. The bottom circumferential edge of the hemisphere is free. The material parameters are $E = 6.825 \times 10^7$ N/m^2 and $\nu = 0.3$. The sphere radius is $R = 10.0$ and the thickness $t = 0.04$.

This problem is also a challenging benchmarking problem to check whether the formulation of the shell can represent inextensional modes of deformation, as it exhibits almost no membrane strains. This problem is a less critical test with regard to inextensional bending, compared to the pinched cylinder problem. However, it is a very useful problem for checking the ability of the present EFG method to handle rigid body rotations normal to the shell surface. Large sections of this shell rotate almost as rigid bodies in response to this load, and the ability to accurately model rigid body motion is essential for any numerical method.

Due to symmetry, only a quarter of the hemisphere is modeled. The results are shown in Figure 12.28. The radial displacements in the direction of loads at loaded points are the same and are found analytically to be 0.0924. This value is used to normalize the numerical results presented in Figure 12.28. The convergence is even better than that of the high-performance results by Simo et al. (1989) using mixed formulation. Among the EFG solutions, the present method also gives better accuracy than that of the thin shell formulation by Krysl and Belytschko (1996b). The results obtained using the quartic polynomial basis functions are better than those using quadratic polynomial basis functions. However, poor accuracy is obtained when very coarse nodes are used in EFG compared with FEM using the same density of nodes. Figure 12.29 shows the effects of the dimension of the support domain on the radial displacement at the loading point on the hemispherical shell. The present EFG results using quartic basis functions do not depend on the size of the support domain and gives an almost exact solution, in the range of our investigation. The results obtained using quadratic basis functions are stable and accurate only in a narrow range (α_s = 4.5 to 6.5). Figure 12.30 shows the effects of the dimension of the support domain on membrane (M) and shear (S) energies in the hemisphere shell. Figure 12.31 shows the same effects but for the bending (B) energy in the hemisphere shell.

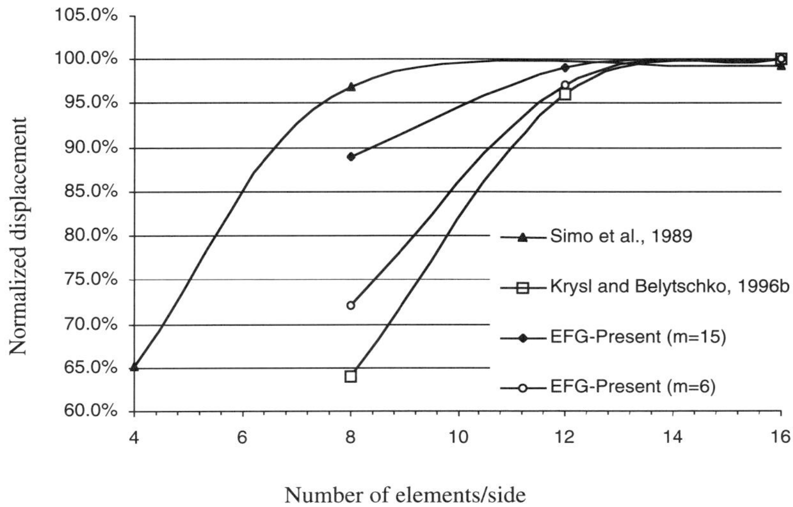

FIGURE 12.28
Convergence the radial displacement at the loading point on the hemisphere shell.

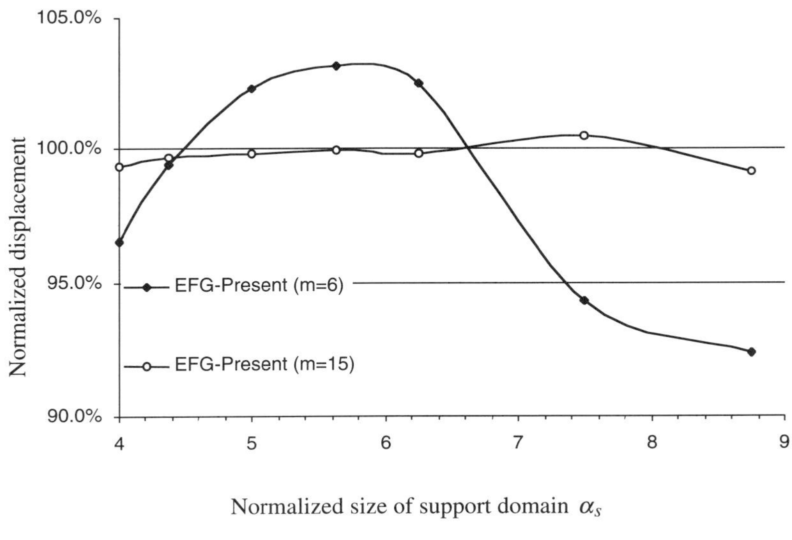

FIGURE 12.29
Effects of the dimension of the support domain on the radial displacement at the loading point on the hemisphere shell.

It can be seen that both the shear and the membrane energy are very small when the quartic basis functions are used. Although the shear energy obtained using the quadratic basis functions is very small, the membrane energy increases with the increase of the support domain. This means that the membrane energy is overpredicted with the increase of field nodes, which is an indication of membrane locking.

Figures 12.32 and 12.33 show the fractions of shear, membrane, and bending energies in the total strain energy calculated using both quartic and quadratic polynomial basis functions.

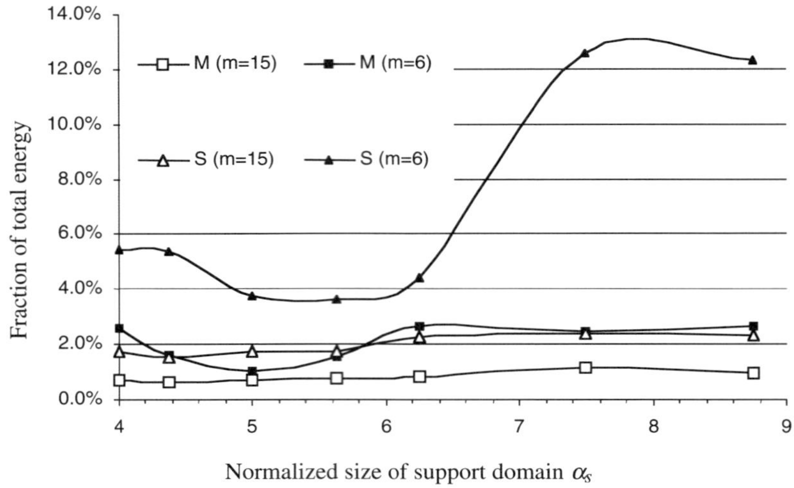

FIGURE 12.30
Effects of the dimension of the support domain on membrane (M) and shear (S) energies in the hemisphere shell.

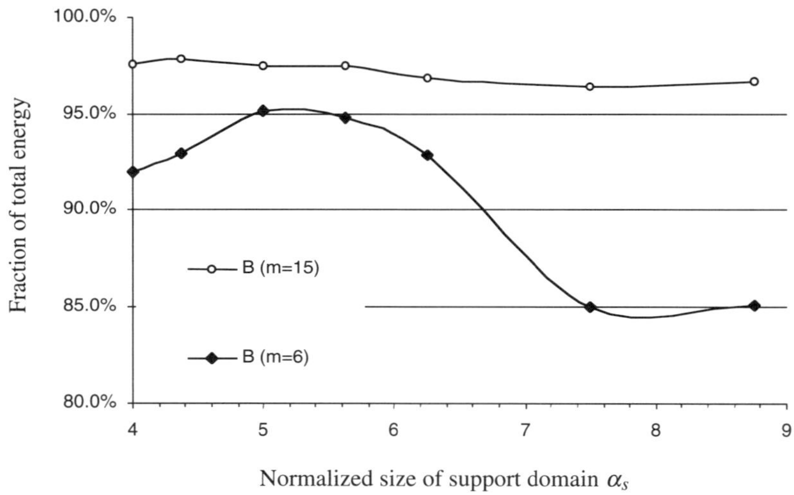

FIGURE 12.31
Effects of the dimension of the support domain on bending (B) energy in the hemisphere shell.

It is found that when quartic polynomial basis functions are used, both membrane and shear energy tend to zero as the field nodes increase. This implies that the shear stress and membrane stress play trivial roles in the deformation of the shell, and the bending energy is dominant. It is also found that membrane locking is dominant for coarse field nodes, but is quickly eliminated with the increase of field nodes.

12.2.4 Remarks

The EFG method has been extended to analyze thick spatial shells based on the stress-resultant shell theory, which has five degrees of freedom assigned to every point of the shell.

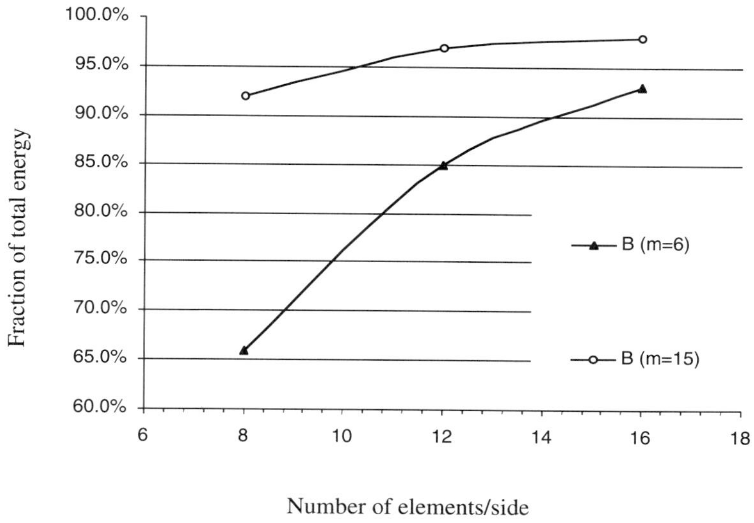

FIGURE 12.32
Convergence of bending energy.

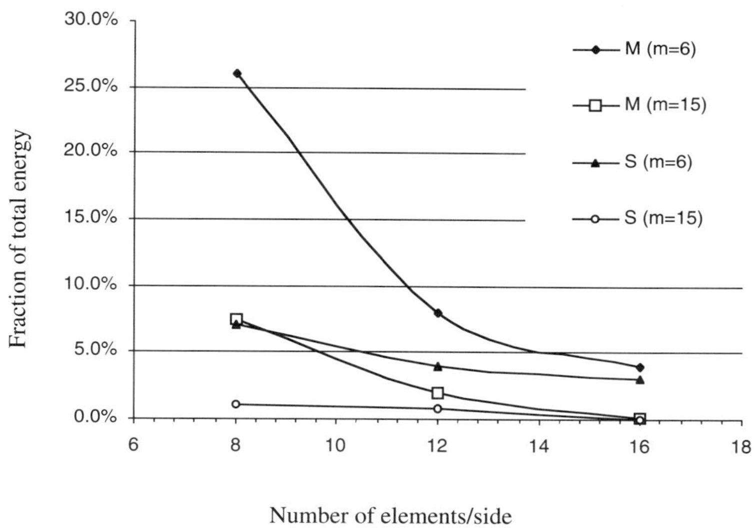

FIGURE 12.33
Convergence of membrane and shear energies.

For thick shells, the formulation allows transverse shear strain and results in the Mindlin plate when the surface becomes flat and the membrane state is negligible. To avoid shear locking and membrane locking, quartic basis function is recommended for MLS approximation. Using high-order basis functions is an effective way to avoid shear locking. A good alternative is to use different orders of shape function for deflection and rotation, as discussed in Chapter 11 for plates (see Example 11.16).

12.3 RPIM for Thick Shells

12.3.1 Formulation Procedure

In the previous two sections, we presented EFG formulations for both thin and thick shells, using MLS approximations. We have shown that special treatment is needed in dealing with the essential boundary conditions, because of the lack of the Kronecker delta function property in the MLS shape functions. The natural progress of our research is then to use PIM shape functions, as we have already seen the advantages of PIM shape functions in Chapters 8 to 11 for solids, fluids, beams, and plates. This section therefore presents RPIM methods that use Galerkin formulation and RPIM shape functions for thick shells. RPIM is then used for static analysis of several benchmark problems to illustrate the validity and efficiency.

Formulation of RPIM was presented in detail in Chapter 5, and the formulation of shells has been detailed in the previous two sections of this chapter. In formulating RPIM for shells, all we need do is replace the MLS shape functions with RPIM shape functions, and ignore all the terms related to essential boundary conditions. Therefore, we need not repeat the process, and go directly to examination of the performance of the RPIM code we have developed. Note that in the following examples nonconforming RPIM or NRPIM is used in the computation. For results of CPIM, readers are referred to the recent work by G.R. Liu et al. (2002c) where excellent convergence of CPIM is reported.

12.3.2 Numerical Examples

Example 12.11 Barrel Vault Roof

Example 12.1 is reexamined using the RPIM formulation for thick shells. The barrel vault roof is shown in Figure 12.2. All the parameters and conditions are exactly the same as those in Example 12.1. The background of this benchmark problem is also mentioned in Example 12.1. This example is used here for investigating the effects of shape parameters in the radial basis functions. The shell roof is loaded by its own weight of uniform vertical gravity load and is supported by diaphragms at the ends, but is free along the sides. Because of symmetry, only one quarter of panel is modeled by the present RPIM method. In presenting the results, we use the reference solution of –3.618 for the vertical deflection at point A to normalize the results calculated using different methods.

Both multiquartic (MQ) radial functions and Gaussian (EXP) radial basis functions are used to construct the RPIM shape function (see Chapter 5). In using radial basis functions, one needs to examine and fine-tune the shape parameters. We therefore use regularly distributed nodes to investigate the shape parameters of these radial basis functions, as defined in Table 5.3.

Effect of Shape Parameters of MQ Radial Basis Functions

When $q = 0.5$ and $q = -0.5$ (see Table 5.3), the MQ radial function becomes the original MQ and the reciprocal MQ (RMQ), which have been traditionally used for many applications. The characteristic length d_c is taken to be the "average" nodal spacing for nodes in the support domain of the quadrature point (see Section 2.10.3).

We now start to investigate the effects of α_C. In our first study, we let $\alpha_C = \alpha_s$, where α_s is the normalized dimension of the support domain. Figure 12.34 shows the effects of the size of the support domain on the results of vertical displacement at the central top of the roof.

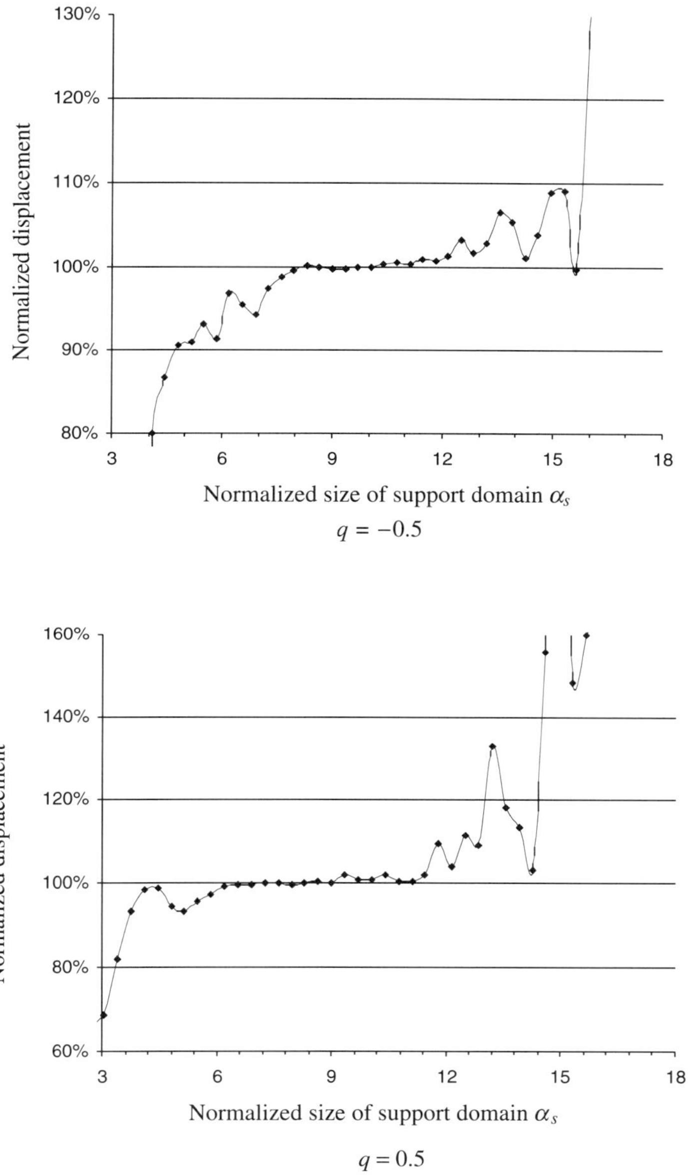

FIGURE 12.34
Effects of the size of the support domain on the results of vertical displacement at the central top of the roof. MQ radial basis function is used with the parameter C the dimension of the support domain ($\alpha_C = \alpha_s$).

Figure 12.35 shows the effects of the size of the support domain on the fractions of shear, membrane, and bending energies in the roof. The membrane-locking phenomenon exists when too small a domain is utilized for approximation. It can be seen that the membrane and bending energies become stable with the increase of the support domain; the difference between the membrane and bending energies is no more than 4.0%. The shear energy is very small and approaches zero with the increase of the support domain, which means that the shear stress plays an almost negligible role in the deformation of the roof. The fractions of membrane and bending energies become oscillating when the normalized

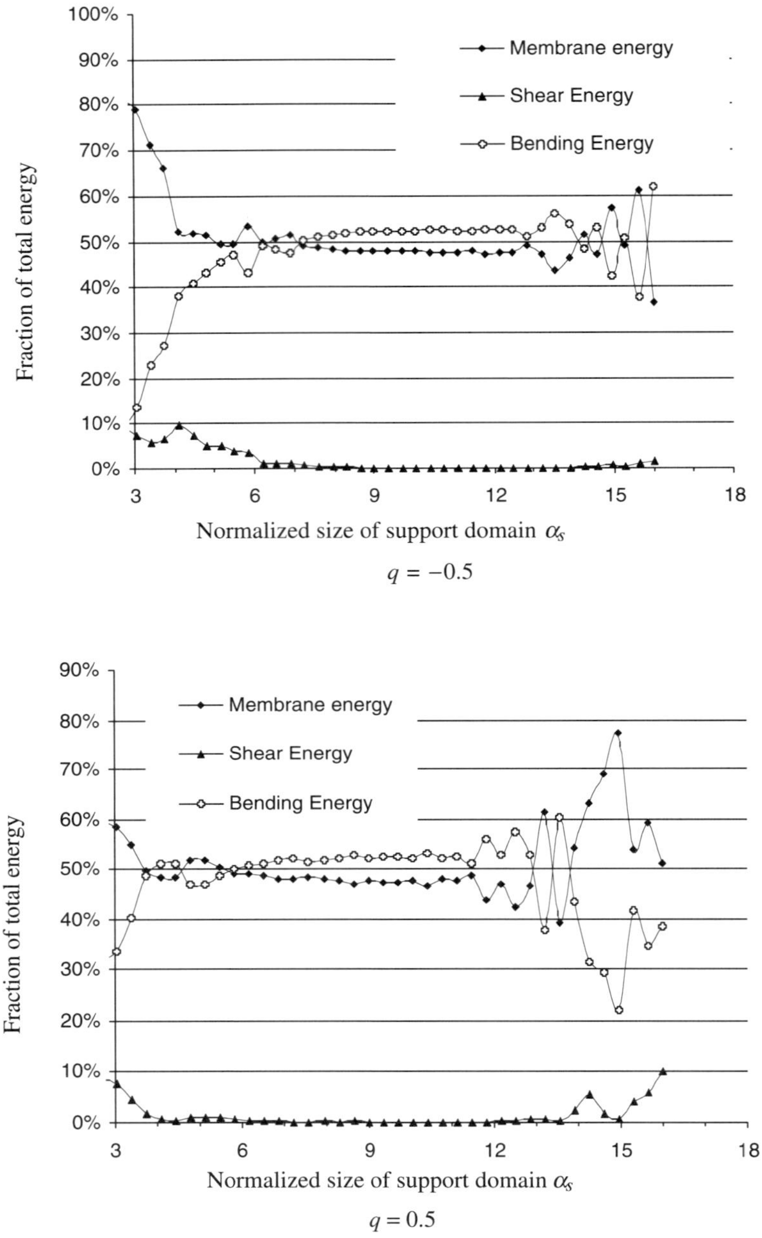

FIGURE 12.35
Effects of the size of the support domain on the fractions of energies in the roof. MQ radial basis function is used with the parameter C the dimension of the support domain ($\alpha_C = \alpha_s$).

support domain is greater than 12.0. Correspondingly, the error displacement of the center point also becomes large and the results are erroneous. Based on the results shown in Figures 12.34 and 12.35, we state that when the shape parameter α_C is chosen the same as the dimension of the support domain, i.e., $\alpha_C = \alpha_s$, the recommended dimension of the support domain for the thin shell is approximately 8.3 to 11.4 and 6.2 to 11.2 times the smallest nodal spacing for $q = -0.5$ and $q = 0.5$, respectively.

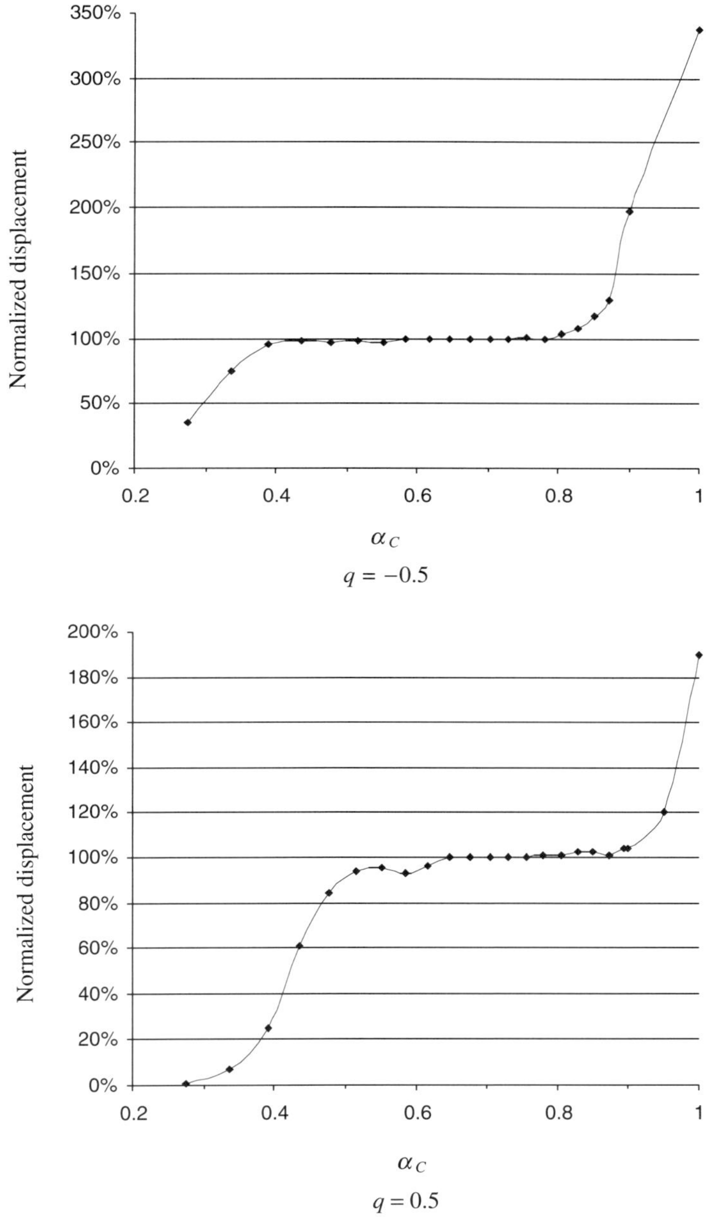

FIGURE 12.36
Effects of the shape parameter α_C of MQ on the displacement results.

Next, we fix the support domain at $\alpha_s = 10.0$, and vary the dimensionless shape parameter α_C. Figure 12.36 shows the effects of the dimensionless shape parameter α_C on the displacement results. This figure shows that to obtain good results, the range of parameter α_C should be 0.65 to 0.85 and 0.45 to 0.78 for $q = -0.5$ and $q = 0.5$, respectively. For convenience, $\alpha_C = 0.7$ is used in the following studies.

Then, we fix $\alpha_C = 0.4, 0.6, 0.8$, and 1.0, and vary α_s, the dimension of the support domain. Figure 12.37 shows the effects of the shape parameter α_C on the displacement results for

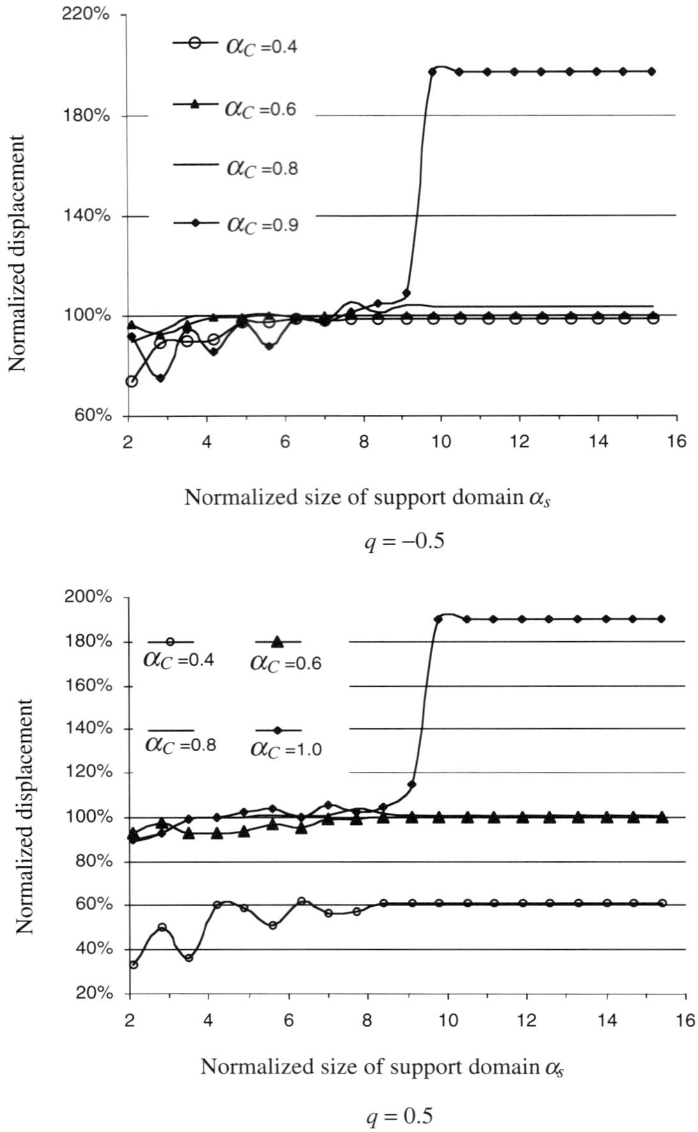

FIGURE 12.37
Effects of the shape parameter C of MQ on the displacement results for different dimensions of support domains used in RPIM.

different dimensions of support domains used in RPIM. We found that the best value of α_C is between 0.6 and 0.8, which supports the finding of 0.7 from Figure 12.36. For an α_C between 0.6 and 0.8, the support domain α_s should be approximately 8.4 to 12.0 and 6.3 to 12.0 for $q = -0.5$ and $q = 0.5$, respectively.

Then the variation of displacement with respect to parameter q in the MQ radial basis function is shown in Figure 12.38. When $q = 0$ and $q = 1.0$, the moment matrix may be singular and the construction of the RPIM shape function may fail, and therefore they should not be considered.

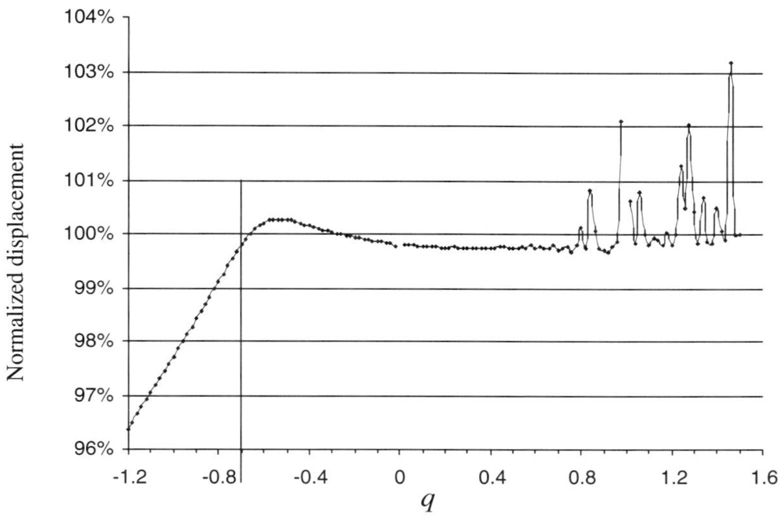

FIGURE 12.38
Variation of displacement with respect to parameter q in the MQ radial basis function.

Based on the above analyses, the following remarks can be made.

- When $q < -0.7$, the results increase smoothly with the increase of q and approach the accurate solution. The error in this range is too large, and should not be considered.
- When $-0.7 \le q < 0$, the results decrease slowly and smoothly, and the error in the results is no more than 0.5%. A q in this range can be used. The best results are observed at $q \cong -0.38$.
- When $0 < q \le 0.7$, the results slightly oscillate, but the results are also very good. The errors are less than 0.4%. A q in this range can also be used.
- When $0.7 \le q < 1.0$ and $1.0 < q < 1.5$, the results are unstable, and oscillate violently. A q in this range should not be considered.

In the previous chapters, a number of results from different studies for different problems have suggested using $q = 0.98$ and 1.03. These two values do not fail in the above-recommended stable range, and we find they perform well for shells. It would be interesting to try out $q = -0.38$ for other problems.

Effect Shape Parameter of EXP Radial Function

There is only one parameter c in the EXP radial basis function, as shown in Table 5.2. The variation of the displacement at the center point with respect to the size of the support domain is shown in Figure 12.39. The results are computed using RPIM-EXP with different shape parameters, c. It can be seen that the results decrease and approach the accurate solution with the increase of c in the range 0.003 to 0.03 and the result is almost equal to the convergent solution when $c = 0.003$. The results are very stable and accurate when parameter c is 0.03 to 0.07. The error increases with the increase of the parameter c. When $c = 0.12$, the error in the result is more than 12%, which is unacceptable.

It is also observed that when a smaller support domain is used, say, $\alpha_s = 4.0$ to 7.0, a small c value between 0.02 and 0.009 performs better. This finding supports the use of

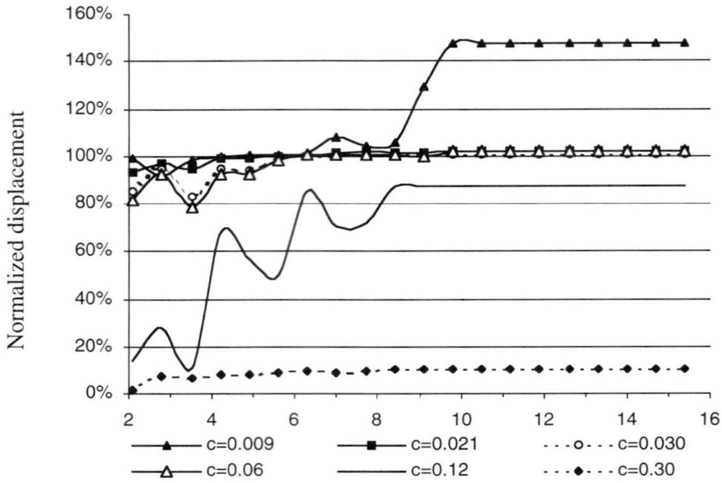

FIGURE 12.39
Variation displacement with respect to size of the support domain. Computed using RPIM-EXP with different shape parameters.

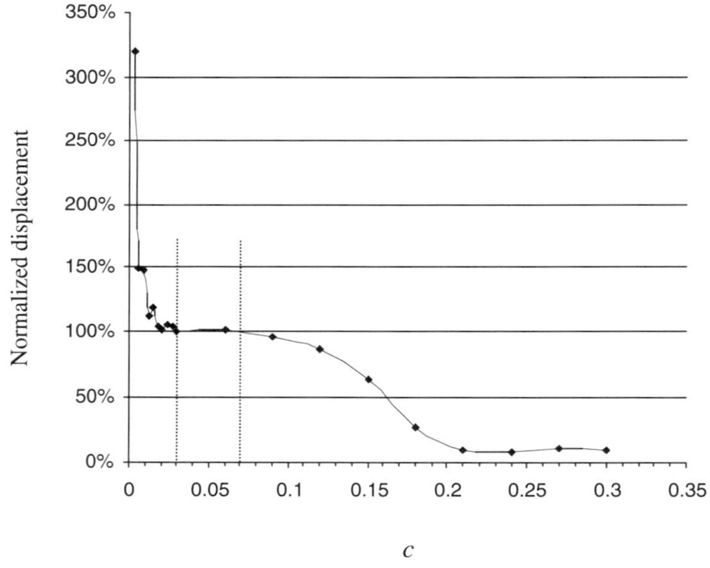

FIGURE 12.40
Variation displacement with respect to parameter c of the EXP radial basis function.

0.003 in the previous chapters where smaller support domains were used. Note also that using too small a b value incurs a risk of instability.

Figure 12.40 shows the variations in displacement with respect to parameter c of an EXP radial for a fixed support domain. This figure also suggests using a c between 0.03 and 0.07.

Based on the results from the above discussion, the recommended parameter c is approximately 0.03 to 0.07 and the results have acceptable accuracy, if the support domain is larger than 5.5 times the nodal spacing. A smaller c, say, 0.003, should be used if a small support domain is used. In the following examples, $c = 0.03$ is used.

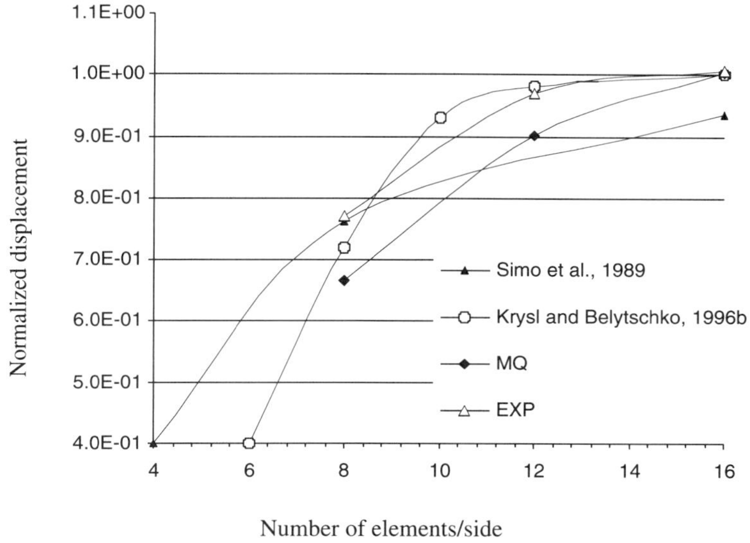

FIGURE 12.41
Convergence of vertical displacement at point A computed using RPIMs and compared with other available results.

Example 12.12 Pinched Cylindrical Shell

The pinched cylindrical shell studied in Example 12.9 is used again to test our RPIM. Configuration of the shell is schematically drawn in Figure 12.22. This problem is one of the critical benchmark tests to evaluate numerical methods in representing inextensional bending modes and membrane states of stress. All the parameters for the shell are exactly the same as those in Example 12.9. Because of the double symmetry, only 1/8 of the cylinder is modeled by the present method. Figure 12.41 shows the radial displacement at loaded points, normalized using 1.8248e-5, which is regarded as the "true" value. The convergence by the present method is excellent in comparison with the results by FEM (Simo et al., 1989) and by another EFG formulation for shells (Krysl and Belytschko 1996b). It can be seen for this particular case that the convergence rates of the results by MQ are relatively slow in comparison with those by EXP.

Figure 12.42 shows the distribution of the vertical displacement along the edge AB computed using both RPIM and FEM. The FEM solution is obtained by commercial software (ANSYS) using a rather fine mesh with 441 nodes. It can be seen that the results of the present method agree very well with the FEM solution.

Example 12.13 Pinched Hemispherical Shell

The pinched hemispherical shell studied in Example 12.10 is reexamined here using RPIM. The configuration of the shell is shown in Figure 12.27. All the parameters used here are exactly the same as those used in Example 12.10. The only difference is we use RPIM to solve the problem.

Due to symmetry, only a quarter of the hemisphere is taken into consideration for the analysis. Figure 12.43 plots the convergence of displacement at load point B computed using RPIM-MQ and RPIM-EXP. The results are normalized using the analytical value of 0.0924, and are compared with the results by FEM (Simo et al., 1989) and by another EFG

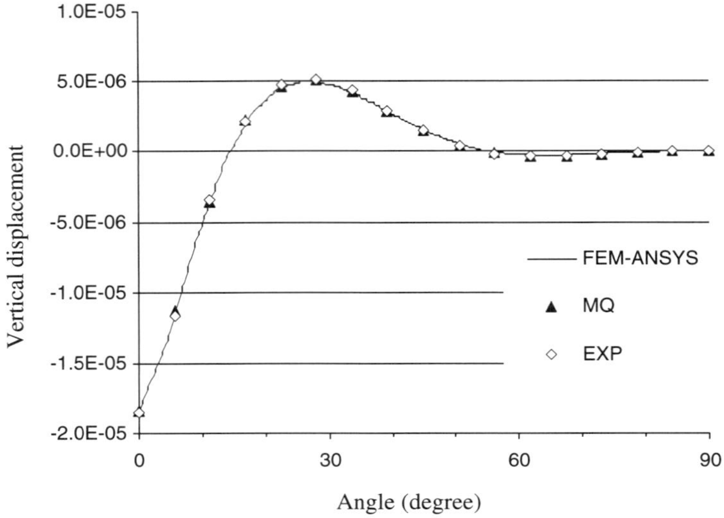

FIGURE 12.42
Distribution of the vertical displacement along AB.

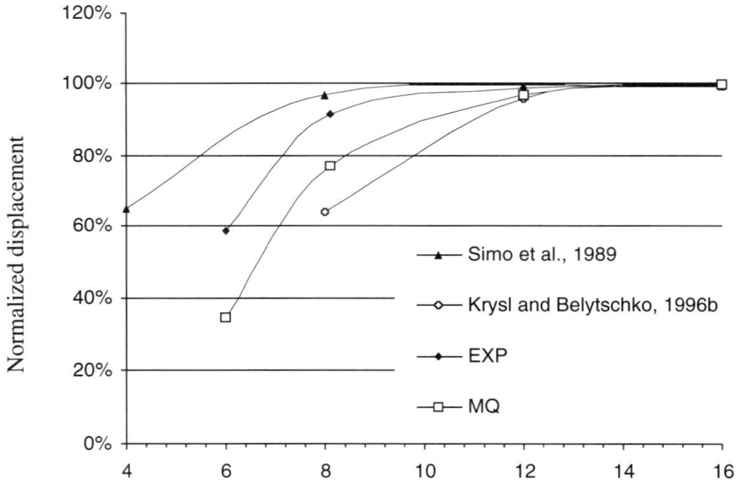

FIGURE 12.43
Convergence of displacement at load point B computed using RPIM-MQ and RPIM-EXP. Other available results are also plotted for comparison.

formulation for shells (Krysl and Belytschko, 1996b). The results obtained by the RPIM-MQ and RPIM-EXP methods are in very good agreement with those obtained using FEM and the EFG methods for thin shell. It also can be seen that the convergence rate by MQ and EXP methods is better than that by the EFG method given by Krysl and Belytschko (1996b).

The distribution of the displacement along the edge AB in the x direction is shown in Figure 12.44, where the FEM solution is obtained by FEM software (ANSYS) using a fine

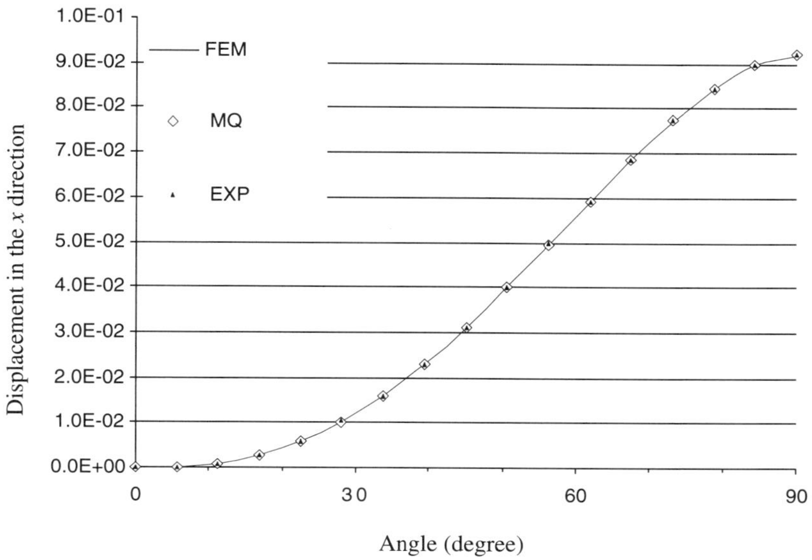

FIGURE 12.44
Distribution of the displacement in the x direction in the hemispherical shell along AB.

mesh with 503 nodes. It can be seen that the results of the MQ and EXP methods agree with the FEM solution very well.

12.3.3 Remarks

In this section, RPIM methods using both MQ and EXP radial basis functions are examined for analyzing 3D shell structures based on the stress-resultant shell theory, which has only five degrees of freedom assigned to every point of the shell.

Both the geometry of the neutral surface of the shell and the field variables are interpolated using RPIM shape functions, which are constructed using nodes in the support domains. The RPIM shape functions possess the Kronecker delta function property. Therefore, the essential boundary conditions can be easily implemented with ease as in FEM. Some important parameters on the performance of the present method are investigated in detail. The findings are summarized as follows:

- The dimensionless shape parameter α_C should be around 0.7.
- The shape parameter q should be in the range of $-0.7 \le q < 0$ and $0 < q \le 0.7$.
- The shape parameter b for EXP radial basis function should be approximately 0.03 to 0.07, if the support domain is larger than 5.5 times the nodal spacing. A smaller c, say, 0.003, should be used if a small support domain is used. However, there is a risk of instability in using too small a value for c.
- RPIM-EXP outperforms RPIM-MQ for our benchmark examples.
- Our study is still ongoing, and currently is investigating the effects of adding polynomial terms. We hope adding polynomials can improve the accuracy and reduce the sensitivities of the results to the shape parameters.

12.4 Summary

In this chapter, we formulated the EFG method for both thin and thick shells and RPIM for thick shells. We have shown through a number of benchmarking problems that they work very well. In all cases, the methods performed much better than standard commercially available FEM packages. There are cases where some fine-tuned FEM results reported in the literature performed better than some of the MFree results. In any case, the author is confident that the MFree method is a better choice for simulating shell structures.

We formulated MFree methods here based only on Galerkin formulation; however we can also use Petrov–Galerkin formulation. This is an ongoing project of the author's research group.

13

Boundary Mesh Free Methods

The boundary element method (BEM) is a numerical technique based on the boundary integral equation (BIE), which has been developed since the 1960s. For many problems, BEM is undoubtedly superior to the "domain discretization" types of methods, such as the finite element method (FEM) and the finite difference method (FDM). BEM has a well-known advantage of dimension reduction for linear problems. For example, only the two-dimensional (2D) bounding surface of a three-dimensional (3D) body needs to be discretized. BEM is applicable to all the problems for which the fundamental solution (Green's functions) is known in a reasonably simple form, preferably in a closed form.

The mesh free idea has also been used in BIE, such as the boundary node method (BNM) (Mukherjee and Mukherjee, 1997a; Kothnur et al., 1999; Chati et al., 1999; Chati and Mukherjee, 2000) and the local boundary integral equation (LBIE) method (Zhu and Atluri, 1998). These boundary type mesh free methods are formulated using the moving least squares (MLS) approximations and approaches and techniques of BIE. In BNM, the surface of problem domain is discretized (presented to be exact) by properly scattered nodes. The BNM has been applied to 3D problems of both potential theory and the theory of elastostatics (Chati et al., 1999; Chati and Mukherjee, 2000). Very good results have been obtained in these problems. However, because the shape functions based on MLS interpolants lack the delta function property, extra efforts are needed to satisfy the boundary condition accurately in BNM. This problem becomes even more serious in BNM because of the large number of boundary conditions, compared with the total number of nodes for the problem. The method used by Kothnur et al. (1999) to impose boundary conditions resulted in a set of system equations that are doubled in number. The advantage of the boundary-type method is therefore eroded and discounted to a certain degree, making BNM computationally much more expensive than conventional BEM.

G. R. Liu and Gu (2001a,e) have developed a boundary-type MFree method called the boundary point interpolation method (BPIM), where the point interpolation method (PIM) proposed by G. R. Liu and Gu (1999; 2001c) is used to construct shape functions. It is confirmed that there is no need at all to use MLS in boundary-type MFree methods, especially for 2D problems. PIM works much more efficiently in constructing shape functions, and all the PIM shape functions possess the Kronecker delta function property. This removes the issue of treatment of boundary conditions, which is especially beneficial to methods based on BIE. The dimension of the equation system of BPIM is equivalent to that of BEM. For 2D problems, BPIM works perfectly well without any special trick and is superior to BNM in simplicity, accuracy, and computational efficiency. For 3D problems for which 2D shape functions need to be constructed, there could be an issue of singular moment matrices. In such cases, the special techniques discussed in Section 5.5.5 should be applied.

An effective way of constructing PIM shape functions is to use radial functions as the basis. The advantage of using a radial function basis is that it guarantees the existence of the inverse moment matrix (see Section 5.7). The methods formulated and coded by Gu and Liu (2001a) are termed boundary radial PIM (BRPIM).

This chapter focuses on the introduction of boundary MFree methods formulated using both MLS shape functions (BNM) and PIM shape functions (BPIM and BRPIM). These boundary MFree methods can be formulated in the same manner, except that in formulating BNM using MLS shape functions proper treatment of essential boundary conditions (see Mukherjee and Mukherjee, 1997b; Kothnur et al., 1999) is required.

Only 2D problems are discussed in this chapter. In all these boundary MFree methods, only the boundary of the problem domain is represented by properly scattered nodes. BIE for 2D elastostatics is then discretized using MFree shape functions based only on a group of arbitrarily distributed boundary nodes that are included in the support domain of a point of interest.

For 3D BNM, readers are referred to the work by Chati et al. (1999).

13.1 BPIM Using Polynomial Basis

A BPIM using polynomial basis in the construction of shape functions is first presented in this chapter for solving boundary value problems of solid mechanics. The method was originally proposed by Gu and Liu (2001e). The PIM shape functions are constructed in a curvilinear coordinate system, and possess the delta function property. The boundary conditions can be implemented with ease as in conventional BEM. In addition, the rigid body movement can also be utilized to avoid some singular integrals. For 2D problems, BPIM with polynomial basis will have no singularity problem of interpolation as we have seen in the domain type of PIMs, as the boundaries are curves, and the interpolation is basically one dimensional. Therefore, there is no reason to use MLS approximations in this case.

Detailed formulation of BPIM using polynomial basis is presented, and several numerical examples are presented to demonstrate the validity and efficiency of BPIM. A comparison study is carried out using BPIM, BNM that uses MLS shape functions, conventional BEM, and analytical methods.

13.1.1 Point Interpolation on Curves

Consider a 2D domain Ω bounded by its boundary Γ, as shown in Figure 13.1. In using boundary MFree methods, only the boundary Γ of the problem domain is represented using nodes. The point interpolants in BPIM are constructed on the 1D bounding curve Γ of 2D domain Ω, using a set of discrete nodes on Γ. As in the conventional BEM formulation, the displacement and traction can be constructed independently using PIM shape functions. The displacement $u(s)$ and traction $t(s)$ at a point s on the boundary Γ is expressed using the surrounding nodes and polynomials:

$$u(s) = \sum_{i=1}^{n} p_i(s)a_i = \mathbf{p}^{\mathrm{T}}(s)\mathbf{a} \tag{13.1}$$

$$t(s) = \sum_{i=1}^{n} p_i(s)b_i = \mathbf{p}^{\mathrm{T}}(s)\mathbf{b} \tag{13.2}$$

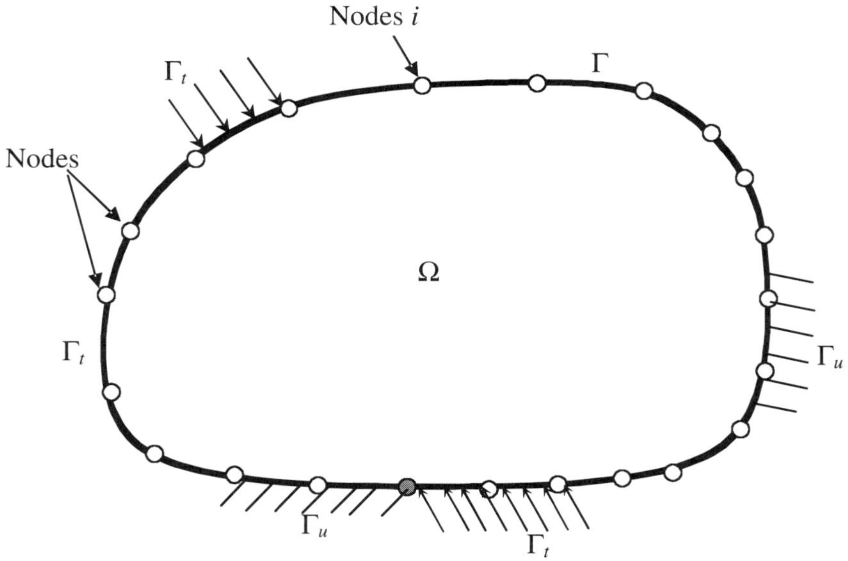

FIGURE 13.1
Domains and their boundaries: problem domain Ω boundary bounded by its boundaries Γ including essential (displacement) boundary Γ_u, natural (force or free) boundary Γ_t. In boundary MFree methods, only the problem boundary is represented using nodes.

where s is a curvilinear coordinate (the arc-length for a 2D problem) on Γ, n is the number of nodes in the support domain of a point of interest at s_Q, which is often a quadrature point of integration, $p_i(s)$ is a basis function of a complete polynomial with $p_1 = 1$ and $p_i = s^{i-1}$, and a_i and b_i are the coefficients that change when s_Q changes. In matrix form, we have

$$\mathbf{a}^{\mathrm{T}} = [a_1, a_2, a_3, \ldots, a_n] \tag{13.3}$$

$$\mathbf{b}^{\mathrm{T}} = [b_1, b_2, b_3, \ldots, b_n] \tag{13.4}$$

$$\mathbf{p}^{\mathrm{T}}(s) = [1, s, s^2, s^3, s^4, \ldots, s^{n-1}] \tag{13.5}$$

The coefficients a_i and b_i in Equation 13.1 can be determined by enforcing Equation 13.1 to be satisfied at the n nodes surrounding the point s_Q. Equation 13.1 can then be written in the following matrix form:

$$\mathbf{u}_n = \mathbf{P}_Q\mathbf{a} \tag{13.6}$$

$$\mathbf{t}_n = \mathbf{P}_Q\mathbf{b} \tag{13.7}$$

where $\mathbf{u}_n$ and $\mathbf{t}_n$ are the vectors of nodal displacement and traction, given by

$$\mathbf{u}_n = [u_1, u_2, u_3, \ldots, u_n]^{\mathrm{T}} \tag{13.8}$$

$$\mathbf{t}_n = [t_1, t_2, t_3, \ldots, t_n]^{\mathrm{T}} \tag{13.9}$$

and $\mathbf{P}_Q$ is the moment matrix formed by

$$\mathbf{P}_Q^{\mathrm{T}} = [\mathbf{p}(s_1), \mathbf{p}(s_2), \mathbf{p}(s_3), \ldots, \mathbf{p}(s_n)] \tag{13.10}$$

Solving **a** and **b** from Equations 13.6 and 13.7, and then substituting them into Equation 13.1, we obtain

$$u(s) = \mathbf{\Phi}^{\mathrm{T}}(s)\mathbf{u}_n \tag{13.11}$$

$$t(s) = \mathbf{\Phi}^{\mathrm{T}}(s)\mathbf{t}_n \tag{13.12}$$

where the matrix of shape function $\mathbf{\Phi}(s)$ is defined by

$$\mathbf{\Phi}^{\mathrm{T}}(s) = \mathbf{p}^{\mathrm{T}}(s)\mathbf{P}_Q^{-1} = [\phi_1(s), \phi_2(s), \phi_3(s), \ldots, \phi_n(s)] \tag{13.13}$$

The shape function $\phi_i(s)$ obtained from above procedure satisfies

$$\phi_i(s = s_i) = 1 \quad i = 1, n \tag{13.14}$$

$$\phi_j(s = s_i) = 0 \quad j \neq i \tag{13.15}$$

$$\sum_{i=1}^{n} \phi_i(s) = 1 \tag{13.16}$$

Therefore, the interpolation functions constructed have the delta function property, and the boundary conditions can be easily imposed as in traditional BEM. The procedure to prove these properties can be found in Chapter 5.

The matrix $\mathbf{P}_Q$ is an $n \times n$ matrix. It needs to be invertible for the construction of the shape functions in Equation 13.13. Fortunately, in the curvilinear coordinate system, the matrix $\mathbf{P}_Q$ is, in general, reversible for the 2D problem (interpolation along a 1D boundary).

It can be found that the accuracy of interpolation depends on the nodes in the support domain of a quadrature point. Therefore, a suitable support domain should be chosen to ensure a proper area of coverage for interpolation. To define the support domain for a point s_Q, a curvilinear support domain is used. The arc-length of the curvilinear domain d_s is computed by

$$d_s = \alpha_s d_c \tag{13.17}$$

where α_s is the dimensionless size of the support domain and d_c is a characteristic length that relates to the nodal spacing near the point at s_Q. If the nodes are uniformly distributed, d_c is the distance between two neighboring nodes. In the case where the nodes are nonuniformly distributed, d_c can be defined as an "average" nodal spacing in the support domain of s_Q. The procedure of determining d_c can be performed following the procedure in Section 2.10.3 for the 1D case based on our current curvilinear coordinate system.

As discussed in Section 5.11, the PIM approximation could be incompatible. Similarly to the domain type in PIM methods, we can also formulate nonconforming and conforming BPIMs.

In using nonconforming BPIM, the support domain is determined for each and every Gauss point. In conforming BPIM, however, we use so-called one-piece shape functions for an integration cell to ensure the compatibility of the field function approximation with the cells. Therefore, the support domain is defined for the cell, meaning all the Gauss points in the cell share the same support domain and hence the same PIM shape functions. The support domain of a cell is determined in a manner similar to that discussed in the previous paragraph. The only difference is that the domain is centered at the geometrical center of the cell and not at the Gauss point.

Note that for boundary type PIMs, because the integration cells are connected at points (not lines), the compatibility is automatically enforced, as long as the one-piece PIM shape functions are used. There is no need to use the penalty method to enforce compatibility as we have done in Chapter 8 for two-dimensional cases.

The number of nodes, n, can be determined by counting all the nodes in the support domain. The dimensionless size of support domain α_s should be predetermined by the analyst. We study this in detail using numerical examination to determine the proper value to use. It must be observed that choosing $\alpha_s = 2.0$ to 3.0 (which includes $n = 3$ to 6) leads to acceptable performance for BPIM.

13.1.2 Discrete Equations of BPIM

The well-known BIE for 2D linear elastostatics, presented by Brebbia (1978), is given by

$$\mathbf{c}_i\mathbf{u}_i + \int_\Gamma \mathbf{u}\mathbf{t}^* \mathrm{d}\Gamma = \int_\Gamma \mathbf{t}\mathbf{u}^* \mathrm{d}\Gamma + \int_\Omega \mathbf{b}\mathbf{u}^* \mathrm{d}\Omega \tag{13.18}$$

where $\mathbf{c}_i$ is a coefficient dependent on the geometric shape of the boundary, $\mathbf{b}$ is the body force vector, and $\mathbf{u}^*$ and $\mathbf{t}^*$ are the fundamental solution for linear elastostatics. The fundamental solution (Brebbia et al., 1984) for a 2D plane strain problem can be used and it is given by

$$u_{ij}^* = \frac{1}{8\pi G(1-\nu)}\left\{(3-4\nu)\ln\frac{1}{r}\Delta_{ij} + r_{,i}r_{,j}\right\} \tag{13.19}$$

$$t_{ij}^* = \frac{-1}{4\pi(1-\nu)r}\left\{\frac{\partial r}{\partial n}[(1-2\nu)\Delta_{ij} + r_{,i}r_{,j}] - (1-2\nu)(r_{,i}n_{,j} - r_{,j}n_{,i})\right\} \tag{13.20}$$

where G is the shear modulus, ν is the Poisson's ratio, Δ is the Kronecker delta function, r is the distance between the source point and the field point, n is the normal to the boundary, and a comma designates a partial derivative with respect to the indicated spatial variable.

Substituting Equations 13.11 and 13.12 into Equation 13.18 yields the BPIM system equation for all the nodes on the boundary of the problem domain.

$$\mathbf{HU} = \mathbf{GT} + \mathbf{D} \tag{13.21}$$

where

$$\mathbf{H} = \mathbf{c}_i + \int_\Gamma \mathbf{t}^*\mathbf{\Phi}^{\mathrm{T}} \mathrm{d}\Gamma \tag{13.22}$$

$$\mathbf{G} = \int_\Gamma \mathbf{u}^*\mathbf{\Phi}^{\mathrm{T}} \mathrm{d}\Gamma \tag{13.23}$$

$$\mathbf{D} = \int_\Omega \mathbf{b}\mathbf{u}^* \mathrm{d}\Omega \tag{13.24}$$

13.1.3 Implementation Issues in BPIM

Singular Integral

To evaluate the integrals given in Equations 13.22 to 13.24, background integration cells that can be independent of the nodes are required. The cells should be created on the boundary of the problem domain with proper dimension to ensure an accurate integration. From Equations 13.19 and 13.20, it can be seen that the integrands in Equations 13.22 to 13.24 consist of regular and singular functions. The regular functions in Equations 13.22 to 13.24 can be evaluated using the usual Gaussian quadrature based on the integration cells. In Equation 13.23, the matrix **G** contains a log singular integral. This type of singular integral can be evaluated by log Gaussian quadrature as follows:

$$I = \int_0^1 \ln(1/x) f(x) \mathrm{d}x \cong \sum_{i=1}^{m} f(x_i) w_i \tag{13.25}$$

where the required points x_i and weights w_i are presented by Brebbia et al. (1984) for conventional BEMs.

In matrix **H** defined by Equation 13.22, **c** is a coefficient dependent on the geometric shape of the boundary, which is easy to obtain for a smooth boundary. However, it is more complicated to obtain **c** for nonsmooth boundaries. In addition, **H** contains a $(1/r)$ type singular integral. Therefore, it can be a nontrivial task to directly evaluate the diagonal terms of **H**. The same difficulty has been experienced in conventional BEMs. Note that shape functions of BPIM possess the delta function property; therefore, the rigid body movement can be utilized in this work to obtain the diagonal terms of **H** (Brebbia et al., 1984).

Application of Boundary Conditions

There are two types of boundary conditions in BPIM

$$\mathbf{t} = \bar{\mathbf{t}} \quad \text{on the natural boundary } \Gamma_t \tag{13.26}$$

$$\mathbf{u} = \bar{\mathbf{u}} \quad \text{on the essential boundary } \Gamma_u \tag{13.27}$$

Because the shape functions of BPIM have the delta function property, the boundary conditions can be imposed in the same way as the traditional BEM. Note that if MLS shape functions were used, proper treatments are needed (Mukherjee and Mukherjee, 1997b; Kothnur et al., 1999).

After applying the boundary condition, the system (Equation 13.21) has $2N_B$ equations and $2N_B$ unknowns for N_B boundary nodes. The system equation can be solved using standard routines of an algebraic equation solver to obtain the displacement and traction.

Handling of Corners with Traction Discontinuities

In handling traction discontinuities in corners, special care should be taken. Double nodes and discontinuous elements at corners are used to overcome this problem in traditional BEM. Because there are no elements used in BPIM, a simple method proposed by Gu and Liu (2001e) to solve this difficulty is by displacing the nodes from the corner. The support domain for interpolation is truncated at the corner. The method is very easy to implement and is used in the following numerical examples. The simple method is proved to be very accurate.

13.1.4 Numerical Examples

The BPIM is coded and used to solve a number of problems of mechanics. A detailed comparison study is carried out using the present BPIM, BNM, conventional BEM, and analytical methods. In BNM, a weight function must be used for constructing MLS shape functions. The exponential weight function (Kothnur et al., 1999) given below is used for the examples in this section:

$$\widehat{W}_i(s) = \begin{cases} \dfrac{e^{-(d_i/c)^{2k}} - e^{-(d_{sQ}/c)^{2k}}}{1 - e^{-(d_{sQ}/c)^{2k}}} & d \le d_{sQ} \\ 0 & d > d_{sQ} \end{cases} \tag{13.28}$$

where d_i is the arc-length, d_{sQ} is the size of the support of the weight function $\widehat{W}_i$, which is the dimension of the support domain in BNM, and k and c are constants. In this section, $k = 1$ and $d_{sQ}/c = 0.75$ are used (Kothnur et al., 1999).

The formulation procedure of BNM is very much similar to that for the nonconforming BPIM except (1) an MLS shape function is used to replace the PIM shape function; (2) Lagrange multipliers, or the method proposed by Mukherjee and Mukherjee (1997b), or any other methods to treat the essential boundary conditions are used.

The following presents some examples analyzed using nonconforming BPIM.

Example 13.1 Cantilever Beam

BPIM is first applied to analyze the displacement and stress field in a cantilever beam, which is shown in Figure 6.4. A plane stress problem is considered. The parameters for this example are

Young's modulus for the material: $E = 3.0 \times 10^7$

Poisson's ratios for two materials: $\nu = 0.3$

Thickness of the beam: $t = 1$

Height of the beam: $D = 12$

Length of the beam: $L = 48$

Load: $P = 1000$

The beam is subjected to a parabolic traction at the free end as shown in Figure 6.4. The analytical solution is available; it can be found in the textbook by Timoshenko and Goodier (1977) and is listed in Equations 6.28 to 6.33.

A total of 120 uniform boundary nodes are used to discretize the boundary of the beam. To evaluate the integral of matrices, 120 uniform integration cells are used. The parameter α_s in Equation 13.17 for the support domains is fixed at 2.0. Therefore, three to six nodes are included in the support domain for constructing shape functions.

Figure 13.2 illustrates the comparison between the shear stress calculated analytically and by BPIM at the section of $x = L/2$. The plot shows a good agreement between the analytical and numerical results. The conventional linear BEM results of this problem are

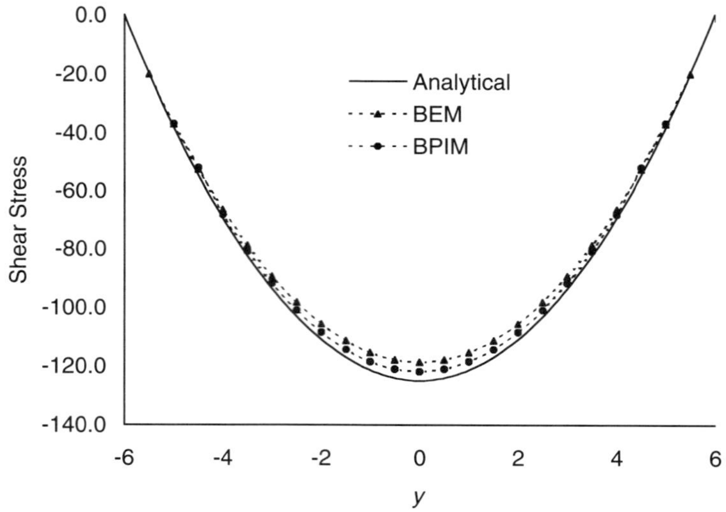

FIGURE 13.2
Shear stress τ_{xy} at the section $x = L/2$ of the beam. (From Gu, Y. T. and Liu, G. R., *Comput. Mech.*, 28, 47–54, 2002. With permission.)

also shown in the same figure for comparison. The density of the nodes in BEM and BPIM is exactly the same. It is clearly shown that the BPIM results are more accurate than the BEM results. This is because BPIM uses more nodes for the interpolation of the displacements and tractions. Therefore, the order of the interpolant in BPIM is higher than the order of the conventional linear elements in BEM.

For the detailed error analysis, we define the following norm as an error indicator using the shear stress, as the shear stress is much more critical than other stress components in the cantilever beam to reflect the accuracy.

$$e_t = \frac{1}{N}\sqrt{\sum_{i=1}^{N}(\tau_i - \bar{\tau}_i)^2 \Big/ \sum_{i=1}^{N}\bar{\tau}^2} \tag{13.29}$$

where N is the number of nodes investigated, τ is the shear stress obtained numerically, and $\bar{\tau}$ is the analytical shear stress.

The convergence for the shear stresses at the section of $x = L/2$ with mesh refinement is shown in Figure 13.3. The convergences of BNM and conventional linear BEM are also shown in the same figure, where h is a characteristic length related to the nodal spacing. Three kinds of nodal arrangement of 40, 240, and 480 uniform boundary nodes are used. It can be observed that the accuracy of BPIM and BNM using MLS approximation is nearly the same, but both BPIM and BNM have higher accuracy than BEM. The convergence of BPIM seems to be the best among these three methods.

As mentioned above, the dimensionless size of support domain α_s defined in Equation 13.17 needs to be chosen such that a "reasonable" number of nodes lie in the support domain of an evaluation point. The results of e_t for different sizes of support domain are shown in Table 13.1. In this analysis, the boundary is modeled by 40 uniformly distributed nodes, and 40 uniformly spaced integration background cells are used. It is found that the accuracy of the results of BPIM changes slightly with the dimension of the support domain when the nodal density is fixed. Although the choice of the support domain may

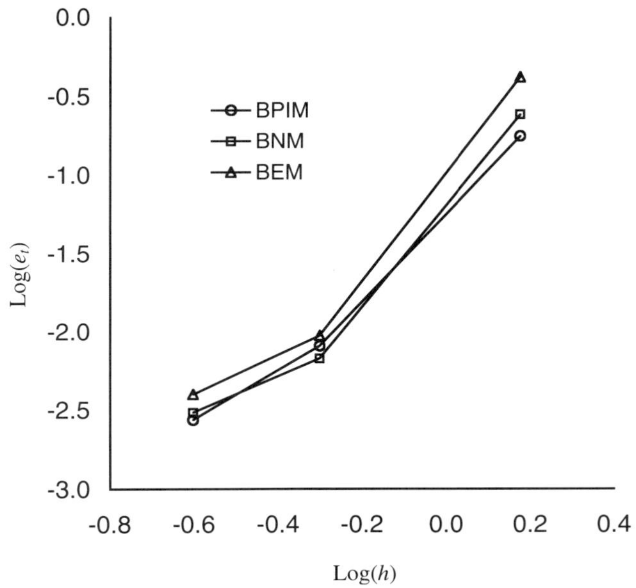

FIGURE 13.3
Convergence in e_t norm of error. (From Gu, Y. T. and Liu, G. R., *Comput. Mech.*, 28, 47–54, 2002. With permission.)

TABLE 13.1
The Effects of the Dimension of the Support Domain on the Error of Energy Norm

Dimension of Support Domain (α_s)	e_t[a]
1.0	0.1688
2.0	0.1410
3.0	0.1775
4.0	0.1812
5.0	0.1794

[a] Defined by Equation 13.29.

also depend a little on the type of the problem, it is found that α_s = 2.0 to 3.0 works well for most of the problems investigated in this section.

It may be mentioned here that the use of a large support domain does not necessarily lead to more accurate results. Similar results were seen for mechanics problems of 2D solids in Chapter 8.

Example 13.2 Plate with a Hole

Consider now an infinite plate with a central circular hole subjected to a unidirectional tensile load of p = 1.0 in the x direction. As a large finite plate can be considered a good approximation of an infinite plate, a finite square plate of 20 × 20 is considered. Making use of the symmetry, only the upper right quadrant of the finite plate is modeled as shown in Figure 13.4. Plane strain condition is assumed, and the material properties are $E = 1.0 \times 10^3$, ν = 0.3. Symmetry conditions are imposed on the left and bottom edges, and the inner

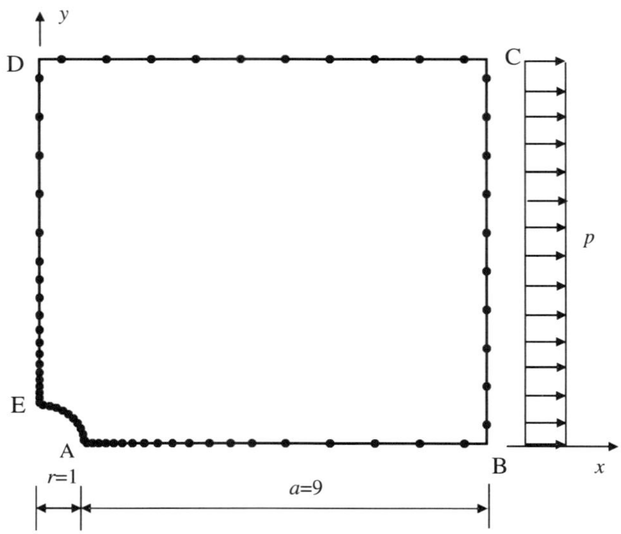

FIGURE 13.4
Nodes in a plate with a central hole subjected to a unidirectional tensile load in the x direction.

boundary of the hole is traction free. The tensile load p is imposed on the right edge in the x direction. The exact solution for the stresses of an infinite plate is given in the textbook by Timoshenko and Goodier (1977) and is listed in Equations 7.59 to 7.64.

A total of 68 nodes are used to discretize the boundary (with 10 uniformly distributed nodes on BC, CD, and AE, and 19 nonuniformly distributed nodes along AB and DE). The same number of integration background cells are used. The support domain of an evaluation point is determined as in Equation 13.17 (with $\alpha_s = 2.0$, and the characteristic arc-length: $d_Q = 1.0$ on AB, BC, CD, and DE, $d_Q = 0.2$ along AE). If the number of nodes in the support domain is more than six, only six nodes with shorter arc-length to the integration point are used in the interpolation.

As the stress is more critical in assessment of the solution accuracy, detailed results of stress are presented here. The stress σ_x at $x = 0$ obtained by BPIM is given in Figure 13.5 together with the analytical solution for the infinite plate. It can be observed from this figure that BPIM gives very good results for this problem. The BNM results of this problem are also shown in the same figure for comparison. It is clearly shown that the BPIM and BNM results possess nearly the same accuracy.

Example 13.3 A Rigid Flat Punch on a Semi-Infinite Foundation

As BEM methods have a clear advantage over domain-type methods for problems with infinite domain, BPIM is used to obtain a solution for an indentation produced by a rigid flat punch in a semi-infinite soil foundation, as shown in Figure 13.6. In this case, Green's functions for a half-plane are employed (Brebbia et al., 1984), and only the contact surface between the punch and half space needs to be discretized.

Consider a rigid punch of length $L = 12$ subjected to a uniform pressure of $p = 100$ on the top surface. The parameters of the soil foundation are taken as $E = 3.0 \times 10^4$, and $\nu = 0.3$. The punch is considered to be perfectly smooth, and does not result in any fraction force in the interface between the punch and the foundation. An indentation is measured by the vertical displacement of the punch. A plane strain condition is assumed. Due to symmetry, only the right half of the contact surface is discretized by 31 distributed boundary

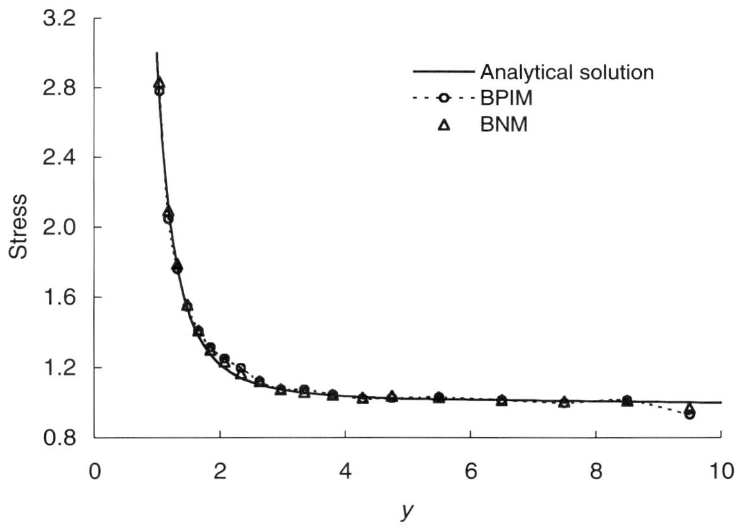

FIGURE 13.5
Stress distribution in the plate with a central hole subjected to a unidirectional tensile load (σ_x, at $x = 0$). (From Gu, Y. T. and Liu, G. R., *Comput. Mech.*, 28, 47–54, 2002. With permission.)

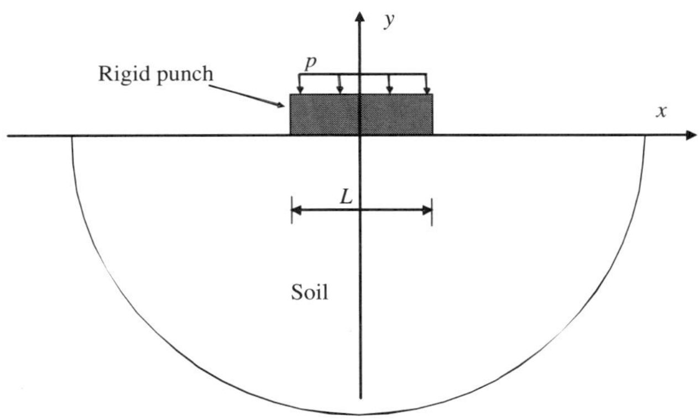

FIGURE 13.6
Rigid punch forced on a semi-infinite soil foundation. (From Gu, Y. T. and Liu, G. R., *Comput. Mech.*, 28, 47–54, 2002. With permission.)

nodes; 31 nonuniformly distributed integration background cells are used. Coordinates of these boundary nodes are obtained using the following formula:

$$x_m = \frac{6.2(m-1)}{m} \tag{13.30}$$

where m is the node number, and $m = 1$ to 31.

The vertical surface displacements of the foundation are assumed to be the same of that of the punch (perfect contact). This assumption is often proved true for a rigid punch. A prescribed vertical displacement of the punch is imposed on the contact surface as a boundary constraint. The prescribed displacement of the punch can be obtained using the approximate method presented by Poulos and Davis (1974); i.e., the vertical displacement of a vertically loaded rigid area in contact with the rigid punch may be approximated by the mean vertical displacement of a uniformly loaded flexible area of the same shape. The approximation is expressed as follows:

$$v_{\text{rigid}} = \frac{1}{2}\lfloor v_{\text{center}} + v_{\text{edge}} \rfloor_{\text{flexible}} \tag{13.31}$$

where v_{rigid} is the vertical displacement of the rigid area in contact with the rigid punch, v_{center} and v_{edge} are vertical displacements at the center and edge, respectively, of the contact area subjected to uniform load, when the contact area is considered flexible. The analytical solution of v_{center} and v_{edge} can be obtained in the textbook by Timoshenko and Goodier (1977). The exact solution (Poulos and Davis, 1974) of contact stresses along the contact surface is

$$\frac{\sigma_y}{p} = \frac{2}{\pi}\frac{1}{\sqrt{1-(2x/L)^2}} \tag{13.32}$$

BPIM is utilized to obtain the contact stresses along the contact surface. The support domain of a quadrature point is determined by Equation 13.17 (with $\alpha_s = 2.0$, and the characteristic arc-length of $d_Q = 3.0$). If the number of nodes in the support domain is more than six, only six nodes with shorter arc-length to the quadrature point are used in the interpolation. When these six nodes are all on one side of the quadrature point along the boundary, one more node nearest to this evaluation point in the other side is purposely added to the support domain to avoid an extrapolation.

Figure 13.7 plots the comparison between the contact stresses calculated analytically and by the BPIM along the contact surface. The plot shows excellent agreement between the analytical and numerical results.

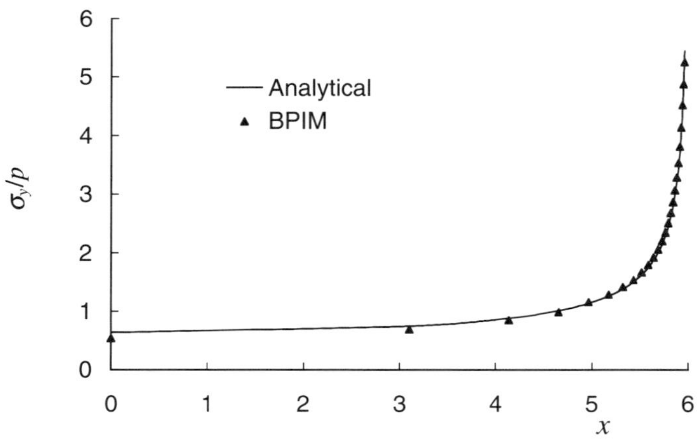

FIGURE 13.7
Contact stresses on the contact surface between the punch and the half-space. (From Gu, Y. T. and Liu, G. R., *Comput. Mech.*, 28, 47–54, 2002. With permission.)

13.2 BPIM Using Radial Function Basis

For 2D problems, the boundaries of the domain are curves. Therefore, the PIM shape function using polynomial basis will have no problem, and BPIM works perfectly well without any special efforts. For 3D problems for which 2D shape functions need to be constructed, there could be an issue of singular moment matrices. The special techniques discussed in Section 5.6 should be applied. One effective way is to use RPIM shape functions. This section introduces the boundary MFree method using RPIM shape functions. This method was formulated and coded by Gu and Liu (2001a) and termed as BRPIM. Although BRPIM performs no better than BPIM for 2D problems, its full advantages are expected to be seen for 3D problems.

The formulation procedure for BRPIM is largely the same as that of BPIM, except for the formulation of the shape function. This section, therefore, focuses only on the portion of the formulation that is different from BPIM. As the radial function is used, a study on the effects of these parameters of the radial function is performed. The performance of BRPIM is discussed using example problems of 2D elastostatics.

13.2.1 Radial Basis Point Interpolation

In BRPIM, the radial basis functions $R_i(s)$ listed in Table 5.1 are used. The point interpolation for displacement $u(s)$ and traction $t(s)$ at a point s on the boundary Γ from the surrounding nodes uses radial basis functions, i.e.,

$$u = \sum_{i=1}^{n} R_i(s)\alpha_i = \mathbf{R}^{\mathrm{T}}(s)\boldsymbol{\alpha} \tag{13.33}$$

$$t = \sum_{i=1}^{n} R_i(s)\beta_i = \mathbf{R}^{\mathrm{T}}(s)\boldsymbol{\beta} \tag{13.34}$$

where s is a curvilinear distance (the arc-length for a 1D curve boundary) on Γ, n is the number of nodes in the support domain of a point of interest s_Q, which is usually the quadrature point, $R_i(s)$ is a radial basis function, and α_i and β_i are the coefficients. In matrix form, we have

$$\boldsymbol{\alpha}^{\mathrm{T}} = [\alpha_1, \alpha_2, \alpha_3, \ldots, \alpha_n] \tag{13.35}$$

$$\boldsymbol{\beta}^{\mathrm{T}} = [\beta_1, \beta_2, \beta_3, \ldots, \beta_n] \tag{13.36}$$

The following multiquadrics (MQ) radial function is used in this section:

$$R_i(s) = (s_i^2 + C^2)^q \tag{13.37}$$

Two parameters, q and C, need to be determined in an MQ radial function. Detailed investigations of these parameters are given in the following numerical examples.

The coefficients α_i and β_I can be determined by enforcing Equations 13.33 and 13.34 to be satisfied at the n-node support domain of point s_Q. Equations 13.33 and 13.34 can then be written in the following matrix form:

$$\mathbf{u}_n = \mathbf{R}_Q\boldsymbol{\alpha} \tag{13.38}$$

$$\mathbf{t}_n = \mathbf{R}_Q\boldsymbol{\beta} \tag{13.39}$$

where $\mathbf{u}_n$ and $\mathbf{t}_n$ are the vectors of nodal displacements and tractions in the support domain, given by

$$\mathbf{u}_n = [u_1, u_2, u_3, \ldots, u_n]^{\mathrm{T}} \tag{13.40}$$

$$\mathbf{t}_n = [t_1, t_2, t_3, \ldots, t_n]^{\mathrm{T}} \tag{13.41}$$

and $\mathbf{R}_Q$ is the moment matrix of radial basis functions

$$\mathbf{R}_Q = \begin{bmatrix} R_1(s_1) & R_2(s_1) & \ldots & R_n(s_1) \\ R_1(s_2) & R_2(s_2) & \ldots & R_n(s_2) \\ \mathrm{M} & \mathrm{M} & \mathrm{O} & \mathrm{M} \\ R_1(s_n) & R_2(s_n) & \ldots & R_n(s_n) \end{bmatrix} \tag{13.42}$$

Solving $\boldsymbol{\alpha}$ and $\boldsymbol{\beta}$ from Equations 13.38 and 13.39, and then substituting them back into Equations 13.33 and 13.34, we obtain

$$u(s) = \boldsymbol{\Phi}^{\mathrm{T}}(s)\mathbf{u}_n \tag{13.43}$$

$$t(s) = \boldsymbol{\Phi}^{\mathrm{T}}(s)\mathbf{t}_n \tag{13.44}$$

where the shape function $\boldsymbol{\Phi}(s)$ is defined by

$$\boldsymbol{\Phi}^{\mathrm{T}}(s) = \mathbf{R}^{\mathrm{T}}(s)\mathbf{R}_Q \qquad \mathbf{R}_Q^{-1} = [\phi_1(s), \phi_2(s), \phi_3(s), \ldots, \phi_n(s)] \tag{13.45}$$

Similar to the polynomial basis shape functions, the shape function $\phi_i(s)$ obtained through the above procedure satisfies

$$\phi_i(s = s_i) = 1 \quad i = 1, n \tag{13.46}$$

$$\phi_j(s = s_i) = 0 \quad j \neq i \tag{13.47}$$

$$\sum_{i=1}^{n} \phi_i(s) = 1 \tag{13.48}$$

Therefore, the RPIM shape functions constructed have the Kronecker delta function property, and the boundary conditions can be easily imposed as in traditional BEM.

The moment matrix $\mathbf{R}_Q$ is an $n \times n$ matrix. It must be invertible for the construction of the shape functions in Equation 13.45. The existence of $\mathbf{R}_Q^{-1}$ has been proved for arbitrary scattered nodes (Kansa, 1990a,b). Therefore, in BRPIM, interpolation using the radial basis function is stable and flexible for arbitrarily distributed nodes on the boundary of the problem domain. This advantage will be very beneficial when using BRPIM to solve 3D problems.

13.2.2 BRPIM Formulation

The formulation of system equations in BRPIM is exactly the same as that for BPIM, except that the PIM shape function given by Equation 13.13 is replaced by the RPIM shape function defined by Equation 13.45.

TABLE 13.2
Comparison of BPIM, BRPIM, BNM, and BEM

	BPIM	BRPIM	BNM	BEM
Mesh	No	No	No	Yes
Interpolants	Polynomial PIM	Radial PIM	MLS	Element based polynomial
Interpolation area based on	Distributed nodes	Distributed nodes	Distributed nodes	Element
Number of basis m and interpolation nodes n	$m = n$	$m = n$	$m \neq n$	$m = n$
Overlapping of interpolation area	Overlapping	Overlapping	Overlapping	No overlapping
Shape function	Simple	Simple	Complicated	Simple
Delta property of shape function	Yes	Yes	No	Yes
Application of boundary conditions	Easy	Easy	Difficult	Easy
Number of system equations	$2N_B$	$2N_B$	$4N_B$	$2N_B$

Source: Gu, Y. T. and Liu, G. R., *Comput. Mech.*, 28, 47–54, 2002. With permission.

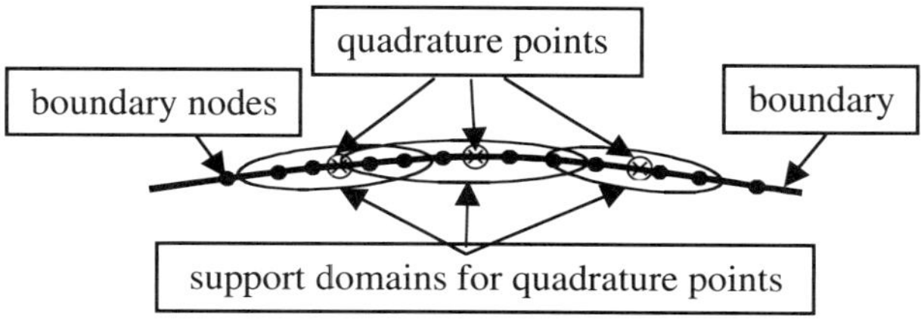

FIGURE 13.8
Interpolation in BPIM and BRPIM.

13.2.3 Comparison of BPIM, BNM, and BEM

A comparison of BPIM, BNM, and BEM is summarized concisely in Table 13.2. It can be found that BPIM, BNM, and BEM are all based on the BIE. The difference is essentially in the means of implementation.

BPIM vs. BEM

Both BPIM and BEM use polynomial interpolants, in which the number of monomials used in the base functions, m, is the same as the number of nodes, n, utilized. Therefore, the interpolant functions possess the delta function property. The boundary conditions can be implemented with ease.

However, BPIM is a boundary-type MFree method, whereas BEM is a boundary-type method based on boundary elements. As other MFree methods (e.g., EFG, BNM, MLPG), the interpolation procedure in BPIM is based only on a group of arbitrary distributed nodes. The interpolation at a quadrature point in BPIM is performed over the support domain of the point, which may overlap with the support domains of other quadrature points, as shown in Figure 13.8. BEM defines the shape functions over predefined regions called elements, and there is no overlapping or gap between the elements.

BPIM vs. BNM

Both BPIM and BNM are boundary-type MFree methods. The difference between these two methods comes from the different interpolants utilized. As discussed above, BPIM uses PIM shape functions, in which the coefficients a and b in Equation 13.1 are constant. The MLS interpolants are used in BNM, in which a and b are also functions of curvilinear coordinate s. Therefore, the shape function of BNM is more complicated than PIM. In addition, the shape function of BNM constructed using MLS approximation lacks the delta function property. It takes extra effort to impose boundary conditions.

13.2.4 Numerical Examples

Example 13.4 Cantilever Beam

Example 13.1 is used here again to examine BRPIM. All the parameters are exactly the same.

Effects of Radial Function Parameters

Two parameters, α_C and q, in the MQ radial function defined in Table 5.3 are investigated to reveal their effects on the performance of BRPIM. The characteristic length d_c is taken to be the "average" nodal spacing for nodes in the support domain of the quadrature point.

The parameter q is first investigated, and q is taken to be –1.5, –0.5, 0.5, and 1.5. Shear stresses for different q are obtained and compared with the analytical solution. Errors for different q are plotted in Figure 13.9a. This figure shows that $q = 0.5$ leads to a better result in the range of studies. Hence, 0.5 is used in following studies.

Errors in shear stresses of different α_C are plotted in Figure 13.9b. It is found that α_C can be chosen from a wide range of $\alpha_C = 1.0$ to 1.6, where steadily accurate results can be obtained. For convenience, $\alpha_C = 1.2$ is used in the following studies.

From studies for 2D interpolation, it has also been understood that α_C has a wider range of choices, but parameter q is very critical and has to be very precise to obtain good results. To determine a more precisely tuned q, more detailed study is required.

Effects of Interpolation Domain

The size of the support domain of a quadrature point is determined by the parameter α_s in Equation 13.17. Results of $\alpha_s = 1.0$ to 5.0 are obtained and plotted in Figure 13.10. It is found that the results obtained using a support domain with $\alpha_s = 3.0$ to 4.5, which covers about 6 to 10 nodes, are very good. Too small a support domain ($\alpha_s < 2.5$) will lead to large error. This is because there are not enough nodes to perform interpolation for the field variable. Too big a support domain ($\alpha_s > 4.5$) will also lead to large error. This is because there are too many nodes to perform interpolation, which results in a very complex shape function and hence a complex integrand in computing the system equations. Numerical integral error becomes very large. Therefore, $\alpha_s = 3.0$ to 4.5 is recommended. For convenience and consistency, $\alpha_s = 4.0$ is used in the following studies.

Comparison with the case of using BPIM, for which $\alpha_s = 2.0$ is the optimum (see Table 13.1) reveals that BRPIM requires a larger support domain, and more points for interpolation. This finding agrees with that for domain-type PIM and RPIM.

Figure 13.11 illustrates the comparison between the shear stress calculated analytically and by the BRPIM at the section of $x = L/2$. The plot shows good agreement between the analytical and numerical results of BRPIM. The conventional linear BEM results of this

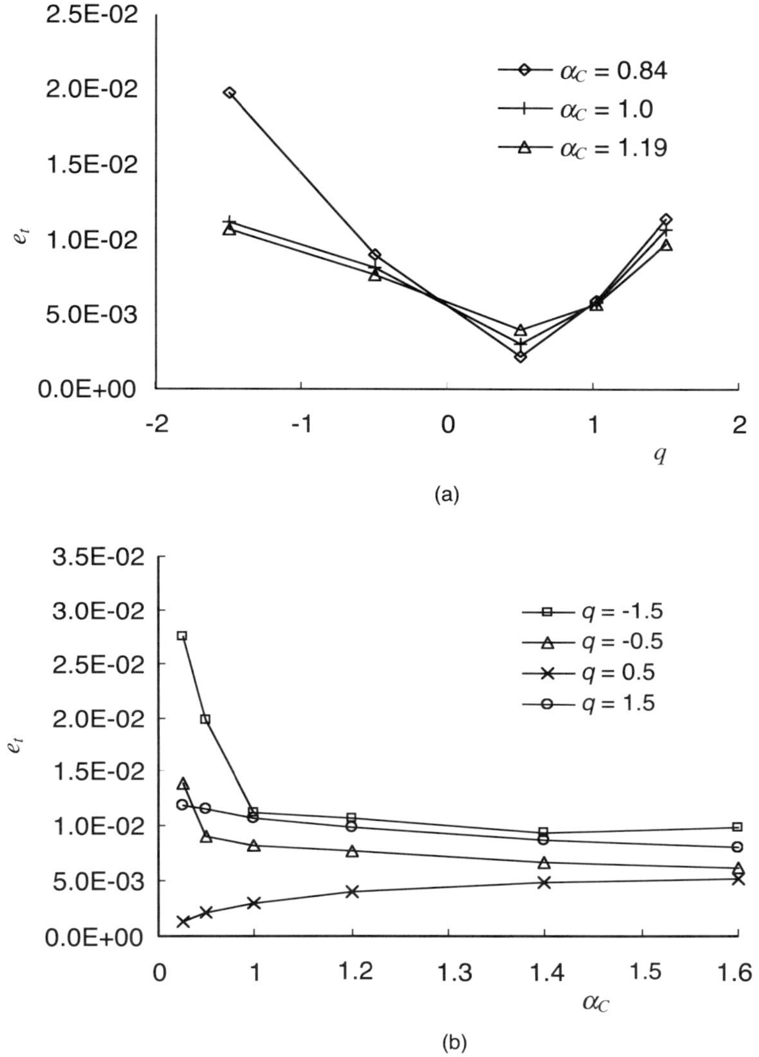

FIGURE 13.9
Effect of parameters q and α_C of MQ radial basis function on the error of shear stress in the cantilever beam computed using BRPIM. e_t is defined by Equation 13.29. (a) Effect of q; (b) effect of α_C.

problem are also shown in the same figure for comparison. The density of the nodes in BEM and BRPIM is exactly the same. It is clearly shown that the BRPIM results are more accurate than the BEM results. This is because the BRPIM uses more nodes for the interpolation of the displacements and tractions.

The convergence for the shear stresses at the section of $x = L/2$ with node/mesh refinement is shown in Figure 13.12, where h is a characteristic length relating the spacing of the nodes. Three kinds of nodal arrangement of 72, 240, and 480 uniform boundary nodes are used. The convergences of BNM and conventional linear BEM are also shown in the same figure. It is observed that the convergence of BRPIM is very good. It can also be observed that BRPIM has higher accuracy than BEM and BNM for this example.

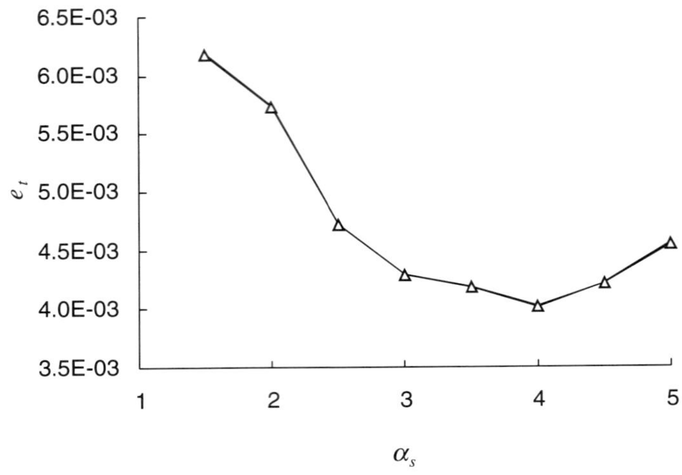

FIGURE 13.10
Influence of the parameter α_s of the interpolation domain.

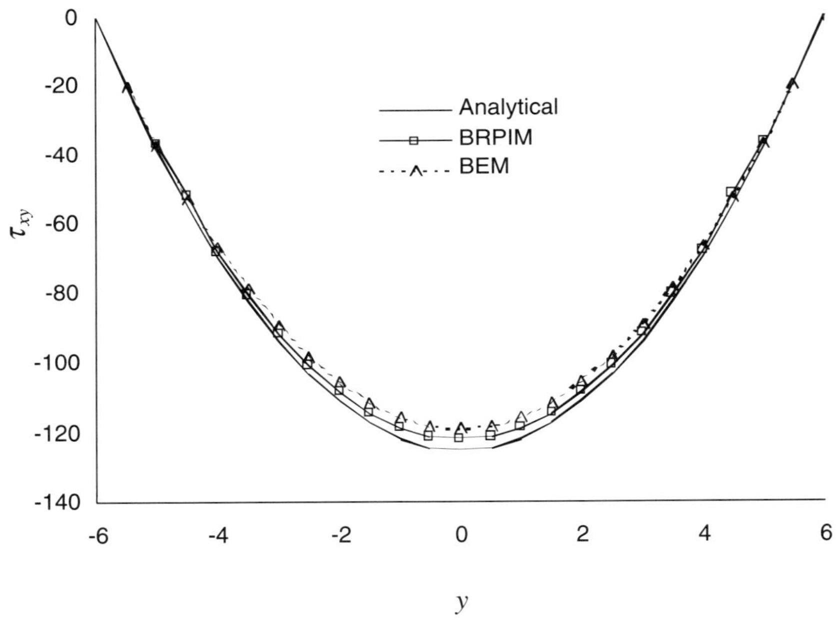

FIGURE 13.11
Shear stress at the section $x = L/2$ of the cantilever beam. Comparison of results obtained by three methods.

Example 13.5 Plate with a Hole

Example 13.2 is also used to examine BRPIM. All the parameters are exactly the same. The nodal distribution is shown in Figure 13.4. The stress σ_x at $x = 0$ computed using the BRPIM code is given in Figure 13.13 together with the analytical solution for the infinite plate. The BEM results of this problem are also shown in the same figure for comparison. It can be observed from this figure that the BRPIM gives very good results. It is clearly shown that the BRPIM and BEM results possess nearly the same accuracy for this problem.

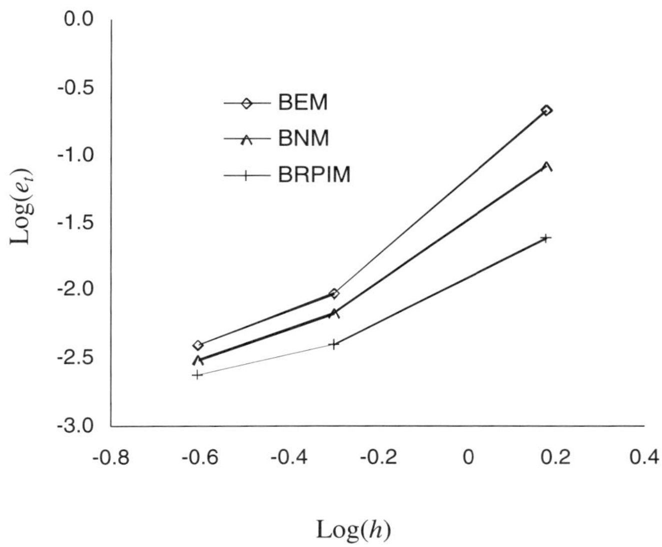

FIGURE 13.12
Convergence in e_t norm of error.

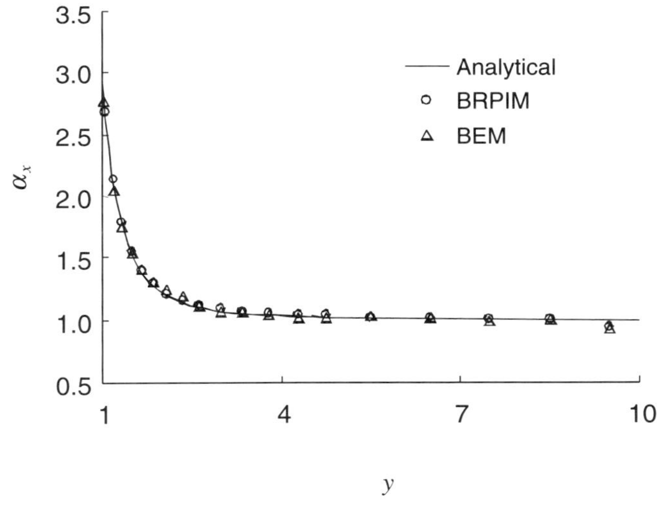

FIGURE 13.13
Stress σ_x distribution in a plate with a central hole subjected to a unidirectional tensile load. Comparison of results obtained using three different methods.

Example 13.6 Internally Pressurized Hollow Cylinder

A hollow cylinder under internal pressure shown in Figure 13.14 is considered. The parameters are taken as $p = 100$, $G = 8000$, and $\nu = 0.25$. This problem has been used by several other authors (see, e.g., Brebbia, 1978) as a benchmark problem, as the analytical solution is available. Due to the symmetry of the problem, only one quarter of the cylinder needs to be modeled. The arrangement of the field nodes is shown in Figure 13.15. The boundary

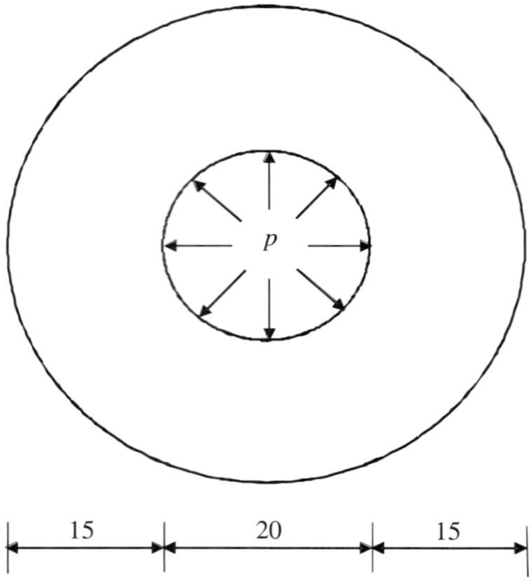

FIGURE 13.14
Hollow cylinder subjected to internal pressure.

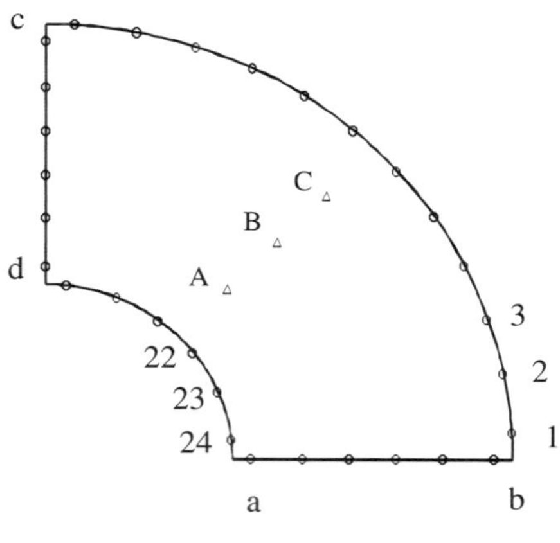

FIGURE 13.15
Arrangement of nodes for a quarter model of the hollow cylinder.

of this domain is discretized by 30 nodes (6 uniformly distributed nodes on *ab*, *cd*, and *ad* and 12 uniformly distributed nodes on *bc*). The same number of background cells are used for integration. Three internal points A, B, and C are selected for examination. The polar coordinates (with the origin at the center of the cylinder) for the three internal points are A(13.75, $\pi/4$), B(17.5, $\pi/4$), and C(21.25, $\pi/4$). The dimensionless size of the support domain of $\alpha_s = 2.0$ is used for all the quadrature points in the integration on the boundary.

The BRPIM and BPIM results are compared with those obtained using BNM, conventional BEM, and the analytical solution. The radial displacements at some of the boundary nodes and the three internal points are listed in Table 13.3. The circumferential stresses

TABLE 13.3

Radial Displacement and Circumferential Stresses in the Pressurized Hollow Cylinder

Nodes	Exact	BPIM	BRPIM	BNM	BEM
Radial Displacements ($\times 10^{-2}$)					
1	0.4464	0.4465	0.4466	0.4462	0.4468
2	0.4464	0.4478	0.4475	0.4463	0.4482
3	0.4464	0.4491	0.4493	0.4498	0.4494
22	0.8036	0.8213	0.8200	0.8220	0.8266
23	0.8036	0.8214	0.8207	0.8215	0.8268
24	0.8036	0.8199	0.8215	0.8223	0.8251
A	0.6230	0.6274	0.6211	0.6256	0.6319
B	0.5294	0.5342	0.5366	0.5353	0.5374
C	0.4766	0.4809	0.4810	0.4826	0.4838
Stress σ_θ					
A	82.0113	81.8947	82.1437	81.8513	82.0192
B	57.9226	58.1285	58.2585	58.1627	58.1691
C	45.4112	45.6471	45.6264	45.4597	45.6575

Source: Gu, Y. T. and Liu, G. R., *Comput. Mech.*, 28, 47–54, 2002. With permission.

σ_θ at points A, B, and C are also listed in the same table. The BRPIM and BPIM results are in very good agreement with the analytical solution. In comparison with the conventional BEM results, the BRPIM and BPIM solution is, in general, more accurate for both displacements and stresses.

13.3 Remarks

Boundary MFree methods are presented in this chapter. Detailed formulations for BPIM and BRPIM are provided for solving 2D problems of elastostatics. The boundary integral equation is discretized using radial MFree shape functions based on a group of arbitrarily distributed points on the boundary of the problem domain. The boundary MFree methods do not require any element connectivity in constructing shape functions, and possess the dimensionality advantage. Numerical examples have demonstrated that boundary MFree methods are superior to conventional BEM in terms of accuracy.

Compared with the BNM, BPIM and BRPIM have the following advantages:

- BPIM is computationally much less expensive than BNM because of its simpler interpolation scheme and smaller system equation dimension. The number of system equations in BPIM is only a half of that in BNM.
- The imposition of boundary conditions is easy in BPIM and BRPIM because the shape functions have the delta property.
- Rigid body movement can be used to avoid some singular integrals.

The parameters for BPIM and BRPIM should be as follows:

- In using BPIM, $\alpha_s = 2.0$ to 3.0 (with $n = 3$ to 6) yields acceptable results.
- In using BRPIM with MQ radial function, $q = 0.5$ and $\alpha_C = 1.0$ to 1.6 lead to acceptable results for most problems studied. $q = 0.5$ and $\alpha_C = 1.2$ are recommended. The dimensionless size of the support domain of $\alpha_s = 3.0$ to 4.5 should work for most problems.

BPIM and BRPIM need to be extended for 3D problems to take full advantages of the meshless concept. For 2D problems, BPIM is the simplest method; it performs the best and is very stable, without any kind of problem.

14

Mesh Free Methods Coupled with Other Methods

In the past few decades, development of finite element methods (FEM) has been accompanied by advances in boundary element methods (BEM). FEM is a domain discretization method, whereas BEM is a boundary discretization method. Both methods have their strengths and weaknesses. FEM is much more flexible for complex structures/domains with high inhomogeneity and nonlinearity, but require intensive computational resources. On the other hand, BEM requires much less computational resources, as discretization of the structure/domain is performed only on the boundary, which leads to a much smaller discretized equation system. BEM, however, is not efficient for inhomogeneous media/domain and nonlinear problems. Efforts to combine these two methods have been made (see, e.g., Liu et al., 1992) and have achieved remarkable results. Commercial software packages have also been developed (e.g., SYSNOISE) and used for solving a wide range of engineering problems.

In previous chapters, we presented both domain-type MFree methods and boundary-type MFree methods. Naturally, attempts have also been made to combine these two types of methods to take advantage of both. There is an additional motivation to couple MFree methods that are formulated using moving least squares (MLS) shape functions and MFree methods are that formulated using point interpolation method (PIM) shape functions or finite element (FE) shape functions. The aim is to simplify the procedure of imposing essential boundary conditions. G. R. Liu and Gu initiated this direction of development and have formulated a number of combined methods including EFG/BEM (Gu and Liu, 2001b), EFG/HBEM (Liu, G. R. and Gu, 2000c), MLPG/FEM/BEM (Liu, G. R. and Gu, 2000a,d), etc. This chapter is devoted to introducing these methods.

14.1 Coupled EFG/BEM

This section focuses on the coupling of the element free Galerkin (EFG) method with the boundary element (BE) method. Techniques for coupling the equation systems of EFG with those of BEM for continuum mechanics problems are presented in detail. This work was originally reported by Gu and Liu (2001b). The major issue is to enforce the displacement compatibility conditions on the interface boundary between the EFG domain and the BE domain. The interface elements, which are analogues of the FE interface element used by Krongauz and Belytschko (1996), are formulated and used along the interface boundary. Within the interface element the shape functions comprise the MLS and FE shape functions. Shape functions constructed in this manner satisfy both consistency and compatibility conditions on the interfaces. A number of numerical examples are presented to demonstrate the convergence, validity, and efficiency of the coupled method. It is shown that the coupled method can take full advantage of both EFG and BEMs. It is very easy to implement, and very flexible for computing displacements and stresses of desired accuracy in solids with or without infinite domains.

14.1.1 Basic Equations of Elastostatics

Consider the following two-dimensional (2D) problem of solid mechanics in domain Ω bounded by Γ:

$$\mathbf{L}^T\boldsymbol{\sigma} + \mathbf{b} = 0 \quad \text{in } \Omega \tag{14.1}$$

where σ is the stress tensor, which corresponds to the displacement field $\mathbf{u} = \{u, v\}^T$, $\mathbf{b}$ is the body force vector, and $\mathbf{L}$ is the differentiation operator defined by Equation 3.28. The boundary conditions are given as follows:

$$\boldsymbol{\sigma} \cdot \mathbf{n} = \bar{\mathbf{t}} \quad \text{on the natural boundary } \Gamma_t \tag{14.2}$$

$$\mathbf{u} = \bar{\mathbf{u}} \quad \text{on the essential boundary } \Gamma_u \tag{14.3}$$

in which the superposed bar denotes the prescribed boundary values and $\mathbf{n}$ is the unit outward normal to domain Ω.

14.1.2 Discrete Equations of EFG

In using a coupled EFG/BEM method, one can use boundary elements to model the portion of the domain that includes the essential boundary and the EFG is used where there is no essential boundary. Following the procedure presented in Chapter 6, without considering the essential boundary, we have the discrete system equation of EFG for all the field nodes in the EFG domain.

$$\mathbf{K}_{\text{EFG}}\mathbf{U} = \mathbf{F}_{\text{EFG}} + \mathbf{P}_{\text{EFG}} \tag{14.4}$$

where the subscript EFG indicates matrices obtained using standard EFG formulation. The vector $\mathbf{f}_{\text{EFG}}$ consists of the equivalent nodal forces contributed from the external force applied on the natural boundary. The nodal force can be obtained using

$$\mathbf{f}_{(\text{EFG})i} = \int_{\Gamma_t} \phi_i \mathbf{t} \, d\Gamma \tag{14.5}$$

The force vector $\mathbf{d}_{\text{EFG}}$ consists of the equivalent nodal forces contributed from the external body force in the form of

$$\mathbf{p}_{(\text{EFG})i} = \int_{\Omega} \phi_i \mathbf{b} \, d\Omega \tag{14.6}$$

Note that if EFG has to be used for the portion of the problem domain containing essential boundaries, formulations using the method of Lagrange multipliers, the penalty method, or any other method discussed in Chapter 6 must be used.

14.1.3 BE Formulation

From Equations 14.1 through 14.3, the principle of virtual displacements for linear elastic materials can be written as

$$\int_{\Omega} (\mathbf{L}^T\boldsymbol{\sigma} + \mathbf{b}) \cdot \mathbf{u}^* d\Omega = \int_{\Gamma_u} (\mathbf{u} - \bar{\mathbf{u}}) \cdot \mathbf{t}^* d\Gamma - \int_{\Gamma_t} (\mathbf{t} - \bar{\mathbf{t}})\mathbf{u}^* d\Gamma \tag{14.7}$$

where $\mathbf{t}$ is the surface traction, $\mathbf{u}^*$ is the virtual displacement, and $\mathbf{t}^*$ is the virtual surface traction corresponding to $\mathbf{u}^*$. The first term on the left-hand side of Equation 14.7 can be integrated by parts to become

$$\int_{\Omega} \mathbf{b} \cdot \mathbf{u}^* d\Omega + \int_{\Omega} \mathbf{L}^T\boldsymbol{\sigma}^* \cdot \mathbf{u} d\Omega = \int_{\Gamma_u} (\mathbf{u}^*\mathbf{t} - \bar{\mathbf{u}}\mathbf{t}^*)\, d\Gamma + \int_{\Gamma_t} (\mathbf{u}\mathbf{t}^* - \mathbf{u}^*\bar{\mathbf{t}}) d\Gamma \tag{14.8}$$

The starting domain integral can be reduced to an integral on the boundary by finding an analytical solution that makes the second integral in Equation 14.8 vanish. The most convenient one is the fundamental solution or Green's function, which satisfies the equation:

$$\mathbf{L}^T\boldsymbol{\sigma}^* + \Delta^i = 0 \tag{14.9}$$

where Δ^i is the Dirac delta function. Substituting Equation 14.9 into Equation 14.8, we obtain

$$c^i u^i + \int_{\Gamma} \mathbf{u}\mathbf{t}^* d\Gamma = \int_{\Gamma} \mathbf{t}\mathbf{u}^* d\Gamma + \int_{\Omega} \mathbf{b}\mathbf{u}^* d\Omega \tag{14.10}$$

The boundary values of $\mathbf{u}$ and $\mathbf{t}$ can now be expressed using interpolation functions and the values at the nodes of the boundary element on the boundary:

$$\mathbf{u} = \boldsymbol{\Phi}^T\mathbf{u}^e \tag{14.11}$$

$$\mathbf{t} = \boldsymbol{\Psi}^T\mathbf{t}^e \tag{14.12}$$

where $\boldsymbol{\Phi}^T$ and $\boldsymbol{\Psi}^T$ can be the conventional FE shape functions constructed based on the boundary elements, or one-piece PIM shape functions. $\mathbf{u}^e$ and $\mathbf{t}^e$ are the values of $\mathbf{u}$ and $\mathbf{t}$ at the boundary nodes. The resulting boundary integral Equation 14.10 can be written in matrix form as

$$\mathbf{HU} = \mathbf{BT} + \mathbf{P} \tag{14.13}$$

where $\mathbf{U}$ and $\mathbf{T}$ are vectors that collect all the nodal values of $\mathbf{u}$ and $\mathbf{t}$ at the boundary nodes, and

$$\mathbf{H} = \mathbf{c}^i + \int_{\Gamma} \mathbf{t}^*\Phi^T d\Gamma \tag{14.14}$$

$$\mathbf{B} = \int_{\Gamma} \mathbf{u}^*\Psi^T d\Gamma \tag{14.15a}$$

$$\mathbf{P} = \int_{\Omega} \mathbf{b}\mathbf{u}^* d\Omega \tag{14.15b}$$

The above integrals are to be carried only on boundaries, and therefore the domain need not be discretized.

To facilitate assembling the system equations of EFG and BE, the BE formulation is expressed in an equivalent form of the EFG formulation. Transforming Equation 14.13 by inverting **B** and then premultiplying the resultant by the distribution matrix **M** (Li and Han, 1986), we have

$$(\mathbf{MB}^{-1}\mathbf{H})\mathbf{U} - (\mathbf{MB}^{-1}\mathbf{P}) = \mathbf{MT} \tag{14.16}$$

where distribution matrix **M** is defined as

$$\mathbf{M} = \int_{\Gamma} \Phi\Psi^{T} \mathrm{d}\Gamma \tag{14.17}$$

Let

$$\mathbf{K}'_{\mathrm{BE}} = \mathbf{MB}^{-1}\mathbf{H} \tag{14.18}$$

$$\mathbf{P}_{\mathrm{BE}} = \mathbf{MB}^{-1}\mathbf{P} \tag{14.19}$$

$$\mathbf{F}_{\mathrm{BE}} = \mathbf{MT} \tag{14.20}$$

Equation 14.17 can then be written in the following equivalent form of the EFG formulation:

$$\mathbf{K}'_{\mathrm{BE}}\mathbf{U} = \mathbf{F}_{\mathrm{BE}} + \mathbf{P}_{\mathrm{BE}} \tag{14.21}$$

Note that matrix $\mathbf{K}'_{\mathrm{BE}}$ derived from the above formulation is in general asymmetric. The asymmetry arises from the approximations involved in the discretization process and the choice of the assumed solution. In the EFG domain, however, the matrix $\mathbf{K}_{\mathrm{EFG}}$ is symmetric. If Equation 14.21 is assembled directly into the EFG matrices Equation 14.4, the symmetry of the coefficient matrix will be destroyed, which leads to inefficiency in solving the system equations. To preserve the symmetry of the system matrix, a symmetrization operation must be performed for $\mathbf{K}'_{\mathrm{BE}}$. One simple method to perform such an operation is to minimize the squares of the errors in the asymmetric off-diagonal terms of $\mathbf{K}'_{\mathrm{BE}}$ (Brebbia et al., 1984). Hence, a new symmetric equivalent BE stiffness matrix $\mathbf{K}_{\mathrm{BE}}$ can be obtained using

$$k_{\mathrm{BE}ij} = 1/2(k'_{\mathrm{BE}ij} + k'_{\mathrm{BE}ji}) \tag{14.22}$$

Equation 14.21 can be rewritten as

$$\mathbf{K}_{\mathrm{BE}}\mathbf{U} = \mathbf{F}_{\mathrm{BE}} + \mathbf{P}_{\mathrm{BE}} \tag{14.23}$$

where $\mathbf{K}_{\mathrm{BE}}$ is now symmetric.

14.1.4 Coupling of EFG and BE System Equations

Continuity Conditions at the Interface

Consider now a problem domain consisting of two subdomains Ω_1 and Ω_2, joined by an interface boundary Γ_I. The EFG formulation is used in Ω_1 and the BE formulation is used in Ω_2, as shown in Figure 14.1. Compatibility and equilibrium conditions on Γ_I must be satisfied. Thus,

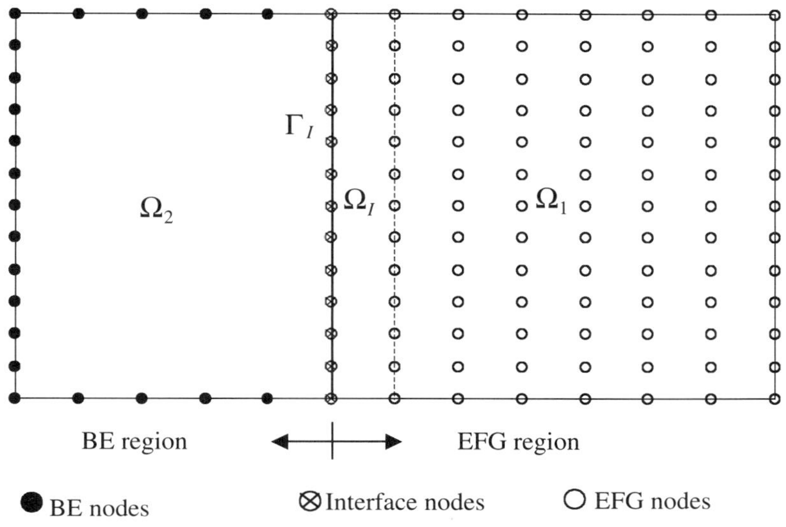

FIGURE 14.1
A problem domain divided into an EFG region and a BE region.

1. The nodal displacements formulated at the Γ_I for Ω_1 and that for Ω_2 should be equal, i.e.,

$$\mathbf{u}_I^{(1)} = \mathbf{u}_I^{(2)} \tag{14.24}$$

2. The summation of the nodal force formulated on the Γ_I for Ω_1 and that for Ω_2 should be zero, i.e.,

$$\mathbf{F}_I^{(1)} + \mathbf{F}_I^{(2)} = 0 \tag{14.25}$$

Because the MLS shape functions used in the EFG method do not possess the Kronecker delta function property, **u** in Equation 14.4 is the parameter of nodal displacement, which differs from the nodal displacement. Proper treatment is required to couple these two equation systems of EFG and BE domains along Γ_I. One simple method is to introduce interface elements in the EFG domain near the interface boundary Γ_I (Figure 14.2). In these interface elements, a hybrid displacement approximation is defined so that the shape functions of the EFG domain along Γ_I possess the delta function property. Therefore, **u** in Equation 14.4 becomes the true nodal displacement on the interface. The system equations for both EFG and BE can be assembled directly.

Shape Functions for the Interface Elements

The detailed characteristics of FE interface elements can be found in Krongauz and Belytschko (1996). Because the nodal arrangement may be irregular in the EFG domain, four to six node isoparametric interface FE elements (Hughes, 1987) are used for the interface elements.

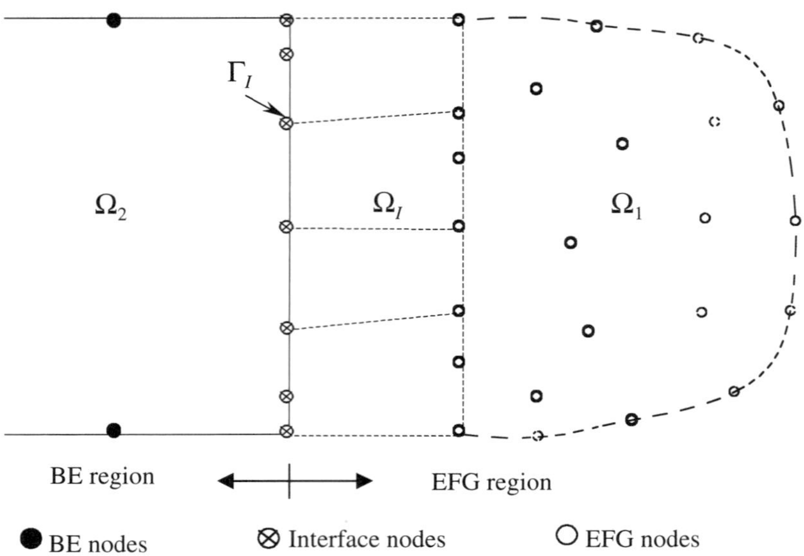

FIGURE 14.2
Interface element used in coupled EFG/BEM. (From Gu, Y. T. and Liu, G. R., *Comput. Methods Appl. Mech. Eng.*, 190, 4405–4419, 2001. With permission.)

A detailed illustration of the interface domain is shown in Figure 14.2, where Ω_I is a layer of subdomain along the interface boundary Γ_I within the EFG domain Ω_1. The modified displacement approximation in domain Ω_1 becomes

$$u_1^h(\mathbf{x}) = \begin{cases} u^{\text{EFG}}(\mathbf{x}) + R(\mathbf{x})(u^{\text{FE}}(\mathbf{x}) - u^{\text{EFG}}(\mathbf{x})) & \mathbf{x} \in \Omega_I \\ u^{\text{EFG}}(\mathbf{x}) & \mathbf{x} \in (\Omega_1 - \Omega_I) \end{cases} \tag{14.26}$$

where u_1^h is the displacement of a point in Ω_1, u^{EFG} is the EFG displacement given by

$$u^{\text{EFG}}(\mathbf{x}) = \sum_{i=1}^{n} \phi_i(\mathbf{x}) u_i \tag{14.27}$$

in which ϕ_i is the MLS shape function given by Equation 5.57. u^{FE} is the FE displacement defined by

$$u^{\text{FE}} = \sum_{i=1}^{n_e} N_i(x) u_i \quad n_e = 3, 4, 5, \ldots \tag{14.28}$$

where $N_i(x)$ is the FE shape function and n_e is the number of nodes in an FE interface element. The ramp function R is equal to the sum of the FE shape functions of a interface element associated with the interface element nodes that are located on the interface boundary Γ_I, i.e.,

$$R(x) = \sum_{i}^{k} N_i(x), \quad x_i \in \Gamma_I \tag{14.29}$$

where k is the number of nodes located on the interface boundary Γ_I for an interface element. According to the property of FE shape functions, R will be unity along Γ_I and will vanish from the interface domain, i.e.,

$$R(x) = \begin{cases} 1 & x \in \Gamma_I \\ 0 & x \in \Omega_1 - \Omega_I \end{cases} \tag{14.30}$$

The new displacement approximation in EFG domain Ω_1 can be rewritten as

$$\mathbf{u}_1^h(x) = \sum_i \tilde{\Phi}_i(x) u_i \tag{14.31}$$

where

$$\tilde{\Phi}_i(x) = \begin{cases} (1 - R(x))\Phi_i(x) + R(x)N_i(x) & x \in \Omega_I \\ \Phi_i(x) & x \in \Omega_1 - \Omega_I \end{cases} \tag{14.32}$$

The derivatives of the interface shape functions are

$$\tilde{\Phi}_{i,j} = \begin{cases} (1 - R)\Phi_{i,j} - R_{,j}\Phi_i + RN_{i,j} + R_{,j}N_i & x \in \Omega_I \\ \Phi_{i,j} & x \in \Omega_1 - \Omega_I \end{cases} \tag{14.33}$$

The approximation using the above modified shape functions will be compatible (or continuous) and reproduce the linear field exactly, which has been proved by Krongauz and Belytschko (1996).

The regular EFG and modified shape functions in one dimension are shown in Figure 14.3. It can be seen that the displacement approximation is continuous from the purely EFG domain passing to the interface domain. The derivative of it is, however, discontinuous across the boundary. These discontinuities do not adversely affect the overall results as they only affect a small number of nodes.

Using the above approximation, the shape functions of the EFG domain along Γ_I possess the Kronecker delta function property, and the system equations of the EFG domain, Equation 14.4, and the system equations for the BE domain, Equation 14.23, can be assembled together directly using the continuity condition on the interfaces of these two domains, which are defined in Equations 14.24 and 14.25.

Coupling Algorithm

The flowchart of coupled EFG/BEM is given as follows:

1. Loop over in EFG domain Ω^1
   a. Determine the nodes in the support domain of point $\mathbf{x}$
   b. Compute the EFG shape functions
   c. If point $\mathbf{x}$ is in the interface element:
      Compute FE shape functions in the element, and $R(\mathbf{x})$
      Compute the interface shape functions
      End if

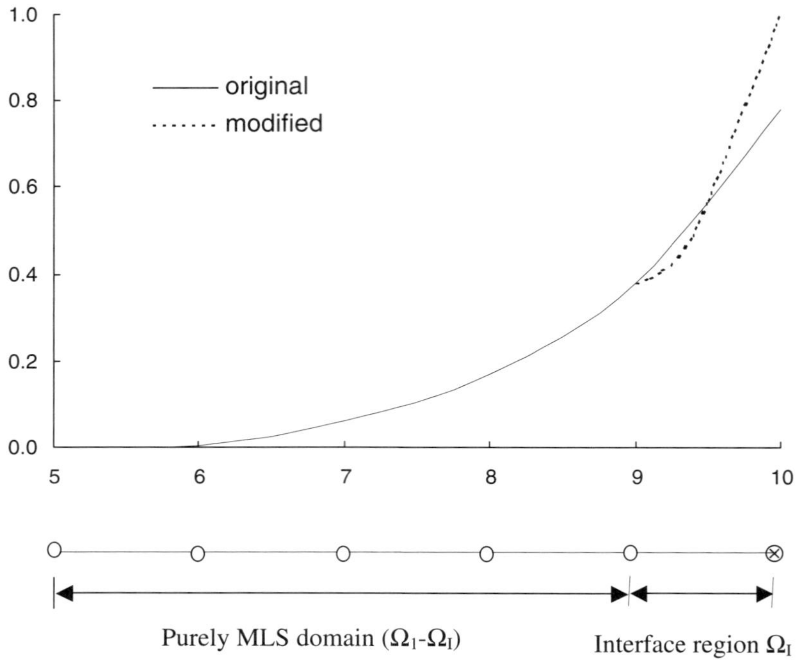

FIGURE 14.3
Comparison of original and modified shape functions in the EFG region where MLS approximation is employed; 1D case. (From Gu, Y. T. and Liu, G. R., *Comput. Methods Appl. Mech. Eng.*, 190, 4405–4419, 2001. With permission.)

   d. Assemble contributions to nodes to get the stiffness matrix $\mathbf{K}_{EFG}$
   e. End loop of EFG domain
2. Loop in boundary elements domain to obtain the matrix **H**, **B**
3. Compute **M**, $\mathbf{K}'_{BE}$, and symmetrize the $\mathbf{K}'_{BE}$ to obtain $\mathbf{K}_{BE}$
4. Assemble $\mathbf{K}_{EFG}$ and $\mathbf{K}_{BE}$ to get the global system equations
5. Solve the system equations for displacements
6. Postprocess to obtain displacement, strain, and stress

14.1.5 Numerical Results

Cases are run to examine coupled EFG/BEM in 2D elastostatics. The programs are developed to combine constant, linear, and quadratic BE with EFG. Interface elements with four to six nodes are used.

Example 14.1 Cantilever Beam

Coupled EFG/BEM is first applied to study the benchmarking problem of the cantilever beam shown in Figure 14.4. A plane stress problem is considered. The parameters for this example are as follows:

Young's modulus for the material: $E = 3.0 \times 10^7$

Poisson's ratios for two materials: $\nu = 0.3$

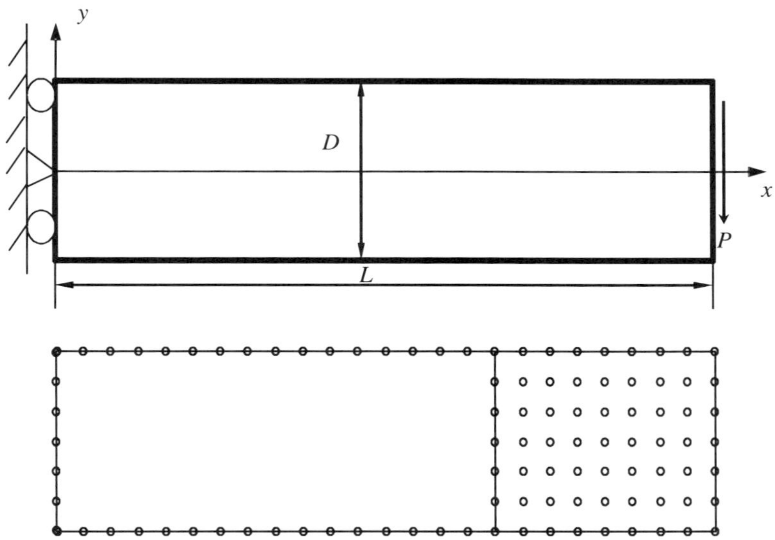

FIGURE 14.4
Nodal arrangement for the cantilever beam subjected to downward traction force on the right end of the beam. (From Gu, Y. T. and Liu, G. R., *Comput. Methods Appl. Mech. Eng.*, 190, 4405–4419, 2001. With permission.)

Thickness of the beam: $t = 1$
Height of the beam: $D = 12$
Length of the beam: $L = 48$
Load: $P = 1000$

The beam is subjected to parabolic traction at the free end as shown in Figure 6.4. The analytical solution is available; it can be found in the textbook by Timoshenko and Goodier (1977) and is listed in Equations 6.28 to 6.33.

The beam is artificially divided into two parts as shown in Figure 14.4. Boundary elements are used to model the left part of the beam in which the essential boundary is included. EFG is used in the right part. The nodal arrangement is also shown in Figure 14.4. Background integration cells of 6×8 are used in the EFG domain. In each integration cell, 4×4 Gauss quadrature is used to evaluate the stiffness matrix of the EFG. Linear boundary elements are employed in the BE domain. Rectangular elements are employed as interface elements. Only 100 nodes in total are used in the entire coupled model. The total number of nodes determines the size of the final assembled system equation, and directly affects the computation time for solving this problem.

Figure 14.5 plotted the shear stress distribution on the cross section of the beam at $x = L/2$ calculated using the present coupled EFG/BEM. Results obtained using analytical formulae and an FEM/BEM (an in-house code developed by G. R. Liu's group) are also plotted in Figure 14.5 for comparison. When FEM/BEM is used, the right portion of the beam is modeled using linear FEs instead of EFG modes. In this case, there is no need to use transition elements, as the shape functions for both FEM and BE are of FE type, which possess the Kronecker delta function property. The plot shows excellent agreement between the results obtained using these three methods. It can be also found that coupled EFG/BEM yields a more accurate result than the FE/BE method. This is because the EFG performs better than the FEM of linear elements.

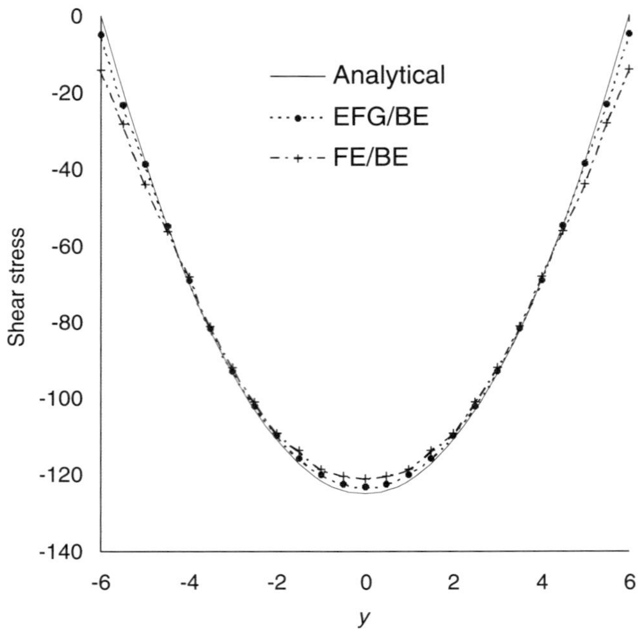

FIGURE 14.5
Shear stress τ_{xy} at the section $x = L/2$ of the cantilever beam computed using three different methods. (From Gu, Y. T. and Liu, G. R., *Comput. Methods Appl. Mech. Eng.*, 190, 4405–4419, 2001. With permission.)

For quantitative error analysis, we define the following norm using shear stresses as an error indicator, as the accuracy in shear strain or shear stress is much more critical for the beam problem:

$$e_t = \frac{1}{N}\sqrt{\sum_{i=1}^{N}(\tau_i - \bar{\tau}_i)^2 \Big/ \sum_{i=1}^{N}\bar{\tau}^2} \tag{14.34}$$

where N is the number of nodes investigated, τ is the shear stress obtained by the numerical method, and $\bar{\tau}$ is the analytical shear stress used as a reference.

The convergence with mesh refinement is shown in Figure 14.6, where h is the nodal spacing or the element size in FEM. It is observed that the convergence of the coupled method is very good. The convergence of the coupled FE/BE method is also shown in the same figure. This figure shows that the accuracy of coupled EFG/BEM is higher than the FE/BE method because of the higher accuracy of EFG. However, the convergence rate of these two coupled methods is nearly same, and is found to be about 2.3 for this problem. This is because that the accuracy of BEM plays a part in the convergence rate of the coupled EFG/BE and FE/BE methods.

Example 14.2 Hole in an Infinite Plate

Consider now an infinite plate with a central circular hole subjected to a unidirectional tensile load of $p = 1.0$ in the x direction. As a large finite plate can be considered a good approximation of an infinite plate, a finite square plate of 20×20 is considered. Making use of the symmetry, only the upper right quadrant of the finite plate is modeled, as shown in Figure 14.7. A plane strain problem is considered, and the material properties are $E = 1.0 \times 10^3$, $\nu = 0.3$. Symmetry conditions are imposed on the left and bottom edges, and the

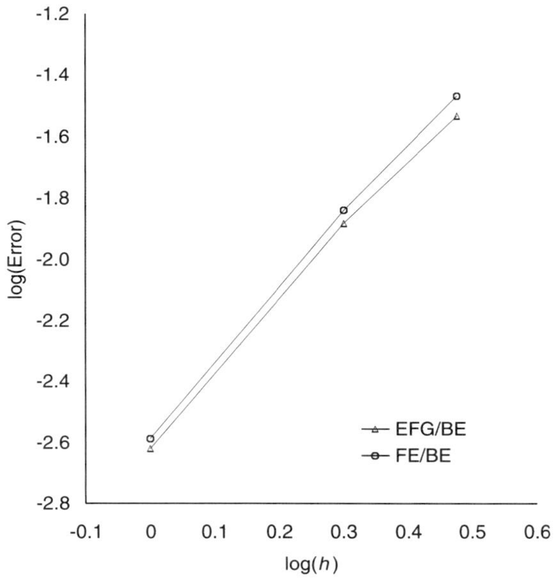

FIGURE 14.6
Convergence in energy norm of error e_t. (From Gu, Y. T. and Liu, G. R., *Comput. Methods Appl. Mech. Eng.*, 190, 4405–4419, 2001. With permission.)

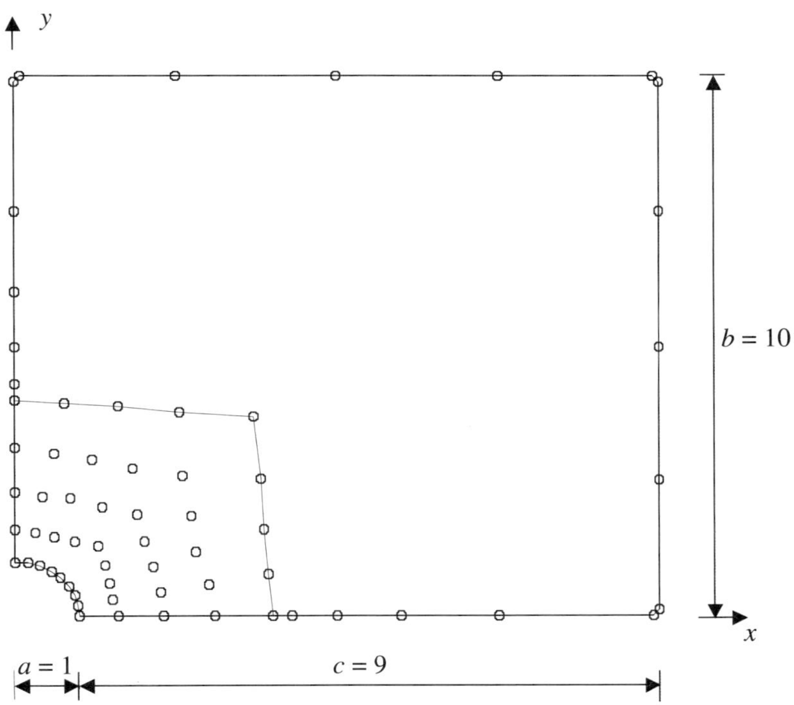

FIGURE 14.7
Nodes in a plate with a hole at its center subjected to a unidirectional tensile load in the x direction. A quarter of the plate is modeled. (From Gu, Y. T. and Liu, G. R., *Comput. Methods Appl. Mech. Eng.*, 190, 4405–4419, 2001. With permission.)

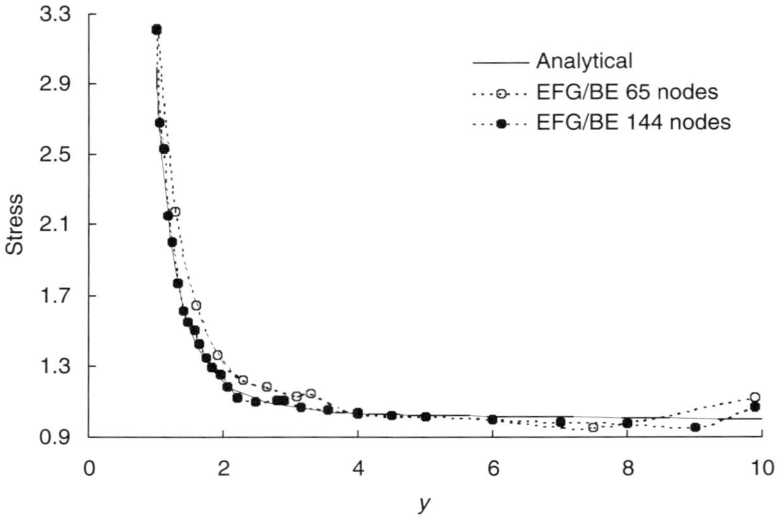

FIGURE 14.8
Stress distribution (σ_x, at $x = 0$) obtained using EFG/BEM together which analytic results for a square plate with a hole at its center. (From Gu, Y. T. and Liu, G. R., *Comput. Methods Appl. Mech. Eng.*, 190, 4405–4419, 2001. With permission.)

inner boundary of the hole is traction free. The tensile load p is imposed on the right edge in the x direction. The exact solution for the stresses of an infinite plate is given in the textbook by Timoshenko and Goodier (1977) and is listed in Equations 6.121 to 6.126.

A quarter of the plate is modeled, symmetry conditions are imposed on the left and bottom edges, and the inner boundary of the hole is traction free. The tensile load in the x direction is imposed on the right edge. The plate is divided into two domains. In the area near the hole, EFG is employed. For the rest of the area of the problem domain BEM is applied.

It is found that the results obtained for displacements are almost identical. As the stresses are much more critical, detailed results of stresses are presented here. The stresses σ_x at $x = 0$ obtained by the coupled method using two kinds of nodal arrangement are given in Figure 14.8. The figure shows that the coupled method yields satisfactory results for the problem when 144 nodes are used. For comparison, the results obtained using EFG/BE, FE/BE, and EFG methods are shown in Figure 14.9. It can be found that EFG/BEM yields better results than the FE/BE method. The accuracy of EFG/BE and EFG methods is nearly the same. However, many fewer nodes are used in coupled EFG/BEM (144 nodes) than are used in the EFG method (231 nodes).

There exist oscillations in the solution of the corner nodes in the BE domain, as shown in Figure 14.8. This is because the tractions are discontinuous in these corner nodes. Special care should be taken in handling traction discontinuities at the corner nodes, as discussed in Chapter 13. One method to overcome this difficulty is simply to split the corner node into two nodes with each node on one side of the corner. These two nodes are very close to the original corner node. A constant BE is then used between these two nodes. Because these two nodes belong to different sides of the corner, the discontinuity of the traction on the corner can be modeled without difficulty. The method is very simple, works very well, and is widely used in BEM. It is also used in Chapter 13 for boundary-type MFree methods.

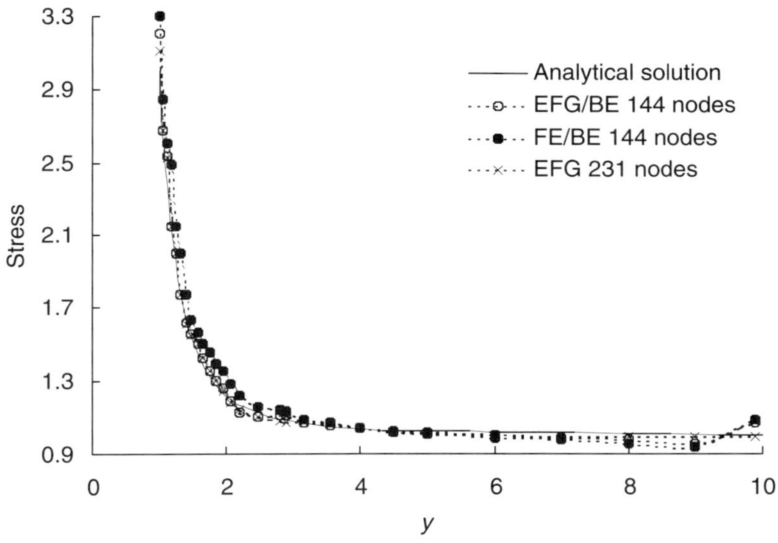

FIGURE 14.9
Stress distribution (σ_x, at $x = 0$) obtained using EFG/BEM, FE/BEM, and EFG together which analytical results for a square plate with a hole at its center. (From Gu, Y. T. and Liu, G. R., *Comput. Methods Appl. Mech. Eng.*, 190, 4405–4419, 2001. With permission.)

Example 14.3 A Structure on a Semi-Infinite Elastic Foundation

In this example, the coupled method is used to solve a foundation–structure interaction problem, illustrated schematically in Figure 14.10. A structure stands on a semi-infinite elastic foundation. The problem has been investigated using coupled FE/BEM by some researchers (Brebbia and Georgiou, 1979). The infinite elastic foundation can be modeled in one of the following three ways:

1. Truncating the plane at a finite distance—approximate method
2. Using a fundamental solution corresponding to the semi-space problem rather than a full-space Green's function in BEM
3. Using an infinite element in FEM

Method 1 is used in this section because it is convenient to compare the coupled method solution with the EFG, FE, and FE/BE solutions.

As shown in Figure 14.10, Region 2 represents the semi-infinite foundation and is given a semicircular shape of very large diameter in relation to Region 1, which represents the structure. Boundary conditions to restrain rigid body movements are applied. The EFG method is used in Region 1, and BEM is used in Region 2. The nodal arrangement of coupled EFG/BEM is shown in Figure 14.11. The problem is also analyzed using FEM, EFG, and FE/BEM. The nodal arrangement of EFG is shown in Figure 14.12. Two load cases shown in Figure 14.13 are analyzed: case 1 considers five concentrated vertical loads along the top, and case 2 considers an additional horizontal load acting at the left corner.

The results of displacement in the y direction (vertical) on top of the structure are listed in Table 14.1. The results obtained using FEM, EFG, and FE/BEM are also included in the same table for comparison. The results obtained using the present EFG/BEM are in very good agreement with those obtained using FE, EFG, and FE/BEM. However, it is interesting to note that the foundation is adequately represented using only 30 BE nodes in coupling cases as compared to 120 nodes for the EFG and FE cases. The saving is considerable.

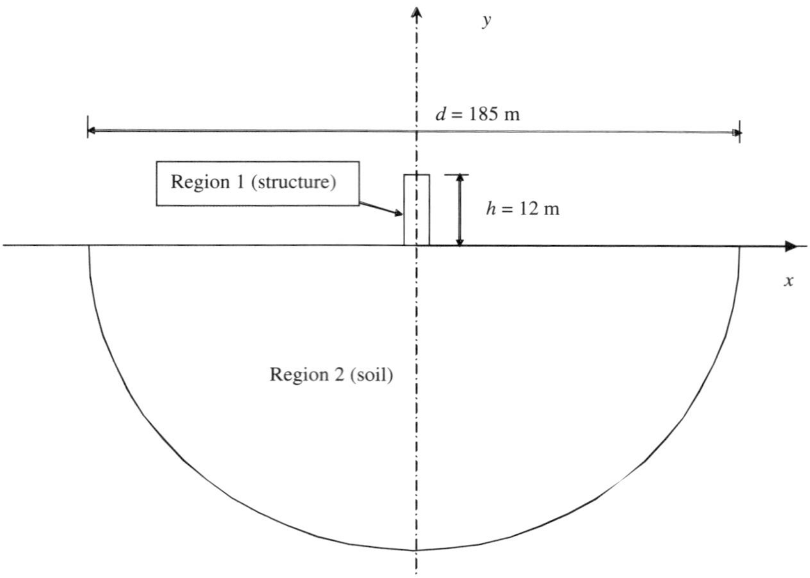

FIGURE 14.10
A structure standing on the top of a semi-infinite soil foundation. (From Gu, Y. T. and Liu, G. R., *Comput. Methods Appl. Mech. Eng.*, 190, 4405–4419, 2001. With permission.)

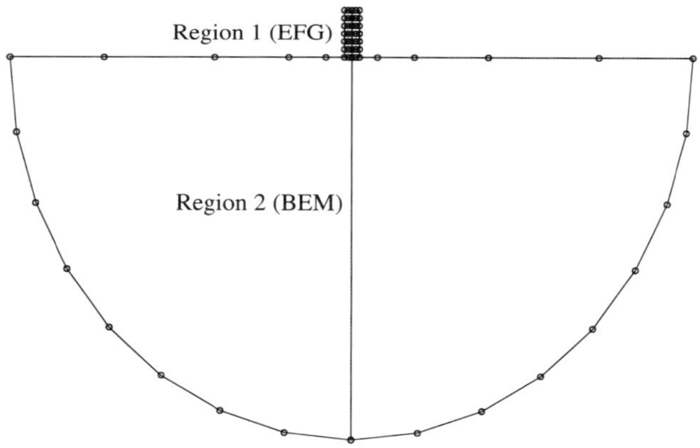

FIGURE 14.11
Nodal arrangement for the coupled EFG/BEM model. (From Gu, Y. T. and Liu, G. R., *Comput. Methods Appl. Mech. Eng.*, 190, 4405–4419, 2001. With permission.)

14.2 Coupled EFG and Hybrid BEM

In the previous section, we demonstrated coupling of the system equations of EFG and BEM. We have seen that a symmetrization of the BE stiffness matrix must be performed before the assembly of the EFG system equations with the BE system equations. This can lead to a loss of accuracy and efficiency of the coupled method. In this section, we present an alternative approach to avoid this disadvantage in coupled EFG/BEM.

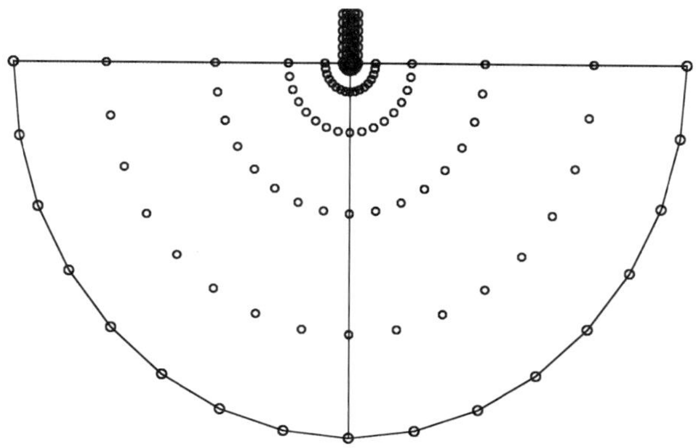

FIGURE 14.12
Nodal arrangement for the coupled EFG model. (From Gu, Y. T. and Liu, G. R., *Comput. Methods Appl. Mech. Eng.*, 190, 4405–4419, 2001. With permission.)

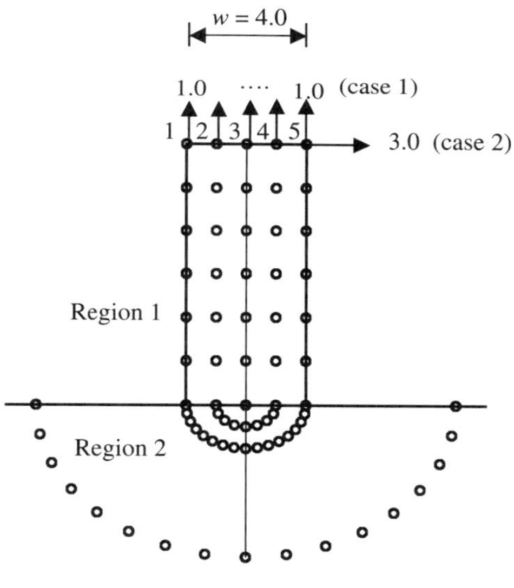

FIGURE 14.13
Nodal arrangement in the structure portion where EFG is used. The structure is loaded by, case 1, a uniformly distributed normal traction in the y direction or, case 2, a concentrate force in the x direction at the top right corner. (From Gu, Y. T. and Liu, G. R., *Comput. Methods Appl. Mech. Eng.*, 190, 4405–4419, 2001. With permission.)

In the late 1980s, alternative BE formulations were developed based on generalized variational principles. Dumont (1988) proposed a hybrid stress BE formulation based on the Hellinger–Reissner principle. DeFigueiredo and Brebbia (1989), DeFigueiredo (1991), and Jin et al. (1996) presented a hybrid displacement boundary element (HBE) formulation. The HBE formulation led to a symmetric stiffness matrix. This property of symmetry can be an added advantage in coupling the HBE with methods that produce symmetric system matrices.

This section presents a coupled EFG/HBE method for continuum mechanics problems. This work was originally reported by G. R. Liu and Gu (2000c). The method of Lagrange multipliers is employed to impose the compatibility conditions on the interface boundary of the EFG and HBE domains. Coupled system equations are derived based on variational

TABLE 14.1

Vertical Displacements along the Top of the Structure

Nodes	Displacements ($\times 10^{-4}$)			
	FE	EFG	FE/BE	EFG/BE
Load case 1				
1	1.41	1.42	1.40	1.42
2	1.34	1.34	1.33	1.33
3	1.32	1.32	1.32	1.32
4	1.34	1.34	1.33	1.33
5	1.41	1.42	1.40	1.42
Load case 2				
1	−3.39	−3.43	−3.55	−3.58
2	−0.97	−1.01	−1.05	−1.04
3	1.35	1.35	1.35	1.34
4	3.61	3.67	3.70	3.68
5	6.00	6.04	6.17	6.13

Source: Gu, Y. T. and Liu, G. R., *Comput. Methods Appl. Mech. Eng.*, 190, 4405–4419, 2001. With permission.

formulation. Several numerical examples are examined using the EFG/HBE to demonstrate the convergence, validity, and efficiency of the coupled EFG/HBE method.

Compared with coupled EFG/BEM discussed in the previous section, the present EFG/HBE method makes the following advances:

1. The coupled system equations are formulated in a different but more general manner.
2. System matrices obtained by EFG/HBE are symmetric without the need for an operation of symmetrization to the BE matrix.
3. There is no need for interface elements; therefore, mesh generation becomes much simpler and there is no special treatment needed on the interface.

The trade-off would be

1. The system matrix is larger than that of EFG/BEM.
2. The system matrix becomes nonpositive.

These drawbacks are similar to that of EFG using the method of Lagrange multipliers. Detailed formulation of the EFG/HBE is presented as follows.

14.2.1 EFG Formulation

Discrete Equations of EFG

Consider again the 2D problem of solid mechanics defined in Equations 14.1 through 14.3. The constrained functional can be written as

$$\Pi_1 = \int_{\Omega} \frac{1}{2}\boldsymbol{\varepsilon}^T \cdot \boldsymbol{\sigma} d\Omega - \int_{\Omega} \mathbf{u}^T \cdot \mathbf{b} d\Omega - \int_{\Gamma_t} \bar{\mathbf{u}}^T \cdot \bar{\mathbf{t}} d\Gamma + \int_{\Gamma} \boldsymbol{\lambda}^T \cdot (\bar{\mathbf{u}} - \mathbf{u}) d\Gamma \tag{14.35}$$

where the fourth term of the integral is for the essential boundary condition, and $\boldsymbol{\lambda}$ is a vector of Lagrange multipliers. Following the procedure in Chapter 6, the discrete system equation of EFG for the EFG domain can be written in the form:

$$\begin{bmatrix} \mathbf{K}_{\text{EFG}} & \mathbf{G}_{\text{EFG}} \\ \mathbf{G}_{\text{EFG}}^T & 0 \end{bmatrix} \begin{Bmatrix} \mathbf{U} \\ \boldsymbol{\lambda} \end{Bmatrix} = \begin{Bmatrix} \mathbf{F}_{\text{EFG}} + \mathbf{P}_{\text{EFG}} \\ \mathbf{q}_{\text{EFG}} \end{Bmatrix} \tag{14.36}$$

where the subscript EFG indicates matrices obtained using standard EFG formulation. The components in vectors $\mathbf{f}_{\text{EFG}}$ and $\mathbf{d}_{\text{EFG}}$ are defined in Equations 14.5 and 14.6, and the reset matrices have been defined in detail in Chapter 6.

14.2.2 Hybrid Displacement BE Formulation

The constrained functional for the hybrid displacement BEM can be written as

$$\Pi_2 = \int_{\Omega} \frac{1}{2} \boldsymbol{\varepsilon}^T \cdot \boldsymbol{\sigma} \, d\Omega - \int_{\Omega} \mathbf{u}^T \cdot \mathbf{b} \, d\Omega - \int_{\Gamma_t} \tilde{\mathbf{u}}^T \cdot \bar{\mathbf{t}} \, d\Gamma + \int_{\Gamma} \boldsymbol{\lambda}^T \cdot (\tilde{\mathbf{u}} - \mathbf{u}) \, d\Gamma \tag{14.37}$$

where $\tilde{\mathbf{u}}$ is the displacement on the boundary, and $\mathbf{u}$ is the displacement in the domain. The fourth term of the integral is for the compatibility of the displacements on the boundary with that near the boundary in the domain, and $\boldsymbol{\lambda}$ is a vector of Lagrange multipliers. As the Lagrange multipliers $\boldsymbol{\lambda}$ represent the traction on the boundary, it is therefore denoted explicitly by $\tilde{\mathbf{t}}$. Hence, Equation 14.37 can be rewritten as

$$\Pi_2 = \int_{\Omega} \frac{1}{2} \boldsymbol{\varepsilon}^T \cdot \boldsymbol{\sigma} \, d\Omega - \int_{\Omega} \mathbf{u}^T \cdot \mathbf{b} \, d\Omega - \int_{\Gamma_t} \tilde{\mathbf{u}}^T \cdot \bar{\mathbf{t}} \, d\Gamma + \int_{\Gamma} \tilde{\mathbf{t}}^T \cdot (\tilde{\mathbf{u}} - \mathbf{u}) d\Gamma \tag{14.38}$$

The first term on the right-hand side can be integrated by parts to become

$$\Pi_2 = \int_{\Gamma} \frac{1}{2} \mathbf{t}^T \cdot \mathbf{u} d\Gamma - \int_{\Omega} \mathbf{u}^T \cdot \mathbf{b} d\Omega - \int_{\Gamma_t} \tilde{\mathbf{u}}^T \cdot \bar{\mathbf{t}} \, d\Gamma + \int_{\Gamma} \tilde{\mathbf{t}}^T \cdot (\tilde{\mathbf{u}} - \mathbf{u}) d\Gamma - \int_{\Omega} \frac{1}{2} \mathbf{L}^T \boldsymbol{\sigma} \cdot \mathbf{u} d\Omega \tag{14.39}$$

The starting domain integral in Equation 14.38 can be reduced to an integral on the boundary using the fundamental solution for Equation 14.9, which is Green's function.

The displacement vector within the domain is approximated as a series of products of $\mathbf{U}^*$, which are formed using the fundamental solutions (DeFigueiredo, 1991) and unknown parameters $\mathbf{s}$, i.e.,

$$\mathbf{u} = \mathbf{U}^* \mathbf{s} \tag{14.40}$$

The displacement vector on the boundary is written as the product of known interpolation functions by the nodal displacement at the boundary nodes, i.e.,

$$\tilde{\mathbf{u}} = \boldsymbol{\Phi}^T \mathbf{u}^e \tag{14.41}$$

Similarly, the traction vector is approximated as a series of products of $\mathbf{T}^*$ that are also formed using the fundamental solutions and unknown parameters $\mathbf{s}$.

$$\mathbf{t} = \mathbf{T}^* \mathbf{s} \tag{14.42}$$

The traction vector on the boundary is written as the product of known interpolation functions and the nodal traction at the boundary nodes, i.e.,

$$\tilde{\mathbf{t}} = \mathbf{\Psi}^T \mathbf{t}^e \tag{14.43}$$

Substituting Green's function and Equations 14.40 through 14.43 into Equation 14.39, we can obtain

$$\Pi = -1/2\mathbf{s}^T\mathbf{A}\mathbf{s} - \mathbf{t}^T\mathbf{B}^T\mathbf{s} + \mathbf{t}^T\mathbf{L}\mathbf{U} - \mathbf{U}^T\mathbf{f} - \mathbf{s}^T\mathbf{b} \tag{14.44}$$

where

$$\mathbf{A} = \int_\Gamma \mathbf{U}^*\mathbf{T}^* d\Gamma \tag{14.45}$$

$$\mathbf{B} = \int_\Gamma \mathbf{\Psi}\mathbf{U}^* d\Gamma \tag{14.46}$$

$$\mathbf{L} = \int_\Gamma \mathbf{\Psi}\mathbf{\Phi}^T d\Gamma \tag{14.47}$$

$$\mathbf{F}_{\mathrm{HBE}} = \int_\Gamma \mathbf{\Phi}\bar{\mathbf{t}}\, d\Gamma \tag{14.48}$$

$$\mathbf{g} = \int_\Omega \mathbf{U}^*\mathbf{b}\, d\Omega \tag{14.49}$$

The stationary conditions for Π can now be found by setting its first variation of Π to zero. As this must be true for any arbitrary values of $\delta\mathbf{s}$, $\delta\mathbf{u}$, and $\delta\mathbf{t}$, one obtains

$$\mathbf{K}_{\mathrm{HBE}}\mathbf{U} = \mathbf{F}_{\mathrm{HBE}} + \mathbf{P}_{\mathrm{HBE}} \tag{14.50}$$

where

$$\mathbf{K}_{\mathrm{HBE}} = \mathbf{R}^T\mathbf{A}\mathbf{R} \tag{14.51}$$

$$\mathbf{R} = (\mathbf{B}^T)^{-1}\mathbf{L} \tag{14.52}$$

$$\mathbf{P}_{\mathrm{HBE}} = \mathbf{R}^T\mathbf{g} \tag{14.53}$$

It can be proved that matrix **A** is symmetric; hence, matrix **K** is symmetric. Equation 14.50 shows that this hybrid displacement boundary formulation leads to an equivalent stiffness formulation. The matrix **K** may be viewed as a symmetric stiffness matrix, but the above integrals are needed to perform only on boundaries, and the domain need not be discretized.

14.2.3 Coupling of EFG and HBE

Continuity Conditions at Coupled Interfaces

Consider a problem consisting of two domains of Ω^1 and Ω^2, as schematically shown in Figure 14.14. These two domains are joined by an interface Γ_I. The EFG formulation is

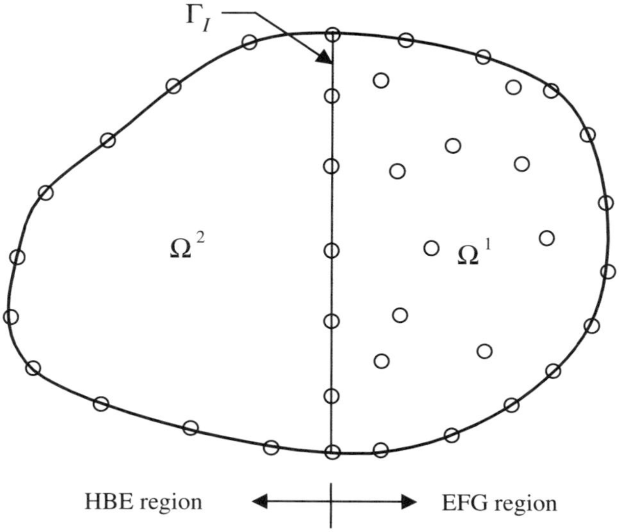

FIGURE 14.14
Domain division into EFG and HBE regions. (From Liu, G. R. and Gu, Y. T., *Comput. Mech.*, 26, 166–173, 2000. With permission.)

used in Ω^1 and the HBE formulation is used in Ω^2. Continuity conditions that must be satisfied on Γ_I are given by

$$\tilde{\mathbf{u}}_I^1 = \tilde{\mathbf{u}}_I^2 \tag{14.54}$$

$$\mathbf{F}_I^1 + \mathbf{F}_I^2 = 0 \tag{14.55}$$

where $\tilde{\mathbf{u}}_I^1$ and $\tilde{\mathbf{u}}_I^2$ are the displacements on Γ_I for Ω^1 and Ω^2, $\mathbf{F}_I^1$ and $\mathbf{F}_I^2$ are the forces on Γ_I for Ω^1 and Ω^2, respectively.

Because the shape functions of EFG are derived using MLS, the displacement vector in Equation 14.36 differs from the true nodal displacement. Proper treatments are needed to assemble these equations of EFG and HBE.

Coupling EFG with HBE via Modified Variational Form

A subfunctional is introduced to enforce the continuity condition, Equation 14.54, by means of Lagrange multiplier λ on the interface boundary

$$\Pi_I = \int_{\Gamma_I} \boldsymbol{\gamma}^T(\tilde{\mathbf{u}}_I^1 - \tilde{\mathbf{u}}_I^2)\, \mathrm{d}\Gamma = \int_{\Gamma_I} \boldsymbol{\gamma}^T \tilde{\mathbf{u}}_I^1\, \mathrm{d}\Gamma - \int_{\Gamma_I} \boldsymbol{\gamma}^T \tilde{\mathbf{u}}_I^2\, \mathrm{d}\Gamma = \Pi_I^1 - \Pi_I^2 \tag{14.56}$$

In Equation 14.54, Π_I^1 and Π_I^2 are the boundary integrations along the EFG side and the HBE side, respectively. Introducing Π_I^1 and Π_I^2 separately into functions, Equations 14.35 and 14.37, generalized functional forms can be written as

$$\Pi_{\mathrm{EFG}} = \int_\Omega \frac{1}{2}\boldsymbol{\varepsilon}^T \cdot \boldsymbol{\sigma}\mathrm{d}\Omega - \int_\Omega \mathbf{u}^T \cdot \mathbf{b}\mathrm{d}\Omega - \int_{\Gamma_t} \mathbf{u}^T \cdot \bar{\mathbf{t}}\,\mathrm{d}\Gamma - \int_{\Gamma_u} \boldsymbol{\lambda}_{\mathrm{EFG}}^T \cdot (\mathbf{u} - \bar{\mathbf{u}})\,\mathrm{d}\Gamma + \int_{\Gamma_I} \boldsymbol{\gamma}^T \cdot \tilde{\mathbf{u}}_I^1\,\mathrm{d}\Gamma \tag{14.57}$$

$$\Pi_{\mathrm{HBE}} = \int_\Omega \frac{1}{2}\boldsymbol{\varepsilon}^T \cdot \boldsymbol{\sigma}\mathrm{d}\Omega - \int_\Omega \mathbf{u}^T \cdot \mathbf{b}\mathrm{d}\Omega - \int_{\Gamma_t} \tilde{\mathbf{u}}^T \cdot \bar{\mathbf{t}}\,\mathrm{d}\Gamma + \int_{\Gamma} \boldsymbol{\lambda}_{\mathrm{HBE}}^T \cdot (\tilde{\mathbf{u}} - \mathbf{u})\,\mathrm{d}\Gamma - \int_{\Gamma_I} \boldsymbol{\gamma}^T \cdot \tilde{\mathbf{u}}_I^2\,\mathrm{d}\Gamma \tag{14.58}$$

In these variational formulations, the domains of EFG and HBE are connected via Lagrange multiplier $\boldsymbol{\gamma}$.

In the EFG domain, $\mathbf{u}$ is given by Equation 5.56 $\boldsymbol{\gamma}$ is given by production of the interpolation function $\boldsymbol{\Lambda}$ and value of $\boldsymbol{\gamma}^I$

$$\boldsymbol{\gamma} = \boldsymbol{\Lambda}^T \boldsymbol{\gamma}^I \tag{14.59}$$

$\boldsymbol{\Lambda}$ can consist of shape functions of the FE type. Substituting Equations 5.56 and 14.59 into Equation 14.57, and using the stationary condition, we can obtain the following EFG equations:

$$\begin{bmatrix} \mathbf{K}_{\text{EFG}} & \mathbf{G}_{\text{EFG}} & \mathbf{B}_{\text{EFG}} \\ \mathbf{G}_{\text{EFG}}^T & 0 & 0 \\ \mathbf{B}_{\text{EFG}}^T & 0 & 0 \end{bmatrix} \begin{Bmatrix} \mathbf{U} \\ \boldsymbol{\lambda} \\ \boldsymbol{\gamma} \end{Bmatrix} = \begin{Bmatrix} \mathbf{F}_{\text{EFG}} + \mathbf{P}_{\text{EFG}} \\ \mathbf{q}_{\text{EFG}} \\ 0 \end{Bmatrix} \tag{14.60}$$

where the subscript EFG indicates the EFG matrices, and $\mathbf{B}$ is defined as

$$\mathbf{B}_{\text{EFG}} = \int_{\Gamma_I} \boldsymbol{\Lambda} \mathbf{F}_{\text{EFG}}^T \, d\Gamma \tag{14.61}$$

The stationary condition of Equation 14.58 leads to the following HBE equations:

$$\begin{bmatrix} \mathbf{K}_{\text{HBE}} & -\mathbf{H}_{\text{HBE}} \\ -\mathbf{H}_{\text{HBE}}^T & 0 \end{bmatrix} \begin{Bmatrix} \mathbf{U} \\ \boldsymbol{\gamma} \end{Bmatrix} = \begin{Bmatrix} \mathbf{F}_{\text{HBE}} + \mathbf{P}_{\text{HBE}} \\ 0 \end{Bmatrix} \tag{14.62}$$

where $\mathbf{K}_{\text{HBE}}$, $\mathbf{f}_{\text{HBE}}$, and $\mathbf{d}_{\text{HBE}}$ are defined by Equations 14.48, 14.50, and 14.53. $\mathbf{H}$ is defined as

$$\mathbf{H}_{\text{HBE}} = \int_{\Gamma_I} \boldsymbol{\Lambda} \boldsymbol{\phi}_{\text{HBE}}^T \, d\Gamma \tag{14.63}$$

Because two domains are connected along the interface boundary Γ_I, assembling Equations 14.60 and 14.63 yields a linear system of

$$\begin{bmatrix} \mathbf{K}_{\text{EFG}} & 0 & \mathbf{G}_{\text{EFG}} & \mathbf{B}_{\text{EFG}} \\ 0 & \mathbf{K}_{\text{HBE}} & 0 & -\mathbf{H}_{\text{HBE}} \\ \mathbf{G}_{\text{EFG}}^T & 0 & 0 & 0 \\ \mathbf{B}_{\text{EFG}}^T & -\mathbf{H}_{\text{HBE}}^T & 0 & 0 \end{bmatrix} \begin{Bmatrix} \mathbf{U}_{\text{EFG}} \\ \mathbf{U}_{\text{HBE}} \\ \boldsymbol{\lambda} \\ \boldsymbol{\gamma} \end{Bmatrix} = \begin{Bmatrix} \mathbf{F}_{\text{EFG}} + \mathbf{P}_{\text{EFG}} \\ \mathbf{F}_{\text{HBE}} + \mathbf{P}_{\text{HBE}} \\ \mathbf{q}_{\text{EFG}} \\ 0 \end{Bmatrix} \tag{14.64}$$

The continuity conditions on Γ_I given in Equations 14.54 and 14.55 are satisfied via the above variational formulation. Note that the system matrix is symmetric, but enlarged and nonpositive.

14.2.4 Numerical Results

Three examples in 2D elastostatics that were examined in the previous section are reexamined using the present coupled EFG/HBE method.

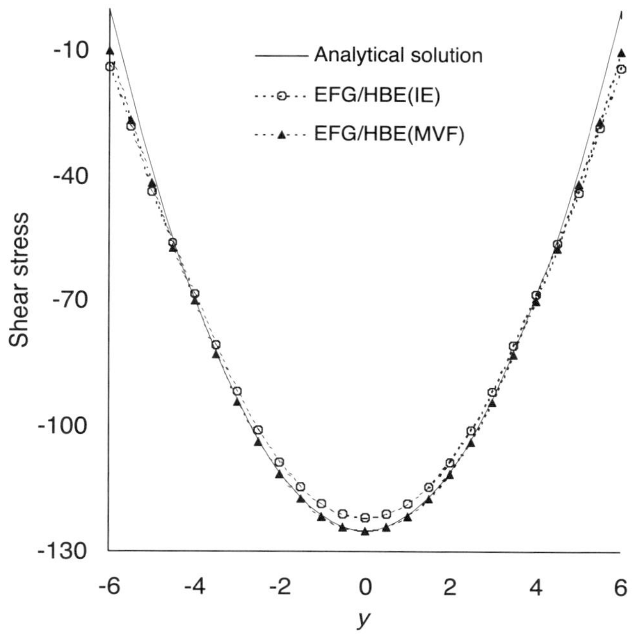

FIGURE 14.15
Shear stress τ_{xy} at the section $x = L/2$ of the beam. (From Liu, G. R. and Gu, Y. T., *Comput. Mech.*, 26, 166–173, 2000. With permission.)

Example 14.4 Cantilever Beam

All parameters and conditions are exactly the same as those in Example 14.1. The nodal arrangement is shown in Figure 14.4, and a background mesh of 6×8 is used in the EFG domain. In each integration cell, 4×4 Gauss quadrature is used to evaluate the stiffness matrix of the EFG. Only 100 nodes in total are used in the coupled method.

Figure 14.15 illustrates the comparison between the shear stress calculated analytically and using the coupled method at the section of $x = L/2$. The plot shows excellent agreement between the analytical and numerical results. The computational result by the present coupled method with interface elements (IE) is also shown in the same figure. There is clear evidence that the accuracy of the coupled method using the modified variational formulation (MVF) is higher than that using the IE method.

The displacement along the interface boundary is shown in Table 14.2. It is shown that the continuity of the displacement is satisfied accurately using the present modified variational formulation method.

Example 14.5 Hole in an Infinite Plate

All parameters and conditions are exactly the same as those in Example 14.2. The nodal arrangement is shown in Figure 14.7. The plate is divided into two domains: in the area near the hole, EFG is employed; in the rest of the domain, the HBE method is applied.

The stresses σ_x at $x = 0$ obtained by the coupled method are plotted in Figure 14.16. The results are obtained using two kinds of nodal arrangement with 65 and 144 nodes. The nodal arrangement of 65 nodes is shown in Figure 14.7. Figure 14.16 shows that the coupled method yields satisfactory results for the problem considered. The convergence of the present method can also be observed from this figure. As the number of nodes increases, the results obtained approach the analytical solution. Compared with the EFG method,

TABLE 14.2

Vertical Displacement along the Interface Boundary (cantilever beam)

Node (y)	EFG/HBE (IE)[a]	EFG/HBE (MVF)[b] EFG Side	HBE Side	Exact
5.75	−4.73203E-03	−4.73090E-03	−4.73093E-03	−4.68750E-03
5.00	−4.72797E-03	−4.72617E-03	−4.72619E-03	−4.68302E-03
4.00	−4.72344E-03	−4.72050E-03	−4.72059E-03	−4.67802E-03
3.00	−4.71970E-03	−4.71664E-03	−4.71670E-03	−4.67414E-03
2.00	−4.71704E-03	−4.71419E-03	−4.71422E-03	−4.67136E-03
1.00	−4.71542E-03	−4.71257E-03	−4.71261E-03	−4.66969E-03
0.00	−4.71488E-03	−4.71199E-03	−4.71203E-03	−4.66914E-03

[a] EFG/HBE (IE): coupled EFG/HBE method using interface element.
[b] EFG/HBE (MVF): coupled EFG/HBE method using modified variational formulation.

Source: Liu, G. R. and Gu, Y. T., *Comput. Mech.*, 26, 166–173, 2000. With permission.

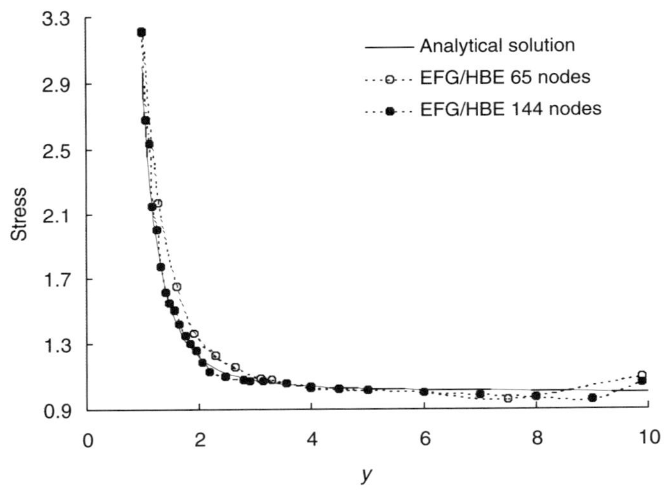

FIGURE 14.16
Stress distribution (σ_x, at $x = 0$) obtained using EFG/HBE method together with analytical results for a square plate with a hole at its center. (From Liu, G. R. and Gu, Y. T., *Comput. Mech.*, 26, 166–173, 2000. With permission.)

fewer nodes are needed in the present coupled method. Comparison with Figure 14.9 reveals that 231 nodes are needed in the EFG method to obtain results of the same accuracy as those obtained by the present EFG/HBE method, where only 144 nodes are required.

Example 14.6 Structure on a Semi-Infinite Foundation

All parameters and conditions are exactly the same as those in Example 14.3, which is schematically illustrated in Figure 14.10. The nodal arrangement is shown Figures 14.11 to 14.13. The only difference is that HBE is used to model the semi-infinite half space instead of BEM.

The results of displacement in the y direction on the top of the structure are given in Table 14.3. The FEM result obtained by Brebbia and Georgiou (1979) is also included in the table. The results obtained by the present method are in very good agreement with

TABLE 14.3

Vertical Displacements along the Top of the Structure on the Semi-Infinite Foundation

	Displacements (× 10^{-4})			
Node No.	**FE**	**EFG**	**EFG/BE (IE)**[a]	**EFG/HBE (MVF)**[b]
Load case 1				
1	1.41	1.42	1.42	1.41
2	1.34	1.34	1.33	1.33
3	1.32	1.32	1.32	1.32
4	1.34	1.34	1.33	1.33
5	1.41	1.42	1.42	1.41
Load case 2				
1	–3.39	–3.43	–3.58	–3.41
2	–0.97	–1.01	–1.04	–1.03
3	1.35	1.35	1.34	1.35
4	3.61	3.67	3.68	3.69
5	6.00	6.04	6.13	6.11

[a] EFG/BE (IE): coupled EFG/BE method using interface element.
[b] EFG/HBE (MVF): coupled EFG/HBE method using modified variational formulation.

Source: Liu, G. R. and Gu, Y. T., *Comput. Mech.*, 26, 166–173, 2000. With permission.

those obtained using other methods, including FEM and EFG for the entire domain. The present method uses many fewer nodes to model the foundation. Only 30 nodes are used in the HBE method compared to 120 nodes used in EFG for the foundation.

14.3 Coupled MLPG/FE/BE Methods

This section presents a method that couples the MLPG method with FEM and BEM. Techniques for the coupled MLPG/FE method and the coupled MLPG/BE method for continuum mechanics problems are presented. This work was performed originally by G. R. Liu and Gu (2000a). The major difficulty with the coupling is to enforce the displacement compatibility conditions on the interface boundary between the MLPG domain and the FE domain or the BE domain. The interface elements are formulated and used along the interface boundary. Within the interface element, the shape functions comprise the MLPG and FE shape functions. Shape functions constructed in this manner satisfy both consistency and compatibility conditions. However, the derivative of the modified interface shape function is discontinuous across the boundary between the pure MLPG domain and the interface domain. In addition to the difficulty of integration in the MLPG method, the use of an interface element presents an additional difficulty in obtaining an accurate numerical integration. A technique is presented for numerical integration to divide the local integration domain into integration subcells using boundaries of FE interface elements. A number of numerical examples are presented to demonstrate the convergence, validity, and efficiency of the coupled methods.

14.3.1 MLPG Formulation

Consider again the 2D problem of solid mechanics defined in Equations 14.1 through 14.3. The problem domain Ω is represented by properly scattered nodes. Using the local weak form, Equation 7.11, and MLS approximation, Equation 5.56, leads to the following discrete system equations:

$$\mathbf{K}_{\mathrm{MLPG}}\mathbf{U} = \mathbf{F}_{\mathrm{MLPG}} \tag{14.65}$$

where the subscript MLPG stands for matrices that are formulated following the standard MLPG procedure, which was described in Chapter 7. Stiffness matrix $\mathbf{K}_{\mathrm{MLPG}}$ and nodal force $\mathbf{f}_{\mathrm{MLPG}}$ vector are defined in Equations 7.20 and 7.21. It is noted that the system stiffness matrix $\mathbf{K}_{\mathrm{MLPG}}$ in the present method is banded but usually asymmetric.

14.3.2 FE Formulation

The weak formulation of FEM is posed as follows:

$$\int_{\Omega} \delta(\mathbf{L}\mathbf{u})^T \cdot \boldsymbol{\sigma}\,\mathrm{d}\Omega - \int_{\Omega} \delta\mathbf{u}^T \cdot \mathbf{b}\,\mathrm{d}\Omega - \int_{\Gamma_t} \delta\mathbf{u}^T \cdot \bar{\mathbf{t}}\,\mathrm{d}\Gamma = 0 \tag{14.66}$$

The interpolation form of FEM for the displacement component u can be written as

$$u = \sum_{i=1}^{n_e} N_i(x)u_i \quad n_e = 3, 4, 5, \ldots \tag{14.67}$$

where n_e is the number of nodes in an element, and the N_i is the FE shape function of node i of the element. Equation 14.67 applies also to displacement component v.

Substituting the expression for u and v given in Equation 14.67 into the weak form, Equation 14.66, yields

$$\mathbf{K}_{\mathrm{FE}}\mathbf{U} = \mathbf{F}_{\mathrm{FE}} \tag{14.68}$$

where $\mathbf{K}_{\mathrm{FE}}$ is the stiffness matrix assembled using the following components computed using FE shape functions:

$$\mathbf{K}_{\mathrm{FE}(ij)} = \int_{\Omega} \mathbf{B}_i^T \mathbf{D} \mathbf{B}_j\,\mathrm{d}\Omega \tag{14.69}$$

where

$$\mathbf{B}_i = \begin{bmatrix} N_{i,x} & 0 \\ 0 & N_{i,y} \\ N_{i,y} & N_{i,x} \end{bmatrix} \tag{14.70}$$

In Equation 14.68, $\mathbf{f}_{\mathrm{FE}}$ is the nodal force vector that is assembled using the following components:

$$\mathbf{f}_{\mathrm{FE}(i)} = \int_{\Gamma_t} N_i \bar{\mathbf{t}}\,\mathrm{d}\Gamma + \int_{\Omega} N_i \mathbf{b}\,\mathrm{d}\Omega \tag{14.71}$$

The discrete system equation of BEM has already been given in Equation 14.23.

14.3.3 Coupling of MLPG and FE or BE

Continuity Conditions at Coupled Interfaces

Consider a problem domain consisting of two subdomains Ω_1 and Ω_2, joined by an interface boundary Γ_I. The MLPG formulation is used in Ω_1 and the FE or BE formulation is used in Ω_2, as shown in Figure 14.17. Compatibility and equilibrium conditions on Γ_I given in Equations 14.54 and 14.55 must be satisfied.

Modified Shape Functions of Interface Elements

Because the shape functions of the MLPG method are derived using MLS, interface elements are introduced in the MLPG domain near the interface boundary Γ_I (Figure 14.17). Ω_I is a layer of subdomain along the interface boundary Γ_I within the MLPG domain Ω_1. In these interface elements, a hybrid displacement approximation is defined so that the shape functions of MLPG domain along Γ_I possess the delta function property. The shape functions and the derivatives for the interface elements are given in Equations 14.32 and 14.33.

The modified displacement approximation in domain Ω_1 becomes

$$\mathbf{u}_1^h(\mathbf{x}) = \begin{cases} \mathbf{u}_{\text{MLPG}}(\mathbf{x}) + R(\mathbf{x})(\mathbf{u}_{\text{FE}}(\mathbf{x}) - \mathbf{u}_{\text{MLPG}}(\mathbf{x})) & \mathbf{x} \in \Omega_I \\ \mathbf{u}_{\text{MLPG}}(\mathbf{x}) & \mathbf{x} \in \Omega_1 - \Omega_I \end{cases} \tag{14.72}$$

where $\mathbf{u}_1^h$ is the displacement of a point in Ω_1, $\mathbf{u}_{\text{MLPG}}$ is MLPG displacement approximated in the form of

$$u_{\text{MLPG}}(\mathbf{x}) = \sum_{i=1}^{n} \phi_i(\mathbf{x}) u_i \tag{14.73}$$

where ϕ_i is the MLS shape function given by Equation 5.56. The expression for the displacement in the FE and the ramp function R are given in Equations 14.28 and 14.29, respectively.

By using the above approximation, the shape functions of the MLPG domain along Γ_I possess the Kronecker delta function property. The equations for the MLPG domain and the FE or BE domain can now be assembled directly.

The regular MLS and modified shape functions in one dimension are shown in Figure 14.3, where a two-node linear interface element is used. It can be seen that the displacement approximation is continuous from the purely MLPG domain (Ω_1–Ω_I) passing to the interface domain Ω_I. The derivative of it is, however, discontinuous across the boundary. These discontinuities do not adversely affect the overall results as they only affect a small number of nodes, but they do increase the difficulties in performing local integration for MLPG.

Numerical Integration in the MLPG Domain

Insufficiently accurate numerical integration may cause a deterioration in the numerical solution. Difficulties in performing numerical integration in MLPG was discussed in Chapter 7. The numerical integration errors result from the complexities of the integrand that is formed using MLS shape functions. To ensure the accuracy of the numerical integration, the Ω_Q should be divided into some regular small partitions. In each small partition, additional Gauss quadrature points should be used.

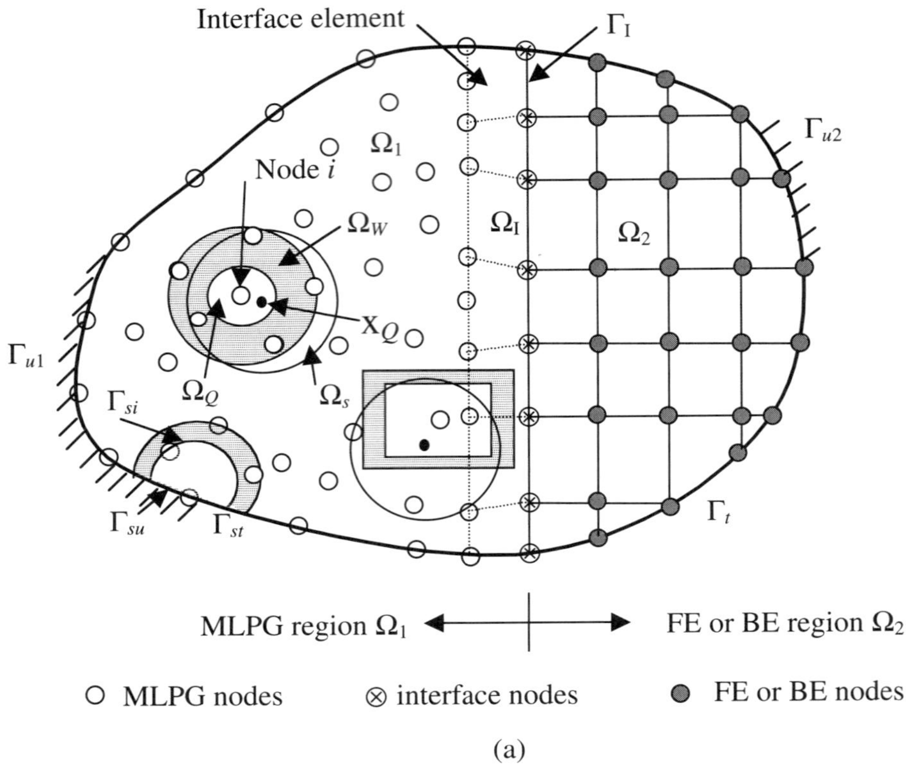

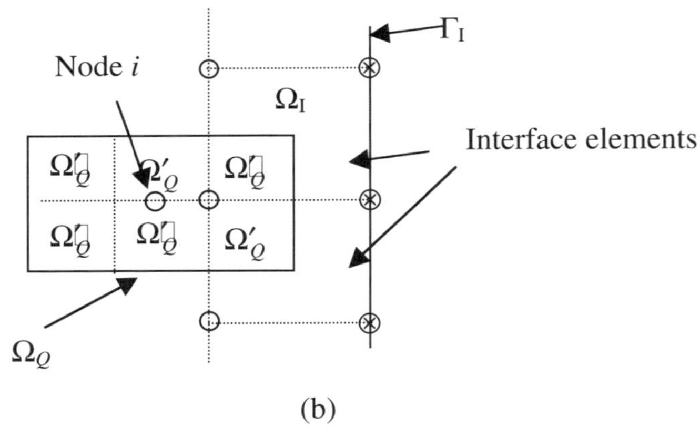

FIGURE 14.17
Domain division into MLPG and FE or BE regions. (a) A layer of interface elements; the weight function domain Ω_W and quadrature domain Ω_Q for node i; the support domain Ω_s for quadrature point x_Q. (b) Detailed integration subdomain Ω'_Q of Ω_Q for node i. (From Liu, G. R. and Gu, Y. T., *Comput. Mech.*, 26, 536–546, 2000. With permission.)

Additional difficulty will be caused in numerical integration when the local integration domain Ω_Q is inside or intersects with the interface domain Ω_I. From the property of the interface shape function, it can be found that the derivative of the modified shape function is discontinuous across the boundary between the purely MLPG domain (Ω_1–Ω_I) and the interface domain Ω_I. In addition, the derivative of the shape functions may be discontinuous across the boundary between two FE interface elements in terms of the property of the FE shape function. The Gauss quadrature can fail to give the exact result for such discontinuous integrand regardless of how many Gauss points are used. The difficulty can be overcome if the domain is divided into integration subdomains by the boundaries of the interface elements, as shown in Figure 14.17b. This division of the subintegration domains can significantly improve the property of the integrand in the subintegration domains, and the accurate integration can be obtained using Gauss quadrature.

14.3.4 Numerical Results

Interface elements with four nodes are used in MLPG. Constant, linear, and quadratic boundary elements are used in the BE domain. Cases presented in the previous sections are used to examine MLPG/FE and MLPG/BE for benchmark problems of 2D elastostatics. In the MLPG part, rectangle local quadrature domains are used for establishing weight functions and obtaining numerical integration. The dimensionless size of the quadrature domain α_Q is chosen as 0.5 to 2.0 in the following examples; i.e., the dimension of the quadrature domain is 0.5 to 2.0 times the average nodal spacing. The dimensionless size of the support domain, α_s is chosen as 1.5 to 3.0 in the following examples; i.e., the dimension of the quadrature domain is 1.5 to 3.0 times the average nodal spacing.

Example 14.7 Cantilever Beam

All parameters and conditions are exactly the same as those in Example 14.1. The beam is divided into two parts. BE or FE is used, respectively, in the part on the left where the essential boundary is included. MLPG is used in the part on the right. The nodal arrangement is shown in Figure 14.18. Four-node isoparametric rectangular FEs are used in the FE part, and linear BEs are used on the BE boundary. In the MLPG part, 63 nodes are used. The results are computed using $\alpha_Q = 1.5$ and $\alpha_s = 3.0$.

The significance of the interface element is first investigated. In the absence of the interface elements, i.e., MLPG region are combined with the FE or BE region directly along the interface boundary Γ_I. The vertical displacements at the right end of the beam are computed and listed in Table 14.4. The same results computed using interface elements are also listed in Table 14.4. It can be found that the absence of interface elements causes

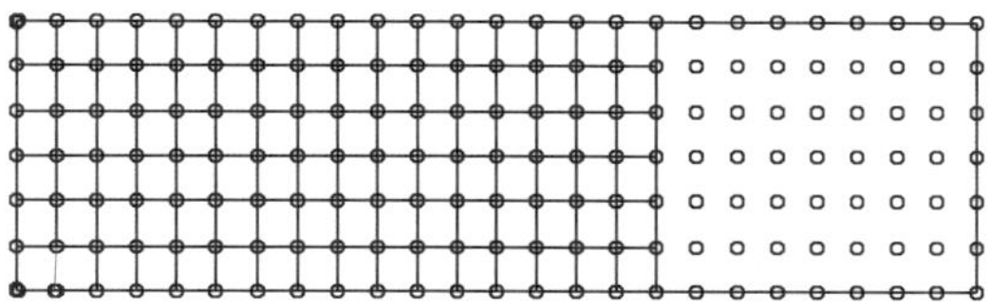

FIGURE 14.18
FE mesh and nodal arrangement of the cantilever beam.

TABLE 14.4
Vertical Displacement at the Right End of the Beam ($\times 10^{-2}$)

	Analytical Solution	MLPG/FE		MLPG/BE	
	u_y	u_y	Error (%)	u_y	Error (%)
With interface elements	0.89	0.8605	−2.81	0.8712	−2.11
Without interface elements	0.89	0.7285	−18.15	0.7232	−18.74

Source: Liu, G. R. and Gu, Y. T., *Comput. Mech.*, 26, 536–546, 2000. With permission.

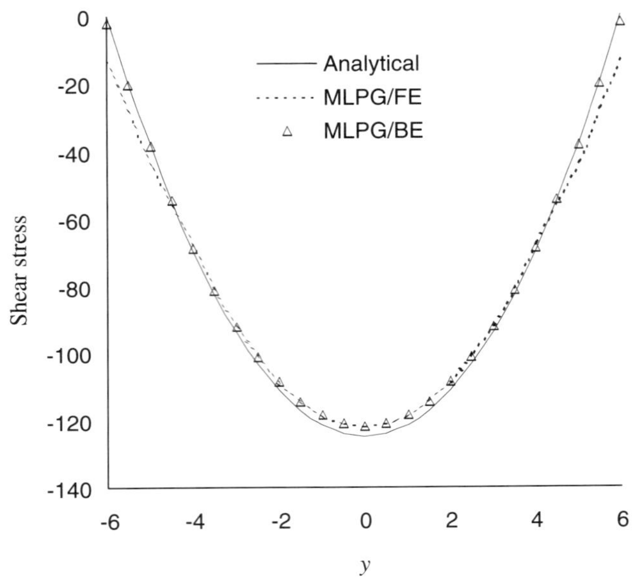

FIGURE 14.19
Shear stress τ_{xy} at the section $x = L/2$ of the cantilever beam. (From Liu, G. R. and Gu, Y. T., *Comput. Mech.*, 26, 536–546, 2000. With permission.)

severe errors. Thus, it is clear that interface elements are imperative in combination MLPG that uses MLS shape functions with FE or BE.

It is found that for displacement, results obtained using different methods are almost identical. As the stress is most critical, detailed results of shear stress are presented here. Figure 14.19 illustrates the comparison between the shear stress at the section of $x = L/2$ calculated analytically and using two coupled methods, MLPG/BE and MLPG/FE. The plot shows excellent agreement between the analytical and numerical results.

The shear stress error defined in Equation 14.34 can be computed for different density mesh, and these results are shown in Figure 14.20, where h is the nodal spacing or the element size in FEM. It is observed that the convergences of the coupled methods are very good. The convergence of using the MLPG for the whole domain is also shown in the same figure. This figure shows that the accuracy of MLPG is the best of these three methods. The accuracy of MLPG/BE is higher than that of MLPG/FE because BE is more accurate than FE in obtaining stresses. However, the convergence rate of these two coupled methods is nearly the same.

Example 14.8 Hole in an Infinite Plate

All parameters and conditions are exactly the same as those in Example 14.2, except the dimension of this example is changed to (10 × 10). Due to symmetry, only the upper right quadrant (5 × 5) of the plate is modeled, as shown in Figure 14.21. The quarter plate is divided into two parts, where MLPG is used in one part, and FE and BE are applied in the other part, respectively. The nodal arrangements of the coupled methods are also shown in Figure 14.21. $\alpha_Q = 1.0$ and $\alpha_s = 2.0$ are used in the MLPG domain. Four-node FEs and linear BEs are used, respectively, in the FE or BE domain.

As the stress is most critical, detailed results on stress are presented here. The stresses σ_x at $x = 0$ obtained by the coupled methods are plotted in Figure 14.22. The result obtained

TABLE 14.5

Radial Displacement for Hollow Cylinder ($\times 10^{-2}$)

Nodes	Exact	MLPG/FE	MLPG/BE	MLPG
1	0.4464	0.4461	0.4468	0.4463
2	0.4464	0.4462	0.4473	0.4466
3	0.4464	0.4478	0.4488	0.4470
4	0.8036	0.8021	0.8120	0.8026
5	0.8036	0.8062	0.8116	0.8068
6	0.8036	0.8101	0.8112	0.8091

Source: Liu, G. R. and Gu, Y. T., *Comput. Mech.*, 26, 536–546, 2000. With permission.

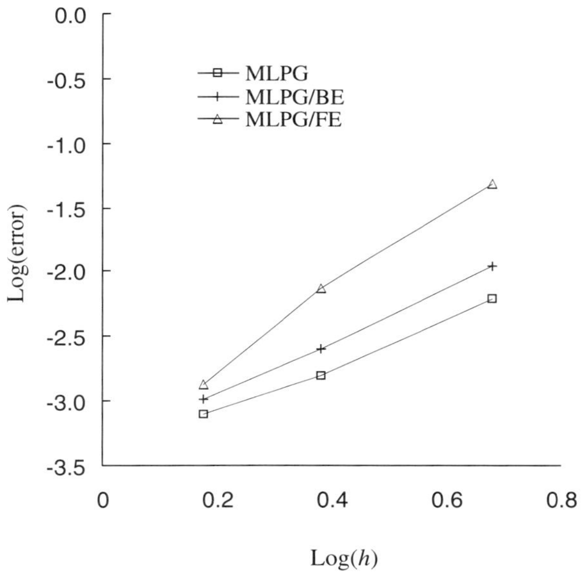

FIGURE 14.20
Convergence in energy norm of error e_t. (From Liu, G. R. and Gu, Y. T., *Comput. Mech.*, 26, 536–546, 2000. With permission.)

by MLPG is shown in the same figure, which shows that the coupled methods yield satisfactory results.

Example 14.9 Internal Pressurized Hollow Cylinder

A hollow cylinder under internal pressure is shown in Figure 13.5. The parameters are taken as $p = 100$, $G = 8000$, and $\nu = 0.25$, which is the same as in Example 13.3. Due to the symmetry of the problem, only one quarter of the cylinder needs to be modeled. The cylinder is divided into two parts, where MLPG and FE (four-node elements) or BE (linear elements) are applied, respectively. As shown in Figure 14.23, 96 nodes are used to discretize the domain and boundary in MLPG/FE. A total of 78 nodes are used to discretize the domain and boundary for the MLPG/BE, as shown in Figure 14.24. In the computation, $\alpha_Q = 1.0$ and $\alpha_s = 2.0$ are used.

The MLPG/FE and MLPG/BE results are compared with the MLPG and the analytical solutions. The radial displacements of boundary nodes are presented in Table 14.5. It can

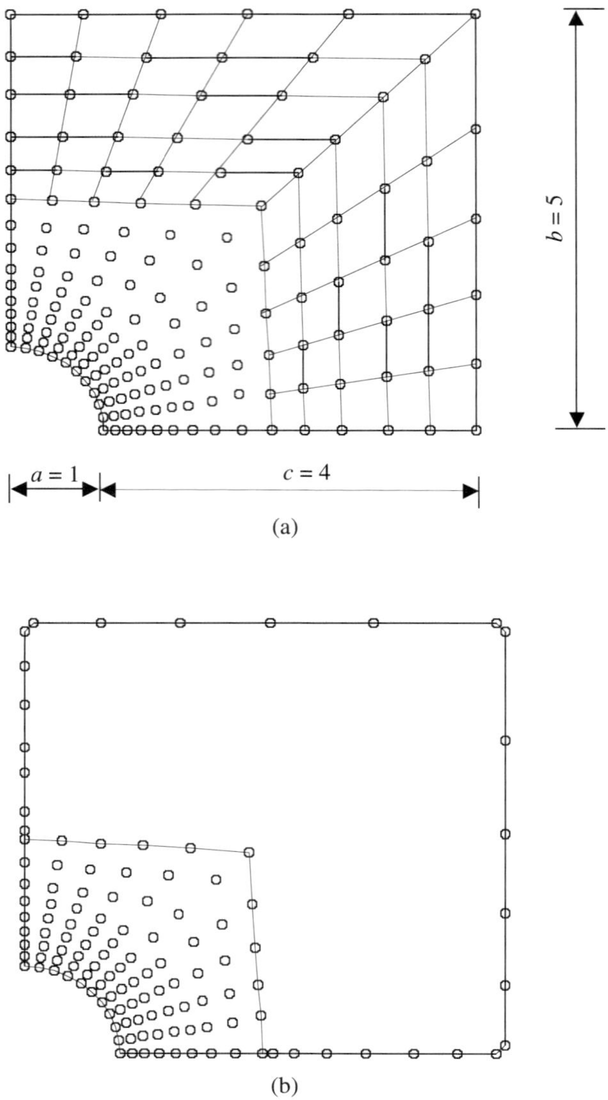

FIGURE 14.21
Mesh and nodal arrangement. (a) MLPG/FEM; (b) MLPG/BEM. (From Liu, G. R. and Gu, Y. T., *Comput. Mech.*, 26, 536–546, 2000. With permission.)

be found that the MLPG/FE and MLPG/BE results are in very good agreement with the analytical solution.

Example 14.10 A Structure on a Semi-Infinite Foundation

All parameters and conditions are exactly the same as those in Example 14.3, which is schematically illustrated in Figure 14.10. The infinite foundation is truncated at a large distance from the structure.

As shown in Figure 14.10, Region 2 represents the semi-infinite foundation and is given a semicircular shape of very large diameter in relation to Region 1, which represents

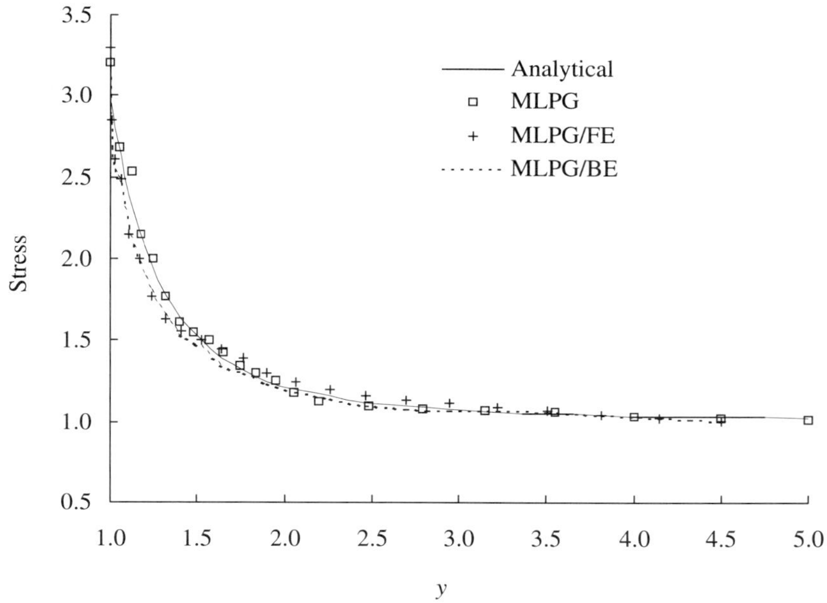

FIGURE 14.22
Stress distribution obtained using MLPG/FEM and MLPG/BEM (σ_x, at $x = 0$). (From Liu, G. R. and Gu, Y. T., *Comput. Mech.*, 26, 536–546, 2000. With permission.)

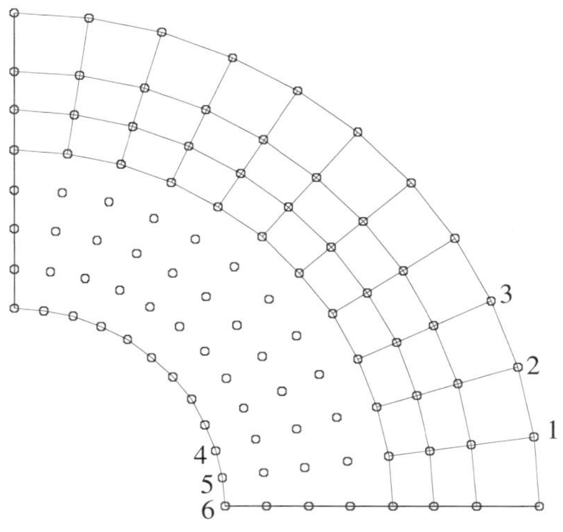

FIGURE 14.23
FE mesh and nodal arrangement for the hollow cylinder. A quarter of the domain is modeled. MLPG/FEM is used for computation. (From Liu, G. R. and Gu, Y. T., *Comput. Mech.*, 26, 536–546, 2000. With permission.)

the structure. Boundary conditions to restrain rigid body movements are applied. MLPG is used in Region 1, and FEM and BEM are used, respectively, in Region 2. The nodal arrangements of the coupled MLPG/BEM and the coupled MLPG/FEM are shown in Figures 14.11 and 14.25. The problem is also analyzed using FEM, MLPG, and FE/BEM. Two loading cases shown in Figure 14.26 are analyzed: case 1 considers five concentrated

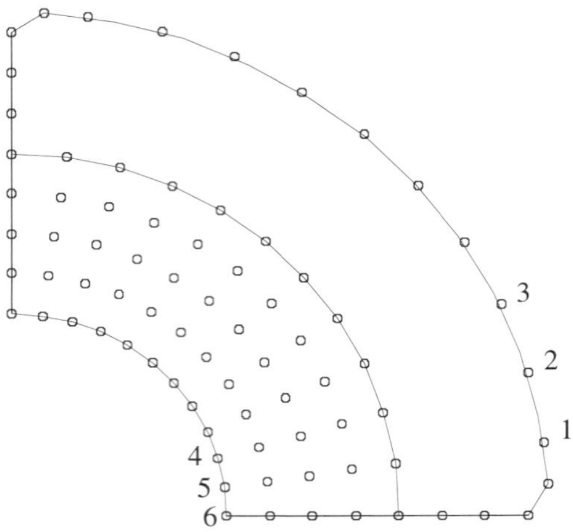

FIGURE 14.24
BE mesh and nodal arrangement for the hollow cylinder. A quarter of the domain is modeled. MLPG/BEM is used for computation. (From Liu, G. R. and Gu, Y. T., *Comput. Mech.*, 26, 536–546, 2000. With permission.)

vertical loads along the top and case 2 considers an additional horizontal load acting at the right corner.

The displacement results of the top of the structure are given in Table 14.6. The results obtained using FEM, MLPG, and FE/BEM are included in the same table. The results obtained using the present MLPG/FEM and MLPG/BEM are in very good agreement with those obtained using FEM, MLPG, and FE/BEM. However, it is interesting to note that the foundation is adequately represented using only 30 BE nodes in the coupling MLPG/BE case as compared with 120 nodes for the MLPG cases. The saving is considerable.

TABLE 14.6
Vertical Displacements along the Top of the Structure

	Displacements ($\times 10^{-4}$)				
Nodes	**FE**	**MLPG**	**FE/BE**	**MLPG/FE**	**MLPG/BE**
Load case 1					
1	1.41	1.43	1.40	1.42	1.43
2	1.34	1.34	1.33	1.34	1.35
3	1.32	1.32	1.32	1.32	1.32
4	1.34	1.34	1.33	1.34	1.35
5	1.41	1.43	1.40	1.42	1.43
Load case 2					
1	−3.39	−3.58	−3.55	−3.53	−3.62
2	−0.97	−1.12	−1.05	−1.00	−1.07
3	1.35	1.36	1.35	1.35	1.34
4	3.61	3.72	3.70	3.59	3.69
5	6.00	6.15	6.17	6.14	6.14

Source: Liu, G. R. and Gu, Y. T., *Comput. Mech.*, 26, 536–546, 2000. With permission.

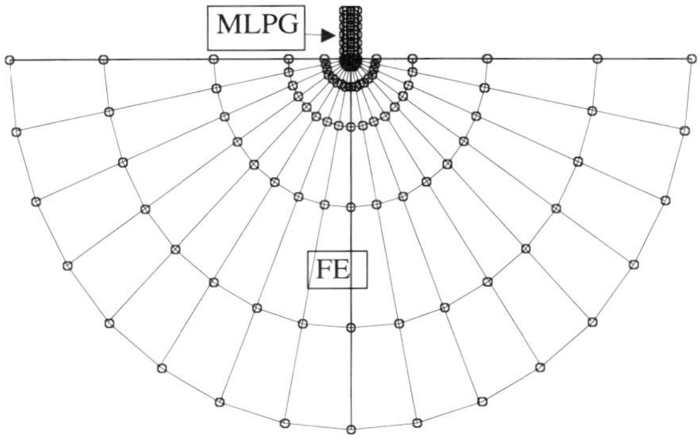

FIGURE 14.25
FE mesh and nodal arrangement of the coupled MLPG/FEM. (From Liu, G. R. and Gu, Y. T., *Comput. Mech.*, 26, 536–546, 2000. With permission.)

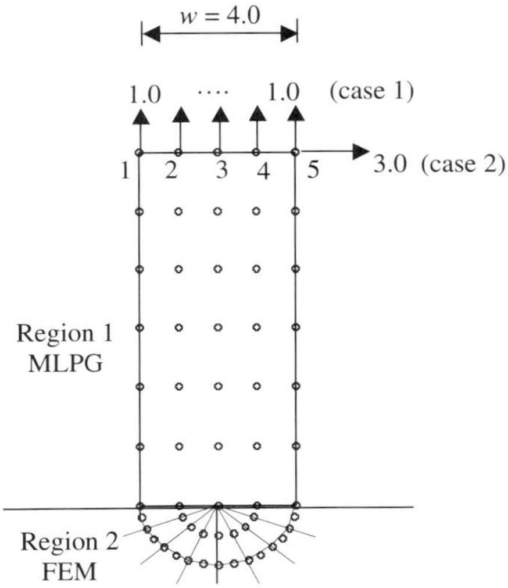

FIGURE 14.26
FE mesh and nodal arrangement of the coupled MLPG/FEM. Closed view of the structure portion. (From Liu, G. R. and Gu, Y. T., *Comput. Mech.*, 26, 536–546, 2000. With permission.)

14.4 Remarks

Methods that couple domain-type MFree methods and boundary-type MFree methods have been presented in this chapter. A number of benchmark examples have demonstrated the feasibility, efficiency, and effectiveness of these coupling approaches. The following important remarks should be made before leaving this chapter:

1. The primary motivation is the same as that for coupling FEM and BEM, which is that domain discretization using finite elements should be confined within the areas of inhomogeneity, nonlinearity, or complex geometry, and that boundary

discretization should be used wherever Green's function is available. This kind of coupling significantly reduces the computational cost, because of the drastic reduction of the number of nodes used for modeling the problem, as well as the reduction of area integration in constructing system matrices.

2. The additional reason for performing such a coupling in MFree methods is that the difficulty of imposing essential boundary conditions for the domain type of MFree methods that use MLS shape functions can be overcome by modeling the portion of the domain with essential boundaries using methods that use shape functions with the Kronecker delta function property, such as FEM, BEM, PIM, and BPIM.
3. Coupling with boundary-type MFree methods works particularly well for problems with infinite domains.
4. BEM can be replaced by any boundary-type MFree method, such BNM, BPIM, radial BPIM, etc. However, if the boundary-type method uses MLS shape functions (e.g., BNM), special treatment of the essential boundary conditions should be performed.
5. Boundary-type methods that produce symmetric system matrices are preferred to produce a symmetric system matrix for the entire problem.
6. On the interface of different methods, there could be an issue of compatibility (not only the continuity condition for nodes), when different orders of shape functions are used on two sides of the domain. The issue can be handled simply by ensuring that the order of the shape function used on the interface is the same as that used by the other method in the domain attached to the interface. Interface or transition elements can be used if necessary.

15

Implementation Issues

Previous chapters have described the detailed formulation and procedures of MFree methods; this chapter focuses on issues related to code implementation of MFree methods. Naturally, different MFree methods contend with different issues related to coding implementation. Here we discuss only those issues that are common to most MFree methods:

- Definition of the support domain or influence domain
- Need for a background mesh for integration
- Node searching
- Dealing with domains of complex and irregular boundaries
- Error estimation
- Adaptive algorithm

15.1 Definition of the Support Domain or Influence Domain

In previous chapters, we used the concept of a support domain for a point of interest (usually the quadrature point $\mathbf{x}_Q$ or the center of the integration cell) in the problem domain, as shown in Figure 4.2. The function of the support domain is basically to determine the field nodes used for constructing shape functions. It is used not only in the process of establishing system equations, but also in the process of retrieving solutions for field variables at locations other than the nodes or derivatives of the field variables at any point in the problem domain. Using the concept of support domain, we simply say that all the nodes that fall into the support domain of a point will be used for constructing shape functions. This works well, if the nodal density does not vary too drastically in the problem domain, but often fails for domains with drastic nodal density variations, such as problems with stress singularity, where the nodal density can vary drastically. For this kind of problem, we suggested the use of the concept of influence domain (see Figure 2.9) to select nodes for construction of shape functions. MFree2D©, which is the focus of the next chapter, uses this approach to select nodes for constructing shape functions. The difference between the support domain and the influence domain is that the support domain is defined for the quadrature points, and the influence domain is defined for the field nodes.

Another advantage of using an influence domain is that we can assign a factor to each node to control its dimension. This factor can always be updated in the process of computation as it accompanies the node. In contrast, if the support domain is used, one would have to assign a factor to all the Gauss points, which can far outnumber the field nodes. The factors need to be changed very often with the change of quadrature method. Hence, it is more difficult to implement.

15.2 Triangular Mesh and Size of the Influence Domain

In using MFree methods that are based on the global Galerkin method, such as the EFG method, a background mesh is required. Any mesh similar to the finite element mesh is applicable, as long as it satisfies the criteria that are stated in Chapter 6 to ensure accurate integration. Ideally, the mesh should be adaptive to the changes in nodal density that may occur within the problem domain in the process of adaptive analysis. The author recommends the triangular background mesh generated based on the well-established Delaunay (1934) triangulation technique. The mesh vertices should be used as the field nodes in the problem domain. We understand that the cells of the integration mesh can be independent of the field nodes. There is also no harm in having them linked together, as long as there is no technical difficulty involved in creating them together. It is actually helpful in adaptive analysis: updating of nodes naturally leads to an update of the cells of the integration mesh. The advantages of using the cells of the triangular integral mesh with its vertices as field nodes are as follows:

- Mature algorithms available in the public domain can be used for the construction of integration cells and field nodes together.
- Triangulation can always be performed in a *fully automated* manner for domains with any complexity.
- Triangular mesh can always be constructed in a *fully automated* manner for both two-dimensional (2D) and three-dimensional (3D) domains.
- Triangular mesh can be easily adapted, which is best suited for adaptive analyses.

One may ask now, why do we not simply use the conventional finite element method (FEM) with triangular elements, which has been fully developed? The answer is simply the well-known fact that the FEM results using triangular elements are very poor. This is due to the poor quality of the linear shape functions constructed based on three-node element confined interpolation schemes. In MFree methods, however, the interpolation/approximation is not cell/element confined. Basically, it can "freely," "automatically," and "dynamically" choose surrounding nodes to construct the shape functions. This freedom of node selection enabled by MFree technology is truly powerful. Based on this argument, the difference between FEM and MFree methods is essentially the difference in the interpolation scheme, and this is also the reason the interpolation technique is the key to the success of the MFree methods.

A mesh generator called MFreePre has been developed for MFree2D for automatic background cell and field node generation. The vertices of the initially generated triangular cell serve as the initial field nodes. When new nodes are added, the triangular mesh is then modified automatically. The following describes the techniques used in MFreePre.

In the process of background mesh generation, each node is assigned a density factor that is used to control local nodal density in the vicinity of the node. Local modification of a mesh is accomplished simply by adjusting the density factor at a node—this is particularly useful in an adaptive analysis.

The triangular background mesh is also employed to determine the size of the influence domain for selecting nodes to be used for constructing shape functions. This ensures the automatic determination of the dimension of the influence domain, and guarantees its flexibility for nodes of any arbitrariness. In the MFree2D 1.0 implementation, a local patch of a node is defined using the triangular cells that surround the node. The formula to compute the dimension of the influence domain becomes

$$r_I = c\sqrt{\frac{2}{n_I}\sum_{i=1}^{n_I} a_i} \tag{15.1}$$

where r_I is the radius of the influence domain of node I, n_I is the number of surrounding triangular cells, a_i is the area of the ith cell, and c is a constant to scale the domain size. The default value is $c = 2$. The size of influence domain by this approach varies reasonably in accordance with local nodal density and is therefore able to represent the influence domain of a node.

To ensure a successful MLS shape function construction, a minimum number is set for the nodes contained in an influence domain. The size of the influence domain can be adjusted automatically in the code if the minimum number of nodes is not achieved.

Once the background cells have been defined, the integration of the system equation is then carried out using conventional Gauss quadrature schemes. This ensures a sufficient number of Gauss points for integrating the system matrices accurately. In addition, the density of the Gauss points is automatically tied to the density change of the nodes. In general, one Gauss point in a cell works well for many problems and three Gauss points in a triangular cell are sufficient for all the problems tested by our research group, which gives a $\alpha_n \approx$ b based on Equation 6.65.

15.3 Node Numbering and Bandwidth of the Stiffness Matrix

The stiffness matrix **K** is sparse and can also be banded. The bandwidth of the stiffness matrix **K** will depend on the numbering system of the fields. It is of little concern if such equation systems as Equation 6.37 are to be solved, as the bandedness of the system matrix has already been destroyed by the introduction of the Lagrange multipliers. It is, however, very important when other methods are used for handling the essential boundary, such as the penalty method, where one can take advantage of the banded feature.

The considerations and techniques of reducing the bandwidth are similar to those in FEM. In FEM, the bandwidth depends on the largest number difference of the elements. All one needs do is determine which element has the largest number difference of all the nodes in the element. Optimization tools for renumbering nodes are routinely available and widely used to ensure the minimum bandwidth. Most of the FEM commercial software packages include some of these optimization tools.

In the MFree method, we do not have elements, and the bandwidth is determined by a much more complex mechanism. Here we suggest the following simplest way to minimize the bandwidth of the system matrix. The principal consideration of the "minimizer" is to minimize the maximum difference in the nodal numbers for all the nodes within a local support domain of any point in the problem domain. This simple scheme is schematically shown in Figure 15.1. The procedure is as follows:

1. Choose a reference point A that is far from the problem domain, i.e., $D \gg H$. Point A has also to be in the longitudinal direction of the problem domain.
2. Calculate the distances between point A and all the field nodes.
3. Rank all the field nodes by the shortest distance.
4. Renumber all the nodes following the rank of the nodes.

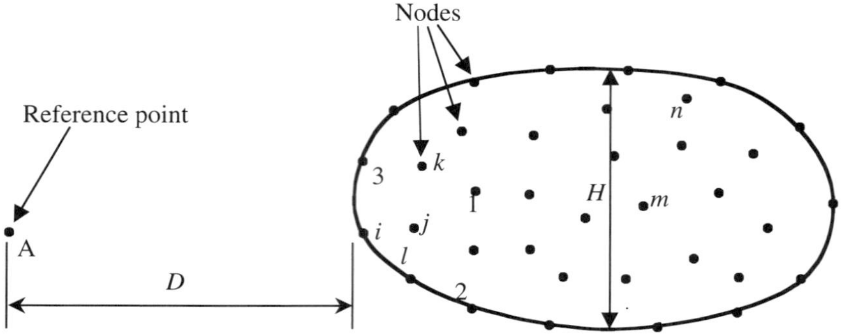

FIGURE 15.1
Reference point chosen at a large distance from the problem domain for optimizing the numbering of nodes.

An index can be easily generated to recode the original numbering of the nodes, which may be useful in retrieving the solution originally assigned to the nodes. This simple procedure is very easy to implement and very effective. The bandwidth of the system matrix should be the minimum or at least very close to the minimum. The procedure is also implemented in MFree2D, as well as in some of our in-house FEM codes. The simplest often work best.

15.4 Bucket Algorithm for Node Searching

To construct the influence domain of a node, a procedure is required to search all the nodes that fall into the influence domain. It is very expensive if every node in the entire problem domain has to be checked against the node, especially when the number of nodes is large. To reduce the cost of node searching, many algorithms have been developed. One very simple approach is to divide a problem domain into buckets, each containing domain nodes up to a predefined number limit. The number limit is defined according to the problem size and the maximum number of nodes allowed in an influence domain. The range of node searching can thus be reduced from the entire problem domain to a number of buckets that have overlaps with the domain of influence under construction. In numerical implementation, the structure of the bucket is defined. The structure contains information about the bucket range and the nodes contained in the bucket. An example of the structure in C language is given as:

```
struct bucket {
int num; //number of nodes contained in bucket
double xmin, ymin, xmax, ymax; //range of bucket
int node[BUCKETNUMLIMIT]; //domain nodes in bucket
};
```

where `BUCKETNUMLIMIT` is the predefined number limit. The bucket defined in the above assumes a rectangular shape.

The next undertaking is to formulate all buckets in a problem domain; this is fulfilled by a recursive procedure. The procedure first sets a bucket to contain the entire domain.

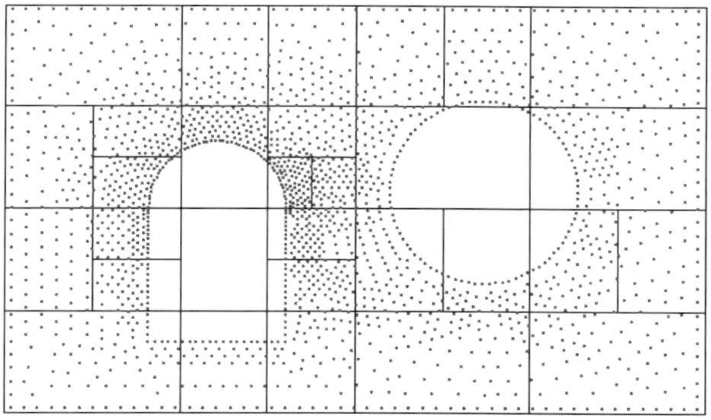

FIGURE 15.2
Problem domain represented by division into buckets. (From Liu, G. R. and Tu, Z. H., *Comput. Methods Appl. Mech. Engrg.*, 191, 1923–1943, 2002. With permission.)

If the number of nodes exceeds the predefined number limit, the bucket is split at its larger dimension into two equal-sized sub-buckets. The same procedure is then applied to each sub-bucket, and the sub-buckets are further divided. The procedure terminates when the number of nodes in every bucket is less than the predefined number limit. In detail, the recursive procedure is as follows:

1. Define the maximum number of domain nodes a bucket is allowed to hold.
2. Create a list to hold the buckets.
3. Set a bucket to hold the entire problem domain.
4. If the number of nodes in the bucket is less than the predefined number limit, append the bucket to the list of buckets and go to 8.
5. Split the bucket into two equal-sized buckets—`BucketOne` and `BucketTwo`.
6. Load `BucketOne` and go to 4.
7. Load `BucketTwo` and go to 4.
8. End.

An application result of this procedure is shown in Figure 15.2, where the rectangles define the bucket ranges. The list of buckets is maintained dynamically with the evolution of the problem domain. In the case of domain refinement or node exchange between adjacent buckets, the disturbed buckets are examined and adjusted accordingly.

15.5 Relay Model for Domains with Irregular Boundaries

15.5.1 Problem Statement

With the use of node-based interpolation techniques, MFree methods offer great opportunities to handle problems of complex geometry. However, with the constraint that no connectivity of nodes be used, difficulties arise in determining the domain of influence of a node. The influence domain usually takes a simple shape, for example, circular

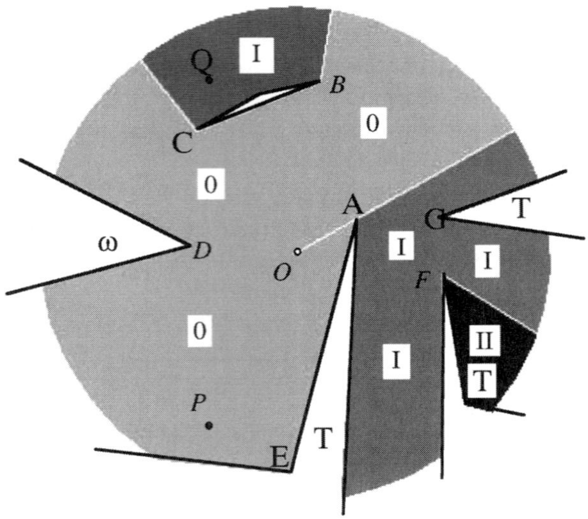

FIGURE 15.3
Influence domain containing numerous irregular boundary fragments.

or rectangular. For simple problem domains, the influence domain can be determined simply by drawing a circle of radius r_I defined by Equation 15.1. Its influence on any other point is directly computed by the distance between the node and the point via the use of the weight functions given in Chapter 5. For complex problem domains involving multiple discontinuities, such as multiple cracks, an influence domain may contain numerous irregular boundary fragments, as shown in Figure 15.3, and computation of nodal weights simply based on physical distance can be erroneous as the discontinuity of the field variables caused by the complex boundary are not accounted for. For example, in Figure 15.3, points P and Q are located at different positions within the influence domain of node O and have the same distance to source node O. A question is: shall O impose the same influence on Q as that on P? If not, how shall the weight of influence on Q be determined, and further on an arbitrary point within the influence domain? Shall an influence domain retain the regularity in shape as in simple problem domains and, if not, how shall the profile of the influence domain be defined? These questions must be addressed properly as the determination of nodal influence is vital to the accuracy of solution using an MFree method.

In view of the aforementioned problems, this section is specifically devoted to these aspects of defining the profile of the influence domain and computing the weight of influence in meshless approximations with irregular boundaries. By starting with a review of the existing techniques in these respects, a relay model aiming to provide a general solution for domains bounded by arbitrary boundaries is proposed. The essence of this model is to construct a hierarchical network of relay points to transmit nodal influence. Moreover, expressions derived from the circle involute curve are employed to define the profile of the portions of the influence domain that are not visible to the source node. Various numerical examples based on the EFG method are presented to verify the effectiveness of the proposed model.

A number of techniques have been reported on the construction of meshless approximations with discontinuities and nonconvex boundaries. The following describes three typical methods: the visibility, diffraction, and transparency methods.

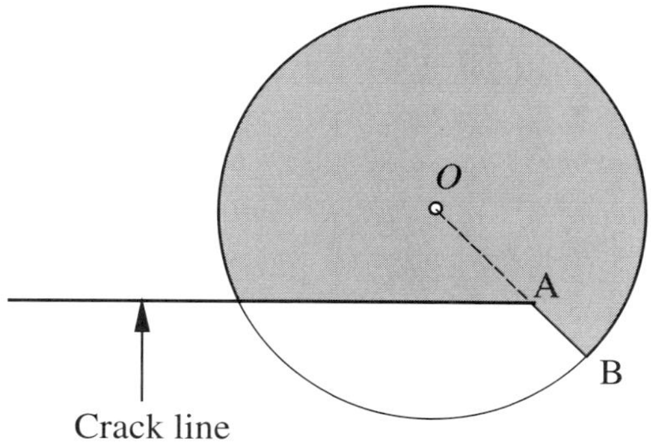

FIGURE 15.4
Domain of influence of a node in the vicinity of a crack tip determined by the visibility method (blank region is excluded by a visibility criterion).

15.5.2 Visibility Method

This is the earliest method used for domains with discontinuities (Belytschko et al., 1994b). Detailed descriptions of the method are available in papers by Belytschko et al. (1996a). The essence of this approach is that the domain boundaries and any interior lines of discontinuities are treated as opaque when constructing weight functions. In this approach, the line from a point to a node is imagined to be a ray of light. If the ray encounters an opaque surface, it is terminated and the point is excluded from the domain of influence. An example of using the visibility criterion is illustrated in Figure 15.4, where the blank region is excluded from the original influence domain of a generic node I in the vicinity of the tip of a straight crack that is represented by the horizontal line in Figure 15.4. This method is simple and straightforward. It is noted, however, that the weight function by the visibility criterion is discontinuous within the influence domain. Along the ray that grazes the tip of the crack, e.g., line AB in Figure 15.4, the weight function is nonzero to the right side of the line, but vanishes to the left. As a consequence, the shape functions constructed are also discontinuous, which is undesirable in a meshless approximation. Moreover, the visibility criterion may not be very easy to adapt to concave boundaries.

15.5.3 Diffraction Method

The diffraction method, proposed by Belytschko et al. (1996a) and Organ et al. (1996), was motivated by the way light diffracts around a sharp corner. This technique applies only to polar-type weight functions where the weights are defined as a function of a single distance-related weight parameter. The mechanism of the diffraction method is illustrated in Figure 15.5, where the straight horizontal line represents a crack. By the diffraction method, the crack line is treated as opaque, but the length of the ray is evaluated by a path that passes around the crack tip C. According to Belytschko et al. (1996a), the weight parameter s associated with A is computed by

$$s = \left(\frac{s_1 + s_2(x)}{s_0(x)}\right)^{\lambda} s_0(x) \tag{15.2}$$

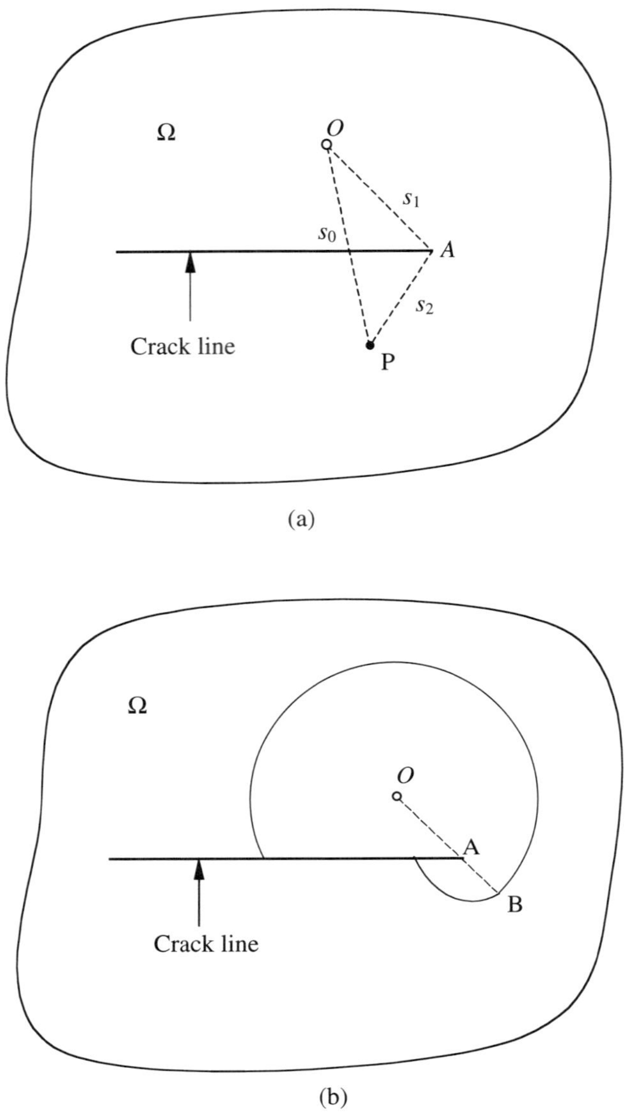

FIGURE 15.5
Schematic illustration of the diffraction technique to determine the influence domain. (a) Definitions of parameters s_0, s_1, and s_2; (b) domain of influence.

where s_0, s_1, and s_2 are distances between O and P, O and A, and A and P, respectively, and λ is a user-defined parameter. As a result, the domain of influence contracts around the crack tip. The diffraction method can also be applied to nonconvex boundaries. Figure 15.6 illustrates such an example, where the parameter s is constructed from the lengths of two line segments that just graze the boundary.

The major benefit of the diffraction method is that the weight function and shape function are continuous within the influence domain. This is, however, achieved at the cost of lower efficiency in computing the derivatives of these functions. Care should also be exercised because, when points O and P are very close, i.e., $s_0(x)$ is close to zero, the weight parameter s becomes infinite, causing difficulty in computation.

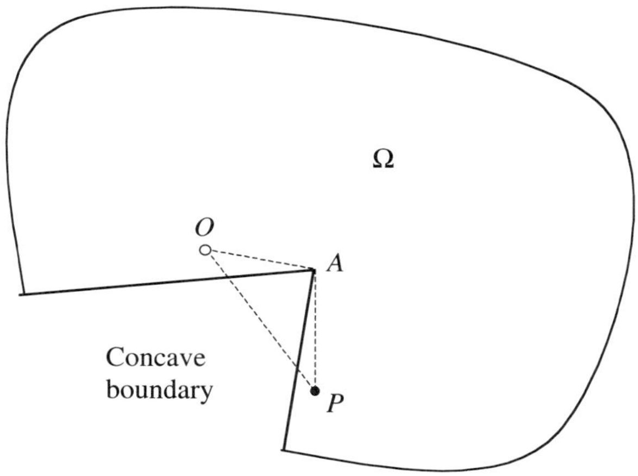

FIGURE 15.6
Diffraction technique applied to concave boundary of a domain Ω.

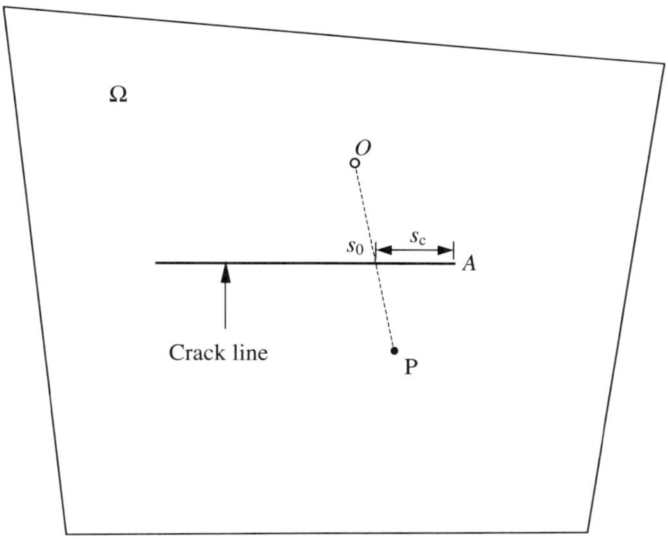

FIGURE 15.7
Schematic illustration of the transparency method for a node near a crack tip.

15.5.4 Transparency Method

The transparency method (Belytschko et al., 1996b; Organ et al., 1996) was developed to smooth a meshless approximation around the tip of a discontinuity by endowing a discontinuity line with a varying degree of transparency—from completely transparent at the tip of the discontinuity to completely opaque at a distance away from the tip. The procedures to compute the weight parameter s are shown in Figure 15.7; the parameter is given by

$$s(x) = s_0(x) + s_{\max}\left(\frac{s_c(x)}{\bar{s}_c}\right)^{\lambda}, \quad \lambda \geq 2 \tag{15.3}$$

where $s_0(x)$ is the distance between O and P, s_{max} the radius of the nodal support, and $s_c(x)$ the distance from the crack tip to the intersection point. The parameter s_c sets the intersection distance at which the discontinuity line is completely opaque. Care should be exercised because, for nodes very close to a boundary, the angle enclosed by the crack and the ray from the node to the crack tip is very small, causing a sharp gradient in the weight function. Therefore, it is required that all nodes have a minimum distance to the crack surface.

15.5.5 The Relay Model

The techniques described above have been applied for many problems of crack propagation. However, these methods are effective for problems of relatively simple domains (e.g., with one or two cracks, or with few nonconvex portions on the boundaries). They are not sufficient for problem domains featuring highly irregular boundaries. A highly irregular domain is a domain containing an arbitrary number of cracks, discontinuities, nonconvex boundary portions, and so forth. Further, there is no restriction to the orientations and distributions of these geometric features. In FEM, this situation causes difficulties mainly in the preprocessing work—meshing—that requires manual operation. Once the domain has been discretized into elements, shape functions are constructed readily with the aid of nodal connectivity and no further consideration is required for the geometric details in the course of computation unless refinement of the domain is needed. However, in MFree methods, the absence of meshes requires a rather sophisticated algorithm to automatically examine the boundary details of an irregular problem domain during a computation to determine the influence domains for a node. This kind of automatic algorithm can be developed, because there is no need to provide connectivity for the nodes. All that has to be done is to determine the nodes that should be included in the influence domain. The relay model is one such algorithm for determining the domain of influence of a node in a complex problem domain. The relay model, developed by G. R. Liu and Tu (2001), is as follows.

The relay model proposed is motivated by the way a radio communication system composed of networks of relay stations works. Consider an influence domain containing a large number of irregular boundary fragments, as depicted in Figure 15.3. O is the source node and the solid lines depict the boundaries. The source node first radiates its influence in all directions equally, just as a radio signal is broadcast at a radio station, until the contained boundaries are encountered. Under the relay model, the influence from the source node is conveyed to the blocked regions via a network of relay points. The following subsections give detailed descriptions about the principles, mechanisms, and computational implementations of the relay model. The descriptions are facilitated by some definitions and notations.

Control Point and Connection

Control Point

A control point is a boundary point at which the angle formed by the boundary on the nonmaterial side (e.g., ω in Figure 15.3) is less than a predefined value. The predefined value is not larger than π; i.e., the boundary at the control point is nonconvex. In the discussion that follows, π is used as the predefined angle value, which means that every boundary node (not only crack tips) is qualified to be a control point if the boundary at this point is concave. (There are cases where a smaller predefined value will greatly simplify the analysis without losing the accuracy of results.) For all control points, the gradient of the boundary changes abruptly in the vicinity of their locations. In Figure 15.3, points A, B, C, D, F, and G are control points, whereas E is not. It is stressed that this definition is purely geometric and contains no physical or mechanics meaning. Therefore, it does not imply a point of singularity.

Connection

A connection is the shortest path that links two points within an influence domain. The path lies completely within the influence domain and is continuous without interruption by boundaries. If such a path does not exist, there is no connection between the two points within the influence domain considered. The connection is said to be direct if the path is straight; otherwise, it is indirect. To denote a connection from one point to another, an arrow is employed in the following notation. For example,

$$\text{direct connections: } P \rightarrow O,\ F \rightarrow A$$
$$\text{indirect connections: } Q \xrightarrow{\{C\}} O,\ T \xrightarrow{\{F,A\}} O$$

The arrows for direct connections are clean, signifying that these connections require no intermediate points. For indirect connections, an assembly of the points via which the connections are established is placed over the arrow. From the left to right in the assembly, the intermediate points are from the nearest to the farthest in terms of their equivalent distances (this will be defined later) to the point on the left side of a connection. The sequence holds if the connection is viewed in reverse order.

Within the original influence domain of a node, points that have no connection with the source node are excluded from the influence domain—this is the criterion of node inclusion in the present model.

Relay Point, Relay Region, and Network of Relay Points

Relay Point

A relay point is a control point with only one boundary segment visible to the source point. It is so named because it is responsible for conveying the influence from the source to the blocked region. This definition is relative—whether a control point is a relay point depends on the position of this point relative to the source point. For example, to source node O, A, B, and C are relay points, whereas D is not. If at another time, P becomes the source node, D is a relay point to P. For a complex influence domain, the relay points are also ranked with the rank order assigned according to their relations with the source node. To illustrate this, the influence domain depicted in Figure 15.3 is again referenced. The relay points that have direct connections with O are the primary relay points; they are ranked level one. Specifically, the source node O is also treated as a relay point and is ranked level zero. Some relay points may possess subrelay points. For example, F is a subrelay point of A; A is called the parent (or master) of F. As F connects indirectly with O via A, F is ranked level two. Subsequently, relay points of higher order are ranked in a similar way. It should be noted that the rank order conferred to a relay point is valid only in the current influence domain; it may change if the same point is in an influence domain of a different source node.

Relay Region

A relay region is a fraction of the influence domain to which a relay point has an exclusive right to transmit the influence from the source node. Two conditions are set for this definition: every point within this region has a direct connection with the relay point and every

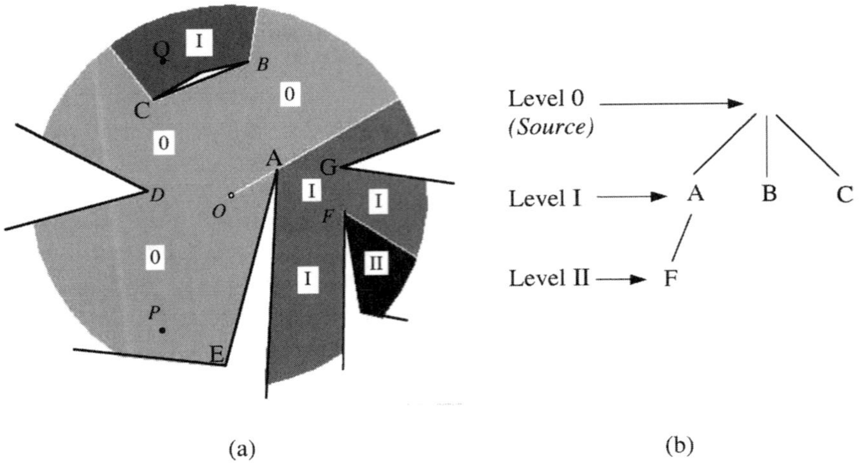

FIGURE 15.8
Relay model for an irregular influence domain. (a) Illustration of relay regions of different levels; (b) tree-representation of the hierarchical network of relay points. (From Liu, G. R. and Tu, Z. H., *Comput. Methods Appl. Mech. Engrg.*, 191, 1923–1943, 2002. With permission.)

point within this region has an indirect connection with the source node via the relay point. The relay point is called the master of the relay region. Relay regions are also ranked with their rank orders inherited from their respective mastering relay points. For the source node, it governs a relay region of level zero. Figure 15.8a shows three levels of relay regions differentiated by gray colors of different degrees—the lightest corresponds to level zero, the intermediate gray to level one, and the darkest to level two. From the above definition, a relay region will have no overlap with another; the interface between two neighboring regions is generally a straight line (Figure 15.8a).

The potential boundary of a relay region is defined by the criterion that the weight function vanishes. By this definition, the potential boundary differs from the real boundary of a relay region in that the former is purely imaginary whereas the latter may comprise portions of the physical boundaries of a problem domain. For the level-zero relay region, the potential boundary coincides with the boundary of the original influence domain. For a relay region of higher level, it is assumed that the radial distance from the mastering relay point to the potential boundary of the relay region decreases with the radial line deviating from the extension of the line connecting the relay point and its parent. As shown in Figure 15.9a, the assumption leads to

$$\rho = \rho(\theta), \quad \text{while } \rho \downarrow \text{ with } \theta \downarrow \tag{15.4}$$

where ρ is the radial distance, decreasing with θ. The preceding expression defines the profile of a relay region that contracts around its mastering relay point (i.e., C_1 in Figure 15.9a). This concept is identical to the diffraction technique described in the preceding section. However, the present model uses expressions derived from the form of circle involute instead of Equation 15.2 this will be presented later.

Network of Relay Points

This is a network formed by all the multitiered relay points within an influence domain. The network is a hierarchical system and corresponds to a tree structure (Figure 15.8b).

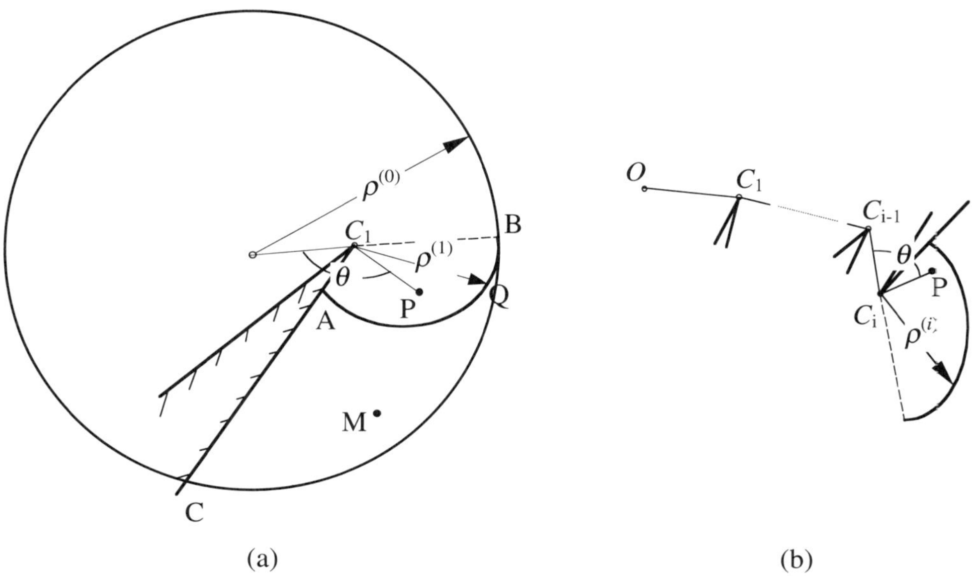

FIGURE 15.9
Profile of relay regions and schemes for computation of equivalent distance of a point using the relay model. (a) Point in a level-one relay region; (b) point in an *i*th-level relay region.

It functions to transmit the influence from the source node to the entire range of the influence domain. This resembles the working principles of a radio communication network system except that the influence (which is the "signal" in radio communication) is conveyed without being enhanced at a relay point. The network formed by relay points is the core of the relay model.

Effective Relay Radius and Equivalent Distance

Effective relay radius is a parameter associated with a relay point; it defines the maximum radial distance from this relay point to the potential boundary of the relay region associated with it. In the present description, it is equal to $\rho(\pi)$. The effective relay radius is calculated by

$$r_{\text{eff}}^{(i)} = \rho^{(i-1)}(\theta) - r = \rho^{(i)}(\pi), \quad i = 1,2,\ldots \tag{15.5}$$

where r followed by a subscript "eff" denotes the effective relay radius and r is the distance from a relay point to its parent relay point. $\rho^{(i-1)}(\theta)$ is the radial distance defined in the relay region associated with the parent relay point, with the radial line passing the relay point considered. The superscript i in parentheses denotes the rank order. In particular, the source node, i.e., the level-zero relay point, has an effective relay radius of

$$r_{\text{eff}}^{(0)} = r_0 \tag{15.6}$$

where r_0 is the original radius of the influence domain.

Under the relay model, an influence domain is divided into numerous relay regions of different levels. Points in different relay regions have different connections with the source node. It will be unfair if the details of a connection are not accounted for in the computation of the weight parameter. This model employs the equivalent distance instead of the physical distance as the basis to measure the weight parameter. For a point in the level-zero relay region, the equivalent distance is identical to the physical distance from this point to the source node, i.e.,

$$r_{\text{eq}} = r \tag{15.7}$$

where r_{eq} and r are the equivalent distance and physical distance from the point to the source node (or the level-zero relay point), respectively.

For points in relay regions of higher level, the equivalent distance is evaluated in a progressive manner. To illustrate this, a point in a level-one relay region is first considered. Figure 15.9a shows a single relay point case where O is the source node, C_1 is a relay point of level one, and P is located in the relay region associated with C_1. According to Equations 15.5 and 15.7, the effective relay radius and equivalent distance associated with C_1 are

$$r_{\text{eff}}^{(1)} = \rho^{(0)} - r = r_0 - r \tag{15.8}$$

and

$$r_{\text{eq}}^{(1)} = r \tag{15.9}$$

respectively, where r is the distance between C_1 and O. Following the notation defined above, the connection between P and O is depicted as

$$P \xrightarrow{\{C_1\}} O \tag{15.10}$$

The equivalent distance of P is computed by

$$r_{\text{eq}} = \frac{r_{\text{eq}}^{(1)} + r}{r_{\text{eq}}^{(1)} + \rho^{(1)}(\theta)} r_0 \tag{15.11}$$

where r represents the distance from P to its mastering relay point C_1. θ is the smaller angle formed by $\overrightarrow{C_1P}$ and $\overrightarrow{C_1O}$ with the direction from $\overrightarrow{C_1P}$ to $\overrightarrow{C_1O}$.

If P is located in an ith-level relay region as shown in Figure 15.9b and has a connection with the source node as described by

$$P \xrightarrow{\{C_i, C_{i-1}, \ldots, C_1\}} O \tag{15.12}$$

the equivalent distance from P to O is computed in the same fashion, i.e.,

$$r_{\text{eq}} = \frac{r_{\text{eq}}^{(i)} + r}{r_{\text{eq}}^{(i)} + \rho^{(i)}(\theta)} r_0 \quad i = 0, 1, 2, \ldots \tag{15.13}$$

where $r_{eq}^{(i)}$ is the equivalent distance associated with the mastering relay point C_i, and $\rho^{(i)}$ defines the potential boundary of the ith-level relay region controlled by C_i. θ is the smaller angle formed by $\overrightarrow{C_iC_{i-1}}$ and $\overrightarrow{C_iP}$; one has:

$$\cos\theta = \mathbf{v}_{\overrightarrow{C_iC_{i-1}}} \cdot \mathbf{v}_{\overrightarrow{C_iP}} \tag{15.14}$$

where $\mathbf{v}_{\overrightarrow{C_iC_{i-1}}}$ and $\mathbf{v}_{\overrightarrow{C_iP}}$ are unit vectors defining the directions of $\overrightarrow{C_iC_{i-1}}$ and $\overrightarrow{C_iP}$, respectively, i.e.,

$$\mathbf{v}_{\overrightarrow{C_iC_{i-1}}} = \frac{\mathbf{x}_{C_{i-1}} - \mathbf{x}_{C_i}}{|\mathbf{x}_{C_{i-1}} - \mathbf{x}_{C_i}|} = \begin{pmatrix} (x_{C_{i-1}} - x_{C_i})/r_{C_iC_{i-1}} \\ (y_{C_{i-1}} - y_{C_i})/r_{C_iC_{i-1}} \end{pmatrix} \tag{15.15}$$

$$\mathbf{v}_{\overrightarrow{C_iP}} = \frac{\mathbf{x} - \mathbf{x}_{C_i}}{|\mathbf{x} - \mathbf{x}_{C_i}|} = \begin{pmatrix} (x - x_{C_i})/r \\ (y - y_{C_i})/r \end{pmatrix} \tag{15.16}$$

Equation 15.13 presents a progressive way to compute the equivalent distance of P sequentially from $C_1, C_2, \ldots, C_i$, to P. The expression also shows that the equivalent distance of a point only needs the information of its mastering relay point; this makes the computation very easy to manage for a very complex influence domain. To save computation expense, $r_{eq}^{(i)}$ and $r_{eff}^{(i)}$ associated with all levels of relay points are computed in advance. There are occasions that a point may have various links to the source node (e.g., Q in Figure 15.8a can connect with O via either B or C). The rule practiced here is that the shortest equivalent distance among all the possible paths is used.

Weight Parameter and Its Derivatives

Weight functions play an essential role in the construction of shape functions. It rules the influence of the source node over the entire influence domain. A general polar-type weight function takes a form of

$$\widehat{W}(s) = \begin{cases} >0 & \text{for } s < 1 \\ =0 & \text{for } s \geq 1 \end{cases} \tag{15.17}$$

and its derivatives are calculated by

$$\frac{\partial \widehat{W}}{\partial x} = \frac{\mathrm{d}\widehat{W}}{\mathrm{d}s}\frac{\partial s}{\partial x} \tag{15.18}$$

$$\frac{\partial \widehat{W}}{\partial y} = \frac{\mathrm{d}\widehat{W}}{\mathrm{d}s}\frac{\partial s}{\partial y} \tag{15.19}$$

where the weight parameter $s = r/r_0$ is a normalized distance. Several commonly used weight functions are given in Chapter 5 (note that variable is now changed to s).

In the relay model, the weight parameter s is measured by the equivalent distance. For example, the weight parameter for a point in an ith-level relay region is

$$s = \frac{r_{eq}}{r_0} = \frac{r_{eq}^{(i)} + r}{r_{eq}^{(i)} + \rho^{(i)}(\theta)}, \quad i = 0, 1, 2, \ldots \tag{15.20}$$

The derivatives of s with respect to x and y are, therefore,

$$\frac{\partial s}{\partial x} = \frac{1}{r_{\text{eq}} + \rho(\theta)} \frac{\partial r}{\partial x} - \frac{r}{(r_{\text{eq}} + \rho(\theta))^2} \frac{\text{d}\rho}{\text{d}\theta} \frac{\partial \theta}{\partial x} \tag{15.21}$$

$$\frac{\partial s}{\partial y} = \frac{1}{r_{\text{eq}} + \rho(\theta)} \frac{\partial r}{\partial y} - \frac{r}{(r_{\text{eq}} + \rho(\theta))^2} \frac{\text{d}\rho}{\text{d}\theta} \frac{\partial \theta}{\partial y} \tag{15.22}$$

where the superscript (i) is dropped for simplicity of description. The derivatives of θ with respect to x and y are computed from Equation 15.14:

$$\frac{\partial \theta}{\partial x} = -\frac{y - y_{C_i}}{r^2} \tag{15.23}$$

$$\frac{\partial \theta}{\partial y} = \frac{x - x_{C_i}}{r^2} \tag{15.24}$$

if θ is counterclockwise. When the direction of θ is clockwise, the derivatives are

$$\frac{\partial \theta}{\partial x} = \frac{y - y_{C_i}}{r^2} \tag{15.25}$$

$$\frac{\partial \theta}{\partial y} = -\frac{x - x_{C_i}}{r^2} \tag{15.26}$$

Numerical Implementation

To implement the relay model, sophisticated algorithms that take into account the details of the boundary fragments are required. This necessitates, at least, two tasks: detection of all boundary fragments contained in an influence domain and construction of the network of relay points. The procedures to construct the relay model within the influence domain of an arbitrary node O are as follows:

1. Construct the influence domain associated with node O.
2. Extract all boundary fragments of the problem domain contained in the influence domain.
3. Filter out the nodes that have no connection with the source node.
4. Construct a list of all control points within the influence domain.
   a. Loop over all boundary nodes within the influence domain.
   b. Append a boundary node to the list if it is a control point.
5. Construct the network of relay points.
   a. Create a new list for relay points.
   b. Set the source node as the level-zero relay point and append it to the list.
   c. Set $L = 1$ (L denotes the level of relay points to be constructed).
   d. Loop over the list of control points.

e. Set the relay points of level L and append them to the list of relay points. This is done by checking the control points against the relay points of level $L - 1$.
f. Update the list of control points by removing the control points that are already relay points.
g. $L = L + 1$.
h. Go to c.

6. Compute the effective relay radius and equivalent distance associated with each relay point.
7. End the construction of the relay model within the influence domain of O.

These procedures are complex and expensive in terms of computation time. This is the cost arising from the absence of meshes and is likely unavoidable for a very complex problem domain. The price has to be paid to relieve the labor involved in meshing. It is not difficult to imagine how much time is required to mesh a domain of complexity shown in Figure 15.3. Compared with the price for labor to perform the work, the cost of the relay model is extremely small. Moreover, if adaptive analysis is required, it is not possible to manually perform the meshing over and over again.

Profile of a Relay Region

Circle Involute Approach

The present model defines the profile, or potential boundary, of a relay region using forms derived from the circle involute (Figure 15.10a). By mathematical definition, the involute of a circle is a curve orthogonal to all the tangents to this circle. For a better understanding, one can imagine there is a circle with a rope wound around its perimeter. If one holds tightly to one end of the rope to unwind the rope from the circle, the circle involute is given by the track of this end. In Figure 15.10a, Q is the end of the rope and T is the point where the extracted portion of the rope makes tangential contact with the circle. The circle involute can be mathematically expressed by

$$x = a(\cos\alpha + \alpha\sin\alpha) \tag{15.27}$$

$$y = a(\sin\alpha + \alpha\cos\alpha) \tag{15.28}$$

where a is the radius of the circle and α is the angle from OA to OT. Alternatively, the circle involute described in polar form is

$$\rho = a(1 + \alpha^2)^{1/2} \tag{15.29}$$

$$\theta = \alpha - \arccos\left\lfloor (1 + \alpha^2)^{-1/2} \right\rfloor \tag{15.30}$$

where (ρ, θ) is the polar position of Q. The preceding equations are used to define the profile of the relay region as shown in Figure 15.9a, with the value of θ confined by $(0, \pi)$.

With Equations 15.29 and 15.30, the following derivatives can be obtained:

$$\frac{d\alpha}{d\theta} = \frac{1 + \alpha^2}{\alpha^2} \tag{15.31}$$

$$\frac{d\rho}{d\alpha} = \frac{a\alpha}{(1 + \alpha^2)^{1/2}} \tag{15.32}$$

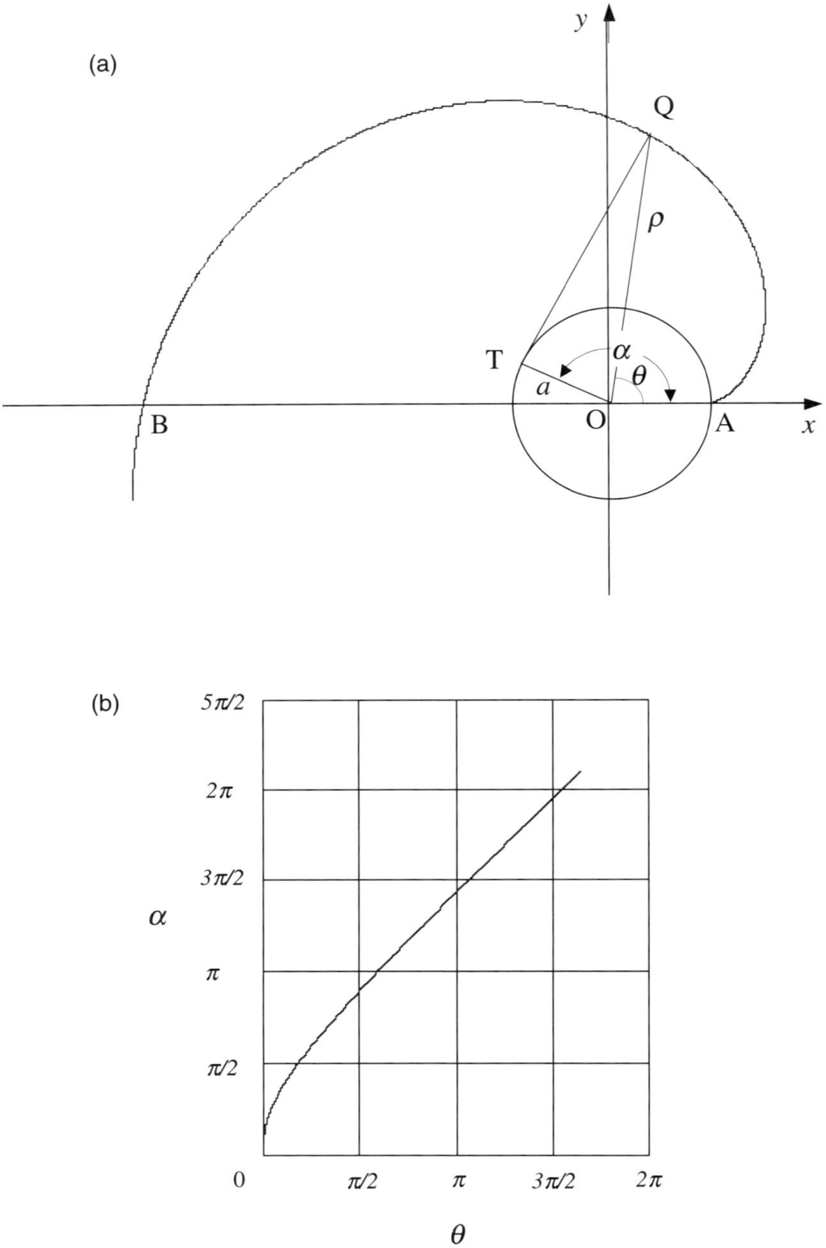

FIGURE 15.10
Circle involute. (a) The involute curve; (b) relationship between α and θ.

Therefore, differentiation of ρ with respect to θ yields

$$\frac{d\rho}{d\theta} = \frac{d\rho}{d\alpha}\frac{d\alpha}{d\theta} = \frac{a(1+\alpha^2)^{1/2}}{\alpha} = \frac{\rho}{\alpha} \tag{15.33}$$

TABLE 15.1

Values of α vs. θ for the Circle Involute Approach

θ	α	θ	α
.0000000000E+00	.0000000000E+00	.1581906633E+00	.8796459430E+00
.8248810247E–04	.6283185307E–01	.1866837799E+00	.9424777960E+00
.6552697876E–03	.1256637061E+00	.2172636968E+00	.1005309649E+01
.2186035144E–02	.1884955592E+00	.2498070719E+00	.1068141502E+01
.5099810682E–02	.2513274123E+00	.2841906568E+00	.1130973355E+01
.9763467994E–02	.3141592654E+00	.4796580601E+00	.1445132621E+01
.1647595382E–01	.3769911184E+00	.7532566507E+00	.1822123739E+01
.2546442135E–01	.4398229715E+00	.1107412823E+01	.2261946711E+01
.3688561209E–01	.5026548246E+00	.1540878884E+01	.2764601535E+01
.5083130149E–01	.5654866776E+00	.2051016866E+01	.3330088213E+01
.6733641460E–01	.6283185307E+00	.2635059918E+01	.3958406743E+01
.8638848454E–01	.6911503838E+00	.3290607926E+01	.4649557127E+01
.1079373643E+00	.7539822368E+00	.4015736576E+01	.5403539364E+01
.1319044111E+00	.8168140899E+00	.4870220170E+01	.6283185307E+01

In a computation, α needs to be solved in terms of θ from the nonlinear Equation 15.30. This adds computation cost if a nonlinear equation solver is used to solve this. To avoid this, one can actually construct a table, such as Table 15.1, to store values of θ vs. α in advance. Figure 15.10b depicts the relationship between α and θ defined by expression 15.30. From Table 15.1, α can be obtained by linear interpolation once θ is calculated from Equation 15.14; i.e.,

$$\alpha = \alpha_1 + \frac{\theta - \theta_1}{\theta_2 - \theta_1}(\alpha_2 - \alpha_1) \tag{15.34}$$

where θ falls into a range defined by two consecutive values θ_1 and θ_2 in the table; α_2 and α_1 are values corresponding to θ_1 and θ_2, respectively.

Next, parameter a is determined. This can be achieved by noting that for ρ there holds:

$$\rho(0) = a \qquad \text{and} \qquad \rho(\pi) = r_{\text{eff}} \tag{15.35}$$

With Equation 15.29, parameter a is obtained as

$$a = r_{\text{eff}}(1 + \alpha(\pi)^2)^{-1/2} = 0.21724 r_{\text{eff}} \tag{15.36}$$

where $\alpha(\pi)$ is 4.449235 by interpolation from Table 15.1.

By combining Equations 15.20 through 15.22, 15.29, 15.30, and 15.33, the weight parameter and its derivatives can be determined; the derived formulae are applicable to any polar-type weight function. To illustrate the effects of the circle involute approach on the weight function, three domains as shown in Figure 15.11 are used for comparison. Figure 15.11a, b, and c represent domains containing no discontinuity, a single crack, and double cracks, respectively. The weight function used is the cubic spline. The calculated weights and derivatives over the influence domain of the source node O are depicted by the surface plots and contour plots in Figures 15.12, 15.13, and 15.14, respectively.

It is noted from Equation 15.33 that the derivative of ρ with respect to θ turns out to be infinite when α is very close to zero. This can be avoided by setting an initial value of α, e.g., $\alpha(0) = \alpha_0 > 0$.

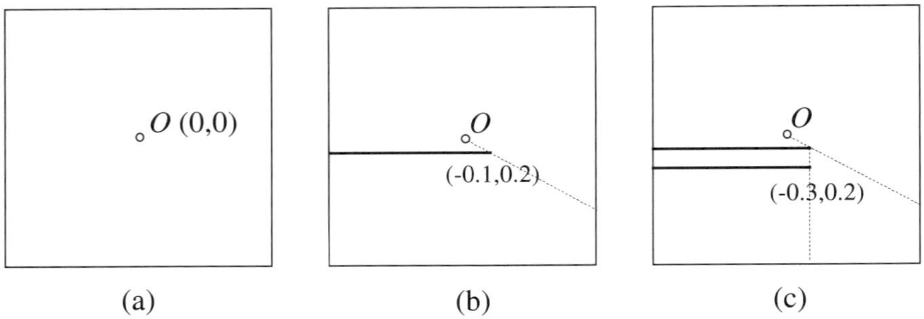

FIGURE 15.11
Three domains: (a) with no discontinuity, (b) with a single crack, and (c) with two cracks.

Modified Circle Involute Approach

Figures 15.13 and 15.14 show that there are small jumps in the derivatives of the weight function across the interfaces between neighboring relay regions; i.e., the derivatives are not smooth. This is not desirable since the smoothness of weight functions is crucial to the continuity of stresses. A modified circle involute approach is therefore proposed to mitigate this problem. This approach uses a different portion of the circle involute curve as shown in Figure 15.15a, where the origin of coordinates, O, moves from the circle center to the starting point of the circle involute with the x axis tangential to the perimeter. The resulting curve in polar form is

$$\rho = a(2+\alpha^2 - 2\cos\alpha - 2\alpha\sin\alpha)^{1/2} \tag{15.37}$$

$$\cos\theta = \frac{a(\sin\alpha - \alpha\cos\alpha)}{\rho} \tag{15.38}$$

The relationship of α and θ defined by Equation 15.37 is depicted in Figure 15.15b. Again, Table 15.2 is constructed to store the values of α vs. θ. For those values not directly given in the table, linear interpolations apply.

The derivative of ρ with respect to θ determined from Equations 15.37 and 15.38 is

$$\frac{d\rho}{d\theta} = \frac{\rho(1+\sin\alpha)}{\alpha - \cos\alpha} \tag{15.39}$$

where θ is defined within $(0, \pi)$ while the bounds of α are given by $\alpha(0)$ and $\alpha(\pi)$. The constant a is

$$a = r_{\text{eff}}/2\pi \tag{15.40}$$

since $\rho = r_{\text{eff}} = 2\pi a$ at $\alpha = 2\pi$.

The cubic spline weight function and its derivatives based on the modified approach are demonstrated by surface plots and contour plots in Figures 15.16 and 15.17. The plots demonstrate that the modified approach gives smooth descriptions for the weight function and its derivatives in the level-one relay region; this can be shown easily from Equation 15.39 mathematically. However, in relay regions of higher level, there are also small jumps across the interface. A more appropriate description of the profile of the relay region is therefore needed; this unfortunately is still under investigation.

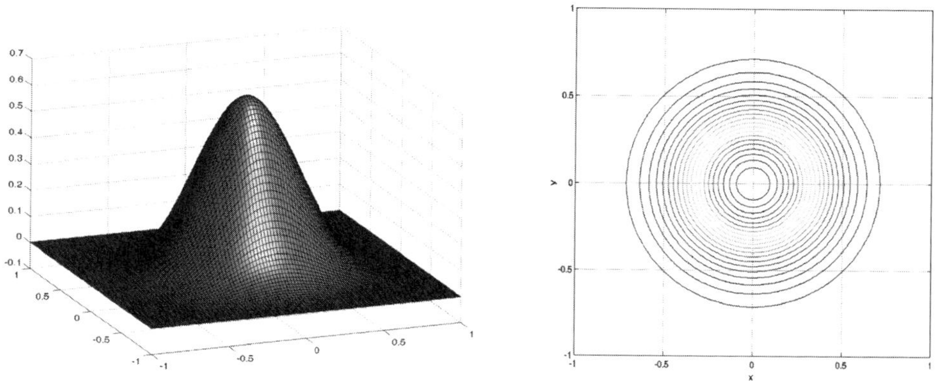

(a)

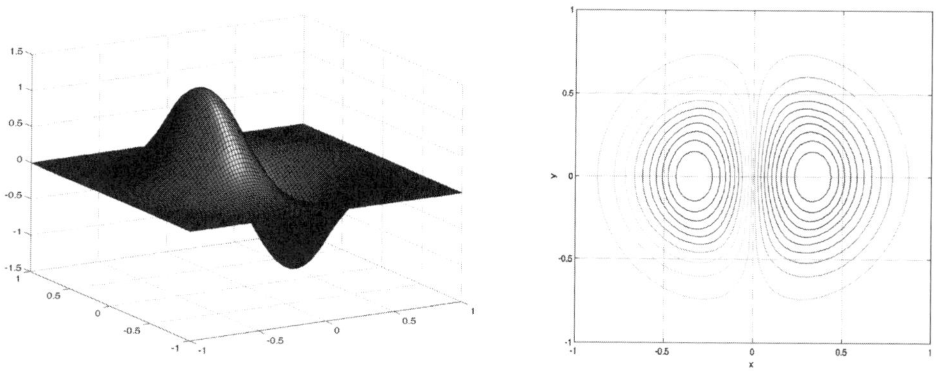

(b)

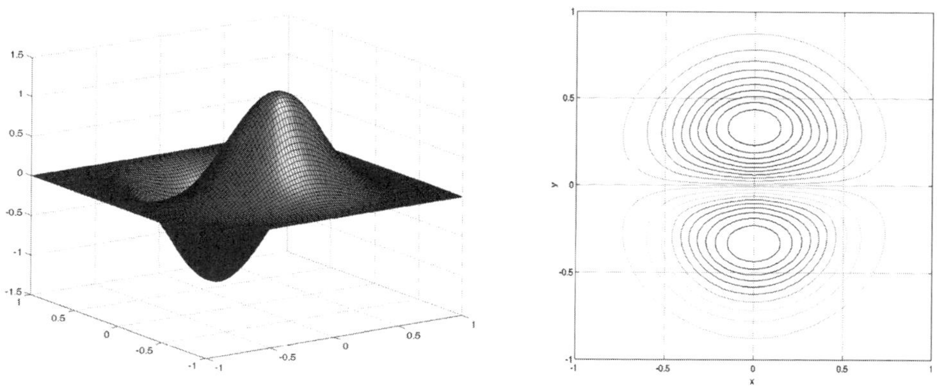

(c)

FIGURE 15.12
Cubic spline weight function and its derivatives over an influence domain with no discontinuity. Surface plots and contour plots for (a) $\widehat{W}(s)$, (b) $\widehat{W}_x$, and (c) $\widehat{W}_y$.

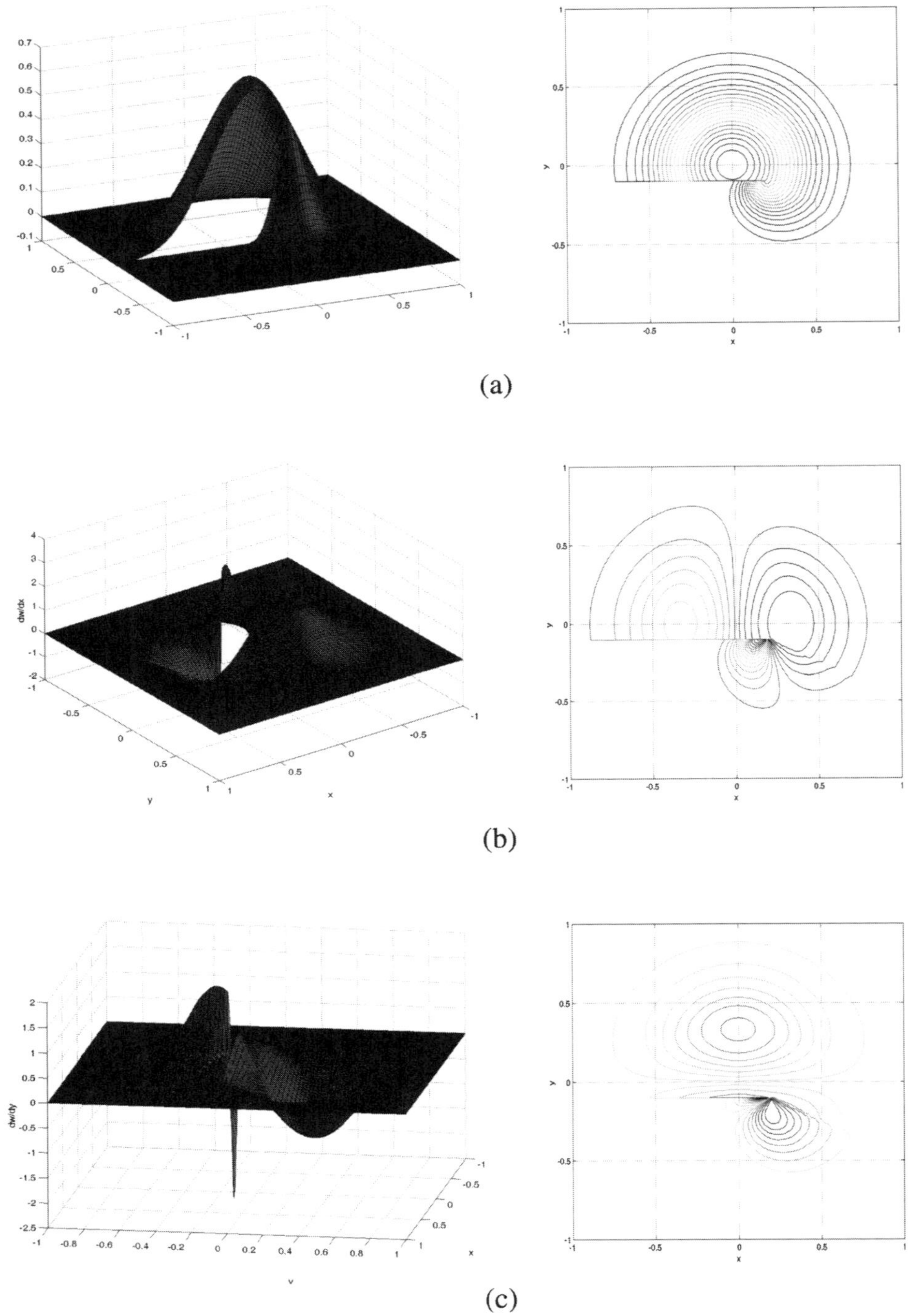

FIGURE 15.13
Cubic spline weight function and its derivatives over an influence domain with a single crack by the circle involute approach. Surface plots and contour plots for (a) $\widehat{W}(s)$, (b) $\widehat{W}_x$, and (c) $\widehat{W}_y$.

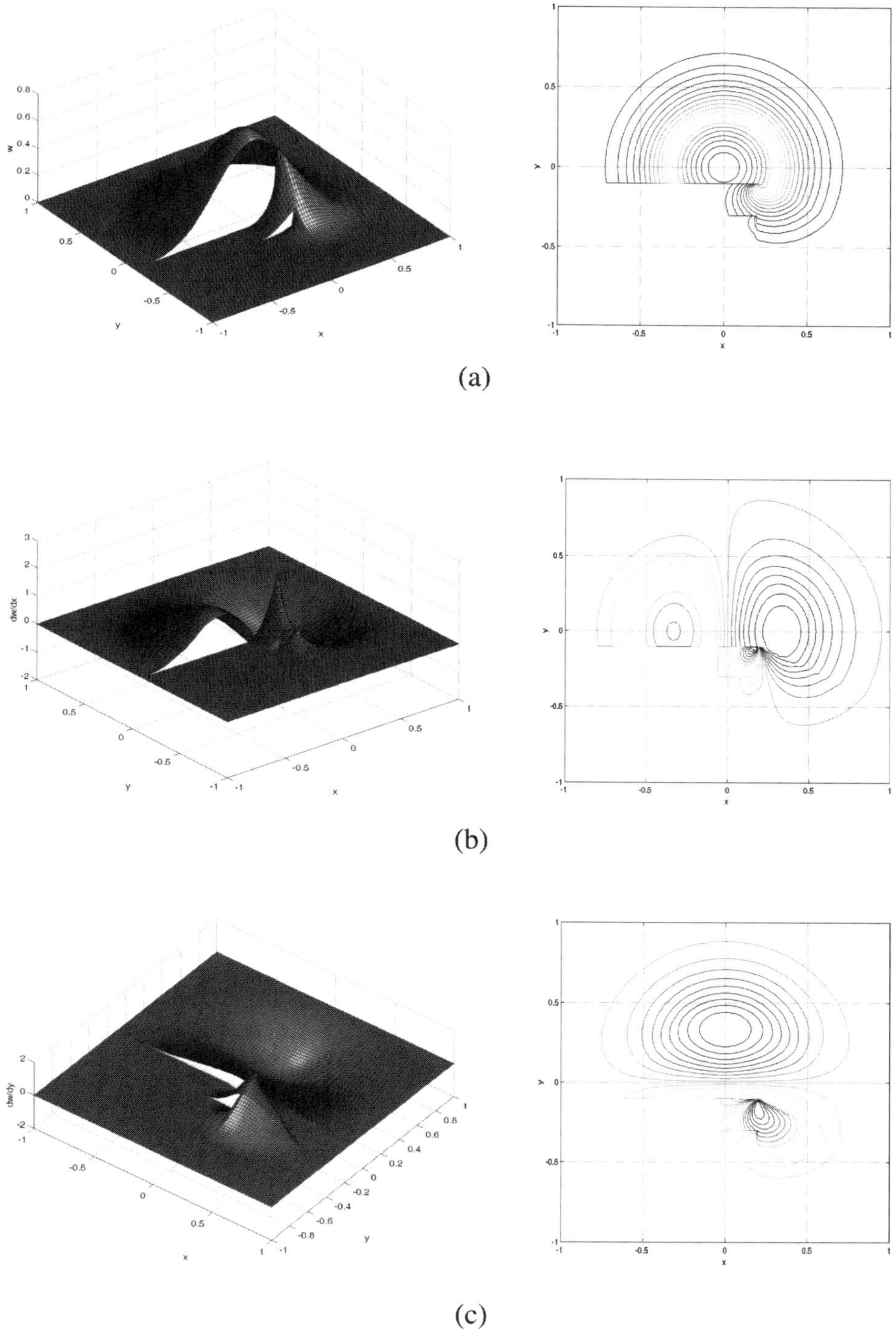

FIGURE 15.14
Cubic spline weight function and its derivatives over an influence domain with two cracks by the circle involute approach. Surface plots and contour plots for (a) $\widehat{W}(s)$, (b) $\widehat{W}_x$, and (c) $\widehat{W}_y$. (From Liu, G. R. and Tu, Z. H., *Comput. Methods Appl. Mech. Engrg.*, 191, 1923–1943, 2002. With permission.)

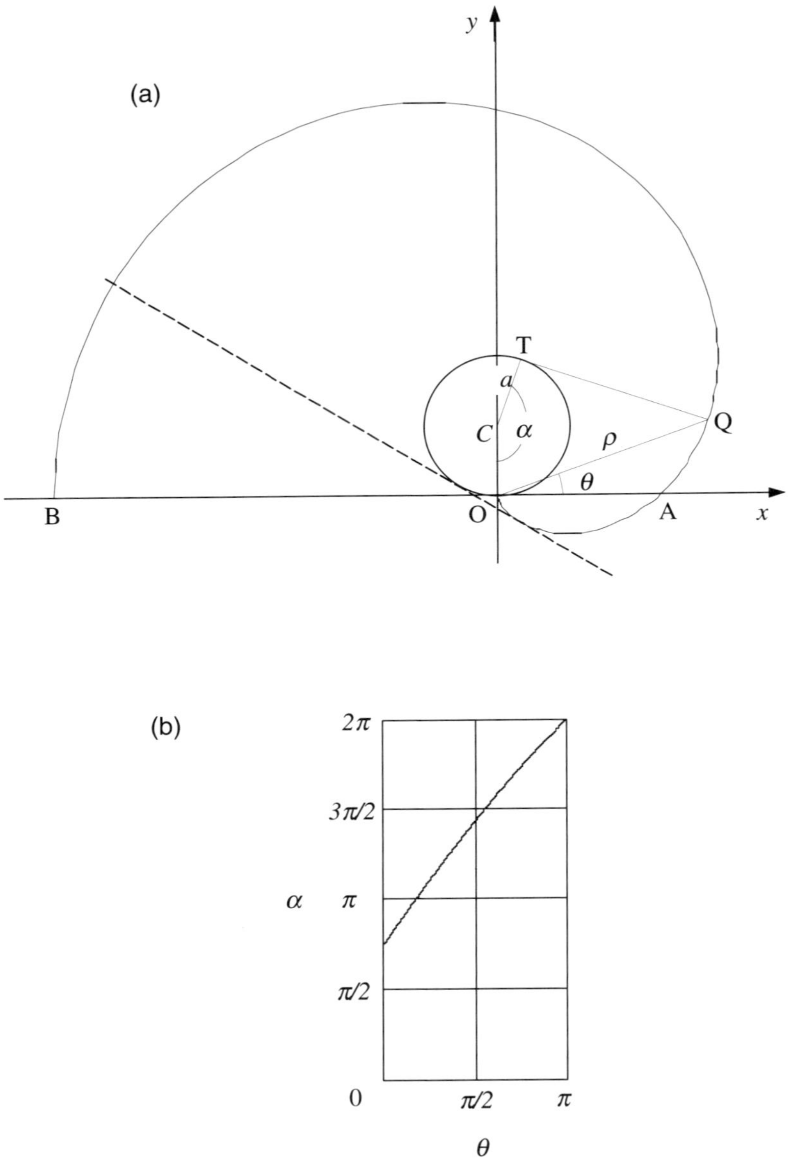

FIGURE 15.15
Modified circle involute approach. (a) The modified curve and (b) the relationship between α and θ.

A more general form for the modified formula is expressed by

$$\rho = a(2 + \alpha^2 + 2\sin(\alpha - \varphi) - 2\alpha\cos(\alpha - \varphi))^{1/2} \tag{15.41}$$

$$\cos\theta = \frac{a(\cos(\alpha - \varphi) + \alpha\sin(\alpha - \varphi))}{\rho} \tag{15.42}$$

where φ is a user-defined value corresponding to an initial rotation angle. The range of φ is recommended to be $[1, \pi/2]$. The extreme case, $\varphi = \pi/2$, corresponds to Equations 15.37

TABLE 15.2

Values of α vs. θ for the Modified Circle Involute Approach

θ	α	θ	α
−.1807289880E−01	.2304887073E+01	.1375632706E+01	.4241150082E+01
.3679373644E−02	.2336460868E+01	.1496629531E+01	.4398229715E+01
.1728543851E−01	.2356194490E+01	.1619396195E+01	.4555309348E+01
.1259017101E+00	.2513274123E+01	.1744098373E+01	.4712388980E+01
.2351114085E+00	.2670353755E+01	.1870915404E+01	.4869468613E+01
.3449731099E+00	.2827433388E+01	.2000040107E+01	.5026548246E+01
.4555502106E+00	.2984513021E+01	.2131677850E+01	.5183627878E+01
.5669115049E+00	.3141592654E+01	.2266044530E+01	.5340707511E+01
.6791318229E+00	.3298672286E+01	.2403363059E+01	.5497787144E+01
.7922927290E+00	.3455751919E+01	.2543857832E+01	.5654866776E+01
.9064832817E+00	.3612831552E+01	.2687746603E+01	.5811946409E+01
.1021800854E+01	.3769911184E+01	.2835229208E+01	.5969026042E+01
.1138352007E+01	.3926990817E+01	.2986472681E+01	.6126105674E+01
.1256253398E+01	.4084070450E+01	.3141592654E+01	.6283185307E+01

and 15.38, whereas $\varphi = 1$ corresponds to the case that the x axis makes tangential contact with the involute curve as shown by the dashed line in Figure 15.15a.

15.6 Adaptive Procedure Based on Background Cells

15.6.1 Issues of Adaptive Analysis

One of the distinctive features of MFree methods is that they evaluate field variables entirely based on a group of discrete nodes and require no predefined nodal connectivity. As nodes need not to be structured, they may be moved, inserted, and deleted freely—this is particularly suitable for adaptive analysis. However, proper algorithms that guarantee a stable, reliable, accurate, efficient, and fully automated adaptive process have to be developed.

In an adaptive analysis, there are essentially two issues—error estimation and domain refinement. The first requires a cheap error estimate to measure the local and global errors, whereby an adaptive procedure determines whether a refinement is required, and if it is required, which part of the domain should be refined. The second is performed based on the information of error distribution provided by the error estimate. The effectiveness and efficiency of these two aspects are critical to the performance of an adaptive procedure. To conduct *a posteriori* error estimation, two values of a quantity—a computed value and a reference value—are usually required. The first is the raw value given by direct computations, and the second is derived from the first via postprocessing (e.g., smoothing or projection). In FEM, the raw stresses/strains (or derivatives) do not possess interelement continuity and have a discrepancy along element boundaries; the improved values are obtained via smoothing the interelement discontinuity on strains. The difference between the raw and improved values formulates a basis for error estimation in FEM; detailed descriptions of this approach are available in the FEM literature (e.g., Zienkiewicz, 1989). There are also other methods (e.g., Radovitzky, 1999) used for error estimation and adaptive meshing in FEM.

In MFree methods, there is no interelement discontinuity of stresses and the resulting stress field is very smooth over the entire problem domain. As a result, error estimates based on stress-smoothing techniques developed for FEM cannot be used for error estimation

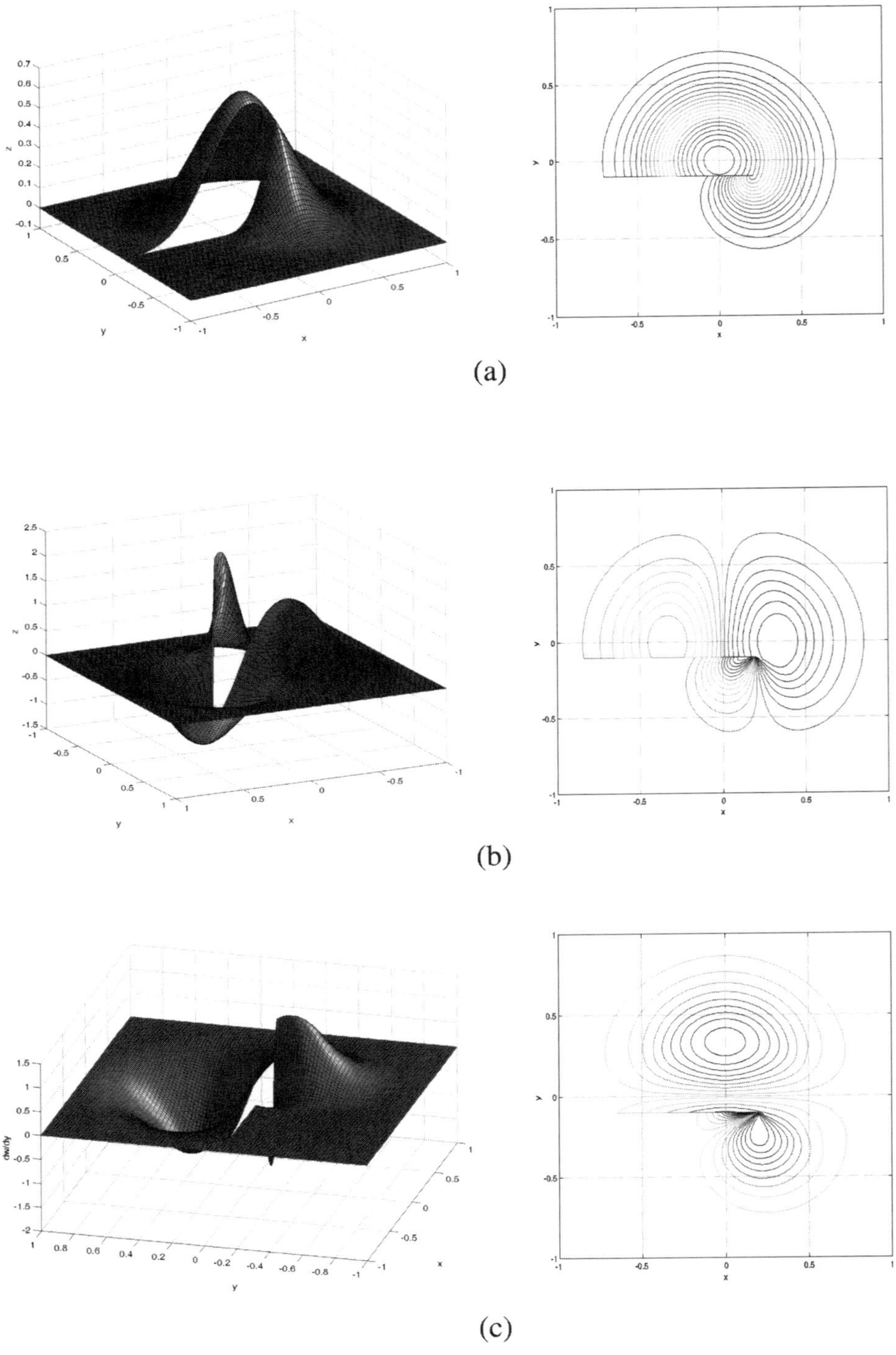

FIGURE 15.16
The cubic spline weight function and its derivatives over an influence domain with a single crack by the modified circle involute approach. Surface plots and contour plots for (a) $\widehat{W}(s)$, (b) $\widehat{W}_x$, and (c) $\widehat{W}_y$.

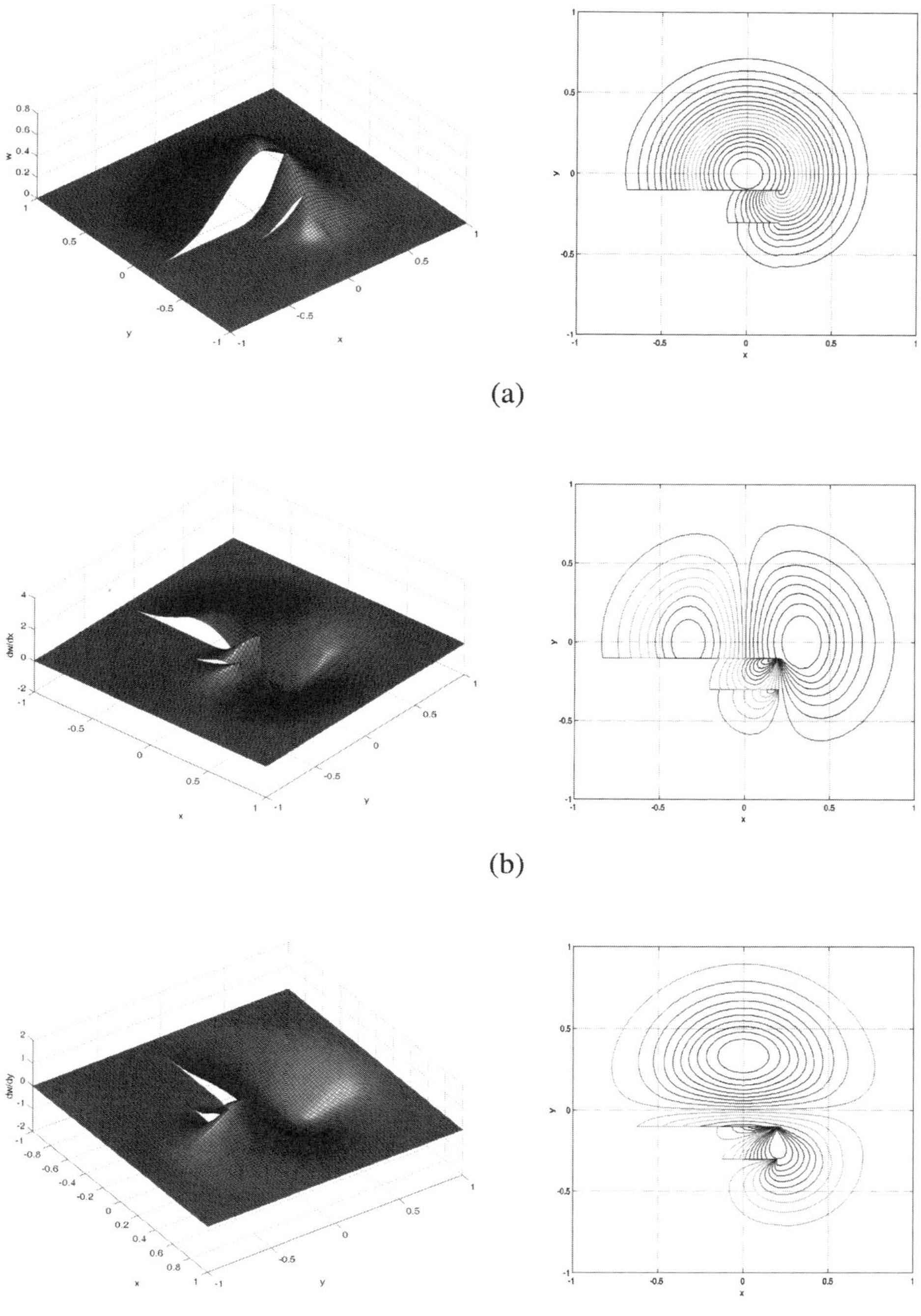

FIGURE 15.17
Cubic spline weight function and its derivatives over an influence domain with two cracks by the modified circle involute approach. Surface plots and contour plots for (a) $\widehat{W}(s)$, (b) $\widehat{W}_x$, and (c) $\widehat{W}_y$.

in MFree methods. There is a need to develop suitable error estimates for adaptive analysis for MFree methods.

15.6.2 Existing Error Estimates

A few error estimates for MFree methods have been reported. One is proposed by Duarte and Oden (1996) for the *h-p* cloud method, which involves the computation of interior residuals and the residuals for Neumann boundary conditions. Chung and Belytschko (1998) adapted the FEM stress projection technique for error analysis in EFG by computing the projected stresses from the raw stresses using a reduced domain of influence. This approach is simple and inexpensive, but its effectiveness depends on the size of the reduced influence domain. Another approach is the strain gradient method proposed by Combe and Korn (1998), who make use of the fact that gradients of stresses and strains may be calculated throughout the problem domain with a high accuracy. The interpolation error is evaluated from the truncated terms in a Taylor expansion of a field variable at a point over its vicinity. This approach was demonstrated to be effective. However, it requires the computations of the second derivatives, which are usually very expensive.

15.6.3 Cell Energy Error Estimate

This section presents a cell energy error (CEE) estimate that can be used for adaptive analysis in MFree methods that use background cells for integration. The material presented here is based on the work done by Tu and G. R. Liu (2000), and G. R. Liu and Tu (2002). The work is based on the fact that many implementations of existing MFree methods rely on a background mesh, global or local, for domain integration of governing equations. The procedure is composed of a cell-based error estimate and a local domain refinement technique. The present error estimate examines the energy error in each cell and uses it as the basis for error estimation.

Error Indicator

An error estimate for an approximation is usually constructed based on the difference between the approximate and exact solutions. For a quantity $\mathbf{q}$ defined over domain Ω and approximated by $\hat{\mathbf{q}}$, a general error indicator of an approximation is defined by

$$e = \mathsf{L}(\mathbf{q}, \hat{\mathbf{q}}) \tag{15.43}$$

where L denotes an operator (e.g., L_1 norm, L_2 norm, etc.) imposed on the exact and approximate values, and e is the error corresponding to the operation. In most problems, $\mathbf{q}$ in exact form is not available and a reference value derived from $\hat{\mathbf{q}}$ is used. In solid mechanics, the quantity can be displacement, strain, stress, or energy.

A conventional implementation of Equation 15.43 is to use the L_2 norm error, i.e.,

$$e = \|\mathbf{q}(x) - \hat{\mathbf{q}}(x)\| = [(\mathbf{q}(x) - \hat{\mathbf{q}}(x))^T \cdot (\mathbf{q}(x) - \hat{\mathbf{q}}(x))]^{1/2} \tag{15.44}$$

This approach is essentially a pointwise approach, as it examines errors at individual points. Errors in local and global domains can be evaluated from pointwise error via integration of the preceding equation. To measure pointwise error, the computed and reference values at a point must be provided. A major task of an error estimate is therefore to formulate the reference value at a point. The error estimate proposed by Chung and

Belytschko (1998) for MFree methods follows this traditional approach, with the reference value obtained by taking a product of the computed value with the shape function constructed on a reduced influence domain. A difficulty in Chung's approach is to minimize the size of the reduced domain while preserving the regularity of the moment matrix in the MLS approximation.

Instead of examining pointwise error, the CEE method uses the error of energy in a cell as the basic measure for error estimation in MFree methods. The quantity $\mathbf{q}$ examined in a cell is the strain energy in the cell. For solid mechanics, the cell energy is

$$E = \int_{\text{cell}} \boldsymbol{\sigma\varepsilon}\, d\Omega \tag{15.45}$$

Note that we disregard the fraction of 1/2 in front of the integration; it is immaterial in our error estimation, because it is the distribution of error that counts for adaptive analysis.

In an MFree approximation, the computed cell energy, E_{comp}, is obtained by using the same Gauss integration scheme as that used for domain integration of the weak-form governing equations, i.e.,

$$E_{\text{comp}} = \sum_{i=1}^{m} c_i \boldsymbol{\sigma}_i \boldsymbol{\varepsilon}_i \tag{15.46}$$

where m is the number of Gauss points used in the cell for integration and c_i the corresponding integration weight. The reference value, E_{ref}, is evaluated using a different Gauss integration scheme:

$$E_{\text{ref}} = \sum_{i=1}^{n} c_i \boldsymbol{\sigma}_i \boldsymbol{\varepsilon}_i \tag{15.47}$$

where n ($n \neq m$) is the number of Gauss points used for the reference value. The energy error in a cell is thus

$$e = |E_{\text{comp}} - E_{\text{ref}}| \tag{15.48}$$

In this approach, the stresses and strains at the n Gauss points are evaluated based on the displacement field given by the original solution and therefore they have a same accuracy as those of the m Gauss points. To reduce computational cost, it is recommended that n is assumed to have a smaller value than m, i.e., $n < m$. Summation of CEE over all cells in the problem domain yields the global error, i.e.,

$$e_{\text{Global}} = \sum_{i=1}^{N} e_i \tag{15.49}$$

where N is the total number of cells. A normalized measure of CEE is

$$e = e/A \tag{15.50}$$

where A is the cell area. The normalized measure is very effective in detection of singular locations.

One prominent feature in the present approach is that it requires only one stress field: both the computed and reference values of cell energy are computed based on values in the same field at different positions. This constitutes a major difference from the conventional pointwise error estimates where a second stress field is required. This is advantageous as the postprocessing for a second field is no longer necessary.

Error Sources

In a numerical approximation, there are mainly two sources of error—interpolation error and integration error. The first arises from the limited order of approximation to the field function, when the interpolation function is performed. For example, a kth-order interpolation results in an error of $(k+1)$th order. Compared with FEM, a meshless approximation (e.g., MLS approximation) generally has a higher order of continuity and therefore a smaller interpolation error. However, the exact order of interpolation error is difficult to determine in MFree methods as the MLS approximation is usually implicit, in contrast to FEM approximation.

The second source of error is introduced in the numerical integration. For example, in a 1D case, m Gauss point integration causes an integration error of $(2m)$th order. The integration error can be minimized if the integration order matches the interpolation order, i.e., when $m = (k + 1)/2$.

Two approaches are usually taken to improve the accuracy of the solution: one is to increase the order of interpolation function and the other is to match the integration order with the interpolation order. In an MLS approximation, the first can be implemented by choosing high-order basis functions and high-order weight functions. Implementation of the second, however, is obscure as there is no prior knowledge about the exact interpolation order. One empirical way is to use a sufficient number of Gauss points in each cell in conjunction with a sufficient number of background cells for integration. It can be expected that because the integration cells are reduced in size, the error evaluated by the proposed error estimate decreases and converges as the stress field in each cell approaches linear (or planar) variation.

The task of an error estimate is to measure the error from every source, whereby an adaptive procedure determines whether to increase the interpolation order or to use a finer background mesh, or both. In the traditional stress-smoothing approach, the smoothed value has a higher order of accuracy than the original, and therefore the interpolation error is measured by the difference. In the CEE approach, the values at the Gauss points in Equations 15.46 and 15.47 are from the same stress field and therefore are of the same order of accuracy. The order of integration accuracy in the two equations, however, is different—it is $(2m - 1)$ in Equation 15.46 but $(2n - 1)$ in Equation 15.47. The CEE evaluated from Equation 15.48 is therefore the integration error between the two integration schemes. It is sensitive to the order of stress field—the higher the order, the larger the energy error. From this point of view, the present approach can be viewed as a variant of the gradient approach as it reflects the gradient change in an approximation field. The major weakness of the CEE estimate is that it does not provide the accuracy of an approximation itself. As a consequence, error estimation by this strategy may be erroneous if a wrong approximation is used as the basis—this is also the case for many of the existing error estimates. Note that the CEE estimate demands that the integration cell sufficiently corresponds to the gradient of the stress/strain. If the order of the Gauss integration for computing the reference cell energy is 3 ($n = 2$), and that for computed cell energy is 1 ($m = 1$), the error estimated by the CEE will approach zero if the cell is small enough so that the stress/strain field can be approximated by a linear function. Otherwise, a large error will be produced that may demand refinement. From this point of view, the CEE performs exactly the job that is required for adaptive analysis. In the case of using an erroneous field variable approximation, it is most likely the adaptive procedure will not converge, unless the erroneous approximation converges.

15.6.4 Numerical Examples

In the following examples, we have adapted the strategy that is used in MFree2D, as follows:

- Triangular cells of background integration are used.
- EFG formulation with the penalty method for essential boundary conditions is followed.
- CEE estimate is used. For the computed cell energy one Gauss point is used, and three Gauss points are used for the reference cell energy.
- Relay model is employed.

Example 15.1 Cantilever Beam (Error Estimation)

The cantilever beam as described in Chapter 6 (Example 6.2) is employed for the first assessment of the error estimate. The beam is schematically drawn in Figure 6.4. For this example, we have an analytical solution (see Example 6.2), so verification can be performed easily. The parameters for this example are as follows:

Loading: $P = 1000$ N

Young's modulus for the material: $E = 3.0 \times 10^7$ N/m^2

Poisson's ratios for two materials: $\nu = 0.3$

Height of the beam: $D = 12.0$ m

Length of the beam: $L = 48.0$ m

At the left boundary ($x = 0$) the displacements are prescribed using the analytical formulae, Equations 6.28 and 6.30, and the left boundary ($x = L$); the traction force is computed from the analytical formula, Equation 6.33.

The "exact" cell energy is computed based on the analytical solutions of stress and strain and integrated over the cell using three Gauss points. Note that the energy is not really exact as the integration is not exact. It can still be used to assess the accuracy of the error estimate.

The MFree model used, shown in Figure 15.18, comprises 737 nodes. The distribution of a normalized estimated energy error is plotted in Figure 15.18b. For the computed cell energy one Gauss point is used while three Gauss points are used for the reference cell energy. The distribution of the exact energy error is shown in Figure 15.18c. The exact error distribution is computed using the exact energy as the reference energy obtained from the analytical solution and integrated over the cells using three Gauss points. Comparison of Figure 15.18b with Figure 15.18c demonstrates close agreement between these two error distributions. The difference in the magnitude of local error, however, is not minor: the maximum estimated value is about 2.52×10^{-4} while the exact is 3.32×10^{-4}. This is also the case in terms of the global error—the estimated and exact values are 2.14×10^{-3} and 2.92×10^{-3}, respectively. This shows that the proposed approach is not sufficient for accurate estimation of the absolute error. It provides a very good indication on the error distribution.

Example 15.2 Infinite Plate with a Circular Hole (Error Estimation)

Example 7.9 is examined. The geometry of the plate is plotted in Figure 7.24. Due to the twofold symmetry, only a quarter of the plate shown in Figure 7.25 is modeled with symmetric boundary conditions applied on $x = 0$ and $y = 0$. The parameters are listed as follows:

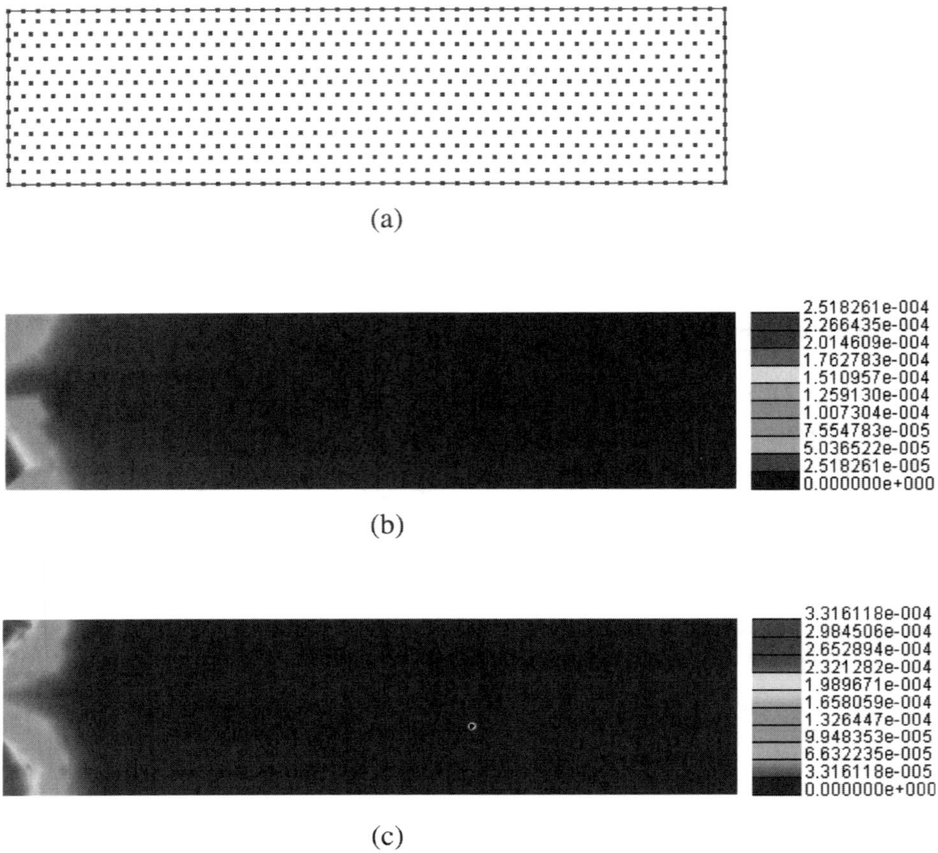

FIGURE 15.18
Error distribution in cantilever beam. (a) MFree model (737 nodes); (b) distribution of estimated error; and (c) distribution of exact error. (From Liu, G. R. and Tu, Z. H., *Comput. Methods Appl. Mech. Engrg.*, 191, 1923–1943, 2002. With permission.)

Loading: $p = 1$ N/m

Young's modulus: $E = 1.0 \times 10^3$ N/m^2

Poisson's ratio: $\nu = 0.3$

Height of the beam: $a = 1.0$ m

Length of the beam: $b = 5$ m

The plate is subjected to a tension in the x direction at the edge of $x = 5$. The boundary condition at $x = 5$ is: $\sigma_{xx} = p$, $\sigma_{yy} = \sigma_{xy} = 0$, and the boundary condition at $y = 5$ is free of all stresses. The analytical solution of displacement and the stress fields within the plate are provided by Equations 7.59 through 7.64 in the polar coordinates (r, θ).

The MFree model (204 nodes) and error distributions are shown in Figure 15.19a, b, and c, respectively. Again, there is very good correlation between the estimated and exact error distributions. Differences in the magnitude of local and global errors are also observed: the maximum estimated and exact local errors are 8.04×10^{-5} and 1.34×10^{-4} and the maximum estimated and exact global errors are 8.44×10^{-4} and 1.07×10^{-3}, respectively.

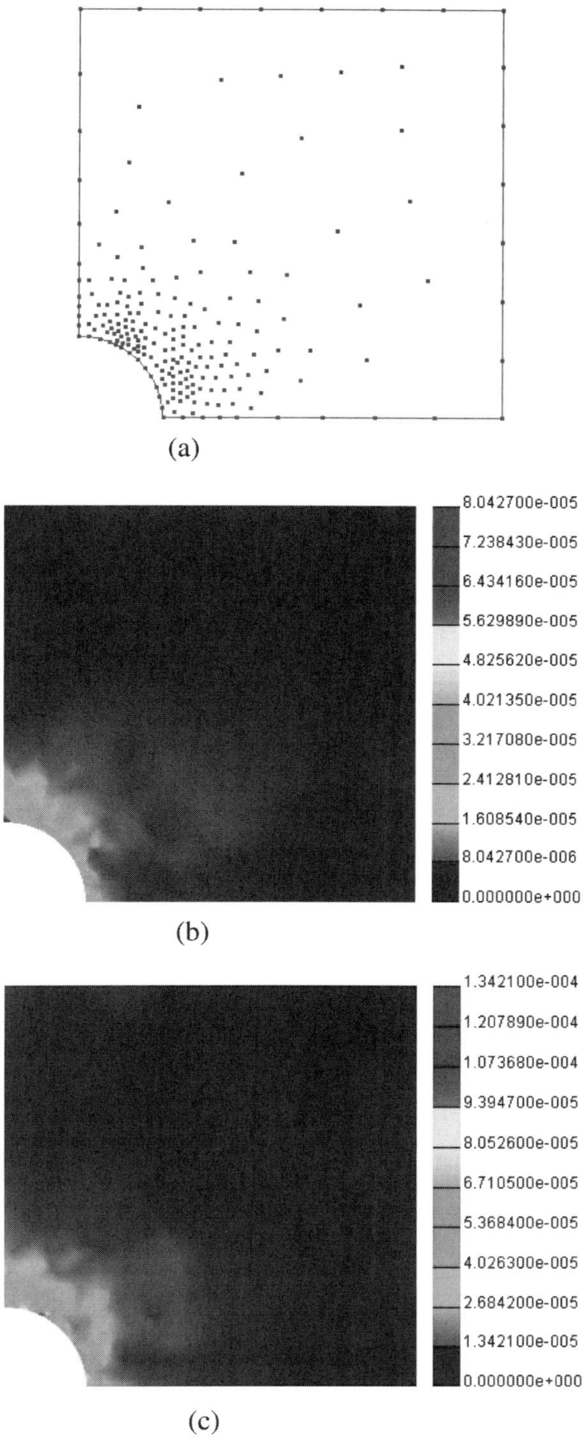

FIGURE 15.19
Error distribution for square plate with a hole problem. (a) MFree model (204 nodes); (b) distribution of estimated error; and (c) distribution of exact error. (From Liu, G. R. and Tu, Z. H., *Comput. Methods Appl. Mech. Engrg.*, 191, 1923–1943, 2002. With permission.)

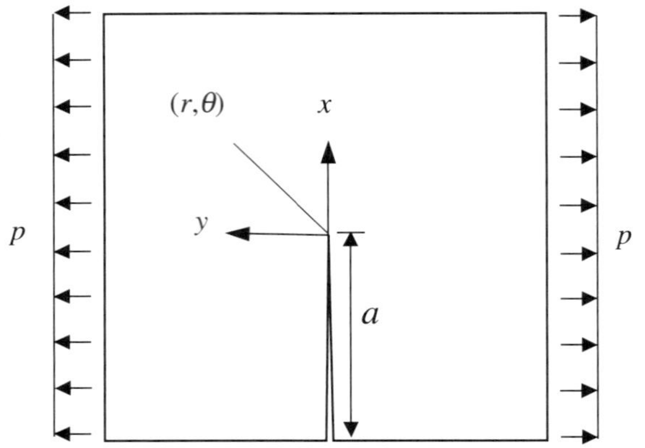

FIGURE 15.20
Square plate with a crack subjected to a horizontal tensile traction. (From Liu, G. R. and Tu, Z. H., *Comput. Methods Appl. Mech. Engrg.*, 191, 1923–1943, 2002. With permission.)

Example 15.3 A Square Plate Containing a Crack

A square plate containing a crack subjected to boundary conditions prescribed by the near crack-tip field solution is shown in Figure 15.20. The material properties are Young's modulus $E = 3.0 \times 10^7$ and Poisson's ratio $\nu = 0.3$. In the crack-tip field problem, the square plate has a side of $2a$ and the crack assumes a length of a. This corresponds to the so-called Griffith mode-I crack problem that has an analytical solution (Anderson, 1991):

$$\sigma_{xx} = \frac{K_I}{\sqrt{2\pi r}} \cos\frac{\theta}{2}\left(1 - \sin\frac{\theta}{2}\sin\frac{3\theta}{2}\right) \tag{15.51}$$

$$\sigma_{yy} = \frac{K_I}{\sqrt{2\pi r}} \cos\frac{\theta}{2}\left(1 + \sin\frac{\theta}{2}\sin\frac{3\theta}{2}\right) \tag{15.52}$$

$$\sigma_{xy} = \frac{K_I}{\sqrt{2\pi r}} \sin\frac{\theta}{2}\cos\frac{\theta}{2}\cos\frac{3\theta}{2} \tag{15.53}$$

with the coordinate system depicted in Figure 15.20. The stress intensity factor K_I is prescribed by $K_I = p\sqrt{\pi a}$. In the MFree model, 397 nodes are used with the majority distributed around the crack tip (Figure 15.21a) to capture the crack-tip field. The relay model is used to handle the discontinuity. As with the previous two examples, the distributions of predicted and exact errors plotted in Figure 15.21b,c show very good agreement.

15.7 Strategy for Local Adaptive Refinement

In adaptive analysis, a problem domain may be refined if the desired accuracy is not achieved. It is usually undesirable to refine the entire domain, as in many cases only a few locations exhibit poor approximations. To achieve high efficiency, it is therefore required to focus on and refine those locations only. A local refinement approach is presented in this section for just this purpose. The approach is also based on a triangular mesh and uses a local Delaunay algorithm with the aid of a density factor. A detailed description of this strategy is given in the following.

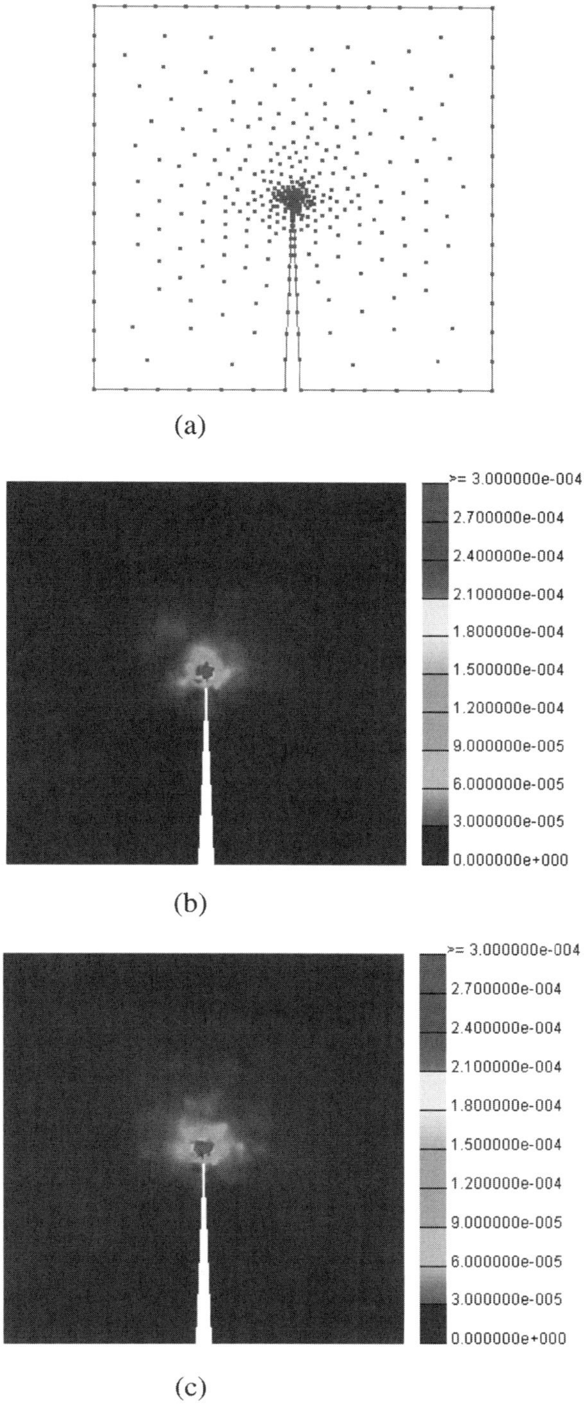

FIGURE 15.21
Error distribution for square plate with a crack problem. (a) MFree model (397 nodes); (b) distribution of estimated error; and (c) distribution of exact error. (From Liu, G. R. and Tu, Z. H., *Comput. Methods Appl. Mech. Engrg.*, 191, 1923–1943, 2002. With permission.)

15.7.1 Update of the Density Factor

The density factor at a node will be changed if refinement is required at its location. The change of density factor is based on the distribution of local error measured by the CEE estimate. This is done by converting the CEE into nodal energy error. The former is equally distributed to cell vertices and the latter is an accumulation of contributions from surrounding cells, i.e.,

$$e_{\text{nodal}} = \sum_{i=1}^{m} \frac{e_i}{n} \tag{15.54}$$

where m is the number of surrounding cells associated with a node, n is the number of cell vertices ($n = 3$ for a triangular cell), and e_i is the energy error of the ith surrounding cell. A relative error measure is then defined for each node

$$R_{\text{nodal}} = \frac{e_{\text{nodal}}}{E_{\text{nodal}}} = \sum_{i=1}^{m} \frac{e_i}{n} \Big/ \sum_{i=1}^{m} \frac{E_i}{n} \tag{15.55}$$

where R_{nodal} is the nodal relative error, E_i the computed energy (given by Equation 15.46) of the ith surrounding cell, and E_{nodal} the nodal energy converted from the cell energy. To determine the locations where refinement is required, a threshold nodal relative error is predefined and the relative error at each node is compared with this value. If the threshold value is exceeded, the density factor at a node will be changed to

$$S^*_{\text{nodal}} = \frac{R_{\text{threshold}}}{R_{\text{nodal}}} S_{\text{nodal}} \tag{15.56}$$

where S_{nodal} and S^*_{nodal} are the old and new values of a density factor, respectively. Refinement of the local domain is based on the change in density factor, and the degree of refinement is manageable as the density factor controls the local nodal density. Upon the change of a density factor, a local Delaunay algorithm is executed.

15.7.2 Local Delaunay Triangulation Algorithm

The Delaunay triangulation technique can be applied to an arbitrary 2D domain. Given a set of nodes and a discretized boundary that encloses the nodes, the technique can generate an optimal triangular mesh for the bounded domain based on the existing nodes. This versatility enables a local domain to be refined easily and forms the basis of the present approach. To illustrate this, consider the example depicted in Figure 15.22a, where the density factor of node I changes from S_I to S^*_I ($S^*_I < S_I$). The procedures for local refinement are as follows:

1. Insert nodes. New nodes are inserted by looping the surrounding triangles: for an acute triangle a node is inserted at the center of its circumcircle whereas for an obtuse triangle, a new node is introduced at the middle of the longest edge (Figure 15.22b).
2. Formulate the local domain. This is done by drawing a circle centered at node I (Figure 15.22c) and then removing all the cell edges inside or intersecting the circle (Figure 15.22d). The circle radius dictates the block size and is used to control the range of mesh revision.

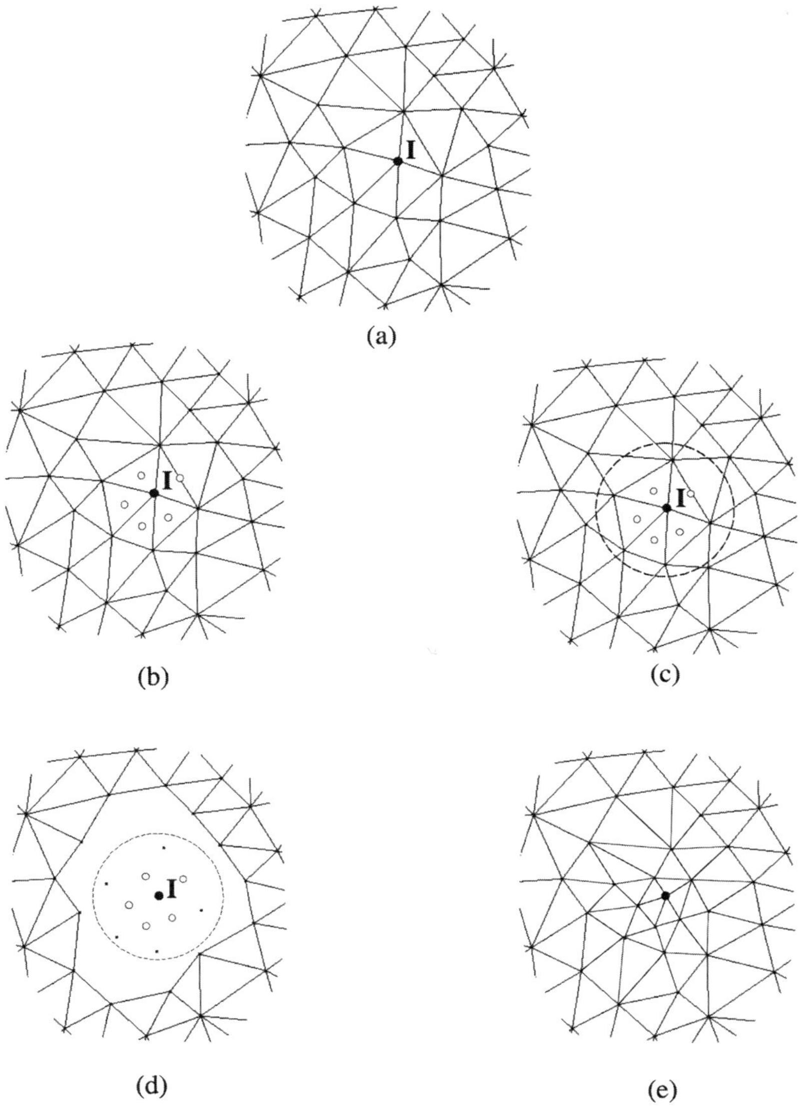

FIGURE 15.22
Procedures for local domain refinement. (a) Original mesh; (b) insertion of nodes; (c and d) formulation of a local block; and (e) refined mesh. (From Liu, G. R. and Tu, Z. H., *Comput. Methods Appl. Mech. Engrg.*, 191, 1923–1943, 2002. With permission.)

3. Triangulate the local domain using the Delaunay algorithm. This regenerates a triangular mesh for the local domain based on the existing nodes inside the block (Figure 15.22e).
4. Recalculate the density factors. Density factors for all affected nodes are updated based on the new mesh.

The refinement procedures will be repeated if the updated density factor of node I is still larger than S_I^*. This local refinement approach is very efficient, especially for problems with a large number of nodes. Other variants of this strategy may be devised. In an iterative solution procedure, the variables at new nodes are evaluated based on the surrounding old nodes, thus giving a starting solution for the next iteration.

Example 15.4 Infinite Plate with a Circular Hole (Adaptive Analysis)
Example 15.2 is reconsidered here for adaptive analysis. The geometry of the plate is plotted in Figure 7.24. Due to the twofold symmetry, only a quarter of the plate shown in Figure 7.25 is modeled with symmetric boundary conditions applied on $x = 0$ and $y = 0$. The parameters are exactly the same as those in Example 15.2.

The local adaptive refinement procedure is used for adaptive refinement. The results are shown in Figure 15.23. The stress concentration occurs around the hole and nodes are automatically added to these locations. The result at the final step is very close to the theoretical solution.

Example 15.5 Square Plate with a Square Hole (Adaptive Analysis)
Adaptive analysis of a square plate with a square hole at its center subjected to a unit tension force is performed. Making used of the symmetry, one quarter of the problem domain, as shown in Figure 15.24, is used for the analysis. This is a classical problem for testing a refinement procedure. The refinement procedure detects that point A (see Figure 15.24) is singular and hence the location at this point is refined. With only three steps, the adaptive procedure yields very accurate stress distributions (Figure 15.25).

Example 15.6 Square Plate with a Crack (Adaptive Analysis)
Adaptive analysis of a rectangular plate with a crack loaded by horizontal unit tractions on the two vertical sides is performed. The problem domain is drawn in Figure 15.26. The domain has dimensions of 100 in length and 50 in height. The material parameters are Young's modulus $E = 3.0 \times 10^7$ and Poisson's ratio $\nu = 0.3$. The results are shown in Figure 15.27. Due to the stress singularity, the vicinity of the crack tip is mostly refined. However, this refinement will never come to an end because of the singularity at the crack tip. Therefore, the computation is terminated at the third refinement step when the crack-tip field is described with a very high resolution. The good numerical results for the three sample problems demonstrate that the proposed local adaptive refinement procedure is effective and efficient.

Example 15.7 Square Plate with Two Parallel Cracks (Adaptive Analysis)
Adaptive analysis of a rectangular plate with two parallel cracks loaded by horizontal unit tractions on the two vertical sides is performed. The domain has dimensions of 100 in length and 50 in height. The material parameters are: Young's modulus $E = 3.0 \times 10^7$ and Poisson's ratio $\nu = 0.3$. Figure 15.28a shows the MFree model. The numerical results give very good descriptions of the stress field around the crack tips as shown in Figure 15.28, verifying the validity of the proposed model.

Example 15.8 Arbitrary Complex Domain (Adaptive Analysis)
A panel carved with "ACES" subjected to a tensile traction loading on the central part of the top side is investigated. The panel is fixed at all four corners. This is a multiconnected body with highly irregular boundaries. Figure 15.29a shows the MFree model for the analysis. The material parameters are: Young's modulus $E = 3.0 \times 10^7$ and Poisson's ratio $\nu = 0.3$. The stress distributions are very well depicted, as shown in Figure 15.29, confirming that the relay model is capable of handling highly irregular problem domains.

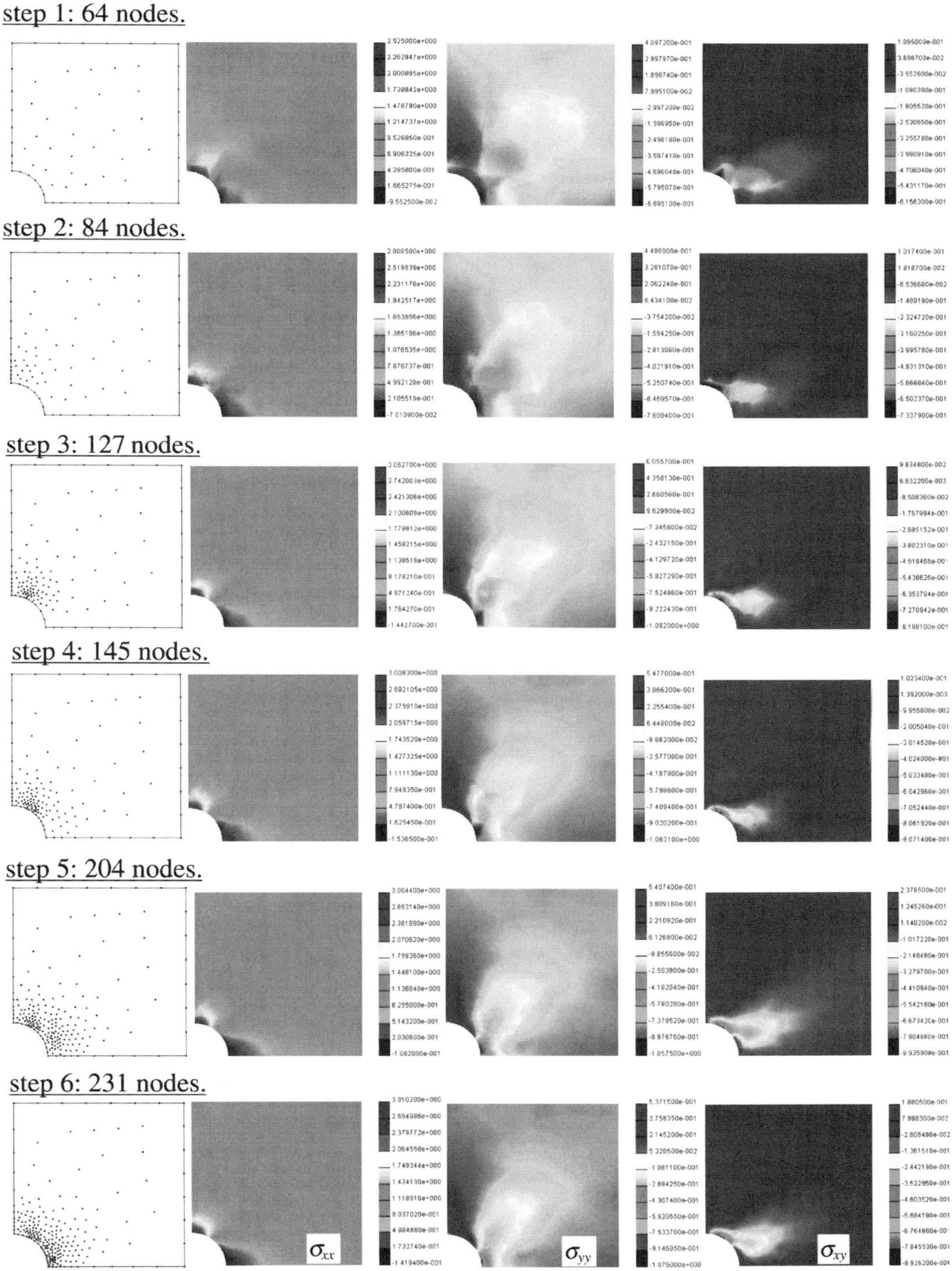

FIGURE 15.23
Adaptive refinement process and distributions of stresses for the square plate with a hole subjected to a unit traction in the x direction.

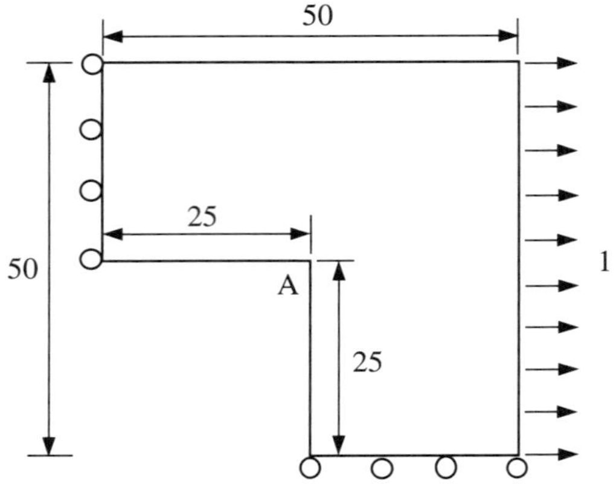

FIGURE 15.24
L-shaped plate subjected to a unit horizontal tensile traction. (From Liu, G. R. and Tu, Z. H., *Comput. Methods Appl. Mech. Engrg.*, 191, 1923–1943, 2002. With permission.)

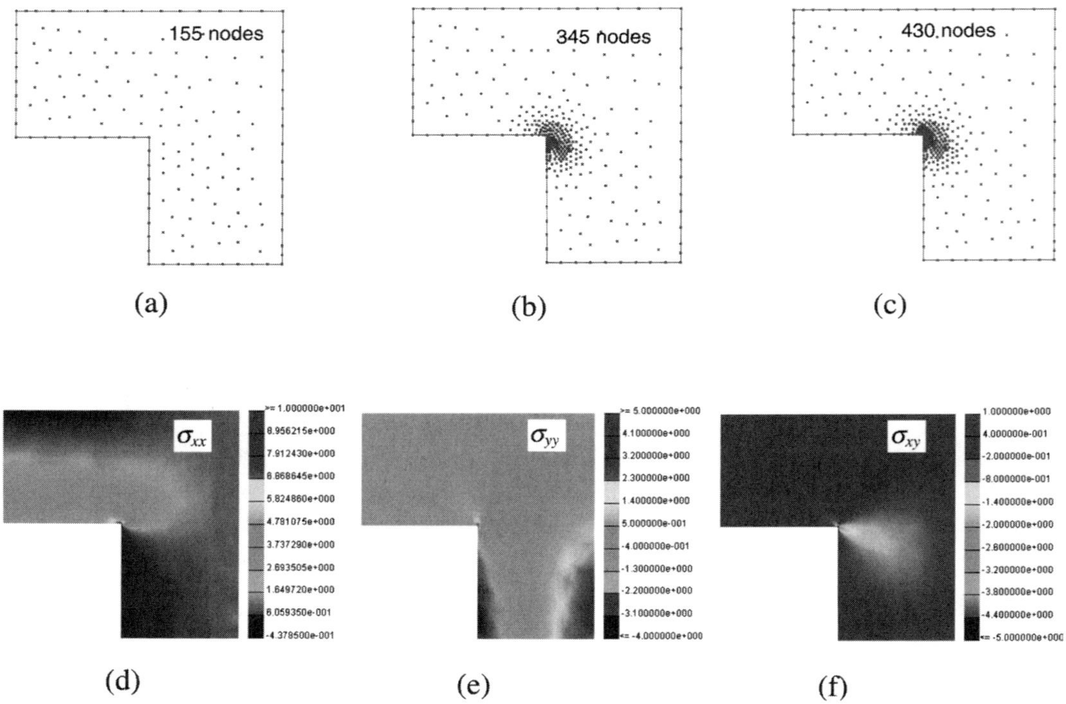

FIGURE 15.25
Adaptive refinement process and the final stress distributions for L-shaped problem. (a to c) Three refinement stages; and (d to f) stress distributions at the final stage. (From Liu, G. R. and Tu, Z. H., *Comput. Methods Appl. Mech. Engrg.*, 191, 1923–1943, 2002. With permission.)

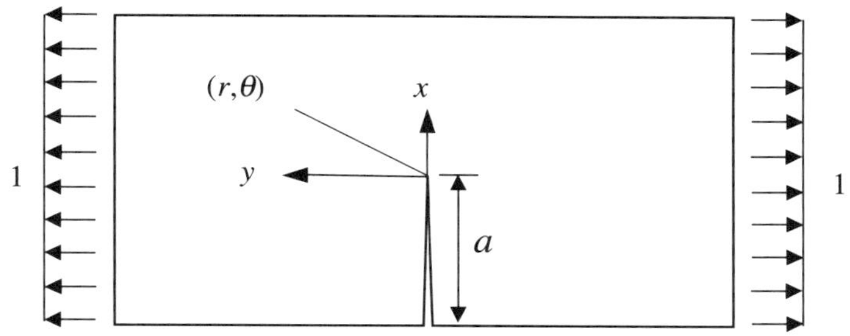

FIGURE 15.26
Rectangular plate with a crack subjected to a unit tensile traction. (From Liu, G. R. and Tu, Z. H., *Comput. Methods Appl. Mech. Engrg.*, 191, 1923–1943, 2002. With permission.)

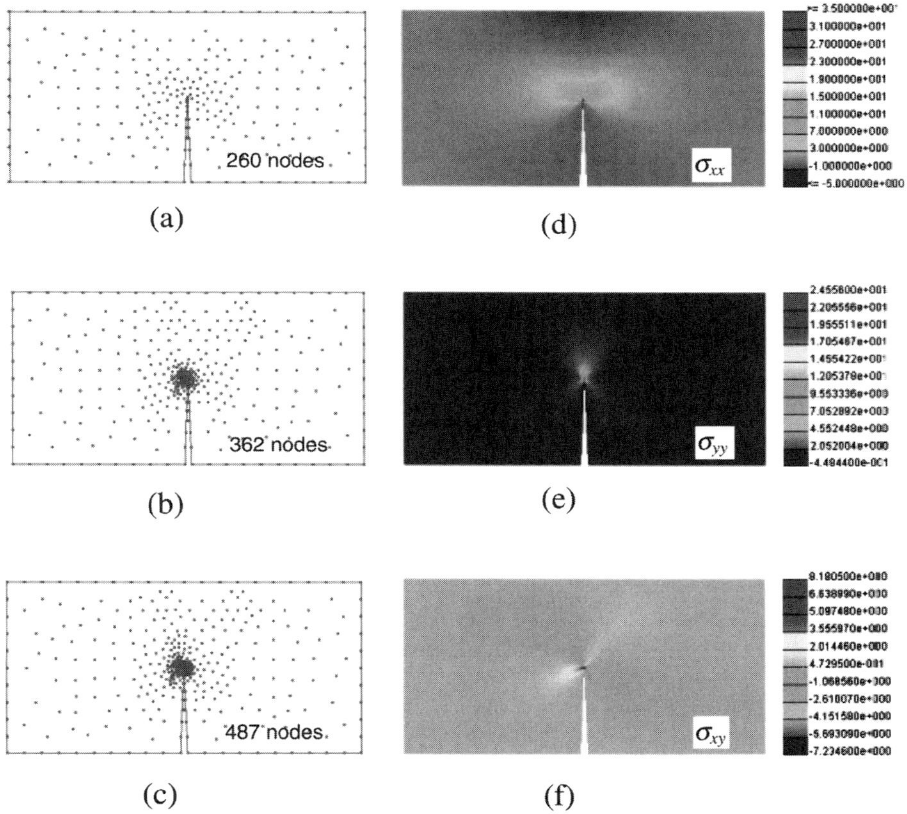

FIGURE 15.27
Adaptive refinement process and the final stress distributions for the rectangular plate with a crack problem. (a to c) Three refinement stages; and (d to f) stress distributions at the final stage. (From Liu, G. R. and Tu, Z. H., *Comput. Methods Appl. Mech. Engrg.*, 191, 1923–1943, 2002. With permission.)

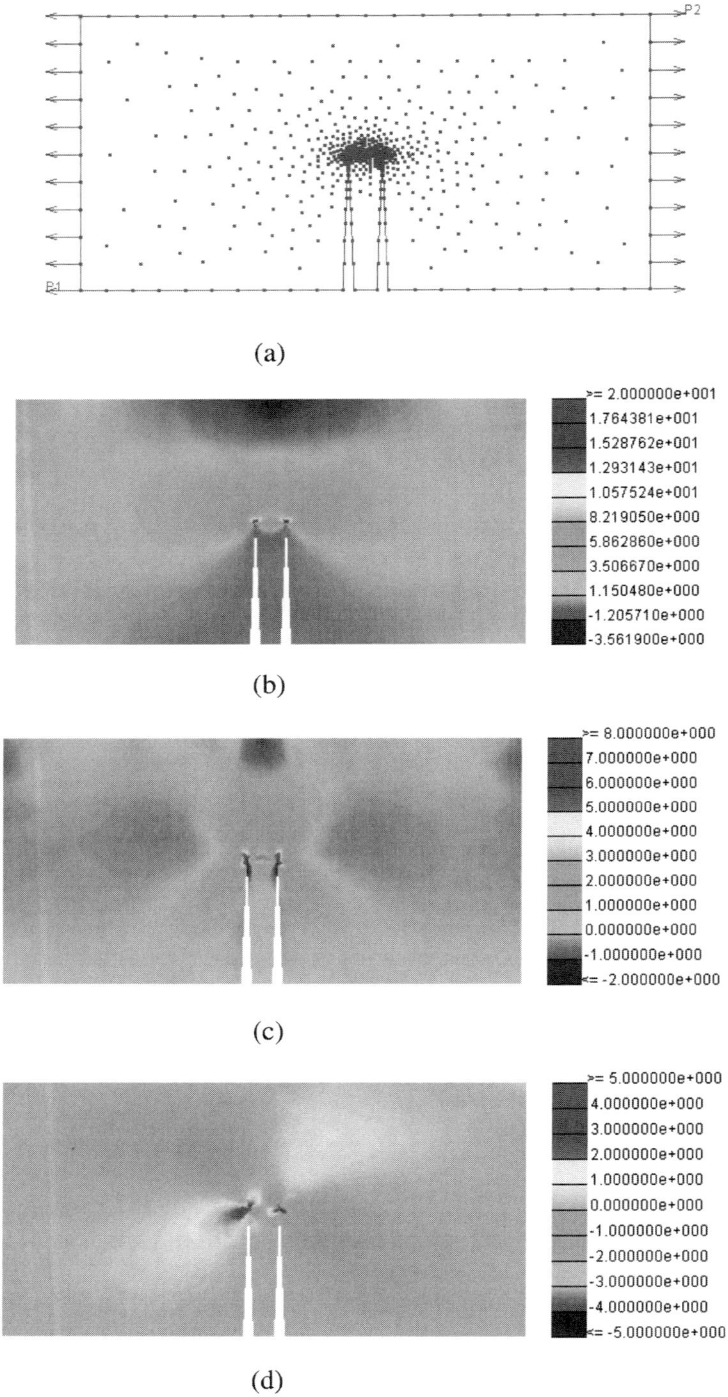

FIGURE 15.28
Interference between two cracks in a rectangular plate subject to a unit tensile traction. (a) Problem model, (b) distribution of σ_x, (c) distribution of σ_y, and (d) distribution of τ_{xy}.

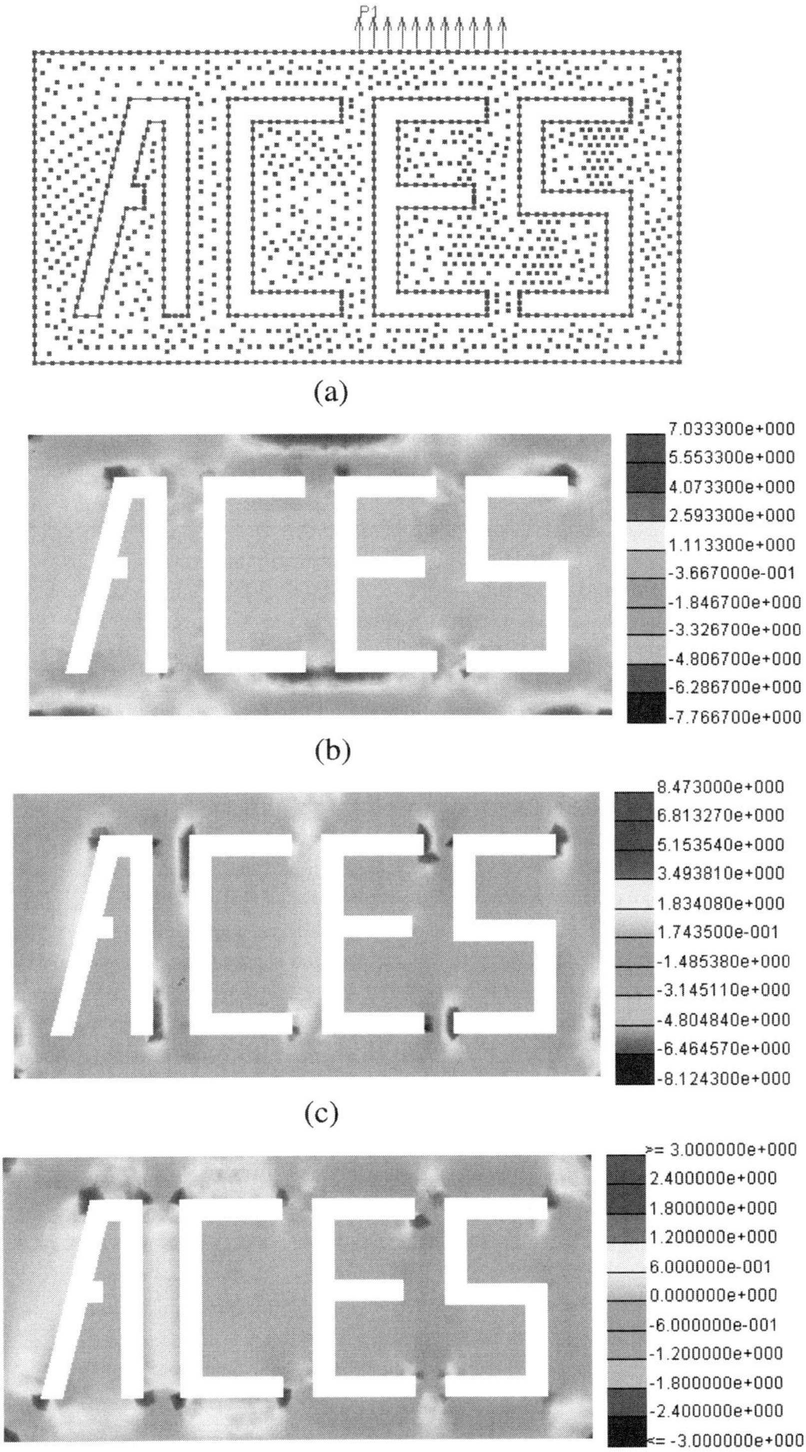

FIGURE 15.29
Panel characterized by "ACES" subjected to a unit tensile traction along the central part of the top side. (a) Problem model, (b) distribution σ_x, (c) distribution of σ_y, and (d) distribution of τ_{xy}.

15.8 Remarks

MFree methods have advantages over conventional mesh-based approaches in dealing with problems involving large deformation and crack propagation. The advantages, however, are undermined by the difficulties in construction of MFree approximations for highly irregular problem domains, which are usually the cases in real applications. A relay model has been proposed in this chapter, aiming to provide a general approach to overcome these difficulties. The essence of this model is to construct a hierarchical network of relay points within an influence domain, via which the influence from the source node is transmitted to the blocked areas. A relay point is said to have a relay region whose profile is described using the circle involute curves. The weight of influence at a point is measured by equivalent distance and is computed in a manner easy to manage—it relates only to its mastering relay point. The approximation functions based on this model are generally well shaped, their smoothness depending on the profile of the relay region. Various numerical examples based on the EFG method have shown that the proposed model is general, effective, and robust. It is envisaged that the result can be improved if a more appropriate description of the profile of the relay region is used.

An adaptive procedure based on background cells is also presented for MFree methods. The core of this procedure is a CEE estimate, which is very different from the conventional pointwise error estimates; it evaluates errors based on cells instead of points and uses CEE as the basic measure. The CEE is measured by the difference between computed cell energy and a reference cell energy given by two different Gauss integration schemes. The stress values at the Gauss points in both integration schemes are from the same stress field. The error estimate is essentially a gradient approach as the CEE can reflect the gradient change in a stress field. Numerical experiments have demonstrated that the proposed error estimate is able to capture the essential features of the error distribution. The limitation of this estimate is that it is formulated without challenging the accuracy of the approximation field, and hence the error estimation may be erroneous if a wrong approximation is used as the basis. In conjunction with the error estimate, a simple, efficient, local domain refinement technique is developed. It is based on a triangular mesh and uses a local Delaunay algorithm with the aid of a density factor. Refinement of a local domain is accomplished by simply adjusting a density factor.

Note that both the error estimate and local domain refinement technique are based on a triangular mesh. From an academic point of view, it is not desirable, as it actually uses a mesh. From a practical point of view, the author sees nothing wrong in using a mesh as long as it can be automatically generated and adapted whenever and wherever needed. From a user's point of view, simplicity, automaticity, stability, and efficiency are always the most important features in any simulation process. The present algorithm, using a triangular mesh, serves well for these practical interests and therefore holds value for practical applications.

16

MFree2D©

16.1 Overview

MFree is a software package being developed by a team at the Centre for Advanced Computations in Engineering Science (ACES) led by G. R. Liu, based on MFree technologies for general-purpose engineering computations. It consists of three components—*MFreePre*, *MFreeApp*, and *MFreePost*. *MFreePre* is a preprocessor to formulate the input required by *MFreeApp*. The latter performs computations and yields the output results, which are then fed to *MFreePost* for postprocessing. These three processors work together seamlessly, and users are not usually aware that they are different processors. However, these three integrated processors can also be separated to perform their work independently. The package is aimed at applying to problems of stress and strain analysis; structural analysis and heat transfer analysis for both static and dynamic situations in two-dimensional (2D) or three-dimensional (3D) space. The first and current version of MFree, ***MFree2D***©, is limited to 2D elastostatics. It runs on a PC in a Window environment. MFree2D was first showcased at the 4th International Asia-Pacific Conference on Computational Mechanics, which was held in Singapore in 1999. The current version is 1.3, which is available for download at http://www.crcpress.com/e_products/downloads/download.asp?cat_no=1238.

MFree2D is *currently* a freeware with a user guide also available on the ACES Web site, but without technical support. Interested readers should make their own arrangements with ACES. Because it is new and still in the development phase, the functions are changing quite frequently and, naturally, there are bugs. ACES is constantly trying to update the new developments and changes via the Web site. All users who are interested trying or using this package are welcome to do so at their own risk, but are currently required by ACES to register and to agree on the terms and conditions set by ACES.

Main features of MFree2D

- The problem domain is discretized using scattered nodes, and no mesh is required for field variable interpolation. The domain discretization process is fully automatic.
- Automatic adaptive refinement techniques are implemented to ensure results of desired accuracy.
- Operation is performed on graphic manual-based user-friendly interfaces.

Advantages over FEM packages

- Meshless—cost of labor for meshing is reduced significantly.
- Higher accuracy—results are accurate as desired.
- There is no problem relating to mesh distortion.

TABLE 16.1

Techniques Used in MFree2D 1.1 and Ongoing Development

Techniques	MFree2D 1.1	Ongoing Development at ACES
Shape function	MLS (Chapter 5)	PIMs (Chapters 5, 8–14)
Weak form	Global Galerkin weak form, EFG (Chapter 6)	Local Petrov–Galerkin (Chapter 7)
Essential boundary condition	Penalty method (Chapters 4, 6–8, 10–13)	Direct approaches for both local Petrov–Galerkin, and global Galerkin
Background integration	Triangular cells + Gauss quadrature (Section 15.2)	Local triangular cells + Gauss quadrature; without numerical integration and hence no background cells are required
Node searching	Bucket algorithm (Section 5.4); relay model (Section 15.5)	Tree algorithm; improvement on relay model
Error estimation	CEE estimate (Section 15.6)	Developing new estimates
Refinement strategy	Density factor control and Delaunay triangulation (Section 15.1)	Refinement of density factor control algorithm
Influence domain	Node based (Section 15.1); relay model (Section 15.5)	Anisotropic influence domain
Problems	Linear elastostatics	Material nonlinear; vibration analysis

Contact Person

Director
Centre for Advanced Computations in Engineering Science
Tel: (65)6874-4795, Fax: (65)6779-1459
E-mail: acehead@nus.edu.sg

16.2 Techniques Used in MFree2D

The tasks involved in an MFree analysis are choice of the size of the influence domain, search of the nodes for the influence domain, computation of nodal weight, which is a coefficient that controls the dimension of the influence domain, generation of the background mesh, and selection of the integration scheme. MFree2D performs all these tasks. Techniques used in MFree2D 1.0 are listed in Table 16.1. The table includes references to relevant chapters and sections of this book for detailed explanations of each technique.

16.3 Preprocessing in MFree2D

MFreePre is used to define and model a problem for meshless analysis. It creates a geometric model, creates nodes and background cells, defines material properties, initial and boundary conditions, and performs solution control. One salient feature is that troublesome and time-consuming manual mesh generations are no longer necessary. In discretization of a problem domain, users do not need to work on the geometric model part by part. One needs only to define the boundaries of the problem domain by specifying its boundaries. MFreePre automatically identifies the geometry and models the domain using scattered nodes and the background cells simply using the Delaunay triangulation algorithm. This saves significant labor cost in mesh creation by engineers. MFreePre allows an analysis to be customized with its open environmental settings while providing default settings for new users. MFree2D requires only the minimum from users, such as geometry, boundary conditions, and material properties. All the issues related to the mesh can be handled automatically. To ease the effort involved in inputting material properties, the material properties can be stored in a database for future use.

A brief user guide is provided as follows. Note that there may be discrepancies between the brief guide shown here and the current version of code downloaded from the Web site. The guide here serves just to describe how MFreePre works.

16.3.1 Main Windows

MFreePre is developed in the Microsoft Windows environment. MFreePre is driven by an interactive graphical user interface. All geometry data, material properties, and boundary conditions can be input by mouse or keyed in with an interactive window. The main working interface (Figure 16.1) consists of six parts: main menu, toolbars, tree view, text view and command view, graphical display area, and status bar. Note that the window has minimal items, because we do not need to create sophisticated element meshes. All we need is geometry creation and boundary, loading, and material definition.

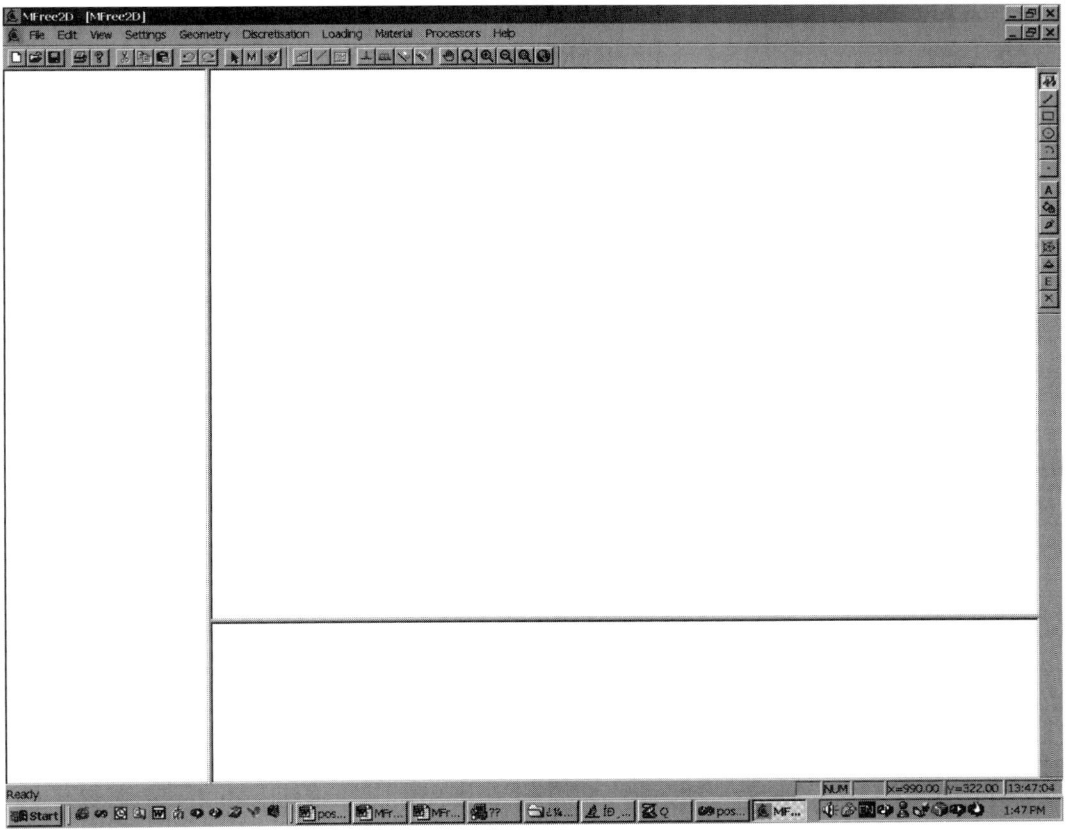

FIGURE 16.1
Main window of MFree2D.

Main menu: The main menu contains all input items and operation facilities. Most of the items are also available as buttons in the toolbars.

Toolbars: Toolbars contain buttons for actions that are related to creating the geometry model, adding loads, defining the boundary conditions, editing, and so on.

Tree view: Tree view is used to show the boundary and to edit the boundary properties.

Text view: Text view is used to show the commands and keyed-in data.

Graphical display area: The graphical display area is where the geometry model is created. All operations of the creation of geometry or postprocessing are mainly done by means of a mouse. In case some option driven by mouse input does not give the desired accuracy, a direct keyboard input is available.

Status bar: The status bar contains three indicators: a line indicator showing the operation of the selected menu item, a cursor position indicator giving the current position of the mouse cursor on the screen, and a time indicator providing the current time.

16.3.2 Geometry Creation

The generation of an analysis model begins with the creation of the geometry model. A geometry model can be formed using points, lines, squares, arcs, and circles. All these entities can be input or edited by means of a mouse or keyboard.

Seed Node

This icon is used to create a *seed node* to control the density of discrete nodes when representing the problem domain with nodes. It is particularly useful when you intend to increase or decrease the density of nodes in the area surrounding the seed node. Use it only when necessary, as for most cases the default works just fine. This item can be selected from the drawing submenu, *source point*, or directly from the toolbar. The point position can be selected with a mouse or keyed in from the keyboard. When the point position has been input, a dialog window (Figure 16.2) will prompt you to key in the values of weight and effect radius. The weight value is used to control the density of discrete nodes, i.e., the average nodal spacing around the seed node. Figure 16.3 shows an example of using a seed node to increase the nodal density at a particular point.

Line

This input icon is for creating a *geometric line*. This item can be selected from the geometry menu as well as directly from the toolbar. You can pick two points using a mouse on the graphical display area to generate a line or key in the data (Figure 16.4) for the desired accuracy. After the first point has been input, a line will be dragged by moving the mouse, and a right click will fix the ending point. Exact coordinates for the ending

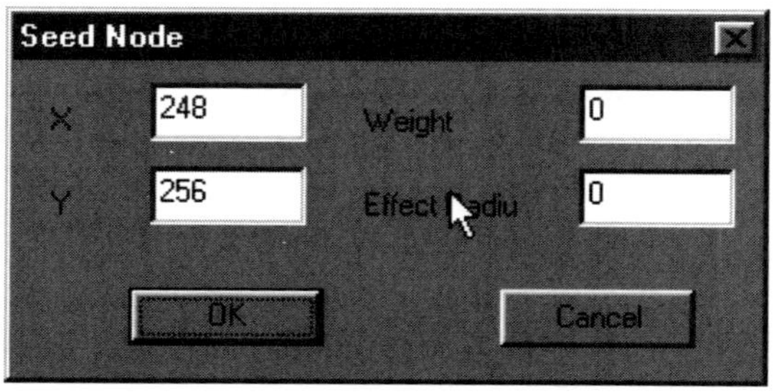

FIGURE 16.2
Edit dialog box for seed nodes.

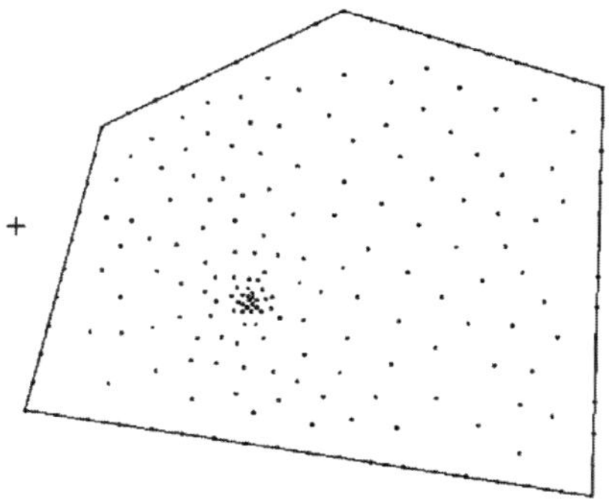

FIGURE 16.3
Example of the effect of a seed node with large density weight.

FIGURE 16.4
Example of generating a line using the keyboard input.

FIGURE 16.5
Example of generating a circle with the keyboard input.

point can be also be specified using the dialog box that is prompted by pressing any of the 10 digit keys.

Square

 This icon is used to create a *square geometry*. This can be done using keyboard input to specify the coordinates for the two diagonal points. The square can be also be created by dragging the mouse while pressing the left button.

Circle

 This icon is used to generate a *circular geometry*. This operation can be carried out by selecting from the geometry menu, or directly from the toolbar. First, the center point of the circle can be selected in the graphical display area using a mouse or by direct input by keyboard. Next, the radius to generate the circle can be keyed in or specified by dragging the mouse to a desired radius of the circle. The key-in boxes are shown in Figure 16.5. When the mouse is used to generate a circle, snap mode can assist you to select particular points conveniently.

Arc

 This icon is used to generate an *arc geometry*. This item can be selected from the geometry menu or directly from the toolbar. In this command, a three-point input method is used to generate an arc. The first point determines the center of the arc, and the second point is

used to determine both the start angle and the radius of the arc. The latter value is calculated by simply taking the distance between the first point, the center of the arc, and the second point. The third point is used only to determine the end angle of the arc. All the point inputs can be done by selecting in the graphical display area with a mouse or by keying in from the keyboard. The second and third point follow the right-hand rule related to the center point of the arc; i.e., the arc should be specified in a counterclockwise direction.

16.3.3 Boundary Conditions and Loads

Boundary conditions are special conditions that are imposed on geometric entities to constrain the displacements. Boundary conditions come in two variants, intensive (point) and distributed constraints.

The user can define two types of loads: intensive (concentrated) forces and distributed tractions. Concentrated forces are applied to geometric points that lie in the geometry. Distributed tractions are applied to geometric entities.

Point Displacement Constraint (intensive displacement)

This icon is used to define a *point displacement constraint* on a certain point. This can be performed from the loading menu, boundary condition → *intensive displacement*, or directly from the toolbar. The user can adjust the point location with a dialog box, as shown in Figure 16.6, which will prompt after the position of the point is defined.

There are seven items in the dialog box. *Location* shows the coordinates of the point. *Displacement* is used to prescribe the fashion of the constraint. The combo box contains five items for selection—*Constraint along X, Constraint along Y, Constraint along X and Y, Surface Normal,* and *Surface Tangent*—which are used to define the direction of the displacement. The last box is used for the prescribed value for this displacement constraint. If the user selects *Constraint along X and Y* and no zero is set for *displacement*, the angle of displacement related to the x direction must be defined. *Discrete Weight* is used to control the density of nodes around the point. The value of *Discrete Weight* is the average nodal spacing around the point load. Figure 16.7 shows an example of specified point constraints along x, y, and x and y.

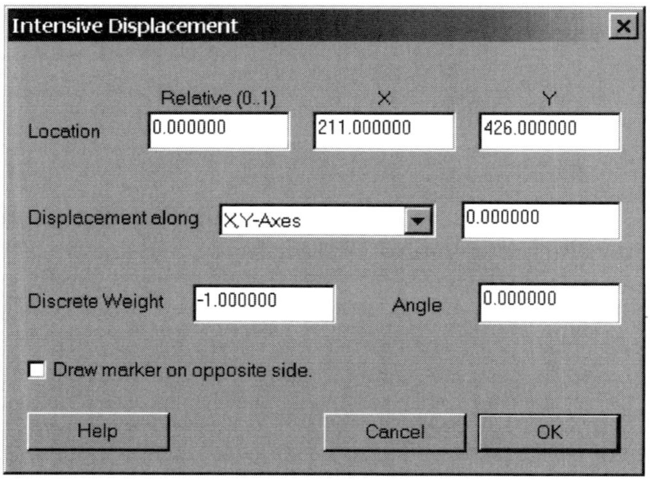

FIGURE 16.6
Dialog box for specifying a point displacement constraint.

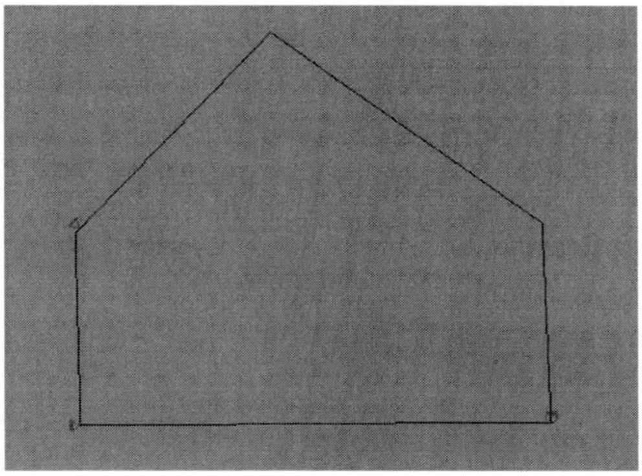

FIGURE 16.7
Example of specified point constraints.

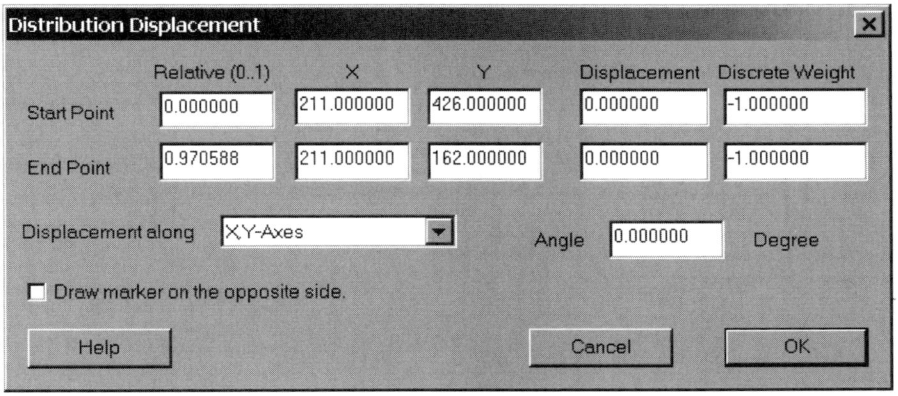

FIGURE 16.8
Dialog box for specifying the distribution displacements.

Distributed Displacement

This icon is used to define distributed constraints along a certain line. This can be performed from the loading menu, boundary condition → *distribution displacement*, or directly from the toolbar. The user can define a linear distribution displacement along a line, circle, and arc, and can input coordinates of the start point and end point by mouse. Figure 16.8 shows the dialog box of distribution displacement. Figure 16.9 shows an example of specified distributed displacement constraints.

Intensive Force (Point Force)

This icon is used to define *intensive (point) force*. Intensive force can be selected from the preprocess menu, *loading* → *intensive force*, or by clicking on the icon button on the toolbar. The position of the point on which the force will be specified can be input by mouse. The load value and direction can be input in the dialog box, Intensive Force, shown in Figure 16.10. In the dialog box, the location shows the coordinates of the point on which the force is to be imposed. The loading value is the magnitude of the force. Angle to x-axis is the force direction related to the x-axis. *Discrete weight* is the average nodal spacing around the point of force. Within the direction box, there are two radio buttons that are

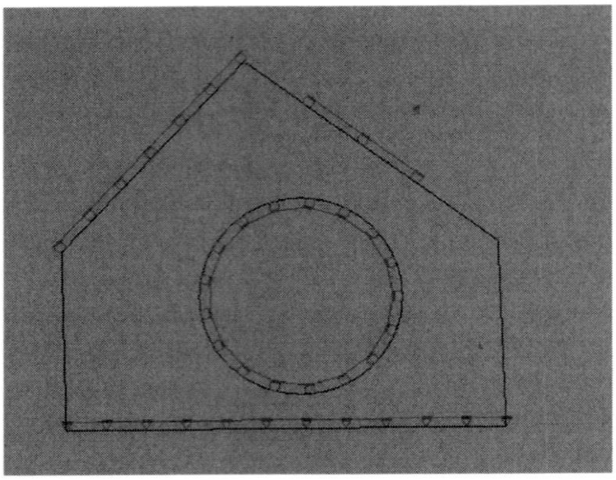

FIGURE 16.9
Example of specified distributed displacement constraints.

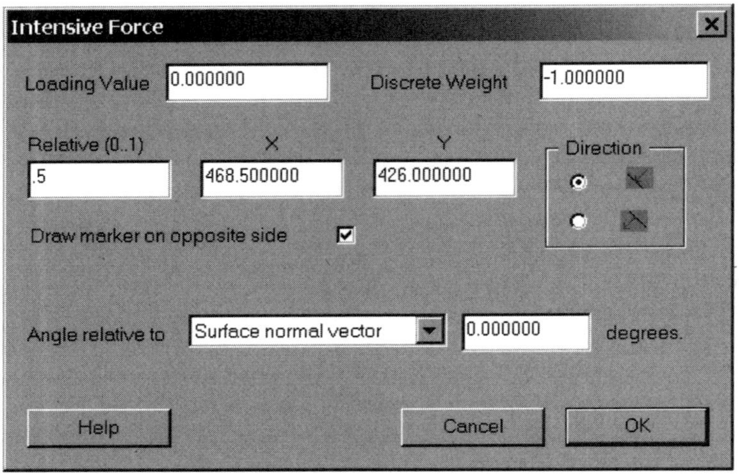

FIGURE 16.10
Dialog box for specifying an intensive force.

used to specify the direction of the force. [icon] means that the force is toward the imposed point. [icon] means that the force is backward from the imposed point. Figure 16.11 shows an example of specified intensive forces.

Traction

[icon] This icon is used to define distributed *traction* on geometric entities. This item can be selected from the preprocess menu, *loading* → *traction*, or by clicking the icon button on the toolbar. The distributed traction represents a distributed force applied on a segment of lines, circles, or arcs. The position of the start and end points of the traction effect can be selected using a mouse. The user can define the linear traction. Figure 16.12 shows a dialog box for specifying a distributed traction. The meaning of all items in the dialog box are the same as that in intensive force. Figure 16.13 shows an example of a specified traction applied on a structure.

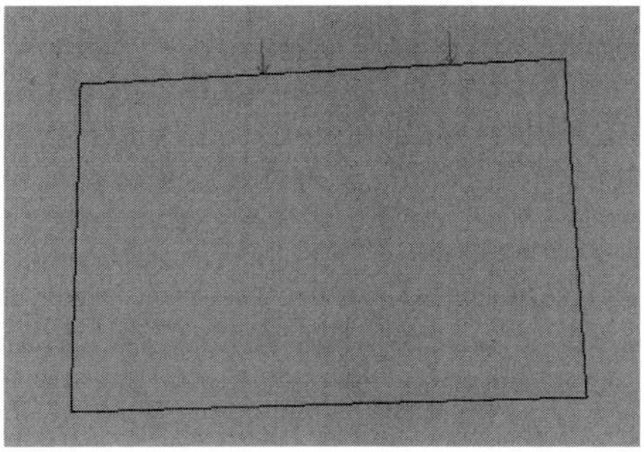

FIGURE 16.11
Example of specified intensive forces.

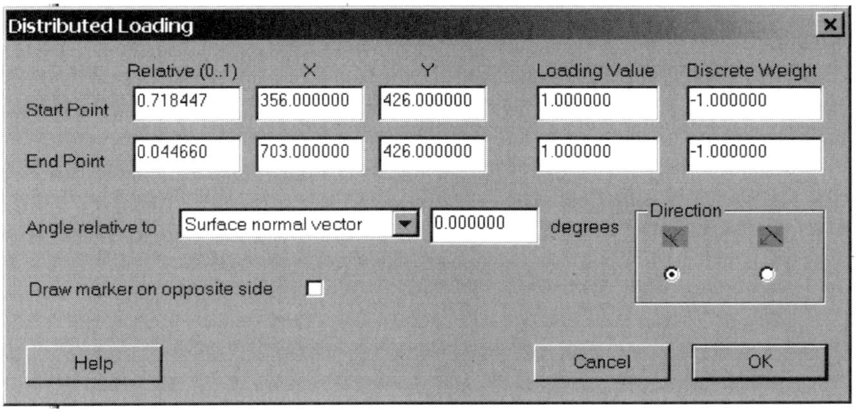

FIGURE 16.12
Dialog box for specifying a distributed traction.

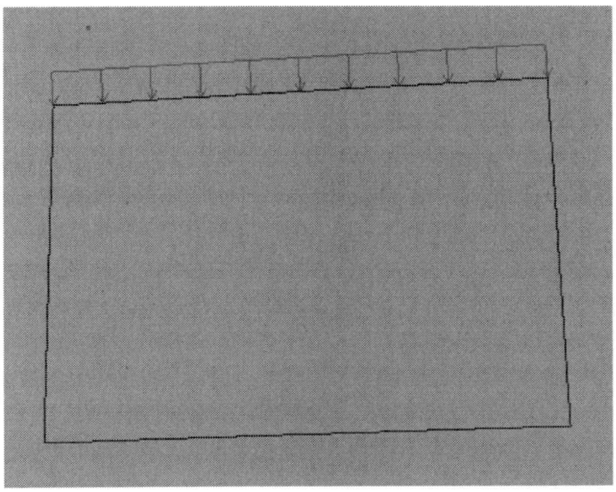

FIGURE 16.13
Example of a specified distributed traction.

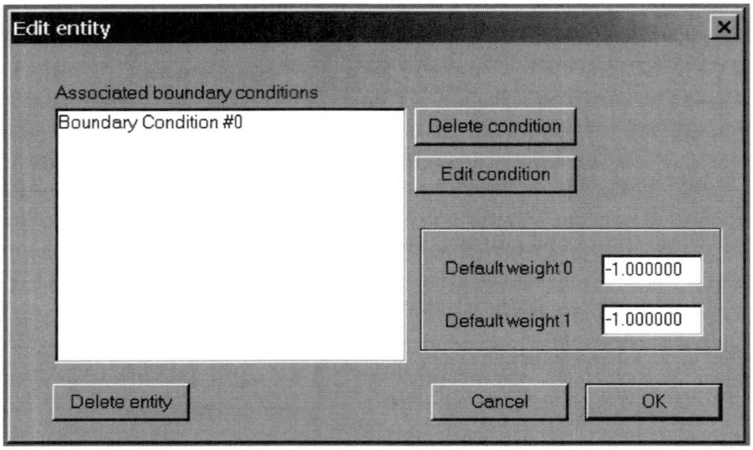

FIGURE 16.14
Example of a dialog box for editing boundary and loading conditions.

16.3.4 Modify and Delete Boundary Conditions and Loads

E This icon is used to edit (*modify* and *delete*) boundary conditions and loads. The user can select the boundary constraint or load applied to the selected entity by double-clicking the entity in the Tree-view. After a dialog window similar to Figure 16.14 appears, the user can edit and delete selected boundary conditions and loads.

16.3.5 Node Generation

In MFree2D, the domain is represented using scattered nodes. All the processes for nodal generation are fully automated based on the boundary of the geometry model created at the initial stage. The density of nodes in the domain can be controlled by the user by setting the *discrete weights* on the end points of each boundary item. Users can conveniently change the local density of nodes by adding seed nodes to obtain a desired density of nodal distribution at specified locations.

Boundary Specification

This icon is used to *specify the boundary* of the geometry of the model. There are two types of boundaries. The first is an external boundary and the other is an internal boundary. The external boundary is the profile that covers the whole domain; internal boundaries are the edges that form holes within the domain of the model. MFreePre will automatically decide which boundary is the external one, and which are the internal ones. The user can specify a boundary by double clicking on the geometric entity that is part of the boundary. All boundaries that form the geometry of the model will be shown in the Tree-view. Figure 16.15 shows an example of specified boundaries (green lines on the screen) for a model.

Modification of Discrete Nodal Density on a Boundary

The user can double-click on the item after selecting it on the tree window (the left window) that needs to be modified and a dialog box will be prompted to show the discrete weight on the item. The user can then modify the parameter in the dialog box as needed, by simply changing the value. Figure 16.16 shows an example of nodal distribution on the boundaries of an MFree model.

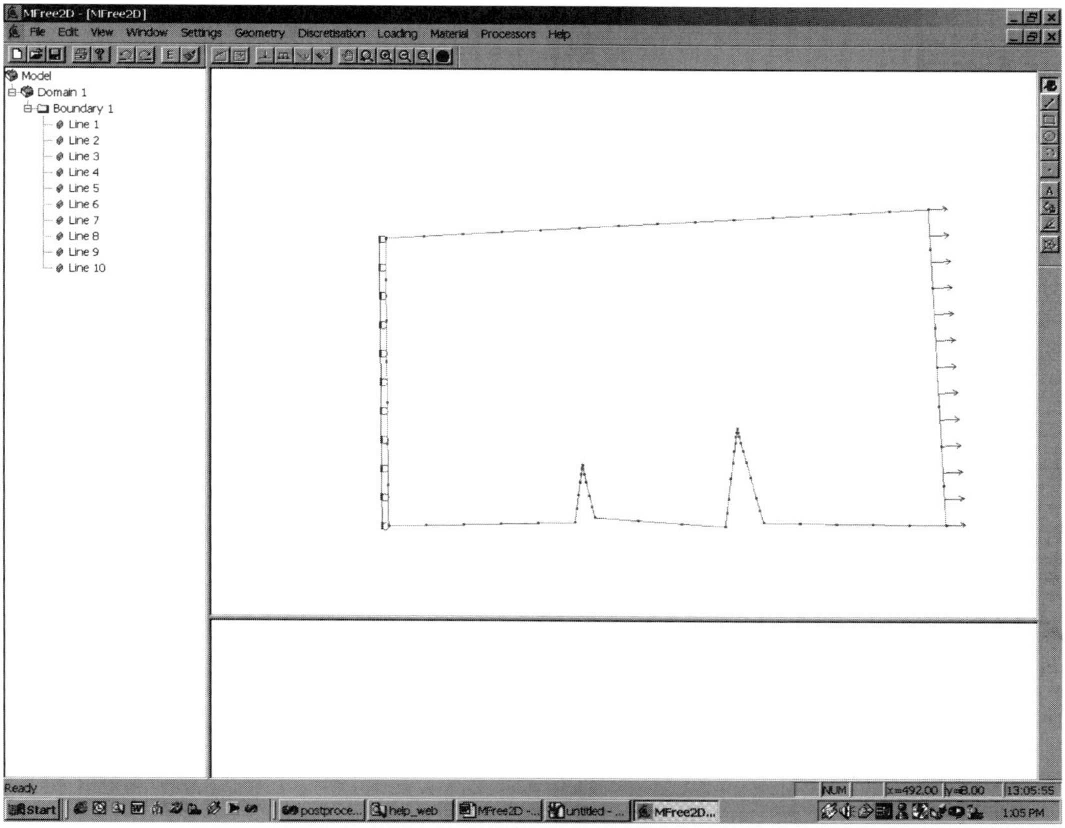

FIGURE 16.15
Example of specified boundaries (show as green lines on the screen) for a model.

The user can also perform the modification by selecting from the menu *settings* → *options* and then modifying the weights on the dialog window. Figure 16.17 shows a dialog box for altering some of the options.

Node Generation

 This icon is used to generate nodes in the domain. It can be selected from the preprocess menu, *discretization*, or by clicking the icon button on the toolbar. This command implements the task of generating nodes in the domain based on the density of nodes specified (or given by default) on the boundaries and at the seed nodes. Figure 16.18 shows an example of nodal distribution generated by MFreePre for a very complex domain. It is not too difficult to imagine what is required if one has created an FEM mode for a geometry of this complexity.

16.3.6 Materials Property Input

The parameters of material can be input using dialog boxes.

Create Material

This item can be selected from *Create* on the *Material* menu. It can be used to create a new material for the material database built together with MFree2D. Figure 16.19 shows a dialog box for inputting material properties. The current MFree2D version released to the

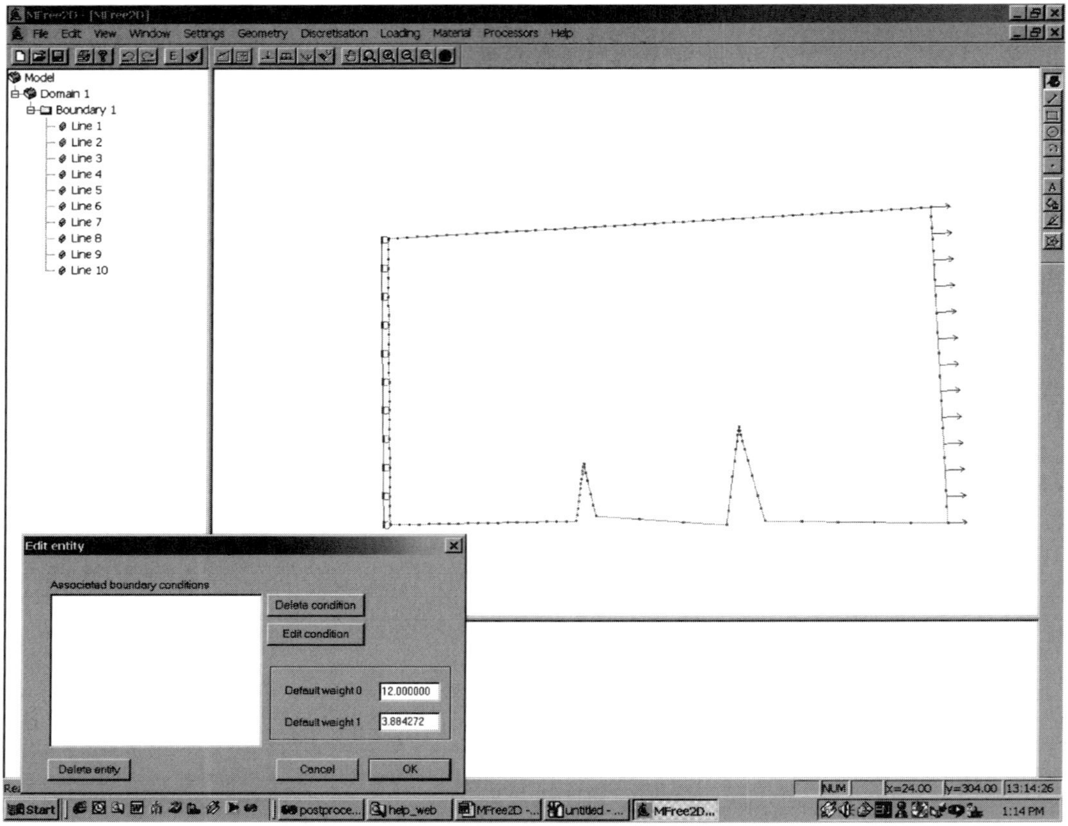

FIGURE 16.16
Example of nodal distribution on the boundaries of an MFree model.

public handles only solids of elastic materials. The material created will be stored in the database for future use.

Show and Update Material

This item can be selected from *show & update* on the *Material* menu. It can be used to show, modify, and delete material from the material database. Figure 16.20 shows a dialog box to display the properties of materials created. Modification is done by simply changing the values. Deleting is done by clicking the delete button.

Material Selection

This item can be selected from *selection* on the *Material* menu. Figure 16.21 shows a dialog box for selecting a material from the material database.

16.3.7 Miscellaneous

MFreePre provides some auxiliary functions to help users view or draw geometry models and change the color of the background or lines.

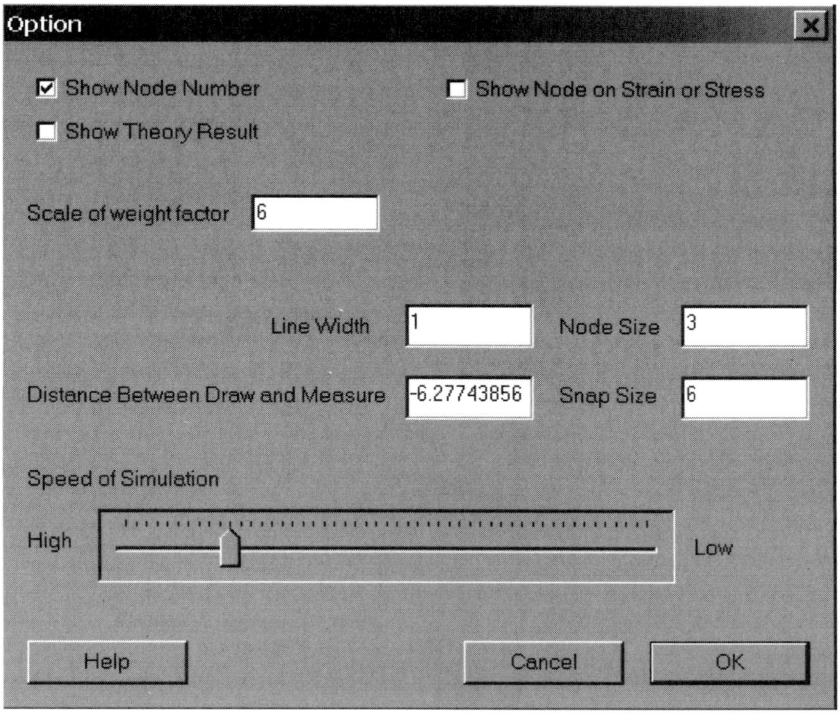

FIGURE 16.17
Dialog box for altering options.

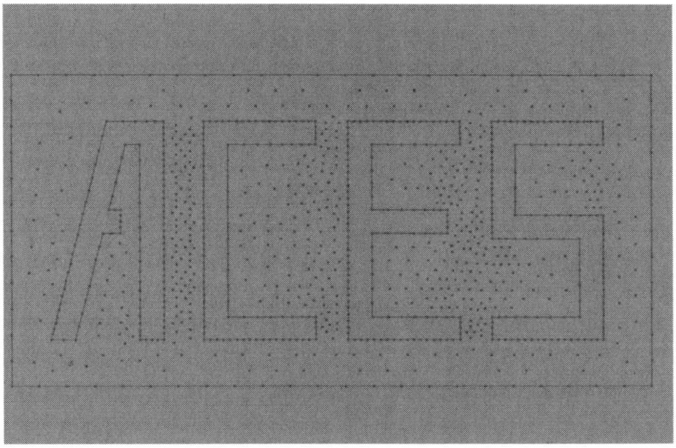

FIGURE 16.18
Example of nodal distribution generated by MFreePre for a very complex domain.

Scale of Graphical Display Area

This icon is used to change the scale of the graphical display area. It can be selected by clicking on the icon button on the toolbar. The default coefficient of scale is 1. Users can change it in the following dialog box. Figure 16.22 shows a dialog box for changing the display scaling factor.

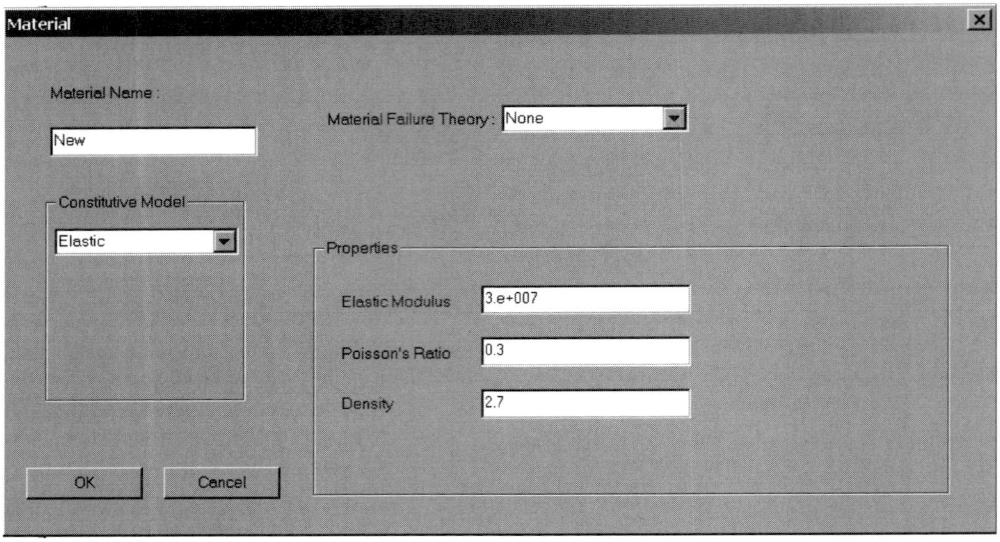

FIGURE 16.19
Dialog box for inputting material properties.

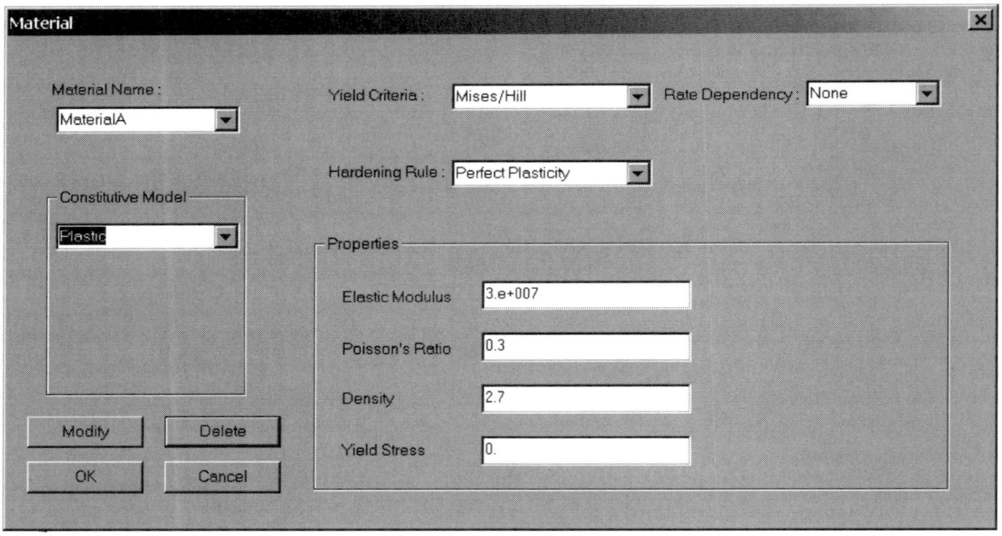

FIGURE 16.20
Dialog box displaying the material properties created. Modification is done by simply changing the values. Deletion is done by clicking the delete button.

Pan Graphical View

This icon is used to pan the graphical view. Users can press the left button of the mouse and drag the drawing to shift it.

Clear

This icon is used to clear the screen. This will not delete the model that you are creating.

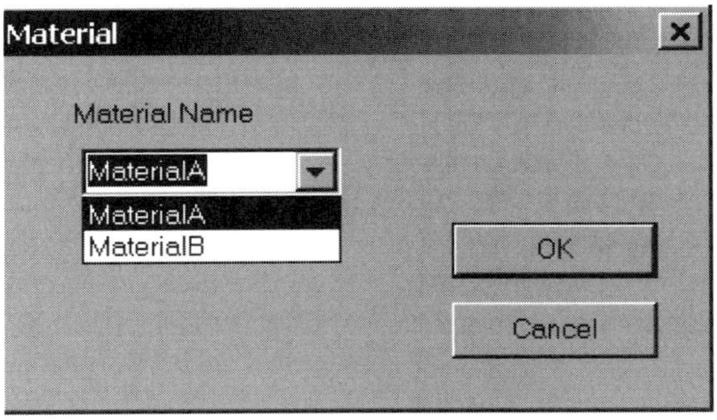

FIGURE 16.21
Dialog box for selecting a material from the material database.

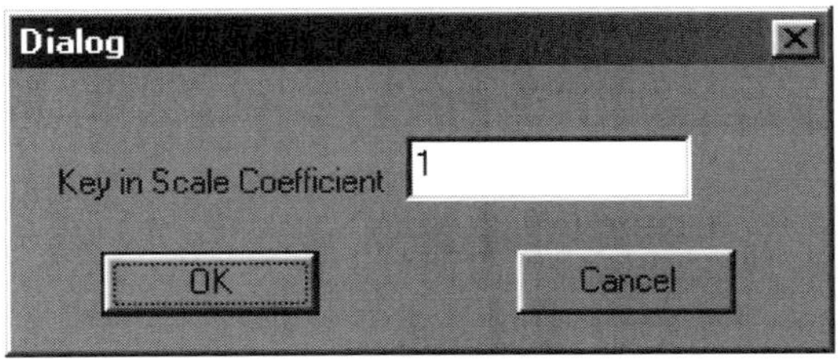

FIGURE 16.22
Dialog box for changing the display scaling factor.

Zoom In

This icon is used to zoom in the view in the display region with a zooming rate of 2.0.

Zoom Out

This icon is used to zoom out the view in the display region with a zooming rate of 0.5.

Zoom Region

To view a detail of a drawing, users can also use the function to magnify a selection area with a window.

Global View

This icon is used to restore the entire view of the model within the screen.

Background Color

This icon is used to change the background color. When this icon is selected, a standard color dialog (Figure 16.23) will appear. You can directly select color from the dialog box or define a custom color for the background.

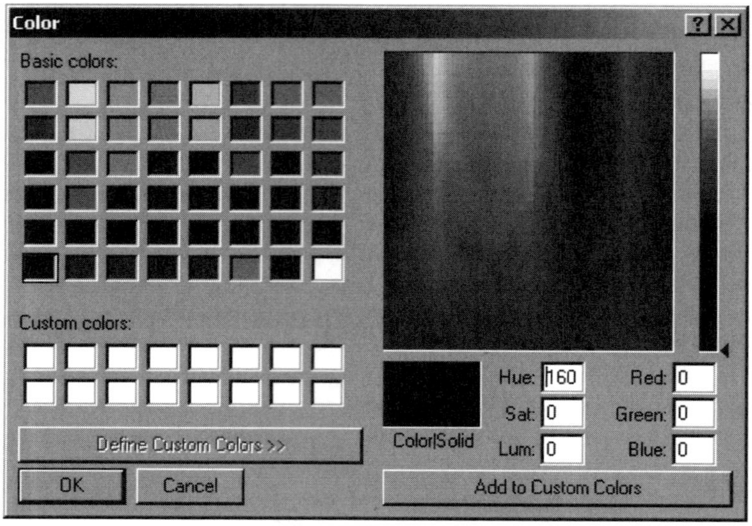

FIGURE 16.23
Color editing dialog box.

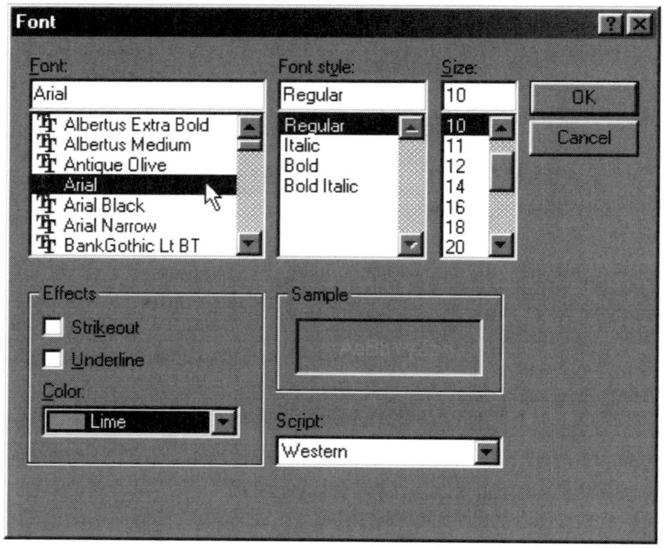

FIGURE 16.24
Dialog box for changing fonts.

Entity Color

This icon is used to change the color of entities that will be drawn on the screen. When this icon is selected, a standard color dialog will appear. You can directly select a color from the dialog box or define a custom color for the selected entities.

Font

This icon is used to change the font, font size, font style, and font color. When this icon is selected, a standard font dialog box (shown in Figure 16.24) will prompt. Users can change the font properties to meet their requirements.

16.4 Postprocessing in MFree2D

MFreePost provides a convenient graphical user interface for visualization of numerical solutions of MFree2D, e.g., initial and deformed domain displaying, field contouring, vector viewing, section projecting, and surface and curve plotting. In addition, it allows animation of the converging and nodal-refining process of the adaptive analysis.

16.4.1 Start of MFreePost

To start MFreePost, choose "Post processor/MFreePost" from "Processors" in the integrated main window shown in Figure 16.1. You can also start MFreePost as an executable file.

16.4.2 Window of MFreePost

Figure 16.25 shows the user graphical interface of the MFreePost. It consists of a title bar, a menu bar, a toolbar, an entity pane (left part of the window), a display region (right area of the window), and a status bar at the bottom of the window.

Main Menu

The menu consists of following menu items: File, Edit, View, Settings, Field, Section, Animation, Export, and Help.

FIGURE 16.25
User–graphical interface of MFreePost.

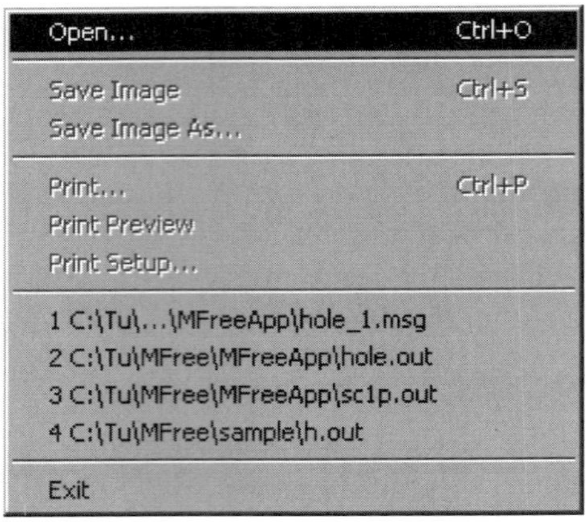

FIGURE 16.26
File submenu in MFreePost.

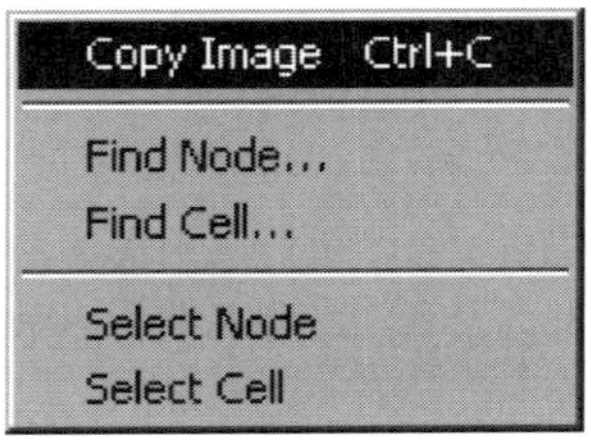

FIGURE 16.27
Edit submenu in MFreePost.

File

Figure 16.26 shows the submenu in the "File" menu. The submenu items are as follows:

Open: To open an MFree result file. The file has an extension of ".out" and is generated by MFreeApp.
Save Image: To save the image displayed in the display region. The file format is Bitmap.
Save Image As: To save the image in the display region with a given file name. The file format is Bitmap.
Print: To print the image in the display region.
Print Preview: To preview the printing effect of the image in the display region.
Print Setup: To set up the printer for use.
List of Recent File: To display the files most recently used. Maximally four files are displayed.
Exit: To end the MFreePost program.

Edit

Figure 16.27 shows the submenu in the "Edit" item. The submenu items and the functions are as follows:

Copy Image: To copy the image in the display region to the clipboard. This enables the image to be pasted in many word/image processing applications.
Find Node: To find the location of a specific node by giving the node ID in the dialog box shown in Figure 16.28.

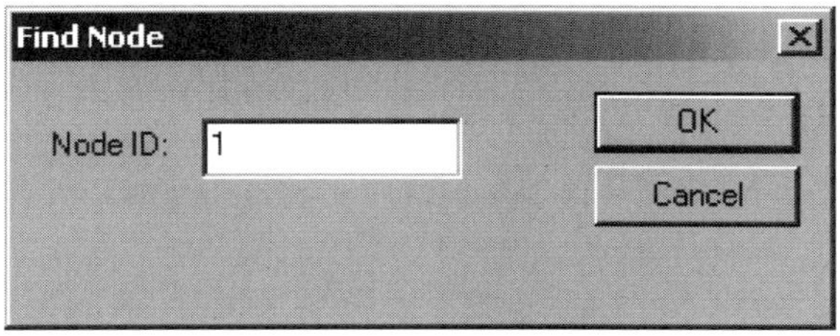

FIGURE 16.28
Find Node dialog box prompts for a node ID.

FIGURE 16.29
Find Cell dialog box requires a cell ID.

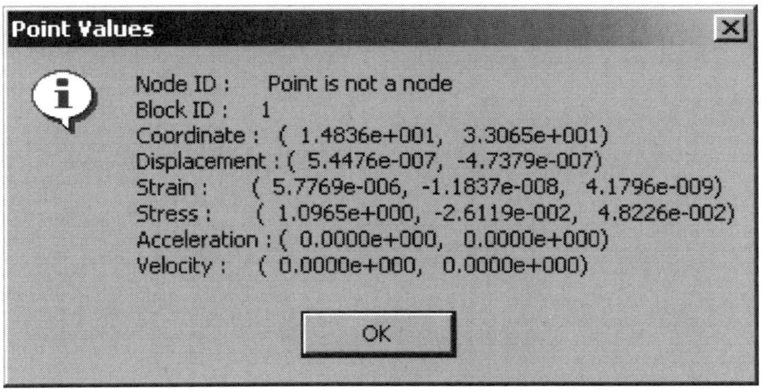

FIGURE 16.30
Output of information of a point.

Find Cell: To find the location of a cell by giving the cell ID in the dialog box shown in Figure 16.29. The cell will be highlighted when it is found. If the cell is out of the display region, it will be shifted to the central location of that region.

Select Node: To enable the node selection mode. By turning on this mode, the information (e.g., ID, coordinates, displacement, strain, stress, velocity, acceleration, etc.) of every point within the problem domain can be displayed simply by clicking the left mouse button. Figure 16.30 shows the output of the information of a point.

Select Cell: To enable the cell selection mode. By turning on this mode, the information (e.g., cell ID, vertices, cell area, etc.) of every cell in the problem domain can be displayed simply by clicking the left mouse button. The output of information related to the clicked cell will be shown as in Figure 16.31.

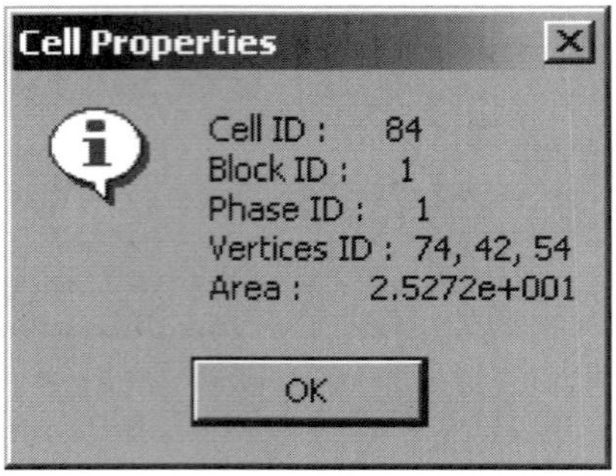

FIGURE 16.31
Output of information of a cell.

FIGURE 16.32
View submenu in MFreePost.

View

The submenu of the "View" item is shown in Figure 16.32. The functions of the subitems are as follows:

Toolbar: To show or hide the toolbar.
Status Bar: To show or hide the status bar.
Visualization Bar: To show or hide the visualization bar.
Node: To show or hide the domain nodes.
Background Cell: To show or hide the background cells.
Domain Border: To show or hide the domain border.
Color Ruler: To show or hide the color ruler.
Vector Scale: To show or hide the vector scale.

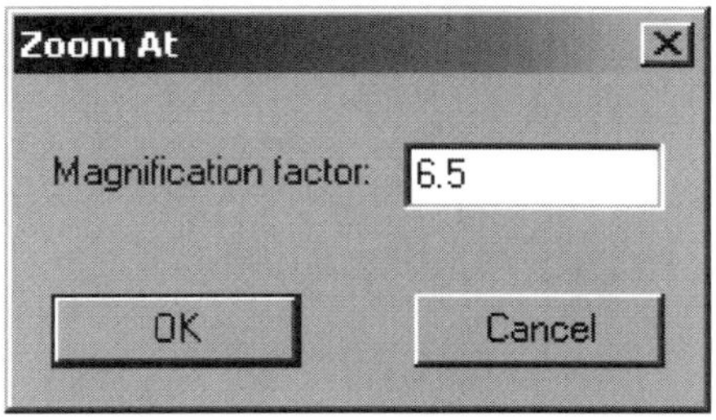

FIGURE 16.33
The Zoom At dialog box requires a magnification factor.

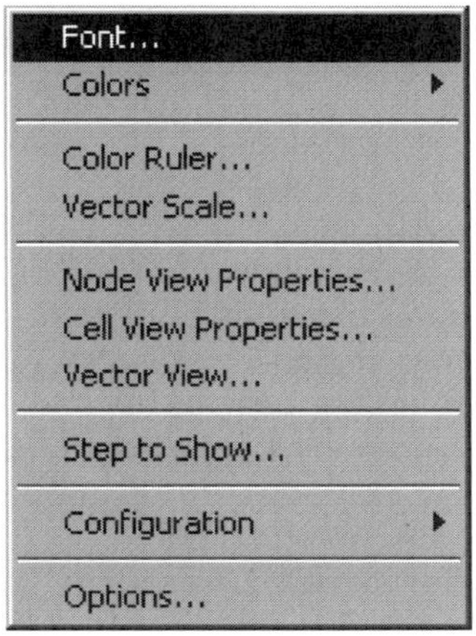

FIGURE 16.34
Submenu of settings in MFreePost.

MFree Logo: To show or hide the MFree logo.
Zoom In: To zoom in the view in the display region with a zooming rate of 50%.
Zoom Out: To zoom out the view in the display region with a zooming rate of 200%.
Zoom At: To zoom at the view in the display region with a given zooming rate. Figure 16.33 shows a dialog box that requires a magnification factor.
Zoom Region: To zoom the view into a selected rectangular region. The region is selected by dragging the mouse and will be projected to the full display region.
Translate (Pan): To shift the view in the display region by moving the mouse with the left button pressed.
Restore All: To restore the image in the display region to its original size and location.
Move Color Ruler: To move the color ruler in the display region.
Move Vector Scale: To move the vector scale in the display region.
Move MFree Logo: To move the MFree logo in the display region.

Settings

The menu item "Settings" is responsible for all view settings in the display region. Figure 16.34 shows its submenu of "Settings" in MFreePost. The functions of the subitems are as follows:

Font: To define the font size, color, and type for the view in the display region (as shown in Figure 16.34).
Colors: To define the background color and drawing color for the display region. The color is selected using the color dialog shown in Figure 16.23.

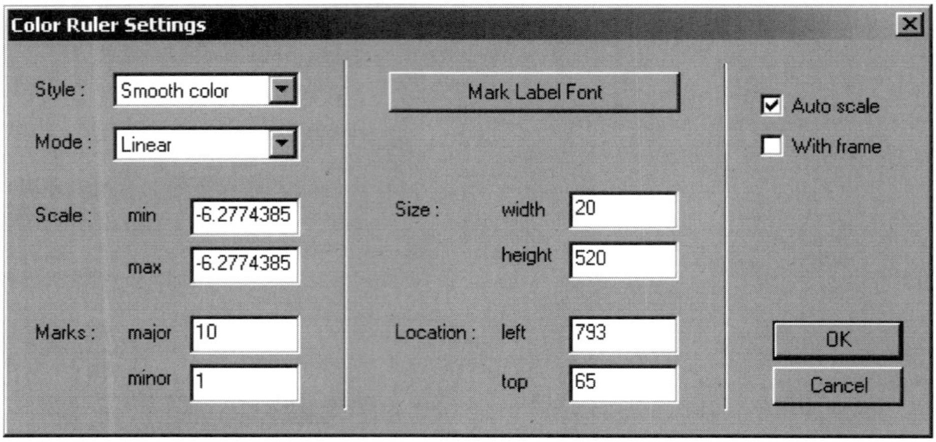

FIGURE 16.35
Dialog box for Color Ruler Settings.

Color Ruler: To define the style, mode, size, location, and other properties of the color ruler. All the settings related to the color ruler are done using the Color Ruler Settings dialog box (Figure 16.35).
Style: To define Line only, Leaped color, Smooth color. Line only is a style that shows contour lines; Leaped color style shows the contour regions and fills each region using a unique color; Smooth color style fills the entire problem domain with smoothly changed colors.
Mode: To define Linear, Logarithmic. The Linear Mode defines that the color along the color scale changes linearly with the value it represents; the Logarithmic Mode defines a logarithmic relationship between the colors and the values.
Scale: To define the minimum and maximum values measured by the color ruler.
Marks: To define the major divisions of the color ruler and the minor divisions in a major division.
Mark Label Font: To define the label font for the color ruler.
Size: To define the size in pixels of the color ruler.
Location: To define the location in pixels of the color ruler.
Auto Scale: To mark the box to set the scale of the color ruler automatically.
With Frame: To mark the box to bound the color ruler with a frame.
Vector Scale: To define the properties of the vector scale. The dialog box for vector scale settings is shown in Figure 16.36.
Reference Value: To define the value represented by the length of the vector scale.
Scale Length: To define the length (in pixels) of the vector scale.
Scale Width: To define the thickness of the vector scale in pixels.
Location: To define the coordinates (in pixels) of the left origin of vector scale.
Label Font: To define the label font of the vector scale.
Auto Scale: To mark the box to set the reference value automatically.
Node View Properties: To define the viewing properties for the domain nodes. The dialog for Setting Node View is shown in Figure 16.37.
Show Node Symbol: To mark the box to show the node symbol.
Show Node ID: To mark the box to show the node ID.
Shape: To choose the nodal shape. The options are none, circle, and rectangle.
Foreground Color: To define the drawing color for the node symbol.
Fill Color: To define the color to fill the node symbol.

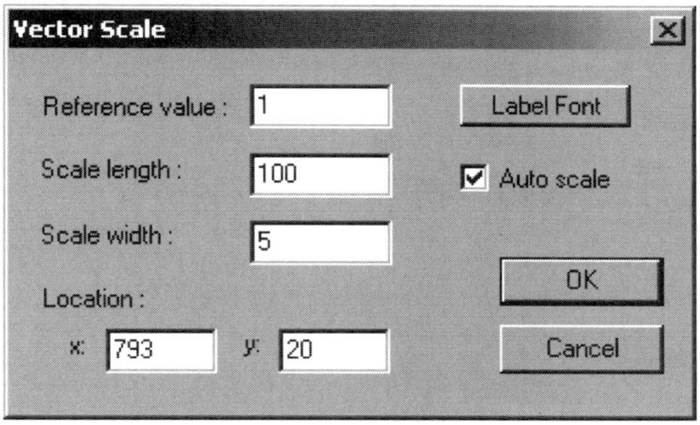

FIGURE 16.36
Dialog box for Vector Scale settings.

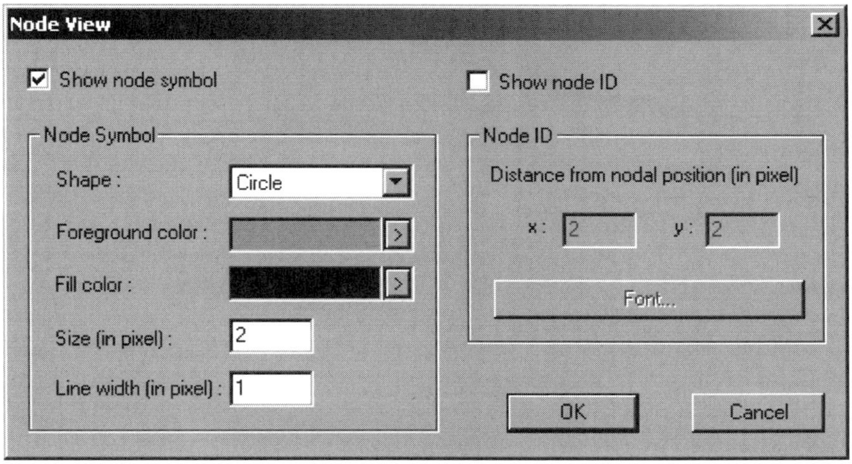

FIGURE 16.37
Dialog box for setting Node View properties.

Size: To define the size of node symbol in pixels.
Line Width: To define the line width for drawing the node symbol.
Distance from Nodal Position: To define the relative location of node label to nodal position.
Font: To define the node label font.
Cell View Properties: To define the cell view properties. The dialog box for setting the cell view properties is shown in Figure 16.38.
Show Cell Edge: To mark the box to show the cell edge.
Show Cell ID: To mark the box to show the cell ID label.
Line Style: To define the line style for drawing the cell edge. The options are solid line, dash line, dot line, and dashdot line.
Line Color: To define the edge color.
Line Width: To define the edge thickness in pixels.
Set Font: To set the label font for the cell ID.
Vector View: To define the vector properties for a vector field. The dialog box for setting the vector properties is shown in Figure 16.39.

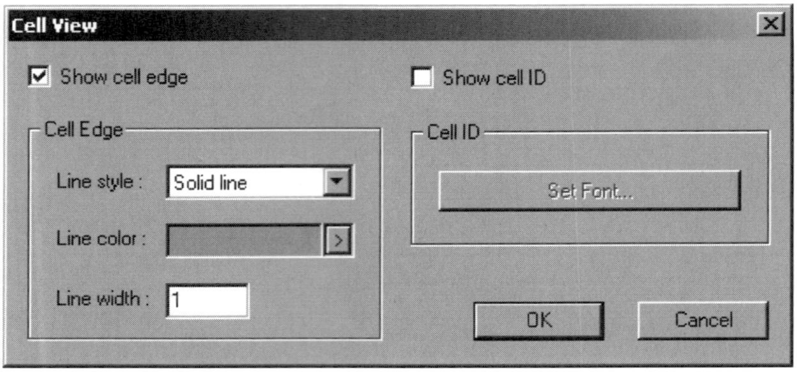

FIGURE 16.38
Dialog box for setting Cell View properties.

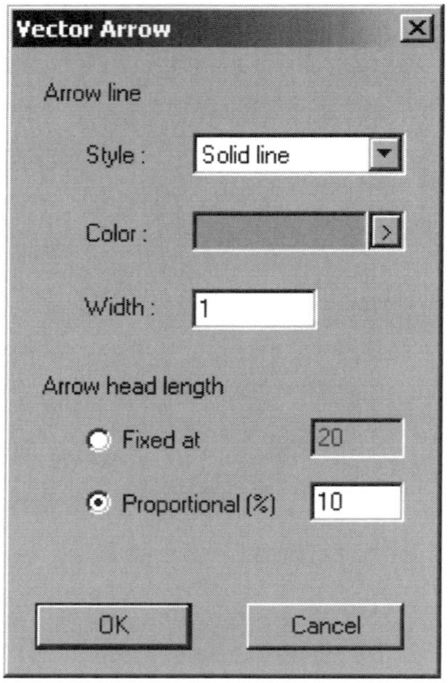

FIGURE 16.39
Dialog box to define Vector Arrow properties.

Style: To define the line style of the vector arrow. The options are solid line, dot line, and dash line.

Color: To define the vector arrow color.

Width: To define the thickness of the vector arrow in pixels.

Arrowhead Length: Fixed: to define an arrow with a fixed head length. Proportional: to define arrow with head length proportional to the arrow length. The length is measured in pixels.

Step to Show: To define the step to be shown in the display region. The dialog box is shown in Figure 16.40.

Show Mesh Refinement Step: To show a specified refinement step.

Show Load Increment Step: To show a specified load increment step.

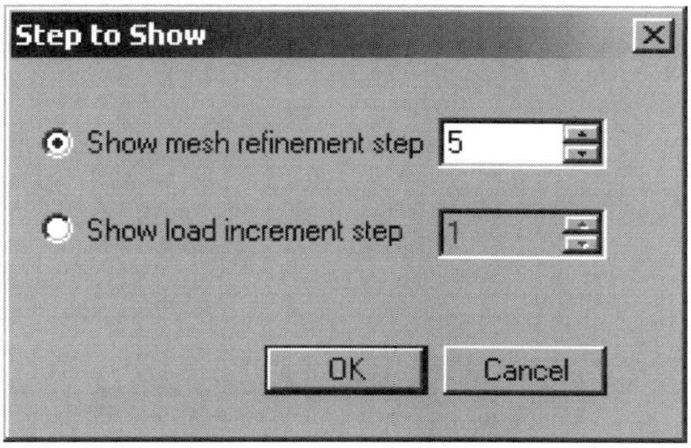

FIGURE 16.40
Dialog box to define Step to Show.

FIGURE 16.41
Submenu in the Configuration item.

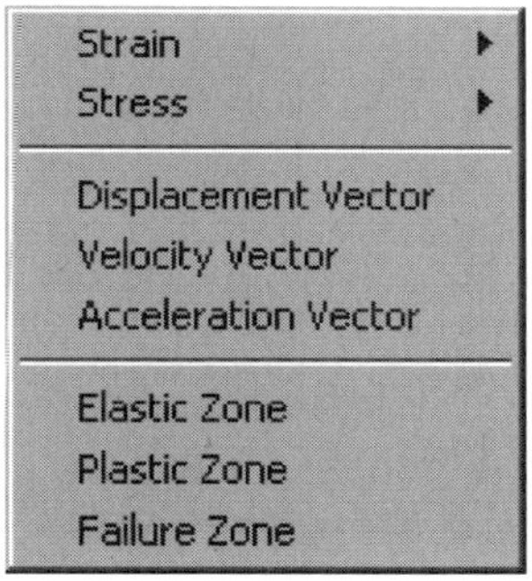

FIGURE 16.42
Submenu of field.

Configuration: To set the configuration of the problem domain. There are two options of configuration: initial configuration and current configuration (Figure 16.41). The former is based on the original undeformed problem domain whereas the latter uses the deformed problem domain at the current step of iteration.

Options: To define other general options.

Field

This item consists of functions related to field variables. The submenu is shown in Figure 16.42. The functions of the subitems are as follows:

Strain: To display the strain field. The strain components consist of strains in the x, y directions (strain_xx and strain_yy), the shear strain (strain_xy), principal strains, the equivalent strain, etc. Figure 16.43 is the submenu of Strain.

Stress: To display the stress field. The stress components consist of stresses in the x, y directions (stress_xx and stress_yy), shear stress (strain_xy), principal stresses, Mises strain, etc. Figure 16.44 is the submenu of Stress.

Displacement Vector: To show the displacement field using vector arrows. A vector arrow starts at the location of a node and points to the displacement direction. The arrow length represents the magnitude of the nodal displacement.

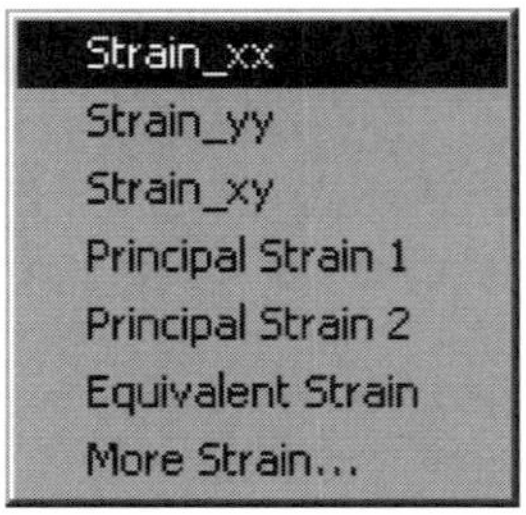

FIGURE 16.43
Strain components that can be displayed.

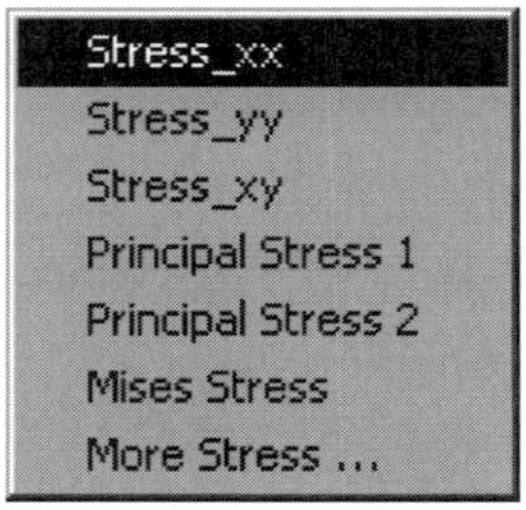

FIGURE 16.44
Stress components that can be displayed.

FIGURE 16.45
Submenu of Create Section.

Velocity Vector: To show the velocity field using vector arrows. A vector arrow starts at the location of a node and points to the velocity direction. The arrow length represents the magnitude of the nodal velocity.

Acceleration Vector: To show the acceleration field using vector arrows. A vector arrow starts at the location of a node and points to the acceleration direction. The arrow length represents the magnitude of the nodal acceleration.

Elastic Zone: To paint the elastic region in the problem domain with a specified color (green by default).

Plastic Zone: To paint the plastic region in the problem domain with a specified color (yellow by default).

Failure Zone: To paint the failure region in the problem domain with a specified color (red by default).

Section

This is a collection of commands related to the section view. The submenu items are shown in Figure 16.45.

Create Section: To create a section inside the problem domain. This is done by moving the cursor to the starting point, pressing and holding the left mouse button, dragging the cursor across the problem domain to the ending point, and releasing the button. A section line is then shown on the screen. Note that only one section line can be created at a time and creation of a new section will replace the existing one.

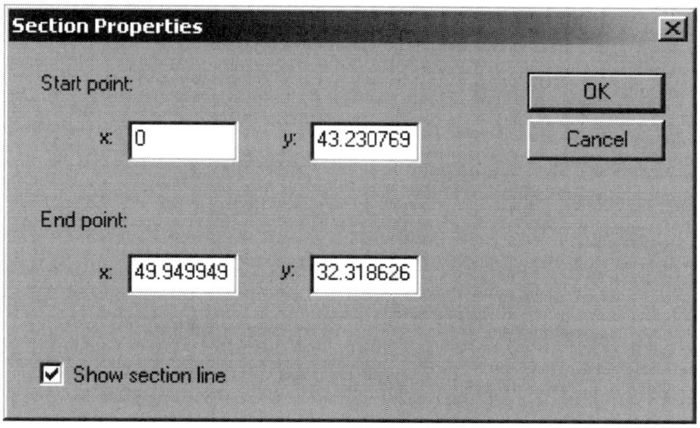

FIGURE 16.46
Dialog box for setting Section Properties.

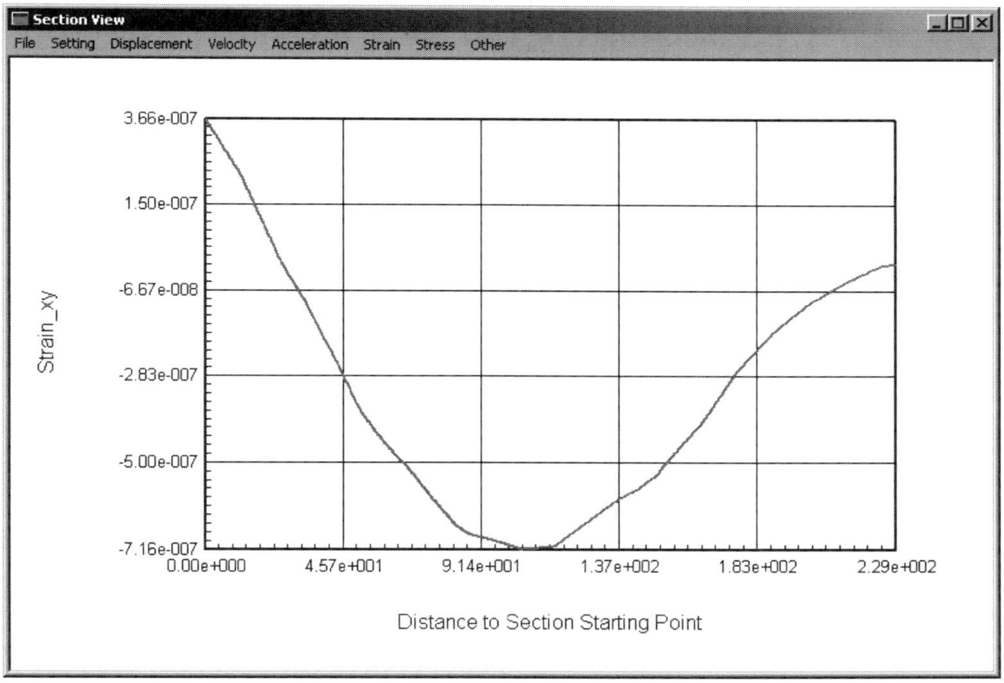

FIGURE 16.47
Section View window to display curve along a section.

Remove Section: To remove the existing section if any.

Section Properties: To define the starting and ending points of a section precisely. The dialog box to set section properties is shown in Figure 16.46.

Show Section Line: To show or hide a section line.

Show Curves: To show curves of field variables along the section line. Choice of this item initiates a separate Section View window (Figure 16.47) where the curves are displayed.

Export Data: To export variable values along a section to a text file.

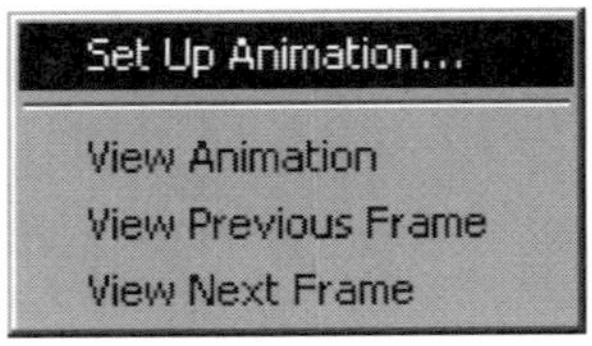

FIGURE 16.48
Submenu of Animation.

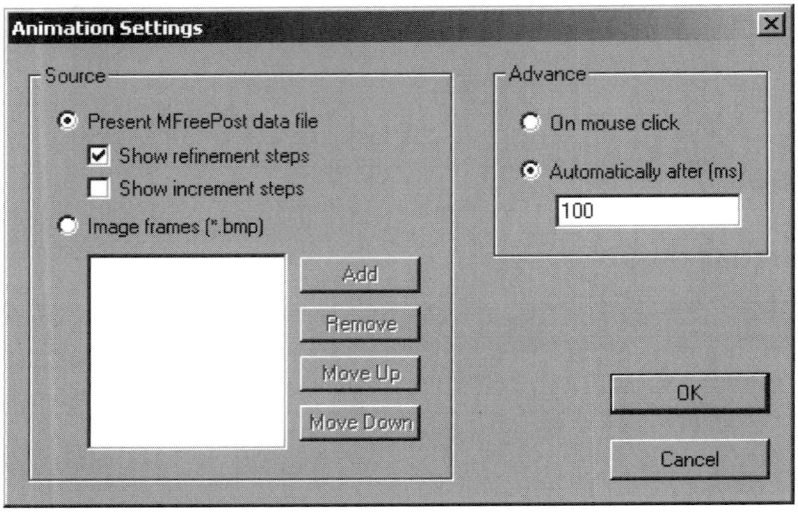

FIGURE 16.49
Dialog box for setting up an animation.

Animation: To define animation. This comprises all commands related to animation. The submenu of animation is shown in Figure 16.48.
Set Up Animation: To set up an animation using a dialog box shown in Figure 16.49. There are two panels in the dialog: the source panel is for defining the source data for an animation; the advance panel defines the animation advancing mode.
Present MFreePost Data File: To define the source data from the currently opened data file. Further, whether to show refinement steps or to show increment steps can also be defined.
Image Frames: To define the source data from a list of image frames. The list can be maintained using four commands: Add, Remove, Move Up, and Move Down.
On Mouse Click: The advancing mode is by clicking the left mouse button and frame by frame.
Automatically After: The advancing mode is automatic with a specified interval time.
View Animation: To display an animation with specified advancing mode.
View Previous Frame: To show the previous frame in an animation.
View Next Frame: To show the next frame in an animation.

Export

This item is specifically for exporting values to text files. The submenu is shown in Figure 16.50.

FIGURE 16.50
Submenu of export.

FIGURE 16.51
Submenu of Help Topics.

Node Value: To export the values of a specified node to a text file. The values include coordinates, displacement, velocity, acceleration, strain, stress, and deformation status at all steps.
Section Value: To export the variable values along a specified section to a text file.

Help

This item comprises help contents and description of the software package. The submenu is shown in Figure 16.51.

Help Topics: To search for a specific help topic.
About MFreePost: To describe the credit and copyright information of MFreePost.

References

Abbassian, F., Dawswell, D. J., and Knowles, N. C., *Free Vibration Benchmarks*, Atkins Engineering Sciences, Glasgow, 1987.

Anderson, T. L., *Fracture Mechanics: Fundamentals and Applications*, CRC Press, Boca Raton, FL, 1991.

Armando, D. C. and Oden, J. T., Hp Clouds—A Meshless Method to Solve Boundary Value Problems, TICAM Report 95-05, University of Texas, Austin, 1995.

Atluri, S. N. and Zhu, T., A new meshless local Petrov–Galerkin (MLPG) approach in computational mechanics, *Comput. Mech.*, 22, 117–127, 1998.

Atluri, S. N., Cho, J. Y., and Kim, H. G., Analysis of thin beams, using the meshless local Petrov–Galerkin method, with generalized moving least squares interpolations, *Comput. Mech.*, 24, 334–347, 1999a.

Atluri, S. N., Kim, H. G., and Cho, J. Y., A critical assessment of the truly meshless local Petrov–Galerkin (MLPG), and local boundary integral equation (LBIE) methods, *Comput. Mech.*, 24, 348–372, 1999b.

Babuška, I. and Melenk, J. M., The partition of unity finite element method, Technical report technical note BN-1185, Institute for Physical Science and Technology, University of Maryland, April 1995.

Barnhill, R. E., Representation and approximation of surfaces, in *Mathematical Software III*, Academic Press, New York, 1977, 69–120.

Bathe, K. J., Wilson, E. L., and Paterson, F. E., SAP IV, a structural analysis program for static and dynamic response of linear systems. University of California, Berkeley. Rep. EERC 73–11, 1973.

Beissel, S. and Belytschko, T., Nodal integration of the element-free Galerkin method, *Comput. Methods Appl. Mech. Eng.*, 139, 49–74, 1996.

Belytschko, T., Gu, L., and Lu, Y. Y., Fracture and crack growth by element free Galerkin methods, *Modeling Simulations Mater. Sci. Eng.*, 2, 519–534, 1994a.

Belytschko, T., Guo, Y., Liu, W. K., and Xiao, S. P., A unified stability of meshless particle methods, *Int. J. Numer. Methods Engng.*, 48, 1359–1400, 2000.

Belytschko, T., Lu, Y. Y., and Gu, L., Element-free Galerkin methods, *Int. J. Numer. Methods Eng.*, 37, 229–256, 1994b.

Belytschko, T. et al., Stress projection for membrane and shear locking in shell finite elements, *Comput. Methods Appl. Mech. Eng.*, 51, 221–258, 1985.

Belytschko, T. et al., Smoothing and accelerated computations in the element free Galerkin method, *J. Comput. Appl. Math.*, 74, 111–126, 1996a.

Belytschko, T. et al., Meshless method: an overview and recent developments, *Comput. Methods Appl. Mech. Eng.*, 139, 3–47, 1996b.

Benz, W. and Asphaug, E., Explicit 3D continuum fracture modeling with smoothed particle hydrodynamics, in *Proceedings of Twenty-fourth Lunar and Planetary Science Conference*, Lunar and Planetary Institute, 1993, 99–100.

Benz, W. and Asphaug, E., Impact simulations with fracture. I. Methods and tests, *Icarus*, 107, 98–116, 1994.

Benz, W. and Asphaug, E., Simulations of brittle solids using smoothed particle hydrodynamics, *Comput. Phys. Commun.*, 87, 253–265, 1995.

Beskos, D. E., Dynamic analysis of plates and shallow shells by the D/BEM, in *Advances in the Theory of Plates and Shells*, Elsevier, Oxford, 1990, 177–196.

Biot, M. A., General theory of three-dimensional consolidation, *J. Appl. Phys.*, 12, 155, 1941.

Bonet, J. and Kulasegaram, S., Correction and stabilization of smoothed particle hydrodynamics methods with applications in metal forming simulations, *Int. J. Numer. Methods Eng.*, 47, 1189–1214, 2000.

Brebbia, C. A., *The Boundary Element Method for Engineers*, Pentech Press, London, Halstead Press, New York, 1978.

Brebbia, C. A. and Georgiou, P., Combination of boundary and finite elements in elastostatics, *Appl. Math. Modeling*, 3, 212–219, 1979.

Brebbia, C. A., Telles, J. C., and Wrobel, L. C., *Boundary Element Techniques,* Springer Verlag, Berlin, 1984.

Breitkopf, P., Touzot, G., and Villon, P., Consistency approach and diffuse derivation in element free methods based on moving least squares approximation, *Comput. Assisted Mech. Eng. Sci.*, 5, 479–501, 1998.

Campbell, J., Vignjevic, R., and Libersky, L., A contact algorithm for smoothed particle hydrodynamics, *Comput. Methods Appl. Mech. Eng.*, 184, 49–65, 2000.

Campbell, P. M., Some new algorithms for boundary value problems in smoothed particle hydrodynamics, DNA Report, DNA-88-286, 1989.

Carlson, R. E. and Foley, T. A., The parameter R^2 in multiquadric interpolation, *Comput. Math. Appl.*, 21, 29–42, 1991.

Carter, R. L., Robinson, A. R., and Schnobrich, W. C., 1. Free vibrations of hyperboloidal shells of revolution, *J. Eng. Mech.*, 93, 1033–1053, 1969.

Chati, M. K. and Mukherjee, S., The boundary node method for three-dimensional problems in potential theory, *Int. J. Numer. Methods Eng.*, 47, 1523–1547, 2000.

Chati, M. K., Mukherjee, S., and Mukherjee, Y. X., The boundary node method for three-dimensional linear elasticity, *Int. J. Numer. Methods Eng.*, 46, 1163–1184, 1999.

Chen, J. K., Beraun, J. E., and Carney, T. C., A corrective smoothed particle method for boundary value problems in heat conduction, *Int. J. Numer. Methods Eng.*, 46, 231–252, 1999a.

Chen, J. K., Beraun, J. E., and Jih, C. J., An improvement for tensile instability in smoothed particle hydrodynamics, *Comput. Mech.*, 23, 279–287, 1999b.

Chen, J. S. et al., Reproducing kernel particle methods for large deformation analysis of nonlinear structures, *Comput. Methods Appl. Mech. Eng.*, 139, 195–228, 1996.

Chen, X. L., Liu, G. R., and Lim, S. P., Deflection analyses of laminates using EFG method, in *International Conference on Scientific and Engineering Computing*, March 19–23, Beijing, P.R. China, 2001, 67.

Chen, X. L., Liu, G. R., and Lim, S. P., A mesh-free method for free vibration and stability analyses of shear deformable plates, *J. Sound Vib.* (in press), 2002a.

Chen, X. L., Liu, G. R., and Lim, S. P., An element free Galerkin method for free vibration analysis of composite laminates of complicated shape, *Composite Structures* (in press), 2002b.

Chen, Y. Z., Evaluation of fundamental vibration frequency of an orthotropic bending plate by using an iterative approach, *Comput. Methods Appl. Mech. Eng.*, 161, 289–296, 1998.

Cheng, M. and Liu, G. R., A finite point method for analysis of fluid flow, in *Proceedings of 4th International Asia-Pacific Conference on Computational Mechanics*, December, Singapore, 1999, 1015–1020.

Cho, J. Y., Kim, H. G., and Atluri, S. N., Analysis of shear flexible beams, using the meshless local Petrov–Galerkin method based on locking-free formulation, in *Proceedings of Advances in Computational Engineering and Science*, Los Angeles, 2000, 1404–1409.

Chow, S. T., Liew, K. M., and Lam, K. Y., Transverse vibration of symmetrically laminated rectangular composite plates, *Compos. Struct.*, 20, 213–226, 1992.

Chung, H. J. and Belytschko, T., An error estimate in the EFG method, *Comput. Mech.*, 21, 91–100, 1998.

Cleary, P. W., Modeling confined multi-material heat and mass flows using SPH, *Appl. Math. Modeling*, 22, 981–993, 1998.

Cleveland, W. S., *Visualizing Data,* AT&T Bell Laboratories, Murray Hill, NJ, 1993.

Coleman, C. J., On the use of radial basis functions in the solution of elliptic boundary value problems, *Comput. Mech.*, 17, 418–422, 1996.

Combe, U. H. and Korn, C., An adaptive approach with the element-free-Galerkin method, *Comput. Methods Appl. Mech. Eng.*, 162, 203–222, 1998.

Cordes, L. W. and Moran, B., Treatment of material discontinuity in the element-free Galerkin method, *Comput. Methods Appl. Mech. Eng.*, 139, 75–89, 1996.

Davis, G. de Vahl, Natural convection of air in a square cavity: a benchmark numerical solution, *Int. J. Numer. Methods Fluids,* 3, 249–264, 1983.

DeFigueiredo, T. G. B., *New Boundary Element Formulation in Engineering*, Springer-Verlag, Berlin, 1991.

DeFigueiredo, T. G. B. and Brebbia, C. A., A new hybrid displacement variational formulation of BEM for elastostatics, in *Advances in Boundary Elements*, Vol. 1, Brebbia, C. A., Ed., Springer-Verlag, Berlin, 1989, 33–42.

Delaunay, B., Sur la sphere vide, *Bul. Acad. Sci. URSS Class. Sci. Nat.*, 793–800, 1934.

Dolbow, J. and Belytschko, T., Numerical integration of the Galerkin weak form in mesh free methods, *Comput. Mech.*, 23, 219–230, 1999.

Donning, B. M. and Liu, W. K., Meshless methods for shear-deformable beams and plates, *Comput. Methods Appl. Mech. Eng.*, 152, 47–71, 1998.

Duarte, C. A. and Oden, J. T., An *hp* adaptive method using clouds, *Comput. Methods Appl. Mech. Eng.*, 139, 237–262, 1996.

Dumont, N. A., The hybrid boundary element method, in *Boundary Elements*, IX, Vol. 1, Brebbia, C. A., Ed., Springer-Verlag, Berlin, 1988, 117–130.

Duncan, J. M. and Chang, C. Y., Nonlinear analysis of stress and strain in soils, *Proc. ASCE J. Soil Mech. Foundation Division*, 96(SM5), 1970.

Duncan, J. M. et al., Strength, Stress Strain and Bulk Modulus Parameters for Finite Element Analyses of Stress and Movements in Soil Masses, Rep. VCB/GT/78-02, University of California, Berkeley, 1978.

Fasshauer, G. E., Solving partial differential equations by collocation with radial basis functions, in *Surface Fitting and Multiresolution Methods*, Mehaute, A. L., Rabut, C., and Schumaker, L. L., Eds., 1997, 131–138.

Flebbe, O. S. et al., Smoothed particle hydrodynamics—physical viscosity and the simulation of accretion disks, *Astron. J.*, 431, 754–760, 1994.

Fleming, M. et al., Enriched element-free Galerkin methods for crack tip fields, *Int. J. Numer. Methods Eng.*, 40, 1483–1504, 1997.

Franke, C. and Schaback, R., Solving partial differential equations by collocation using radial basis functions, *Appl. Math. Comput.*, 93, 73–82, 1997.

Franke, R., Scattered data interpolation: tests of some method, *Math. Comput.*, 38(157), 181–200, 1982.

Gingold, R. A. and Monaghan, J. J., Smooth particle hydrodynamics: theory and applications to non-spherical stars, *Mon. Not. R. Astron. Soc.*, 181, 375–389, 1977.

Gingold, R. A. and Monaghan, J. J., Kernel estimates as a basis for general particle methods in hydrodynamics, *J. Comput. Phys.*, 46, 429–453, 1982.

Golberg, M. A., Chen, C. S., and Bowman, H., Some recent results and proposals for the use of radial basis functions in the BEM, *Eng. Anal. Boundary Elements*, 23, 285–296, 1999.

Golberg, M. A., Chen, C. S., and Karur, S. R., Improved multiquadric approximation for partial differential equations, *Eng. Anal. Boundary Elements*, 18, 9–17, 1996.

Gontkewitz, V. C., Natural vibration of plates and shells, *Nayka Dymka* (Kiev), 1964 [in Russian].

Griffel, D. H., *Linear Algebra and Its Applications*, Ellis Horwood Limited, New York, 1989.

Gu, Y. T. and Liu, G. R., A boundary point interpolation method (BPIM) using radial function basis, in *First MIT Conference on Computational Fluid and Solid Mechanics*, June 2001, MIT, 2001a, 1590–1592.

Gu, Y. T. and Liu, G. R., A coupled element free Galerkin/boundary element method for stress analysis of two-dimensional solids, *Comput. Methods Appl. Mech. Eng.*, 190, 4405–4419, 2001b.

Gu, Y. T. and Liu, G. R., A meshless local Petrov–Galerkin (MLPG) method for free and forced vibration analyses for solids, *Comput. Mech.*, 27(3), 188–198, 2001c.

Gu, Y. T. and Liu, G. R., A local point interpolation method for static and dynamic analysis of thin beams, *Comput. Methods Appl. Mech. Eng.*, 190, 5515–5528, 2001d.

Gu, Y. T. and Liu, G. R., A boundary point interpolation method for stress analysis of solids, *Comput. Mech.*, 28(1), 47–54, 2002.

Gu, Y. T. and Liu, G. R., A meshless local Petrov–Galerkin (MLPG) formulation for static and free vibration analyses of thin plates, *Comput. Modeling Eng. Sci.*, 2(4), 463–476, 2001f.

Haggblad, B. and Bathe, K. J., Specifications of boundary conditions for Reissner/Mindlin plate bending finite elements, *Int. J. Numer. Methods Eng.*, 30, 981–1011, 1990.

Hardy, R. L., Theory and applications of the multiquadrics—biharmonic method (20 years of discovery 1968–1988), *Comput. Math. Appl.*, 19, 163–208, 1990.

Hernquist, L. and Katz, N., TreeSPH—a unification of SPH with the hierarchical tree method, *Astrophys. J. Suppl. Ser.*, 70, 419–446, 1989.

Huang, W. X., *The Engineering Property of Soil*, Hydraulic & Electric Press, Beijing 1983 [Chinese version].

Hughes, T. J. R., *The Finite Element Method*, Prentice-Hall, Englewood Cliffs, NJ, 1987.

Jensen, P. S., Finite difference techniques for variable grids, *Comput. Struct.*, 2, 17–29, 1980.

Jin, Z. L., Gu, Y. T., and Zheng, Y. L., Hybrid boundary elements for potential and elastic problem, in *Proceedings of the 7th Japan-China Symposium on Boundary Element Methods*, Japan, 1996, 143–152.

Johnson, G. R. and Beissel, S. R., Normalized smoothing functions for SPH impact computations, *Int. J. Numer. Methods Eng.*, 39(16), 2725–2741, 1996.

Johnson, G. R., Stryk, R. A., and Beissel, S. R., SPH for high velocity impact computations, *Comput. Methods Appl. Mech. Eng.*, 139, 347–373, 1996.

Kanok-Nukulchai, W., Barry, W., Saran-Yasoontorn, K., and Bouillard, P. H., On elimination of shear locking in the element-free Galerkin method, *Int. J. Numer. Methods Engng.*, 52, 705–725, 2001.

Kansa, E. J., A scattered data approximation scheme with application to computational fluid-dynamics—I & II, *Comput. Math. Appl.*, 19, 127–161, 1990a.

Kansa, E. J., Multiquadrics—a scattered data approximation scheme with applications to computational fluid dynamics, *Comput. Math. Appl.*, 19(8/9), 127–145, 1990b.

Kardestuncer, H., Ed., *Finite Element Handbook*, McGraw-Hill, Boston, 1987.

Kothnur, V. S., Mukherjee, S., and Mukherjee, Y. X., Two-dimensional linear elasticity by the boundary node method, *Int. J. Solids Struct.*, 36, 1129–1147, 1999.

Krongauz, Y. and Belytschko, T., Enforcement of essential boundary conditions in meshless approximations using finite elements, *Comput. Methods Appl. Mech. Eng.*, 131(1–2), 133–145, 1996.

Krongauz, Y. and Belytschko, T., EFG approximation with discontinuous derivatives, *Int. J. Numer. Methods Eng.*, 41(7), 1215–1233, 1998.

Krysl, P. and Belytschko, T., Analysis of thin plates by the element-free Galerkin method, *Comput. Mech.*, 17, 26–35, 1996a.

Krysl, P. and Belytschko, T., Analysis of thin shells by the element-free Galerkin method, *Int. J. Solids Struct.*, 33, 3057–3080, 1996b.

Kuehn, T. H. and Goldstein, R. J., An experimental and theoretical study of natural convection in the annulus between horizontal concentric cylinders, *J. Fluid Mech.*, 74, 695–716, 1976.

Lam, K. Y. et al., Smoothed particle hydrodynamics for fluid dynamic problems, presented at *Int. Symp. Supercomputing and Fluid Science*, August, Institute of Fluid Science, Tohoku University, Sendai, Japan (Plenary Lecture), 2000.

Lancaster, P. and Salkauskas, K., Surfaces generated by moving least squares methods, *Math. Comput.*, 37, 141–158, 1981.

Leissa, A. W. and Narita, Y., Vibration studies for simply supported symmetrically laminated regular plates, *Compos. Struct.*, 12, 113–132, 1989.

Li, H. B. and Han, G. M., A new method for the coupling of finite element and boundary element discretized subdomains of elastic bodies, *Comput. Methods Appl. Mech. Eng.*, 54, 161–185, 1986.

Li, S., Hao, W., and Liu, W. K., Numerical simulations of large deformation of thin shell structures using meshfree methods, *Comput. Mech.*, 25, 102–116, 2000.

Libersky, L. D. and Petscheck, A. G., Smoothed particle hydrodynamics with strength of materials, in *Proceedings of the Next Free Lagrange Conference*, Vol. 395, Trease, H., Fritts, J., and Crowley, W., Eds., Springer-Verlag, New York, 1991, 248–257.

Libersky, L. D. and Petscheck, A. G., High strain Lagrangian hydrodynamics—a three-dimensional SPH code for dynamic material response, *J. Comput. Phys.*, 109, 67–75, 1993.

Liszka, T. and Orkisz, J., The finite difference method at arbitrary irregular grids and its application in applied mechanics, *Comput. Struct.*, 11, 83–95, 1980.

Lin, H., and Atluri, S. N., The meshless local Petrov–Galerkin (MLPG) method for solving incompressive Navier-Stokers equations, *Comput. Model. Engrg. Sci.*, 2, 117–142, 2001.

Liu, G. R., A point assembly method for stress analysis for solid, in *Impact Response of Materials & Structures*, Shim, V. P. W. et al., Eds., Oxford University Press, Oxford, 1999, 475–480.

Liu, G. R. and Chen, X. L., Static buckling of composite laminates using EFG method, in *Proc. 1st Int. Conf. on Structural Stability and Dynamics*, December 7–9, Taipei, Taiwan, 2000, 321–326.

Liu, G. R., and Chen, X. L., A mesh-free method for static and free vibration analyses of thin plates of complicated shape, *J. Sound Vib.*, 241(5), 839–855, 2001.

Liu, G. R. and Gu, Y. T., A point interpolation method, in *Proc. 4th Asia-Pacific Conference on Computational Mechanics*, December, Singapore, 1999, 1009–1014.

Liu, G. R. and Gu, Y. T., Meshless local Petrov–Galerkin (MLPG) method in combination with finite element and boundary element approaches, *Comput. Mech.*, 26, 536–546, 2000a.

Liu, G. R. and Gu, Y. T., Vibration analyses of 2-D solids by the local point interpolation method (LPIM), in *Proc. 1st International Conference on Structural Stability and Dynamics*, December 7–9, Taiwan, 2000b, 411–416.

Liu, G. R. and Gu, Y. T., Coupling of element free Galerkin and hybrid boundary element methods using modified variational formulation, *Comput. Mech.*, 26(2), 166–173, 2000c.

Liu, G. R. and Gu, Y. T., Coupling of element free Galerkin method with boundary point interpolation method, in *Advances in Computational Engineering & Science*, Atluri, S. N. and Brust, F. W., Eds., ICES'2K, Los Angeles, August, 2000d, 1427–1432.

Liu, G. R. and Gu, Y. T., A local point interpolation method for stress analysis of two-dimensional solids, *Struct. Eng. Mech.*, 11(2), 221–236, 2001a.

Liu, G. R. and Gu, Y. T., A local radial point interpolation method (LR-PIM) for free vibration analyses of 2-D solids, *J. Sound Vib.*, 246(1), 29–46, 2001b.

Liu, G. R. and Gu, Y. T., A point interpolation method for two-dimensional solids, *Int. J. Numer. Methods Eng.*, 50, 937–951, 2001c.

Liu, G. R. and Gu, Y. T., A matrix triangularization algorithm for point interpolation method, in *Proc. Asia-Pacific Vibration Conference*, Bangchun, W., Ed., November, Hangzhou, China, 2001d, 1151–1154.

Liu, G. R. and Gu, Y. T., On formulation and application of local point interpolation methods for computational mechanics, in *Proc. First Asia-Pacific Congress on Computational Mechanics*, 20–23 November, Sydney, Australia, 2001e, 97–106 [Invited paper].

Liu, G. R. and Gu, Y. T., Truly meshless methods based on the local concept and its applications, presented at *International Congress on Computational Engineering Science (ICES'01)*, 19–25 August, Puerto Vallarta, Mexico, 2001f [Keynote lecture].

Liu, G. R. and Gu, Y. T., Comparisons of two meshfree local point interpolation methods for structured analyses, *Comput. Mech.* 29(2), 107–121, 2002.

Liu, G. R. and Quek, S. S., *Finite Element Methods for Readers of All Backgrounds*, Academic Press, San Diego, 2003.

Liu, G. R. and Tu, Z. H., MFree2D©: an adaptive stress analysis package based on mesh-free technology, in *First MIT Conference on Computational Fluid and Solid Mechanics*, June, MIT, 2001, 327–329.

Liu, G. R. and Tu, Z. H., An adaptive procedure based on background cells for meshless methods, *Comput. Methods Appl. Mech. Eng.*, 191, 1923–1943, 2002.

Liu, G. R., Wu, Y. L., and Gu, Y. T., Application of meshless local Petrov–Galerkin (MLPG) approach to fluid flow problem. In *Computational Mechanics—New Frontiers for New Millennium*, Valliappan, S. and Khalili, N., Eds., Elsevier, Amsterdam, 2001.

Liu, G. R. and Wu, Y. L., A meshfree formulation of local radial point interpolation method (LRPIM) for impressible flow simulation, *Comput. Mech.*, 2002 (in press).

Liu, G. R. and Xi, Z. C., *Elastic Waves in Anisotropic Laminates*, CRC Press, Boca Raton, FL, 2001.

Liu, G. R. and Yan, L., A study on numerical integrations in element free methods, in *Proc. APCOM '99*, Singapore, 1999, 979–984.

Liu, G. R. and Yan, L., A modified meshless local Petrov–Galerkin method for solid mechanics, in *Advances in Computational Engineering and Sciences*, Atluri, N. K. and Brust, F. W., Eds., Tech Science Press, Palmdale, CA, 2000, 1374–1379.

Liu, G. R. and Yang, K. Y., A penalty method for enforce essential boundary conditions in element free Galerkin method, in *Proc. 3rd HPC Asia'98*, Singapore, 1998, 715–721.

Liu, G. R. and Yang, K. Y., A new meshless method for stress analysis for solids and structure, in *Proc. Fourth Asia-Pacific Conference on Computational Mechanics*, 15–17 December, Singapore, 1999.

Liu, G. R., Yang, K. Y., and Cheng, M., Some recent developments in element free methods in computational mechanics, in *Proc. Fourth Asia-Pacific Conference on Computational Mechanics*, 15–17 December, Singapore, 1999.

Liu, G. R., Tu, Z. H., and Wu, Y. G., MFree2D©: an adaptive stress analysis package based on mesh free technology, presented at *3rd South Africa Conference on Applied Mechanics*, 11–13 January, Durban, South Africa, 2000.

Liu, G. R., Liu, M. B., and Lam, K. Y., Smoothed particle hydrodynamics for numerical simulation of high explosive explosions, presented at First International Symposium on Advanced Fluid Information (AFI-2001), Institute of Fluid Science, Tohoku University, Sendai, Japan, 2001a.

Liu, G. R. Simulation of the high explosive detonation process using SPH methodology, presented at First MIT Conference on Computational Solid and Fluid Mechanics, June, MIT, Cambridge, MA, 2001b.

Liu, G. R., Liu, M. B., and Lam, K. Y., A general approach for constructing smoothing functions for meshfree methods, presented at *Ninth International Conference on Computing in Civil and Building Engineering*, April 3–5, Taipei, Taiwan, 431–436, 2002a.

Liu, G. R. Numerical simulation of the high explosive detonation process using SPH methodology, *Shock Vib.*, 2002b (in press).

Liu, G. R., Liu, L., Tan, V. B. C., and Gu, Y. T., A Conforming Point Interpolation Method for Analyzing Spatial Thick Shell Structures, 2nd Int. Conf. on Structural Stability and Dynamics, December 16–18, (accepted), 2002c.

Liu, G. R., Liu, M. B., and Lam, K. Y., Computer simulation of shaped charge detonation using meshless particle method, (submitted).

Liu, G. R., and Tu, Z. H., An adaptive procedure based on background cells for meshless methods, *Comput. Method Appl. Mech. Engrg.*, 191, 1923–1942, 2002.

Liu, G. R., Yan, L., Wang, J. G., and Gu, Y. T., Point interpolation method based on local residual formulation using radial basis functions, *Struct. Eng. Mech.*, 14(6), 713–732, 2002.

Liu, L., Liu, G. R., and Tan, V. B. C., Element free analyses for static and free vibration of thin shells, in *Proc. Asia-Pacific Vibration Conference*, Bangchun, W., Ed., November, Hangzhou, China, 2001.

Liu, G. R. et al., A combined finite element method/boundary element method for V(z) curves of anisotropic-layer/substrate configurations, *J. Acoust. Soc. Am. U.S.A.*, 92(5), 2734–2740, 1992.

Liu, M. B. et al., Numerical simulation of underwater explosion by SPH, in *Advances in Computational Engineering & Science*, Atluri, S. N. and Brust, F. W., Eds., Tech Science Press, Palmdale, CA, 2000a, 1475–1480.

Liu, M. B. et al., Numerical simulation of underwater shock using SPH methodology, presented at Computational Mechanics and Simulation of Underwater Explosion Effects—US/Singapore Workshop, November, Washington, D.C., 2000b.

Liu, M. B. et al., A new technique to treat material interfaces for smoothed particle hydrodynamics, in *Proc. First Asia-Pacific Congress on Computational Mechanics*, 20–23 November, Sydney, Australia, 2001a, 997–982.

Liu, M. B. et al., Numerical simulation of incompressible flows by SPH, Presented at *International Conference on Scientific & Engineering Computing*, March, Beijing, China, 2001b.

Liu, M. B. et al., Computer simulation of the high explosive explosion using SPH methodology, *Comput. Fluids*, 32(3), 305–322, 2003.

Liu, M. B., Liu, G. R., and Lam, K. Y., An SPH formulation for simulating discontinuous phenomena, *Shock Waves*, accepted.

Liu, M. B., Liu, G. R., and Lam, K. Y., Investigations on water mitigations by using meshless particle method, *Shock Waves*, 12(3), 181–195, 2002.

Liu, M. B. et al., Smoothed particle hydrodynamics for numerical simulation of underwater explosions, *Comput. Mech.*, 30(2), 106–118, 2003.

Liu, W. K., Adee, J., and Jun, S., Reproducing kernel and wavelet particle methods for elastic and plastic problems, in *Advanced Computational Methods for Material Modeling*, Benson, D. J., Ed., 180/PVP 268 ASME, 1993, 175–190.

Liu, W. K., Jun, S., and Zhang, Y., Reproducing kernel particle methods, *Int. J. Numer. Methods Fluids*, 20, 1081–1106, 1995.

Liu, W. K. et al., Advances in multiple scale kernel particle methods, *Comput. Mech.*, 18, 73–111, 1996.

Liu, W. K. et al., Multiresolution reproducing kernel particle method for computational fluid dynamics, *Int. J. Numer. Methods Fluids*, 24, 1–25, 1997a.

Liu, W. K. et al., Multiresolution reproducing kernel particle method for computational fluid dynamics, *Int. J. Numer. Methods Fluids*, 24, 1391–1415, 1997b.

Liu, W. K., Li, S. F., and Belytschko, T., Moving least-square reproducing kernel methods. I. Methodology and convergence, *Comput. Methods Appl. Mech. Eng.*, 143, 113–154, 1997c.

Lu, Y. Y., Belytschko, T., and Gu, L., A new implementation of the element free Galerkin method, *Comput. Methods Appl. Mech. Eng.*, 113, 397–414, 1994.

Lucy, L., A numerical approach to testing the fission hypothesis, *Astron. J.*, 82, 1013–1024, 1977.

Melenk, J. M. and Babuška, I., The Partition of Unity Finite Element Method: Basic Theory and Applications, TICAM Report 96-01, University of Texas, Austin, 1996.

Monaghan, J. J., Why particle methods work, *Siam J. Sci. Sat. Comput.*, 3(4), 423–433, 1982.

Monaghan, J. J., SPH Meets the Shocks of Noh, Monash University Paper, Melbourne, Australia, 1987.

Monaghan, J. J., An introduction to SPH, *Comput. Phys. Commun.*, 48, 89–96, 1988.

Monaghan, J. J., On the problem of penetration in particle methods, *J. Comput. Phys.*, 82, 1–15, 1989.

Monaghan, J. J., Smoothed particle hydrodynamics, *Annu. Rev. Astron. Astrophys.*, 30, 543–574, 1992.

Monaghan, J. J., Simulating free surface flows with SPH, *J. Comput. Phys.*, 110, 399–406, 1994.

Monaghan, J. J., Heat Conduction with Discontinuous Conductivity, Applied Mathematics Reports and Preprints 95/18, Monash University, Melbourne, Australia, 1995.

Monaghan, J. J., An introduction to SPH, *Comput. Phys. Commun.*, 48, 89–96, 1998.

Monaghan, J. J. and Gingold, R. A., Shock simulation by the particle method SPH, *J. Comput. Phys.*, 52, 374–389, 1993.

Monaghan, J. J. and Kocharyan, A., SPH simulation of multi-phase flow, *Comput. Phys. Commun.*, 87, 225–235, 1995.

Monaghan, J. J. and Poinracic, J., Artificial viscosity for particle methods, *Appl. Numer. Math.*, 1, 187–194, 1985.

Morris, J. P. and Monaghan, J. J., A switch to reduce SPH viscosity, *J. Comput. Phys.*, 136, 41–50, 1997.

Morris, J. P., Patrick, J. F., and Zhu, Y., Modeling lower Reynolds number incompressible flows using SPH, *J. Comput. Phys.*, 136, 214–226, 1997.

MSC/Dytran User's Manual, version 4, The MacNeal-Schwindler Corporation, 1997.

Mukherjee, Y. X. and Mukherjee, S., Boundary node method for potential problems, *Int. J. Numer. Methods Eng.*, 40, 797–815, 1997a.

Mukherjee, Y. X. and Mukherjee, S., On boundary conditions in the element-free Galerkin method, *Comput. Mech.*, 19, 264–270, 1997b.

Nagashima, T., Node-by-node meshless approach and its application to structural analyses, *Int. J. Numer. Methods Eng.*, 46, 341–385, 1999.

Nayroles, B., Touzot, G., and Villon, P., Generalizing the finite element method: diffuse approximation and diffuse elements, *Comput. Mech.*, 10, 307–318, 1992.

Nelson, R. P. and Papaloizou, J. C. B., Variable smoothing lengths and energy conservation in smoothed particle hydrodynamics, *Mon. Not. R. Astron. Soc.*, 270, 1–20, 1994.

Noguchi, H., Kawashima, T., and Miyamura, T., Element free analyses of shell and spatial structures, *Int. J. Numer. Methods Eng.*, 47, 1215–1240, 2000.

Oden, J. T. and Abani, P., A Parallel Adaptive Strategy for *hp* Finite Element Computations, TICAM Rep. 94-06, University of Texas, Austin, 1994.

Onate, E. et al., A finite point method in computational mechanics applications to convective transport and fluid flow, *Int. J. Numer. Methods Eng.*, 39, 3839–3866, 1996.

Organ, D. et al., Continuous meshless approximations for nonconvex bodies by diffraction and transparency, *Comput. Mech.*, 18(3), 225–235, 1996.

Orkisz, J., Finite difference method, in *Handbook of Computational Solid Mechanics*, Kleiber, M., Ed., Springer-Verlag, Berlin, 1998.

Ouatouati, A. E. and Johnson, D. A., A new approach for numerical modal analysis using the element free method, *Int. J. Numer. Methods Eng.*, 46, 1–27, 1999.

Ozakca, M. and Hinton, E., Free vibration analysis and optimization of axisymmetrical plates and shells—I. Finite element formulation, *Comput. Struct.*, 52, 1181–1197, 1994.

Petyt, M., Vibration of curved plates, *J. Sound Vib.*, 15, 381–395, 1971.

Petyt, M., *Introduction to Finite Element Vibration Analysis*, Cambridge University Press, Cambridge, 1990.

Poulos, H. G. and Davis, E. H., *Elastic Solution for Soil and Rock Mechanics*, Wiley, London, 1974.

Powell, M. J. D., The theory of radial basis function approximation in 1990, in *Advances in Numerical Analysis*, Light, F. W., Ed., Oxford University Press, Oxford, 1992, 303–322.

Radovitzky, R. M. O., Error estimation and adaptive meshing in strongly nonlinear dynamic problems, *Comput. Methods Appl. Mech. Eng.*, 172, 203–240, 1999.

Rao, S.S., *The Finite Element in Engineering*, 3rd ed., Butterworth Heinemann, Boston, 1999.

Randles, P. W. et al., Calculation of oblique impact and fracture of tungsten cubes using smoothed particle hydrodynamics, *Int. J. Impact Eng.*, 17, 1995.

Randles, P. W. and Libersky, L. D., Smoothed particle hydrodynamics—some recent improvements and applications, *Comput. Methods Appl. Mech. Eng.*, 138, 375–408, 1996.

Reddy, J. N., A simple higher-order theory for laminated composite plates, *J. Appl. Mech.*, 51, 745–752, 1984.

Reddy, J. N., *Finite Element Method*, John Wiley & Sons, New York, 1993.

Reddy, J. N., *Mechanics of Laminated Composite Plates: Theory and Analysis*, CRC Press, Boca Raton, FL, 1997.

Reddy, J. N., *Theory and Analysis of Elastic Plates*, Taylor & Francis, Philadelphia, PA, 1999.

Rippa, S., An algorithm for selecting a good value for the parameter c in radial basis function interpolation, *Adv. Comput. Math.*, 1, 193–210, 1999.

Robert, D. B., *Formulas for Natural Frequency and Mode Shape*, Van Nostrand Reinhold, New York, 1979.

Schaback, R., Approximation of polynomials by radial basis functions, in *Wavelets, Images and Surface Fitting*, Laurent, P. J., Mehaute, L., and Schumaker, L. L., Eds., A. K. Peters Ltd. Wellesley, MA, 1994, 459–466.

Schiffman, R. L. and Gibson, R. E., Consolidation of non-homogeneous clay layers, *Proc. ASCE*, JSMFD, 90(SM5), 1964.

Schinzinger, R. and Laura, P. A. A., *Conformal Mapping Methods and Applications*, Elsevier Science, New York, 1991.

Senthilnathan, N. R., A Simple Higher-Order Shear Deformation Plate Theory, Ph.D. thesis, National University of Singapore, Singapore, 1989.

Sharan, M., Kansa, E. J., and Gupta, S., Application of the multiquadric method for numerical solution of elliptic partial differential equations, *Appl. Math. Comput.*, 84, 275–302, 1997.

Shin, Y. S. and Chisum, J. E., Modeling and simulation of underwater shock problems using a coupled Lagrangian–Eulerian analysis approach, *Shock Vib.*, 4, 1–10, 1997.

Shu, C., Application of differential quadrature method to simulate natural convection in a concentric annulus, *Int. J. Numer. Methods Fluids*, 30, 977–993, 1999.

Simo, J. and Fox, D. D., On a stress resultant geometrically exact shell model, Part I: formulation and optimal parameterization, *Comput. Methods Appl. Mech. Eng.*, 72, 267–304, 1989.

Simo, J., Fox, D. D., and Rifai, M. S., On a stress resultant geometrically exact shell model, Part II: The linear theory; computational aspects, *Comput. Methods Appl. Mech. Eng.*, 73, 53–92, 1989.

Sneddon, I. N., The symmetrical vibrations of a thin elastic plate, *Proc. Cambridge Philos. Soc.*, 41, 27–43, 1945.

Song, B. et al., Application of finite point method to fluid flow and heat transfer, in *Proc. 4th Asia-Pacific Conference on Computational Mechanics*, December, Singapore, 1999, 1091–1096.

Srinivas, S., Kameswara Rao, A. K., and Joga Rao, C. V., Flexure of simply-supported thick homogeneous and laminated rectangular plates, *ZAMM*, 49, 449–458, 1969.

Steinmetz, M. and Muller, E., On the capabilities and limits of smoothed particle hydrodynamics, *Astron. Astrophys.*, 268, 391–410, 1993.

Strang, G., *Linear Algebra and Its Application*, Academic Press, New York, 1976.

Subir, K. S. and Gould, P. L., Free vibration of shells of revolution using FEM, *J. Eng. Mech. Div. ASCE EM2*, 100, 283–303, 1974.

Swegle, J. W. and Attaway, S. W., On the feasibility of using smoothed particle hydrodynamics for underwater explosion calculations, *Comput. Mech.*, 17, 151–168, 1995.

Takeda, H., Miyama, S. M., and Sekiya, M., Numerical simulation of viscous flow by smoothed particle hydrodynamics, *Prog. Theor. Phys.*, 92(5), 939–959, 1994.

Terzaghi, K. and Peck, R. B., *Soil Mechanics in Engineering Practice*, 2nd ed., John Wiley & Sons, New York, 1976.

Timoshenko, S. P. and Goodier, J. N., *Theory of Elasticity*, 3rd ed., McGraw-Hill, New York, 1970.

Timoshenko, S. P. and James, M. G., *Theory of Elastic Stability*, McGraw-Hill, Singapore, 1985.

Timoshenko, S. P. and Woinowsky-Krieger, S., *Theory of Plates and Shells*, 2nd ed., McGraw-Hill, New York, 1995.

Tou, S. K. and Wong, K. K., High-precision finite element analysis of cylindrical shells, *Comput. Struct.*, 26, 847–854, 1987.

Tu, Z. H. and Liu, G. R., An error estimate based on background cells for meshless methods, in *Advances in Computational Engineering and Sciences*, Atluri, S. N. and Brust, F. W., Eds., Tech Science Press, Palmdale, CA, 2000, 1487–1492.

Uras, R. A. et al., Multiresolution reproducing kernel particle methods in acoustics, *J. Comput. Acoust.*, 5(1), 71–94, 1997.

Wang, C. M. et al., Buckling of rectangular Mindlin plates with internal line supports, *Int. J. Solids Struct.*, 30, 1–17, 1993.

Wang, C. M., Reddy, J. N., and Lee, K. H., *Shear Deformable Beams and Plates (Relationships with Classical Solutions)*, Elsevier, Singapore, 2000.

Wang, J. G. and Liu, G. R., Radial point interpolation method for elastoplastic problems, in *Proc. 1st Int. Conf. on Structural Stability and Dynamics*, December 7–9, Taipei, Taiwan, 2000, 703–708.

Wang, J. G. and Liu, G. R., Radial point interpolation method for no-yielding surface models, in *Proc. First MIT Conference on Computational Fluid and Solid Mechanics*, 12–14 June, 2001a, 538–540.

Wang, J. G. and Liu, G. R., A point interpolation meshless method based on radial basis functions, *Int. J. Numer. Methods Eng.*, 2001b (in press).

Wang, J. G. and Liu, G. R., On the optimal shape parameters of radial basis functions used for 2D meshless methods, *Comput. Method Appl. Mech. Engrg.*, 191, 2611–2630, 2002.

Wang, J. G., Liu, G. R., and Wu, Y. G., A point interpolation method for simulating dissipation process of consolidation, *Comput. Methods Appl. Mech. Eng.*, 190, 5907–5922, 2001.

Wendland, H., Error estimates for interpolation by compactly supported radial basis functions of minimal degree, *J. Approx. Theor.*, 93, 258–396, 1998.

Whitworth, A. P. et al., Estimating density in smoothed particle hydrodynamics, *Astron. Astrophys.*, 301, 929–932, 1995.

Xu, X. G. and Liu, G. R., A local-function approximation method for simulating two-dimensional incompressible flow, in *Proc. 4th International Asia-Pacific Conference on Computational Mechanics*, December, Singapore, 1999, 1021–1026.

Yagawa, G. and Yamada, T., Free mesh method, a kind of meshless finite element method, *Comput. Mech.*, 18, 383–386, 1996.

Yagawa, G., Yamada, T., and Furukawa, T., Parallel computing with free mesh method: Virtually meshless FEM, in *IUTAM Symp., Solid Mechanics and Its Applications*, Mang, H. A. and Rammerstorfer, F. G. Eds., Kluwer, Dordrecht, 1997, 165–172.

Yang, H. T. Y., Saigal, S., and Liaw, D. G., Advances of thin shell finite elements and some applications—version I, *Comput. Struct.*, 35, 481–504, 1990.

Zhang, S. Z. et al., *Detonation and Its Applications*, Press of National Defense Industry, Beijing, 1976.

Zhu, T. and Atluri, S. N., A modified collocation and a penalty formulation for enforcing the essential boundary conditions in the element free Galerkin method, *Comput. Mech.*, 21, 211–222, 1998.

Zienkiewicz, O. C., *The Finite Element Method*, 4th ed., McGraw-Hill, London, 1989.

Index

G

H

I

J

K

L

Q

R

W

Y